NASA'S FIRST 50 YEARS
HISTORICAL PERSPECTIVES

Library of Congress Cataloging-in-Publication Data

NASA's first 50 years : historical perspectives / Steven J. Dick, editor
p. cm
NASA SP-2010-4704.
Includes bibliographical references and index
1. United States. National Aeronautics and Space Administration--
History. 2. Astronautics--United States--History. I. Dick, Steven J. II.
Title: NASA's first fifty years.
TL521.312.N383 2009
629.40973--dc22
2009015085

Speakers onstage at the NASA Headquarters Auditorium during NASA's 50th anniversary conference, 28 October 2008. Left to right: John Krige, Maura Mackowski, Michael Neufeld, Edward Goldstein, Michael Meltzer, Joseph Tatarewicz, Stephen Johnson, Andrew Butrica, Steven Dick, Linda Billings, Richard Hallion, John Logsdon, Tony Springer, Rob Ferguson, Erik Conway, David DeVorkin, Jennifer Ross-Nazzal, and James Fleming. Not shown: NASA Administrator Michael Griffin, Howard McCurdy, W. H. Lambright, J. D. Hunley, and Laurence Bergreen.

To the Employees of the
National Aeronautics and Space Administration
During Its First 50 Years
and Their Predecessors at the
National Advisory Committee for Aeronautics

Oh! I have slipped the surly bonds of Earth
And danced the skies on laughter-silvered wings;
Sunward I've climbed and joined the tumbling mirth
of sun-split clouds—and done a hundred things
You have not dreamed of—wheeled and soared and swung
High in the sunlit silence. Hov'ring there,
I've chased the shouting wind along, and flung
My eager craft through footless falls of air . . .
Up, up the long, delirious, burning blue
I've topped the wind-swept heights with easy grace
Where never lark, nor even eagle flew—
And, while with silent lifting mind I've trod
The high, untrespassed sanctity of space,
Put out my hand and touched the face of God.

"High Flight"
by Pilot Officer John Gillespie Magee, Jr.
No. 412 squadron, Royal Canadian Air Force (RCAF)
Killed 11 December 1941 at age 19

A poem beloved by aviators and astronauts alike

Table of Contents

PART I: **NASA at 50**

PART II: **Crosscutting Themes**

PART V: Space Science

PART VI: Earth Science and Applications

PART VII: NASA's Role in History

Acknowledgments

I wish to thank the organizing committee for the NASA 50th anniversary conference on which this volume is based, including Michael Neufeld, Roger Launius, Linda Billings, and the staff of the NASA History Division in the Office of External Relations: Stephen Garber, Nadine Andreassen, Jane Odom, Colin Fries, John Hargenrader, and Liz Suckow. Henry Fingerhut provided important support to the production process, carried out by the staff of the Communications Support Services Center (CSSC) at NASA Headquarters.

I wish to thank Administrator Michael Griffin for his opening remarks, which constitute chapter 1 of this volume, as well as all the authors for their contributions to this landmark conference and volume. In the Office of External Relations, I would like to thank Assistant Administrator for External Relations Michael F. O'Brien, Deputy Assistant Administrator Al Condes, and Tina Alexander for their support.

It has been an honor and a privilege to serve as NASA's Chief Historian during its 50th anniversary. Let us hope 50 years hence NASA will look back on a record of accomplishment as astonishing as its first half century.

Steven J. Dick
NASA Chief Historian
April 2009

Introduction

After considerable discussion in the wake of the Soviet launch of Sputnik, in the spring of 1958 President Dwight D. Eisenhower ordered that a bill be drafted to create a civilian space agency for the United States. The bill was sent to Congress on 2 April 1958, was passed after lengthy congressional debate, and was signed into law 29 July. When NASA began operations on 1 October 1958, no one could have foreseen the full scope of the adventures and accomplishments, the triumph and tragedy that would occur under its auspices over the next 50 years as humans and robots advanced into "the new ocean" of space.

The National Aeronautics and Space Act of 1958 provided for research into the problems of flight, both within Earth's atmosphere and in space.[1] NASA began by absorbing the earlier National Advisory Committee for Aeronautics (NACA), established in 1915, including its 8,000 employees, an annual budget of $100 million, three major research laboratories—Langley Aeronautical Laboratory, Ames Aeronautical Laboratory, and Lewis Flight Propulsion Laboratory—and two smaller test facilities. It quickly incorporated other organizations, or parts of them, notably the space science group of the Naval Research Laboratory that formed the core of the new Goddard Space Flight Center (GSFC); the Jet Propulsion Laboratory (JPL) managed by the California Institute of Technology for the Army; and the Army Ballistic Missile Agency (ABMA) in Huntsville, Alabama, where Wernher von Braun's team of engineers was developing large

1. The National Aeronautics and Space Act of 1958, as amended, with legislative history showing changes over time, is available at *http://history.nasa.gov/spaceact-legishistory.pdf.*

rockets, soon to form the core of the Marshall Space Flight Center (MSFC). Over the next 50 years, each of these NASA Centers—as well as the Kennedy Space Center (KSC), Dryden Flight Research Center (DFRC), Johnson Space Center (JSC), and Stennis Space Center (SSC)—would bring its own expertise to NASA's goals, whether in aeronautics, rocket technology, launch facilities, space science, or human spaceflight.

Fifty years after the founding of NASA, from 28 to 29 October 2008, the NASA History Division convened a conference whose purpose was a scholarly analysis of NASA's first 50 years. Over two days at NASA Headquarters, historians and policy analysts discussed NASA's role in aeronautics, human spaceflight, exploration, space science, life science, and Earth science, as well as crosscutting themes ranging from space access to international relations in space and NASA's interaction with the public. The speakers were asked to keep in mind the following questions: What are the lessons learned from the first 50 years? What is NASA's role in American culture and in the history of exploration and discovery? What if there had never been a NASA? Based on the past, does NASA have a future? The results of those papers, elaborated and fully referenced, are found in this 50th anniversary volume. The reader will find here, instantiated in the complex institution that is NASA, echoes of perennial themes elaborated in an earlier volume, *Critical Issues in the History of Spaceflight*.[2]

The conference culminated a year of celebrations, beginning with an October 2007 conference celebrating the 50th anniversary of the Space Age and including a lecture series, future forums, publications, a large presence at the Smithsonian Folklife Festival, and numerous activities at NASA's 10 Centers and venues around the country.[3] It took place as the Apollo 40th anniversaries

2. Steven J. Dick and Roger D. Launius, eds., *Critical Issues in the History of Spaceflight* (Washington, DC: NASA SP-2006-4702, 2006), pp. 7–35, available at *http://history.nasa.gov/SP-2006-4702/frontmatter.pdf*.
3. Steven J. Dick, ed., *Remembering the Space Age: Proceedings of the 50th Anniversary Conference* (Washington, DC: NASA SP-2008-4703, 2008), available at *http://history.nasa.gov/Remembering_Space_Age_A.pdf*, *http://history.nasa.gov/Remembering_Space_Age_B.pdf*, and *http://history.nasa.gov/Remembering_Space_Age_C.pdf*. Among other 50th anniversary publications are a book of iconic images, *America in Space: NASA's First Fifty Years*, ed. Steven J. Dick, Robert Jacobs, Constance Moore, Anthony M. Springer, and Bertram Ulrich, foreword by Neil Armstrong (New York, NY: Abrams, 2007); a 400-page 50th anniversary magazine, spearheaded by Edward Goldstein in the Office of Public Affairs, with reminiscences, articles, and thematic ads, *NASA: 50 Years of Exploration and Discovery* (Tampa, FL: Faircount Publishing, 2008), available at NASA's 50th anniversary Web site *http://www.nasa.gov/50th/50th_magazine/index.html*; and Steven J. Dick, Stephen J. Garber, and James J. Deutsch, "NASA: Fifty Years and Beyond" in *Smithsonian Folklife Festival 2008* (Washington, DC: Smithsonian

began, ironically still the most famous of NASA's achievements, even in the era of the Space Shuttle, International Space Station (ISS), and spacecraft like the Mars Exploration Rovers (MERs) and the Hubble Space Telescope. And it took place as NASA found itself at a major crossroads, for the first time in three decades transitioning, under Administrator Michael Griffin, from the Space Shuttle to a new Ares launch vehicle and Orion crew vehicle capable of returning humans to the Moon and proceeding to Mars in a program known as Constellation. The Space Shuttle, NASA's launch system since 1981, was scheduled to wind down in 2010, freeing up funds for the new Ares launch vehicle. But the latter, even if it moved forward at all deliberate speed, would not be ready until 2015, leaving the unsettling possibility that for at least five years the United States would be forced to use the Russian Soyuz launch vehicle and spacecraft as the sole access to the ISS in which the United States was the major partner.

The presidential elections a week after the conference presaged an imminent presidential transition, from the Republican administration of George W. Bush to (as it turned out) the Democratic presidency of Barack Obama, with all the uncertainties that such transitions imply for government programs. The uncertainties for NASA were even greater, as Michael Griffin departed with the outgoing administration and as the world found itself in an unprecedented global economic downturn, with the benefits of national space programs questioned more than ever before. There was no doubt that 50 years of the Space Age had altered humanity in numerous ways ranging from applications satellites to philosophical world views.[4] But NASA was still forced to justify its programs alongside all other federal agencies.

Throughout its 50 years, NASA has been fortunate to have a strong sense of history and a robust, independent, and objective history program to document its achievements and critically analyze its activities. Among its flagship publications are *Exploring the Unknown: Selected Documents in the History of the U.S. Civil Space Program*, of which seven of eight projected volumes

---

Institution, 2008), pp. 34–53. For a unique international perspective, written by representatives from each of the major space agencies, including NASA, see P. V. Manoranjan Rao, ed., *50 Years of Space: A Global Perspective* (Hyderabad, India: Universities Press, 2007), published on the occasion of the International Astronautical Congress held in Hyderabad near the 50th anniversary of the Space Age.

4. Steven J. Dick and Roger Launius, eds., *Societal Impact of Spaceflight* (Washington, DC: NASA SP-2007-4801, 2007), available at *http://history.nasa.gov/sp4801-part1.pdf* and *http://history.nasa.gov/sp4801-part2.pdf*.

were completed at the time of the 50th anniversary.[5] The reader can do no better than to turn to these volumes for an introduction to NASA history as seen through its primary documents. The list of NASA publications at the end of this volume is also a testimony to the tremendous amount of historical research that the NASA History Division has sponsored over the last 50 years, of which this is the latest volume.

Steven J. Dick
NASA Chief Historian
April 2009

5. John M. Logsdon, ed., *Exploring the Unknown: Selected Documents in the History of the U.S. Civil Space Program*, seven volumes (Washington, DC: NASA SP-4407, 1995–2008), available at *http://history.nasa.gov/series95.html*. On the NASA history program see Steven J. Dick, Stephen J. Garber, and Jane H. Odom, comps., *Research in NASA History*, Monographs in Aerospace History, No. 43, 3rd ed. (Washington, DC: NASA SP-2009-4543, 2009).

Chapter 1

NASA at 50

Michael D. Griffin

I want to thank NASA Chief Historian Steven Dick and his colleagues for organizing this conference, which I think is a very positive addition to our industry. It allows us to step back for a moment and view NASA and its contribution to society from a more strategic perspective. I believe that such a perspective, and the guidance it can provide in regard to our contribution to society, is our most pressing need as we embark on our next half century. It is too easy to become mired in the day-to-day tactics of budget defense or program execution, too easy to lose sight of the larger goal. A look back at history can provide the context to look forward at what we are doing and why. When I consider NASA and the nation's space program in this way, I am drawn again and again to the overriding need for constancy of purpose in our enterprise, if we are to obtain anything useful from it.

Of course, our purpose must be the right purpose! Prior to the loss of *Columbia*, NASA had a steady purpose for several decades. But I believed then, and believe now, that our space program was guided by the *wrong* purpose. We were doing the wrong things. We were limiting our horizon for human space exploration to low-Earth orbit, with nothing but indefinite promises of future programs without timing, funding, or programmatic content.

In the aftermath of *Columbia*, the Columbia Accident Investigation Board, and especially Chairman Hal Gehman and Professor John Logsdon, who is here with us today, recognized and called attention to this lapse. They recognized the need for an overarching strategic purpose for what we do, a guiding vision for the nation's civil space program. Responding to this need, President Bush put forth the Vision for Space Exploration, now the nation's civil space policy.

The goals of that policy were supported, indeed expanded, in two subsequent NASA authorization acts, the first by a Republican Congress in 2005, and the second by a Democratic Congress just this month. This strong bipartisan—actually nonpartisan—support for NASA and our nation's space program is very satisfying. From a policy perspective, in terms of having a clear statement of national purpose, I think that NASA has not been better positioned in decades. We have rational, cogent, well-balanced priorities for aeronautics, scientific discovery, and expansion of the human range of action and exploration, taking appropriately into account the layout and geography of the solar system. The policy also respects the nation's overall funding priorities, setting goals consistent with the amount of money that can be reasonably made available for civil space programs.

So we have a good policy. I'd like to see us maintain it. We at NASA cannot produce results acceptable to anyone—ourselves, the tax-paying public, our congressional and executive branch overseers, our international partners—if we churn our portfolio on a regular basis, determining anew after every congressional, or presidential, or senatorial election cycle what NASA's purpose is to be. If NASA is to be successful, the Agency must enjoy the stability associated with planning on decadal timescales. I hope that we can achieve that goal and maintain it in the future.

Turning to another subject, I am often asked (and especially so as my tenure comes to its probable end) what my major goals and accomplishments have been. I must leave any assessment of accomplishments, major or minor, to others. I hold firmly to the belief, endemic among credible technical professionals, that one cannot self-assess. That is why independent peer review is such an important part of the work of engineers, mathematicians, and scientists. You who are historians will have to judge my work.

But I can state what my goals were. When I was offered this job, we at NASA simply did not have technical and managerial credibility with the White House, Congress, or the public. Now, in my opinion some of that was unfair. There is always an overreaction to traumatic events, and none is more traumatic than losing a Space Shuttle and seven lives in full public view. But without regard to the mixture of substance versus perception, it is simply a fact that, three and a half years ago, NASA lacked the full measure of technical and managerial creditability that the nation expects of us and that we expect of ourselves.

So to restore that was my first priority, because nothing good can happen without it. After that, as I have stated publicly many times, I wanted to complete the safe return of the Space Shuttle to flight. That was a policy decision made by the President and supported by Congress, and it was stalling. It fell to me to oversee it. It was not in a good place when I joined the Agency, and we needed very rapidly to get it on track if it was to be done at all.

To fly the Shuttle safely, we needed to redevelop a management team, procedures, and methods for doing so. We then needed to do the same thing to retire it, to bring the program to an orderly and disciplined close. Anyone can stop flying the Shuttle; to do it in a disciplined and orderly manner is what NASA and the nation needed and still need. We are working hard on that task every day.

The purpose of returning the Space Shuttle to flight, as stated by the President and, again, supported by Congress, was to use it to finish the ISS. At this point the ISS represents a multidecade international commitment, as well as a commitment to our nation and ourselves. It is a commitment large enough in scale and scope that it was judged to be worth the risk of flying the Shuttle almost 20 more times to finish the job. Doing that job well, efficiently, and safely was my next highest priority after returning to flight.

If we are ever going to do anything in space beyond the ISS—and I began this speech by saying how important I thought that was—then it falls to us in this time to craft a credible human spaceflight architecture that can support operations in low-Earth orbit, as well as to take us back to the Moon and lay the groundwork for eventual voyages to Mars. To go to Mars, we will certainly need much more than is being developed today, but what we develop today should be designed with an eye toward Mars. In my view, we must create systems that enable a logical path to the establishment of a permanent base on the Moon, to a Mars mission, to voyages to the near-Earth asteroids, and to the servicing of large telescopes and other instruments at the Lagrange points, as well as other purposes we might not presently envision. We should work today with an eye toward becoming a permanently spacefaring nation, a permanently spacefaring society, to do things that build on what has been accomplished before.

In planning our next spaceflight architecture, I wanted to plan also for the incorporation of commercially supplied goods and services to the maximum extent possible. Again in my view, it is long past time to incorporate into our spaceflight activities the same policy framework that underpinned the development of aeronautics in the United States throughout most of the 20th century. There was extensive government sponsorship of aeronautics, and there was private development of aeronautics, and they fed each other quite synergistically. Looking back, it seems to me that few things were more central to the rise of the United States as a world power than the lead we forged in the development of aeronautics. It allowed us to project power and influence, commerce and culture, throughout the world in a fashion never seen before our time. The analogies to spaceflight are not, of course, exact. Spaceflight is not aeronautics. But I believe that there are analogies, and that we have not taken proper advantage of them as a matter of government policy. I wanted to do whatever I could to stimulate the commercial development of space as Administrator.

We must understand the proper relationship between governmental and commercial space endeavors. I see important roles for both. But with the history of space development coming about as it did, as a response to Cold War tensions, I think we had what I sometimes call an "excess of government." But actually, that's the wrong term. We certainly do not have an excess of government activity in space, but we do have an insufficiency of private enterprise.

I believe that a key role of government in the development of space is to define, occupy, and extend the frontier of human action and scientific discovery. That is an inherently governmental role; industry cannot make a profit doing it. It's not a productive area for free enterprise, yet. And yet, societies that do not define, occupy, and extend the frontier of human action and scientific discovery will inevitably wither and die. So in my opinion it is a public responsibility, one in which we share the risk as a society.

Now government activity is often inefficient, while properly regulated capitalism is one of the best mechanisms we have found to allocate the resources of a society efficiently. So, I think an important role of commercial enterprise in the development of the space frontier is to help meet government policy goals efficiently. Government's role should be to help bring about the development of space commerce by providing a stable market for service and stable requirements to be met by industry. If industry can meet those requirements, it will almost certainly do so more efficiently than can government. But industry cannot work in an environment where the market lacks stability over the development and sales life cycles of the products and the services they wish to furnish. It cannot be done. So we need a stable policy environment on the part of government in order to enable the kind of space commerce that I believe we would all like to see.

Similarly, international cooperation in spaceflight offers many advantages to the United States as well as to our partners. As a world power, there are things we must do that don't make other people happy, and yet we must do them. Leadership in great enterprises is a hallmark of a great nation, but leaders need allies and partners. We cannot function in the world if every hand is turned against us, or even if others are indifferent to us. So it behooves us to look proactively for things we can do with others to bind us together in common cause. And it is my observation that every society in the world, when it reaches the stage of technical maturity where it can begin to do something in space, does so. It is an arena which everyone seems to find uplifting, exciting, and appealing.

We live in a time, possibly the last time, when only the United States has the technical and financial wherewithal to provide the leadership of great activities in space. I wanted to capitalize on that fact and to take advantage of the hard-won partnerships that have been developed in the course of the Space Station program, where we were really learning how to do these large-scale enterprises in a manner that worked for everybody. I wanted to take that partnership forward to the Moon and to add new members to it. I wanted to keep faith with our partners on the ISS today and return with them to the Moon, establish a research base there, and eventually go on to the near-Earth asteroids and to Mars. Bringing together that collaboration was a major priority for me.

I wanted to do all this while maintaining the scientific excellence of our space science program today. I'm often asked why I've put so much emphasis into human spaceflight in my tenure as Administrator. And the answer has always been easy—I love everything we do, but when I showed up at the Agency our science program wasn't broken and our human spaceflight program was. I frankly didn't have enough hours in the day to do all that needed to be done, and I think most of our management team here could say the same. So, I spend my time and that of our management team where it is most needed. Now, our science program will always have important issues, and we need to work hard to keep it the best in the world. But it wasn't broken, and so I felt that I would do well if we could simply avoid creating collateral damage to our science program while trying to fix things that were damaged.

Finally, I wanted to restore the standing of NASA's aeronautic research program.

If these were the goals, then what have been the main difficulties in reaching them? The biggest of these arises from what I call "democracy in action." I think most of you know that I have spent a good deal of my career in the Department of Defense (DOD) space program, and there is a saying that I picked up from some of my military acquaintances. When frustrated by "the system," they will point out that we are here to protect democracy, not to practice it. That analogy is not completely applicable to a civilian organization such as NASA, but it conveys an important thought in a clever manner.

Winston Churchill noted that democracy was the worst form of government, except for all the others. I will add that in a democratic society there is an inherent tension between the undemocratic autocracy of expertise and the plain fact that the universe doesn't care about the niceties of the democratic process. Technical problems do not yield to majority opinion or produce results on schedules compatible with electoral cycles. Nature punishes technical mistakes, whether they are made democratically or not. It is important to be right.

It is very difficult to manage a large, visible government program efficiently, because far too many people claim the right to a voice in decisions in which they may admittedly have a stake, but for which they lack the expertise necessary to make a useful contribution. When industry is more efficient than government, it is not because it employs better people, but because decisions can be made; actions can be taken; results can be assessed; and corrections can be made, all without engaging anyone not needed for the task. There is personal authority, responsibility, and accountability in the system, all driven by the need to produce a profitable result in a competitive environment. When everyone has a voice, these things are diffused or lacking entirely.

These issues are compounded by any lack of clarity in regard to policy. What should the goals of the civil space program be? To expand the human range of action? To explore, to "go boldly where no one has gone before?" To do more science? To do more technology development? Or are the goals less noble, such as maintaining full employment at major centers? Or is the goal just "don't make waves," to avoid controversial things like retiring the Space Shuttle? Or is the goal even more ignoble—just see to it that whatever you are doing doesn't fail, doesn't make a mess?

Actually, all of these things, at one level or another, for one stakeholder or another, are Agency goals. None of them are entirely compatible, some are completely inconsistent with others, and in any case there is never enough money to accomplish them all. There is no single authority in government to prioritize them. The Administrator isn't allowed to do it—he can recommend, but he cannot act alone. Each of the various stakeholders expects his goal to be the one on top. It's a difficult environment in which to work.

It is always interesting to me that when a crisis looms—a war, the space race, a financial collapse—we nearly always decide to invest resources and authority in what we believe and hope will prove to be expert leadership. We judge the performance of these leaders on outcome, not process. President Lincoln replaced a lot of generals before he found his man, but he didn't deploy White House staff to the field, and he didn't give up on the idea that it took a general to run the army.

We somehow need to balance the tension between the autocracy of expertise and the need for transparent, democratic processes in government. It is very difficult. I think it is useful at times to remind ourselves that we live in a representative democracy, not a direct democracy, not a plebiscite. In a representative democracy such as the framers of our Constitution established, the people do not decide issues directly. The people decide who will decide. Now, through their delegated authority, it is NASA people who decide issues concerning the execution of civil space programs. I think you obtain the best results, the best compromise in the tension of which I spoke, when the leaders of the enterprise possess both demonstrated character and clear expertise.

Expertise without moral character is without value, and good intentions are no substitute for knowing how things work. We need both in the leadership of NASA. If we look at NASA and don't clearly see those traits at all levels, then we still have work to do. If there is not a general understanding that the people who are running the space agency know what they are doing, we get a lot of interference in the doing of it. We get more than enough of it even when the Agency's leadership is generally thought to be competent and objective!

Concern over risk is a perennial theme at NASA and among our stakeholders and can be a major impediment to achieving the goals we set. How much risk should be taken in the name of exploration? My view is that it should be considerably more than we're willing to accept today. It is interesting to note that when Captain Cook set sail on his first voyage to the South Seas, where all he did was discover Australia and New Zealand, he started out with 94 sailors. He was praised upon his return, three years later, for losing only 38 of them to the various hazards of the time: disease, accidents, and hostile action. That praise is easier to understand when one realizes that the first world-girdling voyage by Ferdinand Magellan started out with five ships and almost 300 sailors, yet only one ship and 18 sailors made it back to port. Magellan was not among them. By those standards, Cook did really well.

The current odds of not surviving a Mount Everest climb are just about 1 in 60. This is comparable to, but not as good as, our best estimate for the loss-of-crew risk is for the Space Shuttle. And yet I would venture to guess that the average citizen believes that flying in space is more dangerous than climbing Mount Everest. I haven't seen any public calls to limit the climbing of Mount Everest, and yet I see many people who are concerned about the risk of spaceflight. Why the difference?

Now in all candor, spaceflight is dangerous, and we work hard every day to make it safer. But a sense of perspective is necessary. I've often noted that there is a thousand years in time separating the first open ocean voyages by westerners, the Viking expeditions, from the pleasure cruises that depart Port Canaveral, a few miles from where our Space Shuttle crews lift off. When the

Vikings first set sail from Scandinavia, I doubt that anybody envisioned pleasure cruises as a future possibility.

While I doubt that those Viking expeditions were anywhere close to being as safe as flying on the Space Shuttle, we nevertheless have a long way to go in mastering spaceflight. A very, very long way. It is a risky enterprise and likely to remain so for centuries to come. It is not something for which everyone has a taste, nor should they. We fly volunteers. But we cannot—we simply cannot—define, occupy, and extend the human frontier while at the same time claiming that we can do it safely—not without badly misusing the word "safe."

Not terribly long ago I came across an aphorism concerning the settlement of the West: the pioneers were the ones with the arrows in the front.

So extending the frontier has never been a safe activity, and I think we are disingenuous if we claim that it will be. We should make it as safe as we can. We should try not to make the same mistake twice. I often say that our goal should at least be to make a new mistake. But when we are doing something which has not been done before, which we barely know how to do at all, which is just barely within the range of technical possibility, we should not be surprised when we sometimes fail. As tragic as it is, and as much as we want to prevent it, as much as we want to fix it so that the accident never happens again, we shouldn't be surprised.

I cannot leave the subject of risk, failure, and accidents without noting that there never has been any such thing as a smart failure. Every failure that we encounter looks stupid in hindsight. It is. It reflects something we didn't know, and would like to have known, and, by the time that the investigation is complete, feel that we ought to have known. So when we deal with failure by looking for the guilty parties, my usual suggestion is to start with a mirror. As Shakespeare put it, "The fault, dear Brutus, is not in our stars, but in ourselves . . . ."

Speaking of failure reminds me that it comes in many flavors. There are failures bigger even than the loss of a Space Shuttle and lessons to be learned from those as well. I'm fond of the comment by Santayana that those who are ignorant of history are doomed to repeat it. Regarding our own history, I have often said that the Saturn-Apollo transportation system seems to me to be unique in the history of successful transportation systems, in that we spent 80 percent of the budget of the Apollo program developing it, less than 20 percent of the budget using it, and then threw it away.

That seems to me to have been irrational. And yet the decision to terminate Apollo and all that went with it was made during the Nixon administration with very little, if any, public debate. I certainly don't recall much discussion;

if there was, it was lost in that surrounding the Vietnam War. But looking back, there is a lesson to be learned, and I think the lesson is that it is important to conserve the gains we make. To save what we've built, to adapt it, to reuse it, to take what works and shed what doesn't. But we must try very hard not to lose what we've built, because it comes at very high cost. We must not again throw away capabilities crafted at great expense in terms of money, time, and human skill.

I will close by commenting on another of the questions I am often asked when I represent NASA to those outside the Agency, and that is the question of our impact on society. Looking back across 50 years, I can identify any number of specific, easily defined contributions stemming from our nation's investment in space and NASA. But above these, I think, is a more important contribution. NASA is *the* entity which captures what Americans believe are the quintessential American qualities. Boldness and the will to use it to press beyond today's limits. Leadership in great ventures. Those things are better and more visibly combined at NASA than in any other enterprise in our society. I think that if we can hold true to our desire to continue to make that kind of impact, we will have done well.

Chapter 2

Inside NASA at 50

Howard E. McCurdy

Americans often present the image of Project Apollo and the other great accomplishments of the early years of spaceflight as prime examples of the achievements that a properly run government can produce. "If we can send a man to the Moon," one journalist observed in a commonly heard refrain, "why can't we clean up Chesapeake Bay?" When Al Gore challenged Americans "to commit to producing 100 percent of our electricity from renewable energy and truly carbon-free sources within 10 years," he invoked the memory of Project Apollo. In a work of dramatic fiction, the actor portraying Gene Kranz in the movie *Apollo 13* repudiates the pessimism of fellow flight controllers with the announcement that "this is gonna be our finest hour."[1]

The exaltation of NASA's early years has achieved an almost mythical quality. Such recollections mask the challenges NASA employees faced five decades ago. Those challenges were no less difficult than the ones confronting the Agency today.

NASA officials entered the Apollo age with an Agency ill-suited to the scale of the projects they were asked to complete. "NASA had considerable technical depth," said one of the people brought in to help organize Project Apollo,

1. Tom Horton, "On Environment: If America Could Send a Man to the Moon, Why Can't We . . . ?" *Baltimore Sun* (22 July 1984); Al Gore, "A Generational Challenge to Repower America," 17 July 2008, available at *http://blog.algore.com* (accessed 29 November 2008); Ed Harris as Gene Kranz, Ron Howard (producer), *Apollo 13* (Universal Pictures, 1995).

"but almost no program management experience."[2] Institutional difficulties were so severe on the Ranger project that NASA lost the first six spacecraft and Congress launched an investigation into "problems of management" at NASA Headquarters and JPL.[3] NASA executives reorganized the Agency thrice, reformed their program management practices, and brought in outside talent to show incumbent employees how to do their work.[4]

Engineers assigned to the Moon race, NASA's chief undertaking, worked hard to complete Project Apollo within a firm deadline and a fixed budget. In 1964, NASA leaders set the cost of the eight-year effort to land the first humans on the Moon at $19.5 billion. The figure was adjusted to $22.7 billion two years later, largely to account for the degree to which equipment manufactured to achieve President John F. Kennedy's mandate might be used for additional activities. The actual cost through Apollo 11 was $21.3 billion.[5] Part of the mythology of Project Apollo suggests that engineers compensated for the tight time schedule and high project risk by relaxing spending constraints. In fact, the project was both cost and schedule constrained and completed within the parameters established for both.

The eight-year effort did not begin well. Significant cost overruns and technical problems afflicted Project Gemini, the link between Project Mercury and the actual Moon landings. By 1963, Project Gemini was over budget and behind schedule. Officials at JSC (then the Manned Spacecraft Center) told NASA Associate Administrator Robert Seamans that the Titan II launch vehicle selected to power Project Gemini was unsafe for human flight. The paraglider landing system for the Gemini flight capsule did not work. Exacerbating these problems, Congress cut NASA's requested appropriation for fiscal year 1963 by 3 percent and President Kennedy rejected NASA's suggestion that the government provide a supplemental appropriation as a means of working through project difficulties.[6] If cost growth and the other difficulties had spilled over

---

2. Quoted from Howard E. McCurdy, *Inside NASA: High Technology and Organizational Change in the U.S. Space Program* (Baltimore, MD: Johns Hopkins University Press, 1993), p. 92.
3. R. Cargill Hall, *Lunar Impact: A History of Project Ranger* (Washington, DC: NASA SP-4210, 1977), p. 252.
4. Arnold S. Levine, *Managing NASA in the Apollo Era* (Washington, DC: NASA SP-4102, 1982). See also Robert L. Rosholt, *An Administrative History of NASA, 1958–1962* (Washington, DC: NASA SP-4101, 1966).
5. T. O. Paine letter to Clinton Anderson, with attachments, 21 November 1969, NASA Headquarters History Office historical archives (hereafter referred to as NASA History Office). See also Howard E. McCurdy, "The Cost of Space Flight," *Space Policy* 10, no. 4 (1994): 278–279.
6. Barton C. Hacker and James M. Grimwood, *On the Shoulders of Titans: A History of Project Gemini* (Washington, DC: NASA SP-4203, 1977), pp. 55–56, 105–116, 123–130, 173–175.

into the Apollo phase, it would have undercut much of the political commitment maintaining the Moon program. Concern with Project Gemini motivated much of the top-level effort at NASA Headquarters to reorganize the way in which the civil space agency did its work.

Spending on civil spaceflight activities peaked in 1965. With four years left before the Moon landing, NASA executives had to grapple with the first effects of a declining budget. In 1966, NASA Administrator James E. Webb asked President Lyndon Johnson for $5.3 billion for fiscal year 1967 and $6 billion in new obligational authority for fiscal year 1968. Webb warned Johnson that "there has not been a single important new space project started since you became President."[7] NASA's total appropriation, which had broken through the $5 billion level in 1964, fell below $4 billion by 1969. While funding for Apollo remained adequate, the Agency suffered budget cuts that affected many other flight programs, notably the hope for a vigorous post-Apollo agenda.

Technical difficulties distressed Project Apollo. The Apollo space capsule was clearly not ready for the human test flights NASA officials planned to conduct in late 1966.[8] The Saturn V rocket, later a symbol of technical invincibility, drew similar concerns. The first-stage F-1 engines vibrated, producing a form of oscillation like one might encounter while riding a giant pogo stick. The J-2 engines for the Saturn V rocket presented "the inevitable gaggle of problems" and "difficulties in manufacturing."[9] Concern with the capability of contractors to produce flight-ready equipment for the trip to the Moon deepened after the Apollo 204 fire that killed three astronauts in January 1967.

In hindsight, the Apollo era may have been NASA's finest hour, but to the people working on the civil space program at that time it was a period of successive challenges. Those challenges were in many ways analogous to the difficulties affecting the current effort to return to the Moon and explore beyond. Management difficulties, balky rockets, fixed program budgets, flat or declining total appropriations, cost overruns on individual programs, technical difficulties, dissatisfaction with contractors, and accidents and mission failures were as much a part of the Apollo legacy as the landings on the Moon.

---

7. Quoted in Robert Dallek, "Johnson, Project Apollo, and the Politics of Space Program Planning," in *Spaceflight and the Myth of Presidential Leadership*, ed. Roger D. Launius and Howard E. McCurdy (Urbana, IL: University of Illinois Press, 1997), p. 82.
8. Courtney G. Brooks, James M. Grimwood, and Loyd S. Swenson, *Chariots for Apollo: A History of Manned Lunar Spacecraft* (Washington, DC: NASA SP-4205, 1979), pp. 208–209.
9. Roger E. Bilstein, *Stages to Saturn: A Technological History of the Apollo/Saturn Launch Vehicles* (Washington, DC: NASA SP-4206, 1996), pp. 145, 149.

NASA employees and their contractors overcame these challenges during the Apollo era. What characterizes the early years is not the absence of difficulties, but the ability of people to surmount them. This belief that the people working on the civil space program could overcome those difficulties became part of NASA's organizational creed and American mythology, expressed in phrases such as the repetitive versions of "If we can go to the Moon . . . ."

Deep-seated beliefs such as these come to express what observers characterize as a particular organization's culture. Organization culture consists of the most widely recognized values, norms, beliefs, and practices that motivate employees within a specific institution. The culture often manifests itself in the form of assumptions that employees make about "what works here." In that sense, beliefs about what worked for NASA during the 1960s became part of the Agency's overall organizational culture. Large organizations like NASA—especially those with strong field centers—also possess subcultures. The subcultures may typify practices at particular installations, but the general culture consists of those values, norms, beliefs, and practices widely held throughout the entire organization.

In 1988, I surveyed NASA employees as part of a larger effort to identify the principal characteristics of NASA's organizational culture. The NASA History Division replicated that survey in 2006. Not surprising, the results of the second survey mirrored the results of the first. NASA's central beliefs and practices are well established and have not changed a great deal during its history, a conclusion supported by a number of NASA culture studies.[10]

A central element in NASA's overall culture is the belief in Agency exceptionalism. The belief appears in many ways: confidence that NASA is or was different from other government agencies, the conviction that it possesses great technical capability, the idea that it has received a special mandate, and the less praiseworthy assumption that NASA is a "perfect place."[11]

Beliefs that employees hold about their institution need not be true in order to be part of their culture. If a belief is pervasively held, it can influence behavior even if the belief misrepresents actual conditions. One frequently heard statement associated with the doctrine of NASA exceptionalism is the

---

10. See NASA, "NASA Culture Survey Results," NASA Shared Services Center, 2007; Diane Vaughan, *The Challenger Launch Decision: Risky Technology, Culture, and Deviance at NASA* (Chicago, IL: University of Chicago Press, 1997); McCurdy, *Inside NASA*.
11. Garry D. Brewer, "Perfect Places: NASA as an Idealized Institution," in *Space Policy Reconsidered*, ed. Radford Byerly (Boulder, CO: Westview Press, 1989).

belief that NASA recruits exceptional people. When it begins, a new organization like NASA in the years immediately after its founding may attract a disproportionate share of zealots and talented people. Over time, however, the average large organization by necessity is likely to employ a mix of employees who as a whole reflect the general characteristics of the population from which they are drawn. While a few people will be exceptional, the average organization will consist of average people. Very few institutions can consistently beat this tendency over long periods of time. Yet NASA employees consistently express the belief that they do. The belief is not confirmed by an examination of the social and educational backgrounds of the whole class of professional employees. Still, it is pervasively believed and forms the basis for some of the frequently observed behaviors of people within the organization.[12]

The belief in NASA exceptionalism in some ways mirrors the doctrine of American exceptionalism. The latter refers to the conviction that the United States, because of its historical circumstances and distinctive institutions, developed in ways not typical of other nations, especially those in Europe. In both NASA and America, generally, the doctrine is associated with the influence of the frontier, the presence of which is thought to encourage traits like innovation and equality of opportunity. Like the doctrine of American exceptionalism, the belief in NASA exceptionalism is somewhat controversial, but also influential in explaining how the people involved think about themselves.[13]

Elements of this characteristic grew out of NASA's predecessor organizations, most notably the NACA. One of the more curious manifestations of this legacy was the belief that the NACA's research Centers were not exactly part of the government. The NACA employees, including those who became part of NASA in 1958, understood that they received their paychecks from the U.S. Treasury. Yet they expressed the conviction that they were different from the average government employee. They were not federal bureaucrats, many said; they worked for the N-A-C-A. Many viewed the remainder of

12. See McCurdy, *Inside NASA*, pp. 50–60, 183; Peter Drucker, "Managing the Public Service Institution," *Public Interest* 33 (fall 1973): 259; Steven J. Dick, "2006 NASA Chief Historian Survey on NASA Culture," Office of the Chief Historian, NASA Headquarters, 2007.
13. See Seymour Martin Lipset, *American Exceptionalism: A Double-Edged Sword* (New York, NY: W. W. Norton & Company, 1997); Louis Hartz, *The Liberal Tradition in America: An Interpretation of American Political Thought Since the Revolution* (New York, NY: Harcourt, Brace, 1955); Frederick Jackson Turner, "The Significance of the Frontier in American History," in *Rereading Frederick Jackson Turner*, by John M. Farager (New York, NY: Henry Holt, 1994), pp. 31–60.

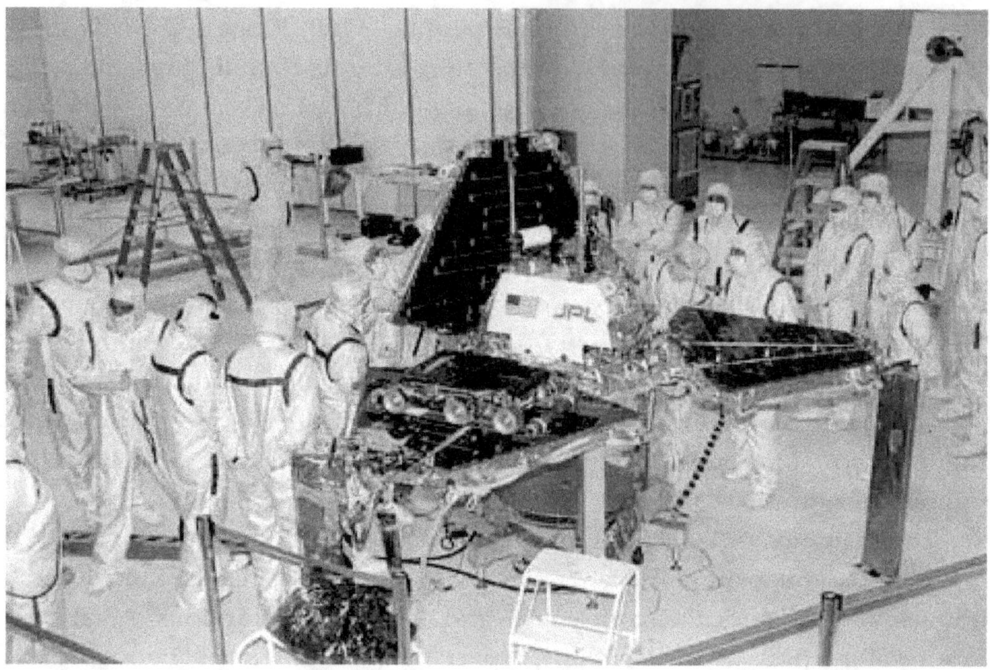

Figure 1: Strong in-house technical capability formed the basis for NASA's original organization culture. Here, Jet Propulsion Laboratory employees close the metal petals of the Pathfinder lander dispatched to Mars in 1996. The Sojourner small rover is visible on one of the three petals. *NASA Image 96PC-1130*

the federal bureaucracy with the same disdain that one might find among purely private citizens.[14]

The acceptance of NASA's exceptional nature grew out of a deeply held respect for the technical capability of employees recruited to work in the NACA, other predecessor organizations like the ABMA and JPL, and the early NASA. Similar convictions can be found in federal organizations such as the National Institutes of Health (NIH), the Centers for Disease Control and Prevention (CDC), and the Federal Bureau of Investigation (FBI), whose workers complete technical tasks within a framework that accords them more independence than the typical government employee.

The source of NASA's early technical capability rested with the reputation of founding groups such as the Space Task Group from the Langley Research Center that directed Project Mercury and the 125-person German rocket team. Speaking of the ABMA and its German rocketeers, one leader of the Apollo

14. See James R. Hansen, *Engineer in Charge: A History of the Langley Aeronautical Laboratory, 1917–1958* (Washington, DC: NASA SP-4302, 1987).

program stated that "their mind set was largely, do it yourself and build it in-house. To do major contracting . . . was foreign to their experience."[15] Comparing NASA's experience with that of the U.S. Air Force, one observer explained that the Air Force relied upon private contractors "because it had neither the depth of competence found in Army laboratories nor the time to recruit engineers."[16] The ABMA and the NACA were different. Their engineering corps and "hands on" approach instilled a level of technical confidence that encouraged early NASA employees to believe that they could do their own work when necessary and exercise strong contractor oversight where required. The relationship between NASA officials and contract workers was much like that of a professor and a student, one of the members of the German rocket team explained. The relationship led to a tradition known as "contractor penetration" in which NASA officials stationed Agency representatives at contractor plants and supervised contractor work to a high degree.

A number of supporting practices amplified the overall faith in the Agency's technical capability. Agency employees placed a great deal of emphasis on research and testing. Defending the extensive testing on the J-2 engine, an MSFC engineer explained that "you would never know for sure [how components] would work until you put them together in the engine."[17] Agency employees defended a culture in which engineers and scientists took risks and learned from the failures that inevitably occurred. Employees embraced the traditions associated with a research and development (R&D) organization in which missions achieved encouraged new challenges. The latter resulted in an innovation mentality favoring the creation of new missions over the repetition of old ones as well as experimentation with new technologies.

The latter point deserves special attention. Surveys and interviews generally confirm the emphasis that NASA employees place on doing new things. So does the Agency's history. "If you want to make progress," said one of NASA's top engineers, "you've got to design things that have not been done before." Yet significant pockets of incrementalism exist with NASA. Incrementalism relies on small, gradual improvements in existing technologies and the use of off-the-shelf components. It means tinkering with the old rather than inventing the new. Change occurs, but it builds upon an established base. The von Braun rocket team, observed one NASA leader, utilized an incremental approach to

---

15. McCurdy, *Inside NASA*, p. 36.
16. Levine, *Managing NASA in the Apollo Era*, p. 70.
17. Quoted from Bilstein, *Stages to Saturn*, p. 150.

Figure 2: Project Apollo joined NASA's technical competence with large-scale systems management, an organizing method imported from the U.S. Air Force. Lieutenant General Samuel Phillips and George Mueller (second and third from the right), who helped install the method, celebrate the launch of Apollo 11 with Wernher von Braun (with binoculars) and other NASA officials. *NASA Image 108-KSC-69P-641*

the improvement of launch vehicles. "They were the world's greatest incrementalists that I have ever seen."[18] The security provided by incremental change existing alongside the desire to undertake new challenges creates a level of tension that is never fully resolved.

These beliefs and practices—the elementary components of the original NASA culture—are confirmed by the words of Agency employees and their attitudes as recorded on various surveys. The beliefs were counterbalanced during the Apollo era by the introduction of management methods from outside the Agency. The development of large-scale systems management, an innovation imported from the U.S. Air Force, has been appropriately characterized by historian Stephen B. Johnson as the "secret of Apollo."[19]

---

18. McCurdy, *Inside NASA*, pp. 76–77.
19. Stephen B. Johnson, *The Secret of Apollo: Systems Management in American and European Space Programs* (Baltimore, MD: Johns Hopkins University Press, 2006).

NASA scientists and engineers schooled in the traditions of their own technical superiority did not readily accept the importation of Air Force management techniques. It would be wrong to say that the great mass of Agency employees willingly looked outside for help or accepted the management reforms when they were introduced. Midlevel officials often resisted the presence of the four dozen Air Force lead managers brought in to reform NASA and what NASA employees viewed as the intrusion of central control. The introduction of large-scale systems management, nonetheless, gave top NASA executives the tools they needed to impose cost, schedule, and configuration discipline over a number of activities that were less tightly organized during the early years. Faith in the exceptional nature of their work, belief in their technical capability, the tradition of in-house activity, and the strength of management systems imported into the Agency characterized NASA's early organizational experience. Such features helped NASA employees attain the capabilities they needed to complete the Moon landings and the other great programs of the formative years. Their ability to do so was expedited by a series of special conditions that helped Agency employees accomplish the tasks they undertook, adroitly summarized by W. Henry Lambright in his writings on the Apollo years.[20] NASA enjoyed an unusually high level of political support. The civil space program was a national priority and received attention from legislators and high-ranking executives, notably the President. When political support for the civil space program waned, especially after the Apollo fire, members of the congressional space committees and President Lyndon Johnson protected the space program. Under James E. Webb and a corps of top NASA executives, the Agency benefited from nearly eight years of stable administrative leadership under an Administrator with access to the highest political circles. The newly created Agency went through a "honeymoon" period during which its budget and workforce expanded rapidly. Agency employees benefited from a national goal that was simple to understand and perceived to be urgent when undertaken. When Richard Nixon became President in 1969, the technical and administrative momentum underlying the Moon program was too strong—and too close to completion—to be undone.

The cumulative force of these conditions produced a climate in which NASA employees enjoyed an unusually high degree of discretion in carrying

---

20. W. Henry Lambright, "Apollo: Critical Factors in Success and Implications for Climate Change" (a paper delivered at the Solutions Summit for Climate Change, Nashville, TN, 14 May 2008). See also Lambright, *Powering Apollo: James E. Webb of NASA* (Baltimore, MD: Johns Hopkins University Press, 1995).

out their activities. Politicians deferred to the technical judgment of NASA employees. The latter believed that they had received a firm presidential mandate and sufficient resources to accomplish it. (Knowledge of the actual budget battles might have shaken that faith.) The first NASA officials carried out their work during an era when the public and most government officials respected scientific and technical judgment. Together, these forces created an environment in which technical criteria superseded political ones and Agency officials had far greater control over the content of their work than they would have in the years to come.

Inevitably, these features changed. The characteristics typifying the early years of any new administrative agency undergo transformation as the agency matures. Historians and social scientists have identified the most common alterations, and NASA was not immune from them. As experience with recurring situations accumulates, so do the number of formal administrative procedures. Rules proliferate and bureaucracy expands, a tendency famously noted by C. Northcote Parkinson in the law that bears his name. Maturing government agencies tend to enter a period of declining flexibility and increasing conservativism, a phenomenon expertly explained by Anthony Down in his classic analysis on the life cycle of bureaus. Such transitions are precipitated by the lessening emphasis given to the agency's mission and the commensurate decline in the resources provided to it. Declining attention and shrinking resources favor the position of leaders who know how to conserve what the agency already has. Other issues ascend to the top of the governmental political agenda. Older agencies remain in place, but their work becomes less important, and they tend to be less flexible and innovative than they were in the period immediately following their creation.[21]

The location of a maturing agency's political base often shifts. A newly created agency can count on the political support of White House and Congressional leaders for a short period of time. As the agency matures, its early benefactors focus their attention on other activities or concerns. Over time, newly established government agencies that once benefited from strong presidential or legislative leadership come to rely more heavily upon constituency groups for support. The process, famously noted by sociologist Philip Selznick in his history of the Tennessee Valley Authority, frequently results in "co-optation," a process in which the needs of the clientele become more important for solidifying

21. C. Northcote Parkinson, *Parkinson's Law, and Other Studies in Administration* (New York, NY: Ballantine Books, 1957); Anthony Downs, *Inside Bureaucracy* (Boston, MA: Little, Brown and Co., 1967).

political support and guiding agency policy. Where constituency groups grow attached to their existing rights and privileges, the result is often a form of "interest group liberalism" in which constituent groups now dependent upon government grants and contracts for their livelihood press to maintain existing advantages. Such behavior further amplifies the tendency of policy-makers in aging government agencies to conserve ongoing programs and resist change.[22]

To encourage change within sluggish administrative agencies, executives and legislators often promote broad administrative reforms. In the latter part of the 20th century, government leaders worldwide embraced what became known as the "new public management." The approach relied upon market forces to alter administrative behavior, propelled by the assumption that the incentives arising from the presence of multiple providers competing to provide government goods and services would enhance performance and reduce costs. For NASA, the most dramatic effects of the new public management emerged from its emphasis upon cost reduction and contracting out.

The full cycle of forces outlined above affected NASA and its workforce as the Agency matured. Bureaucracy and the procedures associated with it grew, a tendency detected in the 1988 survey of NASA employees and reconfirmed in the 2006 History Division survey that replicated it. Only 11 percent of the employees responding to the 1988 survey agreed with the statement that "it is relatively easy to cut through the bureaucracy and get things done within NASA today." Eighty-one percent of the respondents present during the early years agreed that "the amount of paperwork has increased substantially since I came to work for NASA." Workforce statistics also confirm the impression of expanding bureaucracy. The proportion of NASA employees classified as professional administrators increased. The proportion grew when the workforce expanded and grew when the total workforce declined. Parkinson's law suggests that an agency's administrative workforce will proliferate regardless of the amount of work to be done.[23]

In a pattern suggested by Downs in his general description of the life cycle of bureaus, dwindling budgets and declining priorities within NASA favored the position of people who were more cautious and careful in their decisions.

---

22. See Philip Selznick, *TVA and the Grass Roots* (Berkeley, CA: University of California Press, 1949); Theodore J. Lowi, *The End of Liberalism* (New York, NY: Norton, 1969).
23. As reported in McCurdy, *Inside NASA*, pp. 115–116, 137, 179–180. The author conducted a survey of 800 NASA professional employees (engineers, scientists, and professional administrators), randomly selected, in the summer of 1988, with a response rate of 88 percent. See also Dick, "2006 NASA Chief Historian Survey on NASA Culture."

NASA professional employees responding to the 1988 survey uniformly agreed that "risk and failure are a normal part of the business of developing new technologies" but, by a three-to-one margin, agreed with the statement that "at the management level, NASA is dominated by people who are cautious and inclined to avoid risks." Sixty-six percent of long-term Agency employees agreed that such cautious behavior had "increased since I joined the Agency." A commonly referenced turning point was the Apollo 204 fire in 1967, which also coincided with the decline of Agency budgets. "There was a very sharp increase in the bureaucracy immediately following the 204 file," one NASA scientist reported. "After that," he continued, "there was a gradual attrition of the competent risk-takers and a growth of conservatives." Concerns about the diminishing influence of professional judgment and the growing weight allotted the needs of Agency managers appeared frequently after the *Challenger* accident.[24]

NASA's political base shifted. At the commencement of the Space Age, NASA officials drew heavily on national mandates created by political figures like Lyndon Johnson and John F. Kennedy. A pervasive belief arose within the space community that a presidential mandate announced with a major presidential address formed a necessary and sufficient condition for the conduct of any large space initiative. In practice, Johnson and Kennedy did not attend to civil space issues once their initial priorities became established. Interviewed by Walter Cronkite at the point of his retirement from politics, Johnson admitted that he "spent more time in the space field" before 1963 than he did after he became President.[25] Loss of executive attention naturally elevates the importance of groups whose interests are more permanent. In NASA's case, that means the contractors who carry out the bulk of the Agency's work. A statistical analysis of congressional voting behavior completed in 1991 revealed contract distribution to be the second most important factor in predicting legislative support for the civil space program, exceeded only by the ideological disposition of individual members. The influence of presidential support on congressional votes was small.[26]

---

24. McCurdy, *Inside NASA*, pp. 122–123, 176, 180–181; Barbara S. Romzack and Melvin Dubnick, "Accountability in the Public Sector: Lessons From the Challenger Disaster," *Public Administration Review* 47 (May/June 1987): 227–238.
25. Quoted in Dallek, "Johnson, Project Apollo, and the Politics of Space Program Planning," p. 80.
26. John Low, "Economic Benefit, Ideology, and NASA Voting in the U.S. Senate," American University, 1991 (unpublished paper supplemented by an analysis of U.S. House of Representatives voting patterns). See also Launius and McCurdy, "Epilogue: Beyond NASA Exceptionalism," in *Spaceflight and the Myth of Presidential Leadership*, ed. Launius and McCurdy.

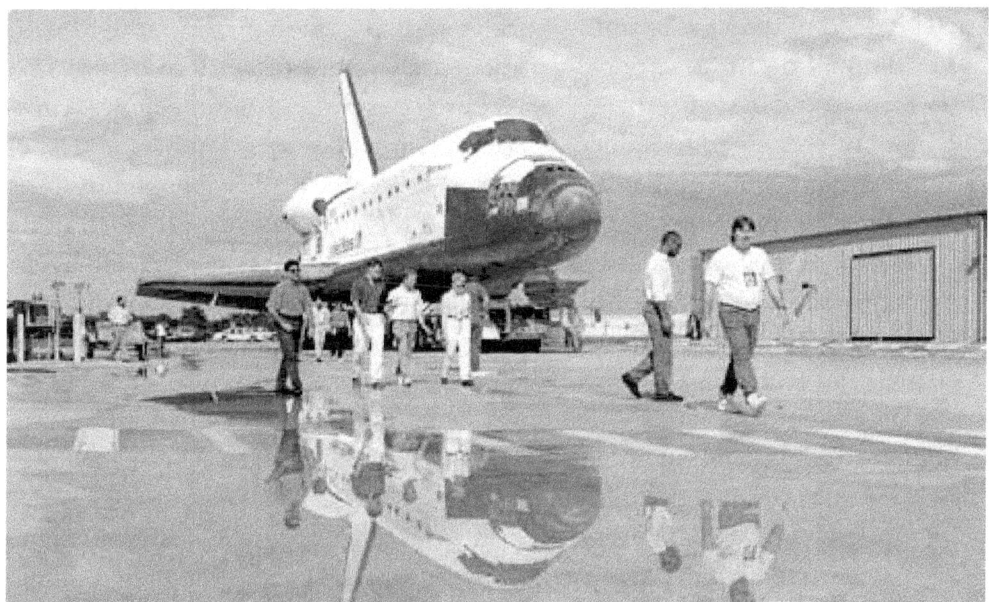

Figure 3: NASA executives sought to maintain sufficient technical strength to effectively supervise contractors. As the space program matured, NASA employees worried that the Agency had transferred too much of its technical competence to outside companies, such as those that worked to launch the Space Shuttle. Here, operations, security, and other personnel escort the orbiter Discovery from the Orbiter Processing Facility to the Vehicle Assembly Building (VAB) for the STS-64 mission. *NASA Image 94PC-0982*

Contractor responsibility grew with the maturing space agency. The first two Administrators, T. Keith Glennan and James Webb, set NASA on a course under which the use of contractors became the preferred method for developing spacecraft and rocketry. The number of contract employees grew from 3 to 10 for every NASA civil servant, and the Agency contracted out nearly 90 percent of its allotted funds. Anticipation of large cost savings, in conjunction with the belief that the Space Shuttle had become "a mature and reliable system," encouraged NASA executives to transfer most Space Shuttle launch service activities to United Space Alliance, an aerospace conglomerate. Political leaders pointed to NASA as an example of how the contractor-agency relationship in the new public management could work. Even if NASA executives had wanted to reduce their reliance upon contractors during the last decade of the 20th century, the existing political situation would not have allowed it.[27]

27. See McCurdy, *Inside NASA*, pp. 134–139; Columbia Accident Investigation Board (Admiral Hal Gehman), *Report*, vol. 1, August 2003, pp. 105–110; E. S. Savas, *Privatizing the Public Sector: How to Shrink Government* (Chatham, NJ: Chatham House, 1982).

The need to direct substantial work to aerospace contractors clashed with the traditional desire to conduct an adequate amount of work in-house. The latter formed the basis for the Agency's much desired in-house technical capability and often helped to reduce risk on missions requiring a substantial level of innovation. Responding to the 1988 survey, NASA professional employees agreed by a margin of six to one that "NASA has turned over too much of its basic engineering and science work to contractors," a ratio that remained essentially the same in 2006. Roughly two-thirds of Agency professional employees wanted the opportunity to do more hands-on work, a proportion that has also remained stable across that time.

Attitudes and factors such as these characterize NASA's overall culture. Based on studies spanning the period between 1988 and 2006, these attributes may be summarized in the following ways:

- NASA professional employees still view the Agency as a special place. They think that the Agency recruits exceptional people and they continue to believe by a ratio of two to one that "NASA has just as much technical capability as its contractors."
- Concurrently, Agency employees express concern that the Agency has turned over too much work to contractors and that the lack of hands-on activity erodes the Agency's technical capability. Significantly, four out of five respondents reported in 2006 that they spent most of their time "working at a desk in an office" without much contact with flight hardware, astronauts, or laboratory facilities. That proportion increased between 1988 and 2006.
- Employees worry about the dominance of administrative procedures, the burden of bureaucracy, and the effect this has on their work.
- By a two-to-one margin, employees are still "very optimistic about NASA's future." They believe that people within the Agency communicate with each other more. They are optimistic about the effect of changes "that I have seen take place within NASA"; that optimism has increased. Movement regarding such beliefs is not large, but it is detectable and possibly the consequence of a number of culture-related activities designed to enhance communications and reform procedures after the 1986 and 2003 Shuttle accidents.[28]

---

28. McCurdy, *Inside NASA*, appendix; Dick, "2006 NASA Chief Historian Survey on NASA Culture."

The persistence of core beliefs during the nearly 20-year period marked by the two surveys illustrates an often observed characteristic of organizational cultures. They tend to be remarkably resilient. Their characteristics change slowly, if at all. Once an organization acquires a particular method of conducting its business that works well, employees continue to use the method even in the presence of efforts to reform. Under such conditions, organizational change is more likely to occur through replacement—the death of one organization followed by the birth of another—than through fundamental change from within.

Persistence can be a source of both opportunity and misfortune. Opportunity flows from the commitment to practices that work, even when those practices might be unpopular with reformers who want to tinker with success. Misfortune can arise when employees come to believe that their organization is a "perfect place," in no need of change even in the presence of shifting circumstances. Garry Brewer warned of the latter in his much-cited article on NASA as a perfect place:

> What do CBS, General Motors, the Bank of America, the U.S. Naval Academy, Yale University, and the horse cavalry have in common with NASA? Each at some time or another came close to being the best organization human beings could create to accomplish selected goals.[29]

Employees in such organizations (Brewer wrote his article in the aftermath of the *Challenger* accident) ignore signals from their surrounding environments telling the employees that the nature of the forces with which they have worked have changed. Simply put, their institutions fail to adapt. Says Brewer, "success reinforces lessons that eventually become obsolete or even harmful."[30]

The institutional challenges faced by NASA employees 50 years after the Agency's founding are similar to those faced by the first employees. Agency missions are hard to organize; cost, schedule, and performance are still difficult to control. Some of the lessons learned—the historic practices developed to respond to those issues—are still applicable. Some of the surrounding circumstances have changed.

Agency employees believe they know how to maintain NASA's technical edge in the presence of bureaucratic procedures, administrative requirements,

29. Gary Brewer, "Perfect Places," p. 158.
30. Ibid.

and accumulating pressures to contract out the bulk of their work. They rely upon in-house technical capability. The practice works well for NASA and the projects it undertakes. It is a central feature of NASA's original culture and one that distinguishes it among government institutions. "The basic policy was that NASA would do enough missions in-house so that it always would have the technical capability in-house to judge a contractor accurately."[31] A commonly stated rule of thumb was 10 percent—the minimum amount of work to be done in-house in order to maintain technical capability. Projects that involved new technologies or untested techniques were prime candidates for in-house development.

Historically, in-house capability has been balanced by a willingness to look outside of the Agency for management techniques that maintain cost, schedule, and performance discipline. During the Apollo era, NASA executives turned to the U.S. Air Force for needed reforms, adopting large-scale systems management as their guiding managerial philosophy. The urgency of mission objectives prompted Agency employees to accept the management reforms—albeit begrudgingly—in spite of the preexisting sense that the Agency already possessed great technical capability.

The cumulative effect of these practices manifested itself as a mindset toward contractors. From this perspective, the job of the contractor was to deliver the products that the government desired—a classic principal/agent relationship. The job of the contractor was to follow instructions, not to modify them.[32] The mindset flowed from the presumption that NASA possessed great technical capability relative to its contractors and other outside groups. This central element in NASA's overall culture persists today. Reported one analyst during the investigation of the Space Shuttle *Columbia* accident: "I have heard repeatedly from several NASA engineers in the months since the accident that NASA is the only organization that flies shuttles and no one else has the expertise to tell it what to do."[33]

At the beginning of the Space Age, many believed that NASA employed the most talented spacecraft engineers and rocket scientists in the world. The facts are hard to judge, but the presence of the belief is incontrovertible.

---

31. Quoted from McCurdy, *Inside NASA*, p. 40.
32. See McCurdy, *Inside NASA*, pp. 40–42. See also Hall, *Lunar Impact*, pp. 253–254.
33. Roger D. Launius, "After Columbia: How We Got Into This Fix and How We Can Get Out of It," 2003 MAPLD International Conference, Washington, DC, 9–11 September 2003, available at *http://klabs.org/richcontent/MAPLDCon03/abstracts/launius_a.htm* (accessed 24 October 2008).

People in NASA believed that they employed the best and brightest in the field. Today, the facts are more apparent. NASA carries out its work within a setting or environment in which technical capabilities for space activities are widely dispersed. Technical capability can be found within the traditional aerospace industry, the commercial space sector, the space entrepreneurial community, research institutes, university laboratories, other nations, other government agencies—and NASA. With respect to NASA's overall culture, this is the most profound way in which the Agency's environment has changed.

As their predecessors did in the first decade of spaceflight, NASA executives recently looked outside their own institution for organizational models that might help them adapt to this new reality. (The desire among top executives to look, it should be said, did not constitute evidence of a widespread desire among Agency employees to accept those models.) Outside experts and Agency executives inspected lessons contained in so-called high-reliability organizations (HROs) and small-scale project teams. Members of the Columbia Accident Investigation Board urged examination of HROs and suggested that NASA executives might look to the U.S. nuclear navy as a possible model.[34] NASA's experimentation with low-cost innovation during the 1990s prompted a revival of organizational techniques associated with the Kelly Johnson "Skunk Works" approach to project management. This model of "too small" project teams also found support in the government's Strategic Defense Initiative ("Star Wars") and a number of entrepreneurial efforts to engage in private-sector spaceflight activities.

The experience with NASA's low-cost initiatives is quite instructive in this regard. During the fifth decade of spaceflight, NASA officials launched (or attempted to launch) a number of low-cost projects including those characterized as "faster, better, cheaper."[35] To reduce cost, the undertakings utilized relatively small project teams often colocated at a single place with extensive responsibilities for assembling, testing, and flying spacecraft. The resulting management style depended upon the potential for frequent communication and intense teamwork found in small, colocated project teams. Frequent face-to-face communication supplemented the use of established systems management techniques. The results were less than satisfactory. A large number of projects failed. Strikingly, low-cost projects developed in-house tended to succeed at a much higher rate than those contracted out. In a study of 23 low-cost projects, about one-third

34. Columbia Accident Investigation Board, *Report*, vol. 1, pp. 182–184.
35. Howard E. McCurdy, *Analysis of Twenty-Three Low Cost Projects 1992–2005*, work in progress, 2008.

of the total number of projects failed to achieve their objectives. None of the projects carried out in-house failed; nearly half of the contracted projects did.

On the surface, this experience seems to reinforce the importance traditionally assigned within NASA's original culture to the maintenance of in-house activity and control of contractors. Closer inspection of the projects reveals more subtle forces at work, ones that respond to the new reality of more dispersed technical capability. Where government officials delegated projects to contractors that utilized NASA in-house practices, the projects did quite well. For example, NASA officials assigned the NEAR-Shoemaker project to the Applied Physics Laboratory (APL) in Laurel, Maryland. Though not part of NASA, APL has a long history of satellite development and a strong culture arising from it. The laboratory conducts its in-house projects much as a NASA Field Center would. NASA officials gave APL employees a great deal of discretion. The latter designed, assembled, and flew the spacecraft in-house. NEAR-Shoemaker exceeded its mission objectives, studying the asteroid Eros from a variety of orbits and eventually landing on it.[36]

In an era of dispersed talent, a low-cost, in-house NASA project has a high probability of success. So does one conducted by a contractor in the manner of an in-house project. For the remaining set of projects, such conditions did not prevail. These contracted projects were not conducted like in-house NASA projects, they did not receive the traditionally close NASA supervision (in part a consequence of the desire to reduce their cost), and they showed a much higher tendency to fail. The projects suffered from a variety of deficiencies. For some, project responsibility was divided between contractors who, for example, might assemble the spacecraft and NASA employees who would attempt to fly it. One such division of responsibility led to the loss of the Mars Climate Orbiter, a classic example of the communication problems likely to appear in projects without a single, unified center of program integration. NASA executives allocated less money to these projects relative to mission complexity than they did to projects completed in-house. The difference likely arose as a consequence of the belief that contracted projects should cost less than similar missions completed by government employees. Finally, some of the contractors lacked experience or what might be characterized as a strong space exploration culture. Together, these factors afflicted their projects to a

36. Howard E. McCurdy, *Low-Cost Innovation in Spaceflight: The History of the Near Earth Asteroid Rendezvous (NEAR) Shoemaker Mission*, Monographs in Aerospace History, No. 36 (Washington, DC: NASA SP-2005-4536, 2005).

much higher degree than the ones done in-house or the ones completed by contractors using in-house practices.

The most important practices supporting mission success were those associated with the traditional NASA culture—a well-integrated center of project authority maintained over the course of the whole mission, adequate funding relative to mission complexity, and a strong culture of technical competence associated with spaceflight. Where those conditions prevailed, it did not seem to matter whether the project was conducted by a NASA Field Center, a contractor organization, or a private entrepreneur. The practice of contracting out influenced the outcome of studied projects much less than the quality of the practices utilized.

To summarize, NASA employees pioneered the use of practices that allowed the completion of very complex space exploration missions. Those practices and the commensurate talent have now spread to the private and nonprofit aerospace sectors. Fifty years of civil spaceflight history suggests that NASA works best when it combines a strong technical culture with a continuing search for organizational methods that fit the challenges posed by the missions it undertakes. For NASA employees, maintaining that internal capability in the presence of growing bureaucracy, administrative proliferation, and the ever-present pressures to contract out has been a continuing struggle. The most pervasive method for maintaining technical capability has been the practice of completing an established proportion of missions with in-house teams. Practices matter. The tendency to divide or weaken centers of integration, weaken the technical culture, or provide margins that are insufficient relative to project complexity create institutional deficiencies that are difficult to surmount. These lessons are especially relevant where NASA employees and government contractors share project responsibilities. The tendency to weaken established practices through excessive decentralization and inappropriate contracting afflicted the human spaceflight program once Project Apollo wound down and contributed to a succession of mission failures on low-cost projects as the 20th century closed.[37] Although the underlying practices are well known, the Agency and its employees seem obliged to constantly relearn them as they enter the next 50 years.

---

37. McCurdy, *Inside NASA*, pp. 130–131; S. C. Phillips, "Summary Report of the NASA Management Study Group: Recommendations to the Administrator," 30 December 1986, Historical Archives, NASA History Office, Washington, DC; McCurdy, *Faster, Better, Cheaper: Low-Cost Innovation in the U.S. Space Program* (Baltimore, MD: Johns Hopkins University Press, 2001).

Chapter 3

Imagining an Aerospace Agency in the Atomic Age

Robert R. MacGregor

Much has been written about the 184-pound satellite lofted into the heavens by the Soviet Union on 4 October 1957. The story is an insidiously seductive one; it is the romantic narrative of a small metal ball usurping the assumed technological authority of the United States. The frenzy of the media and the swift political backlash seem almost comical in light of the diminutive physical size of Sputnik.

The launch of Sputnik was one of the most disruptive singular events in the history of the United States.[1] The temptation to label it a discontinuity is strong. The year following the Sputnik launch saw the formation of the Advanced Research Projects Agency (ARPA), the creation of the new post of Special Assistant for Science and Technology to the President and its associated President's Scientific Advisory Committee (PSAC), the transformation of the NACA into NASA, and the National Defense Education Act. Walter A. McDougall in . . . *The Heavens and the Earth: A Political History of the Space Age* traces the roots of technocracy in America to this "spark":

> Western governments came to embrace the model of state-supported, perpetual technological revolution . . . . What had intervened to spark this saltation was Sputnik and the space technological revolution . . . . For in these years the fundamental relationship between the government and the new technology changed as

1. For a good overview of the Western reaction to Sputnik, see Rip Bulkeley, *The Sputniks Crisis and Early United States Space Policy: A Critique of the Historiography of Space* (London, U.K.: MacMillan Academic and Professional Ltd., 1991).

> never before in history. No longer did state and society react to new tools and methods, adjusting, regulating, or encouraging their spontaneous development. Rather, states took upon themselves the primary responsibility for generating new technology.[2]

McDougall has since revised his original argument by noting that the space technological revolution was an "ephemeral episode in the larger history of the Cold War, rather than the Cold War having been an episode in the larger story of the march of technocracy."[3] This revisionism addresses the eventual fate of the space technological revolution. It is the purpose of the current essay to revise the story of the birth of that technological revolution. Specifically, it will be argued that the conception of the Sputnik launch as a discontinuity that ushered in a technocratic revolution in modern America does not fit the historical record. The environment in which the Sputnik crisis unfolded in the United States was already saturated with preconceived, technocratic notions of the relation of science to the state. The crystallization of the new agency that would become NASA was a process that simultaneously was instigated by a singular event and followed in the footsteps of institutional ancestors. The two are not mutually exclusive; contingency must be embedded in a framework of continuity. The precursor of the space technological revolution was the Atomic Energy Commission (AEC).

"Technocracy" is a contentious term, with definitions running the gamut from a literal etymological interpretation as "the control of society or industry by technical experts"[4] to the idolization of science for propaganda purposes by nonscientific bureaucrats.[5] An attempt at a precise definition is necessarily doomed to failure, but for the purposes of this essay I will adopt McDougall's definition of technocracy as "the institutionalization of technological change for state purposes, that is, the state-funded and -managed R&D explosion of our time."[6] McDougall's definition captures the key features relevant to the

---

2. Walter A. McDougall, . . . *The Heavens and the Earth: A Political History of the Space Age* (Baltimore, MD: Johns Hopkins University Press, 1985), pp. 6–7.
3. Walter A. McDougall, "Was Sputnik Really a Saltation?" in *Reconsidering Sputnik: Forty Years Since the Soviet Satellite*, ed. Roger D. Launius, John M. Logsdon, and Robert W. Smith (Amsterdam, Netherlands: Harwood Academic Publishers, 2000), p. xviii.
4. *The Oxford English Dictionary* (New York, NY: Oxford University Press, 1989).
5. A famous example in space history is Nikita Khrushchev's shrewd tactical use of spaceflight for internal and external political maneuvering. For an overview of Khrushchev's manipulation of the space program, see Asif Siddiqi, *Sputnik and the Soviet Space Challenge* (Gainesville, FL: University Press of Florida, 2003), esp. pp. 409–460.
6. McDougall, . . . *The Heavens and the Earth*, p. 5.

current analysis: massive state funding and intentional control of technological development to serve state purposes. There exist a myriad of other possible definitions, which remain outside the scope of the present argument.[7]

The AEC and NASA are far and away the canonical American institutional examples of technocracy under this definition. The similarities on the surface are obvious. Both the AEC and NASA were characterized by geographically dispersed scientific research laboratories operating as scientific fiefdoms in a confederate framework.[8] Both consolidated to a great extent an entire realm of technology in civilian federal agencies. Unlike other new technologies, such as the microcomputer or early aviation, both were handed over wholesale to civilian agencies created specifically to oversee them rather than entrusting progress to the military or private sector. In introducing the problem the framers of the Atomic Energy Act faced, AEC historians Richard G. Hewlett and Jack M. Holl noted: "How does one best go about introducing a new technology into society? A familiar problem for large manufacturers, the management of technological innovation was hardly a common function for Federal officials, except in the area of regulation . . . in the case of nuclear power, the entire technology was confined within the government."[9] This fundamental historical similarity, domination and encapsulation of an entire area of technology by a civilian government agency, is the basis for the current argument.

---

7. David Noble in *America by Design: Science, Technology, and the Rise of Corporate Capitalism* (New York, NY: Oxford University Press, 1977) inverts the hierarchy and sees this explosion not as state-centric manipulation, but as a "wholesale public subsidization of private enterprise" to serve the ends of technocratic corporate managers working as government contractors (p. 322). John Kenneth Galbraith in *The New Industrial State* (Boston, MA: Houghton Mifflin, 1967) envisions technocracy as having a decision-making mind of its own within a given institutional constellation, the "Technostructure," which operates autonomously from corporate or governmental intentions, often to the detriment of the public good. Don Price argues in *The Scientific Estate* (Cambridge, MA: Belknap Press, 1965) that the fusion of political and economic power seen in the nuclear and Space Age has corrupted market principles by creating corporations solely dependent on government subsidies, resulting in a diffusion of political sovereignty that threatens the American constitutional order. Finally, no discussion of technocracy in America would be complete without mentioning Frederick Winslow Taylor's *Principles of Scientific Management* (New York, NY: Harper & Brothers, 1911), which called for applying scientific principles to the training and management of workers to replace "rule of thumb" factory methods.
8. Peter J. Westwick, *The National Labs: Science in an American System*, 1947–1974 (Cambridge, MA: Harvard University Press, 2003), perhaps borrowing from dialectical materialism, stresses that the systemicity of the labs is central to an understanding of their operation. A single national lab cannot exist in isolation; classified journals and conferences and competition for personnel and research programs were central issues that defined the individual labs.
9. Richard G. Hewlett and Jack M. Holl, *Atoms for Peace and War* (Berkeley, CA: University of California Press, 1989), pp. 13–184.

This paper will examine the links between atomic energy and the processes in the executive and legislative branches that culminated in the signing into law of the National Aeronautics and Space Act on 29 July 1958. While a detailed comparative history of the roles, structures, and functions of NASA and the AEC would immensely contribute to the historical literature, the current analysis will focus more narrowly on the way in which the experience with atomic energy produced unspoken assumptions and shaped the very imagination of politicians of what the new NASA should and could become during the 10-month period from the launch of Sputnik to the passing of the National Aeronautics and Space Act. Specifically, it will be argued that NASA's rise in the 1960s as an engine of American international prestige was rooted in atomic diplomacy and that certain debates in Congress about the new agency were largely approached from within a framework of atomic energy, thereby limiting the range of discourse and influencing the shape of the new agency.

While NASA grew by orders of magnitude in the 1960s, the features that specifically identified NASA as technocratic were frozen into the bureaucracy in this formative period. The sudden influx of money after Kennedy's famous decision to set NASA's sights on a Moon landing merely inflated NASA's existing latent potential.

The Role of Prestige

A large debate in the historiography of NASA Centers is the question of prestige. Is NASA's mission coincident with or even driven by American political imperialism? How did national prestige come to be measured by a cosmic yardstick? These questions are often posed in light of the two temporal sides of the Sputnik rupture. On the one hand, the Eisenhower administration was seemingly caught unawares of the worldwide impact the launch of Sputnik would have on public perceptions of American strength. On the other hand, John F. Kennedy would soon after catapult his career on the program to send humans to the Moon, a program that "transformed NASA from a scientific research agency into a goal-oriented bureaucracy."[10]

In the fall of 1957, high-level officials extrapolated the Sputnik launch into an across-the-board American deficiency in scientific ability. The Democratic majority under Senator Lyndon B. Johnson jumped on the opportunity to place blame on the Republican Eisenhower administration and relaunched hearings

---

10. Giles Alston, "Eisenhower: Leadership in Space Policy" in *Reexamining the Eisenhower Presidency*, ed. Shirley Ann Warshaw (Westport, CT: Greenwood, 1993), p. 117.

by the Preparedness Investigating Subcommittee of the Committee on Armed Services in the Senate in late November. General James H. Doolittle provided one of the early testimonies.[11] In his testimony, Doolittle felt convinced "that the rate of Russian progress is much more rapid than ours; that, in some areas, she has already passed us. If the rate continues, she will pass us in all."[12]

In a meeting of the Office of Defense Mobilization Science Advisory Committee (SAC) with President Eisenhower on 15 October, Edward H. Land explained to the President the reasons for Soviet success:[13]

> The structure of Russian culture and thinking is such that they are learning to live the life of science and its application . . . . Is there a way to tell the country that we should set out on a scientific adventure in which all can participate? If this can be done, with our concept of freedom and the independent, unfettered man, we can move far ahead. We need a scientific community in the American tradition.[14]

Whether or not Land had accurately assessed the Soviet mentality toward science or the true implications of the Sputnik launch is of little importance. The notable point is the reaction produced in the very highest echelons of scientific and military advisory circles. Clearly, the hysteria and "fever" that swept the country in the wake of the Sputnik launch were not limited to an uninformed public. Indeed, the media and public were simultaneously concerned with the integration crisis at Central High School in Little Rock, Arkansas. For those in the government primarily concerned with national security, Sputnik produced a larger effect than in the public at large.

---

11. Doolittle was already famous for his bombing raid on Tokyo shortly after the initiation of hostilities between the United States and Japan in 1942. He later went on to become chairman of the NACA board, a position he held at the time of his testimony.
12. Hearings before the Preparedness Investigating Subcommittee of the Committee on Armed Services, 85th Cong., 1st and 2nd sess., pt. 1, p. 111.
13. At the meeting, I. I. Rabi noted, "most matters of policy coming before President have a very strong scientific component" and "he didn't see around the President any personality who would help keep the President aware of this point of view." Eisenhower concurred and "said that he had felt the need for such assistance time and again." This discussion led to the suggestion by James Killian for the creation of a scientific advisory panel to assist the proposed adviser. This would become the PSAC, which began meeting in November with Dr. Killian as its head. See "Detailed (largely verbatim) notes on a meeting of the ODM Science Advisory Committee with the President on October 15, 1957," folder 012401, NASA Historical Reference Collection, NASA History Division, NASA Headquarters, Washington, DC.
14. Ibid.

The conception of Sputnik as a discontinuity is linked to the conception of scientific prestige as a benchmark for national strength. Since Eisenhower misjudged the impact Sputnik would have on the perception of the United States, so the argument goes, only after the media frenzy and political attacks of fall 1957 did the administration recognize the importance of science to national prestige in the international sphere. Even in the face of Sputnik, Eisenhower seemingly remained steadfast in his dislike of federal bureaucracy and shied away from setting prestige as a goal of space research. On 7 November 1957, Eisenhower announced the creation of the post of Special Assistant to the President for Science and Technology in a televised address on national security. The address summarized American nuclear assets while noting deficiencies in science education in America. The speech concluded with a warning against runaway spending:

> It misses the whole point to say that we must now increase our expenditures of all kinds on military hardware and defense—as, for example, to heed demands recently made that we restore all personnel cuts made in the armed forces. Certainly, we need to feel a high sense of urgency. But this does not mean that we should mount our charger and try to ride off in all directions at once. We must clearly identify the exact and critical needs that have to be met. We must then apply our resources at that point as fully as the need demands. This means selectivity in national expenditures of all kinds.[15]

By analyzing metaphor in his speeches and press conferences, Linda T. Krug notes Eisenhower's "images created a vision of a nation of scientist-generals already hard at work planning how to unlock the secrets of the universe."[16] But the conclusion she draws that "Eisenhower was the only president who saw the space program as a viable entity in and of itself" is based on the assumption that Eisenhower never clothed hidden intentions in crowd-pleasing rhetoric.[17] Such sweeping conclusions about Eisenhower's personal views

---

15. Dwight D. Eisenhower, "Radio and Television Address to the American People on Science in National Security," 7 November 1957, available at *http://www.eisenhowermemorial.org/speeches/19571113%20Radio%20and%20Television%20Address%20on%20Our%20Future%20Security.htm.*
16. Linda T. Krug, *Presidential Perspectives on Space Exploration: Guiding Metaphors from Eisenhower to Bush* (New York, NY: Praeger Publishers, 1991), p. 29.
17. Ibid.

cannot be drawn from televised statements. All presidents must maintain a carefully groomed public persona. While Eisenhower's public proclamations often criticized big government, policy decisions and internal White House discourse did not match his rhetoric.

The National Security Council engaged the question of prestige in relation to the planned American and Soviet satellite launches during the International Geophysical Year (IGY) of 1957–1958. A Technological Capabilities Panel (TCP) was formed in 1954 under James Killian to investigate the satellite question and other technical issues deemed vital to national security.[18] The TCP issued its final report in February 1955, and the National Space Council (NSC), following the TCP's recommendation, concluded in May of that year that the U.S. effort (Project Vanguard) should be given high priority as "considerable prestige and psychological benefits will accrue to the nation which first is successful in launching a satellite."[19] The importance of such benefits was paramount to U.S. foreign policy since "the inference of such a demonstration of advanced technology and its unmistakable relationship to inter-continental ballistic missile technology might have important repercussions on the political determination of free world countries to resist Communist threats, especially if the USSR were to be the first to establish a satellite."[20]

The NSC concluded the U.S. scientific satellite effort should not hinder military missile developments and, therefore, should be vested in a separate, civilian-run program headed by the National Science Foundation (NSF). It is absolutely clear that the Eisenhower administration intended to use the satellite launch to reinforce American scientific prowess in the international arena.

The fact that prestige was an important element after that fateful 4 October and during the formative period of NASA is uncontroversial. In a PSAC meeting

18. The TCP also drew the famous conclusion that establishing freedom of overflight in space, i.e., sovereignty claims of airspace not extending beyond the atmosphere, was in the long-term interests of the U.S. This was motivated by the expectation that the U.S. would have a large lead over the Union of Soviet Socialist Republics (USSR) in electronic satellite reconnaissance capability. For an overview of the TCP and its impact on the freedom of space, see McDougall, . . . *The Heavens and the Earth*, chap. 5. Dwayne A. Day has recently uncovered documents tracing the origin of this principle to a Central Intelligence Agency (CIA) intelligence officer, Richard Bissell, and an Air Force aide working for the CIA. Dwayne A. Day, "The Central Intelligence Agency and Freedom of Space," paper presented at the "Remembering the Space Age: 50th Anniversary Conference," NASA History Division and National Air and Space Museum Division of Space History, Washington, DC, 22 October 2007.
19. "National Security Council Report 5520: Missile and Space Programs." See *A Guide to Documents of the National Security Council*, 1947–1977, ed. Paul Kesaris (Bethesda, MD: University Publications of America, 1980).
20. Ibid.

in March 1958, Hans Bethe commented, "It would be a great mistake for us to oppose popular enthusiasm even though misguided."[21] And in a recently declassified Office of Research and Intelligence Report issued just two weeks after Sputnik on 17 October 1957, it was concluded:

> The technologically less advanced—the audience most impressed and dazzled by the sputnik [*sic*]—are often the audience most vulnerable to the attractions of the Soviet system . . . . It will generate myth, legend and enduring superstition of a kind peculiarly difficult to eradicate or modify, which the USSR can exploit to its advantage, among backward, ignorant, and apolitical audiences particularly difficult to reach.[22]

The report went even further in claiming the United States itself had fanned the flames of the fire in three ways: "first by fanfare of its own announcement of its satellite plans, second by creating the impression that we considered ourselves to have an invulnerable lead in this scientific and technological area, and third by the nature of the reaction within the U.S."

The importance of science to national prestige in the Eisenhower administration existed long before Sputnik; it originated in the experience with atomic energy. Eisenhower had long been an advocate of using atomic energy to further U.S. foreign policy, a fact exemplified by his personal championing of the Atoms for Peace program.

In his 8 December 1953 address to the UN General Assembly, President Eisenhower called for the establishment of an "International Atomic Energy Agency" to serve as a stockpile of nuclear materials for peaceful uses around the world. The proposal was "enunciated by the President almost as a personal hope," with few advisers and only one of the five Atomic Energy Commissioners, Lewis Strauss, aware of the proposal ahead of time.[23] The original proposal was devoid of details but is significant in that Eisenhower displayed a personal

21. PSAC Meeting, 12 March 1958. The transcribed notes of the PSAC are spotty at best, and the argumentative logic is nearly incomprehensible. They are reproduced in *The Papers of the President's Science Advisory Committee*, 1957–1961, microfilm (Bethesda, MD: University Publications of America, 1986).
22. Office of Research and Intelligence Report, "World Opinion and the Soviet Satellite: A Preliminary Evaluation," declassified 1993, pp. 2–4, folder 18106, NASA Historical Reference Collection, NASA History Division, NASA Headquarters, Washington, DC.
23. Hewlett and Holl, *Atoms for Peace and War*, pp. 210–213.

desire to use science and scientific prestige as a tool of international diplomacy. The policy was consciously constructed around the issue of prestige, e.g., the amount of uranium to be contributed by the United States was set at a high enough figure that the Soviet Union would not be able to match the American contribution.[24] While the implementation of the plan was slow in arriving, the middle of the decade saw tangible, albeit often ineffective, international cooperation in atomic technology with the United States as the international lynchpin and guarantor of atomic security. Science in the Eisenhower administration was part and parcel of foreign policy.

The tendency to employ science in the service of international prestige was expressed early on in the discussions concerning a new space agency. Coincidentally, Eisenhower asked James Killian (then president of the Massachusetts Institute of Technology [MIT]) to become his personal science adviser over breakfast on 24 October, the purpose of the meeting being Killian's briefing of Eisenhower in preparation for the Atoms for Peace award being given to Neils Bohr later that day.[25]

In an ODM memorandum issued in January for Secretary of Health Arthur S. Fleming, the analogy to atomic energy was clearly enunciated: "In addition to the military importance of the scientific satellite one should not overlook the benefits of adequate emphasis on peaceful applications of rocketry just as the atoms-for-peace program has served to divert world attention from nuclear weapons."[26] And in a legislative leadership meeting on 4 February, President Eisenhower cautioned against pouring "unlimited funds into these costly projects where there was nothing of early value to the Nation's security. He recalled the great effort he had made for the Atomic Peace Ship but Congress would not authorize it, even though in his opinion it would have been a very worthwhile project."[27]

The relation of prestige to spaceflight has trickled down to the present day. Political pundits still routinely call the value of human spaceflight into question. NASA is frequently attacked as a wolf in sheep's clothing; that is, NASA's stated peaceful exploratory goals are often argued to be merely a

---

24. John Krige, "Atoms for Peace, Scientific Internationalism, and Scientific Intelligence," *Osiris* 21 (1996): 164.
25. James R. Killian, Jr., *Sputnik, Scientists, and Eisenhower* (Cambridge, MA: MIT Press, 1977), p. 24.
26. Executive Office of the President Office of Defense Mobilization, memorandum to Arthur S. Fleming, "Scientific Satellites," 23 January 1957, folder 012401, NASA Historical Reference Collection, NASA History Division, NASA Headquarters, Washington, DC.
27. Supplementary Notes, Legislative Leadership Meeting, 4 February 1958, folder 18106, NASA Historical Reference Collection, NASA History Division, NASA Headquarters, Washington, DC.

façade covering deeper political and military motives. The origins of this dichotomy can be traced directly back to the emphasis placed on prestige during the conception of NASA in the Eisenhower administration, which was in turn based on the experience of atomic foreign policy. By the time a man-in-space investigatory panel was commissioned in 1959 by George Kistiakowsky, then head of the PSAC, it was clear that putting humans in space was solely a prestige issue:

> In executive session of the panel, we talked about these things and I emphasized the need to spell out in our report what cannot be done in space without man. My opinion is that that area is relatively small and that, therefore, building bigger vehicles than Saturn B has to be thought of as mainly a political rather than a scientific enterprise.[28]

Indeed, it can be concluded that space represented a welcome new opportunity for Eisenhower's continuing desire to demonstrate American technological prowess because of a decline in the perception of atomic energy as a positive international technology, a decline spurred on by rising fears of global nuclear annihilation. Certainly the destructive element of nuclear technology had been publicly decried immediately after the Hiroshima and Nagasaki bombings, but the shift in scale from local (bomber-delivered atomic bombs) to global (intercontinental ballistic missile [ICBM]-delivered hydrogen bombs) damned any hope for an unproblematic public perception of nuclear technology. The first hydrogen bomb tests by the United States in 1952 and the Soviet Union in 1953 were followed by the irradiation of the Japanese fishing boat *Lucky Dragon 5* by the Castle Bravo test in March 1954, leading to a widespread public concern over the effects of nuclear radiation.

An illustrative example of the qualitative transformation of atomic energy in the public imagination can be drawn from science fiction. Isaac Asimov's Foundation trilogy, published between 1951 and 1953, portrayed humanity in the far future as a galactic empire in decline. The Foundation, created by a visionary scientist who foresaw the collapse of civilization using new historical-predictive methods, becomes the sole possessor of knowledge of atomic

28. George Kistiakowsky, *A Scientist at the White House: The Private Diary of President Eisenhower's Special Assistant for Science and Technology* (Cambridge, MA: Harvard University Press, 1976), p. 409.

technology and hence the last hope for humanity's future.[29] But by the end of the 1950s, postapocalyptic novels set in nuclear winter ruled the genre: Nevil Shute's *On the Beach* (1957), Pat Frank's *Alas Babylon* (1959), and Walter M. Miller, Jr.'s *A Canticle for Leibowitz* (1959). Space, then, was a natural avenue into which the Eisenhower administration could expand its policy of scientific prestige in the service of the state while avoiding the stigmas becoming associated with nuclear technology.

The National Aeronautics and Space Act of 1958

Of special importance to the current analysis are the sections of the National Aeronautics and Space Act of 1958 that were inspired by the Atomic Energy Acts of 1946 and 1954. Specifically, these are the relation of DOD to the new agency, the role of international cooperation, and the apportionment of intellectual property.

When President-elect Eisenhower was briefed on AEC activities in November 1952, he took special exception to Gordon Dean's acquiescence to the Air Force's demand for atomic-powered plane research in the face of good evidence that such a program would not produce a viable aircraft. "Looking out the window he declared that this kind of reasoning was wrong. If a civilian agency like the Commission thought that a military requirement was untenable or wasteful in terms of existing technology, there was an obligation to oppose it."[30] This was a prescient moment, for it foreshadowed the problem of divvying up responsibility between competing civilian and military institutions during the formation of NASA.

Analogies to the Atomic Energy Commission were widespread throughout the legislative creation of the new space agency. During the congressional hearings, Eilene Galloway, a national defense analyst at the Library of Congress, was invited by representative McCormack (the chair of the House committee) to write a report on the issues facing Congress in the drafting of the National Aeronautics and Space Act.[31] Her report was widely read and was reprinted in both the Senate and House proceedings and is notable for several reasons. First, Galloway drew the immediate conclusion that a comparison to the issues

---

29. Special thanks to Dan Bouk for pointing out this poignant example from a trilogy I have read four times yet somehow overlooked: Isaac Asimov's *Foundation* (New York, NY: Gnome Press, 1951), *Foundation and Empire* (New York, NY: Gnome Press, 1952), and *Second Foundation* (Gnome Press, 1953).
30. Hewlett and Holl, *Atoms for Peace and War*, p. 14.
31. Galloway also served as special consultant to Lyndon Johnson during the Senate hearings and has since become a noted aerospace historian.

facing the drafters of the Atomic Energy Act of 1946 (informally known as the McMahon Act) would be fruitful. To Galloway, the similarities were obvious:

> Atomic energy and outer space are alike in opening new frontiers which are indissolubly linked with the question of war and peace. They combine the possibility of peaceful uses for the benefit of man and of military uses which can destroy civilization. Both are national and international in their scope. They involve the relation of science and government, the issue of civilian or military control, and problems of organization for the executive branch and the Congress. If only their similarities are considered, the legislative task would appear to be the easy one of following the pattern of our present atomic energy legislation.[32]

According to Galloway, the dissimilarities between the two are centered on the problem of delineating military and civilian aspects of aerospace technology. While the boundaries are reasonably clear in the atomic case (bombs versus reactors), nearly every aspect of aerospace technology overlaps the two sides of the military-civilian divide. This is perhaps an oversimplification in that much effort had gone into the Atomic Energy Act of 1954 to allow the development of a civilian atomic energy industry, and the civilian-military divide in practice was quite problematic. Still, it remains true that, in the case of atomic energy, a relatively clear boundary between civilian and military applications could be established through strict regulation of nuclear materials. In the case of NASA, this was not true; yet still a formal divide was automatically assumed to be of paramount importance. In part this was due to concerns of needless duplication of effort and bureaucratic infighting over jurisdictional matters. However, previous experience with the AEC weighed heavily on lawmakers, particularly in the House of Representatives, who now saw science as intimately tied up with national security and felt a need for such a relationship to be codified in law. The administration favored a more informal relationship, as had been the case with the NACA. Both sides weighed heavily on precedent to reinforce their arguments.

The debate surrounding the obligations of the new space agency to DOD and vice versa has long been the center point of the history of the National

---

32. Eilene Galloway, *The Problems of Congress in Formulating Outer-space Legislation* (Washington, DC: U.S. Government Printing Office [GPO], March 1958).

Aeronautics and Space Act of 1958. This is for the reason that the delineation of the role of military and civilian agencies has obvious current political implications, but it remains true that much of the contemporary debate also surrounded the issue. The wording of §102(b) of the National Aeronautics and Space Act established the following criterion by which specific projects could be judged to be NASA- or Defense-centric:

> The Congress declares that the general welfare and security of the United States require that adequate provision be made for aeronautical and space activities. The Congress further declares that such activities shall be the responsibility of, and shall be directed by, a civilian agency exercising control over aeronautical and space activities sponsored by the United States, *except that activities peculiar or primarily associated with the development of weapons systems, military operations, or the defense of the United States . . . shall be the responsibility of, and shall be directed by, the Department of Defense . . . .*[33]

The Act also established a National Aeronautics and Space Council headed by the President and including the Secretary of State, Secretary of Defense, NASA Administrator, and the Chairman of the AEC. The inclusion of the AEC Chairman here is quite curious. In addition, any disputes between departments and agencies over jurisdictional matters were to be settled by the President under advisement of the council.

The original Bureau of the Budget draft bill was quite different from the arrangement in the AEC, which embodied communication with DOD in its Military Liaison Committee. In his official commentary sent to the Bureau of the Budget on the original bill, Strauss suggested "the act provide for interagency liaison similar to that which has operated so satisfactorily in the case of the Military Liaison Committee in the atomic energy program."[34] The House bill included such a liaison committee and, in addition, another for the AEC. The administration had favored informal cooperation in the form of uniformed seats on the advisory committee in the same style as the NACA had tradition-

33. National Aeronautics and Space Act of 1958, Public Law (P.L.) 95-568, available at *http://history.nasa.gov/spaceact.html*. Emphasis and ellipses added.
34. Lewis Strauss to Maurice Stans, Director of the Bureau of the Budget, 31 March 1958, folder 012405, NASA Historical Reference Collection, NASA History Division, NASA Headquarters, Washington, DC.

ally pursued. The Senate kept the administration's arrangement. In the final compromise bill, the military liaison committee was added, while the AEC liaison was dropped.

An internal Bureau of the Budget memo in May snidely remarked on the House bill that "among the trappings of the Atomic Energy Act inserted in this bill are sections establishing and prescribing the functions of a Military Liaison Committee and an Atomic Energy Liaison Committee. Both Committees are to be headed by chairmen appointed by the President . . . . The Department of Defense as well as NACA has opposed this creation of statutory liaison committees, and every effort should be made to secure their elimination in the Senate."[35] The inclusion of the liaison committees in the House bill suggests a strong tendency to adopt portions of the AEC paradigm wholesale. It is particularly remarkable in this case because the civilian-military boundary proposed for NASA was quite different from the model in the AEC. That is, NASA would by default carry on the bulk of aerospace research; but DOD, by sufficiently justifying its need directly to the President, could develop its own aerospace projects. This is in stark contrast to the complete monopolization of basic atomic research by the AEC, which necessitated a reliable and clear avenue of communication to and from the military.

The differences between NASA's and the AEC's relationships with the military deserve elaboration. From the beginning, the AEC was to encompass all levels of nuclear research, nuclear materials production, reactor design, and bomb construction. This centralization was a result of the realities of atomic energy. First, the Manhattan District was already in place during the establishment of the AEC, and maintaining its internal configuration was necessary for the uninterrupted production of atomic weapons. Second, and more important, atomic energy as a technology is unique for a material reason: the regulation of atomic technology is in large part the regulation of a single element and its derivatives. Indeed, the Atomic Energy Act categorically transferred "all right, title, and interest within or under the jurisdiction of the United States, in or to any fissionable material, now or hereafter produced" to the Commission. In effect, all atoms on U.S. territory with 92 or more protons were declared to be the property of the federal government. In addition, an entire new class of information was created. Termed "Restricted Data," this wide umbrella automatically "classified at birth" any and "all data concerning the manufacture

35. Letter from Alan L. Dean to Wiliam Finan, 2 June 1958, folder 12400, NASA Historical Reference Collection, NASA History Division, NASA Headquarters, Washington, DC.

or utilization of atomic weapons, the production of fissionable material, or the use of fissionable material in the production of power."[36] Regulation of fissionable material was also the assumed primary task of early atomic weapons nonproliferation efforts. Containment of atomic technology was seen as synonymous with ownership of nuclear materials.

From the inception of the AEC, the production and control of nuclear materials was the prime directive of the organization. Fissionable material simultaneously was obviously dangerous, was necessary for national defense, and could be relatively easily collected and controlled. The implication of this material reality was tremendous for the bureaucratization of atomic technology in a central governmental agency. In the case of aerospace technology, such a clear compartmentalization was not a natural outgrowth of the relevant technology. Still, the basic structure of the AEC was to provide a perceived "obvious model" for creating an aerospace agency.

§205 of the National Aeronautics and Space Act provided engagement in "a program of international cooperation . . . and in the peaceful application of the results thereof." The Senate Special Committee had noted in a report entitled "Reasons for Confusion over Outer Space Legislation and how to Dispel It" that "the main reason why we must have a civilian agency in the outer space field is because of the necessity of negotiating with other nations and the United Nations from some non-military posture."[37]

The Act specifically authorized the Administrator to grant NASA employees access to restricted AEC data. This violated long-standing AEC policy, which based access on AEC classified status. Strauss thus raised the concern that the act would allow the President to "disseminate Restricted Data to foreign governments . . . . We think that an extension of this existing authority to the proposed Agency would be undesirable and unworkable."[38] In his testimony before the Senate Special Committee, Strauss stressed his preference for limiting international agreements at the outset and noted that "the history of these new agencies, if the Atomic Energy Commission is a prototype, has been that, in the course of time, the basic law is amended by spelling out in greater detail

---

36. Atomic Energy Act, 1946, P.L. 585, 79th Cong., available at *http://www.osti.gov/atomicenergyact.pdf.*
37. Senate Special Committee on Space and Astronautics Report, "Reasons for Confusion over Outer Space Legislation and how to Dispel It," 11 May 1958, folder 012389, NASA Historical Reference Collection, NASA History Division, NASA Headquarters, Washington, DC.
38. Lewis Strauss, General Manager of AEC, to Maurice Stans, Director, Bureau of the Budget, 31 March 1958, folder 012405, NASA Historical Reference Collection, NASA History Division, NASA Headquarters, Washington, DC.

the extent to which cooperation with other nations may be carried on."[39] The strong ties to the AEC are evident.

The issue of intellectual property centered on the allocation of patents. The House bill patterned itself on the Atomic Energy Act, giving the government exclusive ownership of any intellectual property arrived at due to NASA-related work. The American Patent Law Association lobbied against such a provision, for the obvious reason that long-term profits from owning patents was a prime incentive for firms bidding on contracts.[40] In a letter to William F. Finan, Hans Adler (both were in the Bureau of the Budget) wrote in reference to the patent provision in H.R. 12575 (the bill that became the National Aeronautics and Space Act): "this provision is also based on the Atomic Energy Act. However, we doubt that the Atomic Energy Act should serve as the proper precedent, since inventions in the atomic area have peculiar defense and secrecy aspects which make private ownership difficult."[41] Again, we have an example of the adoption of policies crafted for atomic energy without reasoned analysis of their relevance to an aerospace agency. The final language adopted assigned intellectual property to the government, with the Administrator having the right to waive this right if he so desired.

It cannot be overstated how formative the experience with atomic energy was on the psyche of those determining the shape of NASA. The belief that atomic energy would infuse all aspects of future technology was widely held in 1950s America, and rocketry was no exception. The realities of chemical reactive propulsion dictate a maximum theoretical efficiency (specific impulse) due to limited available chemical enthalpy, but the exit velocity of a thermal nuclear rocket is limited only by material failure at high temperatures. The AEC, for these reasons, launched just such a nuclear rocket research program (ROVER) in 1956. Stanislaus Ulam, testifying before the House Select Committee on Astronautics and Space Exploration, reaffirmed that "it is not a question of conjecture or optimism, but one might say it is mathematically certain that it will be the nuclearly powered vehicle which will hold the stage in the near

---

39. Hearings before the Special Committee on Space and Astronautics, United States Senate, 85th Cong., 2nd sess., p. 50.
40. Richard Hirsch and Joseph John Trento, *The National Aeronautics and Space Administration* (New York, NY: Praeger Publishers, 1973), p. 26.
41. Hans Adler to William Finan, "Subject: HR 12575," 4 June 1958, folder 12400, NASA Historical Reference Collection, NASA History Division, NASA Headquarters, Washington, DC.

future."[42] With historical actors like Ulam making such statements, it becomes clear that the birth of NASA as an institution must be historically analyzed through the lens of the atomic experience. The concept of the stewardship of the state over technological affairs had become ingrained in the imagination in the atomic era and was adopted without serious protest during the formation of NASA. Indeed, a sharp contrast can be drawn to the violent reaction by private interests to the original Atomic Energy Act and the relatively benign reception of the National Aeronautics and Space Act. A profound transformation had occurred in the intervening years.

Conclusion

Under the AEC, technocracy had been introduced to America. Under NASA, it was wedded to the federal framework. There are fundamental differences to the two cases, as in the ability to control nuclear material and the need to enforce atomic secrecy through the curtailment of granting patents. But throughout the whole of the discussions in both the executive and legislative branches during 1957–1958, it remains clear that the framers of the new aerospace agency were profoundly affected by their experience with atomic energy, specifically the AEC. When conceiving of a new agency, bureaucrats and legislators actively reached into the past and cherry-picked elements from their prior experience with atomic energy while passively making unconscious assumptions based on the technological realities of atomic energy. Often the decisions they arrived at were not appropriate for the aerospace case.

NASA represented a form of technocracy that divorced military interests as completely as possible. In the 1960s, NASA would become an agency mobilized for social change. Thomas Hughes argues in *American Genesis* that, during the Great Depression, the Tennessee Valley Authority (TVA) was a push for regional social development by progressive politicians via electrification and the management of water resources.[43] NASA followed in these footsteps. Perhaps not so coincidentally, one of the original commissioners of the TVA, David Lilienthal, would later become the first Chairman of the AEC.

But NASA was technocracy in an evolved form. It combined three trends that had not yet together existed in any American organization: 1) Big Science,

---

42. Hearings before the Select Committee on Astronautics and Space Exploration, 85th Cong., 2nd sess., p. 602.
43. Thomas P. Hughes, *American Genesis: A Century of Invention and Technological Enthusiasm 1870–1970* (New York, NY: Viking Penguin, 1989), pp. 360–381.

i.e., the close cooperation of large numbers of scientists and engineers in a vertically integrated hierarchy organized for the production of massive projects; 2) a mandate that pushed science for social benefits and simultaneously minimized obligations to the military; and 3) science in the service of national prestige abroad.

The AEC took over the operation of the entire American atomic machine, from enrichment, to reactor design, to bomb testing in the South Pacific. NASA, instead, was given a mandate to push the boundaries forward in aerospace technology only insofar as they could be peacefully used. This was, then, a pivotal transformation in the history of American technocratic institutions. Under the presidencies of Kennedy and Johnson, NASA was a juicy target to be expanded, but this was merely opportunism. NASA's form had already been cemented in 1958, a form that had atomic roots.

Acknowledgments: This article first appeared, in substantially the same form, in *Remembering the Space Age*, ed. Steven J. Dick (Washington, DC: NASA SP-2008-4703, 2008), pp. 55–70.

Chapter 4

Leading in Space

50 Years of NASA Administrators

W. Henry Lambright

Introduction

The central task of the 10 men who served as NASA Administrator in the 50 years since 1958 has been strategic leadership, by which is meant the setting of priority goals for the Agency for the periods in which they held office. How well the Administrator performed his task depended on his capacity to match the goals he set with the forces in his environment. James Webb, NASA's Administrator in the 1960s, described leadership as "fusing at many levels a large number of forces, some countervailing, into a cohesive but essentially unstable whole and keeping it in motion in a desired direction."[1] He called this process the creation of a "dynamic equilibrium." It entailed setting and implementing policy.

The various Administrators used rhetorical and coalitional skills to achieve their purposes, building internal and external support, while seeking to neutralize opponents.[2] Their role was daunting. Sometimes an Administrator succeeded and at other times failed; but unless an Administrator prevailed to some degree, the forces in his environment determined NASA's fate rather than the Administrator.

---

1. James E. Webb, *Space Age Management: The Large-Scale Approach* (New York, NY: McGraw Hill, 1969), pp. 135–136.
2. For a discussion of administrative strategies, see Jameson Doig and Erwin Hargrove, eds., *Leadership and Innovation: A Biographical Perspective on Entrepreneurs in Government* (Baltimore, MD: Johns Hopkins University Press, 1990).

In every instance, real administrative influence depended on a match among the Administrator's gifts, organizational capacity, and political times. Even when a particular leader was unable, for one reason or another, to make a significant personal mark on history, he could still be important as part of a "relay" leadership, initiating or steering particular programs he inherited forward. Given the brief tenure of most Administrators vis-à-vis the large-scale, long-term programs with which they dealt, that is itself important in NASA's fate.

Creating the Administrator's Role

In drafting the Space Act for NASA in 1958, the White House and Congress clearly wanted NASA to have a leader with significant authority. The leader was given prerogative over budget and personnel, subject to the President and Congress. It is notable that the Administrator is a single figure. Existing models in 1957 included the AEC and the NACA, which had plural heads. Moreover, the title "Administrator" represented a conscious choice by NASA's creators over "Director," the former being a title considered to carry more executive panache.[3]

The Space Act gave NASA a range of tasks in space and aeronautics, but those were relatively broad and vague, leaving it to the Administrator to use his or her discretion in their interpretation. Congress at the time of NASA's creation organized itself into two authorizing and two appropriations committees dealing with space. This pattern simplified reporting assignments for the Administrator.

The mood of Congress, triggered by Sputnik and Cold War competition with the USSR, favored action; but it was up to the initial incumbents in the office to give substance to the words and mood of the time.

1. T. Keith Glennan, August 1958–January 1961

NASA's first Administrator, T. Keith Glennan, was an engineer and university president with experience in industry. He came to NASA from the presidency of Case Institute of Technology in Cleveland, Ohio. He was 53 years in age at the time he took official charge of the new agency on 1 October 1958. Glennan served under a President who wanted to restrain NASA from engaging in any "race" with the Soviet Union. President Eisenhower intended "leadership" in space to be established through sound scientific and technological principles.

---

3. Eilene Galloway, "Sputnik and the Creation of NASA: A Personal Perspective," in *NASA: 50 Years of Exploration and Discovery*, ed. Rhonda Carpenter and Ana Lopez (Washington, DC: Faircount Media Group, 2008), p. 48.

Congress, led by Democrats, was much more "bullish" on space and wished to charge faster ahead. Glennan steered NASA between these two stances.[4]

His goal was to get NASA started quickly, but in a competent way, building on the NACA base of 8,000 scientists, engineers, and support personnel he inherited. He consolidated entities that were transferred from DOD, including the von Braun rocket-engineer team, which became the nucleus of MSFC in Huntsville, Alabama; the space-science group of NRL, which became GSFC in Greenbelt, Maryland; and JPL, run for NASA by the California Institute of Technology in Pasadena, California.

Glennan organized the basic divisions of NASA—manned spaceflight, space science, Earth-oriented applications, and aeronautics. He began the first human spaceflight program, Project Mercury. Finally, he determined that NASA would get most of its work done by industrial and academic contractors, keeping its government base relatively limited.

Glennan's goals were clear, but limited, and he achieved much that he set out to do. However, NASA was still a relatively weak and much-criticized agency when he left office in 1961, and the Soviet Union was well ahead of the United States in space achievements. The Department of Defense remained a formidable bureaucratic rival for supremacy as the nation's premier space agency.

2. James Webb, February 1961–October 1968

The new President, John Kennedy, appointed James Webb his Administrator. Webb, 54 in age at the time, was neither a scientist nor an engineer, but a professional manager and lawyer. He had extensive government experience, having worked for Congress as a young man and later as Truman's Director of the United States Budget and Undersecretary of State. He also had considerable executive experience in industry. He considered himself a government manager, but what most observers saw was a remarkably savvy bureaucratic politician, one capable of dealing one-on-one with Presidents and senior legislators in ways that made him persuasive in advocating NASA's cause. To help him in his inside role of managing a technical agency, he developed a "triad" concept that had a physicist as his Deputy Administrator and an engineer as his Associate Administrator and de facto general manager. Better than any Administrator in NASA's history, Webb understood power in Washington, DC, and how it worked. He sought to enhance NASA's bureaucratic power and his own as its leader.

---

4. Roger Launius, "Leaders, Visionaries, and Designers," in *NASA: 50 Years*, ed. Carpenter and Lopez, p. 258.

Thanks to Kennedy's Apollo decision of 1961, Webb succeeded in building NASA quickly and substantially in budget and personnel and transforming it into a special organization capable of taking America to the Moon. In doing so, he placed NASA clearly in charge of the space enterprise, neutralizing DOD, the White House science adviser, and internal rivals.[5]

Webb understood he had a two-year "honeymoon," given the nation's frustration with being behind in space and especially with having a Russian, Yuri Gagarin, be the first man in space. He used the President's directive to win the race to the Moon to NASA's advantage, taking his organization's estimate of costs to achieve a lunar landing and doubling the figure while also extending the deadline in Kennedy's speech from the original 1967 to "within the decade." Kennedy made the decision to go to the Moon, but Webb shaped it with an eye to implementation. He also extracted from Vice President Johnson promises of help in maintaining support over the long haul.

As NASA's budget doubled and doubled again and its personnel grew apace from 1961 to 1963, he made virtually all key decisions required about facilities and contractors to get to the Moon. He could justify these decisions on technical grounds but usually made them also with an aim at long-term coalition building. For example, the location of the new Manned Spacecraft Center was in the district of the lawmaker with the most control over NASA's budget. Similarly, Webb's managers wanted Gemini, an interim program between Mercury and Apollo, as a stepping stone for technical learning. Webb saw Gemini in "political rhetoric" or public relations terms, a way to keep NASA before the public eye in the mid-1960s. He knew the early honeymoon would not last, and it didn't. But he maintained Apollo in the mid-1960s, as Kennedy, assassinated in late 1963, gave way to Lyndon Johnson.

It is notable that Webb's goals were broader than Kennedy's or Johnson's. In 1962 he and Kennedy had a confrontation in the White House. Kennedy wanted to concentrate NASA's resources virtually all behind the Moon race with the Soviets and indicated he cared little about space per se.[6] Webb said the goal was more than leadership, but preeminence, by which he meant creating a surpassing capability not only in human spaceflight, but also in science, applications, aeronautics—and education. Building capability included human capability. For a time, Webb's NASA had a substantial university program that

---

5. W. Henry Lambright, *Powering Apollo: James E. Webb of NASA* (Baltimore, MD: Johns Hopkins University Press, 1995).
6. John Logsdon, "Ten Presidents and NASA," in *NASA: 50 Years*, ed. Carpenter and Lopez, p. 229.

spread grants geographically throughout the nation. He wanted NASA to have a national constituency, not just one in states with major NASA facilities and contracts. For Webb, Apollo was a *means* to raise the technical level of the country generally, not just an *end* in itself, although he certainly gave priority to this mission.

In the mid-1960s, under Johnson, the nation turned to the Great Society and Vietnam. For a while, Webb used a rhetorical strategy of showing how NASA was also a "Great Society" agency, stressing "spinoffs" and space-based economic development. But Vietnam intervened, and the NASA budget fell in the latter half of the decade. Webb desperately wanted to sell a post-Apollo program that would put to use the extraordinary capability NASA had built with the Apollo-Saturn system, but Johnson, beset with the war and pressing budget problems, put him off.

In 1967, when Johnson seemed ready to give Webb a go-ahead on an interim post-Apollo plan for keeping this capability going, disaster struck. The Apollo fire took the lives of three astronauts. Webb used most of his remaining political capital to persuade the President and Congress to let NASA investigate itself and have Webb make the necessary managerial and technical adjustments. He did so, protecting the Agency while drawing the media and political spotlight and criticism on himself.

Webb got NASA through the crisis and back into space and on target to the Moon, but he himself was weakened. Realizing his power was eroding and he could not sell the kind of post-Apollo program he wanted, he now did not try but instead put almost all his energy and much of NASA's decreasing resources behind achieving the Apollo goal.

His last act was to leave early so as to leverage the choice of his successor. He wanted his deputy, Tom Paine, to take NASA, under a new President, the final leg of its journey to the Moon. He believed it critical, at a time when the country was tearing itself apart because of Vietnam, civil rights unrest, and the collapse of the Great Society, that the nation have a monument to success and to what it could do when it operated at its best.

3. Thomas Paine, Acting Administrator, October 1968–March 1969; Administrator, March 1969–September 1970

Tom Paine, age 46, was an engineer and technical manager who came to NASA from a high executive position at General Electric. He was a solid, risk-taking manager who made the final decisions that got NASA to the Moon. He was also a visionary, a space enthusiast, who wanted to build on Apollo with a bold program of space exploration. His prime goal as Administrator was to sell a comprehensive post-Apollo space program that would enlarge NASA's budget and allow it to go to Mars and build a Moon base, a large space station, and a space shuttle to go to and from the space station.

Unfortunately, Paine's political skills were such that he misread the nation's mood and the interest of President Nixon in a bold space program. A program that had at its mid-1960s height spent close to 4.4 percent of the federal budget fell to approximately 1 percent in the Nixon era.[7] Paine had little access to Nixon, who was one of the most inaccessible Presidents in history. Paine's expansionary rhetoric was out of sync with what the President or his aides wanted to hear. Paine, to his credit, pushed forward with many of the existing programs he inherited and kept his Mars vision alive through his backing the unpiloted Viking program. However, he could not build a coalition for new large human spaceflight programs that would stop the retrenchment of the Agency. He reluctantly ended the Apollo Moon landings and Apollo-Saturn system. For NASA, political support began with the President, and Paine had no influence on Nixon. He resigned early in disappointment and frustration, with NASA's post-Apollo future still undetermined.

4. James Fletcher, April 1971–May 1977

James Fletcher, age 51, came to NASA from the presidency of the University of Utah. A Ph.D. physicist who had achieved wealth in industry, Fletcher fit the mold of the Administrator Nixon wanted. He was conservative politically and economically. Not a space cadet like Paine, he was willing to compromise. He did not seek to promote goals that had no hope of acceptance by Nixon. He was not a combative bureaucrat like Webb. His primary interest was clear: stop the decline of NASA by obtaining a significant goal in human spaceflight that was large enough to maintain the Agency. The only goal with a chance to be adopted, given the national mood of the early 1970s, was the minimal goal of the Space Shuttle. Selling that goal became his primary aim.

---

7. Budget estimates vary. NASA Administrator Michael Griffin's peak percentage is 4.4 percent. Logsdon, "Ten Presidents and NASA," p. 231, places it at 4 percent; Roger Launius has a peak figure of 5.3 percent of the federal budget. The Griffin figure is in "Reality of Tomorrow," Address to American Astronautical Society (5 March 2008), in *Leadership in Space: Selected Speeches of Administrator Michael Griffin, May 2005–October 2008* (Washington, DC: NASA SP-2008-564, 2008), pp. 295ff. The Launius figure is in an unpublished paper, "Project Apollo: A Retrospective Analysis." All numbers reveal the same point, that NASA's budget was substantial in the Apollo era and fell to a much lower priority subsequently.

Paine had believed the NASA Administrator should be a "swashbuckler," a term he used. Fletcher was instead quiet, persistent, and—compared to Webb and Paine—dull to most observers of NASA. While an able inside manager, especially with the help of the legendary NASA veteran, George Low, as his deputy, Fletcher was not experienced in the ways of Washington, DC.

In selling the Space Shuttle, he acquiesced to the pressures of the Office of Management and Budget (OMB), a more powerful organization vis-à-vis NASA under Nixon than it had been under Kennedy or Johnson. This meant the Shuttle had to be justified in the language of OMB, in cost-benefit or economic terms. The Office of Management and Budget pushed Fletcher hard, ratcheting down the price of the Shuttle. The Columbia Accident Investigation Board would later state that the economic pressures of the early years led to technical compromises and overpromising that contributed to subsequent troubles affecting the Shuttle. But Fletcher may have had little choice but to bend. Without the Shuttle, NASA would likely have continued its downward spiral. Also, in building support for the Shuttle, he entered into alliance with DOD and adopted designs to suit DOD needs. The Shuttle was sold not as a NASA system but as a "national" system. In 1972, Fletcher finally got the President to say "yes." The budgetary decline ended. NASA stabilized. Ironically, Nixon appears to have made his decision not on economic grounds, but primarily on a combination of electoral and geopolitical calculations. He wanted California votes and international prestige.

Roger Launius has written the primary published assessment of Fletcher.[8] He points out that Fletcher's Mormon background was important in several non-Shuttle decisions. For example, Fletcher believed in life beyond Earth, a factor in his making Viking a high priority not only for space science, but for NASA generally. He also backed other unpiloted space probes, such as Voyager to Saturn and Jupiter, and the Hubble Space Telescope. Moreover, Launius writes, the Mormon tradition of "stewardship" for Earth had impacts for Fletcher's goals as Administrator. He gave emphasis in his rhetoric and coalition building to the Earth-applications activity of NASA and told Congress NASA was an "environmental agency" as well as a space agency.

Under Fletcher, NASA's image broadened even though the Shuttle was the dominant program by far. Managerially, he decentralized operations, adopting a "lead Center" concept that gave JSC power over the Shuttle, a decision later

---

8. Roger Launius, "A Western Mormon in Washington, DC: James C. Fletcher, NASA, and the Final Frontier," *Pacific Historical Review* 64, no. 2 (May 1995): 214–217.

criticized after the *Challenger* Shuttle disaster. NASA did survive, thanks to Fletcher, reshaped to fit a far more austere environment than that of Apollo. His legacy was not complete when he left. He would return to NASA in the next decade.

5. Robert Frosch, June 1977–January 1981

Robert Frosch, age 49, was, like Fletcher, a Ph.D. physicist. He had risen to high positions in industry and government. Appointed by President Jimmy Carter, he found that his most important goal had to be sustaining the Shuttle development program he inherited. Shuttle cost overruns and schedule slippages became painfully obvious in the Carter years. Carter was not particularly interested in NASA, but he did have a military space priority. He wished to launch intelligence satellites to monitor weapons development, proliferation, and arms control agreements. The Department of Defense connection helped save the Shuttle from a Carter termination decision.[9] Like Fletcher, Frosch promoted the Shuttle as the nation's prime vehicle for scientific, military, and commercial access to space. Carter endorsed a "Shuttle-only" policy for major launches.

6. James Beggs, July 1981–December 1985

Aged 55, James Beggs was a U.S. Naval Academy graduate with a Harvard M.B.A. He had served in the Navy and held a middle-management position with NASA in the late 1960s. He was the second-ranking administrator at DOT in the 1970s. He was a chief executive officer (CEO) in industry before becoming President Reagan's appointee as NASA's leader in 1981.

Beggs was an astute, seasoned, and persistent Administrator. Quiet, but effective, he knew what his prime goals were when he came to NASA. They were to complete the development and testing of the Shuttle and move it to "operational" status. Then, he intended for NASA to take "the next logical step"—to develop a space station. The comprehensive package of programs that Paine had unsuccessfully tried to promote had been broken up by political reality. Successive Administrators were moving incrementally, one major piloted development program at a time. For Beggs, selling the space station would be his legacy. Like his predecessors, he believed that for starting big programs, coalition building in government began with the President.[10]

---

9. Logsdon, "Ten Presidents and NASA," p. 233.
10. Howard McCurdy, *The Space Station Decision: Incremental Politics and Technical Choice* (Baltimore, MD: Johns Hopkins University Press, 1990).

In the early 1980s, the Shuttle finished its tests and NASA declared it "operational," rhetoric that legitimated the transition of NASA, an R&D agency, to a new development program. Beggs faced many forces that wanted to thwart the station's initiation, including OMB, the White House science adviser, and the Secretary of Defense. Beggs proved to be an able bureaucratic politician. He had piqued Reagan's interest in space by inviting him to initial Shuttle launches. He made an end-run around his opposition, getting to Reagan via an interagency committee and making a presentation that appealed to Reagan's desire to use space to project national power, especially vis-à-vis the Soviets. In 1984, Reagan, in a State of the Union address, gave the go-ahead for the space station. Beggs began building a coalition for the station that was both domestic and international.

Sustaining various space science programs he inherited, providing essential additional funding to the Hubble Space Telescope, Beggs's most important other legacy was arguably his support for NASA to move more seriously into the global environmental field. He backed a gradual, growing involvement in global environmental problems, building on Fletcher's policies. NASA's emerging role reached high visibility in the mid-1980s when NASA provided scientific leadership in determining the causes of ozone depletion in the Antarctic.[11] What had been an "applications" program entailing weather and communication satellites in the 1960s was transformed by successive Administrators into what would eventually be called a "Mission to Planet Earth."

In late 1985, however, Beggs took leave from NASA to fight a false charge of wrongdoing while in industry, prior to becoming Administrator. In January 1986, NASA was devastated when the *Challenger* Shuttle exploded, killing the first teacher in space. NASA needed a strong Administrator at the helm and did not have one, and the Agency suffered accordingly.

Beggs's legacy was mixed. He got the space station decision and helped build a coalition of support that included international partners. In doing so, he initiated not only a new program of flagship proportion for NASA but also a model very different from Apollo in carrying out large-scale technology. It stressed the international dimension. But he used a rhetoric in selling the space station that made the Shuttle seem to be far more routine than it turned out to be in fact.

11. W. Henry Lambright, *NASA and the Environment: The Case of Ozone Depletion* (Washington, DC: NASA, 2005).

Figure 1: T. Keith Glennan, first NASA Administrator, 19 August 1958–20 January 1961, served under President Dwight D. Eisenhower. Figures 1 through 8 are the official portraits of the NASA Administrators, which hang in the Administrator's suite on the ninth floor of NASA Headquarters in Washington, DC. *NASA*

Figure 2: James E. Webb, second NASA Administrator, 14 February 1961–7 October 1968, served under Presidents John F. Kennedy and Lyndon B. Johnson. *NASA*

Figure 3: Thomas O. Paine, third NASA Administrator, 21 March 1969–15 September 1970, served under Presidents Lyndon B. Johnson and Richard M. Nixon. Paine was Acting Administrator beginning 8 October 1968. *NASA*

Figure 4: James C. Fletcher, fourth and seventh NASA Administrator, 27 April 1971–1 May 1977, served under Presidents Richard M. Nixon, Gerald R. Ford, and Jimmy Carter, and again for a second term, 12 May 1986–8 April 1989, under Ronald Reagan and George H. W. Bush. *NASA*

Figure 5: Robert A. Frosch, fifth NASA Administrator, 21 June 1977–20 January 1981, served under President Jimmy Carter. *NASA*

Figure 6: James M. Beggs, sixth NASA Administrator, 10 July 1981–4 December 1985, served under President Ronald Reagan. *NASA*

Figure 7: Richard H. Truly, eighth NASA Administrator, 14 May 1989–31 March 1992, served under President George H. W. Bush. *NASA*

Figure 8: Daniel S. Goldin, ninth NASA Administrator, 1 April 1992–17 November 2001, served under Presidents George H. W. Bush, Bill Clinton, and George W. Bush. Goldin was NASA's longest-serving Administrator, exceeding the combined tenures of James C. Fletcher. *NASA*

Figure 9: Sean O'Keefe, 10th NASA Administrator, 21 December 2001–11 February 2005, served under President George W. Bush. *NASA*

Figure 10: Michael Griffin, 11th NASA Administrator, 14 April 2005–20 January 2009, served under President George W. Bush. *NASA*

7. James Fletcher, May 1986–April 1989

President Reagan called Fletcher back to service in 1986. Fletcher's goal was to heal a wounded Agency battered by *Challenger*'s avalanche of criticism and lead its recovery. He guided NASA's return to flight, which took place in September 1988, and authorized many technical and organizational changes in NASA, including pulling power back to Headquarters for Shuttle management, a reversal of his earlier approach. He helped persuade Reagan to add a replacement for *Challenger*, maintaining the Shuttle fleet at four. He also adapted the space station, a project requiring modification in the face of technical and fiscal reality. Much of Fletcher's decision-making involved choices about priorities in what would launch and when, owing to the Shuttle's absence from service for 32 months. The "Shuttle only" policy he had pushed ended with *Challenger*. The Department of Defense and commercial flights would use expendable rockets. Fletcher still wanted as many NASA flights to go up on the Shuttle as possible. He believed deeply in the Shuttle and its promise.

Fletcher also wanted NASA to begin moving toward a mission beyond the space station and Earth orbit. He had Sally Ride, the first female astronaut, provide him with a menu of options for NASA's future. It included various missions, three of which focused on exploration, including a piloted flight to Mars. Fletcher established a new Office of Exploration. He apparently tried to get Reagan interested in exploration; he got some verbal support but little else. However, he did preside over NASA's return to flight and bolstered the morale of an agency profoundly in need of nurturing after *Challenger*.

8. Richard Truly, May 1989–March 1992

The man who served as Fletcher's Associate Administrator for returning the Shuttle to flight was Richard Truly. Truly was President George H. W. Bush's choice of NASA Administrator. Aged 52, Truly was a naval aviator with an aeronautical engineering degree. He was a former NASA astronaut with considerable experience in space. His prime goals as NASA Administrator were to get a fifth Shuttle orbiter and to further develop the space station. The latter was behind schedule, growing in cost, and fast losing congressional support. Indeed, money was being spent, but little or no hardware was being built owing to continuing design changes. President Bush gave him an additional goal, involving the Moon and Mars, one with which he did not appear to the White House to be fully comfortable.[12]

---

12. Thor Hogan, *Mars Wars: The Rise and Fall of the Space Exploration Initiative* (Washington, DC: NASA SP-2007-4410, 2007).

Bush had restored the National Space Council (NSC), a White House interagency body created when NASA came into existence and abolished by Nixon. It was headed by Vice President Dan Quayle, and its role was to set space policy for his administration. While there had been some human spaceflight planning at NASA for a return to the Moon and a Mars program, the stronger impetus for this new Moon-Mars mission came from the NSC. In 1989, the 20th anniversary of the first Apollo landing, Bush proclaimed his Space Exploration Initiative (SEI), a return to the Moon, followed by piloted missions to Mars.

The announcement seemed to come out of the blue, and there had been minimal political spadework with Congress, which was controlled by the Democrats. Congress was underwhelmed by the decision, as were the media and general public. The mood of the nation was to deal with a substantial budget deficit, not start a big new space program. The Cold War was ending; the nation looked to new priorities, but not any involving space. When NASA came up with the estimated cost for the program, perhaps half a trillion dollars, Congress called it "dead on arrival." Bush could not get the program funded.

Truly sought to start the SEI, raising the Office of Exploration to Associate Administrator level and appointing Mike Griffin its director. Griffin could plan, but not implement. Support further eroded in 1990 when the Hubble Space Telescope went up. It provided blurred images and caused NASA ridicule by late-night television comics. An independent panel headed by Norman Augustine called attention to NASA's many infirmities and called for substantial raises for an agency seeking to do more than it could afford. These large raises did not come.

Truly proved incapable of building coalitions for the SEI, and his personal commitment to the President's goal was doubted by Quayle and the NSC staff. He seemed overwhelmed, and his relations with the Vice President and NSC deteriorated. Bush asked for Truly's resignation in early 1992.

9. Daniel Goldin, April 1992–November 2001

Daniel Goldin was the longest serving and most controversial Administrator in NASA's history. He was 51 at the time of his appointment. He had begun his career at NASA shortly after college and then joined TRW, a large, California-based aerospace firm, where he had risen to vice president for space activities. Most of his work was in classified national security programs. He was an "outsider" to NASA and most space watchers when his name as Administrator was announced. An engineer and technical manager, he was closest to Paine in his sheer passion for space. But whereas Paine could not adjust his bold visions to a shrinking budget, Goldin almost found the constrained budget he encountered helpful. It enhanced his own power. Whereas Webb sought power to enlarge NASA and its role on the national stage, Goldin sought power to move the Agency in another direction. Webb, secure with bureaucracy, made NASA his ally. Goldin, distrustful of NASA managers, seemed at times at war with the agency he led. If Truly was uncomfortable with multiple goals, Goldin sought as many goals as he could establish. A self-described change agent, he made organizational chaos his handmaiden and revolution his byword. "Faster, better, cheaper" became his mantra. It fit the times, and his willingness to embrace this philosophy helped keep him NASA Administrator from Bush through two Clinton terms and briefly into the term of the second Bush.[13]

Goldin took the NASA job because he wanted to lead NASA back to Mars. He learned quickly that the SEI was not in the cards. Nevertheless, there was much else he wanted to do, and he lasted long enough to see both successes and failures in his attempts at innovation. He also held to his piloted Mars goal, and he sought to move NASA in that direction, but indirectly and, in some ways, covertly.

Mars was his love, but the space station was his necessary priority. It was NASA's flagship, whether he liked it or not; and it was in deep trouble when he arrived. In 1993, it almost died in Congress, surviving by one vote in the House. Goldin was not especially politically astute at first, but he learned quickly that he needed the President as an ally. He brought Clinton aboard the space station by linking it with Clinton's foreign policy need to help post-Cold War Russia turn to the United States. The space station became the ISS, with Russia the most significant of several international partners. Throughout the 1990s, Goldin struggled to make the U.S.-Russia partnership work and build a U.S.-Russian Space Station core. When he left in 2001, such a nucleus was in space and inhabited by U.S. and Russian astronauts. Getting the Space Station up was his major achievement.

His second major achievement concerned Mars. In 1993, the $1 billion Mars Observer, the first Martian flight since Viking in 1976, died on its way to the Red Planet. Goldin used the failure as an opportunity, remaking the Mars program into the symbol of his faster, better, cheaper robotic program. Similarly, in the mid-1990s, he used claims of fossil bacteria in a Mars meteorite to rekindle NASA's search for life beyond Earth. This came at a time when new planets were being discovered around extrasolar stars. What was called

13. W. Henry Lambright, "Leading Change at NASA: The Case of Dan Goldin," *Space Policy* 23 (2007): 33–43.

"exobiology" in the 1970s became "astrobiology" under change-agent Goldin. "Origins" was initiated as an exciting new program, building on the meteorite and the discovery of new planets. Having gotten the Hubble Space Telescope repaired in late 1993, Goldin wanted to build even more powerful telescopes capable of finding Earth-like planets beyond the Sun. He also sought to develop a Shuttle successor called the X-33. Goldin did all this on a budget that was stagnant at best, declining in real buying power for much of his tenure. He made NASA and himself poster children for Vice President Gore's "reinventing government" campaign, saying it was possible to get more for less.

For most of his time at NASA, Goldin was the best salesman the Agency had had since Webb. He was adept at rhetorical strategies, as exemplified by faster, better, cheaper. He gave a thousand speeches in his tenure and never had a presidentially appointed deputy. He often seemed a one-man show, one the media found fascinating. He was far better at building outside constituencies than inside support. Although some managers—Associate Administrator for Space Science Wesley Huntress especially—learned how to manipulate him to their advantage, many feared him, and he developed a reputation for shooting the messenger of bad news. He was forgiven by many for his inside harshness when the Pathfinder lander arrived successfully at Mars in 1997 and made faster, better, cheaper appear an outstanding success. Observer had failed, but Pathfinder worked at a fraction of Observer's price. An orbiter soon also performed marvelously at the Red Planet at this time, strengthening faster, better, cheaper's claims. So Goldin pushed harder, and he hit the boundary of faster, better, cheaper when the Mars Polar Lander and Mars Climate Orbiter failed in succession in 1999. Now his enemies, of which he had accumulated many, attacked faster, better, cheaper and Goldin. At the same time, the X-33, the Shuttle successor, one on which NASA had spent $1 billion, was killed by Goldin himself. Evidence of overreach by Goldin was everywhere.

Goldin accepted blame and called for NASA's adaptation to technical and fiscal reality. Before he left, he approved substantial funds for Spirit and Opportunity, two probes of Mars involving roving vehicles. These were funds that were geared to mission success, rather than held to an arbitrary cap. The great success of Spirit and Opportunity came after he departed.

Had Goldin left NASA immediately after Pathfinder in 1997, he might have escaped the rather mixed legacy he subsequently gained. But he stayed and saw his credibility and reputation fall. Like Webb after the Apollo fire, Goldin, after the 1999 failures, saw his administrative power wane. However, he did succeed in leaving NASA an agency with a far better image than it had when he came. Whatever his faults—perhaps in part because of his faults—Goldin made NASA interesting and even exciting again. He saved the Space Station. Through the robotic program and Mars rock, he put NASA on a trajectory to the Red Planet and restored interest in the search for extraterrestrial life. But he turned faster, better, cheaper from a means to an end. He also worsened a problem he inherited—an agency trying to do too much with too little.

10. Sean O'Keefe, December 2001–February 2005

It is not unusual for a President to select an Administrator who embodies characteristics he perceives missing in a NASA predecessor. Goldin, with his creative passion and vision, was seen as different from Truly. But Goldin was subsequently seen to be a bad manager by President George W. Bush, as witnessed by a $4.8 billion cost overrun on the Space Station the new President inherited in 2001. Bush wanted a "competent manager" and chose the Deputy Director of OMB, Sean O'Keefe.[14]

O'Keefe was 45 in age and had a career that made him one of the most government experienced Administrators in NASA history. He had worked on the Hill and been the comptroller of DOD and Secretary of the Navy. He had a master's degree in public administration and had taught the subject at his alma mater, the Maxwell School of Syracuse University. O'Keefe had good inside-the-beltway political skills, possibly the best since Webb. He was not a stirring speaker and certainly not a space cadet.

In his first major address, he appalled many space enthusiasts when he said he wanted NASA to be "science-driven," rather than "destination-driven." His primary goal at the outset was to strengthen the Space Station program financially. What came to be his second goal was to address the Shuttle-successor problem, not through Goldin's technologically revolutionary X-33, but through an interim system called the Orbital Space Plane. The Orbital Space Plane would complement the Shuttle and extend its life several years.

However, in February 2003, the *Columbia* Shuttle disintegrated, killing all aboard. Whatever goals O'Keefe had in 2002 were now on hold and subject to change. *Columbia* came to define his tenure at NASA. He got NASA through the disaster and the investigation that followed with relatively modest organizational damage. However, the report of the Columbia Accident Investigation Board forced him and everyone else to face a reality that the Shuttle was old, flawed, and high-risk. O'Keefe, because of *Columbia*, had a window of opportunity for

---

14. W. Henry Lambright, "Leadership and Change at NASA: Sean O'Keefe as Administrator," *Public Administration Review* (March/April 2008): 230–240.

major policy change. The Columbia Accident Investigation Board helped him think big by stating that NASA needed a significant goal—worthy of risking human life—beyond the Space Station.

O'Keefe's inside-coalitional skills and close relations with Vice President Cheney in particular now came to the fore. He elevated and altered an existing body in the White House that was meeting to consider options for a post-*Columbia* piloted space program. He transformed it into an interagency group with high-ranking members that met periodically and behind closed doors. The group came up with the return to the Moon by 2020 as a goal. Bush himself added Mars, perhaps with a nod to his father's failed Moon-Mars program. O'Keefe then hammered out a budget agreement with his old colleagues at OMB that appeared plausible at the time: the Shuttle's expenses would go down as expenditures for a successor went up. In 2010, the Shuttle would complete its work on the Space Station and be retired; in 2014, its successor would come on line. A master budgeter, O'Keefe kept projections of expenses for the new technological system vague. The Moon got emphasis, not Mars. The "sticker-shock" price that killed the SEI under the first Bush was not repeated under his son.

It was ironic. O'Keefe had come in eschewing destination-driven goals and now, once Bush announced his decision in January 2004, he was trying to sell a Moon-Mars-and-beyond goal! O'Keefe went about his task and reorganized NASA for implementation. But before he could start a major congressional, media, and public campaign, he found himself sharply on the defensive. In the wake of *Columbia*, he wanted to change the culture of NASA, and that required, in his judgment, changing the Agency's mindset from "prove to me it's unsafe" to "prove to me it's safe." Using that standard, he could not approve a final servicing mission to the immensely popular Hubble Space Telescope. When word of that decision leaked, O'Keefe was tarred as a bean counter who was trading off the Hubble Space Telescope to get money for Moon-Mars. That was not how O'Keefe saw it, but how his opposition framed the decision. O'Keefe considered a robotic approach to Hubble Space Telescope repair, but a National Academy of Sciences and National Research Council panel told him that method would not work in time to save the Hubble Space Telescope.

He soldiered on with respect to Moon-Mars, aided enormously by the Republican majority in Congress and support for funding the startup of the program by powerful lawmakers he and Cheney enlisted. There was thus a White House and Congress political coalition to begin the new program—what was lacking in 1989. The budget strategy, whatever its merit or demerit, conveyed the image of a pay-as-you-go philosophy instead of the SEI's half-trillion price tag. O'Keefe became the progenitor, therefore, of a destination-driven program. He convinced few critics that he was right about the Hubble Space Telescope, one large reason being that Harold Gehman, chair of the Columbia Accident Investigation Board, did not give him the support he had to have from that quarter to help neutralize opposition.

A "competent manager" who was also an experienced bureaucratic politician, O'Keefe succeeded in getting a Moon-Mars program adopted, but he departed before the new mission's implementation was fully under way.

11. Michael Griffin, April 2005–January 2009

Michael Griffin is the Administrator who most recently completed his tenure at the time of writing this profile.[15] Hence, this assessment is somewhat premature, his legacy being dependent on what President Barack Obama decides to do about NASA. However, certain observations are possible for the Bush period. Griffin, aged 55, came to NASA with a long and deep experience in the space field, indeed a lifetime of interest and work. A man with a Ph.D. in aeronautical engineering and many other degrees, Griffin began his career with NASA via JPL. He served as Associate Administrator for Exploration under Truly and Goldin before the senior Bush's Moon-Mars program was killed at the outset of Clinton's administration. He was head of space research and development for APL of the Johns Hopkins University when he was appointed to head NASA. He was coauthor of a book, *Space Vehicle Design*. Like Paine and Goldin, he had a passion for space, but he displayed that passion more quietly than they. It is hard to imagine anyone who could have been better equipped technically or, probably, managerially to implement the Moon-Mars decision. Indeed, Griffin's prime mandate as NASA Administrator was clear: carry out the Vision for Space Exploration, as it was then called. But many an implementer has to worry about politics. Policy decisions do not leave political and bureaucratic conflict at the adoption door. Typically, the struggle goes on. Griffin wished to move the President's decision down the hardware path so solidly that when he left, his successor would seamlessly take the program the next step toward fruition. Like other Administrators, he had to continually sell the lead program and build a support system for it while fending off opponents of his goals.

O'Keefe got the Moon-Mars decision during the window of opportunity created briefly by *Columbia* and the Columbia Accident Investigation Board. Griffin faced a closing window for implementing this initiative. The Bush administration

15. W. Henry Lambright, *Launching a New Mission: Michael Griffin and NASA's Return to the Moon* (Washington, DC: IBM, 2009).

was weaker in 2005 than in 2004 because of an unpopular war, high deficits, and a declining economy. Hurricane Katrina also occurred at a most inopportune time for Griffin, just as he was trying to ramp up Moon-Mars funding in the fall of 2005. His political/budgetary situation grew subsequently worse.

When Griffin became NASA Administrator, he set three major goals: to return the Shuttle to flight, to finish the Space Station, and to implement the Moon-Mars decision. Griffin said at his confirmation hearings he would try to bring the Shuttle successor into service sooner than 2014. He also said he would revisit the Hubble Space Telescope decision. He took office in April 2005. Griffin moved as quickly as he could on all fronts once in office.

He succeeded in returning the Shuttle to flight and made good progress in finishing the ISS. He reversed O'Keefe's Hubble Space Telescope decision. He came up with a design for the Shuttle successor and return-to-the-Moon system, called Orion-Ares. It was Shuttle-derived and looked much like Apollo. The capsule was Orion and the rocket, Ares I. There would be a large Ares V rocket for cargo shipments to the Moon, as well as a lunar lander. The whole system was called Constellation. First came Orion-Ares. Griffin called Orion-Ares "Apollo-on-steroids." The tentative cost for a Moon return was $104 billion.

He hit a roadblock to his personal goal of accelerating implementation in narrowing the Shuttle/Orion-Ares gap in the fall of 2005. He discovered that the Shuttle budget projections he inherited were unrealistic. The Shuttle costs were not going to go down appreciably prior to 2010. Where would money come from to make up the difference? In announcing the design decision (Orion-Ares), he had said he would not take "one thin dime" from space science to pay for the return-to-the-Moon system. But when he found he needed additional money, OMB refused to revisit the budget projections set in 2004. It proposed that Griffin kill the Shuttle early to get the money, a non-option for Griffin.

The Administrator appealed the budget issue to the White House and Bush in late 2005. With the Iraq war raging and Katrina expenses growing, the time to ask for more money was not propitious. A meeting took place, and he came away with a modest raise and instructions to get the rest of the needed money from reprogramming funds. Griffin took the funds from Orion-Ares and space science to pay for the Shuttle costs. Many space scientists subsequently criticized Griffin and the Moon-Mars program. The result of budget constraints was that Griffin's goal to narrow the gap between the Shuttle and Orion-Ares was set back. He regarded this gap as a serious problem not only for NASA but also for the nation, as it meant primary reliance on the Russians to get to the Space Station after 2010. Consequently, he was hard-pressed to keep the gap at the four-year period he inherited.

In January 2007, the Democrats took control of Congress. Bipartisan support for Moon-Mars continued in general, but not so much support as to give Griffin the money he wanted. Also, White House and congressional fighting over the war in Iraq and other policy issues caused Congress to pass a continuing resolution, which froze NASA's budget for fiscal year 2007 to what it had been in fiscal year 2006. That meant a de facto cut below the presidential request for this year. The result was slippage of Orion-Ares to March 2015. The following year, Congress gave NASA the full amount requested for fiscal year 2008, $17.3 billion. However, the fiscal year 2009 appropriation to be voted on in fall 2008 was also subject to a continuing resolution, and that meant no raise until after the new President took office in January 2009, if then.

Thus, the Griffin era under Bush was largely one in which he fought to implement the Moon-Mars decision in a dreadful budget and political environment. Griffin's great asset was his focus, his ability to keep his eye on Orion-Ares as a priority. He therefore made progress in Moon-Mars implementation in spite of the unfavorable situation he faced. But he did not make the progress he would have liked, owing to his inability to persuade the Bush administration and Congress of the seriousness of the gap and need to fund implementation to narrow it. He could not assuage the scientific community. He could not create or use a bureaucratic power that might have come with a strong and united constituency. Without that power, he could not get the resources he needed.

Griffin's problems were not technical or managerial, but political. Political rhetoric and coalition building were not Griffin's strengths, and he operated in a political environment that would be difficult for almost anyone. In the fiscal year 2005–2010 period, NASA was projected to have $3.9 billion less than was originally proposed in 2004, when the Moon-Mars program was adopted. Griffin made progress as an implementer, but he fell short of what he wanted, and what he believed was vital in the national interest. Then-President-elect Obama promised to provide $2 billion extra to NASA's budget to possibly speed the deployment of Orion-Ares. That may or may not happen under President Obama. Also, the Shuttle's service may also be extended, an extension Griffin did not favor. That the new President appreciated the Shuttle-Shuttle successor gap problem is a testament to Griffin's clarity in sounding an alarm. Whether the new President or his NASA appointee as Administrator does anything about it would not be subject to Griffin's control.

Conclusion

All the various NASA Administrators have been significant in one way or another in the history of the space agency. Each has brought a particular background, training, and set of skills to the Administrator role. All have been important, particularly as "relay leaders." Administrators invariably serve briefer terms than it takes to adopt and fully implement the programs they run. The ISS, for example, began under Beggs and was still unfinished when Griffin completed his tenure with President George W. Bush.[16] They make their mark primarily as they affect the course of these large-scale, long-term programs. Some are initiators; most are implementers; some are saviors. It is also notable that a few stand out as especially influential.

Webb, in particular, is larger than life. He shaped and carried out a program, Apollo, in such a way that it stands as an icon for effective government. He made the most of a rare opportunity on an historic stage and exemplifies what is possible when there is a strong match among man, organization, and time. Fletcher, Beggs, and O'Keefe got presidential decisions for flagship programs; and Frosch and Goldin got presidential decisions that saved two of these programs. Griffin is critical as a relay leader, implementing a program that may (if sustained by successors) at last get NASA out of Earth orbit and back to where it was in the 1960s. Glennan got the Agency and a range of new programs, including the beginning human spaceflight effort, under way. Paine completed Apollo, and Truly maintained the Shuttle and Space Station, but both illuminate what can happen when an Administrator is out of sync with the President.

It is not easy to be the NASA Administrator. He (and eventually she) is CEO of NASA but not of the United States government. NASA leaders need help to be successful. To get help, primarily on their terms, they have to blend political and administrative skills. Success in one provides success in the other. Failure in one leads to difficulty, if not failure, in the other. Typically, a leader has to link a mission that is purely discretionary—space exploration—with national interests that appeal to politicians, media, and the public. A book written years ago had a title that summed up prime motivations that have usually applied to one degree or another as space rationales, at least for human spaceflight: *Pride and Power: The Rationale of the Space Program*.[17]

Space is a source of national pride and a visible demonstration of national power and competence by nonmilitary means. Such values matter to politi-

16. W. Henry Lambright, "Leadership and Large-Scale Technology: The Case of the International Space Station," *Space Policy* 21 (2005): 195–203.
17. Vernon Van Dyke, *Pride and Power: The Rationale of the Space Program* (Urbana, IL: University of Illinois, 1964).

cians and the people they represent. Administrators have to couch their arguments for resources in terms politicians can understand and the public can appreciate. The different Administrators have varied in their political skills in using rhetorical strategies to attract broad public support and build coalitions among Washington, DC, power elites based on mutual interests and quid pro quos. Webb and Goldin were the best rhetorical Administrators. Webb understood how important it was for NASA to keep in the public eye and show success to Presidents. He also knew how to build coalitions with Congress through rewards and sanctions. Goldin's rhetoric was attractive, especially to Vice President Gore, but his coalitions did not seem at times to include his own Agency. Webb, Fletcher, and O'Keefe had additional burdens of being disaster-recovery executives. Most have had to make Hobson's choices—decisions with no real positive alternative.

NASA's leaders have to be good managers, for failure in performance of the Agency usually has retribution in the political arena. Politics and administration go together, as they always have and always will.

Whatever their strengths and weaknesses—and all the NASA leaders had both—they, as a group, did their best and have kept the Agency sailing "this new ocean" of space, often under extremely adverse conditions. NASA and the nation have been fortunate to have had such an able and dedicated set of leaders.

Chapter 5

Space Access

NASA's Role in Developing Core Launch-Vehicle Technologies

J. D. Hunley

Over the past 50 years, NASA has played a significant role in developing the nation's core space launch-vehicle technologies, but it has not done so alone. In typical NASA fashion, the Agency has partnered with military services, private industries, and universities to gain access to space. Since many of the key launch-vehicle technologies first appeared in missiles, the Army, Navy, and Air Force oversaw their development. NASA then borrowed and adapted them for use in launch vehicles. Also, because NASA did not exist until 1 October 1958, its predecessor, the NACA, began some developments that NASA then continued; this also happened when NASA absorbed the von Braun group and JPL from the Army and the Vanguard Project from the Navy. Finally, even after NASA's own space launch-vehicle activities were well established, in the Cold War environment down to 1991 NASA and the military services continued to cooperate in the further development of key launch-vehicle technologies.[1]

Viking and Vanguard

One example in this complicated pattern of innovative cooperation was the Vanguard launch vehicle. Vanguard, which was more successful than many

1. With permission from the publishers, this article is adapted from materials in J. D. Hunley, *The Development of Propulsion Technology for U.S. Space-Launch Vehicles, 1926–1991* (College Station, TX: Texas A&M University Press, 2007); J. D. Hunley, *Preludes to U.S. Space-Launch Vehicle Technology: Goddard Rockets to Minuteman III* (Gainesville, FL: University Press of Florida, 2008); and *U.S. Space-Launch Vehicle Technology: Viking to Space Shuttle* (Gainesville, FL: University Press of Florida, 2008).

people think, had its technical beginnings in the Viking project. Milton W. Rosen at NRL provided the technical leadership for both projects and then become the director of launch vehicles and propulsion in NASA's Office of Manned Space Flight Programs. To prepare for development of the Viking sounding rocket, he arranged to work for eight months from 1946 to 1947 at JPL, which was then under contract with the Army. From 1949 to 1955 Rosen's NRL team—with contractors from the Glenn L. Martin Company and Reaction Motors plus advice from Albert C. Hall—launched 12 Viking rockets that pioneered the use of a gimballed engine for steering and prepared Martin engineers and Rosen for the Vanguard project. Also, Viking's early use of aluminum as a structural material showed the way to other rockets that used this lightweight metal.[2]

The Vanguard project began in the fall of 1955 with a contract between the Office of Naval Research and the Martin Company for design, construction, and preflight testing of a vehicle to launch a U.S. satellite for the IGY (from 1 July 1957 to 31 December 1958). On 30 November 1958, close to the end of the project, NASA took over responsibility for Vanguard. Meanwhile, on 6 December 1957, the first launch (intended as a test) with a small satellite on board was a spectacular failure. But between 17 March 1958 and 18 September 1959, Vanguard vehicles successfully orbited three satellites in nine attempts. This was not a high success rate, but to achieve it the Vanguard team had to overcome problems with all three stages of the vehicle, including two different solid-propellant third stages. It placed a satellite into orbit before the end of the IGY and did so despite a low DOD priority and instructions not to interfere with high-priority missile projects. With the Thor missile as a first stage, the Air Force used modified Vanguard second and third stages in the Thor-Able launch vehicle. A variant of the other third stage, designed by the Allegany Ballistics Laboratory (ABL), became a third stage for the Delta launch vehicle and a fourth stage for the Scout launch vehicle. A follow-on version of this third stage became the third stage for Minuteman I. The strap-down guidance

2. Milton W. Rosen, *The Viking Rocket Story* (New York, NY: Harper & Brothers, 1955), pp. 18–23, 26, 28, 58–62, 64, 66, 236–237; comments of Rosen on a draft treatment of Viking, 8 May 2002, and in a telephone conversation with Hunley, 16 and 17 May 2002; interview with Milton William Rosen by David DeVorkin, Washington, DC, 25 March 1983, pp. 8–31, 44, 52–53, copy available through the Space History Division, Smithsonian National Air and Space Museum (NASM); The Glenn Martin Company, "Design Summary, RTV-N-12 Viking, Rockets 1 to 7," January 1954, pp. 36–37, seen in NASM Archives folder OV-550500-05, "Viking Sounding Rocket." At the time Rosen did his apprenticeship at JPL, it was probably the preeminent rocket propulsion organization in the United States. For an overview of its achievements, see Hunley, *Development of Propulsion Technology*, pp. 16–20.

and control system for Vanguard later found extensive use in launch vehicles. And Vanguard improved the gimballing system used on Viking, a technology later employed extensively on missiles and launch vehicles.[3]

These were significant contributions. Taking over the program so late in the game, NASA did not participate significantly in making them, but the space agency inherited many of the NRL engineers involved in the program, and they contributed to other launch vehicles including Delta.

Delta

In January 1959, Rosen suggested to Abe Silverstein, NASA's Director of Space Flight Programs, that the Air Force's Thor-Able launch vehicle be modified to become the Delta launch vehicle (initially called Thor-Delta). Rosen proposed a more reliable electronics control system than the one on Vanguard, a stainless-steel combustion chamber (instead of the aluminum one used on the second stage of Vanguard) and use of the Bell Telephone Laboratories radio guidance system designed for the Titan ballistic missile, among other changes. Silverstein agreed to these suggestions, and the original Delta launch vehicle thus became an amalgam of NRL and Air Force technologies brought together by Rosen as a NASA engineer—an example of the intricate way in

3. "Project Vanguard, a Scientific Earth Satellite Program for the International Geophysical Year," a report to the Committee on Appropriations, U.S. House of Representatives, by Surveys and Investigations Staff [ca. 15 February 1959], pp. 61–62, 65–66, found in NASM Archives, folder OV 106120-01, "Vanguard II Launch Vehicle 4"; Project Vanguard Staff, "Project Vanguard Report of Progress, Status, and Plans, 1 June 1957," Naval Research Laboratory Report 4969 (Washington, DC: Naval Research Laboratory, 1957), pp. 2-6 to 2-7, 2-25 to 2-32, 2-34, 2-37 to 2-38, 2-41 to 2-42, 2-45 to 2-50, folder 006601, "Vanguard Project: Origins and Progress Reports," NASA Historical Reference Collection, NASA History Division, NASA Headquarters, Washington, DC; Kurt R. Stehling, *Project Vanguard* (Garden City, NY: Doubleday, 1961), pp. 24, 64–66, 82–83, 103, 106–122, 128–130, 132–135, 156–242, 269–281, 301; Constance McLaughlin Green and Milton Lomask, *Vanguard: A History* (Washington, DC: Smithsonian Institution Press, 1971), pp. 57–90, 176–182, 196–198, 204, 210, 213–214, 219, 254–255, 283, 285, 287; John P. Hagen, "The Viking and the Vanguard," in *The History of Rocket Technology: Essays on Research, Development, and Utility*, ed. Eugene M. Emme (Detroit, MI: Wayne State University Press, 1964), pp. 127–128, 140; R. Cargill Hall, "Origins and Development of the Vanguard and Explorer Satellite Programs," *Airpower Historian* 11, no. 1 (January 1964): 109, 111; Project Vanguard Staff, "Project Vanguard Report No. 9, Progress through September 15, 1956," 4 October 1956 (Washington, DC: Naval Research Laboratory, 1956), pp. 5, 7, folder 006601, seen in "Vanguard Project Origins and Progress Reports," NASA Historical Reference Collection, NASA History Division, NASA Headquarters, Washington, DC; The Martin Company, "The Vanguard Satellite Launching Vehicle: An Engineering Summary," Engineering Report No. 11022, April 1960, pp. 4, 49–55, 63, cataloged as a book in the NASA History Office; Kurt R. Stehling, "Aspects of Vanguard Propulsion," *Astronautics* (January 1958): 45–46; Milton Rosen letter to J. D. Hunley, 8 May 2002; document titled "Vanguard Vehicle Characteristics," n.d. [after 16 December 1959], seen in NASM Archives folder OV-106015-01, "Vanguard Project History."

which NASA, the military services, and private contractors interacted in their contributions to launch-vehicle technology.[4]

Beginning with the Thor-Able, the Thor series of launch vehicles was essentially developed by the Air Force but sometimes used by NASA, while the Delta was primarily a NASA launch vehicle until the 1990s, sometimes used by the Air Force. Rosen contracted with the Douglas Aircraft Company to develop the Delta. NASA and Douglas used components that had already been proven in flight, reducing the need for developmental flights. Besides Rosen, many of the NASA personnel working on Delta were former Vanguard personnel, now working either at NASA Headquarters or at the new GSFC in nearby Maryland. Among those at GSFC, William R. Schindler headed a small group that provided direction and technical monitoring of Delta development.[5]

Throughout its history, NASA's Delta increased its payload capacity through a series of models, uprating existing components or adopting new ones that had already proven themselves. The first Delta model could launch 100 pounds of payload to geosynchronous transfer orbit (GTO). Beginning in 1962, a series of Delta models including A, B, C, D, E, J, L, M, M-6, N, 900, 904, 2914, 3914, 3910/Payload Assist Module (PAM), 3920/PAM, 6925 (Delta II), and 7925 (also a Delta II, introduced in 1990) successively increased their payload capabilities, with the Delta 3914 (introduced in 1975) able to lift 2,100 pounds to GTO and the 7925, 4,010 pounds.[6]

---

4. Milton W. Rosen, "A Brief History of Delta and Its Relation to Vanguard," enclosure in letter, Rosen to Constance McLaughlin Green, 15 March 1968, folder 001835, "Rosen, Milton W. (Misc. Bio)" NASA Historical Reference Collection, NASA History Division, NASA Headquarters, Washington, DC; Hunley, *Development of Propulsion Technology*, pp. 31–32.
5. Rosen, "Brief History," pp. 1–3; William R. Corliss, draft, "History of the Delta Launch Vehicle," with comments by L. C. Bruno, 4 September 1973, folder 010246, "Delta Development (1959–1972)," NASA Historical Reference Collection, NASA History Division, NASA Headquarters, Washington, DC, pp. 2-6, 2-12 to 2-13, 3-1 through 3-4, 3-9; V. L. Johnson, "Delta," NASA Program Review, "Launch Vehicles and Propulsion," 23 June 1962, pp. 51, 54–56, 67–68; J. D. Hunley, ed., *The Birth of NASA: The Diary of T. Keith Glennan* (Washington, DC: NASA SP-4105, 1993), pp. 336–337, 357; W. M. Arms, *Thor: The Workhorse of Space—a Narrative History* (Huntington Beach, CA: McDonnell Douglas Astronautics Company, 1972), pp. 6-49 to 6-50; Schindler's obituary in the *Washington Post* (29 January 1992); Stuart H. Loory, "Quality Control . . . and Success," *New York Herald Tribune* (21 April 1963): 4.
6. GSFC, "The Delta Expendable Launch Vehicle," NASA Facts, [1992?], folder 010240, "Delta," NASA Historical Reference Collection, NASA History Division, NASA Headquarters, Washington, DC, p. 9; David Ian Wade, "The Delta Family," *Spaceflight* 38 (November 1996): 373; Jyri Kork and William R. Schindler, "The Thor-Delta Launch Vehicle: Past and Future" (paper SD 32 presented at the 19th Congress of the International Astronautical Federation, New York, NY, 13–19 October 1968), p. 4; J. F. Meyers, "Delta II—A New Era Under Way" (paper 89-196 presented at the 40th Congress of the International Astronautical Federation, Málaga, Spain, 7–12 October 1989), p. 1.

During this period, NASA and its contractors enhanced the capabilities of the first and upper stages, increased the size and length of the tanks in the first two liquid-propellant stages, upgraded the third-stage motors and guidance systems, and added incrementally larger and more numerous solid strap-on motors to improve the boost from the launchpad. As suggested already, most of these changes were adapted from other vehicles, so they constituted NASA contributions to launch technologies only in the way they were adapted to the Delta configurations. But that was itself a significant contribution. For the Delta E, notably, the third-stage motor, the FW-4 (also employed as a fourth-stage motor on the Scout Standard Launch Vehicle), used a case made of fiberglass developed by the Owens-Corning Company that included a low-density silica material weighing 35 percent less than materials used by competitors. Built by the United Technology Corporation of United Aircraft, the FW-4 also had an innovative igniter made of aluminum, which itself burned up during combustion and contributed to propulsion.[7]

A more significant improvement came in the Delta 900 when a strapped-down inertial guidance system replaced the radio-inertial guidance system used previously. The inertial measurement unit for the new system derived from the abort gyro package for the Apollo Lunar Module. But the Delta Inertial Guidance System, as the new system was called, adopted a different digital computer from the one used on the Lunar Module, choosing instead a computer Teledyne had developed for the Centaur upper stage. Once introduced about 1970, it furnished navigation, guidance, and control for the first and second stages of the Delta.[8]

Another major upgrade for the Delta was the introduction of the RS-27 as the stage one engine, derived from the Rocketdyne H-1 developed for the Saturn I. In 1971, McDonnell Douglas, contractor for the Thor first stage,

---

7. Meyers, "Delta II," pp. 1–2; Kevin S. Forsyth, "Delta: The Ultimate Thor," in Roger D. Launius and Dennis R. Jenkins, eds., *To Reach the High Frontier: A History of U.S. Launch Vehicles* (Lexington: University Press of Kentucky, 2002), p. 117; Kork and Schindler, "Thor-Delta," pp. 3–4, 6–7; GSFC, "Delta Expandable Launch Vehicle," p. 9; Chemical Propulsion Information Agency, *CPIA/M1 Rocket Motor Manual*, vol. 1 (Laurel, MD: CPIA, 1994), unit 480, FW-4.
8. E. W. Bonnett, "A Cost History of the Thor-Delta Launch Vehicle Family" (paper A74-08 presented at the 25th Congress of the International Astronautical Federation, Amsterdam, 30 September to 5 October 1974), pp. 6, 8, 17, 20; "The NASA/Grumman Lunar Module," pamphlet in NASM Archives, folder OA-69200-01, "Apollo Lunar Module"; information from procurement list for the Lunar Module, by telephone from Joshua Staff, Cradle of Aviation Museum, Garden City, NY; Pat M. Kurten, "Apollo Experience Report—Guidance and Control Systems: Lunar Module Abort Guidance System," TN D-7990 (Washington, DC: NASA, 1975), p. 9; Charles R. Gunn, "The Delta and Thor/Agena Launch Vehicles for Scientific Applications Satellites," GSFC preprint X-470-70-342, 1970, NASM Archives, folder OD-240014-02, "Delta Launch Vehicle, NASA Report," p. 7.

Figure 1: A Delta E rocket launching from KSC on 5 December 1968 carrying Highly Eccentric Orbit Satellite 1 for the European Space Research Organization. Notice the strap-on motors at the base of the Delta. *NASA*

subcontracted with Rocketdyne for development, testing, and production of the RS-27 by modifying the H-1 for compatibility with the Thor airframe. The RS-27 provided about 207,000 pounds of thrust at sea level, 2,000 more than the most powerful H-1 and 37,000 more than the previous MB-3 Block II engine used on the Thor.[9]

To give but one more instance of the improvements to the Delta, in 1980 it featured a new third stage, the PAM with a Thiokol Star 48 motor. Developed starting in 1976 for use with the Space Shuttle, the PAM was derived from the propulsion unit for Minuteman stage 3, which Thiokol began producing in 1970 using an Aerojet design. On the Shuttle, the PAM propelled satellites to higher orbits than the Shuttle's roughly 160-mile range. The PAM used the same hydroxyl-terminated polybutadiene (HTPB)-ammonium perchlorate-aluminum propellant as Thiokol's Antares IIIA rocket motor, itself a third-generation, third-stage propulsion unit for the Scout launch vehicle. These examples show the complicated ways in which different rocket programs influenced and borrowed from one another. Thiokol also used titanium for the PAM's Star 48 motor case and the recently developed carbon-carbon composite for the nozzle's exit cone.[10]

NASA planned to discontinue use of the Delta in the late 1980s in favor of reliance on the Space Shuttle, but with the *Challenger* disaster on 28 January 1986, the Air Force awarded a contract to McDonnell Douglas for the first Delta IIs, with more contracts from both NASA and the Air Force to follow. In 1989, NASA turned over to the Air Force responsibility for managing Delta. As more an Air Force than a NASA launch vehicle after that, it evolved through

---

9. Paul N. Fuller and Henry M. Minami, "History of the Thor/Delta Booster Engines," *History of Liquid Rocket Engine Development in the United States, 1955–1980*, ed. Stephen E. Doyle, American Astronautical Society History Series, vol. 13 (San Diego, CA: Univelt, 1992), pp. 41, 46–47; Chemical Propulsion Information Agency, *CPIA/M5 Liquid Propellant Engine Manual* (Laurel, MD: CPIA, 1994), unit 85, LR79-NA-11, unit 173, H-1 Booster Engines, and unit 196, RS2701A.
10. Thiokol General Corporation, *Aerospace Facts*, house organ, quarterly (spring 1979): 21, 25, NASM archives, folder B-7-820030-02, "Thiokol General Publications, 'Aerospace Facts'." On the Antares III, see *CPIA/M1*, unit 577, Antares IIIA, and John W. R. Taylor, ed., *Jane's All the World's Aircraft, 1977–1978* (London, U.K.: Jane's Yearbooks, 1978), p. 837. On Minuteman III, see Hunley, *Preludes*, chap. 9, and *CPIA/M1*, units 457 (Aerojet's version) and 547 (for Thiokol's similar variant). On the development of HTPB, see Hunley, *Preludes*, p. 324. On carbon-carbon, see S. Luce, "Introduction to Composite Technology," prepared for Cerritos College Composite Technician Course, n.d., p. 6, filed as a book in NASM library; Andrew C. Marshall, *Composite Basics*, 3rd ed. (Walnut Creek, CA: Marshall Consulting, 1993), pp. 1–4; Julius Jortner, "Analysis of Transient Thermal Responses in Carbon-Carbon Composite," in *Thermochemical Behavior of High-Temperature Composites*, ed. Julius Jortner (New York, NY: American Society of Mechanical Engineers, 1982), p. 19.

Delta III and Delta IV configurations to become part of the Air Force's Evolved Expendable Launch Vehicle (EELV) program.[11]

Scout

Even more than Delta, the Scout family was a NASA contribution to launch-vehicle technology. From 1956 to 1957, an imaginative group of engineers, including later spacecraft designer Maxime A. Faget and promoter of spherical motors Joseph G. "Guy" Thibodaux, Jr., conceived of Scout at the Pilotless Aircraft Research Division on Wallops Island, Virginia, then part of the NACA and its Langley Aeronautical Laboratory. Their idea was to combine four solid-propellant motors to create an all-solid-propellant launch vehicle. By the time that NASA had absorbed the NACA, the Air Force had become interested in the concept and reached an agreement with the new space agency to develop the vehicle. Once that had occurred, the military air arm would consider modifying Scout for its own use, calling the result the Blue Scout.[12]

The original NASA Scout was developed from an Aerojet motor called the Jupiter Senior because that firm had produced it as part of the attempt

---

11. Steven J. Isakowitz, *International Reference Guide to Space Launch Systems* (Washington, DC: American Institute of Aeronautics and Astronautics [AIAA], 1991), p. 203; "Delta II Becomes New Medium Launch Vehicle," *Astro News* 29, no. 2 (23 January 1987), microfilm roll K168.03-2849, Air Force Historical Research Agency (AFHRA), Maxwell AFB, AL; Luis Zea, "Delta's Dawn: The Making of a Rocket," *Final Frontier* (February–March 1995): 46; Frank Colucci, "Blue Delta," *Space* 3 (May–June 1987): 42; James M. Knauf, Linda R. Drake, and Peter L. Portanova, "EELV: Evolving toward Affordability," *Aerospace America* (March 2002): 38, 40; Joseph C. Anselmo, "Air Force Readies Pick of Two EELV Finalists," *Aviation Week & Space Technology* (9 December 1996): 82–83; GSFC, "Delta Expendable Launch Vehicle," p. 3.
12. James R. Hansen, "Learning through Failure: NASA's Scout Rocket," *National Forum* 81, no. 1 (winter 2001): 18–23; Abraham Leiss, "Scout Launch Vehicle Program Final Report—Phase VI," NASA Contractor Report 165950, pt. 1, May 1982, pp. xxxiii, 31–32, 53–54, 56, 65, 90, 131 (available through NASA libraries on microfiche X82-10346); Matt Bille, Pat Johnson, Robyn Kane, and Erika R. Lishock, "History and Development of U.S. Small Launch Vehicles," in *To Reach the High Frontier: A History of U.S. Launch Vehicles*, ed. Roger D. Launius and Dennis R. Jenkins (Lexington, KY: University Press of Kentucky, 2002), pp. 204–213; GSFC, NASA Facts, "NASA's Scout Launch Vehicle," April 1992; Jonathan McDowell, "The Scout Launch Vehicle," *Journal of the British Interplanetary Society* 47 (March 1994): 102–107 (henceforth, *JBIS*); James R. Hansen, *Spaceflight Revolution: NASA Langley Research Center from Sputnik to Apollo* (Washington, DC: NASA SP-4308, 1995), pp. 197–200, 209–210, 214–217; Joseph Adams Shortal, *A New Dimension: Wallops Island Flight Test Range, the First Fifteen Years* (Washington, DC: NASA Reference Publication 1028, 1978), pp. vii, 484–573, 702–709, 712, 716–717, 720; "History—Blue Scout," in the archives of the U.S. Air Force Space and Missile Systems Center History Office (SMC/HO), Los Angeles Air Force Base, CA, "Blue Scout Chronology," entries for August 1958, 14 October 1958, 3 February 1959, 24 February 1959, 8 May 1959, 24–29 May 1959, 9 September 1959, 2 December 1959, 7 December 1959, 9 March 1960, 30 June 1960, 13 September 1960, and 21 September 1960 (hereafter cited as "Blue Scout Chronology" and "Blue Scout History")—1 January 1962 through 20 June 1962 contained in the chronology; Andrew Wilson, "Scout—NASA's Small Satellite Launcher," *Spaceflight* 21, no. 11 (November 1979): 449–452.

to convert the Jupiter missile to a solid-propellant version for use at sea; a Thiokol motor was created by changing the binder for JPL's Sergeant motor to a polybutadiene-acrylic acid substance with metallic additives; an ABL third stage was developed from the Hercules Powder Company third stage for Vanguard but enlarged for Scout; and the original ABL third stage for Vanguard was used as the fourth stage for Scout. The first four-stage NASA Scout flew from Wallops on 1 July 1960. With some of the same stages rearranged and some different motors, the Air Force launched the first of its Blue Scouts, known as the Blue Scout Junior, on 21 September 1960.[13]

Like the Delta, the Scout evolved through a great many configurations, at least one of them with five stages. Gradually, the distinction between Blue and NASA Scouts blurred with an agreement between NASA and DOD on 10 January 1970 providing that NASA would contract for Scout launches from Vandenberg Air Force Base for both DOD and itself. More usually, NASA launched its Scouts from Wallops or Cape Canaveral, Florida. After 26 April 1967, under an agreement between NASA and Italy, some Scouts also launched from the San Marcos platform off the coast of Kenya, Africa. Located on the equator, San Marcos enabled Scouts to place satellites into orbits not achievable from the three U.S. launch sites.[14]

Guidance and control for the early NASA Scouts used a strapped-down inertial reference package from Minneapolis-Honeywell. It included miniature integrating rate gyros to detect deviations from the programmed path and an electronic signal conditioner to convert the outputs of the gyros to appropriate control signals. NASA selected the Chance Vought Corporation as airframe contractor. Under a variety of different names, this firm gradually acquired

---

13. Shortal, *A New Dimension*, pp. 706–709, 720; Hansen, *Spaceflight Revolution*, pp. 199–200, 209–210; Wilson, "Scout," pp. 449–450; U.S. President, Executive Office, National Aeronautics and Space Council, *Report to Congress . . . January 1 to December 31, 1960* (Washington, DC: GPO, 1961), p. 91 (subsequent citations of reports in this series omit place of publication and publisher, which are the same as in this citation); Mark Wade, "Blue Scout Junior," pp. 1–2, available at *http://friendspartners.ru/partners/mwade/lvs/blurjunior.htm* (accessed 6 November 2002); Gunter Krebs, "Blue Scout Junior (SRM-91, MER-6)," p. 1, available at *http://www.skyrocket.de/space/doc_lau/blue_scout_jr.htm* (accessed 20 November 2002); "Blue Scout Chronology," entries for 21 September 1960; 8 November 1960; 17 August 1961; 4 December 1961; 24 July 1962; 21 November 1962; 18 December 1962; "Blue Scout History," narrative included in the "Blue Scout Chronology" package.
14. McDowell, "Scout Launch Vehicle," pp. 101–103, 105–106; U.S. President, *Report to Congress . . . , 1967*, p. 127; GSFC, "NASA's Scout Launch Vehicle"; Leiss, "Scout," pp. 2–3, 50, 53–54, 56–57, 63, 447–448; Wilson, "Scout," pp. 449, 453–457, 459; U.S. President, *Report to Congress . . . , 1965*, p. 149; U.S. President, *Aeronautics and Space Report . . . , 1972 Activities*, p. 89; *Aeronautics and Space Report . . . , 1974 Activities*, p. 128; *Aeronautics and Space Report . . . , 1979 Activities*, p. 90; Bille et al., "Small Launch Vehicles," pp. 208–209.

responsibility for systems management and motor procurement under the overall control of Langley Research Center (LaRC) until January 1991, when responsibility for Scout management shifted to GSFC.[15]

After NASA and Vought overcame some initial systems engineering and quality control problems, Scout went on to become a long-lasting and successful small launch vehicle. It continued to develop, with a long series of improvements increasing the payload capability of the vehicle from only 131 pounds into a 300-mile circular orbit in 1964 to 454 pounds by 30 October 1979, when the final G-1 configuration became operational. The final launch occurred on 5 August 1994.[16]

Meanwhile, on 16 February 1961, Scout became the first entirely solid-propellant launch vehicle, as well as the first rocket from Wallops, to achieve orbit. Among the upgrades from the initial Scout was the Antares IIA third-stage motor that first flew on 29 March 1962. Developed by the ABL, this motor featured one of the early composite modified double-base propellants, a major innovation in solid-propellant technology. This launch marked the shift from Scout version X-1 to X-2, with the payload capability growing from the initial 131 pounds to 168.[17]

Among other notable technological achievements in the Scout program was Thiokol's Antares IIIA third-stage motor developed between 1977 and 1979. It was an early example of the use of hydroxyl-terminated polybutadiene—a major new propellant developed by a number of chemists at the Air Force Rocket Propulsion Laboratory, Thiokol, the Army's Redstone Arsenal, Atlantic Research, Hercules, and the Navy, among others. First used on a Scout launch

---

15. Leiss, "Scout," pp. 26, 31–32, 84–88, 128–133; Shortal, *A New Dimension*, pp. 710–711; LTV Astronautics Division, "The Scout," February 1965, pp. 3-10 to 3-15, filed under "Standard Launch Vehicle I," at SMC/HO; Vought Corporation, "Scout User's Manual," 1 June 1977, NASM Archives, folder OS-050000-70, "Scout Vehicle Users Manuals, 1977"; Bille et al., "Small Launch Vehicles," p. 206; A. Wilson, "Scout," p. 457; SMC/HO; "Press Release Information, Solar Radiation Scout Program," attachment to letter, Space Systems Division/SSVXO (Maj. Reed) to DCEP (Maj. Hinds), 6 April 1962, in "Hyper Environment Test System (TS 609A)" folder, SMC/HO; "NASA Awards Scout Contracts," *Aviation Week* (2 March 1959): 22. None of these sources indicates whether the guidance and control computer was analog or digital, or even whether Minneapolis-Honeywell provided it or procured it from another company.
16. McDowell, "Scout Launch Vehicle," pp. 101–103, 105–106; Leiss, "Scout," pp. 2, 53–54, 56–57, 63, 129, 447–448; Wilson, "Scout," pp. 449, 453–457, 459; Bille et al., "Small Launch Vehicles," pp. 208–211; Gunter Krebs, "Scout," pp. 2, 5, available at *http://www.skyrocket.de/space/doc_lau_fam/scout.htm* (accessed 26 November 2002).
17. Leiss, "Scout" pp. 3–4, 53–54, 56, 65, 131, 437–438; Wilson, "Scout," pp. 449–452, 459; McDowell, "Scout Launch Vehicle," pp. 100–102, 105–106; Hansen, *Spaceflight Revolution*, pp. 209–210; Shortal, *A New Dimension*, p. 720; U.S. President, *Report to Congress . . . , 1961*, p. 9; *CPIA/M1*, unit 428, X254A1; GSFC, "NASA's Scout Launch Vehicle." See also Hunley, *Viking to Space Shuttle*, p. 331.

on 30 October 1979, the Antares IIIA also featured a composite case made of Kevlar 49 and epoxy that was lighter than fiberglass and yielded a mass fraction of 0.923. Because of the motor's high chamber pressure and exhaust velocity, it also used a 4-D carbon-carbon nozzle-throat insert.[18]

Scout was notable not so much for the new technologies it introduced, however, as for the niche it filled in the launch-vehicle spectrum. Had it not been useful, it would not have lasted for nearly three and a half decades and placed into orbit many scientific and applications payloads including Transit navigational satellites and Explorer satellites.[19]

Centaur

Centaur was not originally a NASA project, but NASA contributed significantly to its ultimately successful development. Krafft Ehricke—a former member of Wernher von Braun's group in Germany and then the United States who was working for the Convair Division of General Dynamics in 1957 at the time of Sputnik—and other engineers decided that liquid hydrogen and liquid oxygen were the propellants needed for a powerful upper stage for Convair's Atlas space-launch vehicle. Ehricke presented the idea to DOD's new ARPA on 7 February 1958. The Advanced Research Projects Agency was aware that Pratt & Whitney had designed both an engine burning liquid hydrogen and a centrifugal pump to feed the fuel to the engine for an Air Force aircraft project named Suntan. The following August, ARPA issued an order for an upper stage to be developed by Convair and liquid-hydrogen/liquid-oxygen engines to be developed by Pratt & Whitney. Then in October and November, as NASA was coming into being, the Air Force (at ARPA's direction) issued development contracts to the two firms.[20]

---

18. Leiss, "Scout," p. 54; *CPIA/M1*, unit 577, Antares IIIA; Brian A. Wilson, "The History of Composite Motor Case Design" (AIAA paper 93-1782, presented at the 29th Joint Propulsion Conference and Exhibit, 28–30 June 1993, Monterey, CA), slides entitled "The 70's—Kevlar is King" and "Composite Motor Cases by Thiokol"; C. A. Zimmerman, J. Linsk, and G. J. Grunwald, "Solid Rocket Technology for the Eighties" (International Astronautical Federation paper 81-353, presented at the International Astronautical Federation XXXII Congress, 6–12 September 1981, Rome, Italy), p. 9.
19. Hunley, *Development of Propulsion Technology*, pp. 70, 73; Steven J. Isakowitz, *International Reference Guide to Space Launch Systems*, updated by Jeff Samella (Washington, DC: AIAA, 1995), p. 329.
20. John L. Sloop, *Liquid Hydrogen as a Propulsion Fuel, 1945–1959* (Washington, DC: NASA SP-4404, 1978), pp. 113, 141–166, 178–179, 194–195, 200–201; Virginia P. Dawson and Mark D. Bowles, *Taming Liquid Hydrogen: The Centaur Upper Stage Rocket, 1958–2002* (Washington, DC: NASA SP-2004-4230, 2004), pp. 17–20; Joel E. Tucker, "History of the RL10 Upper-Stage Rocket Engine," in *Liquid Rocket Engine Development*, ed. Doyle, vol. 13, p. 125; Space Division, "Space and Missile Systems Organization: A Chronology, 1954–1979," p. 56 (hereafter cited as

On 1 July 1959, the Centaur project transferred to NASA, but Lieutenant Colonel John D. Seaberg, who previously had overseen the Suntan project, remained as the Air Force's Air Research and Development Command project manager. Milton Rosen became his counterpart in NASA, with Ehricke remaining as Convair's project manager, a post he had held since November 1958. These people had to contend with liquid hydrogen's extremely low density, cold boiling point, low surface tension, and wide-ranging flammability—qualities that went along with its major asset: provision of more thrust per pound than any other chemical propellant then in use (about 35 to 40 percent more than the kerosene used as Atlas's fuel).[21]

In contending with liquid hydrogen's peculiarities, Ehricke in particular was hindered by limitations on funding. The Advanced Research Projects Agency had insisted that General Dynamics restrict its spending on Centaur to $36 million. Another stipulation was that the project not interfere with Atlas development and that, where possible, it use off-the-shelf equipment as well as Atlas tooling and technology. Pratt & Whitney's funding was $23 million, for a total of $59 million to cover the first six launches, beginning in January 1961. This amount did not include the costs of a guidance/control system, Atlas boosters, and launch equipment. As of 1962, Ehricke believed that the limited funding had prevented his project engineers from doing necessary ground testing. Another limiting factor was a lower priority than DOD's highest rating: DX. The absence of a DX priority prevented subcon-

---

"SAMSO Chronology"), copy generously provided by the Space and Missile Systems Center History Office; David N. Spires, *Beyond Horizons: A Half Century of Air Force Space Leadership* (Peterson Air Force Base, CO: Air Force Space Command, 1997), pp. 57–58; Convair Division, General Dynamics, "Atlas Fact Sheet," n.d., p. 3, NASM Archives, folder OA-401060-01, "Atlas Launch Vehicles (SLV-3)."

21. Sloop, *Liquid Hydrogen*, pp. 13–14, 20–26, 37–38, 49–58, 200–201; "Statement of Krafft A. Ehricke, Director, Advanced Studies, General Dynamics/Astronautics," in *Centaur Program*, Hearings before the Subcommittee on Space Sciences of the Committee on Science and Astronautics, H. Rep., 87th Cong., 2nd sess., 15 and 18 May 1962, pp. 5, 63–66 (overall document hereafter cited as Hearings, *Centaur Program*); oral history interview (OHI), Krafft A. Ehricke, by John L. Sloop, 26 April 1974, p. 59, folder 010976, NASA Historical Reference Collection, NASA History Division, NASA Headquarters, Washington, DC; Dan Heald, "LH_2 Technology was Pioneered on Centaur 30 Years Ago," in *History of Rocketry and Astronautics, Proceedings of the Twenty-Sixth History Symposium of the International Academy of Astronautics*, ed. Philippe Jung, American Astronautical Society History Series, vol. 21 (San Diego, CA: Univelt, 1997), p. 207; NASA News Release 62-66, "First Launch of Centaur Launch Vehicle Scheduled," 3 April 1962, NASM Archives, folder OA-40107-01, "Atlas Centaur Launch Vehicle"; General Dynamics/Astronautics, "Centaur Primer: An Introduction to Hydrogen-Powered Space Flight," June 1962, folder 10203, "Centaur General (1959–89)," NASA Historical Reference Collection, NASA History Division, NASA Headquarters, Washington, DC.

tractors from providing the same level of service to Centaur as they afforded to higher-priority efforts.[22]

Under these restrictions, Convair and Pratt & Whitney designed Centaur's structure and engines. Ehricke's engineers used the unusual steel balloon structure developed for Atlas, which employed pressure in the propellant tanks to avoid heavy structural support. The resultant light airframe was critical because liquid hydrogen's low density forced use of a larger propellant tank than was needed for the denser liquid oxygen. Convair placed the liquid oxygen tank on the bottom of the stage. Then, to keep weight at a minimum, Ehricke's engineers made the bottom of the liquid hydrogen tank concave so that it fit over the convex top of the oxygen tank, reducing length, hence weight. But the 4 feet of length and 1,000 pounds of weight that were saved led to problems from the smallness of the liquid hydrogen molecules and their extreme coldness. The oxygen tank's –299°F temperature was so much "warmer" than the liquid hydrogen's –423°F that without insulation, the hydrogen would turn to gas, expand, and exit the tank through a pressure relief valve needed to prevent explosion. To solve this problem, the engineers placed a bulkhead between the two tanks with a 0.2-inch cavity filled with fiberglass-covered Styrofoam. When technicians evacuated air from the pores in the Styrofoam, replaced it with gaseous nitrogen, and then filled the upper tank with liquid hydrogen, the nitrogen froze, creating a vacuum because the solid nitrogen occupied less space than when it was liquefied, a process called cryopumping.[23]

As a result of the low funding and consequent limited testing, it was only in mid-1961 that Convair discovered heat transfer through the bulkhead more than 50 times what was expected. Engineers learned that there were extremely small cracks in the bulkhead through which the hydrogen was escaping, eliminating the vacuum and causing the resultant "heat" from the liquid oxygen to gasify the liquid hydrogen in the tank. The gas that escaped from the pressure relief valve left insufficient liquid hydrogen for a second engine burn needed for propelling a satellite from a transfer orbit to a higher one. Besides insufficient testing, the requirement to use Atlas technology

22. Hearings, *Centaur Program*, pp. 4–5, 105–106.
23. Joseph Green and Fuller C. Jones, "The Bugs That Live at –423°," *Analog: Science Fiction, Science Fact* 80, no. 5 (January 1968): 8–41; Richard Martin, "A Brief History of the *Atlas* Rocket Vehicle," pt. II, *Quest: The History of Spaceflight Quarterly* 8, no. 3 [2000 or 2001]: 43; Heald, "LH_2 Technology," pp. 209–210; G. R. Richards and Joel W. Powell, "Centaur Vehicle," *JBIS* 42 (March 1989): 99.

on Centaur had meant that quality control detected cracks down to about 1/10,000 inch. In 1961 Ehricke's engineers found that hydrogen could escape through even finer openings that would not have been a problem in Atlas's liquid-oxygen tanks.[24]

By the time Ehricke's engineers had uncovered this problem, NASA had assigned responsibility for Centaur to MSFC (on 1 July 1960). Hans Hueter, Director of MSFC's Light and Medium Vehicles Office, managed Centaur for the Center, while Navy Commander W. Schubert served as Centaur project chief at NASA Headquarters. But it was John L. Sloop, Deputy Director of the Headquarters group managing NASA's small and medium-sized launch vehicles, who visited General Dynamics/Astronautics (GD/A) from 11 to 14 December 1961, to have a firsthand look at the problem. "GD/A," he wrote, "has studied the problem and concluded that it is not practical to build bulkheads where . . . a vacuum could be maintained." The firm planned to convert to "separate fuel and oxidizer tanks." Sloop urged sticking with the integral tank, and the Centaur team found that adding nickel to welds increased the single-spot shear strength and fixed the problem.[25]

Centaur experienced many other problems in the course of its development. Some involved the engines, forcing Pratt & Whitney into numerous modifications and redesigns. As a result of all of these problems, the first Centaur launch did not occur until 8 May 1962, 15 months past the date originally planned. Even so, the upper stage exploded, splitting open the hydrogen tank. Before this date, both Sloop and Hueter had complained to GD/A about its matrix organization in which most of the engineers on the Centaur team there did not report directly to Ehricke or his project engineer. As a result, Ehricke was reassigned as not "enough of a[n] S.O.B. to manage a program like this," and

---

24. Hearings, *Centaur Program*, pp. 9, 51, 97; Dawson and Bowles, *Centaur*, pp. 19–20, 34, 51, 74; Irwin Stambler, "Centaur," *Space/Astronautics* (October 1963): 74; John L. Sloop, Memorandum for Director of Space Sciences, 18 December 1961, NASA Historical Reference Collection, NASA History Division, NASA Headquarters, Washington, DC; John L. Sloop Papers, box 22, binder, "Centaur Management & Development, Jan. 1961–Mar. 62"; W. Schubert, "Centaur," in NASA, Office of Space Sciences, "Program Review, Launch Vehicles and Propulsion," 23 June 1962, NASA Historical Reference Collection, NASA History Division, NASA Headquarters, Washington, DC, pp. 173–175.
25. Hearings, *Centaur Program*, pp. 7, 33, 47, 66; MSFC, bio, Hans Herbert Hueter, fiche no. 1067, Marshall History Office Master Collection, copy in folder 001055, NASA Historical Reference Collection, NASA History Division, NASA Headquarters, Washington, DC; Schubert, "Centaur," pp. 121–180; Sloop, 18 December 1961 Memorandum, pp. 1–2, quotations; John L. Sloop, Memo for Director of Space Sciences, 20 December 1961, NASA Historical Reference Collection, NASA History Division, NASA Headquarters, Washington, DC, p. 3, John L. Sloop Papers, box 22, binder "Centaur Management & Development, Jan. 1961–Mar. 62"; Sloop's bio, Sloop, *Liquid Hydrogen*, p. 3; Stambler, "Centaur," pp. 73–75.

Grant L. Hansen was placed in charge of a "projectized" organization with about 1,100 employees reporting directly to him.[26]

At the beginning of January 1962, in agreement with DOD, NASA converted existing Centaur contracts from Air Force to NASA agreements. Direct project management shifted from the Air Force to MSFC. By this time, funding had increased from the original $59 million to $269 million, now for 10 instead of the original 6 vehicles. Following the 8 May explosion, congressional hearings had called Centaur management "weak and ineffective." The hearings had brought out a difference in design approach between MSFC, where Wernher von Braun's team tended to be conservative, and GD/A, which was more willing to gamble on design improvements.[27]

Von Braun was uncomfortable with the "pressure-stabilized tanks" of Centaur and quietly sought to cancel the program in favor of a combination of his own Saturn launch vehicle with an Agena upper stage. Not willing to do this, on 8 October 1962, NASA Headquarters began to transfer management of the Centaur program to Lewis Research Center, which had a history of work with liquid hydrogen and to which Silverstein had returned as Center Director in 1961.[28]

Under Lewis management, Centaur continued to have growing pains but went on to become a highly successful heavy-lift upper stage with numerous upgrades. Of equal importance, the success of Centaur led to the use of liquid hydrogen as a fuel on the Saturn launch vehicle's upper stages and in the Space Shuttle. NASA, especially the Lewis (now Glenn) Research Center, and its contractors had made this major advance in launch-vehicle technology possible.[29]

---

26. Tucker, "RL10 Upper-Stage Rocket Engine," pp. 126–137, 139; Richards and Powell, "Centaur Vehicle," pp. 100, 102–103; Green and Jones, "Bugs That Live at –423°," pp. 21–22; Hearings, *Centaur Program*, pp. 12–27, 115–116; Schubert, "Centaur," p. 131; Isakowitz, *Space Launch Systems* (1995), p. 205; correspondence by Deane Davis, General Dynamics (ret.), in *JBIS* 35, no. 1 (January 1982): 17; Heald, "LH_2 Technology," p. 206.
27. Philip Geddes, "Centaur, How It Was Put Back on Track," *Aerospace Management* (April 1964): 25, 28–29; Hearings, *Centaur Program*, pp. 2, 9, 37, 66, 104; Hearings, *Centaur Program: Centaur Launch Vehicle Development Program*, Report of the Committee on Science and Astronautics (hereafter cited as *Centaur* Report), 87th Cong., 2nd sess. H. Rep. 1959, 2 July 1962, pp. 11 (first quotation), 12; Sloop, *Liquid Hydrogen*, p. 208.
28. Hearings, *Centaur Program*, p. 59 for quotation; Dawson and Bowles, *Centaur*, pp. 54–55, 60; Silverstein biography from Virginia P. Dawson, *Engines and Innovation: Lewis Laboratory and American Propulsion Technology* (Washington, DC: NASA SP-4306, 1991), pp. 169–170, 177–178.
29. See Hunley, *Development of Propulsion Technology*, pp. 185–190; Hunley, *U.S. Space Launch-Vehicle Technology*, pp. 107–121; and the sources they cite.

Saturn

The Saturn launch vehicles were a major part of the Apollo program and a significant NASA contribution to core launch-vehicle technologies. Even they, however, did not originate with NASA. Wernher von Braun's group at the ABMA began in April 1957 to respond to DOD projections foreseeing a need for a huge booster to launch weather and communications satellites. The von Braun team, which joined NASA as MSFC only in mid-1960, looked at the possibility of a launch vehicle with 1.5 million pounds of thrust in its first stage. When ARPA came into existence in early 1958, it urged the ABMA to develop the vehicle with existing and proven engines. Von Braun's group then selected eight uprated Thor-Jupiter engines in a cluster to provide the 1.5 million pounds of thrust, leading to an ARPA order on 15 August 1958 for what soon came to be called the Saturn launch vehicle.[30]

Under an 11 September 1958 contract, the Rocketdyne Division of North American Aviation supplied an H-1 engine that significantly exceeded the description "uprated Thor-Jupiter engine." It was a product of research and development on an X-1 engine begun by an Experimental Engines Group at Rocketdyne in 1957. In other ways, the interim Saturn I and Saturn IB launch vehicles were based on technologies developed for the Redstone, Jupiter, Thor, Atlas, Centaur, and other vehicles and stages. For example, the Saturn I second stage (confusingly designated the S-IV) held six RL10 engines originally developed by Pratt & Whitney for Centaur. The Douglas Aircraft Company built the S-IV using information from Centaur contractors Convair and Pratt & Whitney, as well as its own experience.[31]

Development of Saturn I was problematic. Engineers encountered combustion instability in the H-1 engines, sloshing in the first stage's propellant tanks, stripped gears in a turbopump on the H-1, an explosion of the S-IV stage during

---

30. Roger E. Bilstein, *Stages to Saturn: A Technological History of the Apollo/Saturn Launch Vehicles* (Washington, DC: NASA SP-4206, 1980), pp. 25–28; MSFC, Saturn Systems Office, "Saturn Illustrated Chronology (April 1957–April 1962)," 1962, pp. 1–5, in Bellcom Collection, box 13, folder 3, NASM Archives; Charles Murray and Catherine Bly Cox, *Apollo: The Race to the Moon* (New York, NY: Simon & Schuster, 1989), p. 54.
31. Bilstein, *Stages to Saturn*, pp. 28–31, 188–189; MSFC, "Saturn Illustrated Chronology," pp. 2–4, 8–10, 12–13, 17–20, 22; "The Experimental Engines Group," transcript of a group interview with Bill (W. F.) Ezell, Cliff (C. A.) Hauenstein, Jim (J. O.) Bates, Stan (G. S.) Bell, and Dick (R.) Schwarz, [Rocketdyne] *Threshold: An Engineering Journal of Power Technology*, no. 4 (spring 1989): 21–27; Robert S. Kraemer, *Rocketdyne: Powering Humans into Space* (Reston, VA: AIAA, 2006), pp. 122–127; Linda Neuman Ezell, *NASA Historical Data Book*, vol. 2 (Washington, DC: NASA SP-4012, 1988), pp. 56–58; B. K. Heusinger, "Saturn Propulsion Improvements," *Astronautics & Aeronautics* (August 1964): 25; U.S. President, *Report to Congress . . . , 1964*, p. 128.

static testing, and other problems. Rocketdyne engineers fixed the combustion instability by rearranging the injector orifices and adding baffles to the injector face. Different kinds of baffles solved the sloshing problem. Rocketdyne and MSFC engineers redesigned the gearbox with wider teeth in the gears. And Douglas engineers apparently redesigned a shutoff valve to prevent another S-IV explosion. The 10 test flights of Saturn I between 27 October 1961 and 30 July 1965 revealed problems (including the sloshing), but NASA counted all of the flights as successful, a testimony to the thoroughness and extensive ground testing of the MSFC engineers and their contractors.[32]

Guidance and control for the Saturn I (as well as the IB and V) issued from an instrument unit atop the launch vehicle's uppermost stage. More than most other Saturn components, MSFC engineers designed this unit in-house. The core of the system consisted of a stabilized platform that continued an evolution from those on the V-2, the Redstone, Jupiter, and Pershing missiles. The instrument unit itself evolved through the Saturn IB and Saturn V. The system used gyroscopes and accelerometers that were less than half as heavy as those on the Jupiter. Their weight was the same as on the Pershing missile, but the materials changed from aluminum, beryllium, and Monel to beryllium alone, providing better thermal and structural stability. Designed and developed by the von Braun team at MSFC and its Army predecessor organizations, the Saturn inertial platform system was built by Bendix and tested on a high-speed sled track at Holloman Air Force Base in New Mexico. IBM made the lightweight, high-speed digital computer for the instrument unit (IU). Engineers at MSFC built the first four IUs before NASA selected IBM as the prime contractor for the remaining units on the Saturn IB and those on the Saturn V.[33] This system worked well on the six flights to the Moon and was a noteworthy achievement.

---

32. Bilstein, *Stages to Saturn*, pp. 77–78, 98–104, 184–185, 324–337, 414–415; Summary, "Final Report S-IV All-Systems Stage Incident January 24, 1964, May 11, 1964," seen in George E. Mueller Collection, Manuscript Division, Library of Congress, box 91, folder 14, pp. 2–3; Heusinger, "Saturn Propulsion Improvements," p. 25; MSFC, "Saturn Illustrated Chronology," pp. 46–47; NASA/MSFC et al., "Saturn IB News Reference," December 1965 (changed September 1968), p. 12-2, available at *http://www.apollosaturn.com/ascom/sibnews/contents.htm* (accessed 28 March 2008).
33. Bilstein, *Stages to Saturn*, pp. 241–252, 477; NASA/MSFC et al., "Saturn IB News Reference," pp. 1-4, 7-1, 7-3 to 7-7, B-2; Douglas Aircraft Company, "Saturn IB Payload Planner's Guide," n.d., pp. 38–39, Bellcom Collection, box 14, folder 6, "Saturn IB," NASM; Walter Haeussermann, "Developments in the Field of Automatic Guidance and Control of Rockets," *Journal of Guidance and Control* 4, no. 3 (May–June 1981): 232–235; series of news releases and progress reports on microfilm roll 32,270, frames 127–187, 201–236, 293–311, 404–406, 498–500, 506–507, 510–554, 581, 586, and roll 32,273, frames 215–216, all at the AFHRA, Maxwell Air Force Base, Alabama; MSFC, Astrionics Laboratory, "Astrionics

Saturn IB was a modified version of the Saturn I and a step toward the Saturn V, with the two Saturn I stages modified and redesignated S-IB and S-IVB. The S-IVB was the second stage of the Saturn IB and (with modifications) the third stage of Saturn V. S-IB contained uprated versions of the H-1 engine, while S-IVB employed a new and much larger liquid-hydrogen engine than the RL10. The thrust of the new J-2 exceeded that of all six RL10s on Saturn I. On 10 September 1960, Rocketdyne won a contract to develop the J-2. Its engineers had trouble with the injectors for the new engine until NASA facilitated Rocketdyne's borrowing of technology from the RL10 in one of many examples of shared information between competing contractors in launch-vehicle development as well as of NASA's direct involvement with its contractors' development efforts.[34]

With three instead of just two stages, the 363-foot Saturn V was a substantially larger and more powerful launch vehicle than the 224-foot Saturn IB. The Saturn V's five F-1 engines—built by Rocketdyne under a 9 January 1959 contract with NASA—produced 7,760,000 pounds of thrust as compared with a total thrust of 1,600,000 pounds from the eight H-1s on the Saturn IB. Like the H-1, the F-1 burned RP-1 (kerosene) and liquid oxygen, so its basic technology was not new. But the huge increase in thrust on the F-1 necessitated a major advance in the art of engine design and development. In a hot-fire test of the F-1 at Edwards Air Force Base on 28 June 1962, combustion instability produced meltdown of the engine. A large team of engineers (including Jerry Thomson from MSFC and Paul Castenholz and

---

System Handbook, Saturn Launch Vehicle," 2 January 1964, pp. 4-29, 4-33, 4-35, Bellcom Collection, box 13, folders 8 and 9, NASM; NASA, "Saturn V News Reference," MSFC, KSC, and contractors, August 1967 (portions changed December 1968), pp. 7-1 to 7-5, available at *http://history.msfc.nasa.gov/saturn_apollo/saturnv_press_kit.html* (accessed 28 March 2008), and unpaginated Instrument Unit Fact Sheet (accessed 21 March 2008); S. M. Seltzer, "Saturn IB/V Astrionics System," n.d., p. 5-10, Bellcom Collection, box 13, folder 10, NASM.

34. NASA/MSFC et al., "Saturn IB News Reference," pp. 1-3 to 2-5, 3-1, 3-22, 8-1, 12-2, and unpaginated H-1 engine fact sheet; Ezell, *NASA Historical Data Book*, vol. 2, p. 56; MSFC, "Saturn Illustrated Chronology," pp. 46, 50, 58, 60; Bilstein, *Stages to Saturn*, pp. 83, 97, 138, 140–145; "The Apollo Spacecraft: A Chronology," vol. 1, pp. 53–54, folder 013782, NASA Historical Reference Collection, NASA History Division, NASA Headquarters, Washington, DC; "Propulsion J-2"; Hunley, ed., *Birth of NASA*, pp. 149–150, including n. 6; NASA/MLP (Tischler) to NASA/ML (Rosen), "AGC Proposal," 8 July 1962, folder 013782, NASA Historical Reference Collection, NASA History Division, NASA Headquarters, Washington, DC; Rocketdyne, "Data Sheet, J-2 Rocket Engine," 3 June 1975, folder 013782, "Propulsion J-2" NASA Historical Reference Collection, NASA History Division, NASA Headquarters, Washington, DC; NASA/MLP (Tischler) to NASA/ML (Rosen), "M-1 Engine Review at AGC, July 11–12, 1962," 24 July 1962, folder 013782, NASA Historical Reference Collection, NASA History Division, NASA Headquarters, Washington, DC.

Dan Klute from Rocketdyne) found that, as Thomson admitted, the instability started "for reasons we never quite understood." Employing trial-and-error methods together with high-speed instrumentation and thorough analysis, the team tested 40 to 50 different designs before discovering a combination of baffles, enlarged fuel-injection orifices, and changed impingement angles that proved effective.[35]

Multiple other challenges faced NASA and contractor engineers working on the Saturn V. Eleven failures of the turbopump on the F-1 all required redesign or a change in manufacturing procedures. The S-II second stage, designed and built by the Space and Information Systems Division of North American Aviation (NAA), proved especially defiant of solution to its problems. Engineers at MSFC helped those at NAA to solve welding problems with a 2014 T6 aluminum alloy used for the huge propellant tanks. At NASA's insistence, NAA changed managers. Retired Major General Robert E. Greer introduced management techniques he had learned at the Air Force's Ballistic Missile Division. All of these issues delayed the first launch of the Saturn V from August until 9 November 1967. But then Apollo 4 (flight AS-501) was nearly flawless.[36]

---

35. Ray A. Williamson, "The Biggest of Them All: Reconsidering the Saturn V," in *To Reach the High Frontier*, ed. Launius and Jenkins, p. 315; NASA, "Saturn V News Reference," unpaginated Saturn V Fact Sheet, unpaginated F-1 Engine Fact Sheet, and pp. 1-1, 3-1 to 3-5; Bilstein, *Stages to Saturn*, pp. 58–60, 104–113, 115–116, 339–344, 414–416; NASA/MSFC et al., "Saturn IB News Reference," unpaginated Saturn IB Fact Sheets; MSFC, "Saturn Illustrated Chronology (to April 1962)," pp. 4, 13–14; Ezell, *NASA Historical Data Book*, vol. 2, p. 59; William J. Brennan, "Milestones in Cryogenic Liquid Propellant Rocket Engines" (AIAA paper 67-978, delivered at the AIAA 4th Annual Meeting and Technical Display, 23–27 October 1967, Anaheim, CA), pp. 8–9; Vance Jacqua and Allan Ferrenberg, "The Art of Injector Design," [Rocketdyne] *Threshold*, no. 4 (spring 1989): 4, 6, 9; "Experimental Engines Group," *Threshold*, no. 4 (spring 1989): 21; Murray and Cox, *Apollo*, pp. 145, 147–151, 179–180. Initially, the F-1 produced 1,500,000 pounds of thrust, but later versions yielded 1,522,000 pounds per engine. For more details on achieving combustion stability in the F-1, see Fred E. C. Culick and Vigor Yang, "Overview of Combustion Instabilities in Liquid-Propellant Rocket Engines," in *Liquid Rocket Engine Combustion Instability*, ed. Vigor Yang and William E. Anderson (Washington, DC: AIAA, 1995), pp. 8–9 and sources cited; Kraemer, *Rocketdyne*, p. 166.
36. Bilstein, *Stages to Saturn*, pp. 116–119, 191–209, 211–233, 269; George E. Mueller to J. L. Atwood, 19 December 1965, and MA/S. C. Phillips to M/G E. Mueller, "Command and Service Modules and S-II Review," 18 December 1965, both from George E. Mueller Collection, Manuscripts Division, Library of Congress, box 84, folder 3; Andrew J. Dunar and Stephen P. Waring, *Power to Explore: A History of Marshall Space Flight Center, 1960–1990* (Washington, DC: NASA SP-4313, 1999), pp. 86–89, 90; Barton Hacker and E. M. Emme, notes on interview with Milton W. Rosen at NASA Headquarters, Washington, DC, 14 November 1969, folder 001835, "Rosen, Milton (Miscellaneous Biography)," NASA Historical Reference Collection, NASA History Division, NASA Headquarters, Washington, DC; Mike Gray, *Angle of Attack: Harrison Storms and the Race to the Moon* (New York, NY: Norton, 1992), pp. 20, 22–34, 65–72, 153–156, 159–161, 196–199, 202, 208–209, 253–255, 353–355; Murray and Cox, *Apollo*, pp. 166–171, 183, 231–236.

Resolving problems required teamwork to create and perfect the Saturn rockets that ultimately carried 18 astronauts to the vicinity of the Moon, whence 12 of them actually landed on Earth's natural satellite. Both Germans and Americans at MSFC cooperated with other NASA Centers, universities, contractors, and the U.S. military to produce the Apollo launch vehicles. A key ingredient in the success of the Saturn launch vehicles was management systems. As he had done in Germany developing the V-2 missile and at the ABMA with the Redstone, Jupiter, and Pershing, von Braun served as an overall systems engineer. He displayed an uncanny ability to grasp technical details at meetings and explain them in terms that experts from multiple disciplines could understand. He used weekly notes from and to his managers that communicated difficulties and facilitated solutions across organizations at MSFC.[37]

For Saturn and the Apollo program in general, however, these techniques alone were not enough. Samuel Phillips, a general borrowed from the Air Force to direct the Apollo program, and George E. Mueller, who was Phillips's boss as head of NASA's Office of Manned Space Flight (OMSF), provided contributions of their own. Mueller assumed his post in NASA on 1 September 1963, when the Apollo program was well under way. Determining that the program was behind schedule, Mueller quickly changed the way Saturn was flight-tested from launches with only parts of the vehicle "live" to the all-up method used for the Air Force's Minuteman missile in which the first Saturn IB and the Saturn V launches would use all live stages.[38]

All-up testing conflicted with the step-by-step procedures the von Braun group favored. When von Braun presented the idea to his staff on 4 November 1963, it created a "furor." Recalling numerous failed launches in the V-2, Redstone, and Jupiter programs, structures expert William A. Mrazek said the idea was insane; other lab heads and project managers pronounced it a "dangerous idea" and "impossible."

---

37. Dunar and Waring, *Power to Explore*, Bilstein, *Stages to Saturn*, p. 263; Stephen B. Johnson, "Samuel Phillips and the Taming of Apollo," *Technology and Culture* 42 (October 2001): 691; OHI with Ernst Stuhlinger by J. D. Hunley, Huntsville, AL, 20 September 1994, NASA Historical Reference Collection, NASA History Division, NASA Headquarters, Washington, DC, pp. 48–49.

38. W. Henry Lambright, *Powering Apollo: James E. Webb of NASA* (Baltimore, MD: Johns Hopkins University Press, 1995), pp. 116–118; biographical sketch of Mueller in Arnold S. Levine, *Managing NASA in the Apollo Era* (Washington, DC: NASA SP-4102, 1982), p. 308; Murray and Cox, *Apollo*, pp. 152–154; Ezell, *NASA Historical Data Book*, vol. 2, pp. 6, 625, 642; Bilstein, *Stages to Saturn*, p. 349; interview with Mueller by Robert Sherrod, 21 April 1971, in *Before This Decade is Out . . . : Personal Reflections on the Apollo Program*, ed. Glen E. Swanson (Washington, DC: NASA SP-4223, 1999), p. 108; Lt. Gen. Otto J. Glasser, interview by Lt. Col. John J. Allen, seen in AFHRA, K239.0512-1566, 5–6 January 1984, pp. 75–76; AAD-2/Dr. Mueller to A/Mr. Webb through AA/Dr. Seamans, "Reorientation of Apollo Plans, Oct. 26, 1963," Mueller Collection, box 91, folder 9; Johnson, "Samuel Phillips," pp. 694–695.

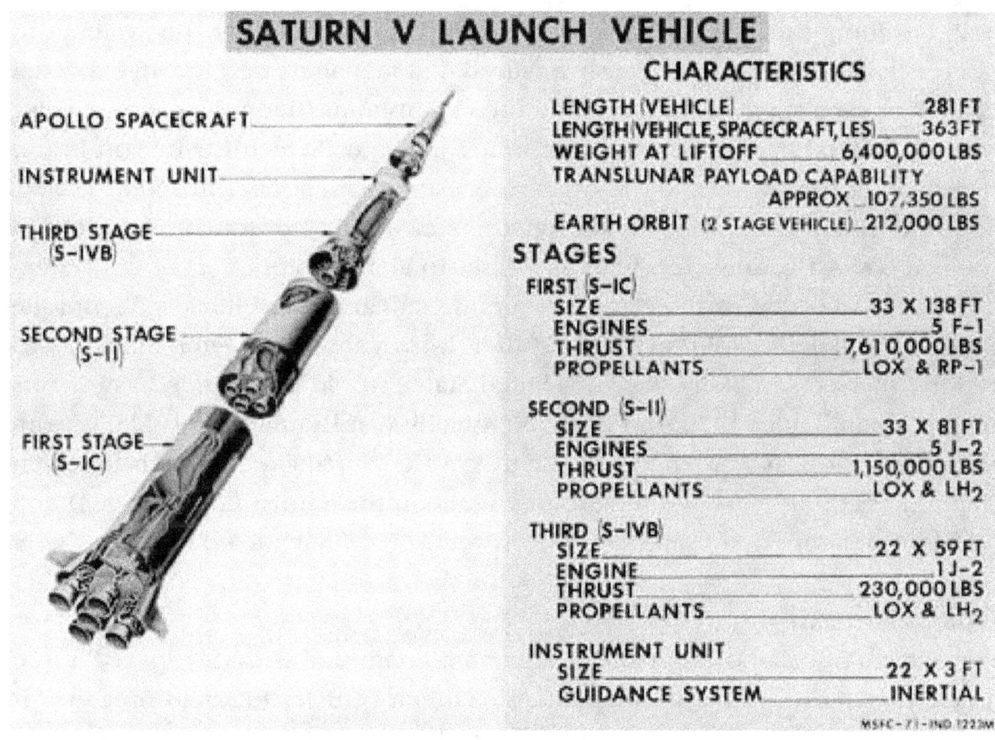

Figure 2: Diagram of a Saturn V launch vehicle's components and characteristics. *NASA*

Von Braun had his own doubts but had to agree with Mueller that launches of individual stages would prevent landing on the Moon on schedule (before 1970).[39] Despite the doubts, all-up testing worked on Saturn as it had on Minutemen.

Another practice at MSFC that did not accord well with the procedures Mueller and Phillips had learned in the Air Force (Mueller as a contractor) was an inclination to base technical decisions on their merits alone without much consideration of schedule or cost. The effects of decisions on time, budget, and configuration control had become critical in the Air Force, and Phillips arranged, soon after his arrival at NASA Headquarters in January 1964, to issue a NASA *Apollo Configuration Management Manual* (May 1964) that was adapted from an Air Force counterpart. In June, Phillips and a subordinate presented configuration management to an Apollo Executive Group of which von Braun was a member. Von Braun complained that costs of developing programs were "very much unknown, and configuration management does not

39. Bilstein, *Stages to Saturn*, pp. 349–351, first quotation from p. 349; Dunar and Waring, *Power to Explore*, pp. 94–95, second and third quotations.

help," arguing the need for flexibility. Phillips explained that the system did not preclude flexibility but simply ensured that managers defined an expected design "at each stage of the game" and then communicated with everyone else when it had to change. It took some time for Center Directors like von Braun to accept the new system, but it was firmly established about the end of 1966.[40]

Mueller and Phillips introduced other management procedures and a control room at NASA Headquarters with data links to Field Centers. Part of the system was a NASA version of the Navy's Program Evaluation and Review Technique used in the Polaris program. Without their borrowings from such military procedures, it seems unlikely that the United States would have landed astronauts on the Moon before 1970. But contributions by von Braun and his MSFC team were also critical, as were those of other NASA Centers (including Lewis Research Center for liquid-hydrogen technology used on the Saturn upper stages) and NASA contractors, among others. Saturn was very much a team effort.[41]

Space Shuttle

The pattern of collaboration and partnership continued with the Space Shuttle, a radical departure from the expendable launch vehicles that had preceded it. Now discredited in many circles by the *Challenger* and *Columbia* disasters, as well as the high costs of Shuttle missions, the reusability of most parts of the Shuttles seemed a good idea in the 1970s. People hoped at the time that like airliners, Shuttles would be reused many times, saving on costs. A key feature of the orbiters was a Space Shuttle main engine (SSME) with a combustion-chamber pressure well above that of previous liquid-hydrogen- and liquid-oxygen-burning engines (including Saturn's J-2), employing something called staged combustion, in which the hydrogen-rich turbine exhaust would flow

40. Dunar and Waring, *Power to Explore*, p. 67; Lambright, *Powering Apollo*, p. 118; Johnson, "Samuel Phillips," pp. 694–695, 697, 700–703, quotations from this source; OHIs of Lieutenant General Samuel Phillips by Tom Ray, 22 July 1970, pp. 8, 12–13, 25 September 1970, folder 001701, NASA Historical Reference Collection, NASA History Division, NASA Headquarters, Washington, DC, p. 22; Levine, *Managing NASA*, p. 309. For a somewhat different perspective on these developments, see Yasuchi Sato, "Local Engineering and Systems Engineering: Cultural Conflict at NASA's Marshall Space Flight Center, 1960–1966," *Technology and Culture* 46, no. 3 (July 2005): 561–583

41. Johnson, "Samuel Phillips," pp. 700–704; MSFC, *Apollo Program Management* 3 (December 1967): 4–29, Mueller Collection, box 64, folder 9; Levine, *Managing NASA*, pp. 156–157; Navy briefing in Samuel C. Phillips Collection, Manuscripts Division, Library of Congress, box 47, folder 12; Swanson, *Before This Decade is Out . . .*, pp. 101–102, 108; Samuel C. Phillips to Dr. Mueller, 3 October 1964, Mueller Collection, box 43, folder 4; OHI of General Samuel C. Phillips by Frederick I. Ordway, Cosmos Club, Washington, DC, 29 January 1988, including extracts from Phillips's Wernher von Braun Memorial Lecture, NASM, 28 January 1988, Washington, DC, pp. 1, 3–4, 7–8, Phillips Collection, box 138, folder 10.

into the combustion chamber to add to the thrust. In July 1971, Rocketdyne won the contract to design and build this complex engine.[42]

Timing for such an engine was difficult and delicate, as was the design of adequate turbopumps. Fires that burned up the evidence of what had caused them and other problems delayed the first flight of the Shuttle from March 1978 to 12 April 1981. "In assessing the technical difficulties that have been causing delays in the development and flight certification of the SSME at full power, it is important to understand that the engine is the most advanced liquid rocket motor ever attempted," stated an ad hoc committee of the Aeronautics and Space Engineering Board in 1981. "Chamber pressures of more than 3,000 psi, pump pressures of 7,000–8,000 psi, and an operating life of 7.5 hours have not been approached in previous designs of large liquid rocket motors," the Board added.[43]

Initial plans for the Shuttle had called for a fully reusable, two-stage vehicle, but budgetary realities in the early 1970s forced NASA and its contractors to compromise and create an only partially reusable Shuttle system with a stage-and-a-half concept in which solid rocket boosters (SRBs) attached to a nonreusable external tank would provide 71.4 percent of the Shuttle's thrust at liftoff and during the early stage of ascent until they separated about 75 seconds into the mission for later recovery and reuse. The SRBs would cost more per launch than reusable liquid boosters, but because of the Air Force's development of the similar, if somewhat smaller, Titan solid rocket motors

---

42. See Dennis R. Jenkins, *Space Shuttle: The History of the National Space Transportation System, the First 100 Missions*, 3rd ed. (Cape Canaveral, FL: D. R. Jenkins, 2001), and T. A. Heppenheimer, *Development of the Space Shuttle, 1972–1981, History of the Space Shuttle*, vol. 2 (Washington, DC: Smithsonian Institution Press, 2002), for (sometimes conflicting) details of Shuttle development and construction. For the specifics of the paragraph in the narrative: Heppenheimer, *Development of the Space Shuttle*, p. 126; Jenkins, *Space Shuttle*, pp. 224–225; Robert E. Biggs, "Space Shuttle Main Engine: The First Ten Years," in *Liquid Rocket Engine Development*, ed. Doyle, vol. 13, pp. 75–76; Al Martinez, "Rocket Engine Propulsion Power Cycles," [Rocketdyne] *Threshold*, no. 7 (summer 1991): 17. The discrediting of the Space Shuttle is at least implied by NASA's plans to replace it with Shuttle-derived hardware in an Apollo-like, nonreusable configuration for Ares I and Ares V, to be discussed at the end of this chapter.
43. Heppenheimer, *Development of the Space Shuttle*, pp. 127–128, 133–134, 148–171; R. Wiswell and M. Huggins, "Launch Vehicle & Upper Stage Liquid Propulsion at the Astronautics Laboratory (AFSC)—A Historical Summary" (unpaginated paper, AIAA-1990-1839, presented at the AIAA/SAE/ASME/ASEE 26th Joint Propulsion Conference, Orlando, FL, 16–18 July 1990); Biggs, "Space Shuttle Main Engine," pp. 80–118; Jenkins, *Space Shuttle*, pp. 225–227; Aeronautics and Space Engineering Board, Assembly of Engineering, National Research Council, Report of Ad Hoc Committee on Liquid Rocket Propulsion Technologies, *Liquid Propulsion Technology: An Evaluation of NASA's Program* (Washington, DC: National Academy Press, 1981), p. 16, Mueller Collection, box 198, folder 9 "NASA, 1981."

(SRMs), SRBs would be cheaper to develop at a time when NASA's budget was most constrained.[44]

Marshall Space Flight Center awarded contracts to Lockheed Propulsion Company, United Technology Center, Thiokol, and Aerojet General to study configurations for the SRB motors. NASA followed these up with Requests for Proposals (RFPs) on 16 July 1973. All four companies responded with technical and cost proposals, but Aerojet ignored a requirement in the RFP and offered a welded case without segmentation, stating that such a case would be less costly, lighter, and safer, with barge transportation to launch sites from Aerojet's production facility. If Aerojet could have won the contract, possibly the *Challenger* accident, caused by a problem with a joint in the segmented motor case, might have been avoided. But Thiokol won the contract on 20 November 1973.[45]

Thiokol acquired some of the technology used to develop the SRBs from participation in the Air Force's Large Segmented Solid Rocket Motor Program, part of which NASA paid for. This included a Lockseal gimballing nozzle, developed by Lockheed, that Thiokol scaled up and called Flexseal. But also included were experience and access to materials, designs, and fabrication methods that Thiokol could apply to the Shuttle SRBs. Thiokol's participation in the Minuteman missile program was also, no doubt, helpful.[46]

As George Hardy, the project manager for the SRB at MSFC from 1974 to 1982, said, the Center tried "to avoid inventing anything new" in the booster's design. Thus the steel case was the same type (D6AC) used on Minuteman and the Titan IIIC, showing once again the interconnectedness of missile and launch-vehicle development. The PBAN propellant was the same type used

---

44. Rockwell International, "Press Information: Space Shuttle Transportation System," January 1984, NASA Historical Reference Collection, NASA History Division, NASA Headquarters, Washington, DC, p. 21; Dennis R. Jenkins, "Broken in Midstride: Space Shuttle as a Launch Vehicle," in *To Reach the High Frontier*, ed. Launius and Jenkins, pp. 358–375; Joan Lisa Bromberg, *NASA and the Space Industry* (Baltimore, MD: Johns Hopkins University Press, 1999), pp. 77–93; T. A. Heppenheimer, *The Space Shuttle Decision: NASA's Search for a Reusable Space Vehicle* (Washington, DC: NASA SP-4221, 1999), pp. 245–290, 331–422; Jenkins, *Space Shuttle*, pp. 139–152, 167–173; John F. Guilmartin, Jr., and John Walker Mauer, *A Space Shuttle Chronology, 1964–1973: Abstract Concepts to Letter Contracts*, 5 vols. (Houston, TX: JSC-23309, 1988), esp. 4: V-5 to V-18, V-23 to V-135, V-193, V-194, V-237 to V-240, and 5: VI-45, VI-46.
45. Jenkins, *Space Shuttle*, pp. 184–186; Heppenheimer, *Development of the Space Shuttle*, pp. 71–78; John M. Logsdon et al., *Exploring the Unknown: Selected Documents in the History of the U.S. Civil Space Program*, vol. 4, *Accessing Space* (Washington, DC: NASA SP-4407, 1999), Document II-17, pp. 269, 271; Bromberg, *NASA and the Space Industry*, p. 100.
46. This paragraph summarizes a longer treatment in J. D. Hunley, "Minuteman and the Development of Solid-Rocket Launch Technology," in *To Reach the High Frontier*, ed. Launius and Jenkins, pp. 271–278.

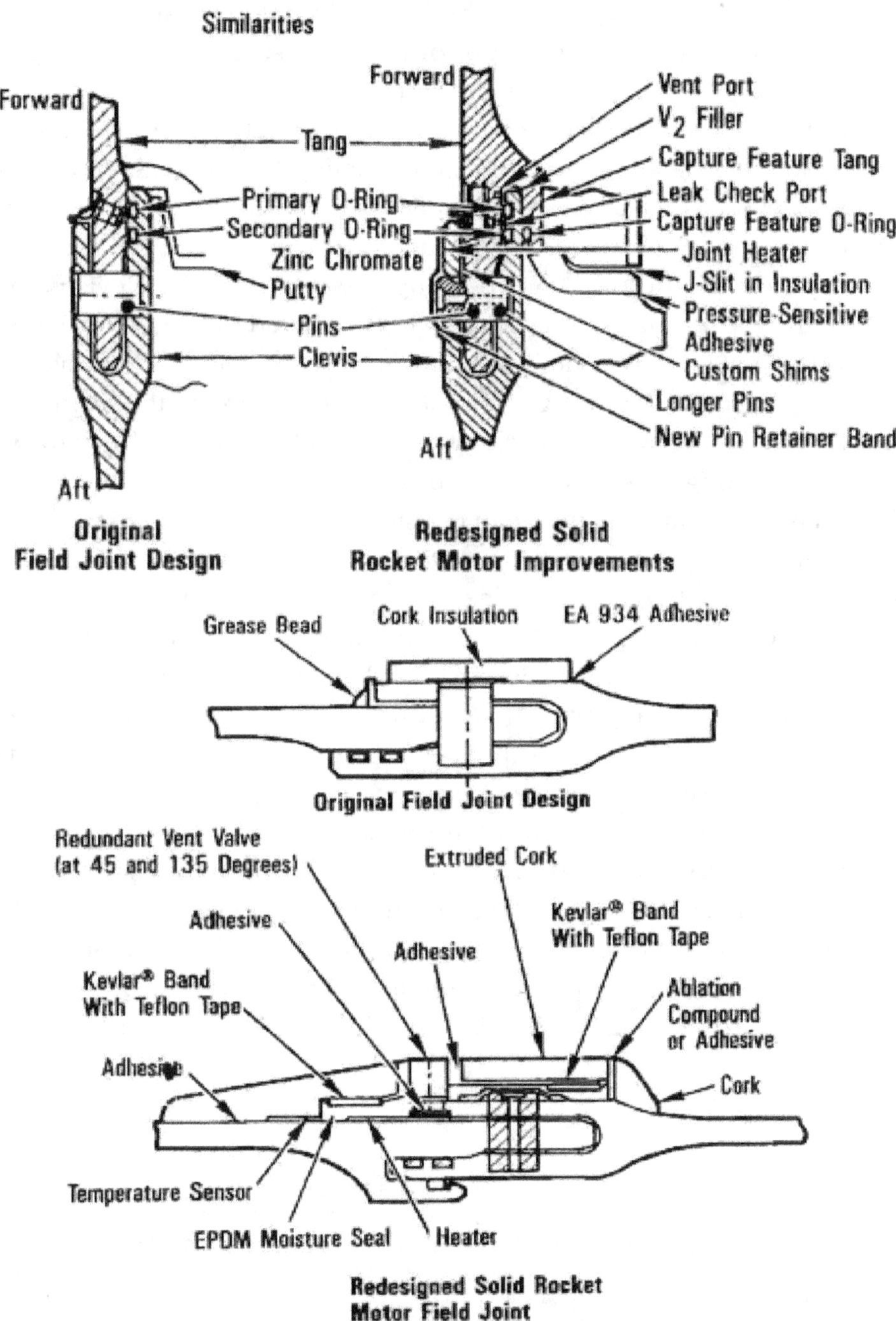

Figure 3: Two sets of cross sections of the original and redesigned field joint for the Space Shuttle solid rocket boosters showing details of both. Taken from NASA, *National Space Transportation System Reference*, vol. 1, *Systems and Facilities* (Washington, DC: NASA, 1988), pp. 33a, 33b.

on the first stage of Minuteman and the Navy's Poseidon missile, even though other propellants provided higher performance. Cost and human rating here were more critical than greater thrust per pound of propellants.[47]

One place where designers departed from MSFC advice "to avoid anything new" was in the tang-and-clevis joints connecting the segments of the SRBs. The Shuttle joints were superficially similar to those in the Titan IIIC but differed in important ways, including orientation. In the Titans, the single tang pointed upward and fit into the two-pronged clevis. With the clevis pointing down, encasing the tang, the Titan joint was protected from rain or dew dripping into it. The Shuttle's joint faced the opposite direction. Additionally, the Titan's joint used only one O-ring; the Shuttle had two. To keep the single O-ring from shrinking in cold weather and then possibly allowing a gas blow-by when the motor was firing, the Titan had heating strips, which the Shuttle lacked.[48]

It seems clear that this change in design, plus bad judgment in launching during cold weather on 28 January 1986, caused the *Challenger* accident. This is perhaps confirmed by the extensive redesign of the field joints following the accident. Instead of having the tang remain a cylindrical piece fitting down into the clevis, a redesign added a tang capture feature, creating in effect a slot in the tang with the capture feature enveloping one side of the clevis. This tang capture feature limited "the deflection between the tang and clevis O-ring sealing surfaces caused by motor pressure and structural loads." A third (or capture-feature) O-ring added to the sealing capability of the new design, with an additional leak check port ensuring that the primary O-ring was in its proper place at ignition and beyond. Custom shims "between the outer surface of the tang and inner surface of the outer clevis leg" compressed the O-rings. The redesign ensured that the seals would not leak under twice the anticipated structural deflection. Also added to the

---

47. Jenkins, *Space Shuttle*, p. 186; Hunley, "Minuteman and the Development of Solid-Rocket Launch Technology," pp. 278–280; Rockwell International, "Press Information: Space Shuttle," pp. 21–46; *CPIA/M1*, Unit 556, Space Shuttle Booster; Heppenheimer, *Development of the Space Shuttle*, pp. 175–176.
48. Comments of Bernard Ross Felix, 22 August 2000, on a draft of Hunley, "Minuteman and Solid Rocket Technology"; see also Hunley, "Minuteman," pp. 278–280; Wilbur C. Andrepont and Rafael M. Felix, "The History of Large Solid Rocket Motor Development in the United States" (paper AIAA-94-3057 presented at the 30th AIAA/ASME/SAE/ASEE Joint Propulsion Conference and Exhibit, Indianapolis, IN, 27–29 June 1994), pp. 7, 14; Heppenheimer, *Development of the Space Shuttle*, pp. 177–178; J. D. Hunley, "The Evolution of Large Solid Propellant Rocketry in the United States," *Quest: The History of Spaceflight Quarterly* 6, no. 1 (1998): 31, 37; NASA, *National Space Transportation System Reference*, vol. 1, *Systems and Facilities* (Washington, DC: NASA, 1988), pp. 33a, 33c.

field joint were external heaters to maintain a temperature of at least 75°F and moisture seals to help maintain the temperature and prevent water from seeping into the joint (see figure 3, p. 103).[49]

A third part of the Shuttle's propulsion system consisted of the external tank, the major nonreusable portion of the launch vehicle. On 6 August 1973, NASA chose Martin Marietta to negotiate a contract for the design, development, and testing of the external tank. Larry Mulloy at MSFC, who worked on the tank, thought it posed no technological challenge despite the aerodynamic heating and heavy loads it faced on ascent. As it turned out, the weight limit MSFC established—75,000 pounds—did present a major challenge, and the external insulation later caused the loss of Space Shuttle *Columbia* on 1 February 2003. There had been a "breach in the Thermal Protection System on the leading edge of the left wing" of the orbiter due to its being struck by "a piece of insulating foam" from the external tank. As the orbiter reentered the atmosphere, this breach produced aerodynamic superheating of the wing's aluminum structure, melting it and causing the breakup of the vehicle as aerodynamic forces increased.[50]

Because it was mainly used as a spacecraft and landing vehicle, the orbiter itself will not be discussed in this account of the Shuttle as a launch vehicle. As the tragic losses of *Challenger* and *Columbia* showed, overall the Shuttle was clearly a flawed launch vehicle. One of the flaws apparently resulted from failure to use the tang-and-clevis joint from Titan III, although there may have been proprietary issues that prevented this. The Aerojet proposal would have eliminated joints altogether, but concerns besides nonresponsiveness to the RFP may have caused that option to be rejected. Other flaws clearly resulted

---

49. Allan J. McDonald with James R. Hansen, "Truth, Lies, and O-Rings: The Untold Story Behind the *Challenger* Accident: An Excerpt from the forthcoming book," *Quest: The History of Spaceflight Quarterly* 14, no. 3 (2007): 6–11; Logsdon et al., *Exploring the Unknown*, vol. 4, *Accessing Space*, Document II-39, "Presidential Commission on the Space Shuttle *Challenger* Accident, 'Report at a Glance,' June 6, 1986," pp. 358–359, 363, 366–368, 386; Stephen P. Waring, "The 'Challenger' Accident and Anachronism: The Rogers Commission and NASA's Marshall Space Flight Center" (paper presented at the National Council on Public History Fifteenth Annual Conference, 22 April 1993), pp. 9–15, 18; NASA, *National Space Transportation System Reference*, vol. 1, pp. 27–33h, quotations from p. 33a.
50. Jenkins, *Space Shuttle*, pp. 186–187; Rockwell International, "Press Information: Space Shuttle," p. 46; Heppenheimer, *Development of the Space Shuttle*, pp. 68–69; Dunar and Waring, *Power to Explore*, pp. 292–294, 301, 323; Columbia Accident Investigation Board, *Report*, vol. 1 (August 2003), p. 9 (for quotations), available at *http://anon.nasa-global.speedera.net/anon.nasa-global/CAIB/CAIB_lowres_intro.pdf* (accessed 26 March 2008). (For the general site of the *Report*, see *http://www.nasa.gov/columbia/home/CAIB_Vol1.html*, accessed same date; henceforth Columbia Accident Investigation Board, *Report*).

from limited funding and negotiations of NASA managers with the Air Force, OMB, and the White House. As the Columbia Accident Investigation Board stated, however, "Launching rockets is still a very dangerous business, and will continue to be so for the foreseeable future as we gain experience at it. It is unlikely that launching a space vehicle will ever be as routine an undertaking as commercial air travel."[51]

Nevertheless, for all of its flaws, the Space Transportation System constituted a notable engineering achievement. It has achieved goals that would be more difficult for an expendable launch vehicle to accomplish. So far, these have ranged from rescue and relaunch of satellites in less than satisfactory orbits to the five repairs and upgrades to the Hubble Space Telescope. These are remarkable feats that make the Space Shuttle an important contribution to launch-vehicle technology, even as NASA engages in developing its successor.

Ares

From 2005 to 2006, NASA began developing two new launch vehicles intended to avoid the hazards of a reusable orbiter that could be damaged by dislodged insulation from an external tank. These vehicles, dubbed Ares I and Ares V, were expected to build upon Shuttle and other existing launch technology to support future missions not just to the ISS, but to the Moon, Mars, and elsewhere in the solar system. By mostly abandoning reusability, the new vehicles will provide greater safety. The Ares I first stage will consist of a five-segment, reusable SRB enlarged from a single Shuttle SRB. The second stage will use a J-2X engine derived from the J-2 used on Saturn and a J-2S version that Rocketdyne developed and tested in the early 1970s but never flew. Ares I is expected to be capable of launching up to six astronauts to the ISS or up to four astronauts to low-Earth orbit for rendezvous of its Orion service module with an Ares V Earth departure stage for travel to the Moon.[52]

Ares V, the heavy-lift vehicle, will consist of two five-and-a-half-segment, reusable SRBs derived from the Shuttle SRB, mounted on each side of a central

51. Roger D. Launius, "NASA and the Decision to Build the Space Shuttle, 1969–72," *Historian* 57 (autumn 1994): 34; Ray A. Williamson, "Developing the Space Shuttle," in *Exploring the Unknown*, ed. Logsdon et al., vol. 4, pp. 161–191, esp. p. 182; Columbia Accident Investigation Board, *Report*, pp. 1–19, quotation from p. 19.
52. NASA, "Overview: Ares I Crew Launch Vehicle," available at *http://www.nasa.gov/mission_pages/constellation/ares/aresI.html* and associated links (accessed 1 July 2008). See also Hunley, *Development of Propulsion Technology*, p. 296. These comments were written in 2008. As of 5 September 2009, the future of Ares I and Ares V seems to be in some doubt.

booster element, itself based in part on the Shuttle external tank with liquid-oxygen and liquid-hydrogen tanks plus six RS-68B rocket engines, the latter modified from engines on the Air Force's Delta IV. An upper stage, also called the Earth departure stage, will be powered by a J-2X. Once the Earth departure stage and its Lunar Surface Access Module separate from the central booster element, they will rendezvous with the Orion module for their journey to the Moon. Planners expect the launch vehicle will be able to lift nearly 414,000 pounds to low-Earth orbit and almost 157,000 pounds to lunar orbit. Ares V will also be able to launch scientific and exploration payloads into space and could take future crews to Mars and beyond.[53]

As can be seen in part from the above description, the two new launch vehicles build upon not only Shuttle and Saturn technologies, but also those from the Air Force and contractors for all three. While it does not appear that the new vehicles will have all of the Space Shuttle's flexibilities in Earth orbit, they will be able to access further reaches of space and return to Earth. The Ares vehicles would reportedly be safer than the Shuttle and an EELV-derived configuration, as well as less expensive than an EELV derivative.[54]

Conclusions

This narrative has not addressed all of NASA's contributions to launch-vehicle technologies, but the major examples discussed here show a fairly consistent pattern. NASA has contributed in major ways to these technologies, but in doing so, it has worked with partners and built upon the work of the military services and industry in particular. Access to space has not come about through the innovations of a few geniuses alone—whether in NASA or elsewhere. It has resulted from an evolutionary process in which many organizations have cooperated in a variety of ways.

---

53. NASA, "Overview: Ares V Cargo Launch Vehicle," available at *http://www.nasa.gov/mission_pages/constellation/ares/aresV.html* and associated links (accessed 1 July 2008).
54. See, e.g., Jeff Foust, "Defending Constellation," *Space Review* (4 February 2008), available at *http://www.thespacereview.com/article/1054/1* (accessed 28 March 2008).

Chapter 6

NASA's International Relations in Space

An Historical Overview

John Krige

"That's one small step for man, one giant leap for mankind." These "eternally famous words," as James Hansen calls them in his biography of Neil Armstrong, expressed both a NASA and an American triumph.[1] They also reached out to the millions watching the spectacle on television screens all over the world, allowing them to make it their own. Elevating the particular to the universal, Armstrong suggested that the awesome technological power embodied in the Moon landing, while indicative of American supremacy, was also a resource that would benefit all—a promise, not a threat. About 30 minutes into the mission, shortly after having been joined by Edwin "Buzz" Aldrin, Armstrong read the words on a plaque attached to one of the ladder legs of the Lunar Module. The *Eagle*—a name deliberately chosen by the astronauts as the symbol of America—had no territorial ambitions: as Armstrong said, "We came in peace for all mankind."[2] "For one priceless moment in the history of man," Nixon told the astronauts as they explored the lunar surface, "all the people on this earth are truly one"—one, that was, under the benevolent American flag that had been erected with some difficulty a few minutes earlier.[3]

The spectacles of the Moon landing and the moonwalk are suffused with quintessentially American tropes: white, athletic males burst the grip of gravity

1. James R. Hansen, *First Man: The Life of Neil A. Armstrong* (New York, NY: Simon & Schuster, 2005), p. 493.
2. Ibid., pp. 393, 503.
3. Ibid., p. 505.

to conquer a new frontier.[4] America's technological superiority in the service of global expansion is affirmed. Feelings of national pride mingle with arrogance, "an arrogance," as Aldrin put it, "inspired by knowing that so many people had worked on this landing, people possessing the greatest scientific talents in the world."[5] The vitality of a dynamic capitalist society imbued with Christian values—Aldrin took Communion soon after the *Eagle* landed on the Moon—is affirmed against the suffocating state socialism of godless communism.[6]

The coupling of national prowess with global leadership was deliberate. For Willis Shapley, Associate Deputy Administrator at NASA Headquarters, the mission would show the world that "the first lunar landing [is] an historic step forward for all mankind that has been accomplished by the United States of America."[7] All the same, we should not be overwhelmed by the political and ideological staging of Apollo 11 as an American-led achievement of transcendent meaning. For the mission also had genuine international components. As everybody knows, beginning with Apollo 11, NASA astronauts collected over 840 pounds of Moon rock and distributed hundreds of samples for public viewing and scientific research all over the world.[8] Less well-known is the fact that the first video images of Armstrong's and Aldrin's steps on the Moon were picked up not in the United States, but by antennas at Honeysuckle Creek and the Parkes Observatory near Canberra in Australia, a tribute to the vast global data and tracking network that supports NASA's missions.[9] Even more pertinent for this article, one of the few scientific experiments conducted on the lunar surface during Armstrong and Aldrin's 160-odd minutes of surface activity on the night of 20 July 1969 had a foreign Principal Investigator.

---

4. For survey of the historical literature, see Roger D. Launius, "Interpreting the Moon Landings: Project Apollo and the Historians," *History and Technology* 22, no. 3 (September 2006): 225–255. On the gendering of the Apollo program, see Margaret A. Weitekamp, *The Right Stuff, the Wrong Sex: The Lovelace Women in the Space Program* (Baltimore, MD: Johns Hopkins University Press, 2004); Margaret A. Weitekamp, "The 'Astronautrix' and the 'Magnificent Male': Jerrie Cobb's Quest to be the First Woman in America's Manned Space Program," in *Impossible to Hold: Women and Culture in the 1960s*, ed. Avital H. Bloch and Lauri Umansky (New York, NY: New York University Press), pp. 9–28.
5. Edwin E. "Buzz" Aldrin (with Wayne Warga), *Return to Earth* (New York, NY: Random House, 1973), p. 231. This feeling was bolstered by the successful management of a last-minute alarm by the astronauts and ground control at Houston as Armstrong and Aldrin were just 6,000 feet above the lunar surface. See also David Mindell, *Digital Apollo: Human and Machine in Spaceflight* (Cambridge, MA: MIT Press, 2008).
6. On the Communion, see Aldrin, *Return to Earth*, pp. 232–233.
7. Quoted by Hansen, *First Man*, p. 495.
8. Ibid., pp. 513–514.
9. Sunny Tsiao, *"Read You Loud and Clear!" The Story of NASA's Spaceflight Tracking and Data Network* (Washington, DC: NASA SP-2007-4232), chap. 5.

Figure 1: Astronaut Edwin E. "Buzz" Aldrin, Jr., Lunar Module pilot during the Apollo 11 extravehicular activity (EVA) on the lunar surface. In the right background is the Lunar Module Eagle. On Aldrin's right is the Solar Wind Composition Experiment already deployed. This photograph was taken by Neil A. Armstrong with a 70-millimeter lunar surface camera. *NASA Image AS11-40-5873*

During their brief sojourn on the Moon the astronauts engaged in six scientific experiments, all chosen by a NASA scientific panel for their interest and excellence. Five of these were part of the Early Apollo Scientific Experiment Package. They included a passive seismometer to analyze lunar structure and detect moonquakes, as well as a device to measure precisely the distance between the Moon and Earth. The sixth was an independent Solar Wind Composition Experiment. To perform this experiment the astronauts had to unroll a banner of thin aluminum metal foil about 12 inches wide by 55 inches long and orient one side of it toward the Sun. The foil trapped the ions of rare gases emitted from the Sun. It was brought back to Earth in a Teflon bag, cleaned ultrasonically, and melted in an ultrahigh vacuum, releasing the gases that were then analyzed in a mass spectrometer.[10] The results provided insights into the dynamics of the solar wind, the origin of the solar system, and the history of planetary atmospheres.

Johannes Geiss, a leading Swiss scientist, was responsible for this experiment. The payload was manufactured at Geiss's University of Bern and was paid for by the Swiss National Science Foundation.[11] What is more, apart from Armstrong's contingency collection of lunar samples immediately on emerging from the Lunar Module, this was the first experiment deployed by the astronauts. Indeed, to ensure that the foil was exposed to the Sun for as long as possible, it was even deployed *before* Armstrong and Aldrin planted the American flag in the lunar surface and spoke to the President. Scientific need trumped political and ideological statement. NASA's commitment to international cooperation could not be expressed by having the flags of many countries, or perhaps just the flag of the United Nations, left on the Moon. Congress decided that this was an American project and that the astronauts would plant the U.S. flag.[12] Instead, NASA's international agenda fused seamlessly with the "universalism" of science to create a niche for flying an experiment built by a university group in a small, neutral European country.

It is striking that even though the Solar Wind Experiment is routinely mentioned in writings on the Apollo 11 mission, the European source of the experiment is not.[13] This is partly because of the iron grip that human space-

---

10. "Experiment Operations During Apollo EVAs. Experiment: Solar Wind Composition," available at *http://ares.jsc.nasa.gov/humanexplore/exploration/exlibrary/docs/apollocat/part1/swc.htm* (accessed 31 August 2008).
11. Thomas A. Sullivan, *Catalog of Apollo Experiment Operations* (Washington, DC: NASA Reference Publication 1317, 1994), pp. 113–116. Geiss's team also measured the amounts of rare gases trapped in lunar rocks: P. Eberhart, J. Geiss, et al., "Trapped Solar Wind Noble Gases, Exposure Age and K/Ar Age in Apollo 11 Lunar Fine Material" (*Proceedings of the Apollo 11 Lunar Science Conference*, vol. 2, ed. A. A. Levinson, Houston, TX, 5–8 January 1970). See also *Chemical and Isotopic Analysis*, pp. 1037–1070.
12. Hansen, *First Man*, p. 395.
13. This is true of scholarly works like Hansen's *First Man*, chap. 29; accounts specifically concerned with lunar science, like William David Compton's *Where No Man Has Gone Before: A History of Apollo Lunar Exploration Missions* (Washington, DC: NASA SP-4214, 1989); autobiographical accounts like Aldrin's *Return to Earth*, chap. 8; and semipopular works like Leon Wagener's *One Giant Leap: Neil Armstrong's Stellar American Journey* (New York, NY: Forge Books, 2004), chap. 14. None of these sources mentions that the Swiss experiment was deployed before the American flag was unfurled. One has to burrow deep into the official records to extract these data (see *Experiment Operations During Apollo EVAs*). I only did so because I was alerted to the existence of Geiss's experiment by Peter Creola, Swiss and European statesman and space enthusiast: see Peter Creola, interview by John Krige, Bern, Switzerland, 25 May 2007, NASA Historical Reference Collection, NASA History Division, NASA Headquarters, Washington, DC. For Creola's own role in space, see anon., *Peter Creola: Advocate of Space* (Noordwijk, Netherlands: ESA SP-1265/E, 2002).

flight has on the imagination, a mindset constructed by enthusiasts whose shrill voices and skillful marketing have capitalized on the frontier myth that is deeply ingrained in America's sense of itself and its destiny, so playing down alternative, less glamorous visions of spaceflight using benign technologies.[14] It is the challenges faced by the astronauts as they conquer new domains, not the scientific content of the Apollo missions, that resonate culturally, that entertain and inspire, that showcase American technological success and project American power abroad.

The European contribution to Apollo 11 is also ignored because so much space history in the United States—as everywhere—is nationalistic and celebratory, a symptom of the high value placed on technological achievement as a marker of national prowess. There is no doubt that NASA's achievements are extraordinary and that they dwarf the efforts of other spacefaring nations. To date, these have only been able to match the American space program in select domains (the Soviets in some aspects of human spaceflight, the Europeans with their civilian launchers and dynamic science program, the French with the Satellite Pour l'Observation de la Terre [SPOT] series of Earth observation/reconnaissance satellites, and so on). But even if the United States is the undisputed leader in space science and technology, it should not be forgotten that "leadership" is relative and that the preeminence it expresses is assessed in relation to what others are doing. Those competitors and collaborators help define the terrain on which key social actors strive to maintain American leadership, not to say dominance, of space. The extraordinary national feats repeatedly celebrated by America-centric space history do not only serve domestic imperatives; they also help the United States situate itself vis-à-vis other space powers, and they lay the groundwork and create the capacity for it to try to shape what others do in line with American objectives and interests. The international dimension is thus not peripheral to NASA's mission to maintain America's leadership in space: it is intrinsic to it.

International Collaboration in the 1958 Space Act

The National Aeronautics and Space Act of 1958 was signed into law by President Eisenhower on 29 July 1958.[15] It distinguished between civilian

14. I owe this point to Howard E. McCurdy, *Space and the American Imagination* (Washington, DC: Smithsonian Institution Press, 1997).
15. The Act is available at *http://www.hq.nasa.gov/office/pao/History/spaceact.html* (accessed 27 January 2005).

and defense-oriented aspects of aeronautical and space activities and called for the establishment of a new agency to provide for the former in parallel to DOD and, although this was not specified in the Act, to the CIA and later to a highly secret covert agency, the National Reconnaissance Office (NRO), established in September 1961.[16] The primary mission of the resulting NASA that formally came into being on 1 October 1958 reflected the dynamics of superpower rivalry and the struggle for leadership with the Soviet Union that had propelled it into existence in the wake of the Sputnik shocks the year before. In particular, the Space Act called on the new agency to ensure "the role of the United States as a leader in aeronautical and space science and technology and in the application thereof to the conduct of peaceful activities within and outside the atmosphere" (Sec. 2 (c) 5). In the fiery political rhetoric of the day, this stress on leadership escalated into a demand for domination. In January 1958, Senate Majority Leader Lyndon Johnson claimed that "Control of space means control of the world, far more certainly, far more totally than any control that has ever or could ever be achieved by weapons, or troops of occupation. Whoever gains that ultimate position gains control, total control, over the earth, for purposes of tyranny or for the service of freedom." John F. Kennedy picked up the refrain in his presidential campaign: "Control of space will be decided in the next decade. If the Soviets control space they can control earth, as in past centuries the nation that controlled the seas dominated the continents . . . . We cannot run second in this vital race."[17] NASA's core mission was thus to preserve American leadership in the mastery of space science and technology, to dominate the new frontier that was outer space so as "to insure peace and freedom" as Kennedy put it.

Other countries, above all from the free world, were to be enrolled in this endeavor. To this end, the Space Act included among NASA's missions "Cooperation by the United States with other nations and groups of nations . . ." (Sec. 2 (c) 7). This objective was developed in a short, separate section headed "International Cooperation." Here it was specified that "The Administration, under the foreign policy guidance of the President, may engage in a program of international cooperation in work done pursuant to the Act, and in the

---

16. Gerald Haines, "The National Reconnaissance Office. Its Origins, Creation and Early Years," in *Eye in the Sky: The Story of the Corona Spy Satellites*, ed. Dwayne A. Day, John M. Logsdon, and Brian Latell (Washington, DC: Smithsonian Institution Press, 1998), pp. 143–156.
17. Both quoted by McCurdy, *Spaceflight*, pp. 75–76.

peaceful application of the results thereof, pursuant to agreements made by the President with the advice and consent of the Senate" (Sec. 205). International collaboration thus went hand in hand with foreign policy: NASA was to be an arm of American diplomacy.

Eisenhower stressed from the outset that this clause was not intended to engage presidential authority for all bilateral or multilateral programs undertaken by NASA. Its aim, rather, was to allow for the rare occasions when cooperation engaged such important questions of foreign policy that it had to be underpinned by international treaties. The *Final Report of the Senate Special Committee on Space and Aeronautics*, dated 11 March 1959, confirmed this interpretation.[18] As a result, as Arnold Frutkin put it, the pace of the cooperative program "was to be faster and its procedures far simpler than would have otherwise been the case." In particular, "NASA's international program was thus immediately distinguished from that of the Atomic Energy Commission which, under its legislation, was required to obtain approval of its international efforts from Congress."[19] The Space Act thus gave NASA considerable latitude to engage in international collaboration as its officers saw fit and to handle the diplomatic dimensions of its policies and practices informally through interagency consultation, above all with the State Department.

The Emphasis on "Peaceful Use"

A commitment to the "peaceful use" of outer space was essential to the successful exploitation of space for civilian scientific and applications programs on both a national and international collaborative level. As Eilene Galloway, who was involved in drafting the Space Act, has put it, the emphasis on peaceful use was intended to preserve space "as a dependable orderly place for beneficial pursuits."[20] It was driven by two main concerns.

First, there was the fear that space would become a military battlefield or provide platforms from which lethal weapons could be launched at targets on

---

18. On the IGY, see Rip Bulkeley, *The Sputniks Crisis and Early United States Space Policy* (Bloomington, IN: Indiana University Press, 1991); Fae L. Kosmo, "The Genesis of the International Geophysical Year," *Physics Today* (July 2007): 38–43; and Allan Needell, *Science, Cold War and the American State: Lloyd V. Berkner and the Balance of Professional Ideals* (Chur, Switzerland: Harwood Academic Publishers, 2000).
19. Arnold W. Frutkin, *International Cooperation in Space* (Englewood Cliffs, NJ: Prentice Hall, 1965), p. 31.
20. Eilene Galloway, "Organizing the United States Government for Outer Space, 1957–1958," in *Reconsidering Sputnik: Forty Years Since the Soviet Satellite*, ed. Roger Launius, John M. Logsdon, and Robert W. Smith (Amsterdam, Netherlands: Harwood Academic Publishers, 2000), pp. 309–325. See also "The Woman Who Helped Create NASA," available at *http://www.nasa.gov/topics/history/galloway_space_act.html* (accessed 20 September 2008).

Earth. Such bellicose scenarios were widespread in the late 1940s and 1950s.[21] Indeed, Wernher von Braun, the most charismatic and persuasive booster of human spaceflight at the time, went so far as to propose the construction of a multipurpose crewed space station that would serve equally as a platform for further exploration, as a reconnaissance tool, and as a base for firing atomic weapons at hostile nations.[22] The thought that the Soviets might also have such ambitions, and indeed might be ahead of the United States in developing space weapons, galvanized stakeholders in space affairs in the United States to advocate peaceful use as a way to "prevent war and ensure peace in this pristine environment," as Galloway puts it.[23] The call for peaceful use thus served both to project a positive image of the United States and to defuse the threat of Soviet space supremacy.

The second major reason was to protect the freedom for satellites to fly over foreign territory. It is well known that national security, and certainly not a space race with the Soviets, was the main driver of Eisenhower's space policy. He was not against the use of space for science and for robotic exploration, but what he wanted above all was to exploit satellite technology to penetrate behind the wall of secrecy that surrounded the Soviet military buildup. The administration's interest in launching a scientific satellite during the IGY was intended to clear the way for this technological development. The ideology of international scientific collaboration was instrumentalized to establish, by setting a precedent, the principle of the freedom of space, i.e., the right of any space power or organization to send a satellite over the territory of another country without being accused of violating national sovereignty.[24]

Spurred on by these concerns, the United States moved rapidly to set up an international regime forbidding the militarization of space. Lyndon

---

21. McCurdy, *Spaceflight.* Major General Bernard Schriever, who played a major role in developing an ICBM for the Air Force, speaking to space enthusiasts in San Diego, CA, in February 1957, remarked that "several decades from now the important battles may not be sea battles or air battles, but space battles, and we should be spending a certain fraction of our national resources to insure that we do not lag in obtaining space supremacy." Quoted by Dwayne A. Day, "Cover Stories and Hidden Agendas: Early American Space and National Security Policy," in *Reconsidering Sputnik*, ed. Launius, Logsdon, and Smith, pp. 161–195.
22. Michael J. Neufeld, "'Space Superiority': Wernher von Braun's Campaign for a Nuclear-Armed Space Station, 1946–1956," *Space Policy* 22 (February 2006): 52–62.
23. Galloway, "Organizing," p. 322.
24. Walter A. McDougall, . . . *The Heavens and the Earth: A Political Economy of the Space Age* (New York, NY: Basic Books, 1985). See also the collection of articles in Launius, Logsdon, and Smith, ed. *Reconsidering Sputnik*, notably the contribution by Dwayne Day, and the special edition of *Quest* 14, no. 4 (2007).

Johnson was invited by President Eisenhower to address the United Nations in November 1958, where he made a stirring plea against unilateral "penetration into space." "Today outer space is free," Johnson said. "It is unscarred by conflict. No nation holds a concession there. It must remain this way." Johnson went on to stress the "orderly course of full cooperation," which, he said, was the only way to avoid "adding a new dimension to warfare" and to "make the substantial contribution yet . . . toward perfecting peace."[25] In the face of considerable Soviet hostility and suspicion, the United States took the lead in establishing an ad hoc Committee on the Peaceful Uses of Outer Space (COPUOS) that became a regular committee of the United Nations General Assembly in December 1959.[26] This body provided the politico-legal framework in which Washington, DC, sought both to permit the ongoing use of satellites for reconnaissance and to outlaw the use of antisatellite weapons. It faced an uphill struggle.[27] The Soviets were stung by the intelligence-gathering capacity of the U-2 spy planes and by the increased potential of satellites to penetrate their closely guarded military secrets. In June 1962, they formally objected to the use of satellites for reconnaissance. They finally dropped their objections in September 1963. Paul Stares explains the timing of this change of attitude as due to three factors: the now-routine use by the Soviets of their Kosmos series of satellites for intelligence gathering, progress with test ban negotiations (in which satellite overflight was a crucial means to verify compliance), and the prospect of successfully banning nuclear and other weapons of mass destruction from space altogether.[28] Indeed, all parties to the negotiations realized what many scientists had been saying all along: that space platforms were no better and considerably worse than Earth-based ballistic missiles for delivering nuclear weapons to terrestrial targets. Recognizing that "neither side could gain a military advantage by placing nuclear weapons in space [the two superpowers] signed a treaty not to do so" in 1967.[29]

25. Galloway, "Organizing," p. 319.
26. Andrew G. Haley, *Space Law and Government* (New York, NY: Appleton-Century-Crofts, 1963), pp. 313–328.
27. See, for example, Homer E. Newell, *Beyond the Atmosphere: Early Years of Space Science* (Washington, DC: NASA SP-4211, 1975), chap. 18.
28. Paul B. Stares, *The Militarization of Space: U.S. Policy, 1945–1984* (Ithaca, NY: Cornell University Press, 1985), p. 71.
29. McCurdy, *Spaceflight*, p. 68.

No clear definition of "peaceful use" was laid down by COPUOS, nor has one been established since. This is because of the immense importance of military space programs and, above all, the role that intelligence and reconnaissance satellites have played since the dawn of the Space Age. As one scholar notes, from the late 1950s, "the legal position of the United States with respect to the meaning of the phrase 'peaceful uses' became crystallized along lines quite dissimilar to the original rhetoric. The term 'peaceful' in relation to outer space activities was interpreted by the United States to mean 'non-aggressive' rather than 'non-military.'" In international law, this entails that all military uses are permitted and lawful as long as they do not engage the threat or the use of force.[30] No state has formally protested the United States' interpretation of peaceful use (or at least had not by 1990). This interpretation has been essential to the preservation of both international stability and the national security of the space powers.[31] It is now a central plank of the military's expanding reliance on space for technological support in the global war on terror.

The treaty on the peaceful uses of outer space was drawn up simultaneously with the Antarctic Treaty and has a close resemblance to it.[32] By coincidence, a key preparatory meeting, which spawned the Antarctic Treaty, took place in Washington, DC, just three days after the launch of Sputnik. It was convened by Paul C. Daniels of the Department of State and attended by representatives from Britain, Australia, and New Zealand, all of whom had a national stake in the region. The abiding fear among those present was that the Soviet Union would place missile bases in the frozen waste. Daniel's idea was to exploit the IGY to override claims to national sovereignty and instrumentalize scientific cooperation to demilitarize the region. Article I of the ensuing treaty, signed on 1 December 1959, declared that

---

30. Ivan A. Vlasic, "The Legal Aspects of Peaceful and Non-Peaceful Uses of Outer Space," in *Peaceful and Non-Peaceful Uses of Space: Problems of Definition for the Prevention of an Arms Race*, ed. Bhupendra Jasani (New York, NY: Taylor and Francis, 1991), pp. 37–55. This is the definition of "non-aggressive" as stipulated in Article 2(4) of the United Nations Charter.
31. John Lewis Gaddis, *The Long Peace: Inquiries into the History of the Cold War* (New York, NY: Oxford University Press, 1987), chap. 7.
32. This paragraph owes much to Simone Turchetti, Simon Naylor, Katrina Dean, and Martin Siegert, "On Thick Ice: Scientific Internationalism and Antarctic Affairs, 1957–1980," *History and Technology* 24, no. 4 (December 2008): 351–376. See also Jacob D. Hamblin, "Masters of Landscapes and Seascapes. Scientists at the Strategic Poles During the International Geophysical Year," in *Extremes: Oceanography's Adventures at the Poles*, ed. Keith R. Benson and Helen M. Rozwadowski (Sagamore Beach, MA: Science History Publications, 2007).

Antarctica was to be used for peaceful uses only; it explicitly prohibited any military activity in the area, including the testing of any kind of weapons.[33] The Treaty respected "previously asserted rights of or claims to sovereignty in Antarctica" (Art. IV.1) but still insisted on the "freedom of scientific investigation in Antarctica and cooperation toward that end" (Art. II.1).[34] Harlan Cleveland, Assistant Secretary of State for International Affairs, claimed that this was the best the United States could hope for and indeed better than making a claim to sovereign territory; such a claim was irrelevant, and indeed restricting, given the techno-scientific power of the United States. As he put it in 1965, "For the United States, as the nation with the greatest capability to mount and support scientific investigations in Antarctica, this Treaty was clearly better than limiting ourselves to one slice of a much-divided pie. As things stand, we are at liberty to investigate anywhere, build anywhere, fly anywhere, traverse anywhere in this vast still mysterious south land."[35] The same logic informed the Space Act's insistence on restricting space to peaceful uses. Claims to national sovereignty were eclipsed by the demand that space be open to all for nonaggressive activities, from science to applications—like telecommunications and meteorology—to intelligence gathering. By avoiding any unambiguous definition of peaceful use, and by roundly rejecting early Soviet demands in COPUOS that reconnaissance satellites be banned from outer space, the United States preserved the possibilities for international collaboration in civilian space projects without impeding the exploitation of space for national defense.

The Scope of International Collaboration

The scope of NASA's international collaboration is truly vast. In 1970, when many countries only had embryonic programs of their own, Arnold Frutkin reported that NASA had already collaborated with scientists in 70 different countries and established 225 interagency or executive agreements with 35 countries.[36] Addressing a congressional subcommittee in 1981, Ken Pedersen remarked that NASA had over 1,000 agreements with 100 countries and that its international programs had resulted in more than $2 billion of economic

33. Vlasic, "The Legal Aspects," p. 43.
34. The Treaty is reproduced in Haley, *Space Law*, appendix I-A.
35. Turchetti et al., "On Thick Ice," 359–360.
36. Arnold Frutkin, "International Collaboration in Space," *Science* 169, no. 3943 (24 July 1970): 333–339.

benefits for the country.[37] In 2005, Roger Launius remarked that NASA had concluded over 2,000 cooperative agreements with other nations for a multitude of various international space activities.[38] In sum, the number of international agreements entered into by NASA ran at an average of 20 per year during its first decade, exploded to a total 1,000 by the end of its second decade, and then doubled again over the next 20 to 25 years. Looking just at scientific collaboration with Europe, we find that it has increased rapidly in recent times. Launius reported that there had been 139 cooperative science agreements with European nations between 1962 and 1997. Twenty years earlier, John Logsdon counted just 33 projects between 1958 and 1983, suggesting an increase by a factor of four or five in the last decades of the 20th century.[39]

Numbers alone cannot capture this vast enterprise. Table 1 surveys the range of international activities that NASA was engaged in for the first 26 years of its existence. These include infrastructural components like tracking and data acquisition and launch provision. They cover collaboration in science using balloons, sounding rockets and satellites, and applications in areas like remote sensing, communications, and meteorology. In addition, NASA sponsored a huge education and training program through fellowships, research associateships, and the hosting of foreign visitors. There is no doubt that the Agency has played a fundamental role in encouraging and strengthening the exploration and exploitation of space throughout the world, or at least among friendly nations. NASA has helped many countries kick-start their space programs and has enriched them once they had found their own feet. More than that, it has helped give thousands of people in over 100 nations some stake in space, some sense of contributing, albeit in perhaps a small way, to the challenges, opportunities, excitement, and dangers that the conquest of space inspires.

---

37. Kenneth S. Pedersen, *Statement to Subcommittee on Science, Technology and Space; Committee on Commerce, Science and Transportation; United States Senate, 97th Congress*, 31 March 1981, Record No. 1669, folder Pedersen, Kenneth S., NASA Historical Reference Collection, NASA History Division, NASA Headquarters, Washington, DC.
38. Roger D. Launius, "NASA and the Attitude of the U.S. Toward International Space Cooperation," in *Les relations franco-américains dans le domaine spatial (1957–1975): Quatrième recontre de l'IFHE, 8–9 décembre 2005* (Paris, France: IFHE Publications, in press), pp. 45–63.
39. John Logsdon, "U.S.-European Cooperation in Space Science: A 25-Year Perspective," *Science* 223, no. 4631 (6 January 1984): 11–16.

Table 1. Cumulative Statistical Summary Through 1 January 1984[40]

Type of Arrangement	A	B
COOPERATIVE ARRANGEMENTS		
Cooperative Spacecraft Projects	8	38
Experiments on NASA Missions		
Experiments with Foreign Principal Investigators	14	73
U.S. Experiments with Foreign Co-investigators or Team Members	11	56
U.S. Experiments on Foreign Spacecraft	3	14
Cooperative Sounding Rocket Projects	22	1,774[a]
Joint Development Projects	5	9
Cooperative Ground-Based Projects		
Remote Sensing	53	163
Communication Satellite	51[b]	19
Meteorological Satellite	44[c]	11
Geodynamics	43	20
Space Plasma	38	10
Atmospheric Study	14	11
Support of Manned Space Flights	21	2
Solar System Exploration	8	10
Solar Terrestrial and Astrophysics	25	11
Cooperative Balloons and Airborne Projects		
Balloon Flights	9	14
Airborne Observations	12	17
International Solar Energy Projects	24	9
Cooperative Aeronautical Projects	5	40
U.S./USSR Coordinated Space Projects	1	9
U.S./China Space Projects	1	5
Scientific and Technical Information Exchanges	70	3
REIMBURSABLE LAUNCHINGS		
Launchings of non-U.S. Spacecraft	15	95
Foreign Launchings of NASA Spacecraft	1	4
TRACKING AND DATA ACQUISITION		
NASA Overseas Tracking Stations/Facilities	20	48

40. Anon., *26 Years of NASA International Programs* (Washington, DC: NASA, n.d.), p. 3. Thanks to Dick Barnes for providing me with a copy of this booklet.

NASA Funded Smithsonian Astrophysical Observatory (SAO) Optical and Laser Tracking Facilities	16	21
REIMBURSABLE TRACKING ARRANGEMENTS		
Support Provided by NASA	5	48
Support Received by NASA	3	12
PERSONNEL EXCHANGES		
Resident Research Associateships	43	1,417
International Fellowships		358
Technical Training	5	985
Foreign Visitors	131	85,177

A: Number of Countries/International Organizations
B: Number of Projects/Investigations/Actions Completed or in Progress as of 1 January 1984
[a] Number of Actual Launches
[b] United States Agency for International Development Sponsored International Applications Demonstration
[c] Automatic Picture Transmission Stations

The Institutional Dimension

NASA's collaborative effort was originally located institutionally in the Office of International Programs. The first Director, Henry E. Billingsley, was quickly replaced by Arnold W. Frutkin in September 1959. Frutkin joined NASA from the National Academy of Sciences. There, he had been the deputy director of the U.S. National Committee for the IGY and had also served as an adviser to the Academy's delegate to the first and second meetings of the International Committee on Space Research (COSPAR). It was at the second COSPAR meeting in March 1959 that the United States representative, Richard Porter, announced that NASA would be willing to fly single experiments from foreign countries as part of larger payloads on American satellites, as well as to launch complete payloads prepared by other countries.[41] This initiative played a major role in stimulating space research with satellites all over the world.

Frutkin's career at NASA, which lasted 20 years, was crowned with many national and international awards. His many notable achievements included the meteorological and Earth resources satellite data reception networks; the advanced technology satellite regional broadcast experiments, including a highly successful educational program in India; the joint U.S.-Soviet Apollo-Soyuz mission; and the Spacelab agreement signed with the European Space Agency

41. Newell, *Beyond the Atmosphere*, chap. 9. The text of the offer is reproduced in H. Massey and M. O. Robins, *History of British Space Science* (Cambridge, U.K.: Cambridge University Press, 1986), Annex 4.

(ESA). Frutkin also served regularly on United States delegations to the United Nations and other international bodies. In short, in the formative years of the space programs, both in the United States and abroad, Frutkin was, as his official biography put it, "personally responsible for an extraordinary successful series of major international space endeavors contributing equally to the nation's foreign policy objectives and to the advancement of human knowledge" as well as to the "prestige the United States space program enjoys today around the world."[42] In 1978, NASA Administrator Robert Frosch appointed Frutkin Deputy Associate Administrator, then Associate Administrator for External Relations. There, he was responsible for the development of external policy with the public, the international community, universities, and local and state governments, as well as DOD and other federal agencies. The post was not to his liking, and Frutkin left government service shortly thereafter in June 1979.[43] Frutkin's activities were taken over by Kenneth Pedersen, Director of the International Affairs Division of the Office of External Relations. Pedersen had been an assistant professor of political science at San Diego State University from 1968 to 1971, before taking on various policy analysis activities in the federal government. Prior to moving to NASA, he had worked for the Nuclear Regulatory Commission that had replaced the AEC in 1975. Pedersen was the director of the Office of Policy Evaluation that dealt with all aspects of nuclear regulation. He also worked closely with the International Atomic Energy Agency in Vienna, Austria.

Frutkin laid down the basic principles that guided NASA's international collaborative projects for two decades in which the United States was the leading space power in the free world. Pedersen frequently remarked that he was dealing with a different geopolitical situation in which the United States' historical rival for space superiority, the Soviet Union, was showing a greater willingness to open out to international partners and in which the space programs in other regions and countries, notably Western Europe and Japan, had matured significantly. The new, neoliberal philosophy of President Reagan also laid greater stress on rolling back the state's engagement in the provision of space technology (notably launchers); private industry

---

42. NASA Release No. 59-210, 3 September 1959; NASA Key Personnel Change, 1 June 1979, Record No. 726, folder 11.2.1, Frutkin, Arnold W., NASA Historical Reference Collection, NASA History Division, NASA Headquarters, Washington, DC.
43. NASA Key Personnel Change, 1 June 1979.

was encouraged to exploit the economic potential of space.[44] Pedersen's programmatic statements stressed the need for NASA to accept these new realities and to adjust its attitudes to collaboration to reflect the fact that its budgets were limited and that it was no longer "the only game in town." In September 1985, Pedersen was named Deputy Associate Administrator for External Relations and was elevated to Associate Administrator three years later in November 1988.[45]

Richard Barnes replaced Ken Pedersen as Director of International Affairs in 1985.[46] Barnes had been with NASA since 1961 after serving with the AEC's Division of International Affairs and being affiliated with the Atomic Industrial Forum. Barnes was Frutkin's right-hand man during the 1960s and 1970s, before moving on to become NASA's European Representative. He was based at the American Embassy in Paris in the early 1980s, a period and a personality fondly remembered by many Europeans who had dealings with him.

During his term of office, Pedersen had taken a year's sabbatical at Georgetown University. In August 1990, Margaret "Peggy" Finarelli took over his duties when he moved definitively into academia; she was elevated to Pedersen's post of Associate Administrator for External Relations in January 1991. Finarelli joined NASA in 1981 after serving in various government agencies including the White House Office of Science and Technology Policy; she also served as a technical adviser at the Arms Control and Disarmament Agency. She was NASA's chief negotiator for the international agreements with Canada, Europe, and Japan regarding cooperation in the Space Station *Freedom* program.[47]

In October 1991, NASA Administrator Richard Truly reorganized NASA's external relations. He created a new Office of Policy Coordination and International Relations at Headquarters to enable NASA, as he put it, "to respond effectively to the growing international and interagency policy aspects of America's civil space and aeronautics activities."[48] It had four divi-

---

44. For one example of this policy and its exaggerated hopes, see John Krige, "The Commercial Challenge to Arianespace. The TCI Affair," *Space Policy* 15, no. 2 (May 1999): 87–94.
45. Special Announcement, 1 February 1979; Release 88-160, 21 November 1988, Record No. 1669, folder Pedersen, Kenneth S., NASA Historical Reference Collection, NASA History Division, NASA Headquarters, Washington, DC.
46. Release 85-132, 20 September 1985, Record No. 000137, folder Barnes, Richard J. H., NASA Historical Reference Collection, NASA History Division, NASA Headquarters, Washington, DC.
47. Release 90-7, 2 August 1990; Release 91-3, 7 January 1991, Record No. 1669, folder Pedersen, Kenneth S., NASA Historical Reference Collection, NASA History Division, NASA Headquarters, Washington, DC.
48. Special Announcement, 3 October 1991, Record No. 640, folder Finarelli, Margaret, NASA Historical Reference Collection, NASA History Division, NASA Headquarters, Washington, DC.

sions. The International Relations Division was led by Peter G. Smith. Smith joined NASA in 1979 as China Desk Officer after a distinguished career in the State Department. His division was institutionally situated alongside the Policy Coordination Division, the Defense Affairs Division, and the Office of National Service. Finarelli was appointed Associate Administrator of the umbrella office. John D. Schumacher, who had joined NASA in 1989 from a New York law firm, was appointed her deputy. In 1995 NASA Administrator Daniel Goldin appointed Schumacher to the post of Associate Administrator for the Office of External Relations, citing his extensive managerial experience and talent to head an office that was dedicated to "international policy formulation, coordination and implementation."[49]

Three points emerge from this brief survey of the organization of international relations inside NASA from the late 1950s to the early 1990s. First, only Frutkin was directly involved in space matters before he joined NASA, notably through his important role in the National Academy of Sciences and the IGY. Second, the officers appointed to these posts had gained extensive international experience though immersion in nuclear matters, either through the civil nuclear energy program (Barnes, Pedersen) or through arms control (Finarelli). Finally, we see a marked shift in profile, beginning with Pedersen, toward people with formal experience in policy formulation and legal affairs. This change is reflected in NASA Administrator Truly's reorganization in the early 1990s, which elevated Finarelli and Schumacher to senior positions in a new office at Headquarters and which placed Smith at the head of the International Relations Division. It is confirmed with the subsequent promotion of Schumacher to head the Office of External Relations. As NASA and its international relations and obligations expanded, as the programs grew in size and in complexity, and as national security agendas promoted the use of antisatellite weapons in space (along with increasing fears of technology transfer), the rather autonomous approach that had marked Frutkin's 20 years in office inevitably yielded to a more formal mechanism for managing the Agency's relations with its domestic and international partners, from policy formulation to implementation.

Frutkin's Guidelines for International Collaboration

There were two original stimuli for international collaboration; both of them were referred to in the episode described at the start of this article, and they

---

49. News release 95-102, 26 June 1995, Record No. 2955, folder Schumacher, John D., NASA Historical Reference Collection, NASA History Division, NASA Headquarters, Washington, DC.

are illustrated in table 1. Firstly, there was the wish, inspired by major international initiatives like the IGY and the exploration of Antarctica, and coherent with an abiding thread in American foreign policy, to engage other countries (especially friendly and neutral countries) in an exciting new scientific and technological adventure where they could benefit from American leadership and largesse.[50] Secondly, there was the practical need for a worldwide tracking and data-handling network to monitor and intervene in NASA's multiple space missions from planetary probes to human exploration. Sunny Tsiao has recently covered the latter dimension in depth.[51] Here I will concentrate on the scientific and technological aspects of international collaboration in scientific and applications satellites and in human spaceflight from the creation of NASA up to the late 1990s.

In 1965, Arnold Frutkin published an important book spelling out the philosophy that he thought should underpin international cooperation in space.[52] It insisted on the need for "A program founded on conservative values, though not necessarily conservative in scope and objectives . . . ."[53] This view was deeply embedded in Frutkin's thinking. It was probably inspired by the emphasis (in the congressional committee hearings that led to the creation of NASA) that international collaboration in space could transform the tense and confrontational international political climate of the day into one of peaceful coexistence. As Don Kash pointed out 40 years ago, this sentiment led Frutkin to stress the differences between the reality of NASA's programs and the broad hopes expressed in Congress and by three Presidents that international space collaboration would create a new political reality.[54] Insisting that space collaboration could not upset the political status quo, Frutkin advised the State Department in February 1959 that "Political commitments regarding

---

50. Marcia S. Smith, "America's International Space Activities," *Society* 21, no. 2 (January/February 1984): 18–25.
51. Tsiao, "Read You Loud and Clear!" For a review, see *NASA News and Notes* 25, no. 3 (August 2008): 1–5. The system comprised four tracking programs: Minitrack, a north-south network through the Western Hemisphere for scientific satellites; the Deep Space Network; the manned spaceflight ground stations; and the Baker-Nunn tracking stations for a Smithsonian astrophysics program. For an account of the last, see Teasel Muir-Harmony, "Tracking Diplomacy: The IGY and American Scientific and Technical Exchange with East Asia, 1955–1973," in *Making Science Global: Reconsidering the Social and Intellectual Implications of the International Polar and Geophysical Years* (proceedings), chap. 16.
52. Frutkin, *International Cooperation in Space.*
53. Ibid., p. 32.
54. Don E. Kash, *The Politics of Space Cooperation* (West Lafayette, IN: Purdue University Press, 1967), p. 126.

U.S. performance or accomplishment in international space matters should be made with the very greatest caution and conservatism."[55]

This is why one of Frutkin's chief concerns in the early 1960s was to puncture the bubble of enthusiasm and misguided optimism surrounding the achievements of the IGY. This huge enterprise, which combined the efforts of as many 60,000 scientists and technicians from about 66 nations in a study of Earth and the upper atmosphere, was rapidly assuming the stature of a myth; its significance was amplified by exaggerated claims made for the possibilities of international scientific cooperation as an instrument to bring governments together. Frutkin deplored both tendencies. The IGY, he noted, was not a unified and integrated program of cooperation between governments. It was a collection of national programs independently working toward purely scientific objectives that were loosely coordinated by a nongovernmental mechanism. Yes, the IGY had built "scientific bridges across political chasms," "but the bridges had no effect on the chasms; these remained and no traffic other than scientific passed between them." As for scientists, his experience had taught him that they were "demonstrably subject to normal, human limitations and nationalist constraints," just like everyone else. Notwithstanding their rhetoric, they had no privileged ability to overcome national rivalry. They cooperated across borders because in some disciplines, including those connected with space, worldwide collaboration was essential if knowledge was to progress. Science was also of "critical value for cooperation because of the critical dangers with which it is associated," typically in the atom and in space, but also in a field like meteorology, where international collaboration was stimulated by the prospect of weather modification. In short, Frutkin was emphatic that in defining policy, one had to discard "sentiment and tenuous history" that misrepresented and exaggerated the possibilities for bringing about closer collaboration between peoples through international space cooperation.[56]

The one-sided emphasis on scientific cooperation as an innovative instrument to reduce global tensions masked the political competition that was intrinsic to the conquest of space. Essential space technologies, wrote Frutkin, "—rockets, radio, guidance, stabilization—were all common to both the military

55. These sentiments are expressed in a NASA and National Academy of Sciences "Advisory Paper for the Department of State on International Cooperation in Space Activities," dated 12 February 1959, p. 12. It was sent by Hugh Odishaw to Homer Newell, and it was intended to guide United States policy in the United Nations. An annotated footnote suggests that Frutkin actually wrote it. I am grateful to John Logsdon for providing me with a copy of this document.
56. Frutkin, *International Cooperation*, p. 19 on the IGY, p. 15 on scientific cooperation.

and to science." Even the scientific results of space research, from a better understanding of the weather to a more precise knowledge of Earth's shape and its magnetic field, straddled the civilian/military divide. Space achievements were also exploited for propaganda purposes in the context of the Cold War, being used to win the battle for people's admiration and allegiance in the politically uncommitted parts of the world.[57] In short, space exploration was necessarily politicized and suffused with national security concerns that were broadly conceived. Quoting NASA Administrator James Webb, Frutkin remarked that space, like Janus, looked in two directions: one emphasizing international cooperation and the other emphasizing international competition.[58] Frutkin did not deny that any international project would have political implications and that these "should serve the political interests of the United States." However, he was convinced that to avoid criticism, "political objectives are best served by solid accomplishment which may then be exploited politically *after the fact*."[59]

In his book published in 1965, Frutkin identified a number of criteria for a successful international collaborative project. Twenty years later, they were presented more or less unchanged as the basic guidelines for NASA's relationship with its partners.[60] In this summary form, they read:

- Designation by each participating government of a government agency for the negotiation and supervision of joint efforts.
- Conduct of projects and activities having scientific validity and mutual interest.
- Agreement upon specific projects rather than generalized programs.
- Acceptance of financial responsibility by each participating agency for its own contributions to joint projects.
- Provision for the widest and most practicable dissemination of the results of cooperative projects.

This list requires some elaboration.

The first requirement was that NASA have just one interlocutor to deal with in the partner country, an interlocutor that had official authority to engage the human, financial, and industrial resources in the collaborative project. Frutkin was aware that at the dawn of the Space Age, many individuals, pressure groups,

57. Ibid., p. 5.
58. Ibid., p. 8.
59. NASA and National Academy of Sciences "Advisory Paper," p. 1, emphasis in the original.
60. In the introduction to *26 Years of NASA International Programs*, signed by the International Affairs Division, then headed by Ken Pedersen, who also wrote the foreword to the booklet.

and government departments would be jockeying for control of the civilian space program, as they had in the United States. He wanted NASA to avoid becoming enrolled in these domestic conflicts or, indeed, unwittingly being used to promote the interests of one party over the other, hence his refusal to negotiate with anyone but a single official representative. This policy, coupled with NASA's offer to fly foreign payloads in March 1959, not only stimulated the creation of space programs in foreign countries; it forced the national authorities to designate one body as responsible for international collaboration and, in some cases, led to thc rapid establishment of a national or regional space agency. Whereas Frutkin originally left the door open for collaborating with "a central, civilian, and government sponsored, if not governmental authority," by 1986 space agencies were so widespread internationally that NASA could simply designate them as its preferred partners.[61]

The second criterion was obviously meant to make scientific exploration, not political exploitation, the core of any collaborative space program. This was consistent with Frutkin's determination to distinguish the technical from the political and make the former the driving force of the effort. At the same time, he was sensitive to the asymmetry in space capability between NASA and any potential partner in the 1960s, the Soviet Union excepted. He did not want the United States to use its advantage to dictate what others did, both so as to encourage local communities to formulate their own programs and to avoid later charges that the United States had "dominated" the space activities of its partners. Hence his demand that each country "poll its scientific community for relevant ideas" and, in consultation with NASA, "develop full-fledged proposals for cooperative experiments having a character of their own."[62]

This concern also informed the criterion that all agreements should be on a project-by-project basis. An open-ended engagement to collaborate could lead to NASA's committing itself to costly projects that were of no interest to United States investigators. By evaluating each proposal on a case-by-case basis, it could be assessed for its novelty and compatibility with the general thrust of the American space effort, so contributing to the knowledge base of both partners. Also for that reason, both would be willing to invest resources in their part of the project without seeking help from the other. This clause, summarized by the slogan "no exchange of funds," was a cornerstone of

---

61. Frutkin, *International Cooperation*, p. 34.
62. Ibid., p. 35.

NASA policy and a touchstone for the willingness of its partners to take space collaboration seriously and invest their (often scarce) resources in a project.

The demand for full disclosure in the fifth and last criterion listed above flows from this. It was also meant to ensure that the joint program did not touch directly on matters of national security at home or in the foreign country. Frutkin, as we have seen, was well aware of the tight interconnection between the civil and the military in space matters. The requirement that the results of any joint effort be disseminated as widely as was practicable was at once a gesture to this commingling and an attempt to carve out a space for the civil alongside the military. The concept of peaceful use, as I stressed earlier, helped define the limits of the civil domain because it restricted the military to the aggressive. These definitions permitted the collaborative exploitation of scientific data on, say, the effect of electric densities in the ionosphere on the propagation of radio waves, a topic of considerable interest to scientists, but also to commercial and government bodies, including the military.[63]

When Frutkin first formulated his programmatic ideals, he focused almost entirely on space science. This was because most nations could not dream of engaging in major joint technological projects with the United States at the time. The exception was the Soviet Union. Indeed, in a famous speech to the United Nations on 20 September 1963, President Kennedy suggested that there was "room for new cooperation, for further joint efforts in the regulation and exploration of space," adding that "I include among these possibilities a joint expedition to the Moon." Kennedy died before he was able to explore these proposals further, but the obstacles posed by technological exchange to any joint lunar venture were obvious to astute observers. As one editorial noted, the United States was too far ahead in design and engineering to have any interest in developing hardware with the Soviet Union. Collaboration would also undermine national security and deaden the competitive drive between, and national support for, the rival programs. It also required levels of trust between the partners that simply did not exist.[64] This is not to say that no col-

---

63. I am referring here to the so-called topside sounder experiment, undertaken in collaboration with the Canadian Defence Research Telecommunications Establishment, who placed the first national satellite built outside the superpowers, Alouette-I, in orbit in September 1962 using a Thor-Agena rocket provided by NASA. For a detailed analysis of this kind of overlap, see David DeVorkin, *Science With a Vengeance: How the Military Created U.S. Space Sciences After World War II* (New York, NY: Springer-Verlag, 1993).
64. "A Lunar Proposal," *Missiles and Rockets* (14 October 1963): 52. See Frutkin, *International Collaboration*, pp. 116–117.

laboration took place between the United States and the Soviet Union in the 1960s; it happened in meteorology, for example.[65] However, as Frutkin stressed, in dealing with the Soviets, the cooperation was "arms length, in which each side carries out independently its portion of an arrangement without entering into the other's planning, design, production, operations and analysis."[66] Put succinctly, the maintenance of "clean technological and managerial interfaces," along with the demand that there be "no exchange of funds," limited the threat to American technological leadership and national security inherent in the transfer of knowledge required by technological collaboration.

The criteria developed by Frutkin necessarily limited NASA's partners to those that posed no serious security risk and who were willing to make a serious commitment to space. It is not surprising, therefore, that of 38 international cooperative spacecraft projects undertaken or agreed on between 1958 and 1983, 33 were with Western Europe. Of a total of 73 experiments with foreign Principal Investigators, 52 were with this region. Canada, Japan, and the Soviet Union, along with several developing countries, made up the balance.[67] This was quite unlike a program like Atoms for Peace, which proliferated research and some power reactors throughout the developed and developing world in the late 1950s and was driven by foreign policy and commercial concerns that had little regard for indigenous capability. This difference was deliberate: Frutkin was emphatic that space collaboration should never become a form of foreign aid, and he effectively restricted the scope of NASA's activities to industrialized or rapidly industrializing countries with a strong science and engineering base.

This also explains the insistence that collaborative experiments should be of "mutual interest" (second criterion above). How could a foreign experiment that had "a character of its own" be of some value to NASA and to American investigators? For Frutkin, it had to dovetail with the broad interests of the American program, if only to justify the expenditure of United States dollars. Thus, each cooperative project had to be "a constructive element of the total space program of the United States space agency, approved by the appropriate program officials and justifying the expenditure of funds for the US portion of the joint undertaking."[68]

---

65. Angel Long, "Making Atmospheric Sciences Global: U.S.-Soviet Satellite Networking and the Development of High Technology" (paper presented at the Georgia Tech School of HTS Seminar Series), 3 November 2008.
66. Frutkin, *International Collaboration*, pp. 101–102.
67. John Logsdon, "U.S.-European Cooperation in Space Science: A 25-Year Perspective," *Science* 223, no. 4631 (6 January 1984): 11–16.
68. Ibid., 33.

Logsdon has put together some of the "constructive" contributions that international collaboration, notably with Western Europe, made between 1958 and 1983, not only to the United States space effort as such, but also to the American economy and to the pursuit of American foreign policy. His findings are summarized in table 2. This table not only shows the concrete ways in which foreign experiments were to be of "mutual interest" scientifically; it also draws attention to the economic and political benefits of space collaboration, including channeling foreign resources down avenues that would not undermine American scientific and technological leadership; creating markets; projecting a positive image of the United States abroad; and promoting foreign policy agendas, including the postwar integration of Europe.

Table 2. Benefits of NASA's international programs, adapted from Logsdon.[69]

SCIENTIFIC/TECHNICAL BENEFITS
Attracts brainpower to work on challenging research problems.
Shapes foreign programs to be compatible with the U.S. effort by encouraging others to "do it our way."
Limits foreign funds for space activities that are competitive or less compatible with the space interests of the United States.
Obtains outstanding experiments from non-U.S. investigators.
Obtains coordinated or simultaneous observations from multiple investigators.
Opens doors for U.S. scientists to participate in foreign programs.
ECONOMIC BENEFITS
Has contributed over $2 billion in cost savings and contributions to NASA's space effort.
Improves the balance of trade by creating new markets for U.S. aerospace products.
POLITICAL BENEFITS
Creates a positive image of the United States in the struggle for the minds of the scientific, technical, and official elite.
Encourages European unity by working with multinational institutions.
Reinforces the image of U.S. openness in contrast to the secrecy of the Soviet space program.
Uses space technology as a tool of diplomacy to serve broader foreign policy objectives.

These putative benefits were not always welcomed by those actually engaged in the practicalities of international collaboration. American scientists and engineers, flush with the enormous success of their own program, feared that their partners were less capable than they and might not fulfill their

69. Ibid., 13.

commitments. They balked at the additional layers of managerial complexity and the assumed added cost of international projects. As resources for NASA's space science program shrunk in the 1970s, they sometimes resented the presence of foreign payloads on NASA satellites, suspecting that they had been chosen less because of merit than because they were free to the Agency. And they noted that by encouraging foreign powers to develop space capabilities, NASA was undermining the American leadership in high-technology industry: it was producing its own competitors.[70] International collaboration was not uncontested at home, particularly as NASA's partners gained in maturity and were competitors as much as collaborators.

The weight of the several factors (scientific and technical/economic/political) that were brought into play in the first two decades of international collaboration varied depending on circumstances. A scientific experiment built with a foreign Principal Investigator and paid for by a national research council—like Geiss's Solar Wind Experiment on Apollo 11—raised few, if any, broader economic or political issues. Complex and expensive projects calling for major technological developments and managerial inputs were at the other end of the spectrum.

The 1975 Apollo-Soyuz Test Project (ASTP) is the best-known example of this. Often reduced to simply a "handshake in space," it involved docking an American Apollo and a Soviet Soyuz spacecraft with each other in orbit 120 miles above Earth. During the two days in which the hatch between Apollo and Soyuz was open, three American astronauts and two Soviet cosmonauts exchanged pleasantries and gifts and conducted a few scientific experiments together. This was above all a political statement, a concrete manifestation of the new climate of détente with the Soviet Union being pursued by President Nixon and his National Security Adviser and Secretary of State, Henry Kissinger.[71]

Political concerns also provided a trigger for two other major projects in the 1960s and 1970s. One was Helios, the $100 million venture to send two probes built in (West) Germany and weighing over 200 kilograms each to within 45 million kilometers of the Sun.[72] Helios was the most ambitious joint

---

70. For these objections, see Logsdon, "U.S.-European Cooperation," 13.
71. For a summary account, see Joan Johnson-Freese, *Changing Patterns of International Collaboration in Space* (Malabar, FL: Orbit Books, 1990), chap. 6; Smith, "America's International Space Activities," 19.
72. NASA's participation in the project is described in Frutkin, "International Cooperation." For the political dimension, see John Krige, "NASA as an Instrument of U.S. Foreign Policy," in *Societal Impact of Spaceflight*, ed. Steven J. Dick and Roger D. Launius (Washington, DC: NASA SP-2007-4801), pp. 207–218. See also John Logsdon, "Astronautical Research in the Transatlantic Perspective," in *Ein Jahrhundert in Flug, Luft- und Raumfahrtforschung in Deutschland, 1907–2007*, ed. Helmuth Trischler and Kai-Uwe Schrogl (Frankfurt, Germany: Campus, 2007).

project agreed to in the 1960s between NASA and a foreign partner. It was the result of an invitation for space collaboration made by President Lyndon Johnson to Chancellor Ludwig Erhard during a state banquet at the White House in December 1965. For Erhard, a major civil space project was one way of reducing German obligations to buy military equipment from the United States, as required by the offset agreements between the two countries. For Johnson, it was a gesture of support for America's most faithful ally in Europe at a time when the Vietnam War was increasingly unpopular and the French were increasingly hostile to the North Atlantic Treaty Organization (NATO). Of course, once the official offer had been made, these political concerns receded into the background (and Erhard was soon punished in domestic elections for being too "pro-American"). Scientific and technical success, however, should not be decoupled from the political will that created the essential window of opportunity for scientists, engineers, and industry to embark on such an ambitious project so early in Germany's postwar space history with NASA's help.

The same can be said of the Satellite Instructional Television Experiment (SITE), another impressive international project that was agreed to with the Indian authorities in 1970. In this experiment, an advanced application satellite (ATS-6) was first placed into geosynchronous orbit to perform some experiments for various U.S. agencies before being shifted further east.[73] From its new position, it could broadcast television programs to village receivers directly or via relay stations provided by the Indian authorities. For India, the satellite was a marvelous way of bringing educational television, produced locally and dealing with local needs like family planning, into otherwise inaccessible rural areas (programs were broadcast in eight languages directly to small receivers in over 2,000 villages), while giving an important popular boost to the indigenous space program. For the United States, it served a variety of political and economic needs. It sealed a bond with an ally deemed unreliable and promoted the modernization of India as an alternative model to China for developing countries. It was part of a broader strategy to channel Indian resources down the path of civilian technologies. And, by withdrawing the satellite from service after a year, NASA successfully encouraged the Indian government to buy additional models from United States businesses. The SITE,

73. For the NASA perspective, see Frutkin, "International Cooperation." For the broader foreign policy dimensions, see Ashok Maharaj, "Regaining Indian Prestige: The Chinese Nuclear Test, NASA and the Satellite Instructional Television Experiment (SITE)" (paper given at the SHOT Conference, Lisbon, Portugal, October 2008).

while being of undoubted benefit to various constituencies in India, also served multiple geopolitical needs for the United States in the region.

In all of three of the cases just described, while political (and economic) motives were part of the broader context inspiring the collaborations in question, they were essentially left behind or bracketed during the scientific and technical definition of the projects and their implementation. Once the programs got under way, the fundamental maxims of clean interfaces and no exchanges of funds dominated development. Perhaps the Soviets learned a good deal about how the United States managed large-scale space programs through the ASTP. However, as far as hardware is concerned, Marcia Smith remarked in 1984 that "it [was] difficult to point to a single example of new space technology being used by the Soviets that might have come from their experience with ASTP (except for the remodeling of the Soviet mission-control center to resemble the one at NASA's Johnson Space Center)."[74] Indeed, the flow of technology facilitated by cooperation of this nature should not be exaggerated: one NASA Task Force insisted in 1987 that "the major paths for Soviet acquisition of US and Western technology are espionage, evasion of export controls, and access to open literature."[75]

Similarly, there was no significant technology transfer in the Helios project. NASA provided two launch vehicles, some experiments, and the use of its deep space network. Germany designed, manufactured, and integrated the two space spacecraft, provided 7 of the 10 experiments, and operated and controlled the two satellites from a center on domestic soil. Once again, there was doubtless a transfer of managerial expertise in the joint working group that, as in all NASA cooperative projects, was involved in the technical implementation of Helios. However, it focused primarily on payload-spacecraft and spacecraft-booster interfaces, so it was not engaged in the industrial development of core hardware on either side of the Atlantic.[76] Finally, in the SITE, the United States provided the space segment (for very little cost to NASA), while India provided the ground segment.[77] I quoted Frutkin earlier as stressing that if there was political advantage to be gained from international cooperation, it should be exploited after the fact. This was possible in these cases because, by enforcing

74. Smith, "America's International Space Activities," 19.
75. Hermann Pollack, "International Relations in Space. A U.S. View," *Space Policy* 4, no. 1 (February 1988): 28.
76. Frutkin, "International Cooperation in Space," 336.
77. Ibid., 333–334.

his criteria for collaboration, NASA could draw a more or less sharp distinction between the technical and the political that mapped onto various phases of the joint ventures. The balance between the two shifted dramatically as one moved from initiation, through technical implementation, and on to operation.

There was a notable exception to this: the major initiative, inspired by NASA Administrator Tom Paine, to engage Europe at the technological core of the post-Apollo program between 1969 and 1973.[78] In a nutshell, with NASA's budget shrinking dramatically after the "golden years" of the Apollo lunar missions, Paine hoped to get Europe to contribute as much 10 percent (or $1 billion) of an ambitious program that initially included a space station and a shuttle to service it. Foreign participation would also help win the support of a reluctant Congress and President for NASA's plans. And it would undermine those who insisted that Europe needed independent access to space— Europeans were told that they were wasting valuable resources by developing their own expendable launcher to compete with a reusable shuttle that, it was claimed, would reduce the cost per kilogram into orbit by as much as a factor of 10. For several years, joint working groups invested hundreds of hours discussing a variety of projects. Some, like having European industry build parts of the orbiter wing, threw clean interfaces to the winds. Others, like the suggestion that Europe build a space tug to transfer payloads from the shuttle's low-Earth orbit to a geosynchronous orbit, a project of interest to the Air Force, touched directly on matters of national security. The entire process was reconfigured soon after President Nixon authorized the development of the Space Shuttle in January 1972. Clean interfaces and no exchange of funds imposed their own logic on the discussion (and were reinforced by anxieties about European capabilities to fulfill commitments and by fears that NASA was becoming entangled in unwieldy and costly joint management schemes). The European "contribution" was reevaluated, and Germany decided to take the lead in building Spacelab, a shirtsleeve scientific laboratory that fitted into the Shuttle's cargo bay and that satisfied all the standard criteria of international collaboration. So too did Canada's construction of the Remote Manipulator System (RMS), a robotic arm that grabbed satellites in space or lifted them from the Shuttle's payload bay prior to deployment. Once built, both Spacelab and the RMS were handed over entirely to NASA to operate.

---

78. For a summary, see Logsdon, "U.S.-European Cooperation."

The debates around technological collaboration in the post-Apollo program threw into relief the limits to international cooperation in space. For the Europeans, it provided the opportunity to share cutting-edge technologies and access to desperately needed project management skills, though at the risk of not acquiring independent access to space. While many in the United States were happy to see Europeans abandon their plans for a powerful expendable launcher, they were concerned about the threat that intimate technological exchange posed to American preeminence and national security. For NASA, the question was whether the financial and domestic political benefits—as well as the enthusiasm of some sectors of U.S. industry to participate in joint ventures with leading British and European aerospace firms—were worth the risks. The decision-making process was complicated by NASA's difficulty in fixing a technical content to the post-Apollo program that would win congressional and presidential support, by Europe's hesitations, and by the multiplicity of stakeholders involved: NASA (of course), but also the State Department, DOD, and the aerospace industry, just to mention the most prominent in the United States. In the event, Germany's decision to build Spacelab (and France's to build the Ariane launcher) reaffirmed and consolidated the criteria of clean interfaces and no exchange of funds. In a single movement, all the anxieties that had accompanied technological transfer from the world's leading space power in a sensitive sector were dispensed with—though not without considerable European resentment.

The willingness to share technology in the post-Apollo program (and also in support of the European Launcher Development Organization in the mid-1960s) was part of a general sentiment in Washington, DC, that something had to be done to close the technological gap that had opened up between the two sides of the Atlantic at the time. Space technology was seen as a crucial sector for closing this gap.[79] Technological sharing would undermine European criticisms of American dominance in high-tech areas while helping to build a European aerospace industry that could eventually serve as a reliable partner sharing costs in civil and military areas: Europe would assume some of the burden for its own defense. Japan also benefited from technological sharing in the domain of rocketry (and, like Europe, was offered a stake in the post-Apollo program, which it declined). The State Department (in the person of

79. See John Krige, "Technology, Foreign Policy and International Cooperation in Space," in *Critical Issues in the History of Spaceflight*, ed. Steven J. Dick and Roger D. Launius (Washington, DC: NASA SP-2006-4702, 2006), pp. 239–262.

U. Alexis Johnson) allowed U.S. firms to transfer rocket technology to Japan in an intergovernmental agreement signed in 1969 and updated in 1976 and 1979. As Japan was forbidden to develop technologies with military potential, the performance of the subsequent N-series of rockets was deliberately constrained and no state-of-the-art technologies were transferred to Tokyo. In addition, the Japanese authorities were not permitted to provide launches for third parties without the explicit approval of the United States government.[80] There was a transfer of technology, but it was under a tight regime that enforced Japan's restricted international status as a technological power and ensured that NASA's monopoly on access to space in the non-Communist world was not yet seriously challenged.

The Changing Context in the 1980s

The context of international cooperation changed importantly in the 1980s. In essence, the technological gap between NASA and its traditional partners began to close in a variety of space sectors. At the same time, the Soviet Union began to open its closed and secretive program to international collaboration. The effective monopoly that NASA had enjoyed for two decades was over, and so was the willingness by foreign partners to accept Washington, DC's constraints on collaboration that they needed to secure access to the most dynamic, technologically advanced, and open space program on the globe.

Launchers were at the cutting edge of this transformation. On Christmas Eve 1979, the ESA successfully tested its first Ariane rocket. After overcoming the normal teething troubles, Ariane soon proved to be a spectacular success. Helped on by the lower than expected launch rate of the U.S. Space Shuttle, Arianespace (the company that commercialized Ariane) had acquired about 50 percent of the commercial market for satellites by the end 1985. A second major new player entered the field of rocketry in the late 1980s. Japan developed its H-series to replace the N-series built under American tutelage. H-I was tested in 1987. The H-2, scheduled for launch early in the 1990s, was able to reach geostationary orbit. It was, the Japanese argued, derived entirely from technology developed at home and so not subject to the restrictions that NASA had placed on the N-I, N-II, and H-I series, notably as regards providing launches for third parties. China's Long March 3 placed a satellite in geostationary orbit in April 1984; the authorities immediately announced that they were keen to

80. For a summary, see Kenneth S. Pedersen, "The Changing Face of International Space Cooperation," *Space Policy* 2, no. 2 (May 1986): 120–139.

find clients abroad. Finally, the Soviet Union was showing a greater willingness to open its previously closed and secretive launcher system for commercial use and was even seeking a contract to launch a satellite for the International Maritime Satellite Organization (INMARSAT), something that had been simply inconceivable several years before. As Ken Pedersen stressed, "It was, after all, America's launch hegemony that was the foundation of its traditional pre-eminence in cooperative enterprises."[81] That hegemony, along with the opportunities it gave the United States to dictate the terms of collaboration and to dominate the global exploitation of satellites, was now crumbling.

Launch technology was not the only area where American leadership was being challenged. Advanced communications satellites and remote sensing satellites with technologies more sophisticated than those available in the United States civil sector were being built in Europe, Japan, and Canada. The French had taken the lead in commercializing images from SPOT, an Earth remote sensing satellite that technologically outstripped the earlier NASA Landsat system, then bogged down in negotiations over privatization. Australia and a number of rapidly industrializing countries—Brazil, China, India—had constructed solid national space programs; and many third world countries, along with the Soviet Union (in a reversal of its historic policy), were clamoring for a greater say in international bodies like Intelsat, which governs the global satellite telecommunications system. In space science as well, America was becoming just one partner among others. In March 1986, an armada of spacecraft surveyed Halley's Comet on its regular 75- to 76-year sweep though the inner solar system. Giotto was the first satellite sent by the ESA into deep space, and it came within about 600 kilometers of Halley's nucleus. Other spacecraft were supplied by the Soviet Union (Vega I and Vega II) and by Japan (Suisei and Sagikake). The mission was conceived as a joint NASA/ESA venture, and although NASA cooperated by providing support through its Deep Space Tracking Network and a number of American scientists were involved in foreign experiments, the U.S. agency did not have a spacecraft of its own in the fleet. Summing up the situation, a special task force of the NASA Advisory Council reported in November 1987 "that there is in process an accelerating equalization of competence in launching capability, satellite manufacturing and management for communications, remote sensing and scientific activity, and in the prospective use of space for commercial purposes."[82] For Pedersen,

81. Pedersen, "The Changing Face," 124.
82. Pollack, "International Relations," 24.

this meant that NASA had to learn to operate in a pluralistic world in which its historic dominance was diluted along with the flexibility and freedom of action it had long enjoyed. "For NASA today," he wrote in 1986, "'power' is much more likely to mean the power to persuade than the power to prescribe."[83]

The end of the Cold War forced yet another reassessment of NASA's role. The rigidity that had marked 40 years of United States and Soviet rivalry, along with the framework for collaboration that it had defined, had now collapsed. The space program "lost an enemy." The political and military rationales for collaboration with Western allies—and the subordination of economic considerations to geostrategic concerns during the Cold War—would come back to haunt the United States: the technological gap was no more, and previous allies were now economic competitors. Most dramatically, President Reagan, the father of both the International Space Station *Freedom* and of a defensive shield in space popularly known as "Star Wars," suggested "recapitalizing" the former Soviet Commonwealth of Independent States (CIS) republics through large-scale purchases of space hardware and systems. Subsequently, "the Bush administration, in a sharp reversal of prior practice, . . . announced that it will henceforth review license applications to export dual-use technology to the CIS countries with a 'presumption of approval'."[84] The hallowed principles of no exchange of funds and clean interfaces to restrict technology transfer were being overturned. Efforts were made to retain the infrastructure and institutional memory of the major Soviet space programs in Russia and later the Ukraine, though technology transfer was restricted through the Missile Technology Control Regime. As a report for the Office of Technology Assessment pointed out in 1995, Russian industrialists involved in the ISS would be obliged to abide by Western nonproliferation rules, e.g., by not selling sensitive booster technology to unreliable partners.[85] Scientists and engineers were given strong incentives to ally themselves with United States and Western-style reforms in an effort to stem "the flow of indigenous high-risk technologies and expertise from those locations [the CIS states] to outside destinations, principally Third World Nations."[86]

---

83. Pedersen, "The Changing Face," 130. See also Johnson-Freese, *Changing Patterns*, chap. 9, and Smith, "America's International Space Activities."
84. Kenneth S. Pedersen, "Thoughts on International Space Cooperation and Interests in the Post-Cold War World," *Space Policy* 8, no. 3 (1992): 208.
85. U.S. Congress, Office of Technology Assessment, *U.S.-Russian Cooperation in Space*, OTA-ISS-618 (Washington, DC: GPO, April 1995), p. 81. I thank Angel Long for bringing this report to my attention.
86. Pedersen, "Thoughts on International Space Cooperation," 216.

This change in context had palpable effects on the evolution of the plans for the ISS. NASA had already shown a new flexibility in defining this huge technological venture with representatives of the ESA, Canada, and Japan even before the President authorized the scheme in 1984; in recognition of the technological maturity of its partners and the absolute necessity to have them share the cost, NASA's "coordination in the *early planning phases* indicated a consideration of foreign partner interests and objectives unprecedented in space cooperation hitherto" (my emphases).[87] With the inclusion of Russia in the venture beginning in 1993, there was an increased move to multilateralization and interdependence. NASA and American industry could benefit directly by collaborating closely with a partner that had extensive experience in human spaceflight. It was reported in 1995 that United States firms and their counterparts in Canada, Europe, and Japan had entered into Space Station-related contracts and other agreements worth over $200 million. NASA had procured about $650 million of material from Russian suppliers over four years.[88] Russia became functionally integrated into the Station in 1998, providing critical path infrastructure elements on what became a U.S.-Russia core. America's traditional partners in Europe, including Italy, as well as Canada and Japan also made critical path contributions to the overall scheme. And in 1997, an agreement was signed with Brazil for the "design, development, operation and use of flight equipment and payloads for the international space station program."[89]

Ken Pedersen summarized the shift in NASA's policies precipitated by the rapidly changing geopolitical context of the late 1980s, and that was expressed in the collaborative arrangements for the ISS at a conference in Florence, Italy, in 1993. Pedersen began by repeating the mantra that had shaped his approach to international collaboration when he first replaced Frutkin: clean interfaces to minimize technological leakage, no exchange of funds, independent management of projects, "which was really just a somewhat nice way of saying that NASA would continue to stay in charge," and that there was "no idea of joint development of hardware. We would each do our own thing, with our

87. Eligar Sadeh, "Technical, Organizational and Political Dynamics of the International Space Station Program," *Space Policy* 20 (2004): 173. For the early history of the Station, see John M. Logsdon, *Together in Orbit: The Origins of International Participation in Space Station Freedom* (Washington, DC: Space Policy Institute, George Washington University, December 1991); Howard E. McCurdy, *The Space Station Decision: Incremental Politics and Technological Choice* (Baltimore, MD: Johns Hopkins University Press, 2007).
88. *U.S.–Russian Cooperation in Space*, p. 76.
89. Sadeh, "Technical, Organizational," 184.

own money, with our own technology and then bring it together." This was no longer the way to do business. As Pedersen put it:

> If we build long term infrastructures in space with long periods of operation, no exchange of funds is simply not going to work. If we are to build truly global space stations, we have to get used to the fact that each of us is going to be on each other's critical paths. We have to be prepared to share and jointly develop infrastructure in a way in which we must all depend on each other to get to the end of the road. We are going to have to find ways of joint decision making in which conclusions and decisions, as to both the development and operation of joint projects are made in forums in which there is genuine voting or genuine ways of expressing agreements and disagreements and reaching resolution without one actor necessarily imposing its will on another.[90]

Yet even as the physiognomy of collaboration in the Space Station was being redrawn to respond to these new principles, there were other factors at work that would undermine them, and limit their general applicability.

The 1990s and Beyond

In March 1983, President Reagan made his famous speech in which he labeled the Soviet Union an "evil empire" and suggested intercepting and destroying ballistic missiles from space before they reached American shores. "Star Wars," as it became popularly known, was never fully implemented, but it signaled a new emphasis on national security in space matters that generated considerable friction between NASA and DOD. If relationships between the two agencies previously had been relatively smooth and trouble-free, by 1987 they were "neither close nor working well."[91] The Department of Defense feared that NASA was "soft" on technology transfer and not attentive enough to national security considerations, even with its close allies. Already in 1984,

90. Ken Pedersen, "International Cooperation: Past, Present and Future," in *The Implementation of the ESA Convention: Lessons from the Past*, ed. European Centre for Space Law, proceedings of the ESA and European University Institute (EUI) International Colloquium, Florence, Italy, 25–26 October 1993 (Dordrecht, Netherlands: Martinus Nijhoff, 1994), p. 215.
91. Pollack, "International Relations," 26.

NASA Administrator James Beggs had warned his senior staff involved in the Space Station program that they were to be careful to avoid "adverse technology transfer" in international programs, notably where the Soviet Union was involved, and expressed concern about "careless and unnecessary revelation of sensitive technology to our free world competitors—sometimes to the serious detriment of this nation's vital commercial competitive position."[92] As if to confirm the point, DOD intervened in Space Station negotiations with Europe, Japan, and Canada in the mid-1980s, so undermining NASA's authority as the lead American negotiator. In short, as national security concerns (including concerns about threats to American technological and economic leadership) came to the fore in the 1980s, the fears of technological leakage threw an increasingly long shadow over civil space cooperation.

As Beggs's letter made clear, heightened concerns about technological leakage were symptomatic of the economic strength of NASA's partners, a strength that made them both valuable partners and formidable competitors. Economic concerns were now complemented by new military demands. As satellite technology became more sophisticated, the military began to appreciate the importance of space-based hardware as a "force multiplier," i.e., its capacity to enhance *traditional* military operations. Satellites began to be used to improve the effectiveness of battlefield surveillance, tactical targeting, and communications.[93] These advantages, and not the fantasies of "Star Wars," were dramatically demonstrated in Operation Desert Storm, the United Nations-sanctioned, United States-led assault on Iraqi forces that had occupied Kuwait in 1991. NASA Administrator Dan Goldin's 1993 *Final Report to the President on the U.S. Space Program* stressed this dimension of the conflict. "Control of space was essential to our ability to prosecute the war quickly, successfully, and with a minimum loss of American lives." Communications, navigation, weather reporting, reconnaissance, surveillance, remote sensing, and early warning—all these were mentioned by Goldin as essential to United States victory.[94] The defense space budget climbed in line with demand. NASA's budget remained roughly unchanged in constant dollars between 1975 and 1984 (hovering between $8 and $9 billion 1986 dollars). The defense space

92. Quoted in Sadeh, "Technical, Organizational," 174.
93. Stares, *Militarization*, pp. 242–243.
94. *Final Report to the President on the U.S. Space Program, January 1993*, submitted by NASA Administrator Dan Goldin to President H. W. Bush, 7 January 1993, available at *http://history.NASA.gov/33082.pt1.pdf* (accessed 15 December 2008).

budget came from behind to equal NASA's around 1981. By 2000, they were approximately the same at $12.5 to $13 billion current dollars. It was recently reported that in fiscal year 2008, the Pentagon's space program cost about $22 billion, almost a third more than NASA's.[95]

The attacks on American soil on 11 September 2001 accelerated demands for the protection of space as a key asset in America's defensive arsenal. We can get a sense of the outlines of the policy shift by comparing the lessons drawn by the United States administration from the two wars in the Persian Gulf. In 1993, Goldin suggested that the first engagement with Saddam Hussein showed how important it was "to develop and maintain our ability *to deny the use of space to our adversaries during a crisis in wartime*" (my emphasis).[96] Ten years later, Operation Desert Storm was followed by Operation Iraqi Freedom and the global war on terror. Even greater emphasis was placed on the need to secure space as an American military asset. In an unclassified summary of what was almost certainly a National Security Presidential Directive (NSPD) of 31 August 2006, it was stressed that "United States National Security [was] critically dependent upon space capabilities, and this dependence will grow." The document emphasized that "Freedom of action in space is as important to the United States as air power and sea power" and, while stressing that space could be used by all nations for peaceful purposes, made a point of adding that "'peaceful purposes' allow U.S. defense and intelligence-related activities in pursuit of national interests."[97] This point was developed in one of the most controversial clauses of the unclassified document that was released in October 2006:

> The United States considers space capabilities—including the ground and space segments and supporting links—vital to its national interests. Consistent with this policy, the United States will preserve its rights, capabilities, and freedom of action in space; dissuade or deter others from either impeding those rights or developing capabilities intended to do so; take those

---

95. Demian McLean, "Obama Moves to Counter China with Pentagon-NASA Link," *http://www.bloomberg.com*, 2 January 2009, available at *http://news.yahoo.com/s/bloomberg/20090102/pl_bloomberg/aovrno0oj41g/print* (accessed 4 January 2009).
96. *Final Report to the President on the U.S. Space Program, January 1993*, pp. 22, 31.
97. The declassified statement, which is presumably derived from NSPD 49, is available at *http://www.fas.org/irp/offdocs/nspd/space.html* (accessed 29 September 2008). The quotations are respectively from section 5, first paragraph, and section 1, second paragraph.

> actions necessary to protect its space capabilities; respond to interference; and *deny, if necessary, adversaries the use of space capabilities hostile to U.S. national interests* [my emphasis].[98]

Many commentators have noted the continuity in United States space policy from the Reagan years to the present and have insisted that the new directive simply renders more explicit what was left vague and inconclusive in previous policy statements, including those by President Clinton (i.e., there is agreement across party lines on the broad direction of United States space policy for the 21st century). At the same time, it is worth noting the difference between my italicized phrase in Goldin's report in 1993 and that in the August 2006 policy statement. The NASA Administrator suggested the need for denial in times of wartime crisis. The new policy is far broader, and uses "national interest" to justify a range of initiatives—dissuasion, deterrence, and denial—to preserve America's "rights, capabilities, and freedom of action in space." It is this all-encompassing demand that so worries America's partners, all the more so as it is coupled with a recent history of preemptive, unilateral actions by an executive that has refused to be tied down by obstructive international agreements—as reaffirmed in the August 2006 directive: "The United States will oppose the development of new legal regimes or other restrictions that seek to prohibit or limit U.S. access to or use of space." In short, there is a fundamental contradiction in the making between NASA's dependence on foreign partners to pursue its international projects and the military's dependence on space technologies to protect national interests (and to secure civil society's dependence on space technology for the successful functioning of "ATMs, personal navigation, package tracking, radio services, and cell phone use").[99]

For the moment, it is not easy to get a clear picture of how far national security concerns are subverting civilian space collaboration by crippling technological exchange. In a recent assessment of trends, Alain Dupas and John Logsdon noted that President Bush had encouraged international collaboration, but only when it "would support U.S. space exploration goals." They went on to suggest that it seemed that a "unilateral approach [was] emerging as the

---

98. NSPD 49, section 2, item 5.
99. The last was stressed by White House spokesman Tony Snow, as reported in Suzanne Goldenberg's article, "Bush Issues Doctrine for US Control of Space," *Guardian* (19 October 2006), available at *http://www.guardian.co.uk/science/2006/oct/19/spaceexploration.usnews* (accessed 29 September 2008).

preferred U.S. path to shaping international participation."[100] In the 1960s, United States dominance was ensured by virtue of the weakness of its partners and its monopoly on access to space. Collaboration with its allies in the free world was driven as much by generosity as by the exigencies of the Cold War. In 2007, the United States once again seeks dominance, but now for political and military reasons; increasing alienation, rather than grateful admiration, is becoming the hallmark of its international relationships. The last word on this matter will be left to the ESA's Director of Science, David Southwood, who in 2007 deplored the constraints on collaboration that resulted, in his view, from the more or less indiscriminate application of International Traffic in Arms Regulations (ITAR) to any and all space technology. As Southwood put it, "It's not 'this is military space or not military'—anything to do with space is a potential military technology, therefore arms, therefore falls under ITAR." He went on to tell me that "It looks to me as if ITAR is working against the interests of the United States in that . . . . By trying to impose a hegemony, which they can't impose, they're only encouraging others to build up alternative routes to do it . . . . Those of us who want to cooperate with the United States are frustrated by the level of regulation and nonsense we're put through, and indeed the problem we face of trying to explain to people that if we really are cooperating we have to have an understanding of what something does in the partner's piece of equipment."[101] It remains to be seen if Southwood's anger is widely shared and if new presidential policies will remove some of the current obstacles to international collaboration that he has identified.

Early in January 2009, it was announced that President-elect Barack Obama would "probably tear down longstanding barriers between the U.S.'s civilian and military space programs to speed up a mission to the moon amid the prospect of a new space race with China." Pentagon funds could be used for the civilian program in a period of recession. NASA's new Ares I rocket could be scrapped in favor of using an existing military booster. NASA-Pentagon cooperation is also being encouraged to strengthen United States antisatellite technology in the light of China's recent investments in antisatellite warfare. Defense Secretary Gates, who has been kept on by Obama, has recently remarked that these and related Chinese initiatives "could threaten the United States' primary means

---

100. Alain Dupas and John Logsdon, "Space Exploration Should be a Global Undertaking," *Aviation Week & Space Technology* (5 July 2004): 70.
101. David Southwood, interview by John Krige, ESA headquarters, Paris, France, 16 July 2007, NASA Historical Reference Collection, NASA History Division, NASA Headquarters, Washington, DC.

to project its power and help its allies in the Pacific: bases, air and sea assets, and the networks that support them."[102]

At the time of writing, these are merely proposals, and it is difficult to know how much store to lay by them. Yet they are entirely consistent with the general drift of United States space policy over the past 20 or 30 years, a drift that is seeing an increasing militarization of space and a radical rethinking of the relationship between the U.S. civilian and military space programs, alongside the historic determination to use space to project United States power abroad. This blurring of the civilian/military divide can eventually only change the face of NASA and the role and limits of international collaboration in the Agency's mission.

Concluding Remarks

Looking back over NASA's first 50 years, it could be argued that while the rationale for international collaboration has changed, there is an underlying continuity in NASA's ambitions. Those ambitions are driven by a quintessentially American determination to lead in the conquest of space, a determination that has been given additional social and historical traction by defining space as a new frontier to be explored and controlled. These themes appear and reappear in presidential proclamations that characterize the conquest of space as simply the next logical step in that outward dynamic push that is the "manifest destiny" of the United States and intrinsic to American identity and American exceptionalism.[103] Thus when the Shuttle *Columbia* touched down on 4 July 1982, signaling the start of a new era in space transport, President Reagan found it fit to say:

> The quest of new frontiers for the betterment of our homes and our families is a crucial part of our national character . . . . The pioneer spirit still flourishes in America. In the future, as in the past, our freedom, independence and national

102. Robert M. Gates, "A Balanced Strategy: Reprogramming the Pentagon for a New Age," *Foreign Affairs* 88, no. 1 (January/February 2009): 28–41.

103. Jacques Blamont talks provocatively of the "The Wright brothers complex," born with the flight of the *Kitty Hawk*, the conviction that Americans have been chosen by God to be the motors of all scientific and industrial progress in the modern world and that space is their privileged domain of conquest, hence their incredulity at the Soviet firsts in space and the Soviet nuclear test in August 1949; see Jacques Blamont, *Venus devoillée* (Paris, France: Odile Jacob, 1987), p. 245.

> wellbeing will be tied to new achievements, new discoveries, and pushing back frontiers.[104]

Similarly, President George H. W. Bush remarked that "Space is vitally important to our nation's future and . . . to the quality of life here on earth . . . It offers a technological frontier, creating jobs for tomorrow . . . . Space is the manifest destination of a new generation and a new century."[105] America does not choose to go into space and dominate it: it does so because that is its destiny.

Historians like Patricia Nelson Limerick have pointed out that the uncritical celebration of the frontier in remarks like these obscures the violence, failures, corruption, and the near obliteration of Native Americans that were part and parcel of the conquest of the West: hardly a congenial "mission model" for NASA. She emphasizes too that much of the mythology surrounding that conquest has been shown by historians to be downright wrong. No matter. The appeal to the frontier and to "manifest destiny" functions in such contexts not as an appeal to what we now know, but as a metaphor that "guides your decisions—it makes some alternatives seem logical and necessary, while it makes other alternatives nearly invisible."[106] The alternative rendered "invisible" here is a mode of international collaboration that dilutes United States sovereignty in the interests of "genuine" collaboration; instead all cooperation must necessarily be subordinate to the preservation of American leadership and the promotion of American interests.

When NASA was first established and was reaching for the Moon, the metaphor of the frontier, and its tight coupling with American identity and America's role in the world, energized and justified the vast expenditure required for the Apollo program. The associated assumptions of conquest and control did not particularly bother the United States' partners in the free world: their space programs were too new and the need to work with NASA was too urgent for them to see the Agency as anything other than benevolent and generous. Fifty years later the metaphor lives on as the "logical and necessary" framework for thinking about how America should conduct itself in space; its partners, now mature, are finding that framework incompatible with "genuine" cooperation.

---

104. Cited by Patricia Nelson Limerick, "Imagined Frontiers: Westward Expansion and the Future of the Space Program," in *Space Policy Alternatives*, ed. Radford Byerly, Jr. (Boulder, CO: Westview Press, 1992), p. 251.
105. Quoted in Goldin's *Final Report to the President, January 1993*, p. 1.
106. Limerick, "Imagined Frontiers," p. 250.

To ask NASA to change its behavior is, however, to ask far more than that new instruments be established to shape new patterns of collaborative action. It is to ask NASA and the people, Congress, and Presidents who support it (along with American industry, which is being encouraged to capitalize on the economic and military possibilities of space) to decouple space activity from a "manifest destiny" to global expansion and the domination of new frontiers.

NASA Administrator Michael Griffin made the point explicitly in his keynote address opening the conference that celebrated NASA's 50th anniversary. "Societies which do not define, occupy and extend the frontier of human action and scientific discovery will inevitably wither and die," said Griffin. That said, NASA's most important contribution over the past half decade, Griffin added, was not simply a series of spectacular space firsts and successful scientific and technological achievements. What mattered was that NASA was "*the* entity which captures what Americans believe are the quintessential American qualities. Boldness, and the will to use it to press beyond today's limits. Leadership in great ventures"[107]—with international partners willing to dovetail their ambitions with NASA's goals. To ask NASA to rethink its global role and to move toward "genuine" interdependency with its space partners as a matter of general policy is to ask the American stakeholders in space to redefine what it means to be American.

---

107. Michael D. Griffin, "NASA at 50" (NASA's 50th anniversary conference, Washington, DC, 28 October 2008), chap. 1 of this volume.

Chapter 7

Fifty Years of NASA and the Public

What NASA? What Publics?

Linda Billings

The history of the relationship between NASA and the public involves the Agency's approach to informing the public about its activities, public opinion and public understanding about the United States civil space program and efforts to foster public support for it, the evolution of "NASA" and "the public" over time, and the role of political appointees in NASA's public affairs operations, among other things.

This history has unfolded in the context of an evolving cultural environment, shaped by the Cold War, the post-Cold War period, the state of journalism, government-citizen relations, government-journalism relations, and other factors. A half century of public opinion polling about the space program, as well as media coverage of the space program, is a part of this cultural history.

The subject of "50 years of NASA and the public" stretches over a huge research space. The historical record of NASA's relationship with the public is immense, including official records and other archival materials, scholarly research, popular literature, media content, and public opinion. In exploring this research space, one must consider how best to go about interpreting the historical record of the space program. What counts? What, or who, is credible? What motivates official statements? What is missing from the record?[1]

1. The researcher can determine, for example, who has donated their records to the NASA History Office or other archives. But the researcher cannot determine what is missing from these archives. The question is: how do we know what we do not or cannot know?

The analysis approaches the subject of "NASA and the public" from several different perspectives. The history of NASA's Public Affairs Office and operations is reviewed, drawing primarily on official and other archival records. NASA's efforts to fulfill its statutory responsibility, articulated in the 1958 National Aeronautics and Space Act, to effect the "widest practicable dissemination" of information on NASA activities, are examined.

NASA's relationship with various "publics" is also examined: how NASA has dealt with "the public" over its first 50 years and how "the public" has responded to NASA and its programs. That public response encompasses public opinion, public interest, public support, public protest, public ignorance, public apathy, and the permeation of popular culture with images and ideas about space exploration. These are all different aspects of NASA's relationship with "the public." NASA "in" public is scrutinized: its public image, its public face, and public perceptions of and interest in NASA. From a critical perspective, the space program will also be considered as a cultural spectacle.

This review of "NASA and the public" is a mix of scholarly analysis and personal history, or participant observation, in social scientific parlance. In exploring the first 25 years of NASA and "the public," the author draws primarily on archival materials, focusing on the origins of the Agency and its public relations apparatus. For the second 25 years, 1983 to the present, the author draws on her own observations and experiences as a participant-observer[2] along with primary and secondary sources. This review does not offer a panoramic, "god's eye" view of this history, as feminist scholar Donna Haraway[3] would call it. It does offer what feminist scholars call "lived experience," informed by relevant theory and research.

The History of NASA's Public Affairs Office and Operations

Even before NASA was created, U.S. engagement in space exploration was shaped by official concerns about public image. U.S. activities in space were intended to be seen an assertion of scientific and technological expertise, political power, and global dominance. From its inception, NASA was part of larger national political effort aimed at "winning hearts and minds" in a bifurcated world of free and Communist nations. Early records of the NASA

---

2. The author has worked in the Washington, DC, aerospace community since 1983 as a journalist, consultant, and researcher.
3. Donna J. Haraway, *Simians, Cyborgs, and Women: The Reinvention of Nature* (New York, NY: Routledge, 1991).

Public Information Office (later renamed the Public Affairs Office) show that NASA's intent was to establish with "the public" that the United States had a national space program, that NASA was in charge of it—not the Air Force or the Navy or any other military group—and that this program served the purpose of supporting national policy goals. The aim was to make it clear, to U.S. citizens and people around the world, that NASA's space program was open while the Soviet space program was secret.

Early NASA information policy documents cited the Agency's statutory responsibility to "provide for the widest practicable and appropriate dissemination of information concerning its activities and the results thereof."[4] This policy, on paper and in practice, also aimed to control the flow of information to the public, including the mass media. At the same time, the media were invited in, by design, to help tell the story of U.S. leadership and conquest in space.

In a memorandum to NASA Administrator T. Keith Glennan dated 9 September 1958, Walter T. Bonney, Assistant to the Executive Secretary of the NACA and also in charge of public relations, noted that NASA, unlike the NACA, was created to "be employed as an instrument of U.S. policy."[5] NASA's objectives were to preserve space for peaceful purposes, promote international cooperation in space, and advance United States leadership in science and technology. To meet these objectives, NASA must master the art of communication—"to use effectively the techniques of information transmission," said Bonney, who would soon become NASA's first director of public information. "The United States must wage peace not only by what we do but by what we say," he continued. "Our problem is not only to explore outer space for peaceful rather than military purposes but to insure that the world knows what we're doing. We must use the truth to counter the Communist lie."[6]

---

4. National Aeronautics and Space Act of 1958, As Amended, P.L. 85-568, Sec. 203(a)(3).
5. Walter T. Bonney, memorandum to Administrator, "NASA Information Program," 9 September 1958, Office of Public Affairs Files, NASA Historical Reference Collection, NASA History Division, NASA Headquarters, Washington, DC.
6. Ibid. Bonney, a former newspaper reporter and editorial executive, had served as the NACA's public affairs chief since 1951. His goals at the NACA were to establish the group as an equal partner with industry and generate "greater public recognition that the work of NACA represented one of the taxpayers' best investments . . . . The effort was to win and to keep the confidence of press representatives." See Ginger Rudeseal Carter, "Public Relations Enters the Space Age: Walter S. Bonney and the Early Days of NASA PR," Association of Educators in Journalism and Mass Communication, Chicago, IL, 1997.

Elaborating on NASA's statutory responsibility to disseminate information, Bonney said, "NASA must tell the truth, modestly, clearly, and with enough vigor to be heard . . . . [A] positive information policy will provide at least partial control of 'the situation.'"[7] The Agency must not even "permit the appearances of engaging in . . . a competition with the Russians to see who can produce the most spectacular space stunts." At the same time, he said, Congress and "the public" need to know "how much is being accomplished how rapidly by NASA . . . . Here, as in all aspects of its information program, NASA needs to maintain a nice sense of balance." In a 22 January 1959 memo to Glennan, Bonney argued, "There is a need to exercise control over the public statements made by NASA staff." He recommended adopting the policy he had established at the NACA: "No information regarding NACA activities should be imparted to the press without knowledge of, and approval by, Mr. Bonney."[8] Glennan complied.

Bonney wrote to Glennan later that year, "So far as the world is concerned, the nation which first succeeds in" putting a man into Earth orbit "will be credited with having demonstrated a measure of scientific superiority of enormous and incalculable value . . . . Around the world," he wrote, "we are fighting for the minds of men."[9] At the same time, "The distinction between publicity and public information must be kept constantly in mind," he noted. "Publicity to manipulate and 'sell' facts or images of a product, activity, viewpoint, or personality to create a favorable public impression has no place" in NASA. A few months later, Bonney reported to Glennan:

> There is a need . . . for a sharpening of the public focus on the picture of NASA and its activities, thus to assure awareness and understanding that our leadership is hard-driving as well as intelligent, that our staff is talented as well as dedicated, that

---

7. Bonney, "NASA Information Program." Bonney did not specify in this memo what he meant by "the situation."
8. W. T. Bonney to T. K. Glennan, "Dissemination of Public Information," 22 January 1959, Office of Public Affairs Files, NASA Historical Reference Collection, NASA History Division, NASA Headquarters, Washington, DC.
9. Walter T. Bonney, memorandum for the Administrator, "NASA Public Information Program," 20 August 1959, Office of Public Affairs Files, NASA Historical Reference Collection, NASA History Division, NASA Headquarters, Washington, DC. This "winning hearts and minds" approach was then, and still is, popular in military and diplomatic circles. See, for example, Pavani Reddy, "Rapporteur's Report: Winning Hearts and Minds: Propaganda and Public Diplomacy in the Information Age" (presented at the Carnegie Endowment for International Peace, Washington, DC, 27 November 2001), available at *http://www.carnegieendowment.org/events/index.cfm?fa=eventDetail&id=428*, and Edward Bernays, *Propaganda* (with an introduction by Mark Crispin Miller) (New York, NY: Ig Publishing, 2005).

> our planning is boldly imaginative as well as sensible, that our prosecution of the job is vigorous and massive as well as urgent . . . . NASA must show itself to be big enough, lusty enough, and courageous enough to accomplish what must be done in space . . . .[10]

Chris Clausen, JPL Public Affairs Officer, wrote to Bonney in 1959:

> There is a distinct payoff to Russia if it can maintain the fiction that Communism is superior to capitalism simply because Russia can fire larger and heavier payloads than can the U.S. . . . [We are] in this competition certainly not by its choice and generally on terms dictated . . . by the Russians . . . . What we have to do now . . . is stress [the] differences [between the Soviet and U.S. space programs] over and over until everyone understands them.[11]

The most important message for NASA to convey, Clausen wrote, is that the Agency's space program is open while the Soviet program is secret:

> It can be shown that our policy of honesty and candor in reporting our entire program . . . represents . . . one of the basic differences between our philosophy and the Russian doctrine. It is the difference between rubber stamp elections and free elections . . . it is the difference between a civilization that is sure and proud of its strength and a dictatorship whose insecurity must be protected by secrecy.[12]

Another important NASA message, Clausen continued, is that NASA's space program is "a national space program." Getting across this message should minimize "the amount of scrambling different services perform in order to grab public credit for NASA programs. All of this points up the dreadfully difficult task one

---

10. Walter T. Bonney, memorandum to the Administrator, "OPI Staffing," 24 November 1959. Office of Public Affairs Files, NASA Historical Reference Collection, NASA History Division, NASA Headquarters, Washington, DC.
11. Chris Clausen to Walter T. Bonney, n.d., Office of Public Affairs Files, NASA Historical Reference Collection, NASA History Division, NASA Headquarters, Washington, DC. A note from Bonney dated 3 May 1973 and attached to the Clausen memo indicates it was written in early 1959.
12. Ibid.

encounters when one tries to act in a democratic manner. A nice balance must be struck between the attitudes of dictatorial inflexibility and foolish anarchy."[13]

From the start, then, there was a tension between NASA's democratic task of informing the public and its political objective of controlling image and message. Whether the Agency's public affairs officials explicitly recognized this tension is not clear.

By 1960, the Agency had codified the functions and authority of the Office of Public Information in a NASA Management Instruction: disseminating public information, advising NASA officials on "public information matters," reviewing public information "for content and policy adherence," and preparing and distributing information for the media. At this point, the office already had a motion picture section and an art and exhibits section. NASA was well along the way to infusing popular culture with the spectacle of space exploration.

In January 1960, Bonney told Glennan that NASA public information efforts should "avoid selling" the space program.[14] A few months later, Bonney's deputy Joe Stein told the Administrator, "Never should the OPI staff, nor others connected with NASA, attempt to pressure or 'sell' NASA information, nor to play favorites among editors, reporters, writers, broadcasters or publications anywhere." Stein continued, "OPI seeks, not to tell other members of the staff what they can and cannot say, but what is consistent with accuracy and policy, and the effects achieved thereby . . . ."[15]

In December 1960, Administrator Glennan asked Benjamin McKelway, editor of the *Washington Evening Star* and president of the Associated Press, and Russell Wiggins, editor of the *Washington Post* and *Times Herald*, for advice on persuading the media to avoid building up public expectations in advance of Mercury missions. Glennan told the editors that NASA's goal was "no undue limit on reporting of events but rather better informed and more responsible interpretation."[16] Wiggins told Glennan that NASA was not the first

---

13. Ibid. It is worth noting that in this memo, Clausen characterized the United States media as "an ex officio part of the government" with a "valid right to poke its nose into government affairs." The Agency "recognizes and serves this right," he asserted.
14. Walter T. Bonney to the Administrator, "NASA Office of Public Information," 16 January 1960, Office of Public Affairs Files, NASA Historical Reference Collection, NASA History Division, NASA Headquarters, Washington, DC.
15. Joe Stein to the Administrator, "NASA Information Program," 14 October 1960, Office of Public Affairs Files, NASA Historical Reference Collection, NASA History Division, NASA Headquarters, Washington, DC.
16. Shelby Thompson to the Administrator, "Conference with Messrs. McKelway and Wiggins—12/8/60," 14 December 1960, Office of Public Affairs Files, NASA Historical Reference Collection, NASA History Division, NASA Headquarters, Washington, DC.

organization "to find that in spite of its best efforts to make facts available, they are not always reflected in print as might be desired." The editors did not share NASA's view that "pre-launch use of [NASA] background information" could raise public expectations and lead to "a letdown if the experiment were postponed or fell short." The editors suggested that NASA engage with the National Association of Science Writers and other leaders in science news about communicating guidelines to journalists on how to report on "the trial and error nature of the Mercury experimental launches."[17]

Bonney left NASA at the end of 1960. In November 1961, NASA's second Administrator, James Webb, approved a reorganization under which the Agency's Office of Public Information and Office of Technical Information and Educational Programs were merged into a new Public Affairs Office (PAO). Hiden T. Cox was the first Assistant Administrator of the new PAO, serving for six months.[18] In a 1962 memo to Administrator Webb, NASA official Jay Holmes advised that while NASA enjoyed "extremely powerful public support" and "a favorable general public opinion . . . this does not pay off nearly so well as an aggressive, sophisticated lobby."[19] Holmes recommended that Webb and other top NASA officials book speaking engagements "in greater numbers than at present at industrial and technical meetings, around NASA installations, and in states like California and Florida, where space activity is heavy."[20] While NASA focused on industry relations, it did not appear to be as concerned about responding to public queries. A few illustrative examples follow.

---

17. In a subsequent letter to National Association of Science Writers President Earl Ubell, Glennan solicited help in improving "public understanding of the truly experimental nature of our work." He also told Ubell that NASA would be lifting its embargo on the use of prelaunch information, no longer prohibiting media use of the information until after a launch. T. Keith Glennan to Earl Ubell, 23 December 1960, Office of Public Affairs Files, NASA Historical Reference Collection, NASA History Division, NASA Headquarters, Washington, DC.
18. Robert L. Rosholt, *An Administrative History of NASA, 1958–1963* (Washington, DC: NASA Scientific and Technical Information Division, 1966), p. 222.
19. Jay Holmes to Webb, "NASA's Public Position," 12 June 1962, Office of Public Affairs Files, NASA Historical Reference Collection, NASA History Division, NASA Headquarters, Washington, DC.
20. Activity calendars maintained by the Office of Public Affairs through the 1960s indicate that this advice was heeded. For example, the 1964 PAO calendar of events, marking NASA speeches, briefings, exhibits, and conferences, includes appearances at museums and state fairs nationwide; scientific and technical conferences; libraries and universities; Kiwanis Clubs; a "women's study club" in Woodsville, TX; and even a speech by NASA Administrator James Webb to the 23rd men's luncheon of the Texas Rose Festival in Tyler, TX. The author has observed that this practice has continued over the past 25 years into the present.

In 1962, college student Claudia Sperry of Albany, New York, wrote to NASA seeking information on the "policies, programs, and publics" of NASA's "Public Relations program." Sperry asked:

> What is your definition of Public Relations? What prompted the creation of your Public Relations Department? Who do you consider to be your publics? What is your Public Relations Department doing to influence public opinion into thinking that our country needs to spend billions of dollars on space projects instead of concentrating solely on . . . problems we have here on Earth . . . ?[21]

NASA Assistant Administrator for Public Affairs Hiden T. Cox replied to Sperry:

> We do not have a public relations office . . . . I do not believe a Public Relations Department is necessary in this field. I believe we would discharge our responsibilities adequately if we were able to provide the widest practicable dissemination of information concerning NASA's activities . . . . The entire American people constitute [NASA's] public . . . . We do, however, have an Office of Public Affairs, whose function is to help NASA officials cope with . . . enormous demands . . . for information about NASA activities and their results.[22]

"As to what prompted the creation of NASA's public relations department," Cox said, "as phrased, the question does not apply." He concluded:

> You seem to assume public relations activity is in progress to create a favorable image and acceptance of the national space program . . . . Even if we wanted to engage in public relations activities, it would be impossible to do so in view of the other demands on our time.[23]

---

21. Claudia Sperry to Director of Public Relations, 2 April 1962, Office of Public Affairs Files, NASA Historical Reference Collection, NASA History Division, NASA Headquarters, Washington, DC.
22. Hiden T. Cox to Claudia Sperry, 9 May 1962, Office of Public Affairs Files, NASA Historical Reference Collection, NASA History Division, NASA Headquarters, Washington, DC.
23. Ibid.

In another example, a NASA public affairs officer took umbrage at questions about media access to information. In a 1965 letter to the Agency's Manned Spacecraft Center public affairs officer Paul Haney, the president of the Greater Houston chapter of the Texas Civil Liberties Union asked for NASA to "reconsider the restrictive measures . . . imposed on the newspapers in regard to their coverage of the activities of the astronauts."[24] Haney replied, "Your letter came as complete surprise and shock to us . . . . I can only conclude that your letter was based on misinformation."[25] He provided tallies of NASA's interactions with the public: numbers of visitors to the Manned Spacecraft Center, viewers of "film clips and TV presentations" about the Manned Spacecraft Center, press briefings and interviews conducted, and so on. In another letter to Read, Haney wrote that the constitutional freedom of the press "is precisely that—a freedom, not a subpoena." NASA prefers to select its astronauts "by means other than a newspaper publicity contest," he wrote. "[M]ay we know what your policy is to be with regard to freedom of speech, particularly that of an individual?"[26]

Haney's response to Read provides an example of NASA PAO's standard approach to assessing public interest: quantification. The PAO measures its (and NASA's) performance by counting hits on NASA's Web site and stories about NASA in print, broadcast, and online media. Elite media coverage of NASA news is always of particular interest. Audiences for various media may also be counted, though apparently with no attention paid to whether audiences actually receive the information that NASA disseminates and what audiences actually do with that information. For example, in a 1995 activity report, NASA PAO stated that "last year an estimated three million people examined

24. Mrs. Clark P. Read to Paul Haney, 11 May 1965, Office of Public Affairs Files, NASA Historical Reference Collection, NASA History Division, NASA Headquarters, Washington, DC.
25. Paul Haney to Mrs. Clark P. Read, 12 May 1965, Office of Public Affairs Files, NASA Historical Reference Collection, NASA History Division, NASA Headquarters, Washington, DC.
26. Paul Haney to Mrs. Clark P. Read, 25 May 1965, Office of Public Affairs Files, NASA Historical Reference Collection, NASA History Division, NASA Headquarters, Washington, DC. Clearly, Haney and Read differed in their interpretation of NASA's statutory responsibility to provide for the widest practicable dissemination of information about its activities. NASA Assistant Administrator for Legislative Affairs Robert Allnut took a similar approach in responding to a query from United States Representative Charles Goodell about NASA spending on public affairs. "NASA does not have what is commonly designated as a 'Public Relations' program," Allnut told Goodell. Referring to the Agency's statutory mandate, he said NASA's task is to disseminate information to the public, noting that in 1967, "530,000 people toured NASA facilities at Cape Kennedy . . . 500,000 people toured the Manned Spacecraft Center at Houston," and NASA "distributed over 3 million publications . . . loaned over 70 thousand motion picture prints, scheduled 3,000 speakers, participated in 1,000 exhibits and conducted 11,400 spacemobile lecture demonstrations" (Robert F. Allnut to the Hon. Charles Goodell, 16 January 1968, Office of Public Affairs Files, NASA Historical Reference Collection, NASA History Division, NASA Headquarters, Washington, DC).

the Space Camp Exhibition when it tours state fairs."[27] Such estimates do not have a clear meaning. Does this number mean that a total of three million people attended the fairs that hosted the exhibit? Or does it mean that three million people walked by the exhibit? Or does it mean that three million people learned something useful from the exhibit?

In mass communication research, audience studies and critical and cultural studies address what this quantified approach does not. How many readers, listeners, and viewers are paying attention to content? Who receives the messages that content providers are aiming to convey? What do people do with the information they acquire from the media? What does media content mean to all of its various audiences? What do people do with what they learn? How does media content influence public opinion? What do the PAO's tallies say about what people know, or think they know, about NASA, and what they do with what they know? Answering these questions grows more complicated by the day, as the number and kind of media outlets, the volume and type of content they produce, and the technological means of interpersonal as well as mass communication continue to proliferate. Add to this mix the increasing sophistication of marketing campaigns in the public and the private sectors, including NASA and its aerospace contractors, and the task of understanding "NASA and the public" appears daunting. It is important to consider that NASA does not have a single, monolithic "public." It has many different publics, and they are changing all the time.[28] Another important factor to consider in examining NASA's public relations is its longstanding and intensive focus on maintaining good relations with Congress and the White House, which colors its relations with other publics.

Marketing the Space Program

In assessing public opinion about, interest in, and knowledge of the space program, NASA and the space community have typically taken an advertising and marketing approach to the task, performing or commissioning administrative research. NASA has repeatedly turned to the advertising and marketing

27. Laurie Boeder to multiple addressees, "Weekly Report of the Office of Public Affairs," 14 February 1995 and 9 November 1995, Office of Public Affairs Files, NASA Historical Reference Collection, NASA History Division, NASA Headquarters, Washington, DC.
28. In recent years, NASA has given some recognition to the need for serving different publics in different ways. The Agency's approaches to dealing with different audiences can appear to be simplistic, however—for instance, compartmentalization of audiences to buttons on the NASA Web site: "for public," "for educators," "for students," "for media," "for policymakers," "for employees."

sector for help in "branding" and "selling" the space program. The result has been a string of similar studies and similar findings—including the finding that public knowledge of NASA is a mile wide and an inch deep—and a series of attempts to cultivate favorable public opinion, along with the increased public support that is erroneously assumed to accompany that favorable opinion, by "pitching" NASA to the public.

In the early 2000s, NASA Administrator Sean O'Keefe commissioned Harmonic International—"a strategic positioning company"—to help the Agency with "brand equity and message concept development." Harmonic reported to NASA in 2004 that "NASA enjoys a strong favorable attitude" and positive "brand equity," though people who hold these views have "a very weak knowledge foundation" for them. Thus, NASA communications "must help explain NASA, building a knowledge base" and reinforcing "the foundation of NASA's brand equity"—that is, advancing knowledge and understanding the universe.[29] A "cultural analysis" of space exploration conducted as part of the larger Harmonic study expanded upon the advertising-and-marketing approach, exploring "NASA and the public" in a broad social context:

> The general public . . . believe space exploration is not a fantasy, but an achievable possibility . . . a noble endeavor. They have a generally positive view of NASA, based primarily on the success of the manned space Mercury and Apollo programs. But they do not believe the government should spend billions of dollars to achieve it.[30]

In 2004, NASA created a new Office of Communications Planning and an Office of Strategic Communications Planning, headed by political appointees. The Office of Strategic Communications Planning was tasked with "developing a strategic communications approach for guiding the activities of the Offices of Communications Planning, Education, Legislative and Intergovernmental

29. Harmonic International, "Brand equity and message concept development" (presentation to NASA Headquarters, Washington, DC, 24 May 2007). The author attended and obtained a copy of this presentation.
30. Center for Cultural Studies and Analysis, "American Perception of Space Exploration: A Cultural Analysis for Harmonic International and the National Aeronautics and Space Administration," 1 May 2004, p. 3. Harmonic International, "Brand equity and message concept development" (presentation to NASA Headquarters, Washington, DC, 24 May 2007). The author attended and obtained a copy of this presentation.

Affairs, and Public Affairs, including strategies and tactics that support NASA's Mission." The Office of Communications Planning was tasked with advising the Administrator "on new and innovative ways to engage and inform a broader cross-section of the . . . public"; identifying "audiences for . . . a wide variety of specialized and targeted resources, information, and messages"; "developing effective, data-driven strategic messages that can be employed Agency-wide and targeted to specific audiences . . . to provide for the widest practicable and appropriate dissemination of information concerning the Agency's activities and results thereof and to increase public awareness and understanding of NASA and its mission"; and ensuring "message consistency and repetition across the Agency to increase the American public's understanding of science, technology, and NASA's mission."[31]

In 2006, NASA adopted a new public information policy to demonstrate its commitment to open communications.[32] In a 2007 briefing to Agency officials, NASA Strategic Communications Chief Robert Hopkins asserted that NASA "is committed to a culture of openness with the media and the public that values the free exchange of ideas, data and information" and that "scientific and technical information from or about Agency programs and projects will be accurate and unfiltered."[33] Nonetheless, NASA's "open" communications under this policy are subject to a complex, multilevel system of review and concurrence. Thus, "openness" in this policy is a relative term.[34]

Also in 2007, Hopkins distributed a "final NASA Message Construct" to Headquarters officials: "NASA explores for answers that power our future."[35] He advised officials to use the message, verbatim, in their communications, and he steered them to NASA's "Strategic Communications Framework Implementation

31. NASA Policy Directive (NPD) 1000.0A, 1 August 2008, "NASA Governance and Strategic Management Handbook," available at *http://nodis3.gsfc.nasa.gov/npg_img/N_PD_1000_000A_/N_PD_1000_000A_.pdf*.
32. NASA Policy on the Release of Information to News and Information Media, 30 March 2006, available at *http://www.nasa.gov/audience/formedia/features/communication_policy.html*. NASA issued this 2006 policy on releasing information to the media in response to press reports that NASA public affairs officials tried to limit Agency climate-change expert James Hansen's public statements. The policy states: "release of public information concerning NASA activities . . . will be made promptly, factually, and completely" and that "in keeping with the desire for a culture of openness, NASA employees may, consistent with this policy, speak to the press and the public about their work."
33. Robert Hopkins, "NASA Media Communications Policy" (presentation to the NASA Senior Management Council, 11 July 2007).
34. The author's observations regarding openness pertain to NASA communications from the time the Agency announced its new public information policy in 2006 through December 2008, when this paper was completed.
35. Robert Hopkins, Memorandum to Officials-in-Charge, "NASA messages," 1 August 2007.

Plan" and "Strategic Communications Implementation Handbook" for further guidance. A few weeks later, Hopkins advised officials that the intent of his "message construct" memo was not to deliver a "mandate" but "to provide some consistency on how we talk about NASA's work with the public." He said the core message was not intended to be "a slogan or tag line," and he encouraged officials to use the themes of "inspiration, innovation, and discovery" in their communications, "depending on whether they work."[36]

To sum up, during its first 25 years, NASA's desire to control image, message, and the overall flow of information from the Agency to the public was in tension with its need to tend to its statutory obligation of disseminating information. This tension has persisted over the last 25 years. Early on, NASA public affairs officials exhibited a tendency to contain or withhold information that might not serve the purpose of boosting NASA's public image and reinforcing its chosen message. They have continued to do so over the 25 years that the author has been watching.[37] There is a tension between the goals and objectives of these political appointees and the civil servants who work with them on disseminating information. The role of appointees is to make the President look good, by making NASA, headed by a leader of the President's choice, look good. Civil servants have the task of fulfilling the Agency's statutory responsibility to disseminate information on all of its activities. They are also compelled to keep their appointee bosses happy—a tough order on some days.[38]

Over the last 25 years, the author has observed a continued institutional sensitivity at NASA about activities that might be construed as "promotional"—even though the Agency regularly engages in all sorts of activities that could easily be construed as promotional. For example, in 2008 NASA held a series of Future Forums in different cities around the country.[39] NASA designed these events to inform the public about NASA's plans for executing the President's Vision for Space Exploration. NASA's press releases, background information, and official statements about the forums could easily

---

36. Robert Hopkins, Memorandum to Officials-in-Charge of Headquarters Offices Directors, NASA Centers, "Updated Guidance on NASA Messaging," 11 September 2007.
37. It is worth noting that NASA's core message has not changed much since the beginning of the space program. See Linda Billings, "Ideology, Advocacy, and Spaceflight—Evolution of a Cultural Narrative," in *Societal Impacts of Spaceflight*, ed. Steven J. Dick and Roger D. Launius (Washington, DC: NASA SP-2007-4801, 2007), pp. 483–500.
38. The author has not been able to verify precisely when NASA adopted the practice of placing political appointees in charge of public affairs.
39. The author reviewed NASA information and media reports about these forums but did not attend any of the events.

be construed as promotional, a carefully orchestrated sales pitch with the tag line "NASA powers inspiration, innovation, and discovery." According to the Agency, the aim of these forums was to "discuss the role of space exploration in advancing science, engineering, technology, education and the economy that benefits your community and the nation" and to provide "an exciting preview of NASA's Constellation Program—America's return to the Moon and beyond." NASA used these forums to talk about its contributions to what the Agency calls "The Space Economy"—"the full range of activities that create and provide value to human beings in the course of exploring, understanding and utilizing space."[40]

NASA and the Media

A core function of the press, historically and presently, is to mediate the flow of information from government to citizens, and NASA has always depended on the mass media to get the word out about its public performances. Reliance on official sources has long been a standard journalistic practice, and by engaging in this practice, the media reinforce and perpetuate official opinions and worldviews.[41] This practice has served NASA well from the Agency's inception to the present.

The history of Science Service, a news syndicate that operated from 1920 through World War II, provides some insight into the longstanding cozy relationship between government and the press and the role of the media in science and technology boosterism. Newspaperman Edwin W. Scripps created Science Service, the first science news syndicate, in 1921 because he believed that science was the basis of democratic life and that scientists were "so blamed wise and so packed full of knowledge . . . that they cannot comprehend why God has made nearly all the rest of mankind so infernally stupid."[42] The Science Service syndicate was controlled by a board of trustees representing prestigious sci-

---

40. Remarks as delivered by the Honorable Shana Dale, NASA Deputy Administrator (San Jose Future Forum, San Jose, CA, 14 May 2008), available at *http://www.nasa.gov/50th/future_forums/sanJoseWithGallery.html*. NASA executed another carefully orchestrated public performance by participating in the Smithsonian Institution's Folklife Festival in 2008. NASA's Future Forums and its presence at the Folklife Festival are promising material for case studies in "NASA and the public."
41. See, for example, Pamela J. Shoemaker and Stephen D. Reese, *Mediating the Message: Theories of Influences on Mass Media Content*, 2nd ed. (White Plains, NY: Longman, 1996); Wolfgang Donsbach, "Psychology of News Decisions: Factors Behind Journalists' Professional Behavior," *Journalism* 5, no. 2 (2004): 131–157; M. Schudson, *The Sociology of News* (New York, NY: W. W. Norton, 2003).
42. Dorothy Nelkin, *Selling Science: How the Press Covers Science and Technology*, rev. ed. (New York, NY: W. H. Freeman, 1995), p. 81.

ence associations, including the American Association for the Advancement of Science and the National Academy of Sciences, "and its editorial policies were dominated by the values of the scientific community." Scripps chose to operate the syndicate as "a press agent for the associations" rather than an independent news service. In line with the interests of Mr. Scripps, Science Service's stories "cast science as a new frontier and scientists as pioneers and discoverers."[43]

After World War II and throughout the Cold War, the U.S. media continued to serve the cause of science boosterism, and NASA rode this wave. At the same time, broadcast media began to supplant print media as the dominant source of news, highlighting the spectacular quality of space exploration. "More active or visual issues . . . became especially newsworthy."[44] Through the 1980s and 1990s, consolidation of media ownership disturbed the traditional balance between the publishing (advertising and profit-seeking) and editorial (reporting and analysis) components of journalism. NASA has benefited from the related media trend toward producing more infotainment content and less news and analytic content in recent years. At the same time, NASA's public affairs, public outreach, and public education initiatives have been trending toward at least the appearance of infotainment. Today, the media are as dependent as ever on official sources—perhaps increasingly so in an increasingly competitive media environment and more tightly controlled government public affairs operations. Concurrently, NASA's Public Affairs Office has become increasingly proficient at peddling the spectacle of space exploration, showcasing rocket launches and astronauts. As political communication expert Shanto Iyengar has observed, the boundaries between news and political marketing "have virtually vanished. The use—even manipulation—of the mass media to promote political objectives is not only standard practice, but in fact is essential to survival."[45]

The author has observed over the past 25 years that the view of the press as subservient to government is persistent at NASA. So is the one-way transmission or "bullet" conception, or model, of communication, whose goal is to deliver a specific message to a specific target. The rhetorical objective of communica-

---

43. Nelkin, *Selling Science*, pp. 81–82.
44. Shanto Iyengar, "Engineering Consent: The Renaissance of Mass Communication Research in Politics," in *The Yin and Yang of Social Cognition: Perspectives on the Social Psychology of Thought Systems—A Festschrift Honoring William J. McGuire* (New Haven, CT: Yale University, 20–22 April 2001), p. 3, available at *http://pcl.stanford.edu/common/docs/research/iyengar/2001/mcguire.pdf*. The Watergate incident in the 1970s may have made the media more skeptical about official sources, but those effects were not necessarily long-lasting.
45. Iyengar, "Engineering Consent: The Renaissance of Mass Communication Research in Politics," p. 1.

tion by this model is persuasion. This was the model employed in Cold War government propaganda campaigns. NASA's Public Affairs Office has always been expert at knowing how to disseminate information to the media. The Agency is not so expert in understanding how journalism works, as a culture, a practice, a system of values. In addition, evidence is lacking of a matching expertise in understanding what people *do* with the information they receive from NASA. This disconnect may at least begin to explain the gap between NASA's good public reputation and its consistently low ranking as a spending priority.

NASA's Relationship with Its Various Publics

For all of its 50 years thus far, NASA has claimed a high level of public interest and a good reputation with "the public." It is not clear how much of this good feeling among citizens is a product of NASA's public affairs efforts and how much is due to other social factors—that is, the social and cultural context for the space program. Over the past 25 years, the author has observed that when NASA and other members of the space community talk about public interest and understanding and engagement, they are usually talking about their desire to expand public support. Public opinion research and studies of public understanding of science and technology have shown how and explored why public interest does not equate to public understanding and how and why neither interest nor understanding equates to public agreement or support.[46]

Numerous public opinion polls and surveys about NASA and space exploration have revealed this disconnect.[47] Poll and survey results have shown consistently over the years that respondents tend to be interested in the space program and tend to value having one. In addition, results do not reveal wide endorsement of big-ticket human spaceflight programs such as the Apollo lunar-landing program and proposed human missions to Mars. And when asked to rank the space program as a government spending priority, respondents have consistently put NASA at the bottom of their lists. One factor that may contribute to this consistently low ranking is NASA's lack of a meaningful rationale for the space program. For people in the space community, the space program means many things: jobs, money, progress,

---

46. See, for example, Alan Irwin and Brian Wynne, eds., *Misunderstanding Science? The Public Reconstruction of Science and Technology* (New York, NY: Cambridge University Press, 1996); National Science Board, *Science and Engineering Indicators 2008* (Arlington, VA: NSB 08-01, NSB 08-01A, January 2008).
47. Roger D. Launius, "Public Opinion Polls and Perceptions of U.S. Spaceflight," *Space Policy* 19 (2003): 163–175.

political capital, and prestige. For 50 years, NASA and the space community have promoted the economic, political, and security benefits of space exploration. And for 50 years, people outside the space community have not been clear about the purpose of the space program. The rationales that NASA has offered over the years[48] do not appear to be especially meaningful to the Agency's "external" audiences.

Over NASA's first 25 years, the Cold War was NASA's driving rationale for space exploration. Over the last 25 years, NASA has been weak on rationale, despite continual attempts to articulate one. What drove the United States space program in its early years, journalist John Noble Wilford observed, was "the pursuit of national prestige and power by a new means and in a new frontier." The lack of a durable rationale for space exploration "contributed eventually to a serious mid-life crisis for the American space effort," he said, deeming the Apollo lunar landings

> . . . a triumph that failed, not because the achievement was anything short of magnificent but because of misdirected expectations and a general misperception of its real meaning. The public was encouraged to view it only as the grand climax of the space program, a geopolitical horse race and extraterrestrial entertainment—not as a dramatic means to the greater end of developing a far-ranging spacefaring capability. This led to the space program's post-Apollo slump . . . . We had been conditioned to think of the space program in terms of the Cold War . . . . The media no doubt perpetuated this attitude, for editors generally viewed every story in those days in terms of whether it meant we or the Russians were ahead. But NASA also played the game, because that was the surest route to the Treasury.[49]

Sylvia Fries Kraemer has also made note of this problematic lack of rationale. Citing "the relative poverty of . . . intellectual efforts to understand the

48. Linda Billings, "Ideology, Advocacy, and Spaceflight—Evolution of a Cultural Narrative," in *Societal Impacts of Spaceflight*, ed. Steven J. Dick and Roger D. Launius (Washington, DC: NASA SP-2007-4801, 2007), pp. 483–500.
49. John Noble Wilford, "A Spacefaring People: Keynote Address," in *A Spacefaring People: Perspectives on Early Spaceflight*, ed. Alex Roland (Washington, DC: NASA SP-4405, 1985), pp. 70, 72, available at *http://history.nasa.gov/SP-4405.pdf*.

significance of space travel . . . and the relative uncertainty of . . . rationales for a space program as a major, national undertaking," she has observed that a sound rationale must "reflect the genuine needs and aspirations of real and important constituencies. The burden of our space program is that it has had only a marginal audience, and marginal constituencies."[50]

In examining the history of NASA's "public" relations, the Agency's expectation that the mass media will help to foster those relations and generate favorable public opinion deserves attention. It is useful to consider that media discourse does not create public opinion, nor does public opinion create media discourse. They interact with each other and with other social phenomena as well, in a process of social construction.[51] Some interesting insights might be gleaned from mapping out the evolution of interactions among NASA's public information efforts, media discourse, and public opinion.

NASA in Public

In regard to "NASA in public" during the Agency's early years, the power-and-prestige rationale for space exploration "exercised major influence" in national political circles at that time,[52] and astronauts and rockets quickly became the public image of the space program. From those early years into the present, NASA and the media have continually "contrived to present the astronauts as embodiments of the leading virtues of American culture." The mythic astronaut was, and still is, depicted as "everyman," "defender of the nation," "virile, masculine," and heroic.[53] In 1959, NASA introduced its first group of astronauts to the press, and the Mercury 7 became the public face of NASA virtually immediately. The Agency soon cut a deal with *LIFE* magazine to tell their stories. This deal was all about marketing on both sides. NASA Public Affairs Chief Walter Bonney approached Washington, DC, celebrity attorney Leo D'Orsey about helping the astronauts with publicity. D'Orsey agreed to represent them, for free, and peddled the rights

---

50. Sylvia Doughty Fries, "Commentary," in *A Spacefaring People: Perspectives on Early Spaceflight*, ed. Alex Roland (Washington, DC: NASA Scientific and Technical Information Branch, NASA SP-4405, 1985), pp. 75–76.
51. See, for example, William A. Gamson and Andre Modigliani, "Media Discourse and Public Opinion on Nuclear Power: A Constructionist Approach," *American Journal of Sociology* 95, no. 1 (1989): 1–37.
52. Kim McQuaid, "Sputnik Reconsidered: Image and Reality in the Early Space Age," *Canadian Review of American Studies* 37, no. 3 (2007): 371–401.
53. Roger D. Launius, "Heroes in a Vacuum: the Apollo Astronaut as Cultural Icon" (43rd AIAA Aerospace Sciences Meeting and Exhibit, Reno, NV, 10–13 January 2005).

to their "personal" stories. *LIFE* won the bidding at $500,000.[54] In a retrospective report, Time-LIFE commented on the Mercury contract: "In 1959, as the seven original astronauts prepared for their missions in space, LIFE Magazine went along, producing four years of intimate coverage of their training, their historic flights and their heroic achievements. The Mercury Astronauts allowed LIFE into their homes and shared with the magazine's readers their thoughts before and after their journeys into space."[55] NASA signed another, more complicated, contract with Time-LIFE and another partner for reporting the life stories of the Gemini and Apollo astronauts. According to Gemini-Apollo astronaut Michael Collins, media interest in the personal stories of the astronauts was "morbid, unhealthy, persistent, prodding."[56] But even if unwanted, stardom came with the job. Consider this anecdote: Apollo astronaut Gene Cernan escorted two Soviet cosmonauts, on a United States visit after Apollo 11, to a party at the home of actor Kirk Douglas, where "every star in Tinsel Town wanted to glitter for the men from space."[57] Guests included Clint Eastwood, Goldie Hawn, Lee Marvin, Groucho Marx, Yul Brynner, Natalie Wood, and Frank Sinatra. The cosmonauts didn't recognize any of them since they had not been exposed to American media content. Everybody recognized the spacemen.

As this cultural spectacle was unfolding, not everyone in official Washington, DC, thought the astronauts should serve as the public face of NASA. In his NASA transition report to President-elect John F. Kennedy in 1961, adviser Jerome Wiesner wrote:

> We should make an effort to diminish the significance of [the Mercury] program to its proper proportion before the public . . . . We should find effective means to make people appreciate the cultural, public service and military importance of space activities other than space travel.[58]

---

54. Jay Barbree, *"Live from Cape Canaveral": Covering the Space Race from Sputnik to Today* (New York, NY: HarperCollins, 2007); Tom Wolfe, *The Right Stuff* (New York, NY: Farrar, Straus and Giroux, 1979).
55. Seth Goddard, ed., "A Giant Leap for Mankind," *LIFE* online, available at *http://www.life.com/Life/space/giantleap/sec3/intro.html* (accessed 31 December 2008).
56. Michael Collins, *Carrying the Fire: An Astronaut's Journeys* (New York, NY: Farrar, Straus and Giroux, 1974), p. 54.
57. Eugene Cernan and Don Davis, *Last Man on the Moon* (New York, NY: St. Martin's Press, 1999), p. 243.
58. Wiesner Committee, "Report to the President-Elect or the Ad Hoc Committee on Space," 10 January 1961, NASA Historical Reference Collection, NASA History Division, NASA Headquarters, Washington, DC.

In 1969, President Nixon's Space Task Group, assembled to consider options for a post-Apollo space program, reported that it had "found strong and wide-spread personal identification with the manned flight program and with the outstanding men who have participated as astronauts." At the same time, "We have found questions about national priorities" and the cost of human spaceflight. The group recommended that "a decision to phase out manned space flight operations, although painful, is the only way to achieve significant reductions in NASA budgets over the long term."[59] What came next at NASA was the Space Shuttle Program, a transportation system with nowhere to go but Earth orbit. Then came the Space Station program, whose schedule and budget ballooned over time while its functions and purpose narrowed.

In 1985, President Reagan appointed a National Commission on Space to develop a 25-year plan for United States space exploration. As part of its research, the Commission conducted a series of public forums around the country to ask citizens what they wanted in a space program.[60] Among the 1,800 people who participated were "former astronauts, folk singers, lawyers, members of Congress, philosophers, teachers, and students." Most participants "had no direct link to the space program." The Commission reported that it was "overwhelmed by the high caliber of comments obtained, and duly impressed by the commitments of the citizens in attendance to respond intellectually to the call for participation."[61] The result of this exercise is that, more than 20 years later, NASA is still struggling over how to execute the sort of long-term plan for human exploration laid out in the Commission's report.

When Daniel Goldin took charge as NASA Administrator in 1992, he held a series of town meetings nationwide to ask citizens for their views on the space program, "with the goal of developing a shared vision for the future of NASA."[62] More than 4,500 people attended these meetings, with half claiming some affiliation with the space program. The results of this exercise included the finding that meeting participants "were interested in all aspects of" NASA and believed that "NASA should do a much better job of communication with

59. Space Task Group, "The Post-Apollo Space Program: Directions for the Future," September 1969, NASA Historical Reference Collection, NASA History Division, NASA Headquarters, Washington, DC.
60. The author served as the Commission's public affairs officer for these forums.
61. National Commission on Space, *Pioneering the Space Frontier* (New York, NY: Bantam, 1986), p. 174.
62. NASA, *Toward a Shared Vision: 1992 Town Meetings* (Washington, DC: NASA NP-205, 1993), p. 6. The author was a member of the NASA team that planned and executed these meetings, attended all of the meetings, and helped to write this meetings report.

the public, both through the news media and via direct means."[63] NASA committed to improving the quality of its public information, upgrading NASA TV and radio programming. Some changes were made, in fact, though the new and improved NASA TV was short-lived due to budgetary limitations. While Goldin was committed to improving and expanding communication, with special attention paid to science communication, his successors Sean O'Keefe and Michael Griffin appeared to be more comfortable with the conventional control-and-persuasion approach established in NASA's early years and maintained through the 1980s.[64] In its relations with its publics throughout the Bush administration, NASA has continued to take the marketing approach to engagement with its publics, with persuasion the objective.

NASA and Public Opinion

From the beginning of the United States space program to the present, polling firms[65] (commissioned by the aerospace industry, aerospace associations, the mass media, and NASA) have been attempting to gauge public opinion on the space program. As previously noted, in assessing public opinion about, interest in, and knowledge of the space program, NASA and the space community have typically taken an advertising and marketing approach to the task, soliciting what we call administrative research. NASA has repeatedly turned to the advertising and marketing sector for help in "branding" and "selling" the space program. The result has been a string of similar studies and similar findings—public knowledge of NASA is a mile wide and an inch deep—and a continuing series of attempts to cultivate favorable public opinion, and the increased public support that is erroneously assumed to accompany that favorable opinion, by "pitching" NASA to the public. NASA has paid considerable attention—arguably too much—to quantitative indicators of public interest provided by public opinion polls and surveys. But it appears that the Agency has paid little attention to the limits of poll data and the practice of polling itself. NASA and others in the space community

63. NASA, *Toward a Shared Vision*, p. 12.
64. In 2004, President Bush's Commission on Implementation of United States Space Exploration Policy reported, "A new model is needed to expand the role of space exploration in our culture . . . a new model for public engagement built on grass roots support." Building public support "requires sustainable, systematic, effective marketing and communication programs . . . . Industry, professional organizations, and the media [must] engage the public in understanding why space exploration is vital to our scientific, economic, and security interests" (Report of the President's Commission on Implementation of United States Space Exploration Policy, "A Journey to Inspire, Innovate, and Discover," June 2004, p. 44).
65. Including Gallup, Harris, Ipsos, Roper, Yankelovich, and Zogby.

continue to interpret high levels of public interest as indicators of public support, a correlation that poll results themselves show to be spurious. Roger Launius has examined the history of public opinion polling about the space program and pointed out that "consistently throughout the 1960s a majority of Americans did not believe" NASA's Apollo program "was worth the cost."[66] He has also noted that while NASA has consistently earned favorable ratings in public opinion polls, respondents consistently rank the space program low as a national spending priority.

Practitioners like to say that public opinion polling allows "the people" to speak for themselves. Research has shown that this is not necessarily the case. Public opinion polling has been described as "a cultural practice that sustains and affirms deeply held founding mythologies about community, democracy, and vox populi."[67] Research has explored how cultural elites "use public opinion polls to manage and control public opinion." It has been argued that polls "legitimate the authority of the state by appealing to the mythical sovereignty of the people without actually, or in practice, doing so."[68] Weaknesses of public opinion polling and public opinion research include a lack of reporting on survey nonresponse rates and insufficient research on the sources and effects of nonresponse.[69] Survey researchers have also found bias in the other direction—people who are interested in the topic of a survey are more likely to respond to it, and this factor can bias survey results.[70]

While polling methods have improved in some respects over the years, polling is still subject to what practitioners call nonsampling error—that is, nonquantifiable sources of error or uncertainty ranging from "interviewing problems to flawed interpretive theories"; the context and timing of surveys; the gender, race, or class of interviewers and respondents; and the phrasing and order of questions and response options. If they are to be useful, poll data "must be interpreted both in terms of larger historical or social trends, and within the context of public debate and discussion."[71] To better understand the

66. Launius, "Public Opinion Polls and Perceptions of U.S. Human Space Flight," 163–175.
67. Lisbeth Lipari, "Polling as Ritual," *Journal of Communication* (winter 1999): 83.
68. Lipari, "Polling as Ritual," 86.
69. Elizabeth Martin, "Unfinished Business," *Public Opinion Quarterly* 68, no. 3 (2004): 439–450.
70. Robert M. Groves, Stanley Presser, and Sarah Dipko, "The Role of Topic Interest in Survey Participation Decisions," *Public Opinion Quarterly* 68, no. 1 (2004): 16.
71. J. Michael Hogan, "George Gallup and the Rhetoric of Scientific Democracy," *Communication Monographs* 64, no. 2 (1997): 168. See also "A Gold Mine and a Tool for Democracy: George Gallup, Elmo Roper, and the Business of Scientific Polling, 1935–1955," *Journal of the History of the Behavioral Sciences* 42, no. 2 (2006): 109–134.

limits of polling data and the practice of polling itself, it helps to look into the history of the business of public opinion research.

In 1935, George Gallup founded the Gallup Organization[72] to do public opinion research. Gallup had come out of the advertising and marketing business, where he had been head of the marketing department at the New York advertising firm Young & Rubicam. Gallup created and employed a "rhetoric of scientific democracy" in attempting to construct legitimacy for what he called the new "science" of polling. Gallup succeeded in legitimizing polling, in part by deflecting questions about methods and accuracy with "a rhetoric of 'scientific mystification.'"[73] The Roper Center for Public Opinion Research, founded by Elmo Roper, a colleague of George Gallup, after World War II, has maintained an archive of polling data, collected by a variety of organizations, ranging from the 1930s to the present.[74] Louis Harris & Associates—now known as Harris Interactive—was founded in New York City in 1956 by Louis Harris, who served as John F. Kennedy's pollster during his 1960 campaign for the presidency.[75] Harris Interactive bills itself as "one of the largest market research and consulting firms in the world and the global leader in conducting online research."[76] This longstanding marketing bias, which continues to characterize the public opinion business today, is an important factor to consider in interpreting poll data. Another important factor to consider is the considerable difference between political polling and other types of polling.

Early on, NASA enlisted scholars and analysts to help define the Agency's image, message, purpose, and publics. But NASA apparently paid little attention to their findings. Apparently "NASA ignored its own early opinion research . . . . [F]indings which argued against widespread knowledge or interest in NASA programs were ignored."[77] During NASA's first few years, social psychologist Donald Michael pointed out to NASA the importance of "understanding . . . the relation of events to attitudes and values" when considering public opinion about the space program. In the case of public response to the launch of Sputnik I, for example, "for many people everywhere, their own affairs, Little

72. Originally known as the American Institute of Public Opinion, the Gallup Organization is now known as Gallup, Inc.
73. Hogan, "George Gallup and the Rhetoric of Scientific Democracy," 161.
74. *http://www.ropercenter.uconn.edu/about_roper.html.*
75. In 1975, political science professor Gordon S. Black founded the Gordon S. Black Corporation (GSBC) to do public opinion research. In 1996, GSBC acquired Louis Harris & Associates from the Gannett Corporation, and Louis Harris & Associates is now known as Harris Interactive.
76. *http://www.harrisinteractive.com/about/heritage.asp.*
77. McQuaid, "Sputnik Reconsidered," 392. See also Launius, "Public Opinion Polls," 163–175.

Rock, and the World Series took precedence over the Soviet leap into space."[78] Michael urged the space community to consider "the socio-psychological context in which efforts to explore space will evolve," pointing out that space exploration would proceed within a "vast matrix of already existing social and psychological values and beliefs, and behaviors which define our society today."[79] "There is," he said:

> *No* good reason to believe that there will be strong pressure from the public for effort and expenditures in this area, *unless very special efforts are made to elicit it* . . . . The matter is not close enough to most people's way of life to fit in with the values and behavior they have learned are important for successfully coping with day-to-day reality.[80]

Today the range of issues people are thinking about may be different, but the situation is the same. While many people may view the space program as a salient issue, they typically do not put it at the top of their list of things they need to think about. NASA continues to struggle to make space exploration relevant to people's lives. The Roper Center's archive of polling data contains the results of numerous surveys about NASA, and typifying this body of work are *New York Times/CBS News* polls conducted in 1994, 1998, and 2004 that asked respondents about space exploration:

- Is the government spending "too much, too little, or about the right amount" on space exploration? In 1998, 32 percent of respondents answered "too much." In 2004, 40 percent answered "too much."
- Should the United States send astronauts to Mars? In 1994, 55 percent favored and 40 percent opposed human missions to Mars. In 2004, 48 percent favored and 47 percent opposed.
- Would it be worth it to build a permanent base on the Moon? In 2004, 58 percent said "not worth it," while 35 percent said "worth it."[81]

In 2003, for the *Houston Chronicle*, Zogby International polled people on their views about NASA:

---

78. Donald N. Michael, "The Beginning of the Space Age and American Public Opinion," *Public Opinion Quarterly* 24 (1960): 573–582.
79. Donald N. Michael, "Society and Space Exploration," *Astronautics* (February 1958): 20.
80. Michael, "Society and Space Exploration," 88–89, emphasis in original.
81. *New York Times/CBS News* poll, 12–15 January 1994.

- "How would you rate the job being done by the space agency, NASA (the National Aeronautics and Space Administration)?" Sixty nine percent of respondents gave NASA an "excellent" or "good" rating, while 23 percent gave it a "fair" to "poor" rating.
- "Do you feel that the amount of tax dollars the government now spends on the U.S. space program should be increased, kept at the present level, decreased, or ended all together?" Zogby reported that "a plurality of people (44%) feels that the amount of tax dollars the government now spends on the U.S. space program should be kept at the present level. One-third (32%) thinks this amount should be increased."[82]

A poll conducted in 2004 by Ipsos Public Affairs for the Associated Press asked:

- "The United States is considering expanding the space program by building a permanent space station on the moon with a plan to eventually send astronauts to Mars. Considering all the potential costs and benefits, do you favor expanding the space program this way or do you oppose it?" Among respondents, 48 percent favored a human mission to Mars, while 48 percent opposed it.
- "On the whole, do you think our investment in space research is worthwhile or do you think it would be better spent on domestic programs such as health care and education?" Among respondents 42 percent said investing in space research would be "worthwhile" while 55 percent said it would be "better to spend on domestic programs."[83]

A *USA Today*/Gallup poll conducted in 2006 found that 48 percent of respondents deemed NASA's investment in the Space Shuttle "worth it," while 48 percent said the money would have been better spent elsewhere. At the same time, 57 percent of respondents said NASA was doing a good to excellent job, while 37 percent rated NASA "fair" to "poor." In reporting these results, Gallup observed, "The fact that less than a majority endorses the spending on a space program is not a new phenomenon. During the 1960s, when the United States increased spending on sending astronauts to the moon, a higher percentage of Americans consistently said it was not worth spending the money

---

82. Joseph Zogby, "America's Views on NASA and the Space Program," Zogby International for the *Houston Chronicle*, 3 July 2003.
83. "Americans Assess NASA's Price Tag," *Angus Reid Global Monitor* (15 January 2004), available at *http://www.angus-reid.com/polls/view/1474/americans_assess_nasas_price_tag.*

to accomplish the feat." At the same time, "ratings of NASA have generally been positive since Gallup first asked this question in 1990."[84]

A Harris Interactive poll conducted in 2007 asked, "If spending had to be cut on federal programs, which two federal programs do you think the cuts should come from?" Fifty-one percent of respondents put the space program at the top of the "cut" list, followed by welfare at 28 percent.[85] A poll conducted by Rasmussen Reports in 2007, for the University of California-Berkeley's BioMars astrobiology research team, asked, "How important is it for the United States to have a manned [*sic*] space program?" Thirty percent of respondents said it was "very important," 27 percent said it was "somewhat important," 22 percent said it was "not very important," and 13 percent said it was "not at all important."[86]

In the 1980s and 1990s, NASA called on political scientist Jon Miller, an expert in public opinion research and public understanding of science, to study "the information needs of the public concerning space exploration." In a 1994 report to NASA, Miller broke up the bloc of "interested" respondents reported by pollsters for decades into more precisely defined groups. He distinguished between "informed" and "attentive" audiences and also reported on gender- and age-based differences of opinion. And "even among those citizens with a high level of interest in space exploration and who believe themselves to be well informed"—a small percentage of respondents in the surveys he drew on[87]—"there are vast areas of ignorance and misunderstanding."[88] He also pointed out that people who are "attentive" to the space program may not necessarily support new initiatives or budget increases.

Over the past few years, Dittmar Associates has conducted market studies aimed at gauging public interest in and support for NASA. In a 2004 marketing study of space exploration, Dittmar found a widespread public perception that "the space program is disengaged from and uncaring about the public."[89]

---

84. Joseph Carroll, "Public Divided Over Money Spent on Space Shuttle Program, Americans Continue to Rate NASA Positively" (Princeton, NJ: Gallup News Service, 30 June 2006), available at *http://poll.gallup.com/content/Default.aspx?ci=23545&VERSION.*
85. Harris Interactive, "Closing the Budget Deficit: U.S. Adults Strongly Resist Raising Any Taxes Except 'Sin Taxes' or Cutting Major Programs" (The Harris Poll #30, 10 April 2007), available at *http://www.harrisinteractive.com/harris-poll/index.asp?PID=746.*
86. "Support for Space Missions Drops in U.S.," *Angus Reid Global Monitor*, 19 June 2007, available at *http://www.angus-reid.com/polls/view/16175/support_for_space_missions_drops_in_us.*
87. Biennial National Science Board surveys.
88. Miller, "The Information Needs of the Public Concerning Space Exploration: A Special Report to the National Aeronautics and Space Administration," 1 June 1994, p. viii.
89. Mary Lynne Dittmar, "Gen Y and Space Exploration: A Desire for Interaction, Participation, and Empowerment," Third Space Exploration Conference and Exhibition, Denver, CO, 27 February 2008.

Participants in these studies expressed a "desire for a responsive NASA—one that goes out of its way to involve interested citizenry in real and meaningful ways beyond traditional 'outreach and education.'" This desire "emerged repeatedly in response to questions asking about relevance of the space program to their daily lives." Dittmar found strong interest in and endorsement of the space program among Caucasians, Asians, males, and people 45–65 years old, and "little interest and less endorsement among women, Hispanics, and younger adults." Among 18- to 25-year-olds, Dittmar found "very little excitement or interest about NASA or its activities"—including the Vision for Space Exploration—"with the exception of Mars rovers." Participants in this age group expressed "confusion about and lack of interest in what NASA does" and a "strong sense that NASA wasn't about them." In a 2006 market study of "Gen Y" (ages 15–35) and space exploration, Dittmar found an "absence of a relationship with NASA, no participation, no interactivity."

Space Exploration as Spectacle

Another way of examining the history of "NASA and the public" is to consider it as 50 years of spectacle. Author Tom Wolfe wrote of the Mercury astronauts' press debut as a theatrical event, spotlighting not the astronauts' piloting abilities but their relationships with "god, family, country." Overnight, he said, the astronauts became "national heroes."[90] The story of the Mercury 7 provides insights into the role of the mass media in the social construction of reality—in this case, the spectacular hyperreality of the astronauts as superhuman, fearless yet god-fearing, patriotic family men.

In his famous essay, "Society of the Spectacle," published in 1967 at the peak of United States space frenzy, French critic Guy Debord (1931–1994) argued that in contemporary industrialized, commercialized society, image had supplanted reality as our social reality. He observed:

> In societies where modern conditions of production prevail, all of life presents itself as an immense accumulation of spectacles. Everything that was directly lived has moved away into a representation . . . . Spectacle is not a collection of images but a social relation among people, mediated by images . . . . The society which rests on modern industry is not accidentally or

---

90. Wolfe, *The Right Stuff*, p. 94.

> superficially spectacular, it is fundamentally spectaclist . . . . The spectacle presents itself as something enormously positive, indisputable and inaccessible . . . . The attitude which it demands in principle is passive acceptance which in fact is already obtained by . . . its monopoly of appearance . . . . In the spectacle, which is the image of the ruling economy, the goal is nothing, development everything. "The language of the spectacle consists of signs of the ruling production . . . . As information or propaganda, as advertisement or . . . entertainment, the spectacle [is] the omnipresent affirmation of the choice already made in production and its corollary consumption . . . . The spectacle's form and content are identically the total justification of the existing system's conditions and goals.[91]

The spectacle "is the opposite of dialogue," Debord concluded. In today's ever-more-mediated cultural environment, the society of the spectacle continues to thrive, and thanks to increasing numbers and varieties of media outlets and mass communication technologies and techniques, the space program is as spectacular as it ever was, and arguably more so. Debord's thinking offers an interesting way to think about the history of "NASA and the public," in which goals are always changing while "development" always proceeds. One condition of "the existing system" today is the power and influence of the so-called military-industrial complex, whose primary goal is dominance in the global aerospace sector and in outer space itself.

Like Debord, culture critic Jean Baudrillard (1929–2007) argued that in contemporary consumerist, mediated, high-technology-dominated society, people live in a social reality of images, spectacles, and simulacra that is so disconnected from actual reality that "reality" is no longer meaningful.[92] "Abstraction today is no longer that of the map," according to Baudrillard. "Simulation is no longer that of a territory . . . . It is the generation by models of a real without origin or reality: a hyperreal . . . . It is the map that precedes the territory . . . it is the map that engenders the territory."[93]

---

91. Guy Debord, *Society of the Spectacle* (Detroit, MI: Black and Red, 1967), unpaged, available at *http://www.marxists.org/reference/archive/debord/society.htm*.
92. "Jean Baudrillard," *Stanford Encyclopedia of Philosophy*, first published 22 April 2005; substantive revision 7 March 2007, available at *http://plato.stanford.edu/entries/baudrillard/*.
93. Jean Baudrillard, "Simulacra and Simulations: Disneyland," 1983, in *Social Theory: The Multicultural and Classic Readings*, ed. Charles Lemert (Boulder, CO: Westview Press, 1993), pp. 524–529.

In the 21st century, people know NASA by its representations—its spacewalking heroes and their spaceships, the Hubble Space Telescope, and anthropomorphized rovers on Mars. What is missing in this pastiche of spectacles is the meaning of NASA for all of its publics.

Conclusion

Throughout its 50 years, NASA has concerned itself with public opinion and public support for the Agency as an entity, or some specific program of the Agency. What people seem to care about is space exploration, in the broadest possible sense. People care as much about the *idea* of space exploration, the *idea* of human and robotic presence in space, as they do about the mechanics, the reality, of these things. When asked to place a value on the idea of space exploration, people rate it highly. When asked to put a price tag on the reality of space exploration, a different picture results.

President George W. Bush's space commission[94] recommended that the space community adopt "techniques employed by the film industry" to "inspire and educate people." Citizens might ask: Is the goal informing and engaging citizens? Or selling the space program and enlisting new advocates? The "space infotainment" trend in the aerospace community is disturbing, as the emphasis seems to be more on entertaining—the spectacle, the simulation—than on informing and empowering citizens. As NASA official Alan Ladwig has observed, "Basing decisions on thrill factors is fine for Hollywood studios, but it's a dubious performance indicator for space science and exploration." At NASA, "publicity shouldn't be the float leading the parade," Ladwig has said. "The legislative charter that created the agency was quite specific concerning priorities and goals . . . . The agency's charter says nothing about excitement or entertainment."[95]

NASA has always been good at framing stories about the space program to make a favorable public impression. A frame is a social construction used to organize stories and make meaning. Assumptions and beliefs, sponsorship (for example, official sources), and media practices (journalistic norms and conventions—for instance, the convention of balance) are among the factors determining what news frames will be and how they will work. In mass communication

94. Report of the President's Commission on Implementation of United States Space Exploration Policy: "A Journey to Inspire, Innovate, and Discover," June 2004.
95. Alan Ladwig, "The Excitement Myth: Space Exploration Shouldn't Have to Entertain to be Worthwhile," *Space Illustrated* (fall 2001): 16.

research, frames have been explored as functional structures,[96] structural forms of bias,[97] ideological processes,[98] structural *and* ideological forms of bias,[99] and special-purpose constructions of social reality.[100] The foregrounding and backgrounding of issues in a story frame contribute to public agenda setting, as they affect not only what issues audiences think about, but also how they think about the issues.[101] It is not clear whether any in-depth understanding of what framing is and how framing works has undergirded these framing efforts.

Medium theory could also help NASA in fostering relations with its various publics. Medium theory describes how media are not simply means for disseminating information but also "are themselves social contexts that foster certain forms of interaction and social identities." The proliferation of mass media and other types of communication technologies has "altered the nature of social interaction in ways that can not be reduced to the content of the messages communicated through them."[102] NASA continues to focus on message content and delivery, depending on counting how many times and to how many people messages are sent. It might be more useful to study whether and how people actually receive those messages and what they do with them when they receive them. This qualitative sort of research is more difficult to do than the conventional quantitative assessment of Web hits, news clips, and air time. It offers, however, insights that quantitative assessments cannot. Cultivation theory posits that repeated exposure to certain media content or frames can cultivate "adoption of a particular point of view that is more in line with media presentation than with reality."[103] It might be

---

96. Robert M. Entman, "Framing U.S. Coverage of International News: Contrasts in Narratives of the KAL and Iran Air Incidents," *Journal of Communication* 41, no. 4 (1991): 6–26; Z. Pan and Gerald M. Kosicki, "Framing Analysis: An Approach to News Discourse," *Political Communication* 10 (1993): 55–75.
97. Salma Ghanem, "Filling in the Tapestry: The Second Level of Agenda Setting," in *Communication and Democracy: Exploring the Intellectual Frontiers in Agenda-Setting Theory*, ed. Maxwell McCombs, Donald L. Shaw, and David Weaver (Mahway, NJ: Lawrence Erlbaum Associates, 1997), pp. 3–14.
98. Kevin M. Carragee and Wim Roefs, "The Neglect of Power in Recent Framing Research," *Journal of Communication* 54, no. 2 (2004): 214–233.
99. Frank D. Durham, "News Frames as Social Narratives: TWA Flight 800," *Journal of Communication* 48, no. 4 (1998): 100–117.
100. Dietram A. Scheufele, "Framing as a Theory of Media Effects," *Journal of Communication* 49, no. 1 (1999): 103–122.
101. McCombs, Shaw, and Weaver, eds., *Communication and Democracy*.
102. Joshua Meyrowitz, "Shifting Worlds of Strangers: Medium Theory and Changes in 'Them' versus 'Us'," *Sociological Inquiry* 67, no. 1 (1997): 59–71.
103. George Gerbner et al., "Growing Up With Television: The Cultivation Perspective," in *Media Effects: Advances in Theory and Research*, ed. J. Bryant and D. Zillman (Hillsdale, NJ: Lawrence Erlbaum Associates, 1994), pp. 17–41, 93.

useful for NASA to consider what points of view, what attitudes, it has been cultivating, or attempting to cultivate, over time, and what perspectives and attitudes it has actually cultivated over its 50 years of existence.

NASA could benefit from engaging in some critical research on this topic of "NASA and the public." In contrast with conventional administrative research, critical research "has to question existing conditions in terms of their historical preconditions and future possibilities."[104] In contrast to administrative research, critical research takes its social responsibility seriously. Critical researchers take care to define the relevance and validity of their research questions. "The sense of being critical is expressed in sharing responsibility for the future by identifying those critical (empirical) conditions which stimulate or fetter humans and democratic developments and recognizing their historical roots."[105]

NASA exists in a social reality where special interests—political and economic and business interests—will continue to ensure, for better or worse, the continuation of the civilian space program. At the same time, most citizens arguably do not "get" space exploration in the same ways that special interests in the space community do. NASA and its advocates are framing space as a resource-rich environment to exploit for economic gain, as a money-making enterprise, as a guaranteed source of employment for scientists. It has not been established that this approach to space exploration best serves the public interest. To serve the public interest as well as special interests, NASA will need to talk with, listen to, and involve citizens in planning a future in space. It will need to look deeply into its history in contemplating its future. It is likely that United States citizens would not be happy if their government were to abandon the civilian space program. It is reasonable to assume that the space program has meaning for many citizens. By engaging with its citizenry, NASA could begin to find out what space exploration means to different people in different socioeconomic sectors and walks of life. Perhaps this perspective can provide a starting-off point for the next 50 years of "NASA and the public."

---

104. Slavko Splichal, "Why be critical?" *Communication, Culture, and Critique* 1, no. 1 (2008): 20–30.
105. Splichal, "Why be critical?" 29.

Chapter 8

NASA Aeronautics

A Half Century of Accomplishments

Anthony M. Springer

NASA has actively promoted the widest practical dissemination of information concerning its research, a policy that has led to the application for commercial use of many of the technologies first derived from NASA research. Aeronautics research did not begin in 1958 with the Agency's formation. Instead, it was a legacy of work transferred from the National Advisory Committee for Aeronautics (NACA). The NACA supplied its rich traditions, cutting-edge facilities, and experienced personnel to NASA's organizational and scientific core.

In examining the accomplishments of the last half century, it is often uncertain who first developed a technology, or even who developed a given technology. In many cases there are no clear answers because different groups of people and organizations were involved at different points along the way. Often the research of one group served as a springboard to another, which then expanded or adapted the research, leading eventually to a solution to the original problem.

Many technologies described here were derived in this fluid, organic, yet still purposeful way. In many cases, even when NASA was not the first or the end developer of a technology, the Agency contributed significantly to a technology's advancement and operational use. NASA-developed technology or its derivatives can be found on every aircraft in the current United States commercial and military aircraft fleets. This paper is a survey of accomplishments in aeronautics by NASA and, in a few cases, the NACA that were made over the last half century. It is by no means complete, but it is intended to give the reader a foundational understanding of the broad range and significance of these key technological accomplishments and what they contributed to the advancement of flight.

Introduction: The NACA and NASA

NASA's aeronautics research has it roots in the NACA, which was formed in 1915 by the Navy Appropriations Act of 1915.

> . . . That it shall be the duty of the [National] Advisory Committee for Aeronautics to supervise and direct the scientific study of the problems of flight, with a view to their practical solution, and to determine the problems which should be experimentally attacked, and to discuss their solution and their application to practical questions.[1]

The NACA was created out of a need to improve United States aeronautic capabilities and technology in response to the great advances made by European countries and companies prior to and during World War I. During these first decades of the 20th century, the United States was severely lacking in the infrastructure and means to develop and produce its own advanced aircraft. The United States government created the NACA to lead this research effort. Forty-three years later, another worldwide event would lead to the formation of the NACA's successor—NASA. This major event was the launch of the first artificial Earth satellite, Sputnik, by the Soviet Union (USSR) on 4 October 1957. The tiny spacecraft proved to be the catalyst for the United States' formation of a civilian agency to develop and operate a civilian United States space program.

At President Eisenhower's request, the NACA was tapped to form the nucleus of the new agency:

> I recommend that aeronautical and space science activities sponsored by the United States be conducted under the direction of a civilian agency . . . . The responsibilities for administering the civilian space science and exploration program be lodged in a new National Aeronautics and Space Agency, into which the National Advisory Committee for Aeronautics would be absorbed . . . . The new agency would continue to perform the important aeronautical research functions presently carried on by the NACA.[2]

---

1. P.L. 271, 63rd Cong., 3rd sess., passed on 3 March 1915, 38 Stat. 930.
2. Statement by President Eisenhower, Hearings before the Select Committee on Astronautics and Space Exploration, 85th cong., 2nd sess. on HR 11881, 15 April–2 May 1958.

Additional assets were transferred to NASA over the next few years, including the transfer of DOD assets such as the Development Operations Division of the ABMA in 1960. The ABMA employed Dr. Wernher von Braun and his German "Rocket Team" along with the core of the Army rocket program that launched the first United States satellite, Explorer 1, after the initial failure of the Vanguard program.

NASA officially came into being with the passage of the National Aeronautics and Space Act of 1958. Signed by the President on 29 July 1958, it was "An Act to provide for research into problems of flight within and outside the earth's atmosphere, and for other purposes." The legislation stated that "The aeronautical and space activities of the United States shall be conducted so as to contribute materially to one or more of the following objectives: (1) The expansion of human knowledge of phenomena in the atmosphere and space; (2) The improvement of the usefulness, performance, speed, safety, and efficiency of aeronautical and space vehicles . . . the term 'aeronautical and space activities' means (A) research into, and the solution of, problems of flight within and outside the earth's atmosphere, (B) the development, construction, testing, and operation for research purposes of aeronautical and space vehicles, and (C) such other activities as may be required for the exploration of space; . . . [and to] provide for the widest practicable and appropriate dissemination of information concerning its activities and the results thereof."[3] This language would have significant impact on NASA's future activities in space—its primary realm of responsibility—and on aeronautics activities as well.

With the passage of this act, NASA began operations on 1 October 1958. But the absorption of the NACA's work did not magically happen overnight. The NACA personnel and facilities had been involved in a number of far-reaching projects prior to their transfer to NASA. This work was not arbitrarily stopped but instead was transferred to NASA, where it grew into fruition. One of the best-known projects was the X-15 research program.

Throughout its nearly 45-year existence, the NACA and its personnel made numerous significant advancements to the field of aeronautics. A small sample of these accomplishments included: the airfoil studies of the 1930s that resulted in the NACA airfoils (4 and 6 Digit Series); the NACA cowling to

---

3. P.L. 85-568, 72 Stat. 426. John M. Logsdon, ed., *Exploring the Unknown: Selected Documents in the History of the U. S. Civil Space Program*, vol. 1, *Organizing for Exploration* (Washington, DC: NASA SP-4407, 1995), pp. 334–335, available at *http://history.nasa.gov/series95.html*.

reduce drag; aircraft handling quality standards; icing research; NACA Report 1135, the "Standard Compressible Flow Handbook"; the "slotted throat" transonic wind tunnel; engine research; compressibility research; and support of the X series of research aircraft from XS-1 in 1947, which broke the "sound barrier," to name a few.[4]

Aerodynamics

The swept-wing concept was originated by the German aerodynamicist Adolph Busemann and presented at the fifth Volta Conference in 1935 in his paper "Aerodynamic Lift at Supersonic Speeds." In 1947, Busemann would be brought to the United States under "Project Paperclip," where he would work at the NACA's Langley.[5] Busemann's highly mathematical paper introduced the idea of sweeping a wing back to reduce its drag rise beyond the critical Mach number. Many at the Volta conference, including Eastman Jacobs, Theodore von Kármán, and Hugh Dryden, didn't realize the significance of the paper. The German Luftwaffe would later classify swept-wing material in 1936; its first production jet fighter, the ME262, used swept wings. Robert T. Jones at Langley independently began research into wing sweep of missiles in summer 1944. Jones pursued the mathematical theory based on previous work by Ludwig Prandtl and Max Munk. He completed his initial report in April 1945, but Langley management refused to publish it until it was verified. As luck would have it, Jones's theory would be experimentally validated that summer both in flight through the use of models and by wind tunnel tests. Results were widely distributed in 1946. Variable sweep was flight-tested on the X-5 aircraft in 1951.[6]

---

4. Pamela E. Mack, ed., *From Engineering Science to Big Science: The NACA and NASA Collier Trophy Research Project Winners* (Washington, DC: NASA SP-4219, 1998); Booz-Allen Applied Research Inc., "A Historical Study of the Benefits Derived From Application of Technical Advances to Civil Aviation," Joint Department of Transportation (DOT)-NASA Civil Aviation R&D Policy Study, Volume I Summary Report and Appendix A (detailed Case Studies) (NASA CR-1808), and Volume II Appendices B through I (NASA CR-1809), February 1971; Ronald Miller and David Sawers, *The Technical Development of Modern Aviation* (New York, NY: Praeger Publishers, 1970); J. G. Paulisick, "R&D Contributions to Aviation Progress (RADCAP) Volume 1: Summary Report," August 1972, Department of the Air Force (NASA-CR-129672); John D. Anderson, Jr., "The Airplane: A History of Its Technology" (Reston, VA: AIAA 2002).
5. John D. Anderson, Jr., "A History of Aerodynamics and Its Impact on Flying Machines," *Cambridge Aerospace Series* (Cambridge, U.K.: Cambridge University Press, 1997), p. 400.
6. James R. Hansen, *Bird on the Wing: Aerodynamics and the Progress of the American Airplane* (College Station, TX: Texas A&M University Press, 2004), pp. 97–100.

During the early 1950s, multiple factors converged to spur NACA scientist Richard Whitcomb into the creation of the "area rule" theory. In 1950, the NACA had developed the slotted throat wind tunnel to enable transonic wind tunnel testing, which had not been possible up to that time. In 1951, Busemann, now at Langley, made a presentation on transonic flows in which he used for the first time a "pipe fitters" analogy for fluid dynamics. From this chain of events, Richard Whitcomb surmised that transonic disturbances and shock waves produced by aircraft were functions of the longitudinal variation of their cross-sectional area. This theory resulted in the "Coke-bottle" or wasp-waist wing-body interface on aircraft.[7]

Figure 1: Richard Whitcomb, the NACA scientist who developed the "area rule" theory. *NASA Image L-89119*

One of the most dramatic examples of the application of the area rule was to the F-102 Delta Dagger in 1953. Whitcomb later developed anti-shock, wing-mounted bodies on the Convair 990. Some also consider the 747 fairing part of his area rule work.[8] In 1954, Whitcomb received the Collier Trophy. In the decades following, it has been acknowledged that the basic theory behind the area rule was implied in 1947 in the doctoral thesis of Wallace Hayes.[9]

Swept wings offer benefits at high speeds, but they result in stability and control concerns at low speeds and at higher landing and takeoff speeds. Flight

---

7. Joseph R. Chambers, *Concept to Reality: Contributions of the NASA Langley Research Center to U.S. Civil Aircraft of the 1990s* (Washington, DC: NASA SP-2003-4259, 2003), pp. 45–56.
8. John D. Anderson, Jr., "A History of Aerodynamics and Its Impact on Flying Machines," *Cambridge Aerospace Series* (Cambridge, U.K.: Cambridge University Press, 1997), pp. 413–416.
9. Pamela E. Mack, ed., *From Engineering Science to Big Science: The NACA and NASA Collier Trophy Research Project Winners* (Washington, DC: NASA SP-4219, 1998), pp. 135–148.

testing of the Bell X-5 that began in June of 1951 was the first full-scale testing of an aircraft that could change its wing sweep in flight.[10]

The X-5 required an extremely intricate and heavy mechanism to move the wing fore and aft along the fuselage to keep the aircraft within acceptable limits of stability and control. The technology would lie dormant until about 1957, when the concept of a multimission military aircraft came into being that required high-performance goals both at low speeds, which are best met with a straight wing, and at high speeds, which are best met with a swept wing. The breakthrough came during experimental testing in November 1958 that resulted in a method to overcome the instability and uncontrollability of previous swept mechanisms. The idea was to move outboard the pivot points, keeping the center section constant, and only sweeping the outboard sections of the wings to keep the aircraft stable in both configurations. These solutions led to the development of the F-111 and later use of swing wings on the F-14, B-1, British Tornado, and U.S. SST concepts.[11]

During work on advanced subsonic aircraft, Whitcomb hypothesized that the increase in drag-divergence Mach number from blowing through a slot in the upper wing surface was caused by delayed shock-induced separation. He envisioned a solution to this problem that could be applied to swept-wing subsonic transport.

Research on this concept started in 1964, leading to the first supercritical airfoil. NASA and the U.S. Navy used a T-2C Buckeye trainer with a 17 percent chord thickness airfoil for the first flight test of the concept in 1969. The results of the flight test validated the wind tunnel test, but the test configuration had a number of drawbacks in aircraft performance and handling characteristics. A more definitive test was needed than using the simple balsa wing modification performed in the T-2C tests. NASA proposed replacing the wing of an F-8C aircraft with that of a new, specially designed supercritical wing. The first flight of the 86-flight program using the F8-C supercritical wing was on 9 March 1971, with the last on 23 May 1973. Results from these flight tests demonstrated the transonic cruise efficiency of the supercritical wing and a potential theoretical increase of cruise Mach number for transport aircraft from 0.82 up to 0.90. The results of these tests were reported to industry in a classified conference in 1972. As stated in the conference

10. Jay Miller, *The X-Planes: X-1 to X-45* (Surrey, U.K.: Midland Publishing, 2001).
11. James R. Hansen, *Bird on the Wing: Aerodynamics and the Progress of the American Airplane* (College Station, TX: Texas A&M University Press, 2004), pp. 123–137.

Figure 2: F-8 supercritical airfoil. *NASA Image EC73-3468*

summary: "The key F-8 supercritical wing results discussed in the earlier papers may be summarized as follows: I feel the overall performance goals of Richard T. Whitcomb, as demonstrated by delayed drag-rise Mach number and a relatively high lift coefficient for the onset of significant separation, have been achieved."[12] Supercritical wings are now used on most military and commercial aircraft.[13]

The concept of a winglet, or a surface at the end of a wing to increase performance, originated with the work of F. Nagel at McCook Field (now Wright-Patterson Air Force Base) in 1924. Frederick W. Lanchester of England patented the endplate concept in 1897, but it was not a functional solution at the time.[14]

12. NASA Flight Research Center, *Supercritical Wing Technology: A Progress Report on Flight Evaluation* (Washington, DC: NASA SP-301, 1972), p. 122.
13. Joseph R. Chambers, *Concept to Reality: Contributions of the NASA Langley Research Center to U.S. Civil Aircraft of the 1990s* (Washington, DC: NASA SP-2003-4259, 2003), pp. 7–20.
14. James R. Hansen, *Bird on the Wing: Aerodynamics and the Progress of the American Airplane* (College Station, TX: Texas A&M University Press, 2004), p. 199. Nagel's work is in Memo Ref 130, "Wings with End Plates."

NACA scientist Richard Whitcomb, inspired both by an article in *Science* magazine about how soaring birds used tip feathers for control and by past research by other scientists, started analyzing the flow around wingtips in the early 1970s. He later theorized that a winglet or endplate at the wingtips extending above and/or below the wing could reduce the trailing vortex and thus drag. Using the 8-foot wind tunnel at Langley, he and his team performed experimental testing from 1974 through 1976. The design approach for the winglet was published in 1976. Understanding the possible benefits of this technology, NASA and the United States Air Force performed flight tests on a modified KC-135 aircraft between 1979 and 1980. The KC-135 was a good stand-in for a commercial transport aircraft. Today, winglets are used on a number of commercial aircraft from business jets to the large Boeing 747.[15]

Weather Hazards Research and the Airspace System

For more than 50 years, NASA and its predecessor institution have performed research related to the safety of aircraft. It took many forms during the NACA and NASA eras, from better understanding and predicting the fundamental science involved in weather phenomena such as lightning and ice formation to mitigating hazards caused by wind shear and wet pavement. Each technology or knowledge base evolved from analytic models and studies through ground tests and actual flight testing to validate the models, thereby gaining the real-world data required to improve the safety of aircraft and their crews in the air and on the ground.

Beginning in the 1950s, traction problems associated with wet airport runways became even more worrisome with the introduction of jet aircraft and their high takeoff and landing speeds. The powerful aircraft were more difficult to control on wet runways as compared to their piston engine counterparts.

In 1954, the NACA Langley Landing Loads Track facility went into operation to help find a solution to this problem. NASA and the Federal Aviation Administration (FAA) conducted joint studies on hydroplaning during the late 1950s and early 1960s. NASA researchers studying the hydroplaning problems for aircraft and land vehicles now attempted to find a practical solution to the skidding problem.

---

15. Joseph R. Chambers, *Concept to Reality: Contributions of the NASA Langley Research Center to U.S. Civil Aircraft of the 1990s* (Washington, DC: NASA SP-2003-4259, 2003), pp. 35–44; Maurice Allward, "Wingtip Technology," *The Putnam Aeronautical Review* 1 (May 1989): 39–44; Richard T. Whitcomb, "A DESIGN APPROACH AND SELECTED WIND-TUNNEL RESULTS AT HIGH SUBSONIC SPEEDS FOR WING-TIP MOUNTED WINGLETS" (Washington, DC: NASA TN D-8260, July 1976).

Researchers proved that cutting thin grooves across concrete runways created channels that would drain excess water from runway surfaces and reduce the risk of hydroplaning. (The British first tested runway grooves in England in 1956.) In 1962 and 1964, NASA tested the groove concept on the Langley Landing Loads Track facility, now named the Landing Dynamics Facility. Promising results from the tests led to a government industry conference.

In 1965, NASA initiated a study of commercial aircraft skidding incidents, which revealed the root causes of hydroplaning—viscous skidding and reverted-rubber skidding. Then NASA, in cooperation with the FAA, undertook a systematic study of grooving configurations and the process of grooving, including groove durability using a set of test patterns at a number of airfields throughout the United States. These studies resulted in a wealth of knowledge, both theoretical and practical, for airports, including standards for hydroplaning and slush drag equations.[16]

From this relatively simple solution, airports around the world today have safety-grooved surfaces, and all 50 of the United States have grooved portions of some of their main highways. The technology has been shown to restore wet friction performance to worn or smooth pavement surfaces and to extend their service lifetime by 5 to 10 years, resulting in significant maintenance cost savings. In 1966, a two-year study of grooved highways revealed that the grooves resulted in a 98 percent reduction of accidents.

Friction testing using a variety of vehicles and groove patterns continued into the 1980s. In 1968, the runway at NASA Wallops Flight Facility was grooved, and evaluations were made of the effectiveness of grooved runway surfaces for safer wet pavement landings using highly instrumented vehicles and runways. In the mid-1980s, tests were performed on 12 different concrete and asphalt runways, grooved and nongrooved, including dry, wet, snow, slush, and ice-covered surface conditions. Over 200 test runs were made with two transport aircraft, and over 1,100 runs were made with different ground test vehicles. The results of these tests showed the best configurations of grooves for specific sets of conditions.[17]

Starting in the late 1970s and through the 1980s, NASA studied lightning strikes and their potential threat to aircraft structures, avionics, and control systems. The program began by focusing on identifying the characteristics of

---

16. NASA LaRC, *Pavement Grooving and Traction Studies* (Washington, DC: NASA SP-5073-1969).
17. Joseph R. Chambers, *Concept to Reality: Contributions of the NASA Langley Research Center to U.S. Civil Aircraft of the 1990s* (Washington, DC: NASA SP-2003-4259, 2003), pp. 199–208.

lightning and then expanded its scope to the acquisition of aircraft flight data during lightning strikes.

NASA conducted research and flight tests to collect the first comprehensive data on intracloud lightning strikes and the effects of in-flight strikes. A special lightning-protected F-106B aircraft was used for the in-flight strike data. During the flight program of almost 1,500 storm penetrations, the aircraft was struck over 700 times, resulting in an extensive database on lightning effects on both metallic and composite structures, aircraft systems, and the characterization of lighting and when it is most likely to occur. This NASA-developed knowledge base is used to improve standards for protection against lightning for aircraft electrical and avionics systems.[18]

From the 1980s through the 1990s, NASA partnered with the FAA and the airline industry to approach the safety issue of wind shear, which is the violent downdraft of air that often forms with thunderstorms that can drive even the largest airliner into the ground if the downdraft occurs close to takeoff or landing.

First, the research team identified the unique characteristics of this hazard—the signature headwind, downdraft, and tailwind—and how these three components might affect a particular aircraft. The tests led to a detailed understanding of microburst and wind shear hazards. The resulting technology base led to the manufacture of airborne remote sensing technology that looks ahead, providing the ability to predict wind shear situations before encountering them. This forward view allows pilots ample time to avoid, rather than react to, wind shear hazards; airborne wind shear detection was born.

Finally, NASA aided in the creation of flight management systems, developing standard operational procedures for pilots to follow to minimize danger if trapped in a wind shear scenario.[19]

Over the decades, NASA has developed a number of air traffic management simulation tools. Beginning in 1991, NASA and the FAA developed the Center Terminal Radar Approach Control (TRACON) Automation System (CTAS).[20] The CTAS is a suite of three software tools that generates new information for air traffic controllers. These tools are 1) Traffic Management Advisor (TMA),

18. Ibid., pp. 173–184.
19. Ibid., pp. 185–198.
20. Heinz Erzberger, "Design Principles and Algorithms for Automated Air Traffic Management," Mission Systems Panel of the Advisory Group for Aerospace Research and Development (AGARD) and the Consultant and Exchange Program of AGARD, Madrid, Spain, 6–7 November 1995, published in LS-200; Dallas G. Denery and Heinz Erzberger, "The Center-TRACON Automation System: Simulation and Field Testing" (Washington, DC: NASA TM-110366, August 1995).

software created to forecast arriving air traffic to help controllers plan for safe arrivals during peak periods; 2) Descent Advisor (DA), software that generates clearances for en-route controllers handling arrival flows to metering gates; and 3) Final Approach Spacing Tool (FAST), software that provides terminal area controllers with heading and speed advisories for good spacing of aircraft on final approach courses.

The TMA was designed and developed by NASA and the FAA to automate workload. "The TMA is a time-based strategic planning tool that provides Traffic Management Coordinators and En Route Air Traffic Controllers the ability to efficiently optimize the capacity of a demand-impacted airport. The TMA consists of trajectory prediction, constraint-based runway scheduling, traffic flow visualization and controller advisories."[21] The TMA was evaluated in 1996 at the Fort Worth Air Route Traffic Control Center (ARTCC). The resulting data showed a 1- to 2-minute delay reduction per aircraft during peak periods. En route controllers felt the tool reduced their workload and increased their job satisfaction. The TMA was left in place at the Fort Worth ARTCC after the tests and is in daily operation.

During the 2000s, two tools were developed to support air traffic management, including the Surface Management System (SMS) tool[22] and the Future Air traffic management Concepts Evaluation Tool (FACET).[23] "SMS is a decision support tool that provides information and advisories to help FAA controllers and traffic managers as well as National Airspace System users to collaboratively manage aircraft on the surface and in the terminal area of busy airports. SMS has three fundamental capabilities: 1) the ability to predict the movement of aircraft on the airport surface and in the surrounding terminal area; 2) the ability to use this prediction engine to plan surface operations; and 3) the ability to disseminate this information and provide appropriate advisories to

---

21. Harry N. Swenson, Ty Hoang, Shawn Engelland, Danny Vincent, Tommy Sanders, Beverly Sanford, and Karen Heere, "Design and Operation Evaluation of the Traffic Management Advisor at the Fort Worth Air Route Traffic Control Center," 1st U.S.A./Europe Air Traffic Management R&D Seminar (Saclay, France, 17–19 June 1997).
22. Stephen Atkins, Yoon Jung, Christopher Brinton, Laurel Stell, Ted Carniol, and Steven Rogowski, "Surface Management System Field Trial Results," AIAA 4th Aviation Technology, Integration and Operations Forum (Chicago, IL: AIAA 2004-6241, 20–22 September 2004).
23. Karl Bilimoria and Banavar Sridhar, "FACET: Future ATM Concepts Evaluation Tool," 3rd U.S.A./Europe Air Traffic Management R&D Seminar (Napoli, Italy, 13–16 June 2000); Banavar Sridhar, Kapil Sheth, Philip Smith, and William Leber, "Migration of FACET From Simulation Environment to Dispatcher Decision Support System," 24th Digital Avionics Systems Conference, 30 October 2005.

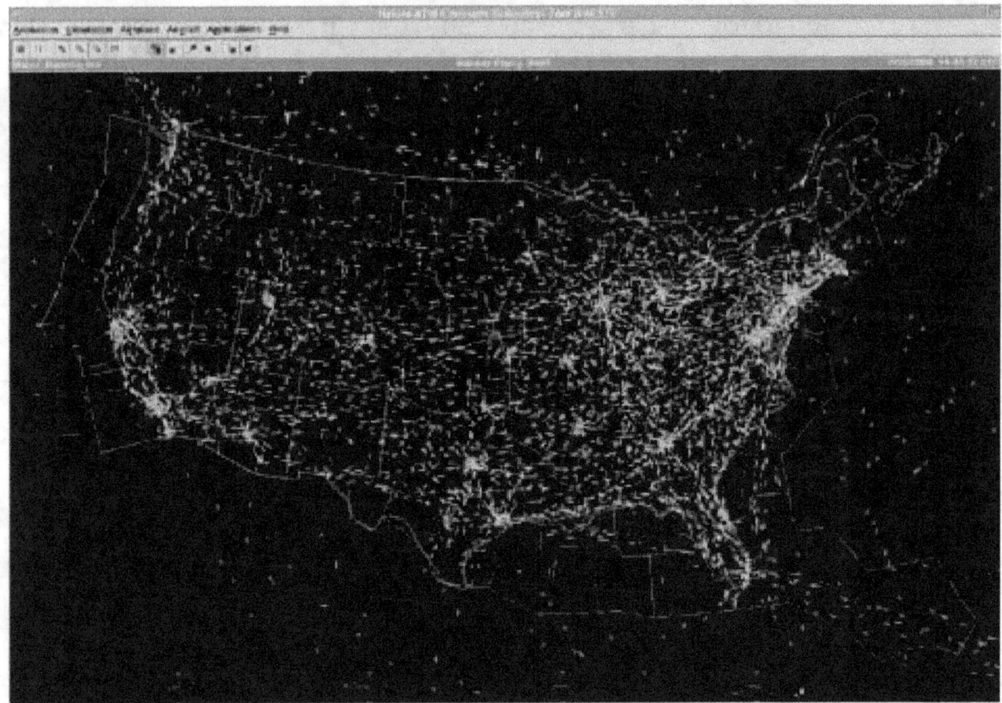

Figure 3: FACET. *NASA*

a variety of users."[24] In general, SMS software provides controllers with data to know when aircraft arrive on the ground or at the gate. NASA and the FAA field-tested the SMS concept at Memphis International Airport during late 2003 and early 2004, which proved to be successful.

To improve traffic flow across the United States, FACET maps thousands of aircraft trajectories. The tool was originally developed as a simulation and analysis tool "to provide a simulation environment for exploration, development and evaluation of advanced Air Traffic Management concepts."[25] As FACET evolved, its uses have increased to a state where FACET is being additionally developed as an air traffic management decision tool for dispatchers at airline operations centers.[26]

---

24. Atkins, Jung, Brinton, Stell, Carniol, and Rogowski, "Surface Management System Field Trial Results," AIAA 4th Aviation Technology, Integration and Operations Forum (Chicago, IL: AIAA 2004-6241, 20–22 September, 2004).
25. Karl Bilimoria and Banavar Sridhar, "FACET: Future ATM Concepts Evaluation Tool," 3rd U.S.A./Europe Air Traffic Management R&D Seminar (Napoli, Italy, 13–16 June 2000).
26. Banavar Sridhar, Kapil Sheth, Philip Smith, and William Leber, "Migration of FACET From Simulation Environment to Dispatcher Decision Support System," 24th Digital Avionics Systems Conference, 30 October 2005.

Aircraft Control

During the 1970s, a number of factors came together that led to the development of a dramatically improved aircraft cockpit that would use flat panel digital displays instead of dials and gauges. Two of these factors were flightworthy cathode-ray tube screens and the increased complexity of the aircraft. The resulting increased number of displays required to provide information were competing for both physical space and pilot attention. The new "glass" instruments gave the cockpit a distinctly different look and suggested the name "glass cockpit."

NASA, working with Boeing and Rockwell Collins, developed and tested electronic flight display concepts, culminating in a series of flights to demonstrate a full glass cockpit system using a NASA Boeing 737 aircraft. The demonstrations showed that a glass cockpit increased safety by reducing pilot workload while maintaining situational awareness. The glass cockpit was introduced commercially on the Boeing 767 in 1982. Today, glass cockpits are used on commercial, military, and general aviation aircraft, as well as on NASA's Space Shuttle fleet.[27]

The F-8 digital fly-by-wire flight research project validated the principal concepts of an all-electric flight control system. As electronics evolved in the 1960s, so did the concept of electronic controls. Neil Armstrong, then Deputy Associate Administrator of aeronautics at NASA, approved the program in 1970. The goal of the program was to have an electronic flight control system coupled with a digital computer to replace conventional mechanical flight controls. A modified F-8C Crusader served as the test bed for the fly-by-wire technologies. Phase I of the program used a computer from an Apollo spacecraft Command Module. The first flight of the 13-year project took place on 25 May 1972, with the last flight on 16 December 1985, for a total of 211 flights.

The electronic fly-by-wire system replaced older hydraulic control systems, freeing designers to design aircraft that would have increased maneuverability but also would be inherently less stable. Increased control provided by the fly-by-wire system allowed designers to compensate for this instability.[28] The F-8 digital fly-by-wire system became the forerunner of current fly-by-wire systems used in the Space Shuttles and on today's military and civil aircraft to make them safer, more maneuverable, and more efficient.

---

27. Joseph R. Chambers, *Concept to Reality: Contributions of the NASA Langley Research Center to U.S. Civil Aircraft of the 1990s* (Washington, DC: NASA SP-2003-4259, 2003) pp. 157–160.
28. James E. Tomayko, *Computers Take Flight: A History of NASA's Pioneering Digital Fly-By-Wire Project* (Washington, DC: NASA SP-2000-4224, 2000); James E. Tomayko and Christian Gelzer, "The Story of Self-Repairing Flight Control Systems" (Edwards, CA: NASA Dryden Historical Study No. 1, October 2003).

Supersonic and Hypersonic Flight

High-speed flight has been a quest since the earliest days of flight. From the NACA's early work on supersonic flight and on breaking the sound barrier, to later work on the vehicles and technologies required to achieve flight at many times the speed of sound, NASA has pushed the limits of flight to hypersonic levels, greater than five times the speed of sound,[29] and is looking to make commercial supersonic flight viable.

Since World War II and the original XS-1 program, exploring high-speed or supersonic flight had been a goal of the NACA and NASA. During various programs, NASA has used high-speed aircraft—A-12s, YF-12s, and SR-71 aircraft—to study the phenomena of sonic booms and ways to reduce sonic boom overpressures, the sharp "thunderclap" sound heard on the ground when an aircraft exceeds the speed of sound.

Two relatively recent programs aimed at the active reduction of the sonic boom were the Shaped Sonic Boom Demonstrator (SSBD) program, led by the Defense Advanced Research Projects Agency, and its follow-on Shaped Sonic Boom Experiment (SSBE) and Quiet Spike program, led by NASA. Each of the concepts explored in these efforts shows promise, but, as of this publication, neither has been implemented on commercial or military aircraft.

The goal of the SSBD and SSBE programs was to demonstrate in flight that incorporating specialized aircraft shaping techniques could substantially reduce sonic booms. The idea of shaping an aircraft to reduce the sonic boom was theorized decades ago but never flight-tested. The concept was successfully demonstrated in flight on 27 August 2003. Pressure measurements obtained on the ground and in the air confirmed that modifications made to an F-5E research aircraft not only changed the shape of the shock wave signature emanating from the aircraft, but also produced a "flat-top" signature whose shape persisted, as predicted.[30]

The Quiet Spike program's goal was to reduce the noise associated with supersonic flight by using a telescoping spike to produce a series of weak

29. T. A. Heppenheimer, *Facing the Heat Barrier: A History of Hypersonics* (Washington, DC: NASA SP-2007-4232, 2007).
30. John M. Morgenstern, Alan Arslan, Victor Lyman, and Joseph Vadyak, "F-5 Shaped Sonic Boom Demonstrator's Persistence of Boom Shaping Reduction through Turbulence," 43rd AIAA Aerospace Sciences Meeting and Exhibit (Reno, NV: AIAA-2005-0012, 10–13 January 2005); Joe Pawlowski, Peter Coen, David Graham, and Domenic Maglieri, "Origins of the Shaped Sonic Boom Demonstration Program," 43rd AIAA Aerospace Sciences Meeting and Exhibit (Reno, NV: AIAA-2005-0005, 10–13 January 2005).

Figure 4: NASA research pilot Bill Dana alongside the X-15 rocket-powered aircraft after a 1966 test flight. *NASA Image EC67-1716*

shocks as compared to a single stronger shock. An F-15B research aircraft was modified with a telescoping three-segment composite structure boom that could vary in length from 14 to 24 feet. Each segment was designed to produce a weak shock. The first flight of the vehicle was on 10 August 2006. Near-field data obtained from the flight experiments validated predictions enabling future studies.[31] Eventually, data from these types of sonic boom studies could lead to aircraft designs that reduce the peak of sonic booms and minimize the "startle" effect they produce on the ground.

The X-15 was an air-launched, rocket-powered, piloted hypersonic demonstrator for advanced technologies, many of which were to be used later for spaceflight. Three of the test vehicles were constructed, with the first flight—an unpowered glide flight—taking place in 1959. More flights that gradually built

31. Robbie Cowart and Tom Grindle, "An Overview of the Gulfstream/NASA Quiet Spike Flight Test Program," 46th AIAA Aerospace Sciences Meeting and Exhibit (Reno, NV: AIAA 2008-123, 7–10 January 2008).

up to higher speeds and altitudes followed for the next 10 years. The gradual nature of expanding the flight envelope was caused in part by delays in the development of the XLR-99 rocket engine that would power the X-15.

The last flight of the X-15's 199-flight test program was in October 1968. During the program, the three test vehicles were modified to increase performance. One was lost, along with its pilot. The fastest flight would exceed Mach 6.70, while the highest would reach 354,200 feet or 67 miles. More than 750 technical reports would be generated over the life of the program, with research results obtained in structures, control systems, life support, aerodynamics, aerodynamic heating, physiological responses, and the vehicle's use as an experimental platform.

The dream of hypersonic flight would then lie dormant for nearly 30 years. Programs would come and go, with some reaching the hardware testing stages (National Aero-Space Plane), but no flight demonstrations of hypersonic flight or a high-speed, air-breathing propulsion system would take place.[32]

The X-43A was a robotic, expendable, air-launched, 12-foot-by-5-foot test vehicle designed to flight-demonstrate the technology of airframe-integrated supersonic ramjet or scramjet propulsion at hypersonic speeds or speeds above Mach 5. A scramjet engine is an air-breathing engine through which the airflow remains supersonic.

The first X-43A dropped from its under-wing position on a NASA B-52B carrier aircraft on 2 June 2001. Shortly after ignition of the Pegasus booster rocket, a failure occurred, and the booster and its mated X-43 had to be destroyed by the range safety officer. The investigation traced the mishap to a failure of the booster flight control system. This was due to incorrect modeling of the forces generated by a launch of the mated vehicle at 20,000 feet instead of the 40,000-foot altitude at which a Pegasus is normally launched.

The second version of the X-43A flew on 27 March 2004. The engine was able to develop more thrust than the drag on the vehicle, accelerating it to a record speed of March 6.83. This was the first time a scramjet engine had ever operated in flight. The third X-43A flight on 16 November 2004 reached

32. Pamela E. Mack, ed., *From Engineering Science to Big Science: The NACA and NASA Collier Trophy Research Project Winners* (Washington, DC: NASA SP-4219, 1998), pp. 149–164; Richard P. Hallion, ed., "Hypersonic Revolution: Case Studies in the History of Hypersonic Technology," Air Force History and Museums Program, 1998, three volumes; Dennis R. Jenkins, *X-15: Extending the Frontier of Flight* (Washington, DC: NASA SP-2007-562, 2007); Dennis R. Jenkins, *Hypersonics Before the Shuttle: A Concise History of the X-15 Research Airplane*, Monographs in Aerospace History, No. 18 (Washington, DC: NASA SP-2000-4518, June 2000).

Mach 9.6, just shy of its planned Mach 10. The X-43A was able to cruise at this speed, meaning engine thrust matched drag and showed that the scramjet engine operated as predicted.

Overall, the program accomplished several important goals, including obtaining the first free-flight data on scramjet engines and validating predictive tools used to design the engine and future engines.[33]

Unconventional Aircraft Configurations

One of the most successful of the vertical takeoff and landing (VTOL) and short takeoff and landing (STOL), VTOL/STOL, programs was that of the XV-15 tilt rotor research aircraft, which combined standard aircraft cruise flight with VTOL and STOL capabilities. The research and experience gained from this program led directly to the military V-22 Osprey and its civilian offspring currently under development.

The development of the XV-15 was initiated in 1973 under a joint Army and NASA "proof of concept" program, with two aircraft built by Bell Helicopter Textron (BHT) in 1977. Bell completed aircraft development, airworthiness testing, and the basic "proof of concept" testing by September 1979. The first NASA flight of the XV-15 tilt rotor occurred in October 1980.[34]

The tilt rotor concept has many advantages over either a helicopter or an aircraft in certain situations. The ease with which the aircraft can be converted from one flight mode to another enhances its maneuverability and permits the aircraft to be configured to meet mission requirements. Airports can be small, needing only a relatively small area for takeoffs and landings, making tilt rotor aircraft ideal for intercity commuter travel. In the STOL mode, tilt rotor aircraft are ideal for long distance transport of heavy cargos into remote areas where only short runways are available. The XV-15 has been the primary influence for Bell's V-22, the first production tilt rotor. Previous to the XV-15 research in the late 1960s, NASA assisted the British government with the testing of the P1127 Kestrel, the forerunner of the Harrier and Harrier II aircraft currently in use by the United States Marines and other military services.[35]

Another nontraditional configuration tested and flown was that of the lifting bodies. A number of lifting body concepts were flown from 1963 to 1975.

---

33. Maurice Allward, "Wingtip Technology," *The Putnam Aeronautical Review* 1 (May 1989): 39–44.
34. Richard T. Whitcomb, "A DESIGN APPROACH AND SELECTED WIND-TUNNEL RESULTS AT HIGH SUBSONIC SPEEDS FOR WING-TIP MOUNTED WINGLETS" (Washington, DC: NASA TN D-8260, July 1976).
35. NASA LaRC, "STOL Technology" (Washington, DC: NASA SP-320, 1972).

The purpose of the lifting bodies was to demonstrate the ability of pilots to maneuver and safely land a wingless vehicle using the vehicle's body shape to generate lift. The lifting body research vehicles were the M2-F1, M2-F2, M2-F3, HL-10, X-24A, and X-24B. Information generated through the lifting body research programs contributed to the database that led to development of the Space Shuttle.

Dr. Alfred J. Eggers, Jr., of the NACA's Ames conceived the original idea for the lifting body around 1957. Eggers found that, by slightly modifying a symmetrical nose cone shape, aerodynamic lift could be produced. This lift enabled the modified shape to "fly" back from space rather than plunge to Earth in a ballistic trajectory.

In 1962, DFRC Director Paul Bikle approved a program to build a lightweight, unpowered lifting body as a prototype to flight-test the wingless concept. Construction was completed in 1963. The first flight tests were over Rogers Dry Lake in California at the end of a towrope attached to a hopped-up Pontiac convertible driven at speeds of up to 120 miles per hour (mph). These initial tests produced enough flight data about the M2-F1 to proceed with flights behind a NASA R4D tow plane at greater speeds.

Success of the M2-F1 tests led to development of two heavyweight lifting bodies based on studies at NASA's ARC and LaRC—the M2-F2 and the HL-10, both built by the Northrop Corporation. The "M" refers to "manned," and "F" refers to the "flight" version. "HL" comes from "horizontal landing," and "10" represents the 10th lifting body model to be investigated by LaRC. The United States Air Force, upon seeing results from the previous lifting body programs, started the joint NASA-Air Force lifting body program with the Martin X-24A, later modified into the X-24B high-speed lifting body program.[36]

Noise, Materials, and Tools

Aircraft engines are a main source for noise heard in the cabin and on the ground. During the last half century, NASA and its industry partners have explored various methods to reduce aircraft noise without degrading engine performance.

During the 1980s, the United States Air Force started looking for ways to reduce aircraft infrared signature by mixing the engine exhaust with free stream air. NASA later observed that the same nozzles reduced noise emissions

36. Jay Miller, *The X-Planes: X-1 to X-45* (Surrey, U.K.: Midland Publishing, 2001).

as well. As an outgrowth of testing derived from noise suppressor concepts for military and civilian aircraft engines, the chevron engine concept was born in the 1990s. The goal was to mix the jet exhaust, or the air that exits from the engine, with the free stream flow in a way to promote suppressing the exhaust noise of the engine. Theoretical, experimental, and then flight validation of chevron nozzles was done on a NASA Lear 25 in the late 1990s.[37] This technology found its way onto a number of business jets. Nearly 10 years later, NASA used computer simulations to improve an asymmetrical scallop design of chevrons, which are now used on the nozzles of some jet engines to reduce the resultant exhaust noise. Ground and flight tests by NASA and its industry partners in 2006 under the Quiet Technology Demonstrator 2 program proved that the new chevron design reduces noise levels both in the passenger cabin and on the ground.[38] Chevrons are implemented on many of today's aircraft, including Bombardier and Embraer regional jets using CF34 engines and the A321 using CFM-56-5B. Many aircraft currently in development are looking into the concept.[39]

NASA did not invent the concept of composite materials but contributed significantly to their advancement and acceptance in aeronautical systems. Composites are high-strength, nonmetallic materials that replace heavier metals in aircraft components to reduce weight and improve durability.

Early work on composites was an outgrowth of German research in the 1930s on fiber-reinforced plastics and an outgrowth of the British laminated aircraft of World War II. It wasn't until the 1960s that researchers at the Royal Aircraft Establishment were able to develop a commercially viable carbon fiber. NASA became involved in the early 1970s through the RECAST project and in 1972 through the Composite Flight Service program, which obtained real-world data on applying composites to commercial aircraft.

In 1975, the Aircraft Energy Efficiency (ACEE) program began and eventually led to the ACEE Composite Primary Aircraft Structures Program that worked with industry on the design, build, test, and flight of larger composite segments on secondary structures.[40] During the 10-year program, more than

---

37. Michael Abrams, "Put a Nozzle on It: Teeth-Like Tabs Are Turning Down the Volume on Jet Cacophony," *Mechanical Engineering* 128, no. 11 (The American Society of Mechanical Engineers, November 2006).
38. William H. Herkes, Ronald F. Olsen, and Stefan Uellenberg, "The Quiet Technology Demonstrator Program: Flight Validation of Airplane Noise-Reduction Concepts," 12th AIAA/CEAS Aeroacoustics Conference (Cambridge, MA: AIAA 2006-2720, 8–10 May 2006).
39. NASA LaRC, "STOL Technology" (Washington, DC: NASA SP-320, 1972), pp. 371–412.
40. Jeffrey L. Ethell, *Fuel Economy in Aviation* (Washington, DC: NASA SP-462, 1983).

600 publications were derived from this research, and the major United States airframe manufacturers gained experience with composite structures.

Another segment of the composite world to which NASA has contributed since the 1980s is that of textile composites (or woven, knitted, or braided composites). NASA has also supported general aviation with a process that allows airframe manufacturers to procure certified composite materials from vendors in the same manner in which they were able to procure metals.

NASA has also supported composites research for rotorcraft through the Advanced Composite Airframe Program. In 1999, NASA and Sikorsky conducted a simulated helicopter crash test at the Impact Dynamics Research Facility.[41]

During most of the last 40 years, NASA, working with industry, has continued its research into composite materials and their applications. Composites have gradually replaced metallic materials on parts of an aircraft's tail, wings, fuselage, engine cowlings, landing gear doors, and finally primary structures. The use of composite materials can reduce the overall weight of an aircraft and improve fuel efficiency.[42]

NASA has made significant contributions to the development of analytical or numeric tools for the analysis of the physical processes and phenomena associated with aeronautical vehicle design and operation. Two of the best-known tools used by government and industry are Computational Fluid Dynamics (CFD) and NASA Structural Analysis (NASTRAN).

Since the 1970s, NASA has developed and partnered for the development of sophisticated computer codes that can accurately predict the complex ways that air flows over and through realistic aircraft and spacecraft designs and their components. Now considered a vital tool for the study of fluid dynamics, CFD greatly reduces the time required to design and test any type of aircraft or spacecraft. NASA has developed numerical methods, flow solvers, and grid generation software and worked to integrate these and other tools to increase computing performance.[43]

---

41. Joseph R. Chambers, *Concept to Reality: Contributions of the NASA Langley Research Center to U.S. Civil Aircraft of the 1990s* (Washington, DC: NASA SP-2003-4259, 2003), pp. 71–88; Eric Schatzberg, *Wings of Wood, Wings of Metal: Culture and Technical Choice in American Airplane Materials 1914–1945* (Princeton, NJ: Princeton University Press, 1999), pp. 223–232; Joseph R. Chambers, *Partners in Freedom: Contributions of the Langley Research Center to U.S. Military Aircraft of the 1990s* (Washington, DC: NASA SP-2000-4519, 2000).
42. Ray Whitford, *Evolution of the Airliner* (Marlborough, U.K.: Crowood Press, 2007).
43. Chambers, *Concept to Reality*, pp. 57–64.

A NASA 1964 structural dynamics review revealed that multiple NASA Centers were each separately developing structural analysis software. After much discussion, it was decided that a single software package would be developed to bring together the individual NASA research and codes. A contract was awarded to Computer Sciences Corporation to develop the new software based on the research and codes NASA had developed. The new software, NASTRAN, was released to NASA in 1968. Over the subsequent years, this integrated software package would become the standard structural analysis code for the industry. The software was a finite element analysis utilizing numerical finite element methods that could perform static response, dynamic response, complex Eigen value, and elastic stability analysis. A commercial version was later developed and made available by the MacNeal-Schwedier Corporation (MSC), entitled MSC.Nastran.[44]

Conclusions

For more than 50 years, NASA has developed aeronautical technologies that have affected all aspects of aeronautics from general aviation to advanced military aircraft to spaceflight. NASA aeronautics remains true to the 1958 Space Act with the goal of the widest practicable and possible dissemination of its research. Innovations developed either solely by NASA or in partnership with industry or academia have benefited the public in their daily lives and in the defense of the United States. It can be safely said that NASA-developed technology or its derivatives can be found on every aircraft in the current United States commercial airliner and military aircraft fleets.[45]

It is a fitting tribute that the main thrust of a statement made more than 30 years ago by former astronaut and research pilot Neil Armstrong, then Deputy Associate Administrator for Aeronautics in the Office of Advanced Research and Technology, remains relevant today. Only the numbers of NASA employees and budget dollars have changed:

> In 1958 the National Aeronautics and Space Administration was brought into being to explore certain broad areas of research

---

44. Richard H. MacNeal, ed., "The NASTRAN Theoretical Manual" (Washington, DC: NASA SP-221(01), December 1972).
45. Anthony M. Springer, "50 Years of NASA Aeronautics Achievements," 46th AIAA Aerospace Sciences Meeting and Exhibit (Reno, NV: AIAA 2008-0859, 7–10 January 2008). Joseph R. Chambers, *Innovation in Flight: Research of the Langley Research Center on Revolutionary Advanced Concepts for Aeronautics* (Washington, DC: NASA SP-2005-4539, 2005).

and development, which included not only the exploration of space but also the continued responsibility in aeronautics which had been the primary function of its predecessor agency, the National Advisory Committee for Aeronautics. It is seldom recognized by the general public that NASA has a vital and necessary role in the advancement of military and commercial aviation in the United States, and that the level of effort while a small fraction of the agency's total program is very substantial. Roughly 2500 NASA employees supported by funding of about $160,000,000 per year are directly engaged in conducting the research described in "Aeronautics." The frontiers of flight have not all been explored and the applications of NASA's advanced research in aeronautics will continue to keep the United States in first place in commercial and military aviation in the years ahead until someday we will be able to travel as casually from New York to Australia at 6000 mph as millions do now from New York to Paris at nearly 600 mph.[46]

Acknowledgments

The author would like to thank the following mission support staff of the NASA Aeronautics Research Mission Directorate for their support in this endeavor: Karen Rugg, Jim Schultz, Lillian Gipson, and Maria Werries. A general overview of NASA contributions can be found on the following lithographs:

- "NASA Aeronautics Research Onboard: Decades of Contributions to Aviation" (GPO, Lithograph NL-2008-10-008-HQ).
- "NASA Aeronautics Research Onboard: Decades of Contributions to General Aviation" (GPO, Lithograph NL-2008-10-009-HQ).
- "NASA Aeronautics Research Onboard: Decades of Contributions to Military Aviation" (GPO, Lithograph NL-2008-10-010-HQ).
- "NASA Aeronautics Research Onboard: Decades of Contributions to Rotorcraft Aviation" (GPO, Lithograph NL-2008-12-011-HQ).
- "NASA Aeronautics Research Onboard: Decades of Contributions to Tilt Rotor Aviation" (GPO, Lithograph NL-2008-12-012-HQ).

---

46. David A. Anderton, *Aeronautics (America in Space: The First Decade)* (Washington, DC: GPO, NASA EP-61, 1970).

Chapter 9

Evolution of Aeronautics Research at NASA

Robert G. Ferguson

While NASA is first associated with space travel, it has always had a mission to study aeronautics. In fact, it was a predecessor agency, the National Advisory Committee for Aeronautics (NACA), that formed the core of NASA when the space agency began in 1958. The NACA, established in 1915, built a formidable reputation in the science and engineering of aeronautics based largely on four decades of rigorous wind tunnel research. Though not the NACA's sole methodology, wind tunnels were nevertheless dominant and, as such, helped guide the substance of the NACA's investigations. Fifty years after the start of NASA, wind tunnels remain an important tool, but they are now joined by alternative methodologies. The goal of this paper is to examine this evolution and the local factors behind the shifts in NASA's aeronautics research methodology.

The first part of this paper provides a brief review of NASA's aeronautics research stretching back to the NACA era. The paper underscores the manner in which wind tunnels were enmeshed within the technical and administrative culture of the laboratories, begging the question of how alternative methodologies could reasonably challenge such organizational momentum. The second part examines the rise of three alternative methodologies: flight test, CFD, and Center-TRACON Automated System research. This latter section argues that three factors account for the rise of alternative methodologies: institutional structure, macro-technological change, and shifting technological frontiers.

One: Tunnel Vision Reprise

A proper history of NASA's research methodologies really goes back to the early years of the NACA and the rise of the wind tunnel as the NACA's central research tool. Wind tunnels, to be brief, are laboratory devices for measuring the flow of air, especially as air moves around solid shapes. Wind tunnels can be used to simulate the performance of aircraft, and parts of aircraft, in a highly controlled environment, thus reducing the need for risky and potentially costly flight testing. There are plenty of aeronautical investigations that do not need or make use of wind tunnels. It is but one experimental device available, and it finds its greatest utility in such areas as aerodynamics (e.g., lift and drag), aeroelasticity, thermodynamics, control, noise, propulsion, and icing. The NACA developed a close association with wind tunnels, being both an innovator in tunnel design and a strong proponent of their use.

The NACA established itself as a lead scientific institution in its field with the construction of the Variable Density Tunnel (VDT) under Max Munk in 1922.[1] Munk was a student of the famous German aerodynamicist Ludwig Prandtl. The VDT was, in essence, a wind tunnel inside of a pressure vessel.

With a high-pressure atmosphere, the VDT gave more accurate results than its contemporaries; high pressure helped counteract some of the inaccuracies that arose from the use of scale models. As with modern physics and the competition for high-energy particle accelerators, more capable wind tunnels bequeathed advantages to the NACA. Exploiting the VDT and its laboratory descendants, the NACA developed a reputation for engineering-oriented wind tunnel research. An excellent example of this style of work from the interwar years is the NACA 4 Digit Airfoil series. Researchers systematically varied four parameters of a wide number of airfoil types and recorded fundamental data for each one. The resulting airfoils, catalogued by four digits (the four variable parameters), served as a basic sourcebook for aircraft designers.[2]

---

1. The history of the NACA and its early wind tunnel research is well covered in two works: Alex Roland, *Model Research: The National Advisory Committee for Aeronautics, 1915–1958* (Washington, DC: NASA SP-4103, 1985); and James R. Hansen, *Engineer in Charge: A History of the Langley Aeronautical Laboratory, 1917–1958* (Washington, DC: NASA SP-4305, 1987).
2. John D. Anderson, Jr., *A History of Aerodynamics and Its Impact on Flying Machines* (Cambridge, U.K.: Cambridge University Press, 1998), pp. 342–352.

Figure 1: The Variable Density Wind Tunnel (VDT). From left to right are Eastman Jacobs, Shorty Defoe, Malvern Powell, and Harold Turner. In this photo taken on 15 March 1929, a quartet of the NACA staff conduct tests on airfoils in the VDT. (In 1985, the VDT was declared a National Historic Landmark.) Eastman Jacobs is sitting (far left) at the control panel. *NASA Image L-3310*

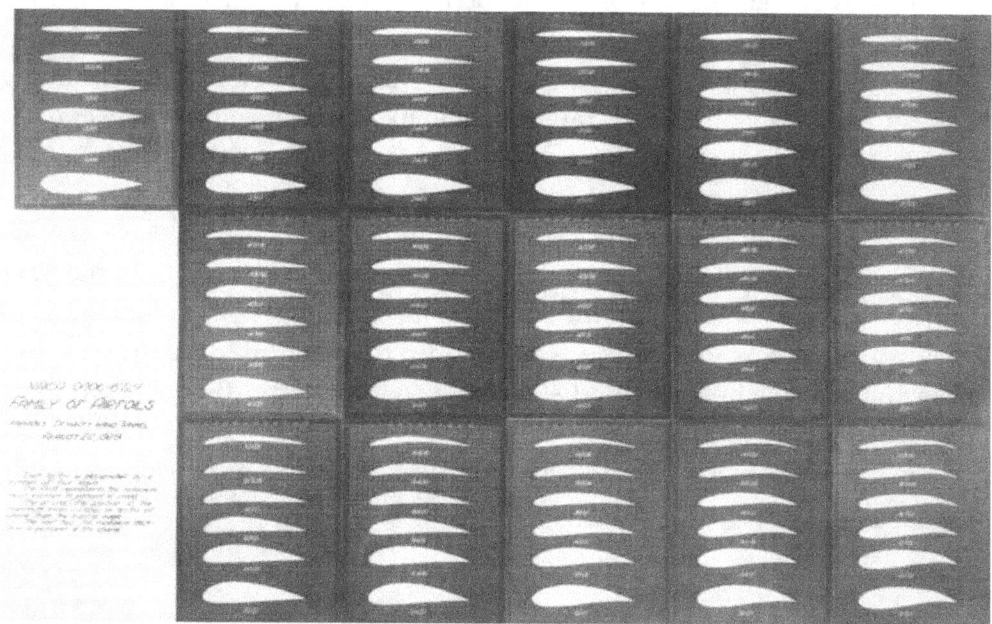

Figure 2: NACA 4 Digit Airfoils. *NASA Image EL-2003-00333*

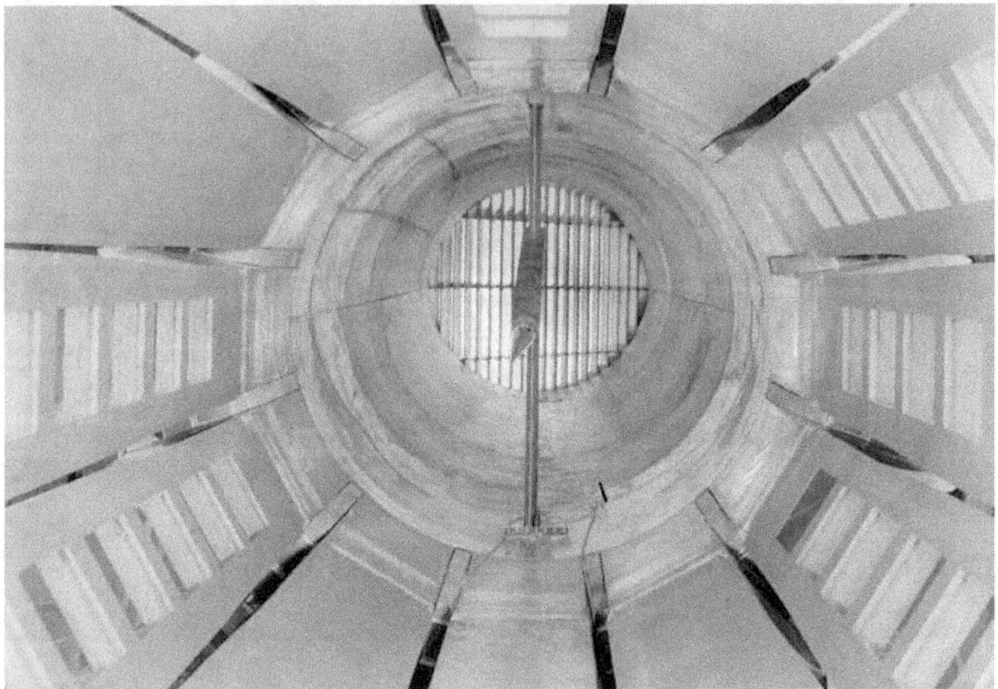

Figure 3: The Langley 8-Foot High-Speed Tunnel. The slotted walls of the test section permitted accurate testing at transonic speeds. *NASA Image EL-2003-00280*

The NACA continued to work at the forefront of aerodynamics into the 1950s, one of its most impressive accomplishments being the development of the slotted wind tunnel. At the time, it was possible to get accurate tunnel results in the subsonic and supersonic speed ranges, but in the transonic range (between the two) shock waves choked the tunnels, corrupting the test data. At the NACA's Langley laboratory, a group under the direction of John Stack solved this problem in the late 1940s by cutting slots in the tunnel's throat. By 1950, their slotted 8-foot High-Speed Tunnel was ready for transonic testing. As with the VDT, advanced test equipment gave the NACA an advantage over its competitors.[3]

Wind tunnels were not the NACA's only research methodology; other important approaches included structural, avionics, and flight testing. But none matched the importance and utility of wind tunnels. They were logical and flexible instruments, useful for theoretical explorations as well as highly applied studies. More crucially, tunnels allowed researchers to shift back and

3. Anderson, *A History of Aerodynamics*, p. 412; Hansen, *Engineer in Charge*, chaps. 9, 11.

forth from mathematical models to flight, thus increasing the reliability of models while also serving to predict aircraft performance. Wind tunnels were, and remain, a critical link between the theory and physical phenomena of flight.

But there were also nontechnical reasons for the prominence of wind tunnels. Tunnels fit the American political context; they were general research tools that, for the most part, were too expensive for the private sector. Only with World War II did American manufacturers have sufficient resources to begin building their own wind tunnel laboratories, but even then they would never attain the breadth of capabilities offered by the NACA.[4] Thus, wind tunnels did not appear to overstep the line between research for the public good and subsidized corporate research. Similarly, wind tunnels allowed the NACA to perform proprietary research for corporate patrons without the outward appearance of subsidy. For many decades, the tunnels were funded regardless of how they were employed. This arrangement worked not only to the benefit of private industry, but of the NACA as well, since the NACA was able to experiment with models provided by the private sector. This truly was an informal, but powerful, form of technology transfer operating in both directions.

Contributing to the institutional importance of tunnels was the fact that laboratory administration usually broke down along tunnel lines and the hierarchical structure of the laboratories was determined by tunnel groupings. For example, an engineering recruit at the Langley or Ames Research Centers in the 1950s would typically find himself part of a branch attached to a particular tunnel or set of tunnels. The principal branches, though they changed over time, were normally associated with speed regimes, as were the tunnels. Researchers cut their teeth in the tunnels and, after earning the respect of their peers, moved up the laboratory hierarchy. It was a system that rewarded intellectual rigor and innovative thinking, but not necessarily managerial skills or fealty to Headquarters. It was not until the early 1970s that the former NACA laboratories began to routinely look outside the branch ladder for laboratory management.[5]

The wind tunnel was thus a bundled sociotechnical package with strong institutional momentum. It was a research community, a continuous research tradition with built-in mechanisms for the transfer of explicit and implicit information from one generation to the next and between government and private researchers. It encompassed a system for professional advancement.

---

4. Robert G. Ferguson, "Technology and Cooperation in American Aircraft Manufacture During World War II" (Ph.D. diss., University of Minnesota, 1996).
5. Robert G. Ferguson, *NASA's First A*, NASA History Series, forthcoming.

It formed the basis for a long-running funding mechanism that dated to the Progressive Era. And it was a very useful laboratory tool.

All of this begs a question: to what extent did the NACA's emphasis on tunnels shape the direction and administration of research?[6] With so much momentum behind a particular methodology, it is easy to see that determining which questions to study quickly becomes a question of what can a tunnel do, and following from that, how to acquire the next generation of wind tunnel. It comes as no surprise that one of the major battles between the NACA and the United States Air Force in the late 1940s and early 1950s was over future wind tunnels (the outcome of which was the Unitary Wind Tunnel system).[7] Similarly, the emphasis on wind tunnels and aerodynamics meant less emphasis on other lines of inquiry. Not until the NACA created the Lewis Engine Laboratory in 1942, for example, did propulsion research begin to approach the priority accorded to traditional tunnel work (and this, perhaps, too late to answer postwar critics who decried the NACA for not having matched the Germans and British in turbine research). Likewise, in the 1950s, aircraft manufacturer Douglas argued that the NACA should do for structures and materials what the agency had done for propulsion: create a separate materials laboratory that was out of the shadow of the aerodynamics core.[8]

Two: What Changed?

Moving ahead to the NASA era, where there was once an agency dominated by wind tunnels, one finds instead an agency embracing a mix of research methodologies. Three explanations come to the fore. The first is institutional

6. Historian Alex Roland, in his history of the NACA, explicitly noted the distorting effects of the agency's attachment to wind tunnels, writing: " . . . research equipment shaped the NACA's program fully as much as did its organization and personnel. The NACA achieved early success and acclaim by developing revolutionary wind tunnels for aerodynamical research. Thereafter the tunnels took on a life of their own, influencing the pace and direction of NACA research; concentrating the Committee's attention on aerodynamics when fields like propulsion, structures, and helicopters had equal merit; and becoming in time a sort of end in themselves" (Roland, *Model Research*, vol. 1, pp. xiv–xv).
7. Roland, *Model Research*, vol. 1, pp. 211–221.
8. Edwin Hartman to Director, NACA, 18 July 1951, "Subject: Visits to the Santa Monica plant of the Douglas Aircraft Company, June 19 and July 2, 1951," Edwin Hartman Memorandums, Langley Archives. On postwar criticism, see Roland, *Model Research*, vol. 1, p. 204.

structure, that the Agency's competing laboratories set the stage for methodological competition. The second is macro-technological change, which generally refers to external inputs that give rise to new methodologies. The third is a shifting technological frontier, which here refers to changes in the laboratories' goals.

Institutional Structure

One of the most important factors in the growth of alternative research methodologies was the establishment of multiple laboratories. At its most simplistic, this argument is about the creation of a competitive market for innovative research programs. The seeds for competition were planted in the NACA era with the establishment of multiple laboratories. The oldest was the Langley Laboratory near Hampton Roads, Virginia. Formally dedicated in 1920, Langley was the NACA's only facility until 1940, and it remained the patriarch among the laboratories for quite some time. At its pre-NASA height in 1952, Langley employed 3,557 people and, from 1920 to 1958, had constructed some 30 wind tunnels (of which about half were still operational in the late 1950s), two towing tanks, and various specialized laboratories.[9] The second laboratory to open was Ames Aeronautical Laboratory next to the Navy's Moffett Field in Sunnyvale, California. Ames was, at least in the beginning, a West Coast Langley, the two labs having similar test equipment and research functions, especially during World War II. Its facilities included 16 wind tunnels constructed from 1940 to 1956.[10] The third laboratory was the Lewis Engine Lab, opened in 1942 next to the Cleveland Municipal (now Cleveland Hopkins) airport. It was in close proximity to the Army Air Force's Power Plants Laboratory in Dayton, Ohio, and relatively close to the nation's aircraft engine industry. The Lewis Engine Lab was a direct descendant of Langley's Aircraft Engine Research Laboratory, begun in 1934. With the founding of NASA, the Lewis Engine Lab devoted itself to the space effort, returning to air-breathing propulsion only in the 1970s.[11] Finally, the NACA operated two test areas: Wallops Island, Virginia (also known as the Pilotless Aircraft Research Station), and the High-Speed Flight Station (HSFS) in Muroc, California.

9. Hansen, *Engineer in Charge*, appendix D.
10. Elizabeth A. Muenger, *Searching the Horizon: A History of Ames Research Center: 1940–1976* (Washington, DC: NASA SP-4304, 1985), appendix C, p. 233.
11. Virginia P. Dawson, *Engines and Innovation: Lewis Laboratory and American Propulsion Technology*, NASA History Series (Washington, DC: NASA SP-4306, 1991).

There are numerous examples of the different laboratories taking different tacks on the same research problem, but the growth of the HSFS as an organic research entity represents a special case. The NACA, of course, had flight-test capabilities prior to the HSFS. But there is a difference between a flight-test group that works in a kind of ancillary supporting role at a wind tunnel-centric laboratory and what evolved on the dry lakebeds of Muroc. The HSFS, later renamed the Flight Research Center (FRC) and eventually renamed the Dryden Flight Research Center, was blessed with a geographic isolation that, over time, nurtured independent research capabilities. This was more than a collection of test pilots and technicians. The FRC worked closely with its peers at the other laboratories, but FRC personnel were also solving problems and acting creatively on their own.[12]

The FRC's capabilities truly flowered in the NASA era, perhaps owing some debt to the intellectual stimulus of the space program (and the freedom conferred to experiment broadly). The early example of this kind of initiative was the FRC's lifting body experiments, begun at a grassroots level and on a shoestring budget. The goal was to create a flyable spacecraft, something that could ride atop a rocket into space and then glide back to Earth. Rather than an expensive and unpredictable ballistic reentry that left heroic astronaut-pilots at the mercy of a flight engineer's trajectory and a flotilla of rescue ships, a lifting body could be flown directly to an airstrip. It was called a lifting body because it generated a small but useful amount of lift from its fuselage shape rather than from wings (which it lacked). And though this was a space vehicle, its distinctive challenges were atmospheric, namely, reentry heating and flight control. Lifting body research was a way to make aerodynamics and the art of aircraft design relevant in an age of ballistic space capsules.

The FRC researchers were not working on this topic in isolation. In fact, the inspiration for lifting bodies came from Ames. In the 1950s, Ames's H. Julian Allen had shown how blunt nose cones could survive the heat of atmospheric reentry, a finding that had an immediate impact on ballistic missile design. Allen and his peers took the idea further and sliced off a side of the cone, creating a shape that produced lift. Here was something that could survive reentry *and* fly. While this worked in the wind tunnel, it was not certain that this was a practical, controllable shape. At the FRC, aeronautical engineer Dale Reed began his own small-scale investigations in 1962 based on the Ames research.

---

12. On the history of the FRC, see Richard P. Hallion, *On the Frontier: Flight Research at Dryden, 1946–1981* (Washington, DC: NASA SP-4303, 1984).

Figure 4: Lifting body research at the FRC. The M2-F1 lifting body is seen here being towed behind a C-47 at the FRC (later DFRC), Edwards, California. The wingless, lifting body aircraft design was initially conceived as a means of landing an aircraft horizontally after atmospheric reentry. The absence of wings would make the extreme heat of reentry less damaging to the vehicle. *NASA Image E-10962*

From paper models to radio-controlled balsa models, Reed grew fascinated by the idea. Drawing together a core of supporters, Reed gained the backing of his director. In four months, Reed's group hand-built a lifting body glider large enough to accommodate a single pilot. By March of 1963, their lifting body glider was in the air, carrying out tests.[13]

---

13. R. Dale Reed, with Darlene Lister, *Wingless Flight: The Lifting Body Story* (Washington, DC: NASA SP-4220, 1997), p. 11.

The FRC's Spartan effort attracted the attention of Headquarters, which agreed to an official program entailing two "heavyweight" vehicles, one using the FRC/Ames blunt body design and another based on a competing idea from Langley. The U.S. Air Force, meanwhile, decided to participate in the program as well, contributing a design based on work that it had done independently of NASA. The FRC airdropped all three aircraft, initially flying them to the ground as gliders and later fitting them out with rockets. Testing continued into the early 1970s, with all three aircraft reaching supersonic speeds (though still well short of the velocities they might experience returning to Earth from orbit). While NASA refused to fund follow-on projects, the tests were influential in contributing to the design of the Space Shuttle, a winged vehicle that glided to a controlled landing from space. Like the lifting bodies, the Space Shuttle had minimal lift and had one chance to land properly on a runway. The FRC showed how this could be done reliably.[14]

What the FRC, now DFRC, became was not simply a competing branch, but a research organization that naturally approached problems differently from the wind tunnel branches. They favored certain questions over others, such as aircraft control. And they championed relatively inexpensive ways of getting vehicles into the air, from remote control models, to remotely piloted vehicles, to gliders, bare-bones prototypes, and reconstructed aircraft. The argument behind DFRC's methodology was that you got, apologies to Tom Wolfe, the "Real Stuff." It offered an integrated platform (i.e., not just an airfoil) operating in free flight and performing maneuvers that were quite difficult in the confines of a tunnel. Additionally, getting a working, flying model into the air drew attention (and potentially funding) in ways that paper designs and small wind tunnel models might not.

From a cynical standpoint, the proliferation of methodologies arising from NASA's institutional structure appears no more than bureaucratic infighting. This is only partially true. At one level, there was a competition between ideas and methods, and not just among the NASA laboratories, but also with outside organizations, such as the U.S. Air Force. As in science, generally, the prestige of a winning idea was, for many at NASA, sufficient incentive. At another level, the competition was sometimes more narrowly about resource allocation and program control. Regardless of the nature of the competition, NASA's decentralized structure has had a lasting impact, encouraging new methods and the reevaluation of traditional

14. Ibid., chap. 5.

Figure 5: The wingless lifting body aircraft sitting on Rogers Dry Lake at what is now NASA's DFRC, Edwards, California. From left to right: the U.S. Air Force X-24A, FRC M2-F3, and Langley HL-10. The lifting body aircraft studied the feasibility of maneuvering and landing an aerodynamic craft designed for reentry from space. These lifting bodies were air-launched by a B-52 mother ship; they then flew powered by their own rocket engines before making an unpowered approach and landing. They helped validate the concept that a space shuttle could make accurate landings without power. The X-24A flew from 17 April 1969 to 4 June 1971. The M2-F3 flew from 2 June 1970 to 21 December 1971. The HL-10 flew from 22 December 1966 to 17 July 1970 and logged the highest and fastest records in the lifting body program. *NASA Image EC69-2358*

ones. It almost goes without saying that NASA's scientific and technological narrative would have been quite different had all of its research been conducted under one roof. To nurture new ideas requires a willingness to support duplicate lines of inquiry (often off-budget ideas), potentially cannibalizing successful teams and methods. Naturally, this would be contrary to a unitary organizational structure with a centralized research infrastructure and administration.

Macro-Technological Change

Macro-technological change refers to external or broader technological innovations that serve as inputs to methodological changes at the laboratories. The prime and most powerful example of this is the rise of electronic computers, which set the stage for CFD (not to mention a revolution in avionics and digital flight control).[15] In turn, CFD was one of the most disruptive of the new research methodologies. This progression suggests a kind of logical unfolding of technological innovation (e.g., automobiles follow horse-drawn carriages). This is somewhat the case, but it is important to understand that the particular shape of CFD, the motivation and timing behind its emergence as well as its political support, could not have been scripted as part of a logical and predictable technical evolution.[16]

In 1958, electronic computers were already a part of NASA's infrastructure. Aeronautics researchers employed them in conjunction with the tunnel branches, helping sift through test data as well as performing some of the less complex mathematical crunching. To use computers as virtual wind tunnels, however, was a different proposition. The mathematical formulas that describe the motion of fluids, the Navier-Stokes equations, are a set of nonlinear, second-order, partial differential equations. The complexity of these equations meant that they could not be programmed into computers. Some researchers, however, realized that simpler mathematical techniques, amenable to programming, could approximate more complex fluid flow equations; indeed, engineers had long made use of such shortcuts in design calculations.[17]

One of the chief advocates of this approach was Harvard Lomax of the Ames Theoretical Division. He wrote his first programs for use on an IBM 650, a 1950s-era computer, and proceeded to make use of newer IBM and Control Data Corporation equipment in the 1960s. Lomax's group focused on converting the Navier-Stokes equations into algebraic approximations that computers could solve through a technique called finite difference. Lomax and his team also reduced the number of calculations that had to be performed by solving for select points in the flow stream.[18]

By the late 1960s, CFD development became another thread in the story of laboratory competition. The leadership at Ames decided to aggressively pursue the technology. The CFD technology offered Ames a way of differentiating itself from Langley (both laboratories relied heavily on wind tunnel research) while

---

15. Interestingly, the FRC put itself at the center of the digital flight control revolution with a program that followed closely on the heels of the lifting body flights. See James E. Tomayko, *Computers Take Flight: A History of NASA's Pioneering Digital Fly-By-Wire Project*, NASA History Series (Washington, DC: NASA SP-2000-4224, 2000).
16. For a classical treatment of technological determinism, see Robert L. Heilbroner, "Do Machines Make History," *Technology and Culture* (8 July 1967): 335–345.
17. For a summary of the Navier-Stokes equations and their place in aerodynamic theory and practice, see Anderson, *A History of Aerodynamics*, pp. 89–93.
18. Ferguson, *NASA's First A*, chap. 4.

diversifying the laboratory's infrastructure and expertise. Lomax's theoretical division became a CFD branch, and the Director of Ames at the time, Hans Mark, successfully lobbied to have a new supercomputer, the Illiac IV, transferred to Ames from the University of Illinois. The Illiac IV did, or attempted, what would later be called vector processing. By the time the supercomputer was operational at Ames, other computers, notably machines designed by Seymour Cray, were outperforming it. But the Illiac IV achieved some key objectives. It signaled Ames's intention to try to capture CFD supercomputing (a political goal) and that it was willing to take a gamble on CFD replacing wind tunnels (both a political and technical goal). And within Ames, the Illiac IV and the 1970s-era CFD work served as a training ground for Ames researchers cutting their teeth on the hardware and software obstacles associated with supercomputers (generally) and parallel processing (specifically).[19]

Researchers at Langley also responded to the twin stimuli of digital computing and laboratory competition. Like Ames, Langley had its own computing facilities and also pursued CFD. It had a Star 100 computer in the 1970s, which was Control Data's first attempt at a vector processing supercomputer. But Langley pursued CFD in a distinct manner, establishing computing branches as complements (rather than as antagonists) of the tunnels. The laboratory set out in two organizational directions. First, in 1972 Langley established the Institute for Computer Applications to Science and Engineering (ICASE). As the name implies, it was more broadly conceived as a general computational center, and unlike Ames, it did not focus on building hardware. Second, Langley established a number of CFD labs that tended to mirror the division of labor among the tunnels; they had CFD laboratories working on transonic, high-speed, low-speed, and aeroelasticity. In the broad scheme of Langley's research, CFD was positioned as a partner to the tunnels, not a replacement. Meanwhile, at ICASE, Langley and visiting researchers created a new mathematical approach to CFD and successfully nurtured this methodology as an alternative to Lomax's finite differences method. Two individuals, David Gotleib (a visiting researcher) and Youssuff Houssaini (one of the directors at ICASE), developed what were called "spectral methods for solving the Navier-Stokes partial differential equations."[20]

Computational Fluid Dynamics did not end the era of wind tunnels. Indeed, accurate CFD modeling still requires both flight testing and wind tunnel

---

19. Ibid.
20. Ibid.

experimentation in order to validate the mathematical models that underlie CFD. In those tasks and conditions (e.g., particular speed ranges and aircraft configurations) where researchers have established reliable models, CFD has replaced some amount of wind tunnel experimentation. But the complexity of aerodynamic phenomena means that numerous qualities, especially those related to cutting-edge aircraft, cannot be reliably predicted with CFD, at least not yet. Until then, researchers will continue to shift between tunnels and computers, searching for new mathematical tricks and exploiting ever more capable computer hardware. Macro-technological change continues to pace CFD. Interestingly, Ames's aggressive pursuit of CFD did not give the Center exclusive ownership of the technology. NASA did choose to centralize supercomputing at Ames (hardware that is accessible to all the laboratories); but persistently lower computing costs have helped democratize CFD, and aerodynamic modeling generally, in a way that remains impossible with high-cost wind tunnels.

Shifting Technological Frontier

The last factor behind changes in research methodology is arguably the least edifying: the technological frontier of aeronautics shifted. What is of interest here is how this played out in the laboratories and in top-level policy. Obviously, the establishment of a civilian space program represented the laboratories' first major shift in technological frontier. That is a topic in and of itself, for many aeronautics researchers took the opportunity to move into space research. But looking more narrowly at aeronautics, the next significant shift occurred during the 1970s when the White House and Congress asked a post-Apollo NASA to better address aviation issues and to make the Agency more responsive to the problems confronting the nation.

NASA's leadership at Headquarters and the Centers replied with plans that dealt with airspace congestion, airport noise, pollution, and energy efficiency. Of course, the NACA and NASA had always responded to more narrowly defined calls for assistance: investigations of structural or control problems in specific aircraft, icing research, wind shear detection, etc.[21] This call to action was more broadly based, and in many cases it involved more

21. William M. Leary, *We Freeze to Please: A History of NASA's Icing Research Tunnel and the Quest for Flight Safety* (Washington, DC: NASA SP-2002-4226, 2002); Lane E. Wallace, *Airborne Trailblazer: Two Decades with NASA Langley's 737 Flying Laboratory* (Washington, DC: NASA SP-4216, 1994), pp. 58–59.

heterogeneous engineering, that is, integrating questions of social behavior alongside technical issues. In materials research, for example, NASA entered into studies about the application of composite materials for fuel efficiency, but the studies involved field-testing different parts with the airlines, partially to establish long-term cost estimates. This was not merely about proving the technical worth of composites that were already in use in military aircraft, but in building public acceptance.[22]

A good example of NASA's shifting technological frontier is air traffic control (ATC) research. In the 1960s, the federal government began to take notice of rising airspace congestion. In 1968, the newly minted DOT created an ATC advisory committee. The committee's findings, known as the Alexander Report, argued for increased automatic communication and control methods. NASA's official response to traffic congestion came in two forms, both of which sought to respect the FAA's purview by working on the vehicle side of ATC (rather than the controller side). The Terminal Configured Vehicle program at Langley sought to increase safety and productivity through advanced avionics and flight procedures.[23] At Ames, researchers explored STOL aircraft as a means of adding airport capacity. There was, buried in the STOL project, a study inspired directly by the Alexander Report, an algorithm for guiding aircraft to land through time-based sequencing. Ames tested the "four-dimensional" guidance system, developed by Heinz Erzberger, as part of the STOL program.[24]

In Washington, DC, DOT was wrestling with the problem of modernizing the ATC system. The Advanced Airways System (AAS), initiated in the 1980s, was one of the most prominent of the FAA's upgrades (along with the Microwave Landing System), and it envisioned a far-reaching system of computer monitoring and control. Ultimately, the AAS did not deliver on these more advanced automated capabilities, which became bogged down in software development and budget overruns. The upgrade was sharply curtailed in the 1990s, with ATC operations hardly more automated than they were two decades earlier.[25]

Erzberger, who had put his four-dimensional guidance system aside in the early 1980s, returned to the topic and chose to examine how it could be integrated into ATC operations on the ground. He put together a small

---

22. Ferguson, *NASA's First A*, chaps. 4, 7.
23. Lane E. Wallace, *Airborne Trailblazer.*
24. A more complete account of Erzberger's research is in Ferguson, *NASA's First A*, chaps. 4, 7.
25. Matthew L. Wald, "Flight to Nowhere: A Special Report; Ambitious Update of Air Navigation Becomes a Fiasco," *New York Times* (29 January 1996).

project called the CTAS; this was to be a laboratory that simulated ATC operations and tested software based on Erzberger's original algorithm.[26] The key ingredient at the CTAS was a live ATC radar feed from the FAA, something that was exceedingly difficult to procure (for bureaucratic, not technical, reasons). The feed allowed simulations to be run against real data and gave researchers access to controller behavior. If there were anomalous situations, researchers could contact the actual FAA controllers and ask them why they had made a particular decision. When NASA and the FAA field-tested the CTAS software at the Denver, Colorado, and Fort Worth, Texas, TRACONs in 1995 and 1996, the tests went sufficiently well that the National Air Traffic Controller Association and the Air Transportation Association asked that the tools be kept in place.[27]

It is important not to overplay the success of the CTAS software. These tools were, against the automation contemplated by the AAS, modest applications. But the key to their success was the CTAS process. This was a laboratory for heterogeneous research, and it taught scientists and engineers that to automate ATC, one had to find solutions that respected the interests and accrued knowledge of human controllers. One might consider the CTAS methodology as a kind of analog to clinical trials in medicine. Further, the point to take away about shifting technological frontiers is not that NASA made a decision to enter into ATC research in a large way (which in fact it did only in the last decade), but that this was part of a larger policy shift to see NASA apply itself aggressively to the sociotechnical problems besetting aviation. This kind of work necessarily involved new methods to accommodate heterogeneous environments. Interestingly, this is the only one of the three factors behind methodological change that derives from top-down research administration, and even here there are important caveats (noted below).

Concluding Observations

The three factors given here for methodological change are not exhaustive. One could argue, for example, that aerodynamics, or at least important segments of it, has reached a kind of maturity. Engineers are now able to predict much of the subsonic aerodynamic performance of conventional aircraft, rendering moot some wind tunnel testing. But this is a more recent shift (that owes much to the spread of CFD) and certainly does not account for changes that were

26. TRACON stands for Terminal Radar Approach Control; it is a radar facility that handles incoming and outgoing traffic around large airports out to a distance of 30 to 50 miles.
27. Ferguson, *NASA's First A*, chap. 7.

taking place through most of the last 50 years. Institutional structure, macro-technological change, and shifting technological frontiers, if not exhaustive, are sufficiently comprehensive to help explain the major shifts. They are also useful in illuminating the murky relationship between research management and scientific and technological innovation. On this, a couple of observations are worth immediate note.

What is remarkable in this history is the degree to which change was undirected. The decentralized, dispersed, and sometimes duplicative structure of the laboratories was not, by design, intended to encourage any kind of disruptive competition. The NACA established multiple facilities largely because they sought capacity growth, or in the case of the HSFS, because of military necessity. NASA inherited this arrangement. Ironically, the omnipresent goal of reducing waste and duplication in federal programs has been simultaneously at odds with the establishment of competing research tracks *and* integral to competition. There is, after all, no bureaucratic drive to differentiate when budgets are assured. And while researchers, such as Harvard Lomax, took on innovative projects because of genuine scientific curiosity, laboratory management understood the value of claiming new research niches.

A second and related facet is the way in which variation grew in a bottom-up fashion, even in cases where a shift in the Agency's technological frontier was a matter of official policy. Erzberger's four-dimensional guidance system was a grassroots response to the Alexander Report, not a laboratory response to Headquarters. Indeed, Erzberger's work was buried within Ames's STOL research and, strictly speaking, was a project more suited to DOT or Langley. Likewise, the laboratories had already contemplated much of the 1970s-era energy efficiency work when the White House and Congress began waving their arms. And it was consistent advocacy from the researchers, not necessarily consistent budgets from the top, that spelled long-term realization of these programs. Erzberger's algorithms took over two decades to mature into usable software tools. This is not to say that the laboratories were unresponsive to policy changes (the creation of the civilian space program being the best example of laboratory realignment), but that policy changes have often tapped into research that was already off topic and/or anticipatory of future directions. It has been to NASA's advantage that researchers have had the ability to try new avenues that are not officially prescribed and that researchers have stuck to their ideas in spite of uncertain and often discontinuous programmatic guidance.

Chapter 10

The NACA, NASA, and the Supersonic-Hypersonic Frontier

Richard P. Hallion

Across the history of flight, adversity and seemingly insurmountable challenges have goaded and inspired aerospace scientists and engineers into producing some of aviation's greatest scientific and technical accomplishments. The advent of the supersonic-hypersonic age[1] and the work of the professional staffs of the NACA and its successor, NASA, certainly exemplify this. Over the first three decades of the NACA, the speed of American operational aircraft rose fourfold, to over 550 mph by August 1945, at the end of World War II. By that time, the anticipated speed of the most advanced American aircraft then under development, the Bell XS-1, was almost double this. Conceived in 1944 and designed and built over 1945, it eventually reached nearly 1,000 mph in 1948. (A derivative of this same design, the X-1A, having greater fuel capacity and thus longer engine-burn time, exceeded 1,600 mph in 1954.)

In 1958, the pioneer era of supersonic flight ended, coincident with the closing of the NACA era, the onset of the NASA era, and the beginning of the Space Age signaled by the launch of Sputnik. That year, the last flying X-1

1. The speed of sound in air is approximately 760 mph, decreasing to approximately 660 mph at 40,000 feet. Because it varies, sonic velocity at any altitude is referred to as Mach 1, honoring the 19th-century Austrian physicist and philosopher Ernst Mach. Though popularly speaking, any flight beyond Mach 1 is supersonic, aerodynamicists classically define three distinctive arenas of flight, reflecting the progression of wind tunnel studies from 1919 onward: subsonic below Mach 0.75, transonic between Mach 0.75 and 1.25, and supersonic from Mach 1.25 to Mach 5. Beyond Mach 5 is the hypersonic realm, with classic aerodynamics receding before aerothermodynamics and magnetohydrodynamics. Mach 5 is a more arbitrary and imprecise demarcation than Mach 1, whose lower velocity is signaled by the formation of standing shock waves, indicating sonic flow formation.

(the X-1E) retired, the Lockheed F-104 Starfighter (the first operational Mach 2 military aircraft) entered service, and airlines began their first transoceanic intercontinental jet transport operations, with de Havilland's Comet IV, Boeing's 707, and the Douglas DC-8. By this time, planners were conceptualizing operational aircraft at speeds over Mach 3, exemplified by a then-highly classified study effort that would spawn the Lockheed A-12/YF-12A/SR-71 Blackbird. The era of piloted hypersonic flight was dawning as the NACA, the U.S. Air Force, the U.S. Navy, North American, and Thiokol put finishing touches on the first of the X-15s, a rocket-powered, air-launched "Round Two" successor to the early "Round One" research airplanes (such as the X-1 and Douglas D-558-2) that had blazed the sonic frontier a decade earlier. Further away, but gestating, were programs for both winged and ballistic orbital vehicles, typified by the "Round Three" study effort leading to the abortive Boeing X-20A Dyna-Soar (an important predecessor to the Space Shuttle), and the Man-in-Space-Soonest (MISS) studies eventually spawning Project Mercury.

From Subsonic to Supersonic

The high-speed breakthrough—from subsonic through transonic and on to supersonic and hypersonic velocities—constituted a singular milestone in the evolution of flight, enabling the achievement of routine rapid global air transport and access to space. The need to understand and resolve some of the challenges of the high-speed regime began as early as World War I, in the era of the open-cockpit, wood-and-fabric, propeller-driven biplane. Although such aircraft typically flew at flight speeds no faster than 120 mph (approximately Mach 0.15), the tips of a rapidly rotating propeller could exceed Mach 0.75. Since a propeller is really a rotating wing, the accelerated flow over the propeller at a tip speed of Mach 0.75 could exceed the speed of sound (Mach 1), producing a standing shock wave and, behind the shock wave, turbulent, separated flows. These flows would seriously degrade the propeller's efficiency. By analogy, if extended to the design of a wing, it was evident to wind tunnel researchers as early as 1919 that the wing would experience both a marked drop in lift and a marked increase in drag.[2] While of largely academic interest at the time (except for propeller designers), such transonic phenomena boded ill for subsequent airplane design as propulsion advances and advances in streamline aerodynamics drove aircraft level-flight speeds above 350 mph.

2. F. W. Caldwell and E. N. Fales, "Wind Tunnel Studies on Aerodynamic Phenomena at High-Speeds," NACA Technical Report (TR) 83 (1920), p. 77.

The practical difficulties of high-speed flight did not become a significant hindrance to safe aircraft operations until the mid-1930s. Nevertheless, the recognition that the compressibility effects of transonic flows at speeds of Mach 0.7 and above could not be ignored drove both efforts to derive new transonic aerodynamic theory and efforts to develop specialized research tools to analyze such phenomena as transonic drag rise. In the course of routine testing of airfoils suitable for propellers, Frank W. Caldwell and Elisha Fales had discovered the characteristic changes in lift and drag and shock wave formation, while testing small airfoils at speeds of up to 450 mph using a small, 14-inch-diameter Army Air Service tunnel at McCook Field, Ohio. Building upon this serendipitous work, Drs. Lyman J. Briggs and Hugh L. Dryden of the Bureau of Standards and Colonel G. F. Hull of the Army Ordnance Department subsequently undertook more detailed investigations using an air jet produced by a turbine-driven, three-stage centrifugal compressor at the General Electric Company's Lynn, Massachusetts, plant. Again, serendipity played a role: a thin coating of oil smeared on the airfoils to prevent them from rusting furnished evidence of airflow separation at transonic speeds, even more proof that as a wing approached the speed of sound, it would exhibit a loss of lift and an increase in drag.[3] Briggs and Dryden continued their studies at the Army's Edgewood Arsenal, testing airfoils drilled to record pressure distribution at various speeds from Mach 0.5 to Mach 1.08 (1,218 feet per second); this series of tests confirmed flow separation, lift loss, drag increase, and shock wave formation and indicated that thin airfoils showed markedly reduced transonic effects compared to thicker ones. From this came the first generalized appreciation that thin symmetrical airfoils showed the best potential for enabling flight at transonic velocities.[4]

By the early 1930s, the global aeronautical community was becoming aware that transonic flow phenomena and the power requirements necessary to overcome attendant high drag rise would seriously constrain aircraft design and performance. Increasingly, aerodynamicists worldwide began to see the speed of sound as a limitation to the expansion of flight, memorably captured

---

3. L. J. Briggs, G. F. Hull, and H. L. Dryden, "Aerodynamic Characteristics of Airfoils at High Speeds," TR 207 (1924), p. 465; H. L. Dryden, "Supersonic Travel Within the Last Two Hundred Years," *Scientific Monthly* 78, no. 5 (May 1954): 289–295.
4. Briggs and Dryden, "Pressure Distribution Over Airfoils at High Speeds," TR 255 (1926), pp. 581–582; Briggs and Dryden, "Aerodynamic Characteristics of Twenty-Four Airfoils at High Speeds," TR 319 (1929), p. 346; Briggs and Dryden, "Aerodynamic Characteristics of Circular-Arc Airfoils at High Speeds," TR 365 (1931).

Figure 1: John Stack, the NACA's foremost and most influential advocate of high-speed aerodynamic research. *NASA Image L-48989*

by British aerodynamicist W. F. Hilton's judgment that Mach 1 loomed "like a barrier against higher speed," the onset of the lurid if not, at the time, altogether inaccurate phrase "sound barrier."[5]

Aerodynamicists in America and Europe used the word "compressibility" as convenient shorthand to encompass all the various problems and challenges associated with transonic flight. One of the most critical was securing accurate, precise, and consistent wind tunnel measurement at speeds just below and above the speed of sound. As flow speed increased within a tunnel and the accelerated flow past a model and its supporting mounts and balance reached Mach 1, shock waves would form on the model and its attachments and reflect across the tunnel, inhibiting accurate measurements. As flow speed increased and the shock waves assumed a more acute angle beyond the speed of sound, researchers could once again take accurate measurements. But this meant a

5. Quoted in James R. Hansen, *Engineer in Charge: A History of the Langley Aeronautical Laboratory, 1917–1958* (Washington, DC: NASA SP-4305, 1987), p. 253.

practically unknown region existed for aerodynamic researchers, and wind tunnels were unable to furnish reliable information between Mach 0.75 and Mach 1.25, precisely the region that aerodynamicists most needed to study.[6]

The NACA's growing interest in transonic aerodynamics reflected not only the state of aviation technology and the deficiencies of test techniques, but also the increasingly pervasive influence within the NACA of a single dominant figure, John Stack. A driving, enthusiastic, insightful, and highly energetic Massachusetts Institute of Technology (MIT)-educated personality, he had joined the NACA in 1928 as a junior engineer in the Langley Memorial Aeronautical Laboratory. Although he did not trigger the NACA's interest in high-speed flight, he subsequently imprinted its high-speed (and international) aerodynamic research in a fashion unmatched by any other single researcher, foreign or American, including fellow agency personages such as Eastman Jacobs, Theodore Theodorsen, and even the legendary Theodore von Kármán at the California Institute of Technology (Caltech). Yet, for all his prominence, he never stinted in recognizing those who had assisted his work or contributed their own.[7]

In 1933, building upon an earlier and less-satisfactory 11-inch tunnel, Stack designed a small, 24-inch "blow-down" tunnel using air drawn from a pressure tank attached to Langley's famed VDT and fitted with an optical Schlieren photographic apparatus to capture flow changes around airfoils at transonic speeds. Subsequent tunnel trials convinced him the best, most reliable means of transonic research (pending development of genuinely transonic tunnels) would be the construction of specialized instrumented "compressibility" research airplanes. That same year, he drew up one such configuration, anticipating a cantilever monoplane powered by a 2,300-horsepower engine, with a symmetrical airfoil having a thickness-chord ratio of 18 percent at the wing root, progressively thinning to a 9 percent section at the tip. According to his calculations, such a design could have a maximum speed of 525 mph, ensuring that it could acquire full-scale "real world" aerodynamic data at speeds where the accelerated airflow around its wings and fuselage would be in excess of Mach 1. In an era when the braced biplane still constituted the "normative" configuration for military and civil aircraft, and even for high-speed racing

6. John V. Becker, *The High-Speed Frontier: Case Studies of Four NACA Programs, 1920–1940* (Washington, DC: NASA SP-445, 1980), pp. 62–63.
7. I was privileged to know John Stack; a copy of a 19 May 1971 interview I did with him on his transonic research is in the archives of the NASA History Division, Washington, DC. Stack's personality is well captured by former colleague John Becker in his previously cited *High-Speed Frontier*, pp. 13–14.

TABLE I.

Fuselage diameter	40 in.
Wing span	29.1 ft.
Wing area	141.2 sq.ft.
Wing chord (average)	4.85 ft.
Aspect ratio	6

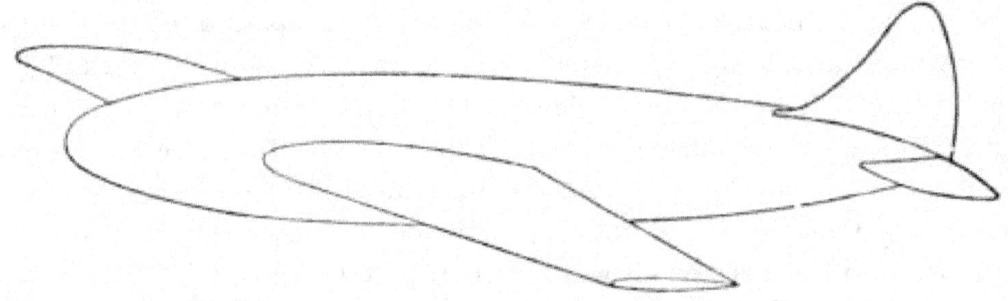

FIG. 1. Hypothetical high speed airplane.

Figure 2: Stack's concept of a compressibility research airplane, from the *Journal of the Aeronautical Sciences*, January 1934. *NASA*

aircraft, Stack's design (published in the inaugural issue of the *Journal of the Aeronautical Sciences*) constituted a remarkably bold and prescient concept.[8] In the short term, however, as engineer John Becker recalled, Stack's concept "had little impact on our outlook" as the NACA's staff contemplated the "enormous challenges" of flight at speeds of over 500 mph. But if seemingly optimistic, such figures were, in fact, reasonable and attained in later years by specialized propeller-driven aircraft.[9]

Compressibility assumed more than academic concern following the advent of the first monoplane fighters capable of exceeding 400 mph in high-speed dives. At that speed, depending upon the degree of streamlining and thickness of a wing, the accelerated airflow around an aircraft could exceed Mach 1, with attendant shock formation and consequent disturbed flow conditions

---

8. John Stack, "Effects of Compressibility on High-Speed Flight," *Journal of the Aeronautical Sciences* 1, no. 1 (January 1934): 40–43.
9. Becker, *High-Speed Frontier*, p. 24. Aircraft achieving such velocities included the Republic XP-47J, XP-72, and, in the late 1960s, a modified Grumman F8F-2 Bearcat.

that robbed a plane of lift, increased its drag, and, in most cases, severely constrained its controllability. In July 1937, a Messerschmitt engineering test pilot perished when his Bf 109 plunged into Lake Muritz during high-speed dive testing at Nazi Germany's Rechlin test center. In November 1941, a Lockheed test pilot likewise died when his experimental YP-38 broke up during a high-speed dive pullout over Burbank, California. Various other compressibility-related accidents claimed examples and airmen flying other new aircraft, such as Britain's Hawker Typhoon and America's Republic P-47.

In response, the NACA's Langley and Ames laboratories acquired loaned examples of representative fighters, instrumenting them and undertaking comprehensive transonic dive trials. It was not something undertaken lightly. In 1940, the NACA had undertaken perilous dive tests to 575 mph (Mach 0.74) of a loaned Navy Brewster XF2A-2 Buffalo at Langley Memorial Aeronautical Laboratory, the trials leaving watching aerodynamicists "with the strong feeling that a diving airplane operating close to its structural limits was not an acceptable way to acquire high-speed research information."[10] Thanks to careful planning, the NACA never lost an airplane or an airman during its dive program, though it had some close calls. The dive tests emphasized the synergy between ground and flight research. Dive recovery flaps, to prevent the formation of shock-producing transonic flows, were wind tunnel tested and then verified through the NACA, military, and contractor flight tests. In the case of the P-38, such flaps transformed it into a safe and operationally effective fighter, flown by America's leading Pacific War aces.[11] Diving tests were undertaken by all the advanced aeronautical nations, particularly Great Britain and Germany. Indeed, in 1943, an instrumented British Spitfire achieved Mach 0.89 (± 0.01) in diving trials at the Royal Aeronautical Establishment, Farnborough, the fastest speed ever recorded by a propeller-driven, piston-powered aircraft.[12]

---

10. Becker, *High-Speed Frontier*, p. 88.
11. Edwin P. Hartman, *Adventures in Research: A History of the Ames Research Center* (Washington, DC: NASA SP-4302, 1970).
12. Squadron Leader A. F. Martindale, "Compressibility Research: Final Dive on Spitfire EN 409," Ref. CTP/K/1/73 (28 April 1944). I wish to acknowledge with grateful appreciation the assistance of the late Air Commodore Allen H. Wheeler, Royal Air Force (RAF), who made this report available to me, and who also put me in contact with Sir Morien Morgan, Professor W. A. Mair, and Mr. R. P. Probert, all of whom furnished further information on the Spitfire trials. See also W. A. Mair, ed., "Research on High Speed Aerodynamics at the Royal Aircraft Establishment from 1942–1945," Aeronautical Research Council *Reports and Memoranda*, no. 2222 (London, U.K.: Her Majesty's Stationery Office [HMSO], 1950). See also Charles Burnet, *Three Centuries to Concorde* (London, U.K.: Mechanical Engineering Society, 1979), pp. 29–52.

Figure 3: A technician readies the recording instrumentation of a P-51D Mustang modified for wing-flow transonic research. The test model (an XS-1 shape with a swept horizontal tail) is at midspan, with mechanical linkages connecting it to the instrumentation installed within the modified gun bay. *NASA Image L-46802*

By 1941, as Theodore von Kármán, Director of the Guggenheim Aeronautical Laboratory at the California Institute of Technology (GALCIT), noted in a seminal paper, aeronautical engineers were "pounding hard on the closed door leading into the field of supersonic motion."[13] The advent of the gas turbine engine in Europe and, subsequently, its adaptation in America promised to revolutionize aircraft flight speeds and make practical the 550+ mph airplane. Under this "technology push" and wartime pressure, the need for reliable transonic information assumed even greater urgency. NACA researchers (and others as well) increasingly adopted stopgap solutions. These included instrumenting and radar-tracking falling bodies dropped from high-flying bombers and firing small rocket-propelled models. In a creative and insightful attempt to take advantage of the flow conditions existing around

---

13. Theodore von Kármán, "Compressibility Effects in Aerodynamics," *Journal of the Aeronautical Sciences* 8, no. 9 (July 1941): 337. Von Kármán intended his paper as an argument to use the tools of mathematical analysis "as a guide for avoiding a premature drop of aerodynamic efficiency."

a diving fighter, engineers installed small free-standing models attached to balances and comprehensive instrumentation located in the gun bay of P-51 fighters, diving the airplanes to subject the models to localized transonic and supersonic flow.

Drawing upon experience with wing-flow models, wind tunnel researchers conceived the transonic tunnel "bump" as a means of replicating the same kind of accelerated flow past a model, though with less success than its airborne predecessor. Though useful, in all of these cases, the information obtained was limited; test duration was short; and, in the case of wing-flow dive trials, the tests involved considerable risk as the aircraft plunged deeper into the dense lower atmosphere where structural loadings and transonic effects would be at their most severe.[14]

From this combination of information need, creative make-do, and limited effectiveness sprang both the postwar X-series research airplanes and the transonic wind tunnel. The story of the NACA's role in the formulation of the X series has been extensively told, and such aircraft certainly constituted the most visible evidence of the agency's research investment in high-speed flight.[15] But its interest was matched as well by significant growing interest in military services and private industry. In 1943, both Nazi Germany and Great Britain had launched national transonic research aircraft study efforts, the former with the Deutsche Forschungsanstalt für Segelflug's DFS 346, a rocket-powered swept-wing vehicle subsequently completed and flown (with indifferent success) in the Soviet Union after the war and the latter with the afterburning turbojet-powered Miles M.52, which was subsequently canceled in 1946, considerably setting back British progress in supersonic design technology. In early 1944, a team working at the direction of Theodore von Kármán at Caltech's JPL undertook performance analysis of a representative transonic research aircraft configuration, a rocket-boosted ramjet-powered configuration remarkably similar in design concept to that of French pioneer Rene Leduc's postwar ramjet test beds and experimental fighters. The study, by America's most prestigious academic aeronautical research establishment, implicitly endorsed the transonic research aircraft concept and offered encouragement

---

14. Robert R. Gilruth to author, 27 January 1972. See also Becker, *High-Speed Frontier*, pp. 84–85.
15. For example, see Kenneth S. Kleinknecht, "The Rocket Research Airplanes," in *The History of Rocket Technology: Essays on Research, Development, and Utility*, ed. Eugene M. Emme (Detroit, MI: Wayne State University Press, 1964), pp. 189–211, and James A. Martin, "The Record-Setting Research Airplanes," *Aerospace Engineering* 21, no. 12 (December 1962): 49–54; both are superb surveys despite their age.

Figure 4: The Air Force-sponsored, rocket-powered Bell XS-1 (X-1), whose fuselage shape copied the .50 caliber machine gun bullet. This is the second X-1 (AAF 46-063), eventually modified as the Mach 2+ Bell X-1E. *NASA Image E49-001*

Figure 5: The Navy-sponsored, turbojet-powered Douglas D-558-1 Skystreak. This is the third D-558-1 (USN 37972); its slender lines earned it the nickname "The Flying Test-tube." *NASA Image E49-090*

that the "sound barrier" might not, in fact, be such a dramatic obstruction to future progress as popular sentiment blithely assumed.[16]

In 1944, coincident with this launching of foreign programs and studies by America's military services, a series of meetings between agency, military, and industry representatives resulted in the launching of what emerged as a joint Army Air Forces (AAF) (later, after September 1947, the U.S. Air Force), U.S. Navy, and NACA research airplane development effort, funded largely by the military, which ran the appropriate aircraft project development offices for each aircraft thus acquired; the effort was overseen technically by the NACA via a multilaboratory Research Airplane Projects Panel (RAPP). To the AAF, which desired even in 1944 to achieve supersonic speeds as quickly as possible—Chief of Staff General Henry H. "Hap" Arnold informed Caltech's von Kármán that in the postwar world, supersonic speed would be a "requirement"—the rocket appeared the most efficient means of first achieving high-Mach flight.[17] Desirous that any such aircraft be immediately applicable to tactical purposes, the Navy favored a turbojet. To the NACA, an agency that, having been founded as a "rider" to a naval appropriations bill, was always more inclined toward the Navy and its Bureau of Aeronautics than to the AAF (and later the Air Force), the jet engine appeared as the best propulsion system, as it would afford longer duration, even though at vastly lower Mach numbers. From this dichotomous (if complementary) approach sprang the first two experimental aircraft developed explicitly for transonic and supersonic research, the AAF/Air Force-sponsored, rocket-powered Bell XS-1 (later X-1) and the Navy-sponsored, turbojet-powered Douglas D-558.

As well, NACA postwar high-speed aeronautical research—as that of virtually every aeronautical research establishment in the postwar world—was dramatically influenced by developments made by the Nazi German aeronautical

16. Theodore von Kármán, F. J. Malina, M. Summerfield, and H. S. Tsien, "Comparative Study of Jet Propulsion Systems as Applied to Missiles and Transonic Aircraft," Memorandum JPL-2 (Pasadena, CA: Caltech JPL, 28 March 1944), fig. 36, p. 76, copy in Malina Papers, file 10.3, box 10, Manuscript Division, Library of Congress, Washington, DC; C. M. Fougère and R. Smelt, "Note on the Miles Supersonic Aircraft (preliminary version)," Tech. Note. Aero. 1347 (Farnborough, U.K.: Royal Aircraft Establishment, December 1943), copy in Wright Field Microfilm Collection, D.52.1/Miles/14, NASM Archives, Paul E. Garber Restoration Facility, Silver Hill, MD; Clark Millikan, "Notes of Visit to the *Deutsche Forschungsanstalt für Segelflug*" (Ainring, Germany: Naval Technical Mission to Europe [NavTechMisEu], June 1945); Clark Millikan, "Technical Report of Visit by C. B. Millikan to British M.A.P. Project E24/43, Miles 52 Transonic Research Airplane," draft (NavTechMisEu, June 1945), last two in Clark Millikan Papers, folder 6-2, box 6, Archives, Caltech, Pasadena, CA.

17. General H. H. Arnold to Theodore von Kármán, 7 November 1944, copy in the office files of the Air Force SAB, Headquarters of the Air Force, Pentagon, Washington, DC.

research community in the crucial decade between 1935 and 1945. Virtually all high-speed aerodynamic configurations and shapes—wings, tail locations, bodies, control surfaces, and inlets, for example—were affected by the discoveries and consequent revelations of Allied technical intelligence teams that ventured to the bleak remains of the Third Reich and sifted through its rubble. The plethora of Nazi German technical developments (some of which, like the ballistic missile and rocket-powered fighter, were quite dramatic) and the grand scale of its aeronautical research effort—visible before the war to prewar NACA officials and staggeringly evident to Allied technical intelligence teams at its end, particularly in its widespread investment in wind tunnels—masked in some cases very poor technical choices (for example, the transonic semi-tailless configuration of the Me 163, which inspired the creation of the British D.H. 108 and American X-4) and a largely inefficient research and acquisition structure, one in which (fortunately for the Allies) energetic effort could not compensate for lack of clear-headed and pragmatic guidance and oversight.[18]

One major German technical development, however, profoundly influenced postwar global aviation: the high-speed swept wing. The swept wing had appeared before World War I, conceived as a means of permitting the design of safe, stable, and tailless flying wings. It imposed "self-damping" inherent stability upon the flying wing, and, as a result, many flying wing gliders and some powered aircraft appeared in the interwar years. In the 1930s, Adolf Busemann postulated a different use of the swept wing as a means of alleviating the deleterious effects of transonic flow as a wing approached and exceeded the speed of sound. Busemann enunciated the concept at the October 1935 Volta Congress on High Speeds in Aviation, held at Guidonia, Fascist Italy's showcase aeronautical research establishment on the outskirts of Rome.[19] (Surprisingly, it went almost unnoticed; its significance was missed entirely by the American delegation, which included no less than GALCIT's Theodore von Kármán, the

---

18. Commander W. E. Sweeney et al. to Chief, U.S. Navy Bureau of Aeronautics, with attached "Resume and Recommendations on High Speed Aircraft Development" (NavTechMisEu, 2 July 1945) folder 6-2, box 6; Diary 6 of Clark Millikan, 28 May–8 July 1945, box 35; both in Millikan Papers, Caltech.
19. G. Arturo Crocco, "Le alte velocità in aviazione ed il Convegno Volta," *L'Aerotecnica* 15, nos. 9–10 (September–October 1935), pp. 851–915; A. Busemann, "Aerodynamische Auftrieb bei Überschallgeschwindigkeit," *Luftfahrtforschung* 12, no. 6 (3 October 1935): 210–220; Adolf Busemann, "Compressible Flow in the Thirties," *Annual Review of Fluid Mechanics* 3 (1971): 6–11; Carlo Ferrari, "Recalling the Vth Volta Congress; High Speeds in Aviation," *Annual Review of Fluid Mechanics* 28 (1996): 1–9; Hans-Ulrich Meier, "Historischer Rückblick zur Entwicklung der Hochgeschwindigkeitsaerodynamik," in Meier et al., *Die Pfielflügelentwicklung in Deutschland bis 1945* (Bonn, Germany: Bernard & Graefe Verlag, 2006), pp. 16–36.

Figure 6: Shock wave pattern formed at Mach 2.5 by the A4 V12/c, a swept-wing variant of the A-4 (V-2) missile, 1940. *Richard Lehnert, "Bericht über Dreikomponentenmessungen mit den Gleitermodellen A4 V12/a und A4 V12/c," Archiv Nr 66/34 (Peenemünde: Heeres-Versuchsstelle, 27 November 1940), Box 674, "C10/V-2/History" file, National Museum of the United States Air Force, Wright-Patterson Air Force Base, Ohio*

Bureau of Standards' Hugh L. Dryden, and the NACA's Eastman Jacobs).[20] His fellow countryman Alexander Lippisch, a developer of self-stabilizing tailless swept-wing gliders, applied it subsequently to the Messerschmitt Me 163, though not (as has been mentioned) with any significant success. The Messerschmitt and Lippisch teams subsequently separated amid increasing acrimony and ill will; Lippisch turned increasingly toward the delta configuration, though with far "fatter" wing sections than desirable, and Messerschmitt to more conventional (but still quite radical for the time) swept configurations having tail surfaces. As a result, at the war's end, Messerschmitt's advanced design project office at Oberammergau had a wide range of experimental fighter projects under way, all exploiting the transonic swept wing. Other German manufacturers, notably the Focke-Wulf and Junkers concerns, had their own swept-wing fighter and bomber projects under way, and the swept wing had even flown (albeit unsuccessfully)

20. Theodore von Kármán, with Lee Edson, *The Wind and Beyond: Theodore von Kármán—Pioneer in Aviation and Pathfinder in Space* (Boston, MA: Little, Brown and Co., 1967), pp. 216–217, 221–222; Eastman N. Jacobs, "Memorandum for Engineer in Charge, Subj.: Trip to Europe," 11 November 1935, p. 1, in Record Group 255, "Foreign Aero Officials Germany 1936–37" folder, box 75, U.S. National Archives and Records Administration, Archives II, College Park, MD.

Figure 7: Robert T. Jones, father of the American swept wing. *NASA Image LMAL 48-705*

on a modified variant of the V-2 missile, the A-4b. (This large swept-wing rocket, representative of an aircraft-type vehicle, successfully accelerated through the speed of sound but then, during reentry, broke up at approximately Mach 4). Among all German technical developments, even including the large rocket, the swept wing constituted the most distinctive element of Nazi-era aeronautics, at once hailed for its significance by virtually all Allied inspectors and, consequently, immediately "exported" to their own nations.[21]

---

21. L. E. Root, "Information on Messerschmitt Aircraft Design," Items No. 5 and 25, File No. XXXII-37, CIOS Target Nos. 5/247 and 25/543, Combined Intelligence Objectives Subcommittee G-2 Division [CIOS G2], SHAEF (Rear), August 1945, pp. 3–4, and Lieutenant Commander M. A. Biot, U.S. Navy, "Messerschmitt Advanced Fighter Designs," Item No. 5, File No. XXXII-41, CIOS G2, July 1945, pp. 4–8 and figs. 1–4, both in Wright Field Microfilm Collection, D52.1/Messerschmitt/143–144, NASM Archives, Garber Facility; Ronald Smelt, "A Critical Review of German Research on High-Speed Airflow," *Journal of the Royal Aeronautical Society* 50, no. 432 (December 1946): 899–934; U.S. Army Air Forces, "German Aircraft, New and Projected Types" (1946), A-1A/Germ/1945 file, box 568, Archives of the National Museum of the United States Air Force, Wright-Patterson Air Force Base, OH; Theodore von Kármán, "Where We Stand: First Report to General of the Army H. H. Arnold on Long Range Research Problems of the AIR FORCES with a Review of German Plans and Developments," 22 August 1945, vol. II-1, copy 13, including Hsue-shen Tsien, "Reports on the Recent Aeronautical Developments of Several Selected Fields in Germany and Switzerland," July 1945; Hsue-shen Tsien, "High Speed Aerodynamics," December 1945; F. L. Wattendorf, "Reports on Selected Topics of German and Swiss Aeronautical Developments," June 1945, copy in Henry H. Arnold Papers, Microfilm Reel 194, Manuscript Division, Library of Congress.

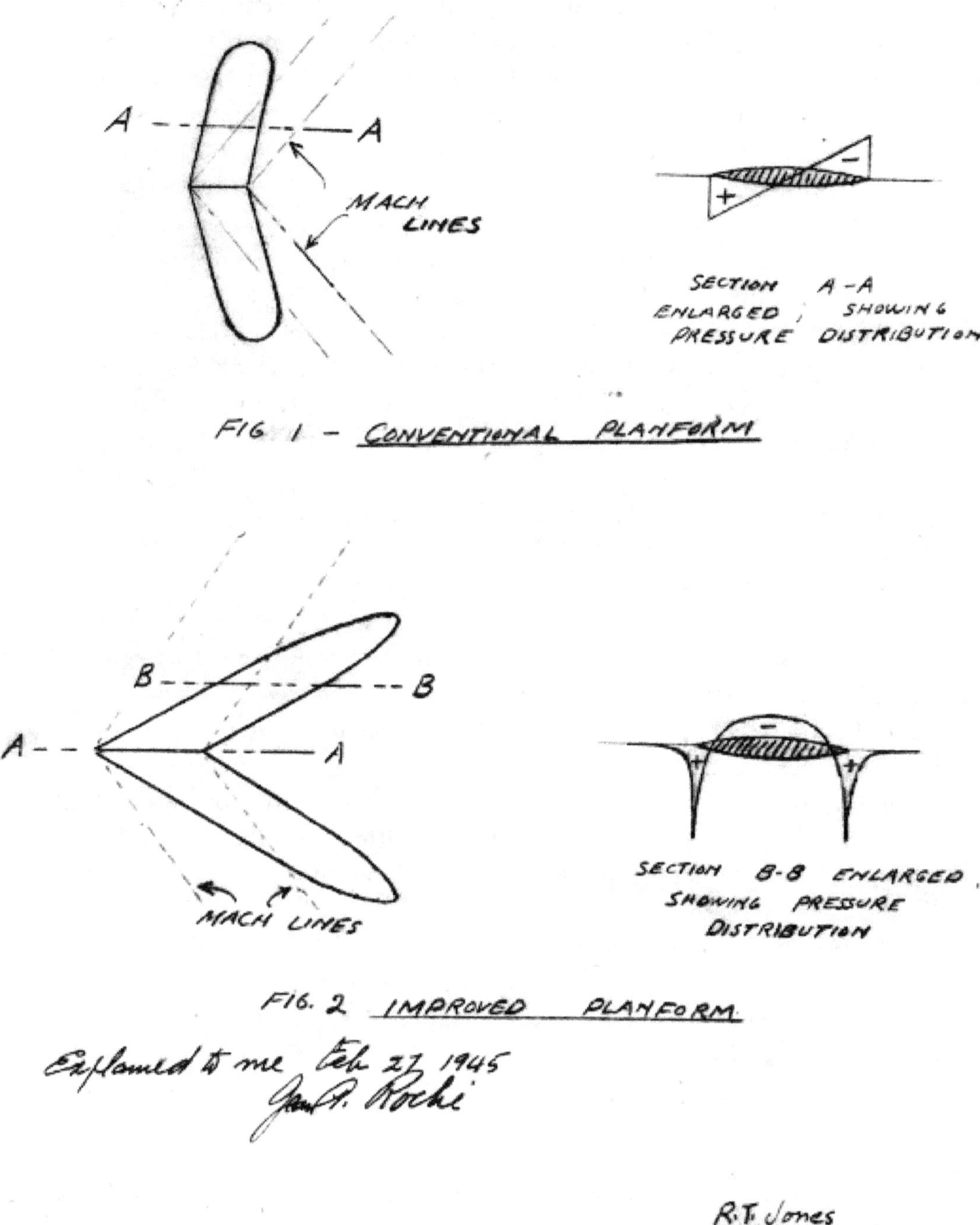

Figure 8: Jones's initial conception of the swept wing, noting his briefing to AAF Representative Jean Roché in late February 1945. *Robert T. Jones biographical file, NASA History Division, Washington, DC*

In the case of the United States, the discovery by technical intelligence teams of the swept wing amid the rubble of Nazi Germany occurred roughly simultaneously with its independent invention in America by NACA researcher Robert T. Jones of the Langley Memorial Aeronautical Laboratory. Assigned to assess the anticipated performance of a proposed glide bomb having large and sharply swept fins, Jones recognized that he could apply aerodynamic theory derived by Max Munk two decades previously to explain the flow field around an inclined airship hull. From this seemingly unlikely source, Jones derived, first, the sharply swept slender delta wing (thinner and more appropriate for high-speed flight than Lippisch's thicker configurations) and the sharply swept transonic wing. Tested in early 1945 at Langley and at the Army's Aberdeen experimental station, the Jones swept wing clearly pointed to the future direction of American aircraft design, though it remained as yet unproven.[22]

In the spring and summer of 1945, the AAF and Navy sent two technical teams to Europe, the AAF Scientific Advisory Group (SAG) and the NavTechMisEu. Composed of academicians, aircraft designers, and military engineers, each swiftly uncovered overwhelming evidence of the Nazi German aircraft industry's overwhelming interest in the swept wing. Through their studies and reports, the swept wing at once became both shorthand and symbol of both the "advanced state" of Nazi aeronautics and the mirror-image "backwardness" of the United States. To assess the swept wing's low-speed behavior, the Navy and the NACA undertook productive low-speed flight tests on a propeller-driven Bell L-39 (a modified P-63 Kingcobra fighter) incorporating a 35-degree swept-wing configuration subsequently adopted for the famed F-86 jet fighter.

The Sabre was one of two major AAF weapon acquisition programs—the other being the XB-47 jet bomber—converted from straight-wing to swept-wing designs. So attractive did the swept wing appear that the NACA briefly came under criticism by the AAF for not having suggested developing the Bell XS-1

---

22. See Robert T. Jones, "Properties of Low-Aspect-Ratio Pointed Wings at Speeds Below and Above the Speed of Sound," NACA TR 835 (11 May 1945), and his "Wing Planforms for High-Speed Flight," NACA Technical Note 1033 (1946, but issued at Langley Memorial Aeronautical Laboratory on 23 June 1945); the dates of both of these confirm the independence of his work. The Munk inspiration came from Max Munk's "The Aerodynamic Forces on Airship Hulls," NACA Report No. 184 (1923); Robert T. Jones, "Recollections from an Earlier Period in American Aeronautics," *Annual Review of Fluid Mechanics* 9 (1977): 1–11. See also Richard P. Hallion, "Lippisch, Gluhareff, and Jones: The Emergence of the Delta Planform and the Origins of the Sweptwing in the United States," *Aerospace Historian* 26, no. 1 (March 1979): 1–10.

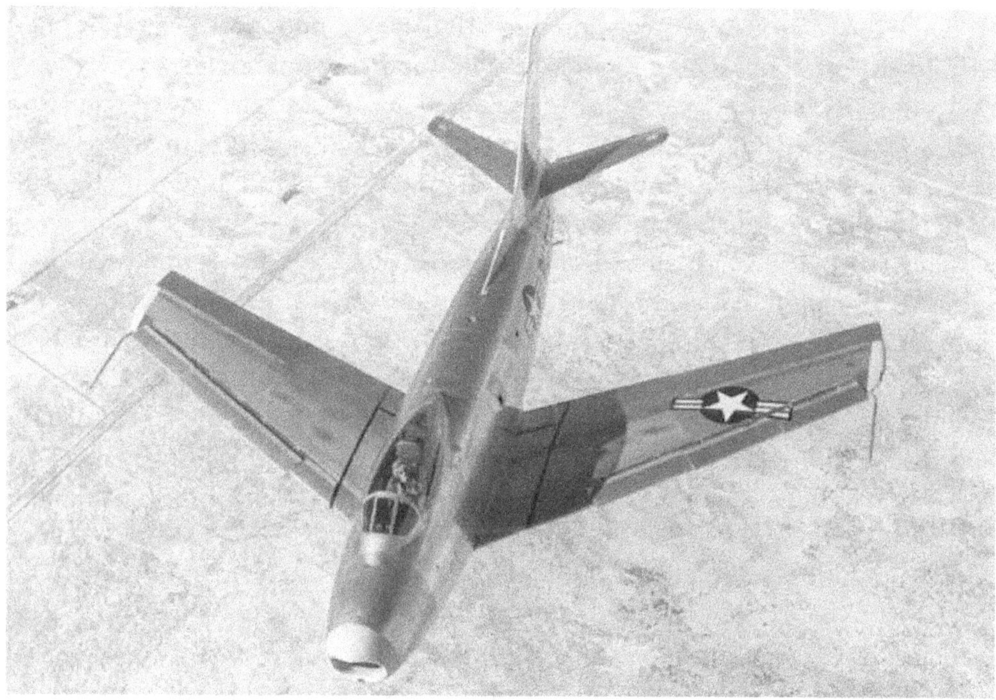

Figure 9: The North American Aviation XP-86 Sabre, the first jet-propelled airplane to exceed Mach 1 (in a dive), 1948. *Air Force Flight Test Center History Office*

Figure 10: The Boeing XB-47, which introduced the design standard of wing-mounted engine pods combined with a swept-wing planform. *Air Force Flight Test Center History Office*

from the outset as a swept-wing aircraft. (The agency, not unreasonably, replied that it had feared burdening the design with too many unknowns.) Accordingly, at AAF urging, the XS-1 program spawned a successor to the swept wing, the Bell XS-2 (later X-2), and the Navy, equally shocked by German swept-wing research, authorized the anticipated six-aircraft D-558 straight-wing program to be split into both a straight-wing variant (the D-558-1) and a swept-wing variant (the D-558-2) of three aircraft apiece. Unrelated to German research, the AAF already had a research program under way for a slender supersonic engine test bed, the Douglas XS-3 (later X-3). Two German design concepts greatly influenced both the Northrop XS-4 (later X-4) and Bell X-5. Despite intelligence reports that clearly enunciated potential difficulties at transonic speeds, Northrop (always interested in semitailless and tailless designs) received a contract to develop a research aircraft to explore the swept, self-stabilizing tailless design Lippisch had employed for the Messerschmitt Me 163 Komet. An abandoned Messerschmitt fighter project, the P 1101, inspired Bell Chief Engineer Robert J. Woods (one of the architects of the XS-1 program) to use its basic design configuration for a variable wing-sweep test bed. This became the Bell X-5, an impracticable if significant precursor to subsequent variable wing-sweep designs.

Together, these and others constituted what subsequent agency authorities termed a "Round One" of research airplanes: the Bell X-1, Bell X-2, Douglas X-3, Northrop X-4, Bell X-5, Consolidated Vultee (Convair) XF-92A, and Douglas D-558-1 and D-558-2. In all cases, the NACA worked with the military services (particularly the Air Force), which bore responsibility for acquiring the aircraft from industry and overseeing contractual execution. Additionally, they were tested at a new NACA research facility established on the shores of Rogers Dry Lake, the Muroc Flight Test Unit (later the NACA High-Speed Flight Research Station, then HSFS, and finally, in the NASA era, the FRC, now, since 1976, the DFRC) at Muroc (later Edwards) Air Force Base, California.[23]

Although used for many research purposes, including some unrelated to their original development, the "Round One" aircraft generally fell into two broad categories of research vehicles. Some were acquired to obtain basic high-speed aerodynamic information (for example, the X-1 and D-558-1). Others were developed to assess the performance, behavior, and handling

---

23. For the history of DFRC and its predecessors, see the author's *On the Frontier: Flight Research at Dryden, 1946–1981* (Washington, DC: NASA SP-4303, 1984), revised in the late 1990s by Michael H. Gorn to cover later years.

Figure 11: The "Round One" Research Aircraft, 1953. From the lower left are the Bell X-1A, Douglas D-558-1, Convair XF-92A, Bell X-5, Douglas D-558-2, and Northrop X-4. At the center is the Douglas X-3. *HSFS Image E-2889*

qualities of a particular aerodynamic configuration, such as the D-558-2 and X-2 (transonic and supersonic swept wing), the X-4 (semitailless design), the X-5 (in-flight variable wing-sweeping), and the XF-92A (the transonic sharply swept delta wing). Despite its seeming "prototype fighter" designation, this latter aircraft was, in fact, a pure aerodynamic test bed originally conceived to support aerodynamic studies for a proposed rocket-ramjet-powered interceptor (the XP-92) that was never built. The Douglas X-3, intended for advanced Mach 2 turbojet propulsion testing, fell largely into the category of configuration explorers, as its performance (due to inadequate engines) never met its original performance goals, though it did furnish useful information on the behavior of slender aircraft having exceptionally high-fineness-ratio fuselages joined to thin, extremely low-aspect-ratio wings.

Flight experience with these aircraft generated a wide range of often-mixed results. On 14 October 1947, flown by Air Force test pilot Charles "Chuck" Yeager, the first Bell XS-1 reached Mach 1.06, the first piloted supersonic flight, following air-launch from a modified Boeing B-29 Superfortress bomber. The flight was considered so risky that, afterward, the Air Force even considered awarding the pilot the Medal of Honor.[24] The first supersonic flights of the first XS-1 (X-1) validated its thin-wing (8 percent thickness-chord ratio) design and adjustable (though not classically single-pivot) "all-moving" horizontal tail. Tests with the thicker-wing (10 percent thickness-chord ratio) second XS-1, and with the equally thick D-558-1, demonstrated that the seemingly small percentage nevertheless triggered early onset of a dramatic and energy-wasting drag rise in the transonic region.[25] While the second X-1, thanks to its rocket engine, could still muscle its way into the supersonic region, the D-558-1 could not, only reaching Mach 1 on a single occasion in a brief dive. The X-1 likewise demonstrated the value of an adjustable horizontal stabilizer at transonic speeds. Though not, strictly speaking, an "all-moving" horizontal tail of the sort used by early aviation pioneers and stipulated by Miles for its abortive M.52 transonic research airplane, the adjustable horizontal tail greatly increased the transonic longitudinal (pitch, that is, nose-up or nose-down) control authority enjoyed by a pilot. It was a lesson taken to heart and applied first to the North American F-86E Sabre in time for Korean combat against the MiG-15 (which lacked such a benefit). Since the adjustable stabilizer greatly improved the Sabre pilot's ability to turn tightly at transonic speeds, critical to "pulling lead" on a hard-maneuvering MiG so as to place the gunsight "pipper" so the Sabre's six .50 caliber machine guns could hose the target, it may be concluded that this NACA-recommended and proven control concept played a major role in ensuring the destruction of at least many of the 800+ MiG-15s shot down by F-86s during the Korean conflict. The adjustable horizontal stabilizer was subsequently supplanted by the single-pivot "slab" all-moving tail incorporated on virtually all American and foreign fighters and other high-performance aircraft beginning with the North American YF-100 Super

---

24. See Lieutenant General B. K. Chidlaw, U.S. Air Force, to General M. S. Fairchild, U.S. Air Force, 7 July 1948; Fairchild to Chidlaw, 7 July 1948; and Lawrence Bell to Chidlaw, 11 June 1948, all in the papers of Muir S. Fairchild, "Air Materiel Command 48-49" file, box 3, Manuscript Division, Library of Congress, Washington, DC. See also "Congressional Medal for Yeager," *Aviation Week* (21 June 1948): 58. Yeager justly received the Mackay Trophy for the flight and shared a Collier Trophy with Stack and Lawrence Bell. He later received a special Congressional Medal of Honor for the flight, akin to that awarded Lindbergh for his pioneering flight across the North Atlantic.
25. For the detailed story of these two programs, see the author's *Supersonic Flight: Breaking the Sound Barrier and Beyond—The Story of the Bell X-1 and Douglas D-558* (New York, NY: Macmillan in association with the Smithsonian Institution, 1972), reissued in an expanded version in 1997 by Brassey's, London, U.K.

Sabre and the Soviet Mikoyan MiG-19, the first supersonic "on the level" jet fighters of the United States and the Soviet Union.[26]

Slightly over six years after the Yeager flight, on 20 November 1953, flying the second Douglas D-558-2 (modified to all-rocket propulsion and air-launched from a Boeing P2B-1S mother ship, a "navalized" B-29), NACA research pilot A. Scott Crossfield extended the piloted supersonic domain past Mach 2, attaining Mach 2.01. The swept-wing D-558-2 and Bell X-2 had no difficulty exceeding Mach 1 but had many quirks of their own. The D-558-2 exhibited pronounced aerodynamic "pitchup" during transonic turns, a combination of swept-wing aerodynamic behavior and the high-fin placement of its horizontal tail. The lesson, as incorporated on subsequent American jet fighters such as the North American F-100, Republic F-105, and Grumman F11F-1 (and many foreign designs as well, such as the MiG-19, Folland Gnat, and English Electric Lightning F.Mk. 1), was to place the horizontal tail low on the aft section of the fuselage, rather than high on the vertical fin.[27]

The X-1 spawned an advanced family of Mach 2+ aircraft (the X-1A, X-1B, and X-1D), which had greater fuel capacity and revised propulsion and cockpit design. These, together with the Bell X-2 and Douglas X-3, exhibited dangerous coupled motion instability at supersonic speeds—combined lateral (rolling) and directional (yawing) motions. This behavior, predicted at Langley Memorial Aeronautical Laboratory by William Phillips in 1948, well before being encountered in flight, reflected both their design (aircraft with relatively short wingspans and long fuselages, thus loaded primarily along the fuselage and not the wing) and deterioration in directional stability characteristic with increasing Mach number. These experiences—nearly fatal with the X-1A and X-3—were fatal with the North American YF-100A Super Sabre, a new fighter aircraft too hastily introduced into service, and with the Bell X-2 rocket-propelled research airplane. On 27 September 1956, the latter claimed Milburn Apt, arguably then the Air Force's most experienced test pilot with coupled motion instability,

26. Walter C. Williams, "Instrumentation, Airspeed Calibration, Tests, Results, and Conclusions," in U.S. Air Force, *Air Force Supersonic Research Airplane XS-1*, Report No. 1 (9 January 1948), copy in NASA Historical Reference Collection, NASA History Division, NASA Headquarters, Washington, DC; W. C. Williams and A. S. Crossfield, "Handling Qualities of High-Speed Airplanes," Research Memorandum (RM) L52A08 (28 January 1952); James O. Young, *Meeting the Challenge of Supersonic Flight* (Edwards Air Force Base, CA: Air Force Flight Test Center [AFFTC], 1997). I also benefited from a 5 December 2008 conversation with Major General Fred J. Ascani, U.S. Air Force (ret.), who test-flew all variants of the F-86 and also the X-1, X-4, X-5, and XF-92A.

27. NACA Flight Test Progress Reports for the D-558-2 #2, dated 15 August 1949 and 14 November 1949; I also benefited from an interview with D-558-2 research pilot Robert Champine, 11 November 1971.

who perished on the world's first Mach 3 piloted flight when his X-2 tumbled out of control after attaining Mach 3.2. Inertial coupling was overcome only by increasing reliance upon stability augmentation technology, increasing the size of vertical and tail surfaces, and, eventually, the near-standardization upon twin vertical fins for high-speed aircraft, evident after 1970 with aircraft such as the F-14, F-15, F-18, MiG-29, and Su-27.[28]

Like the Messerschmitt Me 163 that had inspired it, the X-4 encountered serious longitudinal sine-wave-like "roller-coaster" motions at transonic speeds, which its pilots compared to driving over a washboard road. Fortunately, since its testing was conducted at higher altitudes under conditions of lower dynamic pressure ("low q" in engineering shorthand) and because it had large, rapidly opening aerodynamic speed brakes prudently incorporated in its design, it avoided the dangers Nazi pilots had routinely risked with the Me 163; it also avoided the even more calamitous flight experience of Britain's three de Havilland D.H. 108 Swallows, another Komet-inspired design, one of which disintegrated in September 1946 at Mach 0.87 during an abrupt divergent pitch at low altitude (and hence "high q"), killing pilot Geoffrey de Havilland, son of the firm's founder and one of Britain's most distinguished and experienced test pilots. The experience with the X-4, coupled with this earlier evidence of technical insufficiency leading to in-flight hazard, prevented the semitailless swept-wing configuration from becoming the "mainstream" design element of the transonic era that its proponents had hoped, unlike the conventional tailed swept wing or the triangular delta wing.[29]

The X-5 variable-sweep and XF-92A delta test beds explored wing configurations that did become standard design practice, though neither was as widely adopted as the generic swept wing. Like virtually all of the X series, neither was trouble- or quirk-free. The variable-sweep X-5's "single pivot" technical approach proved miscast—it necessitated the incorporation of a heavy "railroad

---

28. Joseph Weil, Ordway B. Gates, Jr., Richard D. Banner, and Albert E. Kuhl, "Flight Experience of Inertia Coupling in Rolling Maneuvers," RM H55WEIL (1955); HSFS, "Flight Experience With Two High-Speed Airplanes Having Violent Lateral-Longitudinal Coupling in Aileron Rolls," RM H55A13 (1955); Hubert M. Drake and Wendell H. Stillwell, "Behavior of the Bell X-1A Research Airplane During Exploratory Flights at Mach Numbers Near 2.0 and at Extreme Altitudes," RM H55G25 (1955); Hubert M. Drake, Thomas W. Finch, and James R. Peele, "Flight Measurements of Directional Stability to a Mach Number of 1.48 for an Airplane Tested with Three Different Vertical Tail Configurations," RM H55G26 (1955); Walter C. Williams and William H. Phillips, "Some Recent Research on the Handling Qualities of Airplanes," RM H55L29a (1956); Bell X-2 #1 Accident Report, copy in History Office archives, AFFTC, Edwards Air Force Base, CA.
29. See Williams and Crossfield, "Handling Qualities of High-Speed Airplanes."

track" within the airplane over which the wing roots could move to adjust for changes in center of pressure as the wing closed to maximum sweepback (wing root forward) or extended to minimum sweep position (wing root aft). Nevertheless, the X-5 validated the benefits of variable sweep, demonstrating that an airplane could fly with an extensible "straight" wing furnishing good low-speed takeoff and landing performance, adjustable to moderate sweep for good transonic performance, and adjustable to sharp sweepback for supersonic flight. In a fashion, variable wing sweeping was analogous to development of the controllable-pitch propeller in the interwar years, which had enabled the extraction of maximum engine performance across the entire range of the aircraft's performance envelope. Grumman prematurely attempted to develop a single-pivot variable sweep fighter, the XF10F-1 Jaguar, but without success, as it attempted to accomplish too much and had a poorly designed and unreliable power plant. In the mid-1950s, Langley Laboratory's William Alford and Edward Polhamus conceptualized (as did Barnes Wallis independently in Great Britain) the "outboard" wing pivot. This made variable sweep a practicality, as subsequently incorporated on aircraft such as the General Dynamics F-111, the Grumman F-14, the Rockwell B-1, and foreign equivalents, though its inherent complexity and associated weight penalty ensured that it was never adopted as broadly as the "fixed" swept wing.[30]

The XF-92A, built as a flying delta test bed, validated the thin, low-aspect-ratio triangular wing.[31] In contrast to popular myth, this aircraft and all the Convair delta aircraft that followed owed nothing to the delta wing research of Alexander Lippisch. Actually, Convair engineers conceptualized the delta for this design before becoming aware of Lippisch's work, and, in any case,

---

30. Edward N. Videan, "Flight Measurements of the Dynamic Lateral and Longitudinal Stability of the Bell X-5 Research Airplane at 58.7 Deg. Sweepback," RM H55H10 (1955); for its origins, see Robert Perry, "Variable Sweep: A Case History of Multiple Re-Innovation," RAND Study P-3459 (October 1966). I recall with pleasure many conversations with the late Robert Perry and acknowledge with grateful appreciation his contributions to my professional development and thinking on aircraft evolution. For an excellent technical survey, see Robert W. Kress, "Variable Sweep Wing Design," in *Aircraft Prototype and Technology Demonstrator Symposium*, ed. Norman C. Baullinger (Dayton, OH: AIAA, 1983), pp. 43–61. For subsequent adaptation of variable sweep on its most famous—or infamous, in the early years—subject, the "TFX" (F-111), see U.S. Cong., Senate Committee on Government Operations, *TFX Contract Investigation (Second Series)*, Pt. 3, Hearings, 15, 16, 22, 24, and 28 April 1970 (Washington, DC: GPO, 1970), pp. 537–538, and U.S. Cong., Senate, TFX Contract Investigation, Report No. 91-1496 (Washington, DC: GPO, 1970), pp. 5–16, 90. A useful international historical perspective is found in "The Annals of the Polymorph," Part 3, *Air International* 8, no. 5 (May 1975): 249–257.
31. Earl R. Keener and Gareth H. Jordan, "Wing Pressure Distributions Over the Lift Range of the Convair XF-92A Delta-Wing Airplane at Subsonic and Transonic Speeds," RM H55G07 (1955).

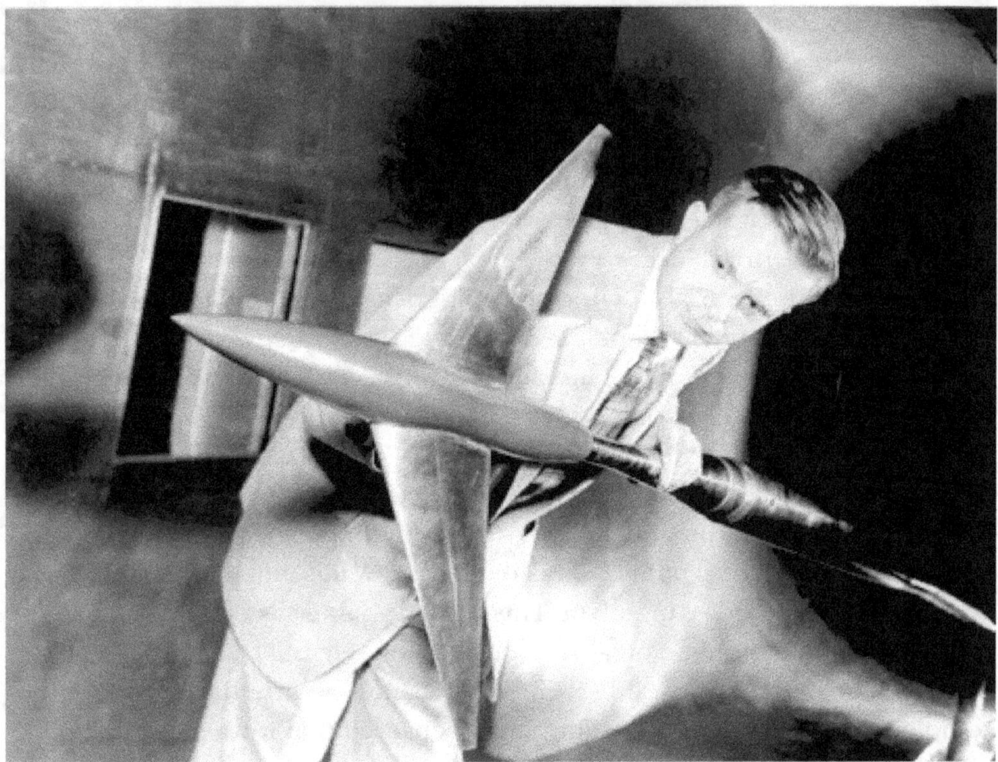

Figure 12: Richard T. Whitcomb with an area-ruled swept-wing research model. *NASA Image L-89119*

his technical approach was dramatically different from their own, with thick-section wings that NACA testing at Langley Laboratory proved completely unsuitable for the requirements of practical transonic and supersonic flight. Rather, the Convair delta was much more like Robert T. Jones's initial concept of the very thin delta in 1944 (indeed, for years, he used his original wind tunnel model of the wing as a letter opener, a testament to its "dagger"-like qualities).[32] However, in one important respect, the XF-92A contributed markedly to transonic design theory, for its drag rise was so high that, despite its refined aerodynamic configuration, it could only marginally attain supersonic speed, and then only in the course of a high-speed dive. Its "operational" successor, the prototype Convair YF-102 Delta Dagger interceptor, experienced an identical problem, as did a prototype naval fighter, the Grumman F9F-9

32. I wish to acknowledge with grateful appreciation a conversation with Dr. Robert T. Jones at ARC, Sunnyvale, CA, on 14 July 1977; at that time, he showed me his original "dagger" tunnel model and graciously donated it for the collections of NASM.

Tiger, disconcerting military planners who recognized that both planes had little prospect of achieving their planned supersonic dash speed. The XF-92A, YF-102, and F9F-9 Tiger exemplified the additive properties of transonic airframe drag rise, in which the summation of drags and their interactions was greater than the summation of the drag of individual components. It was a problem overcome only with development of a new concept of body shaping, the concept of "area ruling."[33]

The area rule stemmed from the work of research engineer Richard T. Whitcomb, who drew upon a new NACA tool, the transonic slotted-throat wind tunnel. The product of Langley's Ray Wright, John Stack, and a team of other researchers, the slotted-throat tunnel largely (though not completely) overcame the problem of tunnel choking that had so plagued prewar investigators and triggered development of the X series. Whitcomb exploited both his experience in the new tunnel and exposure to a seminal lecture to Langley Laboratory staff by Adolf Busemann (who had been brought to America under Project Paperclip, the exploitation program for incorporating Nazi scientists and engineers in American aeronautics and rocketry development) to derive a refined appreciation of transonic drag rise (transonic wave drag) and how it might be mitigated by proper shaping of the wing-fuselage juncture. Whitcomb realized that, for analytical purposes, an airplane could be reduced to a streamlined body of revolution, elongated as much as possible to mitigate abrupt discontinuities and, hence, equally abrupt drag rise. When protrusions occurred—for example, the wing projecting from the fuselage—the fuselage itself could be narrowed, so that the cross-sectional area development of the design remained at a minimum. Whitcomb called his concept "area ruling:" experimentally verified by further tunnel testing, by tests of large-scale rocket-propelled models fired from the Wallops Island Pilotless Aircraft Research Division, and by flight testing with redesigned F-102 and F9F-9 aircraft (the latter redesignated the F11F to signify its extensive reshaping), area ruling's "wasp-waisting" became one of the key visible markers of transonic and supersonic aircraft design, most notably, perhaps, on the Republic F-105 Thunderchief nuclear strike fighter and on the Northrop N-156 prototype that spawned both the lightweight F-5 fighter and the ubiquitous T-38 supersonic trainer. It certainly saved the Air Force's "1954 Interceptor" program: it reduced transonic wave drag by

33. See Lieutenant General L. C. Craigie, U.S. Air Force, to General Hoyt S. Vandenberg, U.S. Air Force, regarding "Engineering Status of Fighter Airplanes," 18 March 1953, "ESOFA" file, box 85, papers of Hoyt S. Vandenberg, Manuscript Division, Library of Congress, Washington, DC.

Figure 13: The Convair YF-102 Delta Dagger interceptor, as originally flown without area ruling. *Convair*

Figure 14: The YF-102A, with area ruling, longer length (hence higher fineness ratio), and shock bodies flanking the tail cone. *AFFTC*

approximately a third (comparing full-scale test results with the initial YF-102 prototype against those of the area-ruled F-102A), turning the troubled F-102 into a success, earning plaudits for the NACA and a richly deserved Collier Trophy for Whitcomb himself.[34]

The "Round One" research airplanes contributed significantly to fundamental understanding in four major areas and thus, as well, to future aircraft design practice. The price for this was nine aircraft destroyed, with five aircrew killed (four pilots and one on-board technician) and others injured. Ironically, only one (the X-2 #1, lost in 1956 from inertial coupling at nearly Mach 3.2) came close to "pushing the envelope" at high Mach. Seven were propulsion-related losses. Four rocket-propelled research aircraft (the X-1 #3, X-1A, X-1D, and X-2 #2) exploded on the ground or in the air from frozen leather seals contaminated with tricresylphosphate detonating under the jolt of pressurization. These catastrophes—the cause of which took far too long to identify—claimed two Boeing EB-50 Superfortress launch airplanes as well. The seventh was a D-558-1 #2 that crashed on takeoff due to turbine disintegration that severed its control lines. The X-5 #2 crashed when its pilot inadvertently entered an unrecoverable (and fatal) spin.

For this price, the program returned vital knowledge and stimulus within the following:

- *Aerodynamics*, including validating and interpreting tunnel test data, aerodynamic heating, lift and drag studies, and inlet and duct studies.
- *Flight loads*, including load distribution with increasing Mach number, the effect of wing sweep on gust loadings, gustiness at high altitudes, buffeting from transonic maneuvering, transonic and supersonic aeroelastic structural effects, and the effect of stability reduction upon flight loads.
- *Stability and control*, including longitudinal (pitch) control over the transonic/supersonic range, the effectiveness of blunt trailing edge control

34. Richard T. Whitcomb, "A Study of the Zero-Lift Drag-Rise Characteristics of Wing-Body Combinations Near the Speed of Sound," NACA RM L52H08 (1952); Richard T. Whitcomb and Thomas C. Kelly, "A Study of the Flow Over a 45 Deg. Sweptback Wing-Fuselage Combination at Transonic Mach Numbers," RM L52D01 (1952); Richard T. Whitcomb, "Some Considerations Regarding the Application of the Supersonic Area Rule to the Design of Airplane Fuselages," RM L56E23a (1956); Edwin J. Saltzman and Theodore G. Ayers, *Selected Examples of NACA/NASA Supersonic Flight Research* (Edwards, CA: SP-513, DFRC, 1995), p. 15, fig. 7; Lane E. Wallace, "The Whitcomb Area Rule: NACA Aerodynamics Research and Innovation," in *From Engineering Science to Big Science: The NACA and NASA Collier Trophy Research Project Winners*, ed. Pamela E. Mack (Washington, DC: NASA SP-4219, 1998), pp. 135–148.

surfaces, the alleviation of aerodynamic pitch-up by wing devices, the effect of the inertial axis upon lateral stability, exhaust jet impingement and directional stability, inertial (roll-yaw-pitch) coupling, directional instability at increasing Mach numbers, and the use of reaction control thrusters in low dynamic pressure ("low q") flight.

- *Operations*, including high-speed flight-test exploration, speed loss in maneuvering flight, high-altitude problems, pressure suit research and use, airspeed measurement, and variable wing-sweep over the transonic range.
- *Aircraft design practices*, including employing the thin, low-aspect-ratio swept-and-delta wing planform as a "normative" configuration for transonic and supersonic flight; using refined aerodynamic "fixes" and design approaches such as vortex generators on wing and tail surfaces and extended "saw-tooth" leading edges; incorporating adjustable horizontal stabilizers for transonic longitudinal control (and the all-moving tailplane subsequently); relocating the horizontal tail to the lower aft fuselage from its traditional position either at the top of the fuselage or, in the early jet era, at the midfin or top-of-the-vertical fin position; increasing vertical fin areas, adding ventral fins, and adopting twin vertical fins; employing variable wing-sweeping to impart good low-speed, transonic, and supersonic aerodynamic performance; employing area ruling on new transonic designs; and stimulating development and application of improved flight control technology, particularly stability augmentation systems to overcome transonic and supersonic inertial coupling, reduced directional stability, and longitudinal trim changes resulting from fluctuating shock wave and center of lift positioning.

Beyond this, by painstakingly evaluating the nuances and quirks of many new aircraft configurations and then assessing them against existing design specification criteria, the NACA identified serious and growing divergences in new aircraft performance away from desired behavior criteria, particularly those dealing with roll (lateral) and yaw (directional) stability.[35] Such documented research led to revised specification requirements, accelerated appreciation for the need for artificial stability augmentation, and resulted in the derivation of new quantitative criteria for evaluating aircraft performance from the

35. For example, see figures 7 and 8 in the previously cited Williams and Phillips, "Recent Research on the Handling Qualities of Airplanes," RM H55L29a, and an earlier Ames investigation, Charles J. Liddell, Jr., Brent Y. Creer, and Rudolph Van Dyke, Jr., "A Flight Study of Requirements for Satisfactory Lateral Oscillatory Characteristics of Fighter Aircraft," RM A51E16 (1951).

"cockpit" standpoint, namely the now-universally employed "Cooper-Harper Rating." Named for Ames Laboratory research pilot George Cooper and Cornell Aeronautical Laboratory research pilot John Harper, the Cooper-Harper scale grew out of Ames and Cornell research with early variable-stability research airplanes, the former with a modified Grumman F6F Hellcat and the latter with a modified Vought F4U Corsair and, subsequently, a Douglas B-26 Invader and Lockheed NT-33 Shooting Star. Translating pilot reaction to modifications to the aircraft's flight control system (and hence performance) into quantifiable data proved so challenging that both Cooper and Harper began searching for a more precise means of ensuring that test pilots, worldwide, could "speak" the same language when evaluating the response and utility of new aircraft designs. Over nearly 20 years of fruitful collaboration, they derived, reviewed, and implemented a scale that NASA released in 1969; it was swiftly and universally adopted across the aerospace community.[36] The Cooper-Harper rating scale, as much a fixture of test pilot training and practice as reliance upon computer-based data reduction, must be counted as one of the NACA and NASA's most significant aeronautical accomplishments in the supersonic-hypersonic era.

As hinted at the beginning of this essay, by the time of the creation of NASA, the "pioneering days" of supersonic flight had passed, and the "macro" performance boundaries of future transonic and supersonic aircraft—to Mach 0.82+ at over 40,000 feet for commercial air transports, and to Mach 2+ at over 60,000 feet for the most advanced military aircraft—were well established. Generally speaking, a half century later, they remain unchanged. From this point on, emphasis would be upon refining explicit aircraft performance parameters and capabilities within these general boundaries, for example, delaying transonic shock formation by tailored supercritical airfoils; improving supersonic lift-to-drag and cruise efficiencies; refining aerodynamic-structural-propulsion integration; enhancing control efficiencies as evidenced by the fly-by-wire revolution; exploring exploitation of advanced electronic stability and control architectures with relaxed-stability (or even inherently unstable) aircraft configurations made possible by the composite revolution to generate previously unattainable designs, such as optimized low observable (LO)

36. George E. Cooper and Robert P. Harper, Jr., "The Use of Pilot Rating in the Evaluation of Aircraft Handling Qualities," NASA TN D-5153 (1969). For an excellent discussion on the background of Cooper-Harper, see George Cooper, Bob Harper, and Roy Martin, "SETP Panel on Pilot Rating Scales," *XLVIII Symposium Proceedings: 2004 Report to the Aerospace Profession* (Lancaster, CA: Society of Experimental Test Pilots, 2004), pp. 319–337. I have also benefited from conversations with George Cooper, Bob Harper, and the late Waldemar "Walt" Breuhaus, father of American V-Stab research at Cornell.

(e.g., "stealth" reduced radar cross-section [RCS]) aircraft, high-aspect-ratio "spanloaders," and highly agile transonic and supersonic aircraft (exemplified by the X-29 and X-31); enhancing the thrust-to-weight and reliability of the gas turbine power plant itself; searching for cleaner, more efficient high-performance engines; and tailoring supersonic aircraft shapes to reduce sonic boom formation and impingement. This continuing refinement defined NASA's aeronautics endeavors in the transonic and supersonic field in the post-1958 period, replacing the "epic" search for solutions and basic knowledge that had characterized the work of the NACA in the "crisis" days of early transonic and supersonic exploration when the transonic slotted throat tunnel was a thing of the future and the transonic-supersonic research airplane the most reliable (if risky) means of securing "real world" data.

While continuing refinement of ground test, flight test, and predictive methodologies (such as the advent of computer-based CFD aerodynamic modeling and structural prediction via programs such as NASTRAN and the Flexible Airplane Analysis Computer System [FLEXSTAB]) greatly assisted this refinement process, the traditional intuitive insight found within NASA's largely empirical research process proved consistently most valuable. Over the 43-year history of the NACA and the 50-year history of NASA that has followed, the Agency has maintained a developmental approach that is, at once, emulative, opportunistic, innovative, integrative, and, at key times, fortunate. NACA and NASA researchers have proven quick to seize ideas and concepts derived elsewhere (emulation); take advantage of circumstances and the ability to be a "fast second" (technological opportunism); and add insightful work, features, or capabilities (innovation), all while maintaining a generally broad "total system" focus and analytical approach (integration). They were fortunate to be able to do this in the 1930s, adapting and exploiting technology available to other nations, but not pursued by them because they generally lacked the readily available resources and wealth then available to the United States. The same situation occurred after World War II and even at the height of the Cold War that succeeded it. Where fortune failed, however—as with the failure to recognize the significance of the swept wing at the Volta Conference, the failure to pursue the gas turbine until confronted with evidence of foreign success, or the failure to anticipate both Sputnik and its likely impact—the price in delayed research and practical implementation, and certainly the price in the Agency's reputation as a center of technical excellence, was high.

Generally speaking, however, the NACA's and NASA's work in the difficult years of the transonic and supersonic era was overwhelmingly excellent, as evidenced by the frequent requests by foreign governments and research establishments for reports and familiarization visits. Continuing a trend found in global aeronautics in the late 1930s, both foreign and American companies in the postwar years generally looked to the NACA as the recognized global authority (certainly in the West) on aircraft design and research. On the other hand, what many of these companies *did* with the products of NACA research was quite something else, with companies and even the military services often entranced with the technologically fanciful at the expense of the militarily and commercially justifiable.

For example, of 25 Air Force fighter development programs contemplated or actually executed between 1946 and 1972, only 3 could be considered unqualified successes; of the rest, 10 required extensive work to be made into satisfactory aircraft, and 12 others were either unrealistic or mission-limited disappointments. (The Navy experience over the same period is equally bleak, with 3 successes, 6 "made to work," and 10 unrealistic or incapable of meeting desired requirements).[37] Among these are Mach 3+ interceptors (too temperamental and complex to meet the "no notice" alert requirement), sled-launched rocket-boosted ramjets (utterly unrealistic), tail-sitting vertical short takeoff and landing (VSTOL) fighters (also unrealistic), and flashy high-performance speedsters of limited range, duration, and weapons carriage. Bombers went down an equivalent path with the Mach 2+ B-58 and the experimental Mach 3+ XB-70A: neither proved adequate for the threat environment each was likely to encounter, and neither outlasted older legacy systems—notably the venerable B-52—that proved more flexible and adaptable. High-speed knowledge and acumen, in short, did not guarantee industry and the military services the ability to develop satisfactory operational aircraft. The need for blending technology with appropriate doctrine and defined operational requirements was ignored

---

37. The three successes in this time period are the F-86, F-4, and F-5. The 10 troublesome "made to work" are the F-84 (both straight and swept wing), the F-86D (an interceptor derivation so different from the F-86 as to be essentially a new airplane [and which was known, albeit briefly, as the YF-95, a separate designation it likely should have retained]), F-89, F-94, F-100, F-101, F-102, F-105, F-106, and F-111A, D, E, and F. The unrealistic or mission-limited include the XF-85, XF-87, XF-88, XF-90, XF-91, XP-92, YF-93, F-103, F-104, F-108, F-109, and YF-12. Of the Navy aircraft over the same time period, the F8U (F-8), F4H (F-4), and F-14 constitute genuine successes (though the latter was never as well developed afterward as it should have been). The F9F (F-9, both straight and swept wing), F3D (F-10), F2H (F-2), F4D (F-6), FJ (F-1), and AV-8 constitute "made to work" aircraft that required extensive developmental work but were generally (some, like the F2H, very) useful. The FR-1, F2R-1, XFY-1, XFV-1, XF2Y-1, XF10F-1, F5D-1, XF8U-3, F3H (F-3), F11F (F-11), and F-111B were unrealistic or operationally deficient aircraft. Designations in parentheses reflect the post-1962 McNamara-era rationalization of AF, USN, and USMC fighter designations, a rationalization that, like many of his policies, likewise proved a disappointment.

in favor of some other value (speed, rate of climb, or exotic propulsion), resulting in wasted effort. The same, of course, was evident in the civilian world, with the great effort expended on developing a Mach 2.7 supersonic transport. Under the aegis of the Supersonic Commercial Air Transport program, the NACA and NASA, building upon Langley Laboratory work with the outboard-pivot variable-sweep wing, Whitcomb's own continued refinement of supersonic wing shaping, and Ames Laboratory development of slender delta and compression lift theory (manifested subsequently in the XB-70 development effort), greatly influenced both federal government and contractor expectations of what such a supersonic transport should be.[38] But a host of factors—fuel economy, performance efficiency, likely utilization, environmental concerns, political circumstances, and antitechnological bias—worked to frustrate both their efforts and those of supersonic transport partisans outside the Agency, derailing the program and not even allowing it to proceed to prototype development. By the late 1970s, NASA was moving in a far different direction: following rising fuel prices—airline jet fuel costs tripled between 1973 and 1975, following the Organization of Petroleum Exporting Countries (OPEC) quadrupling the price of fuel in 1973 and then tripling that price in following years—the Agency embarked on an ACEE program to improve fuel efficiency, one emphasizing economical and firmly transonic performance.[39]

From 1944 to 1945, Jones had enunciated and promulgated the American swept wing; in the late 1940s, Wright and Stack had generated the slotted throat transonic tunnel; in the early 1950s, Whitcomb evolved area ruling. All had materially enhanced the transonic behavior of high-performance aircraft. As well, all spoke to improving the various efficiencies of flight—increasing transonic lift-to-drag values, reducing drag in general, and increasing speed (hence the relative propulsive efficiency of the airplane) and range. By the late 1950s, the United States had emerged as the dominant global jet power: with global-ranging commercial transports, such as the Boeing 707 and Douglas DC-8, reshaping the nature of international air commerce and global-ranging transonic bombers, such as the B-47 and B-52, maintaining one of the three pillars of a nuclear triad—aircraft, land-based missiles, and missile submarines—the United States

---

38. For example, see Richard T. Whitcomb and John R. Sevier, "A Supersonic Area Rule and an Application to the Design of a Wing-Body Combination with High Lift-Drag Ratios," TR R-72 (1960).
39. See Erik M. Conway, *High-Speed Dreams: NASA and the Technopolitics of Supersonic Transportation, 1945–1999* (Baltimore, MD: Johns Hopkins University Press, 2005); Mel Horwitch, *Clipped Wings: The American SST Conflict* (Cambridge. MA: MIT Press, 1982); and Jeffrey L. Ethell, *Fuel Economy in Aviation* (Washington, DC: NASA SP-462, 1983).

came to characterize the "high end" of the spectrum of combat addressed by America's Cold War military forces. A decade later, the fruition of this interest in the transonic had manifested itself in the design of new generations of jet transports to meet a broader range of needs, from business aviation through regional air transport, and on to mass global air movement, the latter exemplified by the Boeing 747, first of the so-called "jumbo" jets; the Boeing 747 carried at times over 400 passengers and would take global air mobility considerably beyond the era of the 140-passenger Boeing 707, though it experienced a protracted period of sluggish sales, delay, and rising costs, coincident with an air transport crisis inflamed, in part, by rising fuel costs.[40]

Again, the creativity of Richard Whitcomb would work to reshape the airplane to meet the challenge of more efficient and less costly flight. In 1941, von Kármán, looking for a method whereby the rigor of mathematical theory could assist "interpretation of experimental research and guidance in design," concluded his essay by noting an international body of research strongly suggested "that careful theoretical and experimental research might be able to push the velocity of flying nearer to the velocity of sound than is possible now. *The mere fact that the air passes over a wing with supersonic velocity does not necessarily involve the occurrence of a compressibility burble and energy loss by shock wave.*"[41] While, as John Becker noted bemusedly years later, the meanings of "these wise words were lost" for the next quarter century, nevertheless, finding a way to delay or minimize shock formation—to raise the critical Mach number of an airfoil, the speed where it forms a shock, with consequent drag rise, loss of lift, and formation of turbulent, separated flows—consumed much creative energy of the NACA's and subsequently NASA's transonic aerodynamicists.[42]

Out of these came three more distinctive Whitcomb contributions: the trailing-edge shock body, the supercritical wing (SCW), and the wingtip winglet. While the former did not become a mainstream design element—though it appeared on Convair's unsuccessful (for reasons other than Whitcomb) 990 jetliner—the latter, both the SCW and wingtip winglet, became (and will undoubtedly remain) standard design elements incorporated in military and civil

40. See Richard P. Hallion, "Commercial Aviation: From the Benoist Airboat to the SST, 1914–1976," in *Two Hundred Years of Flight in America: A Bicentennial Survey*, ed. Eugene M. Emme, vol. 1 of the American Astronautical Society History Series (San Diego, CA: American Astronautical Society, in association with the NASM and the Society for the History of Technology, 1977), pp. 169–171.
41. See von Kármán, "Compressibility Effects in Aerodynamics," 337, 355.
42. See Becker, *High-Speed Frontier*, p. 59.

Figure 15: The NASA F-8 SCW test bed, with a high-aspect-ratio transport-optimized wing planform. *NASA Image ECN 3468*

aircraft alike. Following a four-year tunnel research program, NASA validated the SCW with two series of flight tests. North American Rockwell modified a straight-wing T-2C trainer to incorporate an SCW for initial low-speed testing, in analogous fashion to the Bell L-39's earlier testing of a 35-degree swept wing in anticipation of its being incorporated on the prototype F-86 and B-47 a quarter century before.[43]

Taking advantage of two easily convertible military aircraft, the high-wing Vought TF-8A Crusader and the variable-sweep General Dynamics F-111A

---

43. Richard T. Whitcomb, "Special Bodies Added on a Wing to Reduce Shock Induced Boundary-Layer Separation at High Subsonic Speeds," TN-4293 (1958); John T. Kutney and Stanley P. Piszkin, "Reduction of Drag Rise of the Convair 990 Airplane," *Journal of Aircraft* 1, no. 1 (January–February 1964): 8–12; Richard T. Whitcomb, "Research Associated with the Langley 8-Foot Tunnels Branch," TM 108686 (1970); Flight Research Center, "Supercritical Wing Technology: A Progress Report on Flight Evaluations," SP 301 (1972); Stuart G. Flechner, Peter F. Jacobs, and Richard T. Whitcomb, "A High Subsonic Speed Wind-Tunnel Investigation of Winglets on a Representative Second-Generation Jet Transport Wing," TN D-8264 (1976); Stuart G. Flechner and Peter F. Jacobs, "Experimental Results of Winglets on First, Second, and Third Generation Jet Transports," NASA CP-2036, Pt. 1 (1978).

"Aardvark," NASA undertook extensive transonic investigation of the SCW concept. Technicians first fitted the F-8 with a high-aspect-ratio airliner-style wing, subsequently modifying the F-111 with a lower-aspect-ratio wing (a kind likely to be incorporated on future transonic strike aircraft) after the TF-8A tests from early 1971 into mid-1973 confirmed the benefits of the SCW concept. The F-111A tests, under the aegis of the Transonic Aircraft Technology (TACT) program, a joint NASA-Air Force study effort, demonstrated that the SCW concept could be applied to a strike aircraft, markedly improving its transonic performance—the modified F-111 had twice the lift of a conventional "Aardvark" at transonic speeds—while not hindering its ability to dash to supersonic speeds above Mach 1.3 as well. Whitcomb's winglets, conceived to reduce the drag attending the powerful wingtip vortices created by the wing as it generated lifting forces, underwent their own flight validation from 1979 to 1981; since that time they have appeared on many transonic transport aircraft, reducing the energy required to fly close to the speed of sound. The sum of Whitcomb's creativity—area rule, SCWs, and winglets—marks him as the most influential refiner of high-speed aerodynamic design over the last century of flight.[44]

Into the Hypersonic

As aircraft flew ever faster, a new challenge emerged: aerodynamic heating. It constituted more than merely an issue of finding new materials that could withstand the increased flight loads and temperatures imposed by air vehicles attaining high velocities within the atmosphere. While most attention had focused on the weakening of materials and their eventual melting, as NACA Director Hugh Dryden and Structures Research Chief John Duberg noted as early as 1955, "We now realize that long before a skin temperature is reached at which these effects occur, aerodynamic heating will give rise to serious structural problems."[45] One of the most critical was the combination of aerodynamic heating and thermodynamic effects, or "aerothermodynamics," leading to expansion and distortion of structures and greatly complicating the challenge facing structural designers. Used to configuring structures to withstand the various loads associated with lift, drag, and dynamic pressure, they now had to address the deformation and

---

44. See *On the Frontier* for details on these flight-test validations.
45. Dr. Hugh L. Dryden and Dr. John E. Duberg, "Aeroelastic Effects of Aerodynamic Heating," *Proceedings of the Fifth AGARD General Assembly: The Canadian AGARD Conference, 10–17 June 1955* (Paris, France: Advisory Group for Aeronautical Research and Development, 1955), p. 102.

Figure 16: Ames research scientist H. Julian Allen. *NASA Image A-22664*

stressing of a structure simply because of its expansion in an increasingly hot environment. Understanding and accommodating such heating was a particular challenge on the Lockheed Blackbird flight development program, and it is a remarkable tribute to Clarence "Kelly" Johnson, Ben Rich, and the rest of the Lockheed "Skunk Works" design team associated with this remarkable aircraft that they were able, at a time of very little knowledge, to conceive, produce, and place in service an aircraft facing as daunting and unknown a flight environment as the X-1 had in the 1940s. High-supersonic aerodynamic heating studies constituted one of the key research "targets" for NASA researchers when they had the opportunity to use two early Blackbirds for a decade of concentrated supersonic cruise research beginning in the late 1960s.[46] Heating problems, challenging enough with the Blackbird, became acute with the hypersonic Mach 6+ North American X-15. On one flight, it experienced heating severe enough to shatter a cockpit panel. On another flight, in October 1967, heating nearly led to loss of the X-15 #2 when unexpected localized heating effects seriously damaged its structure, causing a dummy scramjet test article to separate from the craft;

46. A story very well related in Peter W. Merlin's *Mach 3+: NASA/USAF YF-12 Flight Research, 1969–1979*, Monographs in Aerospace History, No. 25 (Washington, DC: NASA SP-2001-4525, 2002).

they also damaged its fuel jettison system, forcing a "heavyweight" landing that might have had—but fortunately did not—disastrous results.[47]

If drag rise was the great challenge of transonic flight, and stability and control challenged supersonic flight, heating posed—and poses still—the great challenge to practical hypersonic flight. Although often considered of recent development, interest in hypersonic flight predated both the invention of the airplane and the beginning of the Space Age; hypersonic systems featured in the speculative ruminations of pioneers such as the Russian Konstantin Tsiolkovskiy, the German-Rumanian Hermann Oberth, and the American Robert Goddard.[48] As a "first step" in 1924, rocket enthusiast Max Valier recommended adding rockets to conventional aircraft and then progressing from these to hypersonic "ether planes" capable of flying around the world.[49] The Austrian Eugen Sänger and Irene Bredt first posited a quasi-realistic hypersonic design, their Silbervogel ("Silver Bird") of 1938, developed as a dual-use space transporter and global strike aircraft, sled-launched off a monorail.[50] The Silbervogel was an extraordinarily influential design study, much read and analyzed after the war. It constituted the first analysis of requirements for a hypersonic single stage to orbit (SSTO) vehicle, the first postulation of the flat-bottom "laundry iron" shape, and lifting reentry theory; and, via multiple translations, it inspired foreign emulation in the United States, Europe, and the USSR.

---

47. Joseph Weil, "Review of the X-15 Program," TN D-1278 (1962); Wendell H. Stillwell, *X-15 Research Results with a Selected Bibliography* (Washington, DC: GPO SP-60, 1965); Albert L. Braslow, "Analysis of Boundary Layer Transition on X-15-2 Research Airplane," TN D-3487 (1966); John V. Becker, "The X-15 Program in Retrospect," 3rd Eugen Sänger Memorial Lecture, Bonn, Federal Republic of Germany, 4–5 December 1968; Milton O. Thompson (with J. D. Hunley), *Flight Research: Problems Encountered and What They Should Teach Us*, Monographs in Aerospace History, No. 22 (Washington, DC: NASA SP-2000-4522, 2000).
48. A. A. Blagonravov, ed., *Collected Works of K. E. Tsiolkovskiy*, vol. 2: *Reactive Flying Machines*, NASA TT-F-237 (Washington, DC: NASA, 1965), pp. 528–530; Hermann Oberth, *Die Rakete zu den Planetenräumen* (Munich, Germany: Verlag von R. Oldenbourg, 1923), pp. 36–39, 49–51, 57–58, 63–64; Robert H. Goddard, "A Method of Reaching Extreme Altitudes," in *The Papers of Robert H. Goddard, vol. 1: 1898–1924*, ed. Esther C. Goddard and G. Edward Pendray (New York, NY: McGraw-Hill Book Company, 1970), pp. 337–406; Frank H. Winter, *Prelude to the Space Age—The Rocket Societies: 1924–1940* (Washington, DC: Smithsonian Institution Press, 1983).
49. I. Essers, *Max Valier: A Pioneer of Space Travel* (Washington, DC: NASA TT-F-664, 1976), pp. 81–97, 130–135, 248.
50. Eugen Sänger, "Neuere Ergebnisse der Raketenflugtechnik," *Flug: Zeitschrift für das gesamte Gebiet der Luftfahrt* 1 (December 1934): 11–16, 19–22; Eugen Sänger and Irene Bredt, *A Rocket Drive for Long-Range Bombers*, Translation CGD-32 (Washington, DC: Technical Information Branch, U.S. Navy Bureau of Aeronautics, 1952); Irene Sänger-Bredt, "The Silver Bird Story: A Memoir," in *Essays of the History of Rocketry and Astronautics: Proceedings of the Third Through the Sixth History Symposia of the International Academy of Astronautics*, vol. 1, ed. R. Cargill Hall (Washington, DC: NASA, 1977), pp. 195–228.

Hypersonics first assumed significance for the problem of ballistic missile warhead reentry; and warhead reentry research, begun by the military with programs such as the Lockheed X-17, naturally benefited the civil human space program of the 1960s and the development of the Mercury-Gemini-Apollo spacecraft families. Crucial to the progression of practical hypersonics was the derivation and exploitation of blunt body reentry theory. In the early 1950s, Ames researcher H. Julian Allen postulated that a blunt shaped reentry body would form a detached shock carrying off much of the heat of reentry. Blunt body theory made possible the ICBM warhead and inhabited spacecraft such as Mercury, Gemini, and Apollo. While Mercury was purely ballistic, Gemini had a modest hypersonic lift-to-drag ratio (L/D) of 0.25, and the Apollo Command Module had a hypersonic L/D of 0.60. In April 1964 and May 1965, anticipating the challenge of Apollo's return to Earth from lunar missions, NASA researchers flew Project Flight Investigation Reentry Environment (FIRE), an Atlas-lofted reentry vehicle that took calorimetric measurements of a spin-stabilized Apollo-like blunt body reentering the atmosphere at over 7 miles per second. In 1968, NASA flew Reentry-F, a slender conical reentry body that completed a Mach 20 reentry on 27 April 1968 furnishing heat transfer and hypersonic boundary layer transition data of value decades later. The greatest contribution of all these programs, whether by NASA or other agencies, was highlighting differences between predicted and actual performance, illustrating (as in the earlier transonic and supersonic era) the great need for more accurate test facilities, simulation techniques, and predictive tools.[51]

H. Julian Allen likewise studied the problems and potentialities of lifting hypersonic craft, postulating various Sänger-like boosted winged vehicles reentering via a "skipping" reentry profile; in December 1957 (just 10 weeks

---

51. Edwin P. Hartman, *Adventures in Research: A History of the Ames Research Center, 1940–1965* (Washington, DC: NASA SP-4302, 1970), pp. 215–218, 266–270, 294–298, 359–363, 451–452; H. Julian Allen, "The Aerodynamic Heating of Atmospheric Entry Vehicles," in *Fundamental Phenomena in Hypersonic Flow: Proceedings of the International Symposium Sponsored by Cornell Aeronautical Laboratory*, ed. J. Gordon Hall (Ithaca, NY: Cornell University Press, 1966), pp. 5–29, esp. pp. 6–10; E. P. Smith, "Space Shuttle in Perspective: History in the Making" (AIAA 75-336, 11th Annual Meeting of the AIAA, Washington, DC, 24–26 February 1975); D. L. Cauchon, *Project FIRE Flight 1 Radiative Heating Experiment* (Washington, DC: NASA TM-X-1222, April 1966); Elden S. Cornette, *Forebody Temperatures and Calorimeter Heating Rates Measured During Project FIRE II Reentry at 11.35 Kilometers Per Second* (Washington, DC: NASA TM-X-1305, November 1966); P. Calvin Stainback, Charles B. Johnson, Lillian R. Boney, and Kathleen C. Wicker, *A Comparison of Theoretical Predictions and Heat-Transfer Measurements for a Flight Experiment at Mach 20 (Reentry F)* (Washington, DC: NASA TM-X-2560, July 1972); W. A. Wood, C. J. Riley, and F. M. Cheatwood, *Reentry-F Flowfield Solutions at 80,000 ft.* (Hampton, VA: NASA TM-112856, LaRC, May 1997); Philippe H. Adam and Hans G. Hornung, "Enthalpy Effects on Hypervelocity Boundary Layer Transition: Experiments and Free Flight Data" (AIAA 97-0764, AIAA 35th Aerospace Sciences Meeting, Reno, NV, 6–9 January 1997).

after Sputnik) he optimistically concluded "the present situation is certainly analogous to that which the Wright brothers faced at the turn of the century. If we give the same painstaking and intelligent treatment to our problems as they gave to theirs a half century ago, our success seems assured."[52] This thinking, accompanied by various industry studies and military efforts, eventually spawned the abortive X-20 Dyna-Soar, "Round Three" in America's postwar research aircraft family. But first was "Round Two," the world's first piloted hypersonic aircraft, the North American X-15. The X-15 sprang from the digestion of the Sänger-Bredt report and subsequent studies resulting from it that led, in June 1952, to a decision by the NACA's leadership to expand the agency's research aircraft program to investigate hypersonic flight. The agency formed a study committee under Clinton Brown; and, at one point, team members proposed modifying the delayed X-2 with reaction controls and strap-on boosters so that it could serve as a high-supersonic/low-hypersonic research airplane, though, for a variety of reasons, the agency did not proceed further with this idea.[53]

In 1954 a team headed by John V. Becker of the NACA Langley Memorial Aeronautical Laboratory (now the NASA Langley Research Center) derived the basic X-15 configuration, stipulating a nickel-alloy Inconel structure, relatively conventional wing configuration, a rocketlike four-surface tail, and "off the shelf" rocket engines, in this case from the Hermes rocket (a V-2 program derivative). This influential study triggered development of the transatmospheric X-15, which first flew in 1959. Powered by a 57,000-pound thrust throttleable rocket engine, the X-15 extended piloted flight through Mach 3 and 4 and on to 5 and 6 and beyond, completing 199 flights by 12 pilots and reaching an altitude of 67 miles in 1963; it reached Mach 6.70 in 1967.[54]

X-15 researchers pursued aerodynamic and structural heating investigations through 1963, following these by using the X-15 to carry experiments into the upper

---

52. H. Julian Allen, "Hypersonic Flight and the Re-Entry Problem," 31st Wright Brothers Lecture (Sunnyvale, CA: Ames Aeronautical Laboratory, 17 December 1957), p. 28, copy in "NACA-NASA" file, box 10, papers of James H. Doolittle, Manuscript Division, Library of Congress, Washington, DC.
53. For example, Hsue-shen Tsien's *Wasserfall*-inspired configuration, detailed in "Instruction and Research at the Daniel and Florence Guggenheim Jet Propulsion Center," *Journal of the American Rocket Society* 1, no. 1 (June 1950): 63; C. E. Brown et al., *A Study of the Problems Relating to High-Speed High-Altitude Flight* (Hampton, VA: NACA Langley Laboratory, 25 June 1953), copy in the files of the historical archives, LaRC, Hampton, VA.
54. John V. Becker, "The X-15 Project," *Astronautics and Aeronautics* 2, no. 5 (February 1964): 52–61; A. Scott Crossfield with Clay Blair, Jr., *Always Another Dawn: The Story of a Rocket Test Pilot* (Cleveland, OH: World Publishing Company, 1960); Milton O. Thompson, *At the Edge of Space: The X-15 Flight Program* (Washington, DC: Smithsonian Institution Press, 1992); Wendell H. Stillwell, *X-15 Research Results* (Washington, DC: NASA SP-60, 1965).

Figure 17: North American X-15 Mach 6+ hypersonic transatmospheric research aircraft. *NASA Image E-7411*

atmosphere or to above Mach 5; many of these supported the Apollo effort. On 3 October 1967, Major William J. "Pete" Knight took a modified X-15, the X-15A-2, to Mach 6.70 (4,520 mph), carrying a dummy supersonic-combustion (scramjet) engine shape. Unanticipated heating, caused by turbulent flows and inadequate dissipation, led to multiple structural failures and the melting of the dummy scramjet from the aircraft; it also damaged its fuel jettison system. Fortunately, Knight landed successfully. Shortly thereafter, Major Michael Adams, U.S. Air Force, was killed when the third X-15 broke up following a combination of instrumentation and control systems failures, aggravated by the pilot's own unusually susceptible vertiginous tendencies. Loss of this aircraft forced NASA to abandon ambitious plans to modify one of the X-15s as a scramjet-powered slender delta, a decision that, in retrospect, was unfortunate. Overall, the X-15 contributed greatly to the understanding of the requirements for practical hypersonic vehicles; the program generated 700 technical reports and demonstrated the value of undertaking repeated flight research missions as opposed to a few "technology demonstrations."[55]

---

55. Joseph Weil, *Review of the X-15 Program* (Washington, DC: NASA TN D-1278, 1962); Johnny G. Armstrong, *Flight Planning and Conduct of the X-15A-2 Envelope Expansion Program* (Edwards Air Force Base, CA: AFFTC-TD-69-4, 1969); Donald R. Bellman et al., *Investigation of the Crash of the X-15-3 Aircraft on November 15, 1967* (Edwards, CA: NASA Flight Research Center, January 1968); Richard P. Hallion, "Flight Testing and Flight Research: From the Age of the Tower Jumper to the Age of the Astronaut," in *Conference Proceedings No. 452: Flight Test Techniques*, by the Advisory Group for Aerospace Research and Development (Neuilly sur Seine, France: NATO AGARD, 1988), p. 24–27; John V. Becker, "The X-15 Program in Retrospect," Third Eugen Sänger Memorial Lecture, *Deutsche Gesellschaft für Luft- und Raumfahrt (DGLR)*, Bonn, Germany, 4–5 December 1968; Dennis R. Jenkins, *Hypersonics Before the Shuttle: A Concise History of the X-15 Research Airplane* (Washington, DC: NASA SP-2000-4518, 2000), pp. 67–81; Milton O. Thompson and J. D. Hunley, *Flight Research: Problems Encountered and What They Should Teach Us* (Washington, DC: NASA SP-2000-4522, 2000), pp. 24–32, 41–46.

The NACA's 1952 decision leading to the X-15 likewise triggered the agency's first examination of hypersonic orbital vehicles.[56] In 1953, engineers Hubert Drake and L. Robert Carman of the HSFS conceived a five-phase evolutionary proposal leading to an orbital air-launched hypersonic boost-glide winged vehicle, the first governmental concept for a "piggyback" multistage orbital concept.[57] In 1956, NACA Ames researchers H. Julian Allen, Alfred Eggers, Clarence Syvertsen, and Stanford Neice presented a Mach 10 air-or-ground-launched rocket-boosted hypersonic design, characterized by a "flat top" delta wing with anhedral tips furnishing both directional stability and compression lift.[58] In contrast, the NACA Langley Laboratory pursued a more traditional delta planform while supporting the Air Force's project for a Mach 15 Hypersonic Weapon and Research and Development System (HYWARDS), this work later influencing the subsequent Project Dyna-Soar (for dynamic soaring, a reference to its Sänger-Bredt-like skipping reentry profile) development effort, conceived in the immediate aftermath of Sputnik. In 1955, supersonic fighters, bombers, and transports accounted for fully 37 percent of the agency's research effort, and space, ICBM, missile defense, and boost-glide aircraft accounted for just 7 percent. In early 1958, in the wake of Sputnik, space, ICBM, missile defense, and hypersonic boost-gliders had risen to 32 percent of the agency's work, and supersonic fighters, bombers, and transports had declined to 18 percent. For the NACA and NASA, the hypersonic era had clearly arrived.[59]

---

56. John V. Becker, "The Development of Winged Reentry Vehicles: An Essay from the NACA-NASA Perspective, 1952–1963," in *The Hypersonic Revolution: Case Studies in the History of Hypersonic Technology*, ed. Richard P. Hallion, vol. 1 (Washington, DC: USAF, 1998); *From Max Valier to Project PRIME (1924–1967)* (Bolling Air Force Base: Air Force History and Museums Program, 1998 ed.), pp. 379–448.
57. Hubert M. Drake and L. Robert Carman, *A Suggestion of Means for Flight Research at Hypersonic Velocities and High Altitudes* (Edwards, CA: NACA High-Speed Flight Research Station, August 1953), copy in the NASA Historical Reference Collection, NASA History Division, NASA Headquarters, Washington, DC.
58. A. J. Eggers, Jr., "Some Considerations of Aircraft Configurations Suitable for Long-Range Hypersonic Flight," in *Hypersonic Flow: Proceedings of the Eleventh Symposium of the Colston Research Society held in the University of Bristol, April 6–April 8th, 1959*, ed. A. R. Collier and J. Tinkler (London, U.K.: Butterworth Scientific Publications, 1960), pp. 369–389; Ames Aeronautical Laboratory, *Preliminary Investigation of a New Research Airplane for Exploring the Problems of Efficient Hypersonic Flight* (Moffett Field, CA: NACA Ames Aeronautical Laboratory, 18 January 1957), copy examined in 1979 in the files of the History Office, NASA Lyndon B. Johnson Space Center, Houston, TX.
59. NACA, "A Review of NACA Research Programs," n.d., Table I-A, and attached letter, Hugh L. Dryden to James H. Doolittle, 21 February 1958, "NACA-NASA" file, box 10, Doolittle Papers, Library of Congress. See also James R. Hansen, *Engineer in Charge: A History of the Langley Aeronautical Laboratory, 1917–1958* (Washington, DC: NASA SP-4305, 1987), pp. 367–378.

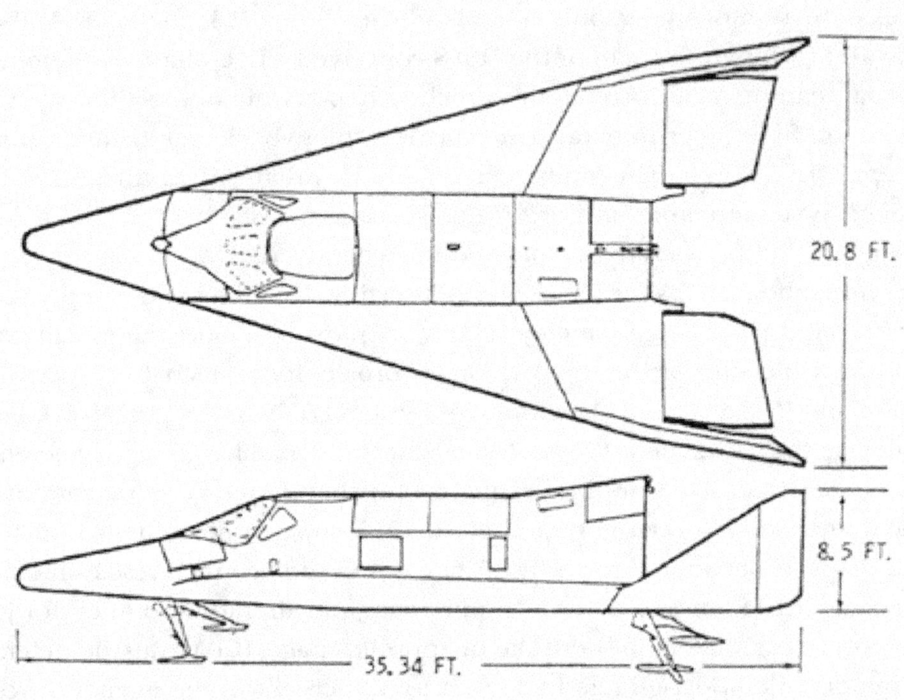

Figure 18: Schematic drawing of the X-20 Dyna-Soar; note the sharply swept delta planform, "toed-in" vertical fins, tilted bow "ramp," flared aft body for hypersonic trim, cockpit aerothermodynamic reentry shield, and wire brush skid landing gear. *Air Force Scientific Advisory Board (SAB)*

Dyna-Soar was "Round Three" of the postwar research aircraft effort and an important step toward developing practical approaches to exploiting the hypersonic frontier. Boeing received a development contract to produce this slender (and attractive) lofted boost-glider, to be launched atop a modified Titan III booster. At the time of its cancellation—again, an ill-considered decision—by Defense Secretary Robert S. McNamara in 1963, it was about two and a half years and an estimated $373 million away from its first flight; $410 million had already been expended. Dyna-Soar served to focus attention on the requirements of an orbital hypersonic vehicle considerably more challenging than the X-15, forcing serious examination of control, structural, crew protection, and light operations issues. It forced investment in specialized test facilities to address problems associated with the emerging fields of aerothermodynamics and high-temperature materials and flight structures. Dyna-Soar's wind tunnel program was three times that of the X-15, involving approximately 30 hypervelocity test facilities. By bringing needed emphasis to facilities development, it constituted an important milestone in the development of a hypersonic technology and design base, despite having never flown.[60]

Hypersonics to this point had strictly involved rocket propulsion. But another form of propulsion, the ramjet, was making spectacular advances into the high-supersonic regime, evidenced by the Mach 4+ Lockheed X-7 test vehicle. In the mid-1950s, NACA ramjet researchers, like others in industry, academia, and the military, envisioned using supersonic combustion flows within the ramjet to enable it to propel vehicles into the hypersonic regime. In 1958, Richard J. Weber and John S. MacKay of the NACA Lewis Flight Propulsion Laboratory concluded: "A number of fundamental problems must be solved before the SCRJ [Supersonic Combustion Ram-Jet] can be considered feasible. The major unknown is whether or not supersonic flow can be maintained during a combustion process. Also, even if a uniform fuel-air mixture can be so burned, there still remains the difficult problem of producing the desired combustible mixtures by fuel injection without causing severe shock losses. Subject to these qualifications, it is concluded from the present preliminary analysis that the SCRJ . . . will provide superior performance at higher hypersonic flight speeds."[61] The scramjet became a powerful impetus for studies on advanced

---

60. Clarence J. Geiger, "Strangled Infant: The Boeing X-20A Dyna Soar," in *Hypersonic Revolution*, ed. Hallion, vol. 1, pp. 185–377; Robert F. Futrell, *Ideas, Concepts, Doctrine: A History of Basic Thinking in the United States Air Force, 1907–1964*, vol. 2 (Maxwell Air Force Base, AL: Air University Aerospace Studies Institute, 1971), pp. 786, 792–795; Aero-Space Division, *Summary of Technical Advances: X-20 Program*, Report D2-23418 (Seattle, WA: The Boeing Company, July 1964); Air Force Scientific Advisory Board, "Some Remarks on the Aircraft Panel on New Technical Developments of the Next Ten Years," 1 October 1954, pp. 2–3, 7–8, in files of the SAB, USAF Headquarters, Washington, DC; John S. Rinehart, "Some Historical Highlights of Hypervelocity Research," in *Proceedings of the National Symposium on Hypervelocity Techniques, Denver, Colorado, October 20–21, 1960*, by the Institute of the Aeronautical Sciences (IAS) and University of Denver—Denver Research Institute (New York, NY: IAS, 1960), pp. 4–10; Harold O. Ekern and Jerry E. Jenkins, *Major High Speed Wind Tunnels in the U.S.* (Dayton, OH: USAF Wright Air Development Division, TM 60-8, July 1960); R. N. Cox, "Experimental Facilities for Hypersonic Research," in *Progress in Aeronautical Sciences*, ed. Antonio Ferri, D. Küchemann, and L. H. G. Sterne, vol. 3 (New York, NY: Pergamon Press, 1962), pp. 139–178; Ronald Smelt, "Test Facilities for Ultra-High Speed Aerodynamics," in *Proceedings of the Conference on High-Speed Aeronautics, Polytechnic Institute of Brooklyn, January 20–22, 1955*, ed. Antonio Ferri, Nicholas J. Hoff, and Paul A. Libby (Brooklyn, NY: Polytechnic Institute of Brooklyn, 1955), pp. 311–333; Julius Lukasiewicz, *Experimental Methods of Hypersonics* (New York, NY: Marcel Dekker, Inc., 1973), quotes from pp. 246 and 250.
61. Richard J. Weber and John S. MacKay, *An Analysis of Ramjet Engines Using Supersonic Combustion* (Washington, DC: NACA TN 4386, September 1958), p. 22; R. R. Jamison, "Hypersonic Air Breathing Engines," in *Hypersonic Flow: Proceedings of the Eleventh Symposium of the Colston Research Society held in the University of Bristol, April 6th–April 8th, 1959*, ed. A. R. Collar and J. Tinkler (London, U.K.: Butterworth Scientific Publications, 1960), pp. 391–408; S. W. Greenwood, "Spaceplane Propulsion," *The Aeroplane and Astronautics* (25 May 1961): 597–599; G. L. Dugger, F. S. Billig, and W. H. Avery, *Hypersonic Propulsion Studies at Applied Physics Laboratory, The Johns Hopkins University* (Silver

hypersonic air-breathing vehicles that could possibly fly into orbit themselves, an interest persisting to the present. From 1965 to 1966, the Aeronautics and Astronautics Coordinating Board of DOD and NASA reviewed a series of concepts for hypersonic logistical vehicles and lofted reentry vehicles of lifting body or winged configurations. Typical of these was a 750,000-pound gross liftoff weight two stage to orbit (TSTO). The first stage had turbofan-ramjet propulsion from subsonic through supersonic and on to Mach 6, then scramjet propulsion from Mach 6 to Mach 12. At that point, the second stage would separate, carrying a 40,000-pound payload into orbit.[62]

As the complexity and impracticality of these systems became ever more apparent, NASA researchers sought more achievable alternative approaches. One involved transforming the classic blunt body into a lifting vehicle. Thus was born the NASA lifting body program, an outgrowth of the work of notable pioneers such as H. Julian Allen, Clarence Syvertsen, Alfred Eggers, George Edwards, George Kenyon, and Eugene Love. NASA produced two notable "rival" lifting bodies, the Ames-sponsored cone-derivative M2 and the Langley-sponsored "fat delta" HL-10, and the Air Force generated another, the SV-5 (which spawned the X-24A and X-24B); but there were many other lifting body study efforts within the military services and industry besides these.[63] Tests up to nearly Mach 2 and descents from 90,000 feet by these rocket-powered research aircraft influenced the decision to have the Space Shuttle complete an unpowered approach and landing, for they gave confidence

---

Spring, MD: the Johns Hopkins University Applied Physics Laboratory, Report TG 405, 14 June 1961), esp. pp. 1–3; Antonio Ferri, "Supersonic Combustion Progress," *Astronautics & Aeronautics* 2, no. 8 (August 1964): 32–37; "Scramjets/Hypersonic Vehicles," in "Report of the USAF Scientific Advisory Board Aerospace Vehicles Panel," by the Air Force Scientific Advisory Board, February 1966, pp. 2–3, in files of the SAB, United States Air Force Headquarters, Washington, DC; Antonio Ferri, "Review of Scramjet Technology," *AIAA Journal of Aircraft* 5, no. 1 (January 1968): 3–10; Frank D. Stull, Robert A. Jones, and William P. Zima, "Propulsion Concepts for High Speed Aircraft" (paper 751092, Society of Automotive Engineers [SAE] National Aerospace Engineering and Manufacturing Meeting, Culver City, Los Angeles, CA, 17–20 November 1975); Paul J. Waltrup, "Liquid Fueled Supersonic Combustion Ramjets: A Research Perspective of the Past, Present and Future" (AIAA 86-0158, AIAA 24th Aerospace Sciences Meeting, Reno, NV, 6–9 January 1986); Edward T. Curran, "Scramjet Engines: The First Forty Years," *Journal of Propulsion and Power* 17, no. 6 (November–December 2001): 1138–1148.

62. NASA-DOD Aeronautics and Astronautics Coordinating Board, *Report of the Ad Hoc Sub-panel on Reusable Launch Vehicle Technology* (Washington, DC: NASA, 14 September 1966), NASA JSC archives.

63. R. Dale Reed, *Wingless Flight: The Lifting Body Story* (Washington, DC: NASA SP-4220, 1997); Alfred C. Draper, Melvin L. Buck, and David R. Selegan, "Aerospace Technology Demonstrators: Research and Operational Options," in *Aircraft Prototype and Technology Demonstrator Symposium, March 23–24, 1983*, ed. Norman C. Baullinger (Dayton, OH: AIAA Dayton-Cincinnati Section in association with the United States Air Force Museum, 1983), pp. 89–102.

Figure 19: The NASA-Air Force lifting body family in 1969. From left to right are the Martin X-24A, Northrop M2-F3, and Northrop HL-10. The X-24 constituted an Air Force initiative and later spawned the X-24B, which had a much higher fineness ratio. The M2-F3 and HL-10 were "rival" concepts, the former a modified half-cone derived by Ames, the latter a fattened delta derived by Langley. *NASA Image E-21093*

that pilot-astronauts could routinely control low L/D during the descent to a pinpoint landing on a runway.[64]

Tests with these complemented what might be considered a "Round Four" of tests of unpiloted vehicles and shapes at hypersonic speeds, including the aforementioned Reentry-F and Project FIRE, and various military and defense projects, particularly the Aerothermo-Structural Systems-Environmental Tests (ASSET) and Precision Recovery Including Maneuvering Entry (PRIME) projects. The McDonnell ASSET resembled the canceled Dyna-Soar, with similar "hot structure" radiative cooling. On 22 July 1964, one attained Mach 15.5 following launch by a modified Thor ballistic missile. The ASSET furnished insight on leeside heating not previously considered a serious design concern. The

64. Milton O. Thompson and Curtis Peebles, *Flying Without Wings: NASA Lifting Bodies and the Birth of the Space Shuttle* (Washington, DC: Smithsonian Institution Press, 1999), pp. 57–62; Johnny G. Armstrong, Flight Planning and Conduct of the X-24B Research Aircraft Flight Test Program (Edwards Air Force Base, CA: AFFTC-TR-76-11, 1977), pp. 12–14, 89–97; Richard E. Day, "Energy Management of Manned Boost-Glide Vehicles: A Historical Perspective" (Edwards Air Force Base, CA: NASA TP-2004-212037, DFRC, May 2004).

Martin SV-5D PRIME was an ablatively cooled lifting body. On 19 April 1967, after being lofted to 400,000 feet and Mach 27 by a modified Atlas booster, it demonstrated a 1,360-nautical-mile maneuvering entry over the Pacific Test Range. The PRIME shape was the first American shape evaluated over the entire subsonic, transonic, supersonic, and hypersonic range, from orbital velocities down to approach and landing, with the X-24A piloted research craft and the pilotless hypersonic SV-5D.[65]

The story of the Shuttle is well known, largely because of the tragic losses of *Challenger* in 1986 and *Columbia* in 2003. The Shuttle offered a vision of cheap, frequent access to space, reflecting an unprecedented (and, in retrospect, unjustified) optimism.[66] Certainly flight-test professionals familiar with the previous history of rocket-powered supersonic and hypersonic vehicles considered the Shuttle a very high-risk system and human spaceflight itself as an inherently dangerous occupation. As Apollo astronaut Michael Collins wrote shortly after the *Challenger* accident, "If someone had suggested to me in 1963, when I first became an astronaut, that for the next 23 years none of us would get killed riding a rocket, I would have said that person was a hopeless optimist, and naive beyond words."[67] The Shuttle's development risked significant unknowns and shortfalls, particularly its complex liquid-fuel rocket propulsion system and tile-based thermal protection system. Even before its first flight, veteran astronauts of the Mercury, Gemini, and Apollo programs criticized its safety, comparing it unfavorably to their earlier spacecraft, regarding its margins as "very low."[68]

---

65. Air Force Flight Dynamics Laboratory (AFFDL), *ASSET Final Briefing*, Report 65FD-850 (Dayton, OH: AFFDL, 5 October 1965); M. H. Shirk, *ASSET: Aerothermoelastic Vehicles (AEV) Results and Conclusions*, Report 65FD-1197 (Dayton, OH: AFFDL, August 1965). See also Richard P. Hallion, "ASSET: Pioneer of Lifting Reentry," in *Hypersonic Revolution*, ed. Hallion, vol. 1, pp. 449–527; J. L. Vitelli and R. P. Hallion, "Project PRIME: Hypersonic Reentry from Space," in *Hypersonic Revolution*, ed. Hallion, vol. 1, pp. 529–745.
66. See, for example, NASA, *Space Shuttle* (Washington, DC: NASA SP-407, 1976), and Jerry Grey, *Enterprise* (New York, NY: William Morrow and Company, Inc., 1979), pp. 164–237.
67. Michael Collins, *Liftoff: The Story of America's Adventure in Space* (New York, NY: Grove Press, 1988), p. 224.
68. "Former Astronauts Criticize Funding for Space Shuttle," *Washington Post* (15 July 1979). See also Eugene S. Love, "Advanced Technology and the Space Shuttle" (10th von Kármán Lecture, 9th Annual Meeting of the AIAA, Washington, DC, 8–10 January 1973); John M. Logsdon, "The Space Shuttle Decision: Technology and Political Choice," *Journal of Contemporary Business* 7, no. 3 (1978): 13–30; Paul A. Cooper and Paul F. Holloway, "The Shuttle Tile Story," *Astronautics & Aeronautics* 19, no. 1 (January 1981): 24–34; A. Scott Pace, "Engineering Design and Political Choice: The Space Shuttle, 1969–1972" (master's thesis, Cambridge, MA: MIT, 1982), pp. 2–3, 103–104, 116, 135–149; Miles Whitnah and Ernest R. Hillje, *Space Shuttle Wind Tunnel Testing Summary* (Washington, DC: NASA Reference Publication 1125, 1984).

Figure 20: The Space Shuttle *Endeavour* landing at Edwards Air Force Base, 1 October 1994. *NASA Image EC94-42789-1*

When STS-1's *Columbia*, crewed by John Young and Robert Crippen, entered the atmosphere on 14 April 1981, it flew down a hypersonic corridor not visited by any American lifting spacecraft since the PRIME lifting body 14 years previously. Fortunately, it survived: particularly since researchers quickly determined that significant differences existed between its actual in-flight performance and ground prediction. These included the timing and magnitude of boundary-layer transition from laminar to turbulent flows, higher roll rates and sideslip excursions during energy management roll reversals, lower motion damping, higher heating loads and rates for the thermal protection system, hot reentry airflow impingement upon the protruding Orbital Maneuvering System pods, and higher-than-predicted wing leading edge heating due to unexpectedly strong wing shock interactions.[69]

The Shuttle inspired global emulation, most notably with the Soviet Buran, French Hermes, and Japanese Hope, and helped generate a climate conducive

69. Kenneth W. Iliff and Mary F. Shafer, "A Comparison of Hypersonic Flight and Prediction Results" (AIAA 93-0311, AIAA 31st Aerospace Sciences Meeting, Reno, NV, 11–14 January 1993), pp. 8–13.

to hypersonic studies of a variety of inhabited and uninhabited systems.[70] In America, post-Shuttle interest ultimately spawned the most ambitious and complex attempt to develop a hypersonic orbital aircraft since the Shuttle: the National Aero-Space Plane Program (NASP), the X-30. Though primarily an Air Force development effort, the NASP involved significant NASA participation from its inception through cancellation. Begun in the mid-1980s and "baselined" in 1991, the SSTO X-30 replicated and/or encountered many of the same problems encountered three decades previously with a similar large air-breathing SSTO program, the Air Force's discredited Aerospaceplane of the early 1960s.[71] At the time of its demise in the early 1990s, its development team had achieved some impressive technical successes involving materials, fuels, and propulsion; but even so, the X-30 remained controversial, having grown in size and complexity, and with an unresolved velocity deficit of approximately 3,000 feet per second that would have prevented it from actually reaching orbit as a single-stage vehicle.[72]

Following a pattern traditional for the hypersonic field, the aftermath of the NASP was one of contraction, frustration, and delayed expectation. Ironically, even as American hypersonics slowed, mastery of the field and foreign interest continued to grow. For NASA, faced with the challenges of maintaining the Shuttle, completing the ISS, and meeting many other ambitious exploration goals, hypersonics was just one of many areas of research interest. Other agencies and organizations, faced with many competing interests, had the same challenge. Accordingly, while many possible hypersonic programs and starts beckoned—

---

70. United States General Accounting Office (GAO), *Investment in Foreign Aerospace Vehicle Research and Technological Development Efforts* (Washington, DC: GAO/T-NSIAD-89-43, August 1989), pp. 5–11; United States GAO, *Aerospace Plane Technology: Research and Development Efforts in Europe* (Washington, DC: GAO/NSIAD-91-194, July 1991), pp. 33–61; United States GAO, *Aerospace Plane Technology: Research and Development Efforts in Japan and Australia* (Washington, DC: GAO/NSIAD-92-5, October 1991), pp. 31–35.
71. United States GAO, *National Aero-Space Plane: A Technology Development and Demonstration Program to Build the X-30* (Washington, DC: GAO/NSIAD-88-122, April 1988); Robert B. Barthelemy, "The National Aero Space Plane Program: A Revolutionary Concept," *Johns Hopkins Applied Physics Laboratory Technical Digest* 11, nos. 2 and 3 (1990): 312–318; United States GAO, *National Aero-Space Plane: Key Issues Facing the Program* (Washington, DC: GAO/T-NSIAD-92-26, March 1992), pp. 4–15.
72. Alan W. Wilhite et al., "Concepts Leading to the National Aero-Space Plane Program" (AIAA 90-0294, 28th Aerospace Sciences Meeting, Reno, NV, 8–11 January 1990); Joseph F. Shea et al., *Report of the Defense Science Board Task Force on National Aero-Space Plane (NASP) Program* (Washington, DC: Defense Science Board, Director of Defense Research and Engineering, November 1992); United States GAO, *National Aero-Space Plane: Restructuring Future Research and Development Efforts* (Washington, DC: GAO/NSIAD-93-71, December 1992), p. 4; R. L. Chase and M. H. Tang, "A History of the NASP Program from the Formation of the Joint Program Office to the Termination of the HySTP Scramjet Performance Demonstration Program" (AIAA 95-6031, AIAA 6th International Aerospace Planes and Hypersonics Technologies Conference, Chattanooga, TN, 30 April 1995).

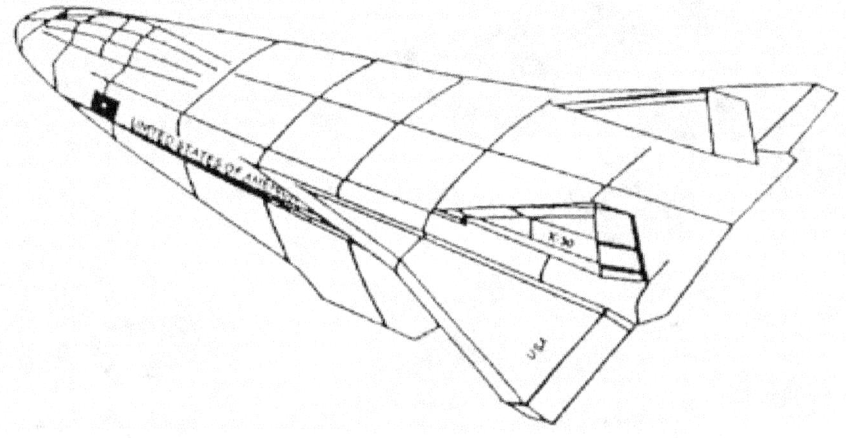

Figure 21: Final general design configuration of the X-30 National Aero-Space Plane. *Air Force SAB*

some, such as the Affordable Rapid Response Missile Demonstrator (ARRMD), X-33, X-34, X-38, and Hyper-X (subsequently designated the X-43), showing real promise—only a handful went ahead, with just one actually flying, the X-43.[73]

NASA's scramjet-powered X-43 achieved the first demonstration of in-flight hypersonic scramjet ignition and operation with an airplanelike configuration. Its flight-test success followed an innovative in-flight hypersonic combustion experiment by a team of Australian researchers from the University of Queensland's Centre for Hypersonics. On 30 July 2002, a team of researchers from the University of Queensland's Centre for Hypersonics launched HyShot over Australia's Woomera test range; it was a small combustor test article lofted by a two-stage booster into the upper atmosphere. HyShot demonstrated 5 seconds of hypersonic combustion at Mach 7.6 as it plunged toward Earth.[74] The stage was now set for a comprehensive demonstration of a true scramjet, the X-43.

The X-43 joined a sophisticated scramjet engine module developed by the General Applied Sciences Laboratory (GASL) to a surfboardlike 100-inch-long, 60-inch-span slender lifting body, lofted to hypersonic velocity by an Orbital Sciences solid-fuel winged Pegasus booster air-launched from a Boeing NB-52B Stratofortress (incidentally, the same launch aircraft that had

73. C. R. McClinton, J. L. Hunt, R. H. Ricketts, Paul Reukauf, and C. L. Peddie, "Airbreathing Hypersonic Technology Vision Vehicles and Development Dreams" (AIAA 99-4978, AIAA 9th International Space Planes and Hypersonic Systems and Technologies Conference, Norfolk, VA, 1–5 November 1999).
74. A. Paull, H. Alesi, and S. Anderson, "HyShot Flight Program and How It Was Developed" (AIAA 02-4939, AIAA/Association Aéronautique et Astronautique de France [AAAF] 11th International Space Planes and Hypersonics Systems and Technologies Conference, Orleans, France, 29 September–4 October 2002).

Figure 22: The Boeing NB-52B Stratofortress "008" at DFRC and the first X-43A Hyper-X scramjet test bed on a "captive carry" test flight over the Pacific Ocean, April 2001. *NASA Image EC01-0126-07*

dropped the X-15, M2-F2/3, HL-10, and X-24A/B lifting bodies). Developers began the Hyper-X program in 1995, drawing upon a Boeing study effort for a Mach 10 global reconnaissance cruiser and space access vehicle for its overall configuration. In October 1996 they completed its preliminary design, and Orbital Sciences subsequently received a development contract for the modified Pegasus booster (the HXLV) in February 1997. The Hyper-X vehicle fabrication contract (the HXRV) went to Microcraft, Inc., of Tullahoma, Tennessee, partnered with the GASL for the engine, Boeing, and Accurate Automation Corporation. The engine underwent comprehensive ignition and combustion stabilization hypersonic testing in Langley Research Center's 8-foot High-Speed Tunnel. Microcraft delivered three X-43A flight-test vehicles to DFRC for launch over the Naval Air Warfare Center's Weapons Division Sea Range. The first flight attempt in June 2001 failed after the Pegasus booster shed a control fin just after launch, forcing its destruction by the Range Safety

Officer. Thereafter NASA undertook a painstaking review before clearing the program for a second flight attempt. This reached Mach 6.8 on 27 March 2004. Although the Agency briefly considered terminating the program following this demonstration, pressure from hypersonic partisans led to a third flight attempt, this reaching Mach 9.7 (around 6,500 mph) at 110,000 feet on 16 November 2004. The third vehicle faced a thermodynamic environment more than 1,000°F harsher than that faced by the second, experiencing airframe temperatures of 3,600°F.[75] Though a planned follow-on, the hydrocarbon-fueled X-43C, had been canceled, NASA research on hypersonics continued with the support of the Air Force's X-51, a hydrocarbon scramjet test bed to be air-launched in 2010 at DFRC. NASA's research on the hypersonic frontier continues, possibly pointing toward a future where hypersonic technology will benefit both the military and civil needs of the nation.

As the 19th century opened, the "normative paradigm" of mass transportation was the horse-drawn wagon; at the outset of the 20th, it was the train; and at the beginning of the 21st, it was the airplane, a progression of 6, to 60, and then to 600 mph. Despite critics who decry supersonics and hypersonics for commercial transportation, both show surprising resilience. While it remains doubtful that a competitive supersonic airliner can be built and introduced into service, certainly a definable market exists for supersonic business jets, and NASA research on both configurations and ways of tailoring their shape to reduce boom impingement may well speed their introduction into service. The hypersonic global transport, boosting out of the atmosphere and returning through it halfway around the world, affords a vision of a "greener" and less environmentally damaging mobility system

75. Thomas J. Bogar, Edward A. Eiswirth, Lana M. Couch, James L. Hunt, and Charles R. McClinton, "Conceptual Design of a Mach 10, Global Reach Reconnaissance Aircraft" (AIAA 96-2894, ASME, SAE, and ASEE 32nd Joint Propulsion Conference, Lake Buena Vista, FL, 1–3 July 1996); C. R. McClinton, Vincent L. Rausch, Joel Sitz, and Paul Reukauf, "Hyper-X Program Status" (AIAA 01-1910, AIAA/NAL-NASDA-ISAS 10th International Space Planes and Hypersonic Systems and Technologies Conference, Kyoto, Japan, 24–27 April 2001); C. R. McClinton, V. L. Rausch, J. Sitz, and P. Reukauf, "Hyper-X Program Status" (AIAA 01-0828, 39th Aerospace Sciences Meeting, Reno, NV, 8–11 November 2001); David E. Reubush, Luat T. Nguyen, and V. L. Rausch, "Review of X-43A Return to Flight Activities and Current Status" (AIAA 03-7085, 12th AIAA International Space Planes and Hypersonics Systems and Technologies Conference, Norfolk, VA, 15–19 December 2003); NASA, "NASA Hyper-X Program Demonstrates Scramjet Technologies," NASA Facts, FS-2004-10-98-LaRC (20 October 2004); "NASA's X-43A Scramjet Breaks Speed Record," available at *http://www.nasa.gov/centers/dryden/home/X-43A_Speed_Record.html* (Washington, DC: NASA, 16 November 2004), accessed 28 December 2004; Jay Levine, "Exploring the Hypersonic Realm," *X-Press* 46, no. 10 (26 November 2004): 1, 8.

Figure 23: The third X-43A accelerates to Mach 9.7 prior to release from its Pegasus booster, 16 November 2004, following air-launch over the Pacific Ocean from its Boeing NB-52B mother ship. *NASA Image EC04-0325-37*

than is the conventional subsonic jetliner today.[76] And so, might not humanity extend the "6, to 60, and then to 600" pace and enter the next century at 6,000 mph, the speed of a hypersonic commercial vehicle? If so, it will constitute the logical outgrowth of NASA's pioneering the supersonic and hypersonic frontier well over a century before.

76. See, for example, Jesse Ausubel, Cesare Marchetti, and Perrin S. Meyer, "Toward Green Mobility: The Evolution of Transport," *European Review* 6, no. 2 (1998): 37–156.

Chapter 11

Fifty Years of Human Spaceflight

Why Is There Still a Controversy?

John M. Logsdon

From the beginning of the U.S. program of human spaceflight until today, there has been controversy with respect to the value of human presence in space compared to its costs and risks. For example, the first science adviser to President Dwight D. Eisenhower, James R. Killian, Jr., said in 1960 that "Many thoughtful citizens are convinced that the really exciting discoveries in space can be realized better by instruments than by man."[1] Yet less than six months later, President John F. Kennedy committed the United States to send Americans to the Moon "before this decade is out." The two men clearly applied different judgments on the value of human spaceflight to their words and actions.

It has been 48 years since the first U.S. spaceflight, the 15-minute suborbital mission of Alan Shepard on 5 May 1961. Since then, 12 Americans have walked on the Moon, and there have been (as of 30 September 2009) 127 launches of the Space Shuttle, several long-duration stays of U.S. astronauts aboard the Russian *Mir* space station, and continuous occupancy of the ISS for over 10 years. As of 30 September 2009, 320 U.S. citizens had gone into orbit or beyond—278 men and 42 women. Many U.S. astronauts have made several trips into space; two have gone seven times each. An additional 176 non-U.S. citizens have also flown in space.[2] It is worth asking: "With all this

---

1. John M. Logsdon, *The Decision to Go to the Moon: Project Apollo and the National Interest* (Cambridge, MA: MIT Press, 1970), p. 20.
2. These statistics are taken from CBS space reporter Bill Harwood's Web site, "CBS Space Place," available at *http://www.cbsnews.com/network/news/space/spacestats.html#GENERALDEMO* (accessed 3 September 2009).

U.S. experience, why is there still controversy in this country regarding the wisdom of continuing a government program of human spaceflight?"

Some Preliminary Observations

As suggested by the quotation from James Killian above, many, if not most, leaders of the scientific community from 1957 to today have questioned the value of humans as opposed to robotic spacecraft in terms of scientific payoff from space missions. Others have pointed out that a human presence is not required to provide the multiple tangible benefits from space such as relaying communications, observing meteorological conditions, remote sensing of Earth's surface, or providing navigation and timing services. Over the same period of time, the U.S. Presidents and other leaders involved in the decisions to carry out a program of human spaceflight have never used scientific output as a primary justification. So a first thing to observe is that the protagonists in the debate regarding humans versus robots have historically talked past one another. To scientists, a space program is first of all about advancing scientific knowledge, and their dismissal of the value of human presence is couched in terms of that objective. To national leaders, the value of human involvement is measured in terms of national power and pride; for example, Project Apollo was about getting Americans to the Moon and back to Earth, thereby demonstrating the technological and organizational power of the United States. It was not about what research astronauts did while on the lunar surface or what we learned about the Moon by going there.

When the debate on whether to approve the Space Shuttle was in full swing in the second half of 1971, there were multiple justifications offered for going ahead with the program. Two apparently were decisive for President Richard Nixon. One was very parochial: starting the Shuttle in 1972 would produce jobs in states key to Nixon's reelection. The other was the argument that not approving the Shuttle Program as a follow-on human flight program to Apollo would send a message that "our best years are behind us, that we are turning inward, reducing our defense commitments, and voluntarily starting to give up our super-power status, and our desire to maintain world superiority."[3]

The U.S. Space Station program was publicly justified both as another demonstration of U.S. leadership and as a demonstration of the research value of humans in orbit, and thus as a means of providing an instrumental and tangible

3. Caspar Weinberger, Deputy Director, OMB, to President Richard Nixon, 12 August 1971, reprinted in *Exploring the Unknown: Selected Documents in the History of the U.S. Civil Space Program*, ed. John M. Logsdon, vol. 1, *Organizing for Exploration* (Washington, DC: NASA SP-4407, 1995), p. 547.

answer to the question of the value of human presence in space. Unfortunately, the checkered history of the program has meant that there still, a full quarter of a century after President Ronald Reagan approved development of a U.S. space station, has been little opportunity to demonstrate the Station's instrumental value as a venue for research; and it is not clear that the U.S. utilization program planned for the ISS will remedy that shortfall. So the jury is still out with respect to whether the results of human-tended research in orbit can produce results that justify the costs and risks of carrying it out, separate from the somewhat questionable value of the Space Station as a tool of U.S. leadership.

It is worth noting that many people, and many governments, are attracted by the opportunity to go into space. China is the newest entry in the human spaceflight club, and India is moving toward becoming the fourth country to acquire membership.[4] The primary motivation behind these recent efforts seems to be similar to that which motivated the Soviet Union and the United States almost five decades ago—the quest for pride and power. Recently, several space agencies have carried out a recruitment campaign for new government-sponsored astronauts. The United States got "only" 3,535 applications for perhaps 15–20 slots. The ESA received 8,413 applications from 17 countries for what were supposed to be 4 positions; ESA ended up selecting 6 candidates due to the overwhelming response. The Japanese Space Agency received 963 applications for 3 opportunities. And Canada received 5,352 applications for 2 slots![5] Countries such as Malaysia and South Korea have paid substantial sums to Russia for the launch of their government-sponsored astronauts. Russia continues to launch fare-paying private citizens to the ISS, and there are many forecasts of a robust market for commercial flights to orbit.[6]

So What Is the Problem?

There are several reasons why the controversy regarding human spaceflight lingers after almost a half century of experience. First of all, human spaceflight is undeniably expensive both in absolute terms and compared to robotic space missions. Thus it is entirely legitimate to continually question whether its benefits

4. Of course, people from many other countries have gone into space on United States and Russian spacecraft, but only three countries currently have the ability to send people into orbit.
5. These numbers come from *http://brianshiro.blogspot.com/2008/09/nasas-2009-astronaut-class-selection.html* (accessed 16 January 2009).
6. This paper will not discuss the wisdom of privately funded human spaceflight. If there is both a demand for such an experience and a socially acceptable means of meeting that demand, there appears to be no fundamental issue with commercial spaceflight.

and risks justify the expenditure of public funds for its support. Calculating the exact amount of the NASA budget devoted to human spaceflight is a somewhat arcane effort and will not be attempted here. Suffice it to say that operating the Space Shuttle and ISS, along with developing the next-generation systems for human spaceflight, is a multi-billion-dollar annual enterprise. Nor will this paper attempt to specify a comparative cost-benefit ratio for human and robotic missions. This is an almost impossible task, given that many of the benefits of human (and to some degree, robotic) missions are intangible and thus very difficult to specify in measurable terms, while it is more feasible—but still difficult—to measure the scientific output of space missions.

In addition to costs, the risks associated with human spaceflight are among the highest of any human undertaking. There have been 1,092 individual trips to space, and 18 people have died during spaceflight.[7] This is a 1.6 percent fatality rate; of every 100 people who have flown into space, nearly 2 did not return alive. There are important ethical questions with respect to government sponsorship of such a risky (peacetime) activity, even given the fact that the astronauts themselves voluntarily accept the risk.

Finally, there are multiple objectives that motivate governments to sponsor human spaceflight. Those skeptical of this activity give priority to different motivations than do its supporters. The result is that advocates and critics use different measures of value, and this is a situation in which it is very difficult to reach a consensus on the question of the absolute value of humans in space.

An excellent example of this point is the long-standing skepticism of the first U.S. space scientist, the late James Van Allen, about the value of human spaceflight. In a 2004 essay titled "Is Human Spaceflight Obsolete?" James Van Allen asked, "Does human spaceflight continue to serve a compelling cultural purpose and/or the national interest?" or "Does human spaceflight simply have a life of his own, without a realistic objective that is remotely commensurate with its costs?" His response places tangible science and application benefits as the highest priority and trivializes intangible benefits: "Almost all the space program's advances in scientific knowledge have been accomplished by hundreds of robotic spacecraft." Also, "In our daily lives, we enjoy the pervasive benefits of long-lived robotic spacecraft." With respect to human spaceflight, a "dispassionate comparison" would, he argues, show that "the only surviving motivation for continuing human spaceflight is the

7. This includes the seven astronauts aboard *Challenger* on 26 January 1986; this was technically not a spaceflight since the vehicle did not achieve orbit.

ideology of adventure."[8] (The use of the rather pejorative phrase "ideology of adventure" suggests that Van Allen's analysis might not be quite as "dispassionate" as he suggests.)

It is worth noting that Van Allen's views are not shared by all space scientists. For example, Steve Squyres, the Cornell professor who is the Principal Investigator for MERs Spirit and Opportunity, observes in his book *Roving Mars* that "The unfortunate truth is that most things our rovers can do in a perfect sol (day's work), a human explorer on the scene could do in less than a minute," and "The rovers are our surrogates, our robotic precursors to a world, as humans, we're still not quite ready to visit." What Squyres really wants, he says, is "boot prints in our wheel tracks."[9]

What Are the "Primary Objectives" of Human Spaceflight?

In a thoughtful white paper published in late 2008, the Space, Policy, and Society Working Group at MIT discussed "The Future of Human Spaceflight." The white paper divides the reasons for undertaking human spaceflight into "primary" and "secondary" objectives. Primary objectives are "those that can only be accomplished through the physical presence of human beings, those whose benefits exceed the opportunity costs, and those worthy of significant risk to, and possibly the loss of, human life." The paper identifies as primary objectives "exploration, national pride, and international prestige and leadership." Secondary objectives "have benefits that accrue from human presence in space but do not by themselves justify the cost or the risk." The paper identifies as secondary objectives "science, economic development and jobs, technology development, education, and inspiration." It adds, "None of this is to say that secondary objectives are unimportant; all have contributing roles to play in justifying government expenditures on space exploration."[10]

If this formulation is accepted, the reasons for the continuing controversy over human spaceflight become clearer. Those advocating the value of humans in space place their priority on the "primary" set of objectives; those critical of human spaceflight evaluate it in terms of the "secondary" objectives. With these

---

8. James Van Allen, "Is Human Spaceflight Obsolete?" *Issues in Science and Technology* (summer 2004), available at *http://www.issues.org/20.4/p_van_allen.html* (accessed 3 September 2009).
9. Stephen Squyres, *Roving Mars: Spirit, Opportunity, and the Exploration of the Red Planet* (New York, NY: Hyperion, 2005), pp. 234, 378.
10. Space, Policy, and Society Working Group, MIT, "The Future of Human Spaceflight," December 2008. The paper is available at *http://web.mit.edu/mitsps/MITFutureofHumanSpaceflight.pdf* (accessed 18 January 2009). The quoted passages are on p. 6 of the paper.

differences in evaluation, it is not surprising that no closure has been reached in the controversy over the value to the nation of a program of human spaceflight.

Viewing human spaceflight as a particularly effective means of enhancing United States national pride and international prestige and leadership has been the underpinning rationale for almost 50 years of support for the activity by the top levels of government. Consider the following:

> To the layman, manned space flight and exploration will represent the true conquest of space. No unmanned experiment can substitute for manned exploration in its psychological effect on the peoples of the world.
>
> —*"U.S. Policy on Outer Space," 26 January 1960*[11]

> Dramatic achievements in space . . . symbolize the technological power and organizing capacity of a nation.
>
> *This nation needs to make a positive decision to pursue space projects aimed at enhancing national prestige.* Our attainments are a major element in the international competition between the Soviet system and our own . . . . [P]rojects such as lunar and planetary exploration are, in this sense, part of the battle along the fluid front of the cold war.
>
> It is man, not machines, that captures the imagination of the world.
>
> —*Memorandum to Vice President Lyndon Johnson from NASA Administrator James Webb and Secretary of Defense Robert McNamara, 8 May 1961*[12]

> We go into space because whatever mankind must undertake, *free* men must fully share.
>
> —*Address by President John F. Kennedy to a joint session of Congress, 25 May 1961*[13]

11. Logsdon, *Exploring the Unknown*, vol. 1, p. 363.
12. Ibid., pp. 444, 446.
13. Ibid., p. 453.

Recent Apollo flights have been very successful from all points of view. Most important is the fact that they give the American people a much needed lift in spirit (and the people of the world an equally needed look at American superiority).

—Memorandum to President Richard Nixon from Caspar Weinberger, Deputy Director, OMB, 12 August 1971

I agree with Cap.

—President Richard Nixon, handwritten note on Weinberger memorandum[14]

Man has learned to fly in space, and man will continue to fly in space. This is a fact. And, given this fact, the United States cannot forgo its responsibility—to itself and to the free world—to have a part in manned space flight . . . . For the U.S. not to be in space, while others do have men in space, is unthinkable, and a position which America cannot accept.

—Memorandum to the White House from NASA Administrator James Fletcher, 22 November 1971[15]

Some people say you can do it all in space with robots. In fact, you must have man. He—and she—are the essential ingredient. The presence of man is the key to leadership in space.

—NASA Administrator James Beggs, Briefing to President Ronald Reagan, 1 December 1983[16]

A fundamental objective guiding United States space activities has been, and continues to be, space leadership . . . . The overall goals of United States Space activities are . . . to expand human presence and activity beyond Earth orbit into the solar system.

—Presidential Directive on National Space Policy, 11 February 1988[17]

---

14. Ibid., p. 547.
15. Ibid., p. 556.
16. Ibid., p. 597.
17. Ibid., p. 602.

Human Spaceflight and Soft Power

These samplings from the historical record suggest the continuity in support at the top level for human spaceflight based on its contributions to national pride, national prestige, and international leadership. In turn, these elements are attributes of what has come to be called "soft power," defined as the ability of the United States to "obtain the outcomes it wants in world politics because other countries want to follow it, admiring its values, emulating its example, aspiring to its level of prosperity and openness."[18]

Contemporary space leaders clearly see the value of human spaceflight in terms of its contribution to U.S. soft power and thus to U.S. national security. This point was articulately made by former NASA Administrator Michael Griffin in a 2006 speech. He suggested that "The most enlightened, yet least discussed, aspect of national security involves being the kind of nation and, doing the kinds of things, that inspire others to want to cooperate as allies and partners rather than to be adversaries. And in my opinion, this is NASA's greatest contribution to our nation's future in the world." He added:

> Today, and yet not for much longer, America's ability to lead a robust program of human and robotic exploration sets us above and apart from all others. It offers the perfect venue for leadership in an alliance of great nations, and provides the perfect opportunity to bind others to us as partners in the pursuit of common dreams. And if we are a nation joined with others in pursuit of such goals, all will be less likely to pursue conflict in other arenas. No enterprise of national scale offers a more visibly attractive and interesting collaboration than does space exploration. This great enterprise threatens no one while enriching everyone. It is about the lure of the frontier; leaders occupy and extend the frontiers of their times.[19]

How might one make an independent assessment of this and the previously cited statements? Is there any way to actually measure the contribution of human spaceflight to U.S. soft power? Or must we depend for this connection on subjective judgments by those with the responsibility for allocating public

---

18. On soft power, see Joseph S. Nye, Jr., *The Paradox of American Power: Why the World's Only Superpower Can't Go It Alone* (Oxford, U.K.: Oxford University Press, 2002), pp. 4–9.
19. Michael Griffin (speech to National Space Symposium, 6 April 2006).

resources to space in the context of competing priorities? It is beyond the scope of this brief paper to address these questions in any depth, but they are crucial to assessing whether human spaceflight is an undertaking that continues to make significant contributions to the pride and power of the United States.

The record of human spaceflight in the United States since the end of the Apollo program may not provide the best basis for assessing its contributions to national power. That record has also raised the question of whether any program of human spaceflight can be effective in this regard, even if it consists of repetitive flights of the same spacecraft carrying out a series of missions that to the layman look very similar. Or is it the case that only spaceflight that results in new experiences can make such a contribution? Unfortunately, neither the Space Shuttle nor ISS program has lived up to its promised performance, and thus it is a fair question to ask whether human spaceflight as carried out by NASA over the past quarter of a century has been a significant contributor to U.S. soft power.

Even so, it may well be that past, and to some degree current, space achievements involving direct human presence do remain a potent source of national pride, and that such pride is the underpinning reason why the U.S. public continues to support human spaceflight. Certainly, space images—an astronaut on the Moon, a Space Shuttle launch—rank only below the American flag and the bald eagle as patriotic symbols, and such patriotism is a foundation of U.S. soft (and hard) power. Most Americans probably cannot name a current astronaut but still have a very engaged reaction once they meet someone who has experienced spaceflight. Young individuals, in particular, remain fascinated by the possibility that they could one day travel to space, and that possibility appears to motivate them toward excellence in their education. The self-image of the United States as a successful nation is threatened when we fail in our space efforts, and catastrophes such as *Challenger* and *Columbia* seem to tap deep emotions. If human spaceflight has the potential to contribute to U.S. soft power, but repetitive flights to and long-duration activity in low-Earth orbit do not tap that potential, what does? The obvious answer is human travel beyond Earth orbit.

Exploration as a Compelling Rationale

Many believe that the only sustainable rationale for a government-funded program of human spaceflight is to take the lead in exploring the solar system beyond low-Earth orbit.[20] The MIT white paper provides an insightful definition of exploration:

---

20. John M. Logsdon, "A Sustainable Rationale for Human Spaceflight," *Issues in Science and Technology* (winter 2003), available at *http://www.issues.org/20.2/p_logsdon.html* (accessed 3 September 2009).

> Exploration is a human activity, undertaken by certain cultures at certain times for particular reasons. It has components of national interest, scientific research, and technical innovation, but is defined by none of them. We define exploration as an expansion of the realm of human experience, bringing people into new places, situations, and environments, expanding and redefining what it means to be human. What is the role of Earth in human life? Is human life fundamentally tied to the earth, or could it survive without the planet?
>
> Human presence, and its attendant risk, turns a spaceflight into a story that is compelling to large numbers of people. Exploration also has a moral dimension because it is in effect a cultural conversation on the nature and meaning of human life. Exploration by this definition can only be accomplished by direct human presence and may be deemed worthy of the risk of human life.[21]

In the wake of the 2003 *Columbia* accident that took the lives of seven astronauts and the report of the Columbia Accident Investigation Board that criticized the absence of a compelling mission for human spaceflight as "a failure of national leadership,"[22] the United States, in January 2004, adopted a new policy to guide its human spaceflight activities. The policy directed NASA to "implement a sustained and affordable human and robotic program to explore the solar system and beyond" and to "extend human presence across the solar system, starting with a human return to the Moon by the year 2020, in preparation for human exploration of Mars and other destinations."[23] This policy seems totally consistent with the definition of exploration provided in the MIT white paper. The issue is whether such a policy and its implementation, focusing on human exploration beyond Earth orbit, can provide an adequate and sustainable justification for a continuing program of government-sponsored spaceflight that will make contributions that will outweigh the costs and risks involved to the "primary objectives" of national pride and prestige, and also to some of the several "secondary objectives."

---

21. MIT, "The Future of Human Spaceflight," p. 8.
22. Columbia Accident Investigation Board, *Report*, vol. 1 (Washington, DC: GPO, August 2003), pp. 209, 211.
23. The White House, "A Renewed Spirit of Discovery: The President's Vision for Space Exploration," January 2004.

In the absence of any agreed-upon metrics to provide an answer to this question, it seems unavoidable that the answer will come from the subjective judgments of national leaders. The past two U.S. Congresses, by strong bipartisan majorities, have made a continued program of human exploration beyond Earth orbit the law of the land. This paper is being completed just as Barack Obama takes the oath of office as the 44th President of the United States. It will be up to President Obama to decide whether to continue along the lines of the current space exploration policy, or whether to return the focus of U.S. space activities, including human spaceflight, to Earth and its near vicinity. As he makes this decision, one hopes that his vision of the future incorporates having the United States remain a spacefaring nation. In this context, it is once again useful to quote the words of former NASA Administrator Michael Griffin:

> Imagine if you will a world of some future time—whether it be 2020, or 2040, or whenever—when some other nations or alliances are capable of reaching and exploring the Moon or voyaging to Mars, and the United States cannot and does not. Is it even conceivable that in such a world America would still be regarded as a leader among nations, never mind *the* leader?
>
> And if not, what might be the consequences of such a shift in thought upon the global balance of economic and strategic power? Are we willing to accept those consequences? In the end, these are the considerations at stake when we decide, as Americans, upon the goals we set for, and the resources we allocate to, our civil space program.
>
> Humans will go to the Moon and Mars; the only questions are which humans, what values they will hold, what languages they will speak.[24]

24. Michael Griffin (speech to the National Space Symposium, 6 April 2006).

Chapter 12

From the Secret of Apollo to the Lessons of Failure

The Uses and Abuses of Systems Engineering and Project Management at NASA

Stephen B. Johnson

In his books, *To Engineer Is Human*[1] and *Success through Failure*,[2] Henry Petroski has documented an interesting and important relationship between an engineering discipline and the reliability of the technical products that it produces. He found that over the course of several generations of engineers in a given discipline, such as civil engineers specializing in bridge building, that the reliability (or conversely, the failure rate) of their products swings back and forth from highly conservative, highly reliable designs to more innovative, less reliable designs. Ultimately the less reliable designs lead to outright failure, such as the famous Tacoma Narrows Bridge failure of 1940. This prompts engineers to determine the causes of the failure and implement more conservative designs on their next projects. Eventually, after many successes, some designers reduce "excessive" design margins to save money, to improve performance, or simply to try new ideas. Eventually someone goes too far and creates a design with inadequate margins, leading once again to failure. Petroski's examples came from civil engineering, but he found this same pattern in other engineering disciplines, including aerospace. He noted that the tragic losses of the Space Shuttles *Challenger* in 1986 and *Columbia* in 2003 follow the same pattern.[3]

1. Henry Petroski, *To Engineer Is Human: The Role of Failure in Successful Design* (New York, NY: St. Martin's Press, 1982).
2. Henry Petroski, *Success through Failure: The Paradox of Design* (Princeton, NJ: Princeton University Press, 2006).
3. Ibid., pp. 163–167.

Petroski's analysis is relevant to NASA because he emphasizes multigenerational knowledge transfer and learning in engineering design and how changes in the perception of risk affect failure rates. NASA has a strong tradition of research and system development, and it also operated these systems, creating organizations focused on launch and mission operations. Understanding NASA's ability to create and operate complex systems requires an understanding of both its large-scale engineering development and its operation of these systems. While academic researchers of "high reliability organizations" have studied operations of complex, high-risk systems such as aircraft carriers and nuclear power plants, there is a relative dearth of research on the dependability of engineering design.[4] Such research is needed, given the emerging understanding that one of NASA's fundamental issues is its culture.

In the 1960s, NASA's Apollo program was a shining example of what humans could accomplish when they set their minds to achieving a difficult goal. As many noted at the time, it was an incredible feat of organization as well as technology. NASA's ability to direct hundreds of thousands of factory workers, engineers, scientists, and managers to achieve multiple lunar landings drew accolades in the United States and abroad.

Yet 31 years after the last astronaut left the lunar surface, the loss of NASA's second Space Shuttle, *Columbia*, and its seven astronauts left the Agency devastated and distraught. The *Challenger* disaster in January 1986 was a shock, shattering NASA's aura of invincibility. The loss of *Columbia* in February 2003 implied more fundamental problems. No longer could the blame for an accident be placed on a few overconfident engineers or managers. Something inherent to NASA as an institution was flawed, something the Columbia Accident Investigation Board identified as NASA's "culture."

"Culture" is a famously holistic and ambiguous term, even for social scientists who use it in their daily work. According to *Webster's Ninth New Collegiate Dictionary*, culture is "an integrated pattern of human knowledge, belief, and behavior that depends upon man's capacity for learning and transmitting knowledge to succeeding generations," or "the customary beliefs, social forms,

4. See, for example, T. R. LaPorte, "High Reliability Organizations: Unlikely, Demanding, and at Risk," *Journal of Crisis and Contingency Management* 4, no. 2 (June 1996): 55–59; K. H. Roberts, "New Challenges to Organizational Research: High Reliability Organizations," *Industrial Crisis Quarterly* 3 (1989): 111–125; and K. E. Weick, "Organizational Culture as a source of high reliability," *California Management Review* 29 (1987): 112–127. Much of this research is a response to Charles Perrow, *Normal Accidents: Living with High-Risk Technologies* (New York, NY: Basic Books, 1984), which argued that accidents are almost inevitable or "normal" in complex systems.

and material traits of a racial, religious, or social group."[5] While accurate, this diagnosis was problematic for NASA. Which beliefs, social forms, or material traits did NASA need to change? The Columbia Accident Investigation Board did not elaborate.

From a historical perspective, was the NASA culture that produced the amazing feats of the 1960s the same culture that also created the disasters of 1986 and 2003? If not, then what changed? Furthermore, can we pinpoint the specific beliefs, social forms, or material traits within NASA's people and organization that cause failure? As NASA celebrates its 50th anniversary in 2008 and embarks on a new journey back to the Moon, can it recreate the magic of Apollo, or will its cultural baggage set the stage for a tragedy in deep space? The fate of America's civilian space agency, and perhaps of humanity's future in space, depends critically on whether NASA understands and can improve its culture sufficiently to make long-term endeavors in deep space viable.

NASA's Original Technical Culture

Between October 1958, when NASA began operations, and July 1960, NASA acquired a number of research and development organizations. From the NACA, NASA inherited three research Centers: Langley, Lewis, and Ames. From NRL, NASA acquired the Vanguard division, which formed the base of GSFC. The Army transferred Caltech's JPL to NASA control as well as the ABMA, which became MSFC. NASA also acquired some projects from the U.S. Air Force, including the F-1 engine used for the Saturn launch vehicle. NASA Headquarters attempted to weld these disparate organizations into a coherent agency.[6]

Despite their differences, these organizations shared some common characteristics, which political scientist Howard McCurdy has identified as NASA's "original technical culture." A strong "in-house" technical competence was shared among all of NASA's original organizations and personnel. They had decades of experience of hands-on technical work and were at least as competent as the contractors that NASA managed as it grew in the 1960s. A crucial part of NASA's technical competence was its insistence on rigorous testing, which grew more elaborate along with NASA's machines. NASA also prided itself on its exceptional personnel. Its original staff members and those it hired in the expansion years of the early 1960s were among the best and brightest that

---

5. *Webster's Ninth New Collegiate Dictionary* (Springfield, MA: Merriam-Webster, Inc., 1991).
6. Howard E. McCurdy, *Inside NASA: High Technology and Organizational Change in the U.S. Space Program* (Baltimore, MD: Johns Hopkins University Press, 1993), chap. 1.

the United States (and its allies!) had to offer. Spaceflight had a glamour and excitement in the 1960s that attracted exceptionally bright and talented staff.[7]

A key organization for Apollo, and unique in its heritage and capabilities, was MSFC's "Rocket Team" under Wernher von Braun. Von Braun himself was one of the founders of rocket technology, as the leader of the V-2 project in Nazi Germany. He was by all accounts a charismatic visionary, an extraordinary manager, a technical leader, and a cultured, charming man. His team of German engineers had been together for decades, working on the same technology throughout that time. They all knew their tasks and how they related to the tasks of their team members. Von Braun, in 1963, described his management style as that of a gardener nurturing and cultivating a capability grown over years, a rather accurate description of the evolution of his team.[8]

Von Braun used simple but effective methods that capitalized on this experienced, "organic" group. He used a policy of "automatic responsibility," whereby division leaders, and even low-level engineers, were required to take responsibility to resolve problems they uncovered, even outside of their own local organizations. If the problem was outside of their area, they were required to alert the relevant organizations of the issue, at which point they were then "automatically responsible" to resolve them. For difficult technical issues, he chaired meetings where the key parties openly debated their views and disagreements. Von Braun would summarize and explain the issues, and then he would make a decision as to how the organization would proceed. By the late 1960s, von Braun also implemented a system of "Monday Notes," whereby all of the division leads would submit a single page of their major issues to von Braun, who would then comment on the full set of notes and circulate them to the entire team. These relatively informal but rigorous techniques worked well due to von Braun's tact and competence and the intimate knowledge that each team member had with other team members.[9]

With the exception of MSFC's unique group, by and large, NASA's extremely experienced and competent engineers and scientists were not particularly

---

7. McCurdy, *Inside NASA*, chap. 2.
8. Michael J. Neufeld, *Von Braun: Dreamer of Space, Engineer of War* (New York, NY: Vintage Books, 2007); Wernher von Braun, "Management of Manned Space Programs," in *Science, Technology, and Management*, ed. Fremont E. Kast and James E. Rosenzweig (New York, NY: McGraw-Hill, 1963).
9. Phillip K. Tompkins, *Organizational Communication Imperatives: Lessons of the Space Program* (Los Angeles, CA: Roxbury, 1993), pp. 62–70; Yasushi Sato, "Local Engineering and Systems Engineering: Cultural Conflict at NASA's Marshall Space Flight Center, 1960–1966," *Technology and Culture* 46, no. 3 (July 2005): 570–575.

good at managing large, complex projects. The NACA's engineers trained on high-technology programs, but these were typically in association with contractors and often with DOD, which managed the truly large-scale programs and manufacturing capabilities. Engineers and scientists from JPL and NRL had similar backgrounds, and they were frequently researchers more than designers or project managers. NASA primarily organized itself by informal committees, ultimately reaching the point where, on the Mercury project, it created a committee to organize the other committees.[10] To remedy this chaotic situation, NASA hired George Mueller in 1963 from TRW's Space Technology Laboratories to head the Office of Manned Space Flight. Mueller quickly realized that he needed to reorganize NASA Headquarters to convert its hands-on engineers into executive managers and that he needed help from outside of NASA to manage the massive Apollo program.[11]

Mueller's most important recruit was Minuteman ICBM Program Manager Samuel Phillips. Phillips had made a name for himself as a manager by bringing this large and complex project to deployment on time and under budget, a rarity for large aerospace projects. The Air Force agreed to assign Phillips to NASA, but only if he became Apollo Program Manager. In January 1964, Phillips submitted a request to his former boss, Air Force Systems Command Chief Bernard Schriever, for further Air Force personnel to be assigned to Apollo to help manage the massive program. Schriever agreed and transferred over 150 senior, middle, and junior officers to NASA.[12]

Mueller, Phillips, and their military cohorts brought to NASA a management system developed during the previous 15 years of ballistic missile development. This included several key elements; the most prominent were concurrency, change control and configuration management, environmental testing, systems engineering, phased planning, and project management.

- *Concurrency* was a method to speed up development by designing, developing, manufacturing, and testing a missile's various pieces and support systems in parallel. This required more detailed planning than serial design and development, since changes in one component often impacted related components, causing simultaneous changes.

10. Stephen B. Johnson, *The Secret of Apollo: Systems Management in American and European Space Programs* (Baltimore, MD: Johns Hopkins University Press, 2002), pp. 116–120.
11. W. Henry Lambright, *Powering Apollo: James E. Webb of NASA* (Baltimore, MD: Johns Hopkins University Press, 1995): 114–118; Johnson, *The Secret of Apollo*, pp. 130–135.
12. Johnson, *Secret of Apollo*, pp. 135–137.

- To handle this problem, engineers developed *change control* so that changes in one component had to be approved by a central systems engineer who coordinated the impacts of those changes on other components.
- *Configuration management* was the use of change control by managers to ensure that cost and schedule estimates were submitted along with each technical change, so as to predict its cost and schedule ramifications. This gained managers some cost and schedule prediction capability through the ability to veto changes that were too expensive or delayed schedules.
- *Environmental testing* improved system reliability by testing a prototype design in a simulated environment in which system components had to operate, such as the projected temperature ranges, vibration levels, and vacuum environment.
- *Phased planning* provided top-level managers with checkpoints in the project's development cycle, at which managers could cancel a project if it was projected to have insurmountable technical, cost, or schedule risks.
- *Systems engineering* encompassed all of these facets, including systems analysis to trade off potential design solutions.
- *Project management* organized a project on the basis of its technical products, as opposed to the disciplines from which the individuals that staffed the project were drawn. Each portion of a project organization was organized around its individual product, such as a structure, a guidance system, or a rocket engine.

These techniques evolved as responses to technical or managerial failings within the Air Force during the 1950s. For example, project management was implemented in response to management issues in early 1950s missile projects, where personnel being yanked from one group to another by line management in charge of many projects left critical projects without needed staff. Change control and environmental testing were responses to ballistic missile test failures caused by mismatched components when design changes had not been communicated between different groups, or components failed due to unexpected environmental factors. Configuration management and phased planning were responses to cost and schedule overruns on a variety of large scale military development programs.[13]

---

13. Ibid., chaps. 2 and 3; Thomas P. Hughes, *Rescuing Prometheus* (New York, NY: Pantheon, 1998), pp. 106–139.

Recognizing Apollo's size and complexity, NASA brought top-level management of the entire program to NASA Headquarters. At Administrator James Webb's insistence, Headquarters hired General Electric and Bellcomm (an offshoot of American Telephone and Telegraph specifically established for the purpose) to provide Apollo program support to Headquarters.[14] By the late 1960s, Headquarters was controlling cost and schedules through Phillips's system of configuration management. To make it work, Phillips needed NASA's unruly designers to define Apollo's actual design. Once defined, this "technical baseline" could be "frozen." This baseline configuration would not be changed unless a change request was made with proper technical, cost, and schedule justification. While Phillips faced a number of objections from MSFC and JSC management who were not eager to be controlled by Phillips's system, through persistence and persuasion by the end of 1966, he was well on his way to full implementation of this system, collectively called "systems management," over Apollo's technical committees.[15]

Failures of the 1960s: Strengthening Systems Management

In the early planetary programs of the late 1950s and early 1960s, a similar evolution from simple committee structures and processes to more sophisticated and bureaucratic methods occurred. The Jet Propulsion Laboratory began in World War II as an Army-funded organization to develop ballistic missiles. During the development of the Corporal missile, JPL ran into the same difficulties with missile failures and cost and schedule overruns as the Air Force had in its ballistic missile programs, and it developed the same kinds of solutions. It implemented these solutions, enumerated in the previous section, in the follow-on Sergeant program, resulting in much higher reliability in Sergeant than Corporal.[16]

After the launch of Sputnik in October 1957 and the subsequent failure of Vanguard's first launch attempt two months later, JPL Director William Pickering gained Army approval to build the satellite for the Army's first attempt to place a spacecraft in orbit, Explorer 1. Its success in January 1958 and JPL's subsequent transfer to NASA the next year put JPL on the path to lead NASA's planetary exploration. In the race against the Soviet Union into space, JPL placed a higher priority on speed than on reliability, and not surprisingly, its

---

14. Johnson, *Secret of Apollo*, pp. 124–125.
15. Ibid., pp. 139–141; Stephen B. Johnson, "Samuel Phillips and the Taming of Apollo," *Technology and Culture* 42, no. 4 (October 2001): 683–709.
16. Johnson, *Secret of Apollo*, pp. 80–99.

early satellites had a high rate of failure, roughly 50 percent in the late 1950s and early 1960s. However, on the Ranger program, which was to take close-up pictures of the lunar surface just prior to crashing into it, a series of six consecutive failures proved more than was politically acceptable. These failures led to congressional investigations of the implementation on Ranger and all of JPL's later spacecraft of the methods evolved from Corporal and deployed on Sergeant. Having already deployed and improved these methods on the Mariner project to send a spacecraft to Venus in 1962, Mariner Project Manager Jack James spearheaded JPL's early efforts to deploy them on other JPL projects and NASA robotic spacecraft programs. These systems engineering methods dramatically improved the reliability of JPL's spacecraft from then on.[17]

In the meantime, Apollo continued rapidly forward in its determination to land an American astronaut on the Moon before the Soviets. By 1966, Samuel Phillips's implementation of systems management techniques was well under way but hardly complete. When three astronauts died on the launchpad on 27 January 1967 during a prelaunch test, the resulting investigation put Apollo and its management methods under the microscope. The accident investigation, run by NASA, concluded that the Agency had severely underestimated the danger of a pure oxygen atmosphere at sea level pressure. The Apollo 204 fire had been caused by a spark in the Apollo Command Module, which ignited the pressurized, pure oxygen atmosphere. Earlier warnings about the potential danger from General Electric safety personnel had been forwarded to NASA, whose safety groups concluded that the risk was acceptable.[18]

Phillips's methods survived the scrutiny unscathed and even strengthened. He had been actively implementing configuration management over NASA's committee structures since his arrival at NASA, and he had uncovered problems with North American Aviation, the prime contractor for the Apollo Command Module and Saturn second stage. By its silence about Phillips's methods, the investigative team and Congress sanctioned Phillips's techniques. In an interesting brief sentence, Congress noted cultural issues played a role: "The committee can only conclude that NASA's long history of testing and launching space vehicles with pure oxygen environments at 15.7 psi and lower pressures led

---

17. Ibid., pp. 92–114.
18. Johnson, *Secret of Apollo*, pp. 145–147; Mike Gray, *Angle of Attack: Harrison Storms and the Race to the Moon* (New York, NY: Penguin, 1992), pp. 232–235; Alexander Brown, "Accidents, Engineering, and History at NASA, 1967–2003," in *Critical Issues in the History of Spaceflight*, ed. Steven J. Dick and Roger D. Launius (Washington, DC: NASA SP-2006-4702, 2006), pp. 379–383.

to overconfidence and complacency." Success bred complacency and created the conditions for future failure.[19]

In the fire's aftermath, NASA made many technical design improvements to Apollo and implemented a new safety system, while Phillips implemented more project reviews and strengthened configuration control.[20] Apollo went on to a series of spectacular successes. These included the first piloted lunar landing of Apollo 11, the near-disaster and heroic recovery of Apollo 13, and several valuable science missions up to Apollo 17. After the successful Apollo 11 landing in July 1969, a congressional hearing and staff study gave NASA the opportunity to showcase its management system, which was widely believed to be one of the primary reasons for Apollo's success. With Apollo, NASA had earned a reputation as an organization capable of incredible technical feats. NASA was an extraordinarily competent and confident institution. However, NASA's competence would soon begin to erode, and its confidence would be ultimately misplaced.[21]

Weakening Systems Management: To *Challenger*, Hubble, and Mars Observer

From its inception in 1958 through the early 1960s, NASA's workforce grew dramatically, up to 36,000 in 1967, and the contractor force working for NASA grew even faster, peaking at roughly 300,000 in 1966. After that time, NASA's workforce slowly declined, and the contractor workforce dramatically shrank, down to 100,000 by 1972. NASA was generally able to reduce its force through regular attrition, though from 1972 to 1975, NASA had to lay off workers. NASA's workforce decline was over by the early 1980s, with roughly 22,000 personnel in 1982. From 1967 through the 1980s, NASA's hiring remained anemic, and the average age of NASA's technical personnel peaked in 1982 at 44.5 years old.[22]

---

19. Johnson, *Secret of Apollo*, pp. 143–150; Nancy G. Leveson, "Technical and Managerial Factors in the NASA Challenger and Columbia Losses: Looking Forward to the Future," in *Controversies in Science & Technology: From Climate to Chromosomes*, ed. Daniel Lee Kleinman, Karen A. Cloud-Hansen, Christina Matta, and Jo Handelsman (New Rochelle, NY: Mary Ann Liebert, 2008), p. 257. Quotation from Senate Committee on Astronautical and Space Sciences, *Apollo 204 Accident*, 90th Cong., 2nd sess., 30 January 1968, with additional views, pp. 9–10.
20. Johnson, *Secret of Apollo*, pp. 143–150. Phillips's centralization went too far by 1968, slowing the program's progress, and Phillips relaxed some of his new overzealous rules that brought the smallest modifications to the attention of executive management.
21. House Committee on Science and Astronautics, *Apollo Program Management*, Staff Study for the Subcommittee on NASA Oversight, 91st Cong., 1st sess., July 1969.
22. McCurdy, *Inside NASA*, pp. 101–106.

In the meantime, in January 1972, President Richard Nixon approved NASA's next major human spaceflight program, the Space Shuttle. The Shuttle Program was sold on the basis that it would provide low-cost access to space. NASA intended it to become the sole transport system for all U.S. payloads and astronauts. To enable this, NASA needed the support of the Air Force. The Air Force needed the Shuttle to have a much larger payload bay able to deploy reconnaissance satellites, and a larger cross-range capability, which required larger wings to maneuver in the atmosphere. However, funding was limited in the 1970s, and Nixon approved a $5.5 billion development program, which was far less than what NASA needed to develop a fully reusable system. This limitation forced design changes on the Shuttle Program, making the Shuttle only partly reusable, using a throwaway external tank and SRBs that could be refurbished between flights. In combination, these conflicting goals and insufficient development funds put strains on the Shuttle Program that would contribute to its later failures.[23]

The Shuttle's development proceeded during the 1970s, and it was organized with the "lead Center" concept, whereby JSC led the program, instead of NASA Headquarters as on Apollo. Because JSC was at the same institutional level as MSFC and KSC, it had less clout for the Shuttle than NASA Headquarters had had for Apollo. Despite a number of technical problems, including the complicated SSME and the novel tiles of the orbiter's thermal protection system, the Shuttle's development proceeded, though with some delays and cost increases. The first flight of the Space Shuttle in April 1981 was perhaps the riskiest mission NASA ever attempted. This was the first, and to date the only, time a new launch vehicle's first test flight had astronauts on board. While successful, it showed NASA's extreme self-confidence at the time. Despite the loss of over one-third of its civil servants and two-thirds of its contract personnel, those that remained were very experienced and were able to pull it off. This further confirmed NASA's confidence in its own abilities.[24]

However, subtle shifts in NASA's engineering and management practices, as well as changes in the attitudes of its personnel, were weakening the Agency's abilities. NASA's goals remained ambitious, yet the sudden drop in funding in the early 1970s and its continued tightness through the early 1980s made

23. T. A. Heppenheimer, *The Space Shuttle Decision: NASA's Search for a Reusable Space Vehicle* (Washington, DC: NASA SP-4221, 1999), chap. 9.
24. T. A. Heppenheimer, *Development of the Space Shuttle 1972–1981: History of the Space Shuttle*, vol. 2 (Washington, DC: Smithsonian Institution Press, 2002).

the achievement of these ambitions problematic. At the same time, the federal government levied more regulations on the Agency to ensure external oversight and compliance with other goals such as workplace and environmental safety and workforce diversity. These new regulations drove an ever-larger burden of paperwork and a corresponding increase in administrative personnel as compared to technical workers. This made NASA a less desirable place to work as compared to its glory days in the 1960s; NASA had more paperwork and less hands-on engineering. The increasing regulations decreased the Agency's decision-making flexibility, with more weight given to cost and schedule factors, along with other regulations. Promotions were harder to come by, and with fewer jobs available, many highly qualified personnel took less demanding positions simply to remain employed. With fewer projects, and those few projects now under greater levels of scrutiny, the management of these projects became more averse to risk, while being required to pay closer attention to schedules and budgets. As a result of all these factors, morale suffered.[25]

Within the Shuttle Program, these factors combined to create conditions that made catastrophic decisions almost inevitable. In particular, MSFC, which was the institution in charge of the SSMEs, SRBs, and the external tank, underwent a number of changes after Apollo that weakened its abilities. The first was the loss of von Braun himself, who left MSFC in 1970. Von Braun's deputy, Eberhard Rees, was MSFC Director until 1973. Rocco Petrone took over until 1974, followed by William Lucas, who was Director from 1974 until the aftermath of the *Challenger* accident in 1986. While Rees understood von Braun's management system, neither Petrone nor Lucas caught its nuances. As the German team retired or were forced out (as many of them believed occurred under Petrone's regime), the informal bonds of von Braun's "organic" team broke down, while a formal system of systems engineering had not really taken hold at MSFC.[26]

Up to and through the 1960s, von Braun's team neither needed nor wanted systems engineering, which is in essence a formal method to ensure proper communications among different engineers and their disciplines in building a product. The German team did not need formal coordination methods, as they knew what to do and when to do it. Von Braun insulated systems engineering and Phillips's centralizing methods from MSFC's core engineering laboratories

25. McCurdy, *Inside NASA*, pp. 90–124.
26. Andrew J. Dunar and Stephen P. Waring, *Power to Explore: A History of Marshall Space Flight Center 1960–1990* (Washington, DC: NASA SP-4313, 1999), pp. 152–169.

and committees by placing systems engineering in the Industrial Operations Directorate (IOD), created in 1962. The Directorate was ultimately headed by Air Force Colonel Edmund O'Connor, whom Samuel Phillips recommended to von Braun to lead the new organization in September 1964. The relationship between Phillips and O'Connor ensured close communication between Headquarters and MSFC, while IOD minimized the impact of Phillips's Air Force-based processes on MSFC's less formal methods, which were the standard techniques in R&D Operations, where the various MSFC laboratories were institutionally housed. Though MSFC began to adopt systems engineering in the 1960s, not all parts of the organization fully accepted it.[27]

Finally, personalities mattered. William Lucas, MSFC's Director from 1974 to 1986, was an extremely intelligent, but difficult, person to work with. Lucas demanded precision from his MSFC managers and engineers, but, unlike von Braun, he did not appreciate hearing of bad news. Whereas von Braun sought out problems and rewarded those that brought problems into the open, Lucas often grew angry when he learned of problems, with a "shoot the messenger" attitude. Engineers at MSFC worked very hard to avoid mistakes, so as not to face Lucas's wrath. However, this also inhibited open discussions, since few were willing to talk about their work until they had done everything possible to prepare for a technical grilling. Lucas also modified the "Monday Notes" system, which became a method of upward communication only, without von Braun's commentary and feedback. Lucas used systems engineering reviews and configuration management to control MSFC's portions of the Shuttle Program, but the underlying attitude of fear and the resulting lack of communication subverted one of the primary goals of systems engineering, which is to enhance communication and ensure proper cross-checks and balances in the engineering design and decision process.[28]

While none of these issues alone caused the *Challenger* accident of 28 January 1986, they all contributed to the continuation of problems with the Shuttle's SRBs and to the fatal decision to launch the Shuttle despite record cold temperatures and a recommendation from the contractor, Thiokol, that the flight should be delayed. Problems with the SRBs had manifested themselves

27. Johnson, *Secret of Apollo*, pp. 150–152; Sato, "Local Engineering and Systems Engineering," pp. 564–578; Tompkins, *Organizational Communication Imperatives*, pp. 76, 88–90. Even as late as the early 1990s, some pockets of resistance to systems engineering remained, as propulsion experts claimed that systems engineering is "what all good engineers do." Author's recollection of meetings at MSFC from that time period.
28. Tompkins, *Organizational Communication Imperatives*, chap. 10.

starting in November 1981 with the flight of Space Transportation System (STS)-2, when the first incident of O-ring erosion (partial burning and charring of the rings) was detected after the flight. Efforts to understand and fix the problem found a number of issues, including problems with the putty that insulated the rings from the SRB flames in flight, rotation of the joint, and some stiffness in the rings in lower temperatures. Further O-ring erosion incidents occurred, but from that time until 1986, engineers and managers ultimately decided that the Shuttle could continue to fly, sometimes citing the fact that the system had a redundant O-ring. Even though tests had shown by 1978 that the second O-ring was ineffective as a backup, not until 1982 was the SRB joint design considered nonredundant, and even after that time, decision-makers continued to treat the design as if its redundancy was effective. Over time, O-ring erosion became classified as typical and acceptable behavior.[29]

On the night of 27 January 1986, with record cold predicted for the next day's launch, a Flight Readiness Review (FRR) teleconference was held to decide whether the Shuttle would fly the next morning. For the first time ever, Thiokol engineers, concerned that the low temperature would stiffen the O-rings sufficiently to cause them to fail in flight (allowing the hot gases to blow through the rings), recommended that the launch be delayed. Inquiring further to understand the recommendation's basis, NASA engineers and managers aggressively questioned Thiokol and concluded that Thiokol's argument, constructed quickly earlier in the day, was technically flawed. This was based largely on the existence of various kinds of problems associated with the SRBs described above and the inability of Thiokol engineers to differentiate temperature effects from other causes. Caucusing privately with the phone on "mute," Thiokol managers and engineers agreed with NASA's point that they could not prove that the SRBs would fail. Thiokol managers decided to reverse their recommendation, and the flight went forward the next morning.[30]

During the course of the discussions that evening, the essential point of an FRR had been unconsciously subverted. The FRR was intended to prove that the Shuttle could fly. Sufficient doubt about this should have been sufficient

---

29. Diane Vaughan, *The Challenger Launch Decision: Risky Technology, Culture, and Deviance at NASA* (Chicago, IL: University of Chicago Press, 1996), chap. 4; Leveson, "Technical and Managerial Factors," pp. 247–251. Vaughan claims that NASA treated O-ring erosion as "normal," while Leveson disagrees, stating that NASA considered it "acceptable," though not normal. I treat this issue in the last section, essentially claiming that this behavior was no longer considered "anomalous" by some members of the community.
30. Vaughan, *Challenger Launch Decision*, chap. 8.

to stop the launch. Unfortunately, NASA's intense questioning raised doubts at Thiokol about their own arguments and gave the impression that NASA disagreed with them. Given that NASA ultimately had technical and funding authority over Thiokol, the contractor decided it could not press the point against its customer when it could not prove its case. Other NASA personnel who had doubts did not speak up. Thiokol should not have had to prove the case, and NASA management fatally accepted the changed position.[31]

In the aftermath of the *Challenger* accident, the Rogers Commission concluded that NASA's decision-making processes had been fundamentally flawed, largely due to communication problems between engineering and management personnel and their differing perspectives. The commission noted that the safety program on the Shuttle was significantly weakened in comparison to Apollo, and it was effectively "silent" regarding the problems leading to *Challenger*. This was due in part to safety functions being part of the program, with little independence from the Program Manager, who had to account for cost and schedule concerns. The commission included indictments of faulty information flows and the fact that executive management did not receive information about O-ring problems over several years prior to the accident or information about the controversial decision the night before the tragedy.[32]

Marshall Space Flight Center was also involved with another major NASA embarrassment, the flawed optics of the Hubble Space Telescope, which was discovered shortly after the telescope's launch on 27 April 1990. The resulting investigation could not determine with absolute certainty the cause of the spherical aberration problem, but it hypothesized how a flawed measurement of the lens's position for grinding led to a 1.3-millimeter error and to the mirror's ultimately being ground too flat at and near its edges. More troubling than the error itself was the fact that optical tests that would have found the problem had been deleted due to a variety of budget issues in the 1980s. As with the Shuttle, NASA had sold the Hubble Space Telescope to Congress and the Gerald Ford administration on the basis of a cost estimate that was far too optimistic for the level of technical complexity the project entailed. NASA

---

31. Vaughan, *Challenger Launch Decision*, chap. 9; Larry B. Rainey, Kevin B. Kreitman, Bradley A. Warner, and Stephen B. Johnson, "Critical Thinking," in *Methods for Conducting Military Operational Analysis*, ed. Andrew B. Loerch and Larry B. Rainey (Military Operations Research Society, 2007), chap. 18, pp. 607–611; Stephen B. Johnson, "Revisiting the *Challenger*," *Quest: History of Spaceflight Quarterly* 7, no. 2 (1999): 18–25.
32. Alexander Brown, "Accidents, Engineering, and History at NASA," pp. 383–388; Leveson, "Technical and Managerial Factors," pp. 251–254.

judged the technical risks of not performing these tests (which were typical on the reconnaissance satellites on which the Hubble Space Telescope optics design was based) to be acceptable.[33]

NASA's deep space projects were not spared the string of disasters. On 21 August 1993, JPL lost communications with its Mars Observer probe as it neared the Red Planet. While the cause of the accident could not be definitively determined, the resulting investigation concluded that the most likely cause was a propellant system rupture that occurred during a monomethyl hydrazine propellant tank repressurization in preparation for the Mars orbit insertion burn. The investigation board speculated that a valve leak allowed nitrogen tetroxide to leak into the propellant system tubing, such that when the repressurization occurred, the monomethyl hydrazine reacted with the nitrogen tetroxide to rupture the tubing, causing the spacecraft to spin up, which then triggered spacecraft fault protection software to stop the command sequence prior to turning on the radio transmitter back to Earth. Despite cost overruns and schedule slips that increased the cost of Mars Observer from $250 to $800 million, NASA could not ensure a successful mission. Contributing to the fiasco was the growth of the spacecraft's complexity because it was the first mission returning to Mars since the mid-1970s Viking missions; it also used a fixed-price contract. These two factors complicated the system and focused the project's attention on cost to the detriment of reliability.[34]

NASA drew three separate and inconsistent lessons from these failures. *Challenger* drew attention to technical and communication problems with regard to safety in the human flight program, and the human flight program took immediate action to fix the technical problems, oust managers that had been directly implicated in the flawed decisions, and improve safety by creating an independent safety organization at NASA Headquarters. The Hubble Space Telescope embarrassment placed emphasis on the need to ensure proper testing for large-scale robotic projects, but in the long term, the Shuttle missions to fix the optics and later to replace and improve its instruments made the Hubble Space Telescope and the Shuttle a heroic and successful combination in the eyes of the public. The embarrassments of 1990 were largely forgotten.

---

33. Robert W. Smith, *The Space Telescope: A Study of NASA, Science, Technology, and Politics* (Cambridge, U.K.: Cambridge University Press, 1993), pp. 399–425.
34. Howard E. McCurdy, *Faster, Better, Cheaper: Low-Cost Innovation in the U.S. Space Program* (Baltimore, MD: Johns Hopkins University Press, 2001), pp. 18–19; "Mars Observer Investigation Report Released," NASA Press Release 94-01-05, 5 January 1994; Mars Observer Mission Failure Investigation Board, *MARS OBSERVER Mission Failure Investigation Board Report*, 31 December 1993.

The loss of Mars Observer contributed to an entirely different dynamic, as pressures on NASA's budget led to a new initiative to make robotic science satellites much smaller and cheaper, so that the loss of any one of them would not be a major problem.

Alternatives to Systems Management

In the 1980s and 1990s, systems management, as it had been practiced from the mid-1960s, came under increasing criticism. To many, NASA's increasingly bureaucratic system appeared unproductive and wasteful. By the early 1990s, systems management had apparently been unable to prevent major disasters and tragedies, despite its perceived high costs and bureaucratic cross-checks. External events were drawing attention to the relative failings of American management in general and to potential new approaches. NASA management, in part because of federal government directives, began to consider alternatives to improve productivity, lower costs, and provide better service to its customers.

By the early 1980s, American competitiveness in certain key industries, most prominently automobiles and commercial electronics, was declining rapidly in the face of foreign, and in particular Japanese, competitors. American management experts began to look to Japan and other nations to search for the secrets of these dramatic and unexpected foreign successes. Japanese culture, with its emphasis on cooperation instead of competition, seemed to uniquely adopt and adapt American statistical quality control methods from World War II; the Japanese created a new and powerful tool: Total Quality Management (TQM). These methods were publicized by journalists, corporate executives, and management experts and became national topics of conversation by 1981.[35]

Experimentation with TQM methods soon began in American corporations and in certain branches of the U.S. government including NASA, which became one of the early adopters of the new management technique. By 1990, TQM activities at NASA were being coordinated by the Safety Mission Quality Office at Headquarters. A report in that year boasted of a number of ongoing TQM initiatives. In 1989, Lewis Research Center won the U.S. government's Quality Improvement Award and was teaching TQM seminars to other government

35. William M. Tsutsui, *Manufacturing Ideology: Scientific Management in Twentieth-Century Japan* (Princeton, NJ: Princeton University Press, 1998); Christopher Byron, "How Japan Does It," *Time* (30 March 1981): 54–60; William Ouchi, *Theory Z: How American Business Can Meet the Japanese Challenge* (Reading, MA: Addison-Wesley, 1981); Ezra F. Vogel, *Japan as Number One: Lessons for America* (Cambridge, MA: Harvard University Press, 1979).

organizations; MSFC executives met with TQM founder Edward Deming; and SSC established a steering committee to implement a TQM program at every Center. This same report also categorized dozens of traditional activities to improve technologies and processes as TQM-related improvements, though it appears unlikely that TQM inspired or controlled many of them. Dan Goldin, who became NASA's Administrator in 1992, was a strong believer in TQM and made it an Agency priority.[36]

While many NASA organizations took TQM seriously and a number of NASA managers gave it executive-level support, NASA's rank and file remained largely skeptical. Total Quality Management's emphasis on work processes and on serving its customers seemed only marginally applicable to NASA. Many of NASA's jobs were one-of-a-kind research tasks, or development tasks that changed over the course of a project, though similar to tasks on other projects. Defining NASA's customer was even more problematic. Was the customer Congress, the President, the American people, or merely other NASA engineers that used NASA test results or analyses? Finally, how did one define the productivity and quality of NASA's products? While quality could be related to NASA's traditional quality assurance functions, productivity was not something easily quantified in NASA's nonprofit, high-creativity environment. In the end, TQM did not take hold; and by the mid- to late 1990s, TQM faded from the NASA scene.[37]

The Jet Propulsion Laboratory was relatively late in using TQM methods, beginning its TQM initiatives in 1991 under its new Director, Ed Stone, to help change JPL's culture to cut through its increasingly cumbersome bureaucracy. By 1993, Stone and his aide, Richard Laeser, recognizing that their initiative was

---

36. NASA Safety and Mission Quality Office, NASA Quality and Productivity Improvement Program, *NASA Total Quality Management 1989 Accomplishments Report*, June 1990, pp. 6–8; Peter J. Westwick, "Reengineering Engineers: Management Philosophies at the Jet Propulsion Laboratory in the 1990s," *Technology and Culture* 48, no. 1 (2007): 74.
37. Historical research on the TQM management fad at NASA has, with a few exceptions, yet to be written. For its impact on JPL, see Peter J. Westwick, "Reengineering Engineers: Management Philosophies at the Jet Propulsion Laboratory in the 1990s," *Technology and Culture* 48, no. 1 (2007): 67–91. A number of the observations made here are from the author's experience. From 1990 to 1991, the author learned TQM at Martin Marietta Corporation Astronautics, which at that time was implementing the TQM Quality Function Deployment technique of generating and assessing requirements on research programs. The author then taught these same methods to some technology R&D groups at MSFC. The problems encountered there were typical of many other attempts to deploy TQM, as the author learned from many student papers on the application of TQM to NASA, the Air Force, and industry while teaching courses on space systems management in the University of North Dakota's Space Studies Department. These observations need to be further fleshed out by research on this subject, but I have no doubt about their general validity when properly qualified by local experiences.

encountering continued resistance among JPL's regular engineering workforce, decided that the key to furthering cultural change was to focus on the lab's processes. Laeser and Stone promoted Mike Hammer's "reengineering" method, which aimed to redefine an organization's processes, starting by charting out the organization's current processes and then redesigning them to eliminate inefficiencies. Stone assigned high priority to the process teams, moving key managers to head them.[38]

The reengineering initiative did not go smoothly. It made certain processes more efficient, such as Voyager mission operations and business processes defined by the International Standards Organization. However, it also proliferated the number of processes, distributed responsibilities in an alarming manner, and intensified rank-and-file resistance to management initiatives, as process ownership increased responsibilities.[39]

The Jet Propulsion Laboratory's efforts at reengineering occurred in parallel with its efforts to respond to Dan Goldin's "faster, better, cheaper" initiatives. Prior to becoming NASA Administrator in March 1992 during the George H. W. Bush administration, Goldin had been an executive at TRW Corporation and was a strong proponent of small satellite technology, including the Space Defense Initiative (SDI) project, Brilliant Pebbles. Upon becoming the head of NASA, his encounters with the current and projected NASA budgets and massive cost overruns on the Space Station program, and the likelihood of limited future funding from Congress, led him to the realization that NASA would do little science unless these missions could significantly reduce costs. In a May 1992 speech at JPL, Goldin discussed the need to reduce spacecraft and mission costs. He elaborated on this theme over the next few months, arguing that NASA needed to build more, but smaller and less expensive, spacecraft, while taking more risks, since the loss of a smaller craft would not be a major disaster to the science program.[40]

Goldin was building on an idea that had been growing in the military and at NASA, that smaller, cheaper spacecraft were appropriate for many robotic missions. The SDI program was studying the launch of hundreds of small spacecraft to intercept and destroy ICBMs. To support this effort, it was miniaturizing a number of technologies to make small, intelligent spacecraft feasible. The $80 million Clementine project was a key demonstrator of the

38. Westwick, "Reengineering Engineers," pp. 73–83.
39. Ibid., pp. 80–84.
40. McCurdy, *Faster, Better, Cheaper*, pp. 48–55.

small satellite philosophy; it started in early 1992 to test ballistic missile defense sensor technologies by performing observations of the Moon from lunar orbit. Built and operated by NRL, its mission to gather data about the lunar surface succeeded in 1994 and provided a concrete example of the "faster, better, cheaper" concept in action. NASA's Earth science community recognized the potential of small spacecraft as well. The Small Explorer program, started in 1988, successfully launched its first spacecraft, the Solar, Anomalous, and Magnetospheric Particle Explorer, in July 1992.[41]

Another inspiration for potential reformers was Lockheed's Skunk Works. This division of Lockheed, based in Burbank, California, had created a host of revolutionary aircraft, including the World War II P-38 Lightning fighter, the P-80 and F-104 jet fighters, the U-2 and SR-71 spy planes, and the F-117 stealth fighter. However, its later fame was based on more than its innovative flying machines; its fame was based on its methods for developing them. Run by Kelly Johnson from World War II until January 1975, and after that by Ben Rich, the Skunk Works had evolved a method of using small teams for its highly secret, high-technology aircraft. Johnson developed a set of 14 rules that defined the constraints and rules to run his Skunk Works projects. These included minimal reporting but critical documentation of "important work," minimizing access by outsiders to the project, delegating authority but retaining a strong project manager, steady funding, and daily interaction with the customer to build trust. Seemingly the opposite of systems management, "faster, better, cheaper" advocates pointed to the Skunk Works approach as a legitimate alternative to systems management.[42]

Ironically, the abortive and potentially massive SEI was also a spur to the development of the "faster, better, cheaper" concept. When the George H. W. Bush administration announced SEI in July 1989, NASA responded with a 90-day study to achieve a human mission to Mars. Its massive costs convinced NSC, which Bush had created in April 1989, that NASA was far too conservative. Many Council members perceived ex-astronaut and NASA Administrator Richard Truly as a member of NASA's old guard that needed to be replaced. In early 1992, they succeeded in their goal and replaced him with Dan Goldin, who they learned was supportive of smaller innovative projects, as NASA's

---

41. Ibid., pp. 46–47, 53–55; Stephanie A. Roy, "The Origin of the Smaller, Faster, Cheaper Approach in NASA's Solar System Exploration Program," *Space Policy* 14 (August 1998): 153–171.
42. Ben R. Rich and Leo Janos, *Skunk Works* (Boston, MA: Back Bay Books, 1994); McCurdy, *Faster, Better, Cheaper*, pp. 90–93.

Administrator. At the same time, the Senate Appropriations Committee directed NASA to develop a plan to "stimulate and develop small planetary or other space science projects." This became the Discovery program, started later that year. By 1993, the Discovery program had two projects in place: Near Earth Asteroid Rendezvous (NEAR) and Mars Pathfinder.[43]

Goldin continued to support and push the "faster, better, cheaper" cause. In 1994, NASA established the New Millennium program to use small spacecraft to flight-test new technologies to enable science missions. Its first mission was Deep Space 1, launched in 1998, which tested ion engines and autonomous navigation technologies. The Small Satellite Technology Initiative also started in the mid-1990s, with the Lewis and the Clark Earth observation satellites. Lewis launched in August 1997, while Clark was canceled due to cost overruns the next year. Finally, the Mars Surveyor program, which consisted of three Mars probes, also used the "faster, better, cheaper" philosophy. Its first launched satellite was the Mars Global Surveyor, which reached Mars in September 1997. All in all, through the 1990s, NASA launched 16 projects related to the "faster, better, cheaper" philosophy in five major programs (Small Explorer, Discovery, New Millennium, Small Satellite Technology, and Mars Surveyor). In addition, other projects, including the proposed Pluto flyby probe, drew Goldin's attention. Under Goldin's watchful eye and direction, it underwent years of studies aimed at reducing costs even further, before it was finally approved as the New Horizons spacecraft and launched in 2006.[44]

Up through 1998, the "faster, better, cheaper" programs had an excellent rate of success, given their lower costs and the higher risks that they assumed. Of the 16 "faster, better, cheaper" projects identified by Howard McCurdy in his book *Faster, Better, Cheaper*, by the end of 1998, 11 had launched, and of these it appeared that only 2 of them had failed: NEAR failed because its orbit insertion burn around asteroid Eros failed in December 1998, and Lewis also failed. Clark was canceled before it ever flew, and thus it failed as a project as well. NEAR's mission ultimately succeeded as it successfully orbited Eros in 2000. Four more "faster, better, cheaper" spacecraft, Mars Polar Lander, Deep Space 2, Stardust, and Wide-Field Infrared Explorer were slated for launches in 1999. Goldin's initiative and prodding to implement "faster, better, cheaper" looked like a stunning success, in particular the very popular Mars Pathfinder

43. Thor Hogan, *Mars Wars: The Rise and Fall of the Space Exploration Initiative* (Washington, DC: NASA SP-2007-4410, 2007); McCurdy, *Faster, Better, Cheaper*, pp. 44–47, 55–56.
44. McCurdy, *Faster, Better, Cheaper*, pp. 6–7, 56–58.

project with its little rover, Sojourner. Unfortunately, the events of 1999 would change that impression dramatically for the worse.[45]

In the meantime, in the human flight program, NASA responded to the *Challenger* accident with several changes. The first was fixing the flawed O-ring design, followed by a variety of other improvements in the safety and reliability of the Shuttle's components and systems. The Rogers Commission also had criticized the communication between NASA's engineers and managers and between NASA organizations. To help remedy this, the management of the Shuttle Program was shifted from JSC to NASA Headquarters. Finally, the Rogers Commission indicted NASA's "silent safety program." NASA's primary response was to create an independent safety organization at NASA Headquarters. However, this seemingly appropriate move was rendered less effective because it never acquired the authority needed to fully discharge its duties. The reporting requirements from the NASA Field Centers remained unclear, and the lines of safety authority and the responsibilities of the safety groups were confused. In addition, within NASA's system, each project purchased safety support, which gave them some latitude and control over the safety function and compromised safety independence in the process.[46]

By the early 1990s, cost-cutting pressures began to affect the Shuttle Program. In the decade from fiscal years 1993 to 2002, NASA's budget declined in real terms by 13 percent. In a period in which space station (soon, the ISS) expenses were taking a larger share of NASA's budget, the budget squeeze hit not only the science programs (where "faster, better, cheaper" was being implemented in large measure due to the funding problems), but the Shuttle Program as well. From 1991, NASA reduced Shuttle operating costs 21 percent by reducing the contractor workforce from 28,394 to 22,387 and the civil service personnel from 4,031 to 2,959. By 1997, contractors and civil servants were down to 17,281 and 2,195 respectively.[47]

These cost reductions were accompanied by organizational changes that some observers believed compromised safety. To reduce costs, Administrator Goldin wanted to take NASA out of repetitive operations such as Shuttle operations, and in 1994, he directed NASA to investigate how to do so. The 1995 Kraft Report claimed that the Shuttle had become a "mature and reli-

45. Ibid., pp. 6–7.
46. Leveson, "Technical and Managerial Factors," pp. 251–252; Columbia Accident Investigation Board, *Report*, vol. 1 (Washington, DC: NASA, August 2003), pp. 99–101.
47. Ibid., pp. 102–107.

able system," that it should "consolidate operations under a single business entity" and should "restructure and reduce the overall Safety, Reliability, and Quality Assurance elements" without compromising safety. These recommendations were accepted and led to the creation of the Space Flight Operations Contract, in which Lockheed Martin and Rockwell created a joint venture called United Space Alliance, to which NASA awarded a sole-source contract for Shuttle operations in 1995. The new managerial arrangement led to a new relationship between NASA and Shuttle safety, known as "insight" instead of "oversight." This meant that instead of directly monitoring and managing the work of NASA and contractor safety personnel, United Space Alliance ran the safety program and provided management with certain contractually agreed information. In 1998, Congress directed NASA to plan for eventual privatization of the entire Shuttle Program. Among other things, this would have made astronauts private employees. Another managerial move was to shift Shuttle Program management from NASA Headquarters back to JSC, which returned the Program to the pre-*Challenger* organizational structure, reversing changes made in response to the Rogers Commission recommendations. Some considered this a safety issue, as the move to Headquarters had been made to improve program communications.[48]

Both in robotic and human flight programs, NASA's emphasis in the 1990s had shifted away from concerns for safety and reliability to pressures to reduce costs. Despite the inherent riskiness of spaceflight, complacency had set in, and it was only a matter of time before it would be shattered.

The End of "Faster, Better, Cheaper," *Columbia*, and the Columbia Accident Investigation Board

For NASA's robotic spacecraft programs, and in particular its Mars science program, 1999 marked the end of an era . . . the "faster, better, cheaper" era. In March, the Wide-Field Infrared Explorer failed shortly after launch when the frozen hydrogen used to cryogenically cool its detectors vented into space after the spacecraft's protective cover was prematurely ejected. It was the fifth spacecraft in the Small Explorer program. In the meantime, three Mars spacecraft were on their way to the Red Planet: Mars Climate Orbiter, Mars Polar Lander, and Deep Space 2. Mars Climate Orbiter was intended to perform observations of the Martian atmosphere from orbit. Mars Polar Lander and Deep Space 2

48. Ibid., pp. 107–110; Leveson, "Technical and Managerial Factors," pp. 256–257.

had been launched together, aiming to land near the poles, with Deep Space 2 containing two subsurface probes to search for water ice.

Hopes were high as Mars Climate Orbiter approached Mars in September 1999. In the week prior to Mars orbit insertion, mission navigators noticed that the spacecraft's trajectory seemed closer to Mars than expected. As it made its closest approach, mission controllers awaited the signal from the spacecraft indicating it had achieved orbit. That signal never came, and attention quickly focused on the odd trajectory. The trajectory problem turned out to hinge on a unit conversion problem. The files delivered from contractor Lockheed Martin had their propulsion maneuvers defined in English units, instead of the specified metric units. The difference was 4.45, the conversion factor of newtons to pounds. The difference in units led to an error conversion factor of 4.45 in the estimated effect of trajectory corrections, and as a result the spacecraft went too close to Mars and burned up in the atmosphere. The operations teams that might have otherwise noticed the error were smaller than on many previous missions, due to the reduced budgets of "faster, better, cheaper."[49]

In December, more bad news, or more accurately, no news at all, came from Mars. Both Mars Polar Lander and Deep Space 2's two probes seemed to be working properly as they entered their entry, descent, and landing phases. None were heard from again. Deep Space 2's failures were never definitively determined, but possibilities ranged from soil being significantly harder than planned when the probes hit the surface, to handling problems at KSC prior to launch that inadvertently sent an electrical pulse that mimicked separation and turned on their batteries, draining them of power.[50]

The investigation of the Mars Polar Lander failure provided a more definitive cause. To detect touchdown, the spacecraft used Hall Effect sensors that detected movement of the spacecraft legs. Leg deployment produced transient signals in these sensors. During deployment, it is almost certain that these transient signals were processed by the software as the real touchdown, turning off the engines while the spacecraft was well above the surface; the spacecraft crashed to the surface and was destroyed. During development, the transient signal problem was known, but the software requirement had been written in such a way that, when ultimately coded, it did not properly meet the intent of

49. McCurdy, *Faster, Better, Cheaper*, pp. 6–7; David M. Harland and Ralph D. Lorenz, *Space Systems Failures: Disasters and Rescues of Satellites, Rockets, and Space Probes* (Chichester, U.K.: Springer-Praxis, 2005), pp. 339–341.

50. Harland and Lorenz, *Space Systems Failures*, p. 239.

the requirement. Testing of the descent and landing did not catch the problem because one of the landing legs was wired incorrectly during this test. After the wiring had been fixed, the test was not rerun due to tight schedules and budgets of the "faster, better, cheaper"-style project.[51]

The failure of all three Mars missions in 1999 drew unwanted attention from both NASA executive management and the press. Mars missions always drew significant interest, and the failure of all three provided strong evidence of programmatic problems. The various failure investigations implicated reductions in testing and in systematic safeguards and cross-checks. In other words, "faster, better, cheaper" had cut more than the fat and into the meat of systems management, leading to failures. Management at JPL reassessed its project management and systems engineering methods, and it found them wanting. Culture change and reengineering distracted management from its core activities and diluted responsibilities. In 2001, Charles Elachi took over for Ed Stone as Director of JPL, and he quickly reinvigorated JPL's historical systems engineering methods and traditions. The lab increased funding for individual projects and reinstituted rigorous design reviews. "Faster, better, cheaper" was out, and systems management was back.[52]

Two years later, on 1 February 2003, a crowd of guests was waiting at KSC for the Space Shuttle *Columbia* to return from its mission to perform a variety of microgravity experiments in low-Earth orbit. Like JPL's mission controllers in 1999, they waited in vain. *Columbia* had broken up over east Texas and was destroyed, along with its crew of seven. The resulting investigation concluded that hot plasma had entered a hole in the leading edge of the Shuttle's left wing, which burned through the structure. The wing fell away, and the Shuttle lost control, tumbled, and broke apart. The hole in the leading edge was created during ascent 17 days before, when insulation foam from the external tank fell off and hit the leading edge at high speed.[53]

Further investigation into the causes of the accident uncovered a trail of events both prior to the fated flight and during the flight itself. Much like the problems leading to the *Challenger* accident 17 years before, foam debris falling off the external tank during ascent was a problem that had been going on from the inception of the Shuttle Program. Also like *Challenger*, this behavior had been reclassified over time from a major safety concern to a minor maintenance

51. Ibid., pp. 330–331.
52. Westwick, "Reengineering Engineers," pp. 84–87.
53. Columbia Accident Investigation Board, *Report*, vol. 1, chaps. 2 and 3.

issue. Foam strikes during ascent had caused minor damage to the Shuttle's thermal protection system on many flights, leading to repairs between flights. Some of the foam pieces, particularly those from the "bipod ramp," were quite large and caused significantly more damage. The real risks of external tank debris hitting the orbiter were misunderstood and underestimated, while the costs to fix the problem were considered too high.[54]

Decision-making during *Columbia*'s flight was also flawed, very much like the decision-making the night prior to *Challenger*'s final flight. During ascent, cameras photographed the foam strike hitting the left wing's leading edge. Shuttle engineers began to assess the potential damage, and even reporters began asking questions about it. Ultimately, the engineers could not determine the actual damage but were worried enough to inquire into the possibility of using a military reconnaissance satellite to photograph the suspect area. Because this request did not go through proper channels, NASA management stopped it. Poor organizational structure inhibited engineering information from making its way to management. Management believed that, even if there was a problem, nothing could be done about it in flight, so it did not make much sense to make extraordinary efforts to determine the amount of damage. Changes in personnel made estimates of the foam strike damage problematic, because the model used to do the estimate was not valid for large pieces such as the one that hit *Columbia*, and the new personnel were unaware of this limitation.[55]

The Columbia Accident Investigation Board noted the many similarities between the organizational and communication problems leading to the *Challenger* and *Columbia* accidents. While finding several managers at fault, the Columbia Accident Investigation Board ultimately found the causes of the accident to be much more insidious than the Rogers Commission had. Some of the major organizational issues the Columbia Accident Investigation Board emphasized included the following:

- Conflicting goals of cost, schedule, and safety, in which safety lost out.
- Overemphasis on bureaucratic procedures, to the detriment of engineering insight and expertise.
- An organization and structure that blocked effective communication of technical problems.

---

54. Ibid., pp. 121–131.
55. Ibid., pp. 140–172.

- Changes to the safety organization that eroded NASA's safety expertise by transferring safety tasks and responsibilities to contractors.
- A lack of resources, independence, authority, and personnel in NASA's safety organization to supply alternate perspectives to developing problems.

Whereas the Rogers Commission cited violations of NASA's procedures, the Columbia Accident Investigation Board concluded that since these issues were common to both the *Challenger* and *Columbia* accidents, the problems leading to the accidents were inherent to NASA itself, part of its "organizational culture." NASA's traditional methods of fixing the technical problems and tightening its procedures was not going to work, since the technical problems were caused by violations of those very organizations, processes, and procedures.[56]

After *Columbia*, NASA quickly went to work to fix the difficult problems with the external tank foam insulation and scrubbed various problematic aspects of the Shuttle's design and operations. It took a number of organizational measures following the Columbia Accident Investigation Board recommendations. It shifted control of the Shuttle Program from JSC back to NASA Headquarters (following the precedents of Apollo and the post-*Challenger* organization). It established the Safety, Reliability and Quality Assurance organization at Headquarters. Flight Readiness Review procedures were modified to allow engineers to participate, and they required astronaut managers to participate and sign off on the launch decision. The Agency established the NASA Engineering and Safety Center at LaRC (and later a second NASA Safety Center at Glenn Research Center), which was tasked to independently review recurring anomalies and act as a resource for connecting engineering to safety issues. Another Columbia Accident Investigation Board recommendation was the creation of an Independent Technical Authority (ITA), which NASA began to implement in November 2004. The ITA funded "Technical Warrant Holders" as technical experts to assess engineering and safety designs and decisions. Ultimately, the ITA was transformed, in February 2006, into a process known as "Process-Based Mission Assurance" (PBMA), which emphasized the development of "technical excellence" as the basis of building safety into NASA's systems. As it evolved, the ITA/PBMA was intended to provide a separate line of communication for technical personnel to air problems. This reinvigorated a matrix manage-

56. Ibid., chaps. 7 and 8, esp. pp. 199–202.

ment system in which personnel reported both to project managers (who controlled the funding and schedules for projects) and functional managers (who controlled the technical content and personnel). Under the new system, the functional managers were tasked with ensuring technical quality and acting as counterbalances to the project managers. The Columbia Accident Investigation Board was far less specific in its recommendations on how NASA should change its culture, and NASA itself had great difficulty trying to determine how to interpret that mandate.[57]

NASA concluded that it needed help with culture change, and in December 2003 it sent out an RFP to perform cultural analysis to pinpoint cultural problems that affected safety and then take measures to fix them. Over 40 bidders responded, and in March, NASA hired Behavioral Science Technology (BST), who then instituted cultural surveys across the Agency; and in a February 2005 report, BST stated that significant progress was being made in cultural change, as measured by its surveys. One new initiative was to train managers to be more open to engineering opinions.[58]

Major changes in NASA personnel and programs quickly began to shift attention away from the Columbia Accident Investigation Board recommendations. *Columbia*'s demise made it clear that the replacement of NASA's Shuttle fleet could no longer be postponed, and this led to a broader assessment of what NASA's goals should be. The result of these discussions was the speech by President George W. Bush in January 2004, announcing the Vision for Space Exploration to complete the Space Station by 2010, retire the Shuttle, conduct the first human mission with a new Crew Exploration Vehicle by 2014, and return to the Moon by 2020. The next month, NASA

---

57. Diane Vaughan, "System Effects: On Slippery Slopes, Negative Patterns, and Learning from Mistake," in *Organization at the Limit: NASA and the Columbia Disaster*, ed. William Starbuck and Moshe Farjoun (Oxford, U.K.: Blackwell, 2005); author's conversation with Michael Griffin, 26 April 2007; "NASA Announces New Safety Center," NASA Press Release, 11 October 2006, available at *http://www.spaceref.com/news/viewpr.html?pid=21031* (accessed 31 January 2009); "Final Report of the Return to Flight Task Group," July 2005, pp. 95–110, available at *http://www.scribd.com/doc/995714/NASA-125343main-RTFTF-final-081705* (accessed 31 January 2009); "Technical Excellence/Technical Authority," 16 March 2006, available at *http://pbma.nasa.gov/index.php?fuseaction=ita.main&cid=501* (accessed 31 January 2009).
58. "NASA Enlists Behavioral Science Technology, Inc. to lead agency-wide culture change," *EDP Weekly's IT Monitor* (26 April 2004), available at *http://findarticles.com/p/articles/mi_m0GZQ/is_17_45/ai_n6264746*; *http://www.FindArticles.com* (accessed 31 January 2009); John Schwartz, "Some at NASA Say Its Culture is Changing, but Others Say Problems Still Run Deep," *New York Times* (4 April 2005), available at *http://query.nytimes.com/gst/fullpage.html?res=9D03EEDB1E3FF937A35757C0A9639C8B63&sec=&spon=&pagewanted=all* (accessed 31 January 2009).

released its initial interpretation of the Vision. A presidential commission gave its assessment in June.[59] NASA established the Exploration Systems Mission Directorate to implement the exciting new program, and NASA's attention quickly shifted to its implementation. Michael Griffin took over from Sean O'Keefe in April 2005. Griffin, who was without question the most technically educated Administrator NASA had had to date, had his own strong opinions about how to address NASA's problems, and BST's cultural surveys were not among them. In June 2005, he terminated the BST contract. He believed that one of the most important things that NASA needed was an organizational structure that provided alternate communication lines for engineering and safety concerns, which was provided by the ITA/PBMA organizational structure and processes. By early 2009, the culture issue, while not forgotten, did not have the priority it had had in the immediate aftermath of the *Columbia* tragedy.[60]

The Social Nature of Failure

Even though much of NASA's attention had shifted from the difficult and uncertain problems of its culture to a new and exciting program of exploration, the culture problem had not gone away. Dealing with the issue remained problematic due to the inherent slipperiness of the concept. The Columbia Accident Investigation Board *Report* described "organizational culture" as "the basic values, norms, beliefs, and practices that characterize the functioning of a particular institution." Explaining further, "organizational culture defines the assumptions that employees make as they carry out their work; it defines 'the way we do things here.'"[61] Something in these basic values, norms, beliefs, and practices led to catastrophic system failure. NASA's problem was, and is, to determine the connection between culture

---

59. President Bush Announces New Vision for Space Exploration Program, "Remarks by the President on U.S. Space Policy," Press Release, 14 January 2004, available at *http://history.nasa.gov/Bush%20SEP.htm* (accessed 31 January 2009); NASA, *The Vision for Space Exploration* (Washington, DC: NASA, February 2004); *A Journey to Inspire, Innovate, and Discover: Report of the President's Commission on Implementation of United States Exploration Policy* (Washington, DC: GPO, June 2004); Frank Sietzen, Jr., and K. L. Cowing, *New Moon Rising: The Making of America's New Space Vision and the Remaking of NASA* (New York, NY: Apogee, 2004); "NASA pulls plug on culture change contract," *Industrial Safety & Hygiene News* (1 August 2005), available at *http://www.highbeam.com/doc/1G1-135467272.html* (accessed 31 January 2009).
60. "Michael Griffin Takes the Helm as NASA Administrator," NASA Press Release, 14 April 2005, available at *http://www.nasa.gov/about/highlights/griffin_admin.html* (accessed 31 January 2009); author conversation with Griffin, 27 April 2007.
61. Columbia Accident Investigation Board, *Report*, vol. 1, p. 101.

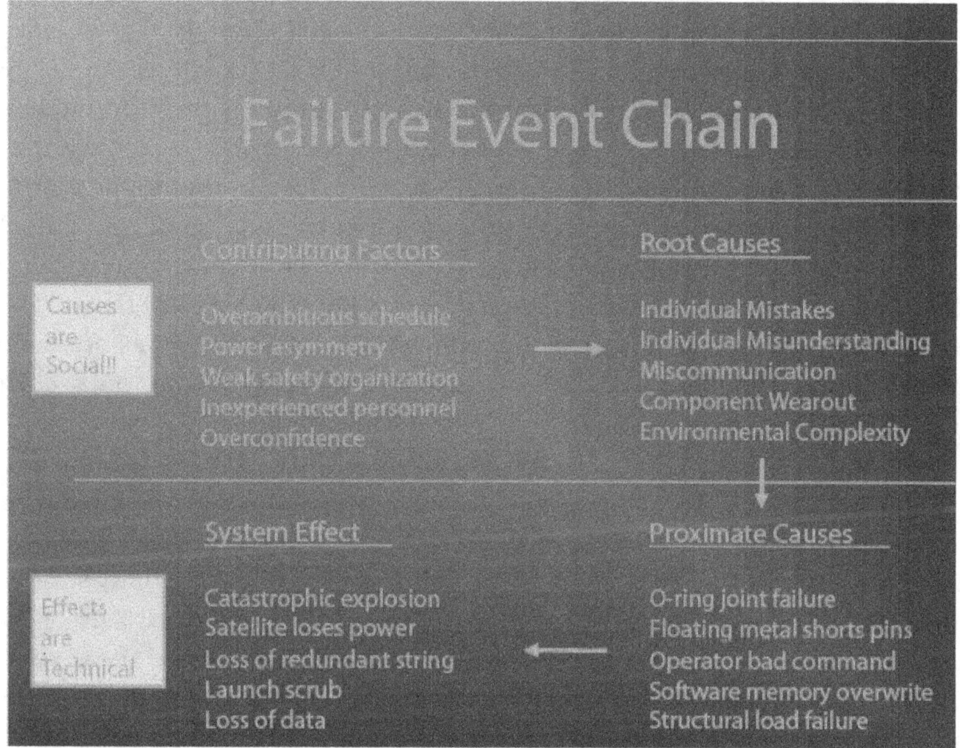

Figure 1: Failure chain of events. *Courtesy of Stephen B. Johnson*

and failure and then to make improvements to culture so as to reduce failure rates and criticality.

To make this connection, we need to understand the nature of failures. In engineering terms, failure is defined as "the unacceptable performance of intended function or the performance of an unintended function."[62] That is, when the system can no longer do what it was designed for, or does things that it was not intended to do, it has failed. Failure is generally the outcome of a chain of events, which are made more likely by various contributing factors. Failure investigations start by assessing the final failure effects, which can include complete system loss, like the Space Shuttle *Columbia*'s burning up in the atmosphere, or can be more benign, such as the scrub of a Shuttle

62. This definition, which is in development on the Constellation program, draws from a variety of engineering sources and has a few improvements to those earlier definitions. See Stephen B. Johnson, "Introduction to System Health Engineering and Management in Aerospace," *Proceedings of the First International Forum on Integrated System Health Engineering and Management* (Napa, CA, November 2005).

launch. The proximate causes of these failures are generally the technical items that malfunctioned and led to the failure effects, such as the O-ring failure of the *Challenger* accident or the foam that fell off the external tank and hit *Columbia*'s wing during ascent. But proximate causes have their genesis in root causes, such as human-induced errors in the application of the foam to the external tank in the *Columbia* case, the decision to launch *Challenger* on a morning when the temperature was lower than rated environmental limits, or human error in creating the Shuttle's original flawed SRB segment joint design. Finally, there are contributing factors, such as pressures to launch the Shuttle on an accelerated schedule, pressures to lower costs, or use of a teleconference instead of a face-to-face meeting contributing to miscommunication.

Frequently the failure effects and the proximate causes are technical, but the root causes and contributing factors are social or psychological. Successes and failures clearly have technical causes, but a system's dependability strongly depends on the human processes used to develop it, the decisions of the funders, managers, and engineers who collectively determine the level of risk. Fallible humans make individual cognitive or physical mistakes, or they make social errors through lack of communication or miscommunication.

Although the statistics have not been studied fully, my sense from experience in the field and from discussions with experienced engineers is that 80 to 95 percent of failures are ultimately caused by individual human errors or social miscommunication between individuals and groups. Most of these are quite simple, which makes them appear all the more ridiculous after the fact when the investigation gets to the root cause and finds, for example, the Mars Climate Orbiter's English-to-metric-unit conversion problem, a nut or bolt left inside the propulsion system (a Centaur failure in 1991), a reversed sign or wiring (for example, the Total Ozone Mapping Spectrometer—Earth Probe), or a single digit left off a command sequence (Phobos 1). Contrary to popular belief, it is the very banality of the causes that makes them so hard to find. We constantly carry out simple daily tasks and communications. Thousands of such tasks and communications happen every day on a project, and any one of them can be the cause of tomorrow's dramatic failure.[63]

63. Harland and Lorenz, *Space Systems Failures*. This book catalogs many types of failures, though it in general discusses the proximate as opposed to root causes.

Failure, then, is caused by a fault, which is defined as "a physical or logical cause, which explains a failure."[64] Faults can be proximate causes or root causes, where the root cause is the first event in the explanatory chain of events. The vast majority of root causes, if pursued far enough, are due to individual or group mistakes by humans. This should be no surprise. Technologies are merely the final products of human knowledge applied to creating useful artifacts, and an artifact merely embodies and incarnates knowledge from its creators. Hence if an artifact has a fault, this is ultimately due to a flaw in the knowledge of its creators or in a mismatch between the knowledge of its creators and that of its users.[65]

Making a system dependable is akin to the problem of reducing the number of needles in haystacks. Most problems are very simple in their causes (the needles), and it is best to prevent them to begin with, as finding them amid all the complexities of the design and how it operates in all possible conditions (the haystack) is very difficult. In essence, dependability is gained by minimizing the number of initial mistakes (fewer needles) and testing the system to find the inevitable mistakes that occur (finding and removing the remaining needles). Skunk Works or "faster, better, cheaper" approaches can succeed because small, experienced teams make fewer mistakes because there are simply fewer people, and with experience they make fewer mistakes as individuals, and also because having fewer people reduces the number of interactions between people where miscommunication may occur. In addition, experienced personnel have the intuition to sense where the remaining mistakes are likely to be found, so they can target their relatively smaller documentation and testing to find them. However, over the long run, small teams cannot provide repeatable results. That is because humans are unable to maintain focus for long periods. Eventually we become lax and forget a

64. This definition is drawn from ongoing work on failure terminology in NASA's Constellation program. It uses many prior engineering sources (from both from academia and industry) and also draws from insights in the philosophy of science that emphasizes that much of science is really about "explanation." From this point of view, a failure is a phenomenon that requires explanation, and a fault is the explanation. B. C. van Fraasen, *The Scientific Image* (Oxford, U.K.: Oxford University Press, 1980), pp. 132–134; Ronald N. Giere, *Explaining Science: A Cognitive Approach* (Chicago, IL: University of Chicago Press, 1990), pp. 104–105. Giere hypothesizes that scientific explanation is characterized by the use of models. This accords well with many explanations of failure, by reference to specific hypothesized failure modes, often backed up by analysis, simulation, and testing.

65. The remaining 5 to 15 percent of faults are caused by a lack of knowledge about the environment in which the system operates. An example of this would be the lack of understanding of the near-Earth space environment and high radiation levels in the Van Allen Radiation belts in the early Space Age, or how the zero-g environment for satellites caused particles to float and short out electronic components.

key detail or skip a critical process because "we know" that we have done the right things and don't need to doublecheck.

By contrast, systems management and systems engineering reduce failure rates by providing formal cross-checks that catch and fix most potential mission-ending faults. Systems management and systems engineering cannot guarantee absolute success, but history shows that they do significantly reduce project failure rates.[66] This should be no surprise, because this is one of the major reasons why they were created to begin with. Systems management is needed when a project gets so large that the simple communication of small teams breaks down. This certainly is the case for huge projects such as Apollo or the Space Shuttle, but also for larger robotic systems and for teams that are distributed or have contracting or other barriers to communication.

The recognition that individual and social (cognitive, communicative, and organizational) factors are critical to system dependability has been slowly growing. The Columbia Accident Investigation Board's *Report*, with sections on organizational culture that largely drew from, and were partly written by, sociologist Diane Vaughan (who had written the authoritative book, *The Challenger Launch Decision*), was a major milestone in documenting and broadcasting this fact to NASA, but, as noted above, it fell short of providing a framework for or specific solutions to the problem. Dozens of failure investigations have concluded that individual (operator or design error) or social (communication) factors are implicated in failure. The demise of "faster, better, cheaper" and the renewed emphasis on systems management in NASA's robotic programs is also an indicator that management and engineering "philosophies" matter. However, an academic and theoretical framework has been lacking, and those that have been developed are currently little known. It is unlikely that approaches driven primarily by the social sciences are going to have much impact on NASA's engineers. What NASA engineers and managers are more likely to understand and implement are ideas couched in engineering terms that draw from social science research, instead of the other way around.

One recent approach to the problem has been developed by Nancy Leveson at MIT. Leveson, a safety engineer and researcher, developed modeling techniques that could begin to address the safety implications of the culture

---

66. On early missile, launcher, and satellite projects without these methods, failure rates of 50 percent were typical. After implementation of systems management, failure rates decreased to around 5 percent. Not all of this was due to systems management, as other learning and design improvements were occurring. Nonetheless, systems management deserves some of the credit. See Johnson, *Secret of Apollo*.

issue. To move forward, she looked backward, resurrecting the decades-old methods of systems dynamics developed at MIT in the late 1950s and 1960s by Jay Forrester.[67]

Forrester, who was one of the leaders of the Whirlwind and Semi-Automatic Ground Equipment (SAGE) real-time computer projects for air defense in the 1950s, became restless with this work and joined MIT's Sloan School of Management in 1956 to apply computer simulation methods to develop new management methods. His 1961 book, *Industrial Management*, showcased his simulations of corporate decision-making, which he and his students modeled as a feedback information system with time-critical information flows for making management decisions. Forrester broadened his approach in 1969, developing models of cities on a similar simulated basis, and published *Urban Dynamics*, which again showcased his interactive simulations that contained multiple interacting feedback loops that indicated to Forrester that decision-makers were unable to make proper decisions without computer-based modeling assistance. Finally, Forrester won funding from the Club of Rome, a small international group of prominent businessman, scientists, and politicians, to apply his methods on a worldwide scale. His models grouped the world into five major subsystems: natural resources, population, pollution, capital, and agriculture. The results of these models led to the controversial but widely circulated *Limits to Growth*, published in 1972, which argued that human civilization would, sometime around 2050, have a catastrophic collapse. Forrester's work was a forerunner of many comprehensive global environmental models and drew from his background in control systems and cybernetics as well as the newly developed techniques and technologies of computing.[68]

Leveson believed that NASA's culture and safety problem was ripe for a similar approach, and she began to model NASA's safety organization and decision-making. Her results, like those of Forrester's in the 1950s and 1960s, showed a periodic roller coaster behavior of concern for safety with a cycle

---

67. Leveson, "Technical and Managerial Factors," pp. 239–245; Nancy G. Leveson, *System Safety: Back to the Future*, unpublished book draft, available at *http://www.sunnyday.mit.edu/book2.html*.

68. Paul N. Edwards, "The World in a Machine: Origins and Impacts of Early Computerized Global System Models," in *Systems, Experts, and Computers: The Systems Approach in Management and Engineering, World War II and After*, ed. Agatha C. and Thomas P. Hughes (Cambridge, MA: MIT Press, 2000), pp. 221–253; Kent C. Redmond and Thomas M. Smith, *From Whirlwind to MITRE: The R&D Story of the SAGE Air Defense Computer* (Cambridge, MA: MIT Press, 2000); Jay Forrester, *Industrial Dynamics* (Cambridge, MA: MIT Press, 1961); Jay Forrester, *Urban Dynamics* (Cambridge, MA: MIT Press, 1969); Donella H. Meadows et al., *The Limits to Growth: A Report for the Club of Rome's Project on the Predicament of Mankind* (New York, NY: Universe Books, 1972).

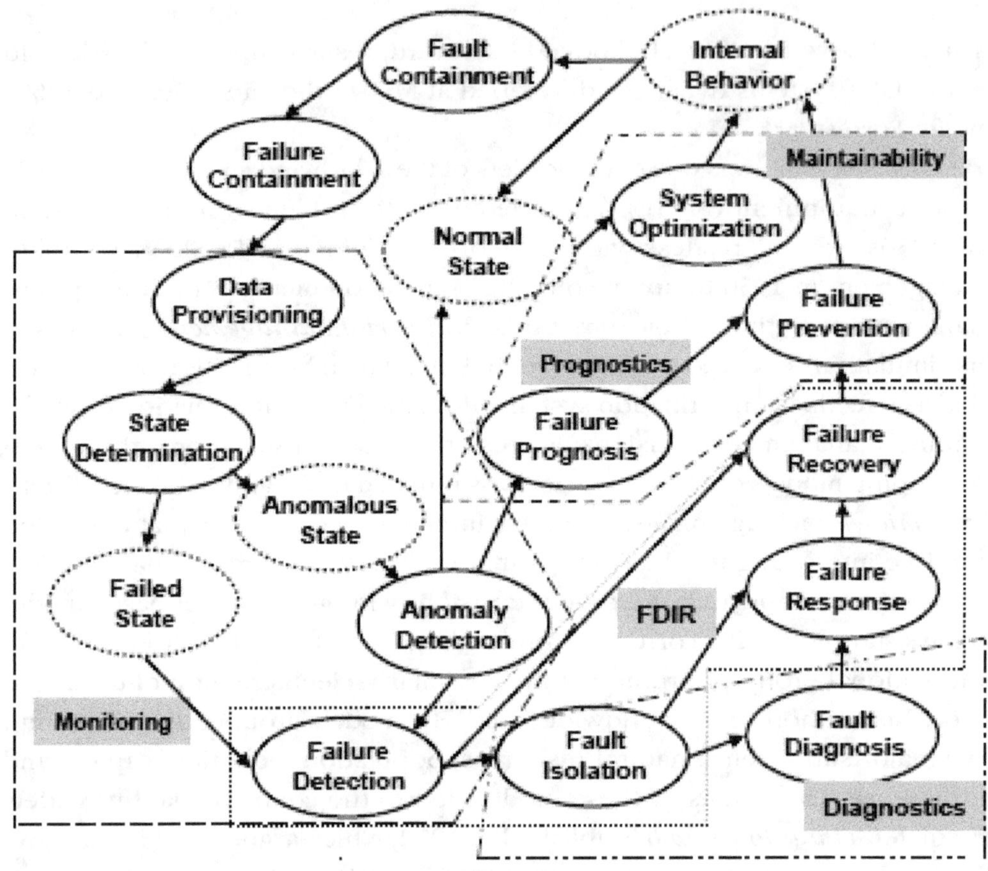

Figure 2: System health management functional flow. *Courtesy of Stephen B. Johnson*

time of roughly 15 to 20 years (NASA's actual human flight accidents showed a 17-year cycle). That is, after spikes of great safety concern immediately following major accidents, NASA quickly reverts to its regular behavior with relatively low, and decreasing, concern for safety. Leveson and her students continue to use these models to hypothesize the impact of potential changes to NASA's organizational dynamics on its safety outcomes.[69]

Another approach, developed primarily by this author with many others contributing since the late 1980s, also uses control theory insights, along with others from systems engineering and from the history, sociology, and philosophy of science and technology. In this approach, the general concern is for system dependability, which is defined as "the ability of a system to function

69. Leveson, "Technical and Managerial Factors," pp. 239–245.

in a manner meeting human expectations." Another term for this budding discipline is "system health management." In this view, as with Leveson's, complex systems such as those required for human and robotic spaceflight are complex mixtures of humans and machines, and from the standpoint of dependable systems, the functions needed to make systems dependable can be allocated to people, software, or hardware. In an operational system, the functions needed to monitor, predict, detect, isolate, respond, and recover from internal failures are arranged into control loops, and those loops are potentially analyzable in terms of time and criticality to create a system architecture that can successfully respond to impending or existing failures. The design of such a system requires new processes that are only partially understood as of yet.[70]

Creating dependable systems requires a proper mix of prevention of failure and the mitigation of internal failures. Humans are ultimately responsible for all dependability functions, but some functions can be placed in hardware or software. Even if placed in hardware or software, these functions are still designed based on human knowledge and intentionality. It is assumed that human designers, operators, and analysts are all fallible, with a certain probability of making mistakes, depending on various "contributing factors" of their social environment. These errors are just as likely to occur in design as in operations.

Certain well-known design principles, such as "clean interfaces," are reconceptualized as principles based on the minimization of communication errors between people, so that reducing the complexity of the functions between system components reduces the needed communications between individuals within differing organizations and their different "cultures." The principle of analytical independence, often seen as crucial for safety purposes, is seen as impossible to achieve in any one person, since complete independence also means no knowledge of the application and hence no ability to constructively say anything about it. Instead of trying to find that mythical single organization or person that can be independent, multiple knowledge overlaps based on differing principles and approaches are needed to achieve plausible results while cross-checking for errors.[71]

---

70. Johnson, "Introduction to Integrated System Health Engineering and Management." The first publication of this "closed-loop operational architecture" appeared in Jeffrey Albert, Dian Alyea, Larry Cooper, Stephen Johnson, and Don Uhrich, "Vehicle Health Management (VHM) Architecture Process Development," *Proceedings of SAE Aerospace Atlantic Conference* (May 1995, Dayton, OH).
71. Ibid.

A reorientation of NASA's thinking is needed, from seeing technical problems as purely technical to understanding that they are primarily flaws in individual knowledge, performance, and social communication. If pursued, these insights may lead to significant improvements in NASA's organizational culture. Diagnosing the culture problem need not be mysterious. One needs only to pursue all failure investigations back to their individual and social roots to identify the individual and organizational flaws that must be addressed. Deciding exactly how to address those problems is more problematic; but by the nature of the problem, it will involve education and training for individuals and changes to institutional structures and processes to improve organizational communication. Technical improvements can also assist, by finding ways to pinpoint which processes correlate with certain kinds of errors and then providing automated means of cross-checking for those error types.

Conclusion

System dependability and system safety, and their inverses, system failure and system hazards, are ultimately functions of individual and social understandings, communications, choices, and actions. Technical systems fail because they embed human failings, mistakes, and misunderstandings. It is unlikely that significant improvements to dependability and safety can be made until engineers and managers learn that ultimately they themselves are the causes of failure and that several individual and social actions must also be taken, along with technical improvements, to improve these qualities.

As Henry Petroski elegantly narrated in his studies of failure, there is an alternating pattern of conservatism and innovation in design over the course of engineering generations, which is rooted in the long-term trends of cultural factors. Failures result more frequently at the end of "innovation periods" as cost cutting and design originality push past reasonable limits. The fact of having pushed too far is generally revealed by the failures themselves. The resulting investigations, if pressed far enough, uncover the individual and social causes of the failures.

NASA has displayed this same dynamic. In JPL's deep space programs, just getting into space was a highly innovative effort that entailed much learning and many failures. In this brief but exciting period from the late 1950s to the early 1960s, JPL's managers, engineers, and contractors discovered many things about the space environment and about how to change its own institutional structures and organizations to operate spacecraft in that environment. A long stable period, with growing conservatism and creeping bureaucracy, ensued from the mid-1960s to the mid-1980s. The encroaching bureaucracy, limited funding, and recognition of alternate methods both within (mainly from the Strategic Defense Initiative) and without (the TQM fad) the space industry bred a growing discontent; and by the late 1980s and early 1990s, JPL was pressed into, and also decided to adopt, new TQM, "faster, better, cheaper," and Skunk Works methods. A short period of institutional change ensued, with a number of successful lower-cost projects at JPL and elsewhere. However, the failures of 1999, with their associated bad publicity, showed the limits of "faster, better, cheaper," and the pendulum swung back to conservative design with systems engineering.

The human flight program showed a similar dynamic, but on a shorter timescale. In the early human flight programs of Mercury and Gemini, NASA successfully navigated the treacherous hazards of space, though sometimes by a hair's breadth (such as Gemini 8, in which Neil Armstrong played a critical role in averting disaster). It accomplished this despite, and perhaps because of, the fact that nearly everything it did was new. However, success bred overconfidence, which was shattered by the Apollo 204 accident of 1967. The resulting investigation uncovered many other design problems besides the one(s) that killed the crew. Many Apollo veterans acknowledged that the enforced pause after the accident probably saved the program by rooting out and fixing many other impending technical and organizational problems. The Skylab and early Shuttle programs also succeeded despite the major new technologies, probably for similar reasons to those of the Mercury and Gemini projects, as engineers paid close attention to every detail for these new systems. But once again, latent problems compounded by reduced attention to safety led to the *Challenger* disaster of 1986. A round of safety improvements, augmented by a few more organizational changes, produced solid results for the next 17 years. However, cost cutting and safety reductions took their toll by the late 1990s, and these in turn contributed to the 2003 *Columbia* tragedy.

After the Columbia Accident Investigation Board *Report*, the resulting investigation went beyond the usual culprits of engineering and management structures, flawed decisions, and technical problems to indict NASA's organizational culture as inherently flawed. While the diagnosis was true, it was vague, and NASA's managers and engineers found it mostly "non-actionable." NASA made some specific and beneficial organizational and process changes, but the broader issue of culture, for which NASA hired BST, for the most part did not get addressed because NASA could not determine exactly what it meant, and BST likely did not understand the uniqueness of NASA's culture and how to develop precise and convincing actions. The NASA culture problem remains largely unresolved.

The culture problem at NASA has not gone away, but cultural change at NASA is no longer a major priority. However, the relationship of culture to dependability and safety has not gone unnoticed, and efforts to make the connection between the two are under way. While NASA's initial efforts at cultural change were stymied by a lack of understanding of the relationship of individual and social factors to failure, significant progress has been made in understanding these relationships. This bodes well for the future of NASA's programs, but only if the Agency both learns from its past and makes use of these growing insights from its own history.

Chapter 13

The "Von Braun Paradigm" and NASA's Long-Term Planning for Human Spaceflight

Michael J. Neufeld

In 1994, political scientist and space historian Dwayne A. Day coined the term "von Braun paradigm" to describe what he saw as an entrenched—and counterproductive—NASA long-term strategy for human spaceflight.[1] Roughly speaking, he boiled that strategy down to: space shuttle → space station → Moon → Mars. Day was responding to the ignominious failure of President George H. W. Bush's Space Exploration Initiative (SEI) from 1989 to 1990, which he, like many others, blamed on the space agency's penchant for gigantomania in its human exploration program. In response to the presidential announcement on the steps of NASM on the 20th anniversary of Apollo 11, NASA's 90-day study group advocated building, on the foundation of the Shuttle and then-projected space station, a lunar base and an ambitious spacefaring infrastructure that within 20 or 30 years would lead to a permanent human foothold on Mars. The cost turned out to be politically suicidal: several hundred billion dollars. The 90-day study reprised the Space Task Group report of 1969, which was an almost equally ignominious political failure. From that earlier proposal for a grand (or grandiose) post-Apollo space program, NASA salvaged only a scaled-back version of its first goal: a winged, reusable Space Shuttle.[2]

---

1. © Smithsonian Institution. Portions of this paper have been excerpted from Michael J. Neufeld, "Von Braun and the Lunar-Orbit Rendezvous Decision: Finding a Way to Go to the Moon," *Acta Astronautica* 63 (2008): 540–550, also © Smithsonian Institution.
2. Dwayne A. Day, "The Von Braun Paradigm," *Space Times* (November–December 1994): 12–15; Day expanded that opinion piece in the AAS newsletter as "Paradigm Lost," *Space Policy* 11 (1995): 153–159. On SEI, see Thor Hogan, *Mars Wars: The Rise and Fall of the Space Exploration Initiative* (Washington, DC: NASA SP-2007-4410, 2007).

Day traced this strategy back to the German-American rocket engineer Wernher von Braun, of course, specifically the series of articles the space visionary wrote or cowrote in *Collier's* magazine between 1952 and 1954. These laid out his grand vision, which Day argued "had its greatest influence on *how the U.S. space community envisions space*. Von Braun convinced those who worked in the field that space was worthy of a concentrated, integrated human exploration effort."[3] In fact, his vision was controversial from the start, drawing vigorous objections from many rocket engineers, but it was undeniably influential over the long run, especially on spaceflight true believers inside the movement and in the general public.[4]

Since Day coined that term, it has gained a certain currency in space history and policy. Particularly my esteemed NASM colleague, and former NASA Chief Historian, Roger Launius, has both popularized and expanded the use of the "von Braun paradigm" as an analytical term for describing a pattern in American space development. In his book *Space Stations*, Roger increases the list of von Braun's essential stages to six, including preliminary ones of robotic satellites and nonreusable piloted vehicles. But in the most recent formulation, in his important book with Howard McCurdy, *Robots in Space*, the two posit *five* stages as the core of von Braun's thought: 1) "Development of multi-stage rockets capable of placing satellites, animals and humans in space; 2) "a large, winged, reusable spacecraft . . . to make space access routine"; 3) "a large, permanently occupied space station" for observing Earth and launching "deep space expeditions"; 4) "human flights around the Moon, leading to the first landings" and eventually to "permanent lunar bases"; and 5) assembling "spaceships in Earth orbit for the purpose of sending humans to Mars and eventually colonizing that planet." Launius and McCurdy also posit the existence of an anti-von Braun, "Rosen/Eisenhower/Van Allen Alternative," for a more measured, and more robotic, space program.[5]

The primary aim of this paper will be to examine von Braun's history of space advocacy carefully, to see how much his ideas actually correspond with the later construct of a von Braun paradigm. My secondary objective is

---

3. Day, "The Von Braun Paradigm," p. 12. See also "Paradigm Lost," p. 154.
4. On the controversy, see esp. Howard McCurdy, *Space and the American Imagination* (Washington, DC: Smithsonian Institution Press, 1997), chap. 2.
5. Roger D. Launius, *Space Stations: Base Camps to the Stars* (Washington, DC: Smithsonian Institution Press, 2002), pp. 26–27; Launius and Howard E. McCurdy, *Robots in Space: Technology, Evolution, and Interplanetary Travel* (Baltimore, MD: Johns Hopkins University Press, 2008), pp. 64–65 (quotes) and chap. 3 (generally).

to look briefly at the history of later NASA human spaceflight planning to try to discern von Braun's influence or at least that of his so-called paradigm. My conclusions are 1) that von Braun was not a systematic and consistent space planner, but rather was often driven by enthusiasm and by a Moon obsession that meant he was as interested in going straight to the lunar surface as using the "logical" steps he laid out in *Collier's*; 2) that although von Braun himself may have been inconsistent, his public advocacy in the 1950s did tend to consolidate a paradigm among space advocates focusing on the *four* main elements of shuttle, station, Moon, and Mars; 3) that while there was always opposition to his plans, the "Rosen/Eisenhower/Van Allen Alternative" is an artificial construct that conflates different ideas from different times; 4) that von Braun's direct influence, still important in the 1960s, diminishes drastically from the 1980s on; but 5) that the classic four-element von Braun paradigm does seem to have been a shaping factor in NASA's planning, from the 1969 Space Task Group to the 2004 Vision for Space Exploration. However, its influence was weakest in the latter case, and its persistence may in part by due to other factors, notably the loss of Venus as a feasible destination and the lack of interest (until the 1990s) in asteroids, which tended to foreclose other possible options for human deep space exploration besides the Moon or Mars.

Did von Braun Have a Paradigm?

Von Braun's career in space advocacy began when he was literally and figuratively in the wilderness in the late 1940s. Relatively underemployed at Fort Bliss, outside El Paso, Texas, when budget cuts forced the U.S. Army to reduce its guided missile projects, including the ones he and his group of about 120 German and Austrian engineers were working on, von Braun sought an outlet for his boundless creative energies. He decided that he needed to sell the American people on spaceflight, so he set out to prove the feasibility of a human expedition to Mars, based on conservative projections of late-1940s technology. Showing how much he was ahead of almost everyone, he felt that it was *too easy* to demonstrate a human Moon landing. But to make his Mars study palatable to the general public, he concluded in 1947 that he had to package it inside a science fiction novel.[6]

---

6. See Michael J. Neufeld, *Von Braun: Dreamer of Space, Engineer of War* (New York, NY: Alfred A. Knopf, 2007), chaps. 9–10, for elaboration; key sections on the novel are excerpted in Michael J. Neufeld, "'Space Superiority': Wernher von Braun's Campaign for a Nuclear-Armed Space Station, 1946–1956," *Space Policy* 22 (2006): pp. 52–62.

By then, von Braun had been making informal plans and back-of-the-envelope calculations for at least 15 years. Arthur Rudolph, later (in)famous as the only member of von Braun's group forced to leave the United States for involvement with concentration camp labor, tells a story of staying up nights at the Kummersdorf officer club in about 1935, calculating trajectories and payloads for a Mars expedition. But von Braun's central obsession was the Moon, specifically leading an expedition to it himself—a dream that seized him as teenager during the German spaceflight fad of the late 1920s. Several anecdotes attest to his continuing fascination with a lunar landing throughout the Nazi period; one or two even speak of a specific proposal, but we have no details as to whether he was speaking of a direct launch from Earth or an assembly in Earth orbit near his space station, another major obsession. He may have contemplated both. Brief comments he made to the press and the public in El Paso, Texas, in winter 1946–47, however, described the station as a "refueling" stop on the way to the Moon.[7]

His plans for a large, rotating, wheel-type space station appear to have developed in parallel to his Moon and Mars ideas. A major influence were the writings of his hero, the German-Rumanian spaceflight theoretician Hermann Oberth, but the wheel format seems likely to have come from the 1929 book by the Slovenian-Austrian Hermann Noordung (pseudonym for Potoçnik), although von Braun never acknowledged the influence. From Oberth, von Braun definitely drew his ideas of the station as a superweapon for observing and dominating Earth. Following the revelation of the atomic bomb and his arrival in the United States, he reconceived it as a battle station controlling co-orbiting nuclear missiles; he became convinced that he had the key to defeating the Soviet Union and winning the Cold War. His ill-fated science fiction novel, originally titled *Mars Project: A Technical Tale*, has a fascinating

7. Arthur Rudolph OHI by Michael J. Neufeld, 4 August 1989, NASM Archives; Daniel Lang, "A Romantic Urge," in *From Hiroshima to the Moon* (New York, NY: Simon & Schuster, 1959), pp. 191–192, originally published in *New Yorker* 21 (April 1951): 75; Peter Wegener, *The Peenemünde Wind Tunnels* (New Haven, CT: Yale University Press, 1996), pp. 41–42; Hans Kehrl, *Krisenmanager im Dritten Reich* (Düsseldorf: Droste, 1973), p. 336; Jak. van den Driesch to Wernher von Braun, 6 January 1969, in U.S. Space and Rocket Center (USSRC), von Braun Papers, file 423-4; Wernher von Braun, "Survey of Development of Liquid Rockets in Germany and Their Future Prospects," in *Report on Certain Phases of War Research*, by Fritz Zwicky (Pasadena, CA: Aerojet, 1945), pp. 66–72; "German Scientists Plan Re-fueling Station in Sky on Route to Moon," *El Paso Times* (4 December 1946), copy in NASA KSC archives, Debus collection; Wernher von Braun (Rotary Club speech, 16 January 1947), in USSRC, Huntsville, AL, Wernher von Braun Papers, file 101-3. For more on the history of von Braun's Moon plans, see Neufeld, "Von Braun and the Lunar-Orbit Rendezvous Decision."

and disturbing opening, "A.D. 1980." It is set after the USSR is destroyed by nuclear strikes from his space station, "Lunetta," a name he treasured from a science fiction story he wrote as a teenage boy. In his early 1950s writings, he discussed using preemptive atomic attacks to protect the station—making the speculation that he was later a model for Dr. Strangelove seem not unjustified![8]

In order for his station to be useful as a reconnaissance and bombing platform against the Soviets, it had to be in a polar or near-polar orbit, which he set at a 2-hour period at 1,075 miles (later shown to be infeasible when the radiation belts were discovered). This orbit was in the wrong plane for his Mars expedition, which needed to depart in the ecliptic plane of the solar system so as to minimize the energy needed to reach the Red Planet. Thus, in the novel, his gigantic fleet of 10 spaceships, each with a mass of 8.2 million pounds and carrying seven men (and only men) apiece, was *not* assembled next to the station. Temporary living quarters for the work crews were set up inside the Mars ships instead. He thus believed a station was not essential to launching a human Mars expedition, but he took it for granted that it would come first as mankind's initial foothold beyond Earth.[9]

To orbit 82 million pounds of hardware and propellants (mostly the latter) required a huge logistics operation he developed at length. In the novel, he alludes to an earlier class of "Jupiter" multistage boosters, but for Mars, the "United States of Earth" develops the "Sirius" class, a huge three-stage rocket much squatter and heavier than the later Saturn V. The first and second stages are recovered at sea and reused. The third is the winged rocket freighter that delivers materials and people into orbit. The assumption that humans would fly in craft with wings was not original to him, of course, as space advocates like the Austrians Max Valier and Eugen Sänger had already argued that the transition was most natural from an atmospheric rocket plane to what we

---

8. Neufeld, "'Space Superiority'"; Wernher von Braun, "Survey . . . ," in *Report on Certain Phases of War Research in Germany*, by Fritz Zwicky (Pasadena, CA: Aerojet, 1945), pp. 66–72; Wernher von Braun, "Questions and Answers on A-9, A-10 and A-11," July 1946, in National Archives, College Park, MD, RG156, E.1039A, file "Ch. II New Material—Revision Material," box 79; "Giant Doughnut is Proposed as Space Station," *Popular Science* (October 1951): 120–121; Wernher von Braun, "Crossing the Last Frontier," *Collier's* (22 March 1952): 24–28, 72–73, and the accompanying magazine editorial on p. 23, "What Are We Waiting For?"; Wernher von Braun "Space Superiority," *Ordnance* (March–April 1953): 770–775.
9. The original "Mars Project" typescripts are in USSRC, Wernher von Braun Papers, file 204-7 (German), 205-1 (English); the failed novel was recently published as *Project Mars: A Technical Tale* (Burlington, ON: Apogee Books Science Fiction, 2006); what originally appeared in print was his revised mathematical appendix, Wernher von Braun, *Das Marsprojekt* (Frankfurt, Germany: Umschau, 1952), and *The Mars Project* (Urbana, IL: University of Illinois Press, 1953).

would now call a space shuttle. To von Braun, wings were needed above all for reentry, as no one had yet conceived of an ablative heat shield for a ballistic return. He pictured a glide halfway around Earth and put active cooling in the wings and nose to prevent them from melting. Of course, landing on a runway also made believable the airlinelike operations needed to fire two giant boosters *per day* and accomplish 950 launches in eight months! Complete reusability and the essential economies it provided were critical to all of von Braun's early spaceflight conceptions, even more critical than a station. It was the only way he could justify the economic feasibility of his monumental space infrastructure.[10]

As is well known, von Braun's often woodenly written novel was rejected by something like 18 publishers, but his revised mathematical appendix did appear in German in 1952 and in English in 1953 as *The Mars Project*. By that time, he had made his great breakthrough in the *Collier's* magazine series, together with several other authors, notably his friend Willy Ley. The first issue, on 22 March 1952, and the first book that came out of the series, *Across the Space Frontier*, introduced the public to aesthetically improved versions of his booster and station as redrawn by artists Chesley Bonestell, Fred Freeman, and Rolf Klep. The magazine endorsed von Braun's militant Cold War argument for using the space station to establish "space superiority" over the Soviet Union. In two more issues in October 1952, and in the spinoff book, *Conquest of the Moon*, von Braun presented his conception of the first lunar expedition, which involved three ships and 50 men and took six months to assemble in the space station's polar orbit. He reveled in imagining huge space voyages, but he was far from committed to it as the only strategy, as he earlier and later discussed small, direct expeditions to the Moon.[11]

The popularity of the space issues caused *Collier's* and its series editor, Cornelius Ryan, to put off the projected Mars number and ask von Braun and some of the other collaborators to generate articles on the training and preparation of "space men" (the word "astronaut" was not then used for that

---

10. Wernher von Braun, *Project Mars*, pp. 23, 113–120, 215–31; Wernher von Braun, *The Mars Project*, pp. 9–36.
11. Neufeld, *Von Braun*, pp. 246–247, 251–269; Neufeld, "'Space Superiority'"; Wernher von Braun, "Crossing the Last Frontier," *Collier's* (22 March 1952): 24–28; Wernher von Braun, "Man on the Moon: The Journey," *Collier's* (18 October 1952): 24–28, 51–58, 60, 72–73; Wernher von Braun (with Fred L. Whipple), "Man on the Moon: The Exploration," *Collier's* (25 October 1952): 38–40, 42, 44–48; Cornelius Ryan, ed., *Across the Space Frontier* (New York, NY: Viking, 1952) and *Conquest of the Moon* (New York, NY: Viking, 1953). On von Braun's other Moon plans, see Neufeld, "Von Braun and the Lunar-Orbit Rendezvous Decision," and footnote 6 in this chapter.

Figure 1: Von Braun in 1955 against the backdrop of a famous Chesley Bonestell painting from the first *Collier's* space issue of 1952. He is holding a model of the Disney version of his winged space shuttle. *U.S. Army photo courtesy of NASM*

purpose). In the spring of 1953, Ryan also asked von Braun to speculate on space exploration before humans went up. Von Braun produced an article on a biological satellite he called the "baby space station," in which several monkeys would spend two months in weightlessness before being euthanized prior to satellite burnup. Even the three-stage expendable booster he proposed for this mission was 150 feet high and 30 feet in diameter at the base.[12] Prior to this time, he had taken little interest in the preliminary stages of space exploration and thought not at all about robotic spaceflight. His Mars expedition is the first mission of any kind to the Red Planet. He had little faith that spacecraft would work without humans on board to fix them, and he was simply uninterested in any other form of exploration. For him, as for other space advocates of his generation, sending humans was the point.

---

12. Wernher von Braun (with Cornelius Ryan), "Baby Space Station," *Collier's* (27 June 1953): 33–35, 38, 40; Neufeld, *Von Braun*, pp. 272–273.

Yet, at almost the same time, von Braun conceived in the classified world an absolute "minimum" satellite booster and craft, dubbed "Orbiter" in late 1954, using his new Redstone ballistic missile as the first stage. The satellite in its initial version would have been only a 5-pound inert ball or balloon.[13] Writing von Braun's biography, I was struck by the apparent contradiction in his character: on the one hand he reveled in gigantism when he conceived of the space future, and on the other he was a very conservative rocket engineer. He and his Army group were *not* on the cutting edge of missile propulsion or structures in the 1950s because of that engineering conservatism. He had two sides: a vivid imagination that led him into romanticism and a deep-seated pragmatism that shaded into naked opportunism; but he did not see it as a contradiction, as he expected that small, practical steps in the near term would lead quickly to the glorious future he imagined.

The *Collier's* series ended in April 1954 with the much-delayed Mars issue. By then von Braun was already moving beyond his original *Mars Project* conceptions. He had come under attack from his colleagues in the American Rocket Society for gigantomania, and his associate Ernst Stuhlinger was studying ion propulsion for interplanetary voyages. But his busy schedule meant that he did not want to rethink his 10-ship expedition in 1954. When von Braun, Ley, and Bonestell's book version, *The Conquest of Mars*, was finally published two years later, however, he cut the expedition back to two ships, in response to criticism, but stuck with chemical propulsion.[14]

In the interim, the Walt Disney Company had become interested in spaceflight for its new television series and hired von Braun and Ley to be its consultants and on-screen spokesmen, along with an ex-German space medicine expert. Disney presented to a television audience of millions between 1955 and 1957 yet another version of von Braun's vision: giant booster with winged spacecraft, orbiting wheel station, Moon exploration (a preliminary circumlunar voyage in this case), and a Mars expedition (using Stuhlinger's solar-powered, ion-engine ships). It is fair to say that Disney helped solidify a "von Braun paradigm" of four main elements (shuttle, station, Moon, Mars) in the minds

13. Michael J. Neufeld, "Orbiter, Overflight and the First Satellite: New Light on the Vanguard Decision," in *Reconsidering Sputnik: Forty Years Since the Soviet Satellite*, ed. Roger D. Launius, John M. Logsdon, and Robert W. Smith (Amsterdam, Netherlands: Harwood Academic Publishers, 2000), pp. 231–257.
14. Wernher von Braun (with Cornelius Ryan), "Can We Get to Mars?," *Collier's* (30 April 1954): 22–29; Willy Ley, Wernher von Braun, and Chesley Bonestell, *The Exploration of Mars* (New York, NY: Viking, 1956); Neufeld, *Von Braun*, pp. 270–272, 275–277, 286; Tom D. Crouch, *Rocketeers and Gentlemen Engineers* (Reston, VA: AIAA, 2006), pp. 134–137.

of the public, which was now more likely to believe that spaceflight would soon become a reality.[15]

One interesting lacuna no one ever talks about is the virtual absence of Venus in the *Collier's*-Disney popularizations of von Braun et al. Here was a planet almost exactly the same size as Earth and actually slightly closer and more accessible than Mars, and one also the frequent subject of science fiction, both written and filmed. Science fiction movies (mostly bad) about landing on Venus continued to be made into the 1960s. One popular speculation was that there must be a steamy swamp world under the impenetrable blanket of clouds, as the planet was closer to the Sun. It was not until Mariner 2's flyby in December 1962 that we knew for certain that the beautiful evening and morning star was actually a hellish world with temperatures hot enough to melt lead. The causes of this neglect by von Braun and his compatriots are not far to seek: the Western cultural obsession with Mars that had flourished since telescopes improved our view in the late 19th century, showing an apparently Earth-like planet with probable life, as opposed to the blank white mystery of Venus. Von Braun's *Mars Project* novel featured an updated version of Percival Lowell's Red Planet with canals designed by an older, superior civilization to move water from the polar caps to its cities. It is unclear how much he still believed in a Lowellian Mars in the 1950s, but he clearly had become deeply fascinated by the idea in his youth and never completely lost it thereafter.[16]

Shortly before the first broadcast of the last Disney program in December 1957, the program about Mars, the Soviets launched two Sputniks. Von Braun immediately proposed a crash project that bore no resemblance to the winged vehicle he had recently depicted as the necessary first step in human space travel. To launch a man (I use the term advisedly) as soon as possible, he argued for using a Redstone to lob a fairly primitive capsule on a brief suborbital flight. Called "Man Very High" and then "Project Adam" (for "first man"), this idea was famously dismissed by NACA Director Hugh Dryden in the spring of 1958 as having "about the same technical value as the circus stunt of shoot-

15. Neufeld, *Von Braun*, pp. 284–290, and for the impact of *Collier's* and Disney on the public, see esp. McCurdy's seminal *Space and the American Imagination*.
16. Robert Markley, *Dying Planet: Mars in Science and the Imagination* (Durham, NC/London, U.K.: Duke University Press, 2005), esp. pp. 2–3, 21–22; Wernher von Braun, *Project Mars*; Neufeld, *Von Braun*, pp. 28–29; Launius and McCurdy, *Robots*, pp. 66, 271n11. I have found a single reference to Venus in the *Collier's* series, in a Wernher von Braun answer to a question in a "Space Quiz" of miscellaneous information put together by the editors, *Collier's* (22 March 1952): 38. He states that a space station would have to be built around Venus before humans could land there.

ing the young lady from the gun." But the idea quickly reappeared, albeit as part of a technically more sophisticated NASA program, Mercury, for putting a man into orbit.[17]

Soon thereafter, von Braun outlined a direct trip to the Moon in a popular magazine. His fame magnified by his central role in launching the first American satellite, he was finally able to realize his frustrated ambitions as a science fiction writer. In the fall of 1958 and spring of 1959, the Sunday newspaper supplement *This Week* published his novella, *First Men to the Moon*, in four parts, detailing a two-man expedition to that body using a huge rocket and a direct launch from Earth. Turning around as it approached the Moon, his spacecraft ignited a landing stage to alight on the lunar surface without going into orbit; that stage provided the launch platform for the two astronauts in their winged reentry vehicle to propel themselves back to Earth. It seems likely that this concept went back to some of his original German ideas, as one anecdote of the Nazi period indicates he was thinking of a two-man expedition. The story was skillfully illustrated by one of his *Collier's* collaborators, Fred Freeman. Padded with popular science material on spaceflight, it appeared as a short book in 1960. That same year, the magazine published a modified excerpt from his failed Mars novel, depicting the encounter between his adventurers and the inhabitants of the Red Planet.[18]

At almost exactly the same time as *First Men to the Moon* was first published, from 1958 to 1959, von Braun and his Army associates developed their first detailed lunar exploration plans. The context was the red-hot space race, interservice rivalry with the U.S. Air Force, and a search for missions for their new Saturn launch vehicle, then going into development. It would combine eight engines in the first stage for an unprecedented 1.5 million pounds (6.67 million newtons) of thrust. Lacking the authority to develop the gigantic launcher needed for direct ascent, which NASA would soon call Nova, and needing to justify Saturn, von Braun and his advanced missions people, Ernst Stuhlinger and H. H. Koelle, favored assembling and fueling the lunar landing vehicle in orbit around the home planet using many launches. This was the conservative approach that von Braun advocated to NASA at the end of 1958 when trying to sell Saturn, and it came up again

17. Dryden quoted in Neufeld, *Von Braun*, p. 329.
18. "First Men" corr. in USSRC, Wernher von Braun Papers, file 200-31; Wernher von Braun, *First Men to the Moon* (New York, NY: Holt, Rinehart and Winston, 1960); Klaus H. Scheufelen, *Mythos Raketen: Chancen für den Frieden. Erinnerungen* (Esslingen, Germany: Bechtle, 2004), pp. 82–83.

in Project Horizon, an Army lunar base study carried out in 1959. These studies helped shape NASA's long-range plan of that year, which rated an accelerated human circumlunar voyage as a goal at least as important as an Earth-orbital station. It was the first step on the road to a rush trip to the lunar surface to beat the Soviets. Von Braun, as a lifelong Moon obsessive, was thrilled at the possibility, and even more so after Kennedy made it real in 1961. He was quite willing to postpone the shuttle and station until later. He was, as we have seen, a romantic not rigidly committed to the plans laid out in *Collier's* and Disney.[19]

With the completion of his group's transfer to NASA in July 1960, von Braun's days as a visionary were essentially over. He spent the next decade as the Director of MSFC and as chief salesman for the Agency's programs, primarily Apollo-Saturn. While he made suggestions for, and critiqued, many NASA-funded studies of space stations and lunar and planetary exploration, the ideas were no longer really his. His influence stemmed largely from the impact of *Collier's* and Disney on a generation of rocket engineers and space enthusiasts. I will explore further the impact of his ideas on NASA planning below.

Thus I agree with Day and Launius; there was indeed a von Braun paradigm that was a product of his popular activities in those two media outlets in the 1950s, and I agree with Day that it consisted of only four fundamental elements. The addition of one or two preliminary stages does not correspond to von Braun's very limited public discussion of the early phases of spaceflight in public (consisting essentially of one solicited article on the "baby space station"), nor to his disinterest in robotic probes and his obsession with monumental human exploration. I might add that he was not rigid either in describing the relationship between the stages, especially in the case of the space station, which did not always have to serve as a base for launching lunar and planetary expeditions and did not necessarily have to come before going to the Moon.

In short, one must distinguish between von Braun and the von Braun paradigm, as there were several von Brauns. One was the pragmatic and

19. Courtney G. Brooks, James M. Grimwood, and Loyd S. Swenson, Jr., *Chariots for Apollo: A History of Manned Lunar Spacecraft* (Washington, DC: NASA SP-4205, 1979), pp. 4–6; Frederick I. Ordway III, Mitchell R. Sharpe, and Ronald C. Wakeford, "Project Horizon: An Early Study of a Lunar Outpost," *Acta Astronautica* 17 (1988): 1105–1121; NASA, "The Long Range Plan of the National Aeronautics and Space Administration" in *Exploring the Unknown*, ed. John M. Logsdon, vol. 1, *Organizing for Exploration* (Washington, DC: NASA SP-4407, 1995), pp. 377–378 (introduction), 403–407 (document).

conservative engineering manager who had a burning desire to accomplish something right now (especially if he could put his name on it) and proposed short-term, "quick fix" programs like Orbiter and Adam. Another was the Moon obsessive fascinated by traveling there, if possible personally, with the result that he privately worked out what it would take to make a direct trip as early as the 1930s. It led him, I think, into a Faustian bargain with the Nazis.

What of Launius and McCurdy's "Rosen/Eisenhower/Van Allen Alternative"?[20] As indicated earlier, I do not believe it is a useful analytical device. It conflates criticism of von Braun, the von Braun paradigm, and large-scale human spaceflight made at different times for different reasons. I will take the three named protagonists in turn.

Milton Rosen served as chief engineer for NRL's Viking and Vanguard programs and made himself famous in October 1952 for debating von Braun at the Hayden Planetarium in New York. At issue was von Braun's March *Collier's* proposals for a giant booster and nuclear-armed battle station. Rosen expressed a widespread feeling among engineers in the American Rocket Society, who thought that the German's grandiose plans were infeasible and would prove a massive distraction from urgent guided-missile work; indeed, von Braun's plans were a threat to national security. Rosen and his American Rocket Society compatriots were in part misled by von Braun's, and the magazine's, disinterest in describing the preliminary stages of spaceflight and by von Braun's willingness to paint a grand picture to sell the public on space, even as he acted in his day job as a rocket engineer every bit as conservative as they were. After Sputnik, Rosen would become as caught up in the space race as anyone else. He advocated building the gigantic Nova launch vehicle for a "direct ascent" mission to land on the Moon, even after von Braun and other NASA engineers had already switched to Earth-orbit or lunar-orbit rendezvous as the way to go during Apollo. So he will hardly serve as the leading name in a united front of anti-human-spaceflight advocates.[21]

President Dwight Eisenhower's secret motives in establishing the first scientific satellite project as a stalking horse for a reconnaissance satellite are now well known, as is his public, post-Sputnik attempt to contain the growth rate of space spending. Motivated by traditional fiscal conservatism, he was worried that human spaceflight programs would grow so large as to add mas-

---

20. Launius and McCurdy, *Robots*, pp. 64–70.
21. Rosen OHI by Michael J. Neufeld, 24 July 1998, NASM Archives; "Journey into Space," *Time* (8 December 1952): 62–64, 67–70, 73; Crouch, *Rocketeers*, pp. 134–137.

sively to a national debt already ballooned by the Cold War and nuclear arms race. He was often exasperated by the now German-American's penchant for loud public speaking on behalf of such programs. Eisenhower's vision for NASA, as Launius, McCurdy, and others have detailed, was of an Agency with a billion-dollar-a-year budget focusing mostly on robotic spacecraft for applications and exploration. It was a vision quickly overthrown by Kennedy and Johnson, who quintupled NASA's budget.[22]

James Van Allen was an Iowa physicist forever linked to von Braun by the iconic picture of the three holding up a replica of Explorer I on the night the first United States satellite was launched (the third was the Director of JPL, William Pickering). Several months later, he became even more famous as the discoverer of the radiation belts because of his Explorer experiment. His opposition to expensive human spaceflight programs grew up as a result of Apollo and the perceived lack of meaningful science return for the money expended, as opposed to the output of robotic exploration of Earth's cosmic environment and deep space. He became the most vocal spokesman, mostly from the 1970s on, for the skepticism about human spaceflight in the scientific community, an attitude still common there today. Van Allen's vision of NASA's ideal program thus bears resemblance to Eisenhower's, but his motivation was rather different.[23]

In sum, there has been opposition to von Braun's ideas, the von Braun paradigm, and large-scale human spaceflight from the beginning until now, but it is more differentiated and complex than is easily encapsulated in a single "alternative." It bears some resemblance to the more complex reality of von Braun and his paradigm, which I have outlined above, and deserves further study. But I will turn my attention back to the last part of my examination of the paradigm thesis, that of its apparent influence on later planning.

---

22. Walter A. McDougall, . . . *The Heavens and the Earth: A Political History of the Space Age* (New York, NY: Basic Books, 1985); David Callahan and Fred I. Greenstein, "The Reluctant Racer: Eisenhower and U.S. Space Policy," in *Spaceflight and the Myth of Presidential Leadership*, ed. Roger D. Launius and Howard E. McCurdy (Urbana/Chicago, IL: University of Illinois Press, 1997), chap. 1.
23. Abigail Foerstner, *James Van Allen: The First Eight Billion Miles* (Iowa City, IA: Iowa University Press, 2007), pp. 250–257. On p. 66 of *Robots*, Launius and McCurdy speak of Eisenhower adopting the "Rosen-Van Allen point of view." I doubt there is any evidence that he paid attention to Rosen's ideas as reported in *Time* back in 1952, and Van Allen had not started campaigning yet at the time of Eisenhower's decision-making on space, from 1955 to 1960. The only Van Allen references they give in the endnotes on p. 271 date to the 1980s. Eisenhower's scientific advisers, notably George Kistiakowsky and James Killian, were likely the sources of his arguments for the superiority of scientific satellites.

The von Braun Paradigm and Long-Term NASA Planning

My analysis of this topic will be briefer, primarily because I have not done the depth of research equivalent to my work on von Braun. The topic is large and sprawling, covering as it does nearly a half century of NASA plans; I will leave detailed examination to others. My primary purpose is to try to test the other part of Day's original thesis, as extended and amplified by Launius and McCurdy, namely, that the von Braun paradigm has exercised a profound influence on NASA's vision for human spaceflight after Apollo, pushing the Agency to build large, expensive programs focusing on the four main objectives: shuttle, station, Moon, and Mars. There have been three milestone events, the Space Task Group (STG) of 1969, the SEI of 1989, and the Vision for Space Exploration of 2004, the first two of which motivated Day's thesis. There have also been less visible proposals and studies, notably two not long before the SEI: the 1986 National Commission on Space and the 1987 Ride Report, neither of which fit neatly the paradigm thesis.

The first of these three major events, the STG, can be interpreted as an attempt to return to the script of the von Braun paradigm after the Moon landing, and it was the only one on which von Braun exercised any direct influence. Shortly after President Richard Nixon's inauguration, which came only weeks after the spectacular circumlunar voyage of Apollo 8, he asked Vice President Spiro Agnew to produce a proposal for a post-Apollo NASA program. The STG's direction and content were largely driven by Administrator Thomas Paine, who had a strong ally in the Vice President, a former Maryland governor with zero space expertise. Paine, an engineer who fondly remembered the *Collier's*-Disney series and was a fan of von Braun, was determined to exploit the Apollo success to get the maximum program he could out of the political system, which he pictured in classic paradigm fashion as a large human spaceflight program culminating in a Mars landing. When the Associate Administrator for Manned Spaceflight, George Mueller, produced an "integrated program plan" that spring for a shuttle, station, and cislunar nuclear shuttle to support continuing Moon exploration, Paine asked von Braun's Center to add a Mars expedition using those elements.[24]

The MSFC Director gave a famous viewgraph presentation to the STG and to a Senate committee in early August 1969, just two weeks after Apollo 11. It

24. Paine OHI by Logsdon, 12 August and 3 September 1970, file 4185, NASA History Division; David S. F. Portree, *Humans to Mars: Fifty Years of Mission Planning, 1950–2000* (Washington, DC: NASA SP-2001-4521, 2001), pp. 47–48; Heppenheimer, *Space Shuttle Decision*, pp. 159–174.

Figure 2: President Richard Nixon announces NASA Acting Administrator Thomas Paine's nomination as Administrator on 5 March 1969. Vice President Spiro Agnew is on the right. Agnew and Paine would push an ambitious shuttle-station-Moon-Mars strategy for the post-Apollo space program that Nixon refused to support. *NASA Image 69-H-225*

audaciously gave an exact date in November 1981 for the departure of such an expedition from Earth orbit. He did so, I believe, in spite of harboring doubts about Paine's risky strategy. Before the Administrator's request, he had not pushed a human Mars program, knowing that public opinion was unlikely to support it. A year later, no doubt influenced by hindsight, he told political scientist John Logsdon: "I have never in the last two or three years strongly promoted a manned Mars project . . . . People . . . have tried to cast me in the image . . . of the Mars or bust guy in this agency, which I am definitely not." A little earlier in the interview, he said: "I, for one, have always felt that it would be a good idea to read the signs of the times and respond to what the country really wants, rather than trying to cram a bill of goodies down somebody's throat for which the time is not ripe or ready."[25]

Like many in NASA's human spaceflight establishment, he thought a space station more salable and was committed to it as necessary infrastructure. A winged shuttle, an idea strongly pushed by Mueller in the late 1960s, von Braun and his counterparts conceived largely as a station adjunct, a logistics vehicle needed to transfer crew and cargo (the station components themselves would be launched on a Saturn V or other heavy-lift vehicle). With the almost instantaneous failure of Paine and von Braun's Mars initiative in the summer of 1969, followed by the slow death of the fall STG report, which laid out shuttle-station-Moon-Mars proposals differing only in timetable, it was the station and shuttle agenda NASA returned to. But the station's purpose was unclear to the politicians and gained no traction with the public. In the end, only the Space Shuttle, oversold as a vehicle that would revolutionize the economics of spaceflight, was politically feasible in the brutal post-Apollo budgetary environment. As the other elements of a big human program faded into a vague and distant future, the Shuttle became for a decade an end in itself—not so much a space policy as an excuse not to have one.[26]

None of what transpired in the STG and its aftermath obviously conflicts with the von Braun paradigm thesis. The NASA human spaceflight establish-

25. Wernher von Braun OHI by Logsdon, 25 August 1970, file 2629, NASA History Division.
26. Joan Hoff, "The Presidency, Congress, and the Deceleration of the U.S. Space Program in the 1970s," in *Spaceflight*, ed. Launius and McCurdy, pp. 98–100, 103–104; STG report, September 1969, in *Exploring the Unknown*, ed. Logsdon, vol. 1, *Organizing for Exploration*, pp. 522–543. The STG report covers actually a much larger spectrum of space policy, including space science and military space projects, so the centrality of the von Braun paradigm is not so easily visible there. Yet the emphasis placed on a big human program leading to Mars is clear in the emphasis Paine put on it that summer of 1969.

ment seems to have accepted the centrality of the four basic elements (the scientific and robotic spacecraft communities were another matter). If I have any second thoughts about the thesis as it applies to this episode, it is that the solar system itself seemed to foreclose other options for near-term human exploration, at least within the framework of discussion before 1975. With the 1962 confirmation that Venus was uninhabitable, there was apparently no place else to go after the Moon. There had been discussion of human flybys of Venus in the 1960s, but only because certain Mars trajectories required a gravity-assist from the second planet in one direction or the other. As robotic missions to the planets succeeded, the scientific return of human planetary flybys seemed scarcely credible for the expense anyway. Human spaceflight advocates were still fixated on the colonization of planetary surfaces, based on the analogy of the voyages of world exploration, and in the United States, especially, of the western frontier. But it would be hard to attribute that exploration and colonization focus primarily to von Braun or the paradigm, as it was embedded in the assumptions of the space travel movement since its origins. Similarly, the winged Space Shuttle had a long prehistory in space advocacy based on the analogy of aeronautics. So if the von Braun paradigm has any analytical meaning, it has to be in the centrality of the four elements, probably in the usual order, but not necessarily rigidly linked to each other.[27]

In the 1970s and 1980s, however, other options for human exploration of the inner solar system appeared. Gerard O'Neill popularized the idea of huge space colonies at the libration points of the Earth-Moon system, based on mining of the Moon and building solar-powered satellites for Earth. Robert Farquhar demonstrated, with the robotic spacecraft ISEE-3, the feasibility of libration-point halo orbits and the possibilities of exotic trajectories when he used lunar swingbys to send it to a comet. The growing concern in the 1980s and after about the threat of asteroid and comet impacts focused attention on those possible targets for human missions.

This changing context had a visible impact on the 1986 National Commission on Space, led by ex-Administrator Paine, which attempted to produce a new space policy for the Reagan administration. In its ill-timed report, which appeared just after the Shuttle *Challenger* accident, Paine's group once again painted a vision of a massive human spaceflight program. Paine even inserted a visual salute to von Braun as the frontispiece: a reproduction of the classic

27. Portree, *Humans*, pp. 20, 26, 32, 37; McCurdy, *Space and the American Imagination*, chap. 6; Launius and McCurdy, *Robots*, pp. 55–61.

Figure 3: On 20 July 1989, at NASM, President George H. W. Bush announces the SEI, a new attempt to implement the von Braun paradigm. Among those present were the Apollo 11 crew, Vice President Dan Quayle, and NASA Administrator Richard Truly. *NASA Image 89-H-380*

Bonestell painting of the shuttle, station, and space telescope from *Collier's*, paired with a modern version of the same trio painted by Robert McCall. But as O'Neill was a member of the National Commission on Space, the report also broke somewhat with the paradigm in discussing space colonies and libration-point missions. That only added to its flavor of impractical utopianism, and it was quickly dismissed in Washington. NASA instead commissioned astronaut Sally Ride to produce a report. Her group also strayed somewhat from the paradigm in 1987 by proposing a robotic "Mission to Planet Earth" as one focus and discussing an option of going straight to human Mars missions without necessarily going back to the Moon, although a lunar base was also an option.[28]

So why did the classic von Braun paradigm apparently reappear only two years later in George H. W. Bush's SEI and the 90-day study NASA produced in response to it? Part of the reason was that in the meantime, NASA had sold

28. United States National Commission on Space, *Pioneering the Space Frontier: The Report of the National Commission on Space* (New York, NY: Bantam Books, 1986); Portree, *Humans*, pp. 67–75; Hogan, *Mars Wars*, pp. 27–32.

the Space Station to President Reagan from 1983 to 1984 at the apogee of early Shuttle optimism. The Station decision might be interpreted as another return to the traditional script, as one of its missions would be to support later lunar and planetary exploration. When President Bush cast about for an ambitious new space policy in 1989, NASA engineers and managers from the human spaceflight side were the primary influences on the staff of the new NSC, which was headed by his Vice President, Dan Quayle. Bush's SEI speech singled out the Station, Moon bases, and Mars outposts explicitly, linking them together as stepping-stones, letting Agency engineers and space planners off the hook for even bothering to think about another strategy. It should be added that if a big human program was viewed as foundational, the lack of other targets—or at least the ability to imagine other targets—remained fundamental. By the end of the 1980s, the fad for O'Neill's space colonies had faded as their utopian character became clear; they would have to follow extensive lunar colonization anyway. The asteroid and comet impact issue had not yet risen to the level of public interest it would in the 1990s, when it resulted in two Hollywood movies. So an entrenched mindset at NASA that might be described as the von Braun paradigm appears central to SEI and its rapid failure, reinforced again by the apparent lack of any other place to go with humans.[29]

In contrast to the Space Task Group, however, von Braun (who had died in 1977) was essentially invisible in this episode. Bush did not mention him, nor did anyone in the Agency invoke his name much.[30] Since it is difficult to prove a negative—why von Braun's name was absent—I can do little but offer speculative explanations. Primarily, I think, he already was a figure from the distant past by 1989, as his last substantive ideas were formulated three decades earlier. The growing controversy about his Nazi past in the late 1980s might also have made his name somewhat "politically incorrect," at least outside his hometown of 20 years, Huntsville, Alabama, where even now denial is the order of the day. Following the October 1984 revelation that Arthur Rudolph had voluntarily left the United States and renounced his citizenship to avoid a

---

29. Hogan, *Mars Wars*, chaps. 3–5; Portree, *Humans to Mars*, chap. 9; copy of Bush speech, 20 July 1989, file 9008, NASA Historical Reference Collection, NASA History Division, NASA Headquarters, Washington, DC.
30. The study NASA did in response to the Bush speech does mention von Braun's name once as a precursor in planning, attributing to him the 1969 plan; see "Report of the 90-Day Study . . . ," November 1989, page 2-1, in NASA History Division, file 17922, and the viewgraph summary by Mark Craig, 18 January 1990, in file 9007. Similarly, von Braun gets a single mention under "Mission Scenarios" in an earlier NASA briefing for NSC and OMB staff: Franklin Martin, "Exploration Background Briefing . . . ," 25 August 1989, file 17923.

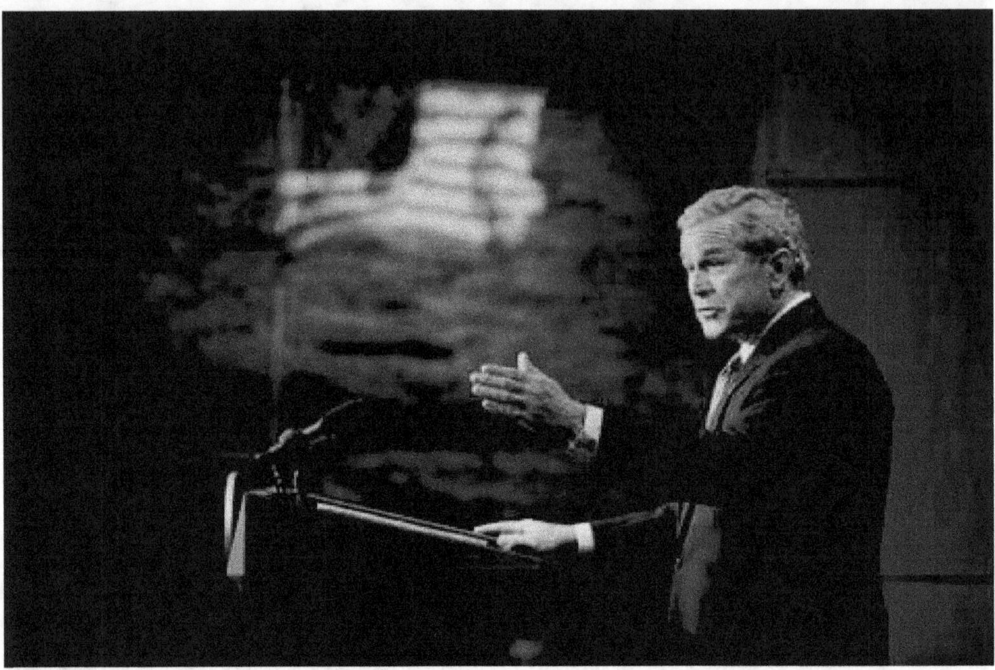

Figure 4: President George W. Bush speaks at NASA Headquarters on 14 January 2004 on the Vision for Space Exploration. It would depart from the classic von Braun paradigm, notably in ending the Space Shuttle Program and marginalizing the Space Station. *White House Official Photo P-37074-33*

denaturalization hearing over his involvement with concentration camp labor, newspapers around the world ran major stories. Subsequently, investigative journalists dug up a lot of dirt on the Third Reich records of von Braun and his key associates. Always a problematic hero, he posthumously became a touchy problem for NASA.[31]

The failure of the SEI put NASA long-term human spaceflight planning once again on the back burner. Faced with the unpopularity of more ambitious objectives, Agency leaders circled the wagons around Space Station

31. The most important pre-1989 publications on the Nazi issue were Linda Hunt, "U.S. Coverup of Nazi Scientists," *Bulletin of the Atomic Scientists* (April 1985): 16–24, and Tom Bower, *The Paperclip Conspiracy: The Battle for the Spoils and Secrets of Nazi Germany* (London, U.K.: Michael Joseph, 1987). Hunt appeared on CNN pursuing von Braun team members in their Huntsville driveways, and Bower had a major special on the PBS-TV program *Frontline* in early 1988. On von Braun's record in particular, see Michael J. Neufeld, "Wernher von Braun, the SS, and Concentration Camp Labor: Questions of Moral, Political, and Criminal Responsibility," *German Studies Review* 25 (2002): 57–78. Monique Laney is studying the history of the Germans in the city and its connection to the memory of the Nazi rocket program; see "'Operation Paperclip' in Huntsville, Alabama," in *Remembering the Space Age*, ed. Steven J. Dick (Washington, DC: NASA SP-2008-4703, 2008).

Freedom, as it was then called, to protect NASA's budget and human spaceflight establishment. New planning began only at the end of the 1990s, late in the term of Administrator Daniel Goldin. In the meantime, Red Planet enthusiasts like Robert Zubrin and his Mars Society had popularized an alternative they called "Mars Direct"—rejecting the Moon as a way station and emphasizing the exploitation of in situ resources to avoid the massive infrastructure of the von Braun approach. Zubrin explicitly criticized the German-American.[32] Inside NASA, the Decadal Planning Team, as Goldin dubbed it, did consider whether to skip the Moon and whether the now-ISS was a worthwhile investment. But it was not until the crisis provoked by the Shuttle *Columbia* disaster in early 2003 that a new space policy could emerge under President George W. Bush.[33] While it is too early to analyze this process in detail as the first historical work is only now being done, what emerged as the Vision for Space Exploration in 2004 was in some ways the anti-SEI. Big budget increases and any global money numbers that might be politically toxic were to be avoided; the Shuttle and Station were de facto rejected. The Columbia Accident Investigation Board had already recommended phase-out of the Shuttle as dangerous; the gigantic cost overruns and lengthy delays of the ISS had left a bad taste in everyone's mouth. One way of financing a new human space program on the cheap was to try to get out of those two obligations as soon as possible. However, the international dimensions and sunk cost of the ISS made it impossible for the United States to get out of the Shuttle quickly or abandon the Station entirely, and sending humans to Mars was simply impractical based on existing technology—targets closer to home were needed to test the new spacecraft and habitation modules. Under the new Administrator, Mike Griffin, the Explorations Systems Architecture Study in 2005 brought the Moon back to the fore as the next critical objective; Mars quickly began fading into the background.[34]

Do all these developments mean that the von Braun paradigm is dead or has little influence any longer? How one answers that question depends a great

32. For a later version, see Robert Zubrin, *The Case for Mars: The Plan to Settle the Red Planet and Why We Must* (New York, NY: Touchstone, 1997), pp. 47, 66.
33. I am indebted to Glen Asner and Stephen Garber, former and current historians at NASA, respectively, for lending me drafts of chapters from their forthcoming history of the Decadal Planning Team and Vision for Space Exploration. What I know of this history depends heavily on them. For Bush's space policy statement, see NSPD 31, 14 January 2004, file 12886, NASA Historical Reference Collection, NASA History Division, NASA Headquarters, Washington, DC, virtually the only primary document available in these files as of October 2008.
34. *Exploration Systems Architecture Study: Final Report* (Washington, DC: NASA TM-2005-214062, 2005), chap. 1, electronic copy courtesy of Glen Asner and Stephen Garber.

deal on how one defines the term, which is slippery when one gets past the fundamentals of a massive human spaceflight infrastructure and shuttle → station → Moon → Mars. If we have already built the Space Shuttle and International Space Station, does it matter if we dump them as long as we follow the allegedly logical order of the four steps? Von Braun would never have expected the first two to be abandoned, seeing them as necessary infrastructure for human spaceflight, but then he did not link the steps rigidly either—at least in the case of the Station, which might or might not serve as an orbital base for construction of interplanetary ships. What one can say is that the paradigm has weakened as an entrenched mindset in the NASA human spaceflight establishment; the disillusionment with the reusable space plane is the most visible sign of that. But not all aspects of the paradigm are dead. Even though I am a Moon buff, having grown up in the 1960s, and think there is much interesting science to be done there, its lingering influence is certainly one possible explanation for NASA's commitment to building a big lunar base, which will likely not be built because it would become another giant money sink like the ISS. And why the determined resistance to discussing the proposals of Bob Farquhar and others for asteroid missions instead, especially in view of our long-term need to build a planetary defense? But here I have strayed from the role of historian into that of commentator, as these events are too recent to provide the historical perspective and research needed to judge them.

In conclusion, I would agree that Dwayne Day's thesis of a von Braun paradigm consisting of *four* main elements remains a plausible interpretative device for analyzing a half century of U.S. human spaceflight planning, especially up to 1989. However, a distinction must be made between von Braun and the paradigm, although he was the one who created it in *Collier's* and Disney. As we have seen, it by no means represented all the dimensions of his enthusiasms, thoughts, and actions.

As for the paradigm he launched, it appears to have flourished in NASA because it offered an alluring vision of the future for human spaceflight enthusiasts and a program of action for engineers and planners that was "logical" yet malleable in its details. Its continuing influence was perhaps aided by the discovery of Venus's inaccessibility, which did nothing to disturb a Western and American cultural obsession with Mars as a possible abode of life. The Red Planet also appeared to be the only habitable, Earth-like objective anywhere in the neighborhood, even as robotic scientific discoveries showed it to be less appealing than hoped. The availability of the Moon relatively close by, and the focus on colonizing planetary surfaces, also tended to reinforce the paradigm while marginalizing other possible destinations like the libration points and the asteroids, not to mention the possibility of rejecting human spaceflight altogether to concentrate on robotic exploration. There are certainly other factors—social, cultural, and professional—that shaped the thinking of NASA engineers and space planners. It could well pay space historians and policy analysts to further test this thesis, as such investigations can cast new light on the fundamental assumptions behind the human spaceflight enterprise in the United States.

Chapter 14

Life Sciences and Human Spaceflight

Maura Phillips Mackowski

Life sciences activities, as carried out by NASA for 50 years, can be roughly divided into two categories: operational biology, meaning mission-related (astronaut health and safety, habitat, environmental life support, psychology, and spacesuits) and fundamental/basic biology. The latter category, in turn, offers three possibilities for study: species to be used for life-support systems or for consumption; "model" species (well-understood life-forms with at least one physiological system that could stand in for humans); and evolutionary biology, including astro- or exobiology, embryology, and genetics. Evolutionary scientists ask where life came from, how living beings might have changed over time, where life exists now, and how it might evolve later or somewhere else besides Earth's surface.

The Space Act of 1958 dictated no role for evolutionary research per se, but it outlined an implicit role for operational components of biology, medicine, and those life sciences with broader applications to all humankind. Its Declaration of Policy and Purpose, Section 102 (c), called for "The development and operation of vehicles capable of carrying instruments, equipment, supplies and living organisms through space" and "The establishment of long-range studies of the potential benefits to be gained from, the opportunities for, and the problems involved in the utilization of aeronautical and space activities for peaceful and scientific purposes." Section 203 authorized the Administrator, in pursuance of these activities, to "arrange for participation by the scientific community in planning scientific measurements and observations to be made through use of aeronautical and space vehicles, and conduct or arrange for the conduct of such measurements and observations; and provide for the widest practicable and appropriate dissemination of information concerning its activities and the results thereof." In doing so, he or she was allowed "to acquire . . . construct, improve, repair, operate, and maintain laboratories, research and testing sites and facilities, aeronautical and space vehicles, quarters and related accommodations . . . and such other real and personal property . . . as the Administration deems necessary within and outside the continental United States." NASA's leader could also "establish within the Administration such offices and procedures as may be appropriate to provide for the greatest possible coordination of its activities under this Act with related scientific and other activities being carried on by other public and private agencies and organizations."[1] Signed when a grand total of four U.S. vehicles had exited the atmosphere, the Space Act may have presciently used the term "living organisms" in the sense of preventing forward contamination by space probes so that future missions did not register false positives in a search for extraplanetary life.[2]

"Vehicles" was the word that garnered the most attention, as space dreamers and visionaries of the early 20th century pictured orbiting or lunar high grounds manned by the military, not biomedical research labs staffed by scientists in lab coats.[3] Associate Administrator Homer Newell, writing in 1980, acknowledged the uneven appreciation at that time by NASA—and the science community at large—for biology in the space program: "One could sense an ambivalence in the life sciences community concerning the space program, a fascination with its novelty and challenge mixed with skepticism on the part of most that space had much to offer for their disciplines . . . ."[4] That lack of integrated support—by researchers, politicians, nonscientists within the Agency, the military, the medical community, and the public—would dog NASA life science research throughout its first half century, resulting in a patchwork of successes and failures across all its endeavors.

---

1. "National Aeronautics and Space Act of 1958 (Unamended)," P.L. 85-568, 72 Stat., 426. Signed by the President on 29 July 1958, RG 255, NARA, Washington, DC.
2. See Steven J. Dick and James E. Strick, *The Living Universe: NASA and the Development of Astrobiology* (New Brunswick, NJ: Rutgers University Press, 2005), pp. 23–24, 29–30. Vanguard 1 and Explorers 1, 3, and 4 all achieved orbit in early 1958, the year after "living organism" Laika, a Soviet dog, had flown aboard Sputnik 2.
3. See, for example, Michael J. Neufeld, *Von Braun: Dreamer of Space, Engineer of War* (New York, NY: Knopf, 2007), pp. 243, 254–255, 257, 266–267, 271, 280.
4. Homer E. Newell, *Beyond the Atmosphere: Early Years of Space Science* (Washington, DC: NASA SP-4211, 1980), p. 274.

The 1960s

In the 1960s, NASA put the first American into space and the only men, ever, onto the Moon. By the end of the decade, one or more U.S. astronauts had flown 44 times, on 22 missions, accumulating more than 2,000 hours in space. Four men had walked on the Moon and returned safely to Earth. Operational medicine was thus the most publicly visible aspect of NASA's space life sciences effort.

Given President John Kennedy's overriding Cold War goal of reaching the Moon, NASA recognized that it would need specialized medical talent and equipment to screen and maintain the health of its astronauts. Pressure to avoid costly duplication of unique personnel and one-of-a-kind equipment led NASA initially to draw on the decade of work already done by the military, primarily the U.S. Air Force at its School of Aviation Medicine at Brooks Air Force Base in San Antonio, Texas, and its Aerospace Medical Laboratory at Wright-Patterson Air Force Base. Along with a contract facility in Albuquerque, New Mexico, the Lovelace Clinic, they had done pioneering research in the post-World War II years that persuaded flight surgeons that piloted spaceflight was survivable.[5] One of the first life sciences activities NASA did after its creation was to employ these aviation medicine specialists to screen a select pool of military jet test pilots to find its first astronauts. Then, from 1961 to 1963, the Mercury years, NASA used flight surgeons, nurses, and medical technicians detailed from the military for clinical work such as postflight exams, in-flight health monitoring, and astronaut-candidate screening.[6] This cooperation continued through the Gemini years, 1965 to 1966, strengthened by the inauguration of the military's Manned Orbital Laboratory (MOL) program, which DOD expected would put a number of soldier-astronauts into low-Earth orbit.[7]

---

5. Loyd S. Swenson et al., *This New Ocean: A History of Project Mercury* (Washington, DC: NASA SP-4201, 1966), pp. 131, 161–163; Jake W. Spidle, Jr., *The Lovelace Medical Center: Pioneer in American Heath Care* (Albuquerque, NM: University of New Mexico Press, 1987), pp. 111, 133; Maura Phillips Mackowski, *Testing the Limits: Aviation Medicine and the Origins of Manned Space Flight* (College Station, TX: Texas A&M University Press, 2006), pp. 3–4.
6. Green Peyton, *Fifty Years of Aerospace Medicine, 1918–1968* (Brooks Air Force Base, San Antonio, TX: U.S. Air Force School of Aerospace Medicine, 1968), pp. 205–210, 215–217, 223–228.
7. Peyton, *Fifty Years of Aerospace Medicine*, pp. 240–243, 250–251.

The duration of Apollo missions (6 to 11 days), combined with two to three times the amount of interior space per astronaut that the Mercury or Gemini crews enjoyed (Command and Lunar Modules combined), allowed NASA to test medical, training, management, and human factors problems in spaceflight. Apollo crews reported candidly on eating, sleeping, and going to the bathroom in space; suit flexibility, cumbersomeness, and temperature control; cabin noise, heat, odor, vibration, and interior design; water potability; and menu appeal. Debriefings also covered the adequacy of training and medical advice, the ability to reach switches or handholds under heavy g-loads, and the abrasiveness of physiological monitors. Frankly discussing their fear and discomfort while slamming forward and back during various flight sequences or while awaiting pickup aboard a capsule taking on seawater, along with describing the ineffectiveness of pharmaceuticals and countermeasures recommended for nausea and the smell of their capsule ablating during a fiery reentry, also helped flight surgeons and spacecraft designers plan realistically for upcoming Shuttle missions and contingencies.[8]

A rising tide lifts all boats, so the broader field of biomedical research also benefited by the missions of the 1960s. The Agency's top management did not seek, let alone demand, in-flight fundamental life sciences research in those tiny early capsules; but Mercury, Gemini, and Apollo astronauts did carry out occasional life sciences experiments. Each mission carried dosimeters to measure the amount of radiation astronauts received in orbit. The Gemini program flew bread mold and human blood samples to examine the effects of space radiation. Apollos 16 and 17 took along pocket mice, microbial samples, and a European experiment, Biostack, which contained numerous biosamples, all to study radiation. After Apollo 10, all crewmembers, except for one, observed light flashes when their eyes were closed, so Apollo 16 and 17 crewmembers wore experimental headgear outfitted with radiation particle detectors during some sleep sessions to look for the cause. It proved to be a type of radiation called HZE particles (for their high atomic number [Z] and energy level [E]). The mouse and Biostack experiments revealed that these particles had passed through the retinas of the astronauts. Adequate controls

8. See the very candid, lengthy, and declassified "Technical Crew Debriefings" for Apollos 7, 8, 9, and 10, stamped "Confidential." These were prepared for the Mission Operations Branch, Flight Crew Support Division, at what was then the Manned Spacecraft Center in Houston, TX. They were published between 17 October 1968 and 2 June 1969, providing rapid feedback from the lunar shakedown mission astronauts.

were not always incorporated into each experiment, and exact repetition was nearly impossible, though, so in-flight studies in this period sometimes produced confounded science and were not considered conclusive.[9]

Early satellites increased interest in life off Earth. NASA opened a Life Sciences Office in 1960 and put biologists to work at Ames. As it grew, Ames built facilities and took on postdoctoral researchers.[10] Gemini capsules had detectors outside that looked for microbial life in orbit.[11] As for satellite studies *dedicated* to fundamental biology of any sort, though, Homer Newell recalled two schools of thought in the 1960s as to the value of such research: either proceed with a program in hopes of gaining insight into designing and operating the carrier and/or system, that is, the satellite itself and/or the rocket, or wait for ground research in biology to catch up and produce more scientifically sound and "definitive" experiments.[12]

An example of how this strategy might (and did) go wrong was the 1962 Biosatellite program, managed by Ames Research Center. Biosatellite I launched in 1966, intended to carry out a three-day science agenda that was specimen dependent. That meant the actual insects, eggs, larvae, etc., had to be retrieved. However, the reentry engines failed to ignite, so while *engineers* might be able to claim having learned from that experience, life scientists got zero data. NASA decided to maintain the program at three flights but to skip a step. Ames would refly the Biosatellite I payload. Leapfrogging over the Biosatellite II agenda of orbiting a rat colony for three weeks, it would then go directly to Biosatellite III's substantially more challenging payload: a monkey.

---

9. Barton C. Hacker and James M. Grimwood, *On the Shoulders of Titans: A History of Project Gemini* (Washington, DC: NASA SP-4203, 1977), pp. 229–230, 541, 546–547, 551–552, 558, 563–564, 568–569; Swenson et al., *This New Ocean*, pp. 385, 467, 497; W. Royce Hawkins and John F. Zieglschmid, "Clinical Aspects of Crew Health," p. 80; Horst Bücker, "Biostack: A Study of the Biological Effects of HZE Galactic Cosmic Radiation," pp. 344, 348; W. Zachary Osborne et al., "Apollo Light Flash Investigations," pp. 355–356, 360–365; Webb Haymaker et al., "The Apollo 17 Pocket Mouse Experiment (Biocore)," pp. 382–383, 391–396, 403; and J. Vernon Bailey, "Radiation Protection and Instrumentation," p. 112, all in *Biomedical Results of Apollo*, ed. Richard S. Johnston et al. (Washington, DC: NASA SP-368, 1975). HZE particles were not available for research on Earth until after Apollo 16.
10. Dick and Strick, *The Living Universe*, pp. 37–39.
11. Hacker and Grimwood, *On the Shoulders of Titans*, pp. 569–570. Finding none, scientists decided that any that might have existed had been killed by radiation.
12. Newell, *Beyond the Atmosphere*, p. 277. For a detailed critique of Ames Research Center's early efforts to become lead Center for fundamental biology research at NASA, see Elizabeth A. Muenger, *Searching the Horizon: A History of Ames Research Center, 1940–1976* (Washington, DC: NASA SP-4304, 1985), chap. 5.

Biosatellite II did not entirely succeed, though, having to end the 1967 flight in less than two days because of recovery area weather.[13]

So, with two-thirds of the program ostensibly "complete" but meeting neither its biological nor engineering goals, and with development costs markedly higher for a primate mission, there was intense pressure to produce *something*. Thus NASA stuck to the original plan for Biosatellite III. Bonnie, a male pigtailed macaque, blasted off 28 June 1969 for what was to be 30 days in orbit. However, NASA terminated Bonnie's mission after eight days, as his health rapidly deteriorated. The monkey died hours after recovery. Bonnie had lost 25 percent of his weight through dehydration, but a necropsy determined the actual cause of death to be "over-instrumentation," that is, too much crammed into one "platform." Thirty-three channels of data came from sensors implanted in his brain; chest leads monitored his heart; sensors in his abdomen monitored Bonnie's temperature; devices in the veins and arteries of his legs measured his blood pressure; catheters collected his urine; and other implants monitored muscle and eye activity. Preflight, Bonnie—and four ground control monkeys—also underwent testicular biopsies, incisor extractions, tail amputations and anal suturing so that, strapped to seats the entire mission, their bodily waste could be collected. Two of those four monkeys died not long afterwards. In essence, the pressure to produce "science" cost NASA not only three payloads—and three very expensive monkeys—but the technology learning experience Homer Newell wrote about.[14]

NASA went ahead in 1970 with the long-planned Orbiting Frog Otolith project. A University of Milan experiment, sponsored by the Office of Advanced Research and Technology, the project put two bullfrogs into orbit for six days, alongside another experiment atop a Scout 1B rocket. An on-board centrifuge alternated their gravity environment from weightlessness to 0.6 g. The frogs wore implanted sensors and had been "demotorized"; they were intentionally not recovered, but usable data were telemetered to the ground.[15]

However, the Agency canceled further development of its own biosatellites. This would have left U.S. researchers with only the three Skylab missions (1973 to 1974) and the one Apollo-Soyuz Test Project flight (1975) until the

---

13. "U.S. Biosatellite Program: Biosatellite I and II," in *Life Into Space: Space Life Sciences Experiments, NASA Ames Research Center, 1965–1990*, ed. Ken Souza et al. (Washington, DC: NASA RP-1372, 1995), available at *http://lis.arc.nasa.gov/lis/index.html*.
14. Colin Burgess and Chris Dubbs, *Animals in Space* (Chichester, U.K.: Praxis, 2007), pp. 280–282; "U.S. Biosatellite Program: Biosatellite III."
15. "Orbiting Frog Otolith Program," in *Life Into Space*, available at *http://lis.arc.nasa.gov/lis/index.html*.

Shuttle was launched 12 years later, had not the Soviets unexpectedly stepped in. They invited visiting NASA life scientists to fly four experiments alongside investigators from five European nations, and they offered to share rat tissue samples, all as part of their 1975 Bion 3 mission (Cosmos 782).[16] Years later, Oleg Gazenko, Director of the Institute for Medical and Biological Problems (IMBP), revealed that one reason for the invitation was that the Soviets wished to fly monkeys, which they would do with Bion 6 (Cosmos 1514) in 1983. NASA had primate experience that the IMBP hoped to tap. However, including CIA, Air Force, and Army launches, the American record between 1948 and 1961 for hominids was five to four *against* survival. NASA did have one feather in its cap, successfully putting the first primate, a chimpanzee no less, into orbit. The feisty Enos rode a Mercury capsule through two orbits on 29 November 1961 and returned cranky but unscathed, paving the way for John Glenn the following February.[17]

The 1970s

With just four crewed missions by middecade, totaling 180 days, the 1970s can be seen as a low point for U.S. crewed spaceflight. In contrast, the Soviet Union launched 25 crewed Soyuz missions, and cosmonauts from the USSR and Warsaw Pact allies visited five different Salyut space stations for a total of over 600 days. However, during this time NASA began recruiting scientists, male and female, no longer just military jet test pilots (who at the time were overwhelmingly white), for anticipated Space Shuttle missions. By thus broadening its astronaut pool, NASA, in contrast with the essentially males-only philosophy of its space rival, increased knowledge in a few pockets of science, mainly operational medicine, including gender differences in anthropometry and physiological reactions to microgravity. NASA also sent the two Viking landers to Mars in 1976 to search for signs of life there. Although the analyti-

---

16. "Mission Information," Biosatellite III, Life Sciences Data Archive, available at *http://lsda.jsc.nasa.gov*; Burgess and Dubbs, *Animals in Space*, pp. 282–286.
17. Burgess and Dubbs, *Animals in Space*, pp. 266–268, 293–294, 228–234, 375–377, 387; Oleg Gazenko, interview by Cathleen S. Lewis, IMBP, 28 November 1989, transcript, p. 10, Smithsonian Archive, Washington, DC. Two French monkeys had made successful ballistic flights in 1967.

cal equipment on board the landers did not find conclusive evidence of even microbial life there, the mission overall was very successful in garnering public awareness of, and support for, NASA and its accomplishments.

The operational medicine subspecialty known as human factors provided NASA engineers with data needed for human-machine integration, for example, designing Shuttle instrumentation that everyone could reach and passageways that everyone could fit through. Suit designers needed to craft garments that fit all sizes and body types among the astronaut corps and allowed wearers to do every imaginable task in relative comfort. The most difficult item proved to be the glove, due to flexibility needs precluding the use of much thermal insulation or padding.

Just prior to World War II, what was then the AAF had begun accumulating anthropometrical (body size) data on military air crew personnel, including females flying for the Nursing Division and the Women's Airforce Service Pilots (WASPs) and black Reserve Officer Training Corps (ROTC) cadets.[18] Military and FAA labs, universities, contractors, and the Japanese space agency all did numerous studies of body size, reach, and so forth during the 1970s and into the early 1980s as the aviation community began to integrate women into jobs previously reserved for men. At NASA, an ad hoc group in Houston, Texas, and Florida composed of astronaut William Thornton, M.D., John T. Jackson, and Gene Coleman, Ph.D., had decided early on that it would be useful to have a similar database for astronauts, and from 1977 to 1979 they gathered statistics on astronauts and applicants alike. NASA (including planetary mission habitat designers in the early 1990s) would use all of these to develop its own human factors standards.[19]

Medical researchers found women to be 15 to 61 percent more susceptible to orthostatic intolerance, that is, passing out when standing after landing, and learned that physical conditioning and pressure suits were less effective countermeasures for females. An early 1990s study would show height to be the predictor in men. The taller the astronaut, the more likely he would be to pass out. Counterintuitively, short women had lower tolerance than tall men, in this case due to a lower blood volume that caused the heart to refill between beats more slowly, thus making it more difficult to maintain sufficient blood pressure. Another observation was that blood pooled in the pelvic area

18. Mae Mills Link and Hubert A. Coleman, *Medical Support of the Army Air Forces in World War II* (Washington, DC: U.S. Air Force, 1955), pp. 238–245.
19. *Man-Systems Integration Standards*, vol. 2, rev. B (Houston, TX: NASA STD-3000, JSC, 1995), appendix A.

among women at as much as six times the rate it did with men, whose blood tended to pool in the legs. Thus pressure suits for females would need to squeeze the pelvic area more and the legs less. Gravity deconditioning after long flights was 50 percent greater among females; however, they recovered more quickly than males.[20]

The three Skylab missions, from 1973 to 1974, carried radiation dosimeters, as the station would be passing regularly through various high-radiation zones. There was also concern over nuclear testing under way back on Earth. Skylab astronauts reported HZE flashes, and the blood of the Skylab 2 and 3 crews was tested for chromosomal abnormalities caused by radiation.[21] Each crewmember was examined several years later to look for evidence of long-term bone demineralization.[22]

NASA investigators flew experiments on board three Soviet Bion/Cosmos missions from 1975 to 1979. Participation was very limited, with NASA contributing only some carrot samples and minor pieces of flight hardware and the passive dosimeters, but they were able to make use of Soviet-supplied tissue samples from rats, fruit flies, and other organisms. Beginning with Bion 4/ Cosmos 936 in 1977, animals (usually rats and quail eggs) were rotated during flight to simulate artificial gravity.[23]

Astrobiology studies at NASA still called for the process of sterilizing planetary probes to avoid contaminating the eventual landing site with Earth microbes that would mislead later investigators. As for life-forms coming *to* Earth, at least from the Moon, NASA dropped the astronaut quarantine requirement after Apollo 12. NASA had canceled a planned exobiology mission to Mars, Voyager, in 1971 but recycled some equipment for the two Viking Mars landers that set down there in 1976. The Jet Propulsion Laboratory funded and coordinated research related to the search for extraterrestrial life in the 1960s and 1970s. Researchers ventured to far-flung deserts, including Antarctica,

---

20. Victor A. Convertino, "Gender Differences in Autonomic Functions Associated With Blood Pressure Regulation," *American Journal of Regulatory Integrative Comparative Physiology* 275 (1998): R1909-20; Mary M. Connors, *Living Aloft: Human Requirements for Extended Spaceflight* (Washington, DC: NASA SP-483, 1985), pp. 29–30.
21. Experiment Information: "Radiological Protection and Medical Dosimetry for the Skylab Crew (SKYRAD)," "Cytogenetic Studies of Blood (M111)," and "Visual Light Flash Phenomena (M106)," Life Sciences Data Archive (LSDA), NASA JSC, data baselined 15 July 2004, available at *http://lsda.jsc.nasa.gov/* (accessed 22 July 2008).
22. Frederick E. Tilton et al., "Long-term Follow-up of Skylab Bone Demineralization," *Aviation, Space, and Environmental Medicine* 51, no. 11 (November 1980): 1209–1213.
23. "The Cosmos Biosatellite Program," in *Life Into Space*, vol. 1, available at *http://lis.arc.nasa.gov.*

for Mars analog locations where they could test life-detection devices for the Viking mission.[24] Ames approved Project Cyclops in 1971, putting researchers at Ohio State University to work with their "Big Ear" radio telescope in 1973, searching for signals from intelligent life.[25]

The 1980s and Beyond

Here the tale of NASA's life sciences research becomes much less straightforward, as the U.S. civilian space program branched out to include an operational Shuttle Program, Space Station, planetary habitat design, and plans for helping private industry make money literally *in* space. Politics and economics, of course, had their roles to play as well.

The Space Shuttle

The focal point of NASA's life sciences research became the ironically named Space Shuttle in the 1980s.[26] An orbiter could carry a much bigger payload than the space capsules of the 1960s and 1970s; in fact, it could be outfitted to carry an entire habitable laboratory in its cargo bay.[27] Its size would also allow the United States, for the first time, to bring along researchers from allied nations, supported on the ground by *their* universities and research labs. However, Shuttle orbiters were confined to just a portion of low-Earth orbit and to missions typically lasting one to two weeks. Plans for extended-duration missions of several weeks and launches from Vandenberg Air Force Base that would have put an orbiter into a polar orbit did not come to fruition. The Shuttle fleet never reached its planned size of five orbiters. Still, the flights of *Columbia*, *Challenger*, *Discovery*, *Atlantis*, and *Endeavour* (the *Challenger* replacement)

---

24. Dick and Strick, *The Living Universe*, pp. 59–61, 82–89.
25. John Kraus, *Big Ear* (Powell, OH: Cygnus-Quasar Books, 1976), pp. 197, 203.
26. Dictionaries define a shuttle as a vehicle that regularly goes back and forth between two places. Since there was no station to serve as the second "place," NASA redefined the destination as simply low-Earth orbit. Also, the Shuttle never achieved a "regular" schedule, and, as *Challenger* proved, it should not have been sold on that basis. The program's formal name was the more accurate Space Transportation System.
27. Asif A. Siddiqi, *Challenge to Apollo: The Soviet Union and the Space Race, 1945–1974* (Washington, DC: NASA SP-2000-4408, 2000), p. 836. Because the Soviets already had stations in orbit, their look-alike shuttle, Buran, apparently was not meant to be an orbiting lab.

added considerably to the *scope* of the Agency's space life sciences capability, especially the dozen-plus Spacelab missions.

Most of the medical research was operational, in particular to reduce or prevent negative effects of microgravity. Vestibular studies by the United States and ESA focused on space adaptation syndrome and the accompanying nausea. NASA researchers, primarily at JSC, designed a suite of experiments, known as the Extended Duration Orbiter Medical Project, that would measure the extent of debilitation that could be expected postflight. The goal was to ensure that returning astronauts would be fit enough to make an emergency egress, but the data were also useful in thinking about future astronauts arriving on Mars after months of weightlessness.[28] Would they be able to get out of their spacecraft, stand, walk, and do any work in even one-third gravity? A key area of interest was bone and muscle loss. Demineralization had been studied as early as Gemini IV, and Gemini astronauts did show significant bone loss in comparison to bedrest patients.[29] However, Apollo and Skylab crews showed mixed results, giving flight surgeons the idea that it was of no real concern, particularly in an operational sense.[30] It became a matter of real concern when astronauts and cosmonauts on longer missions started coming back with significant loss to the weight-bearing bones in their legs, pelvis, and spine. Consequently, animal experiments on Cosmos/Bion and early Shuttle flights began looking at bone loss and bone repair.[31] Flight surgeons at both space agencies had been attempting to prevent or combat the cardiovascular deconditioning and tissue loss since the Apollo program days by prescribing a daily regime of exercise. Twenty years later, they were beginning to question whether that helped at all. Spaceflight seemed to be a speeded-up version of the bone loss experienced by the aging and bedridden, and flight surgeons eventually prescribed the medications used on Earth to treat osteoporosis, rather than creating a medication that was later "spun off" for Earth use.

---

28. Charles F. Sawin, "Introduction to the Extended Duration Orbiter Medical Project," in *Extended Duration Orbiter Medical Project, Final Report 1989–1995*, ed. Charles F. Sawin et al. (Houston, TX: NASA SP-1999-534, JSC, 1999), p. xxiii.
29. Hacker and Grimwood, *On the Shoulders of Titans*, pp. 538, 542, 548. Gemini bone demineralization studies were carried out only on missions IV, V, and VII.
30. Paul C. Rambaut et al., "Skeletal Response," in *Biomedical Results of Apollo*, pp. 303–321; Malcolm C. Smith, Jr., et al., "Bone Mineral Measurement Experiment M078," in *Biomedical Results from Skylab*, ed. Richard S. Johnston and Lawrence F. Dietlein (Washington, DC: NASA SP-377, 1977), pp. 183–190.
31. Kenneth A. Souza and Evgeniy A. Ilyin, "Major Results from Biological Experiments in Space," in *Space Biology and Medicine*, vol. 3, *Humans in Spaceflight*, bk. 1, ed. C. S. Leach Huntoon et al. (Reston, VA: AIAA and Moscow: Nauka Press, 1996), pp. 37–39.

Another area of operational medicine had an easier time making it through the budget and priority tussles over five decades: radiation protection. Radiation shielding was built into *all* vehicles because computer components required protection as much as human beings did. Since the military was putting up many of the satellite payloads, they sponsored a great number of radiation hardening studies. The Department of Energy had been charged with studying human safety at nuclear power reactors and in medical applications such as x rays. Several components of the NIH studied radiation for its risk and curative powers. Weather satellites operated by the National Oceanic and Atmospheric Administration (NOAA) monitored space for solar radiation bursts, allowing commercial and government satellites time to move into "safe" mode. NASA was able to piggyback on the work of these organizations, all of them federally funded, so the need to ask Congress for money was reduced, and everyone felt they had gotten more bang for their bucks.

Each Shuttle mission carried dosimeters on both the crew and the vehicle, and JSC developed an analog human head that was carried along on 11 Shuttle flights between 1989 and 1993.[32] Engineers at LaRC and JPL had worked since the Apollo days to create mathematical models to predict the type and intensity of radiation at various points and times. Complicating things, however, were the discoveries after 1958 of previously unknown types of radiation particles and unrecognized byproducts they created. The radiation belts themselves became better understood, and the inclination of the Shuttle's orbit proved to make a difference in the amount of radiation occupants would be exposed to.[33] NASA developed the Longitudinal Study of Astronaut Health in 1992, partly to track astronauts for the development of cataracts, cancer, and infertility, all potential outcomes of radiation exposure.[34]

Lack of gravity allowed for greater purification of certain biochemical mixtures in space. NASA, industry, and Congress envisioned orbiting factories mass-producing unique medications with no impurities. McDonnell Douglas and the Ortho Pharmaceuticals division of Johnson & Johnson teamed to

---

32. A. Konradi et al., "DSO 469A: Low Earth Orbit Radiation Dose Distribution in a Phantom Head," in *Results of Life Sciences DSOs Conducted Aboard the Space Shuttle 1988–1990* (Washington, DC: NASA, 1991), pp. 59–64; "Experiment Information" and "Experiment Description for: Inflight Radiation Dose Distribution (DSO 469)," LSDA, baselined 15 July 2004, updated 30 July 2008, available at *http://lsda.jsc.nasa.gov/scripts/experiment/exper.cfm?exp_index=200* (accessed 1 August 2008).
33. Life Support Branch, Life Sciences Division, OSSA, "Space Radiation Health Program Plan" (Washington, DC: NASA TM-108036, November 1991), pp. 1–5, 8, 13–14, 25–26.
34. David E. Longnecker et al., *Review of NASA's Longitudinal Study of Astronaut Health* (Washington, DC: National Academies Press, 2004), pp. 1–3.

study erythropoietin purification in space for the treatment of anemia aboard STS flights.[35] Each of the seven flights of Electrophoresis Operations in Space advanced them toward that goal. Electrophoresis Operations in Space also brought about the first of what some expected would be a considerable number of "industrial astronauts," Charles D. Walker, who operated his Continuous Flow Electrophoresis System (CFES) on STS-41D, -51D, and -61B. The *Challenger* accident and the coincidental discovery of a way to make better erythropoietin on Earth combined to dissuade the McDonnell Douglas team from more space electrophoresis research.[36]

The Shuttle era also saw the introduction and evolution of protein crystal experimentation and an attempt at commercial biomaterials processing in orbit. Proteins associated with various diseases (such as diabetes and cancer) were allowed to form crystals for examination by electron microscopy. An intact crystalline structure was almost unattainable on Earth due to gravity and the fragility of these tiny forms. A better understanding of the three-dimensional shape of such proteins would, in theory, allow pharmaceutical manufacturers to tailor-make a cure that would exactly suit the disease. Protein crystallization likewise combined both basic science and technology development, but without the pressures of a rapid commercial application needed to fund it. In recent decades European, Japanese, Canadian, Australian, Brazilian, and U.S. scientists at universities, military labs, and pharmaceutical manufacturers have flown samples inside increasingly sophisticated chambers holding sometimes over 1,000 specimens. In 1986, NASA established a Commercial Space Center, the Center for Macromolecular Crystallography at the University of Alabama-Birmingham, which worked with MSFC to do protein crystal studies.[37] That

---

35. "Feasibility of Commercial Space Manufacturing: Production of Pharmaceuticals," Document III-23 in *Exploring the Unknown*, ed. John M. Logsdon, vol. 3, *Using Space* (Washington, DC: NASA SP-4407, 1998), pp. 534–539.
36. "Biographical Data [Charles D. Walker, MDC Payload Specialist]," Astronaut Biographies—Payload Specialists, Astronaut Office, Flight Crew Operations, JSC, February 1999, available at *http://www11.jsc.nasa.gov/Bios/PS/walker.html*; Charles D. Walker, interview by Jennifer Ross-Nazal, Washington, DC, 19 December 2004, transcript, pp. 15–17; Report from the Payload Specialist Liaison Office, 19 February 1986, folder "Space Operations Directorate Weekly Activity Reports Oct. 1985–Feb. 1986," box 7, Charlesworth files, Center Series, NASA-JSC History Collection, University of Houston–Clear Lake. Per Walker, the Riker Pharmaceutical division of 3M and Japanese and French pharmaceutical firms later joined the collaboration.
37. Since renamed The Center for Biophysical Sciences and Engineering; for a good overview of this program, see "Commercial Protein Crystal Growth-High Density (CPCG-H)," fact sheet number FS-2001-11-186-MSFC, released November 2001, last updated 12 April 2008, available at *http://www.nasa.gov/centers/marshall/news/background/facts/cpcg.html* (accessed 31 December 2008).

much is encouraging; however, the success rate is still far from 100 percent, and no commercial products have resulted after years of research and millions of dollars spent.

Shuttle missions also reminded NASA constituents in the scientific community, Congress, and in the population at large that it was paying attention to *their* needs and to the Agency's relationships on Earth. The STS-90 Spacelab mission, Neurolab, and STS-95, the John Glenn reflight, were touted as having been conceived as medical studies for the general population. Neurolab was a direct response to the 1989 House Joint Resolution 174, declaring the 1990s as "The Decade of the Brain." Its experiments looked at vertigo, eye-hand coordination, prenatal development of neurosensory systems, circadian rhythms, and neural control of the body's other functions.[38] Ex-Mercury astronaut/Ohio Senator Glenn's participation at age 77 was more of a thank-you from the Agency for four decades of participation in, and support of, the space program, but the Agency did put together an agenda of pre-, post-, and in-flight tests and targeted public relations efforts on their applicability to medical issues facing the elderly.

Stunting the Shuttle's potential, however, was the extent to which it became a tool of domestic and international politics, even in relation to the life sciences. NASA, industry, Congress, and the executive branch wrangled over space commercialization; the Shuttle was considered for front-line service in the Cold War; and its seats were seemingly going to the highest political bidder. Especially after the *Challenger* accident and the subsequent investigation that revealed the extent to which internal and external pressures squelched safety concerns, NASA no longer sat atop the world's highest pedestal.[39]

Privatization became a watchword under Ronald Reagan and subsequent Presidents, and NASA made the orbiters the sole platform for it, expecting Space Station *Freedom* to be operational within a decade. The atmosphere was bullish enough on space bioprocessing throughout the 1980s that proposals of industry-sponsored free flyers, orbiting platforms that might be serviced by

---

38. See Jay C. Buckey and Jerry L. Homick, ed., *The Neurolab Spacelab Mission: Neuroscience Research in Space* (Houston, TX: NASA-JSC SP-2003-535, 2003).
39. U.S. Presidential Commission on the Space Shuttle *Challenger* Accident, *Report to the President by the Presidential Commission on the Space Shuttle Challenger Accident* (Washington, DC: Presidential Commission on the Space Shuttle *Challenger* Accident, 1986) offers five volumes of data, analysis, and recommendations. Two subsequent books with useful insights include Diane Vaughan, *The Challenger Launch Decision: Risky Technology, Culture, and Deviance at NASA* (Chicago, IL: University of Chicago Press, 1996) and Howard E. McCurdy, *Inside NASA: High Technology and Organizational Change in the U.S. Space Program* (Baltimore, MD: Johns Hopkins University Press, 1993).

visiting Shuttles, found support in Congress and the space community.[40] The program that came closest to reality was the Industrial Space Facility (ISF), brainchild of Maxime Faget, designer of the Mercury capsule, in partnership with former astronaut Joe Allen, Westinghouse, and a half dozen people from private industry. The tale of the ISF is worthy of a book in its own right, and there is no common agreement about the reasons for its demise. The main suspects, though, are top NASA Administrators guarding their own fledgling Space Station program and some in Congress unwilling to spend on a speculative endeavor without a firm customer base.

Crew safety nearly lost its top priority status in the name of national security. Just months after moving into the White House, an official on Reagan's National Security Council asked other members to give renewed thought to using NASA to fight the Cold War.[41] The unspecified author of a later-declassified secret document posed questions about the Shuttle and its astronauts. "If the conflict is protracted," the writer asked, "do we run the risk of losing all our space launch assets through attrition?" The assets in question were the orbiters and Expendable Launch Vehicles (ELVs). This is clear in the next question: "Will the nation accept a conscientious decision to expose the Shuttle flight crews to anti-satellite attact [*sic*]?"[42]

With a seven-person vehicle, NASA began in 1983 to fly guests and astronauts from some 20 nations, carrying out studies relevant to their national interests: nutrition, agriculture, medical imaging, physiology, and pharma-

---

40. In 1990 the Advisory Committee on the Future of the U.S. Space Program, colloquially called the Augustine Committee, heard from numerous space commercialization promoters, including Max Faget. See also "Johnson Space Center Group 2 Fact Finding Session," folder 5, "Advisory Committee on the Future of the U.S. Space Program, Journal, recording initiation and activities of White House," box 14, the papers of Thomas O. Paine, Manuscript Division, Library of Congress, Washington, DC. Former astronaut and New Mexico Senator Harrison Schmitt testified as an enthusiastic supporter of commercial Helium-3 (He3) mining on the Moon. See also "The Future of NASA," folder 1, "Advisory Committee on the Future of the U.S. Space Program: Presentations Made to Committee," box 18, Paine Papers. The National Commission on Space (The Paine Commission), a similar study under Reagan, had heard similar arguments five years before.
41. The MOL had been canceled in 1969.
42. "Presidential Directive—Space Transportation Policy," item "Space Policy Review, Terms of Reference," Section I, "Future Launch Vehicle Needs," and Section II, "Shuttle Organizational Responsibilities and Capabilities," pp. 1–4, attached to Allen J. Lenz, Staff Director, National Security Council, to Martin Anderson, Edwin Harper, Verne Orr, Richard Darman, George A. Keyworth II, James Beggs, Hans Mark, and William Schneider, Washington, DC, 17 July 1981, file "National Security Directive 144" (accessed online 15 March 2007); "NSDD—National Security Decision Directives, Reagan Administration," *Intelligence Resource Program*, Federation of American Scientists, available at *http://www.fas.org/irp/offdocs/nsdd/index.html*, last updated 25 February 2003 (accessed 22 March 2007); the document was declassified in 1996.

ceuticals manufacturing. Making space accessible to "ordinary" people was touted as an Agency goal until the *Challenger* tragedy in January 1986. The death of teacher Christa McAuliffe made very explicit the difference between a research vehicle and an operational one and halted plans for a journalist, artist, or any other nonscientist or nonengineer in space. In addition to an extreme risk exposure for the "ordinary" person, every Congressman, teacher, writer, or member of royalty in a payload specialist suit meant a seat lost to a bona fide researcher.

Shuttle-*Mir* was a highly controversial program of the late 1990s, criticized in the press and on Capitol Hill for endangering NASA astronauts and giving away U.S. technology in exchange for a glimpse at the Russians' aging station.[43] The Shuttle flew several missions to *Mir*, bringing seven NASA astronauts for stays lasting several months after lengthy training at Star City, near Moscow. Likewise, four cosmonauts were part of Shuttle crews during missions to *Mir*. Astronauts were able to try out Russian Orlon EVA suits slated for use on the ISS, but overall the program did not produce a lot of or particularly novel space life sciences data. It foreshadowed later difficulties in constructing the ISS with the Russians, giving the two space agencies joint operational, management, and emergency experience. As of early 2009, the entire orbiter fleet was expected to retire by September 2010, leaving NASA and the international community only the Russians and their three-person Soyuz capsules to shuttle supplies and personnel to and from the ISS.[44]

Ground-Based Research

It was not until the Soyuz 9 mission in June 1970 that the Soviets noticed problems with cosmonaut deconditioning. Their attitude had been casual to the effects of longer flights on the cardiovascular system until Andrian Nikolayev and Vitaly

---

43. A number of books have been written on the Shuttle-*Mir* program, mostly critical; notable are Bryan Burrough, *Dragonfly: NASA and the Crisis Aboard* Mir (New York, NY: HarperCollins, 1998); James Oberg, *Star-Crossed Orbits: Inside the U.S.-Russian Space Alliance* (New York, NY: McGraw-Hill, 2002); and Jerry M. Linenger, *Off the Planet: Surviving Five Perilous Months Aboard the Space Station Mir* (New York, NY: McGraw-Hill, 2000). Less strident was an account by astronaut Mike Foale's father about his *Mir* experience, which included a collision with a Progress supply ship. (Linenger survived a fire aboard *Mir*.) See Colin Foale, *Waystation to the Stars: The Story of Mir, Michael, and Me* (London, U.K.: Headline Book Publishing, 1999). NASA's official version was Clay Morgan's *Shuttle-Mir: The U.S. and Russia Share History's Highest Stage* (Houston, TX: NASA-JSC SP 2001-4225, 2001). Clay Morgan was the husband of Barbara Morgan, Christa McAuliffe's backup. Twelve years later, Barbara Morgan was selected as a mission specialist. She eventually flew one mission in the summer of 2007.
44. Space Station Consolidated Launch Manifest as of 4 December 2008, available at *http://www.nasa.gov/mission_pages/station/structure/iss_manifest.html*.

Sevastyanov returned feeling dizzy, weak, and ill after 18 days in space. Suddenly there was a lot of concern about lengthy space stays, and even opposition among Soviet scientists, designers, and cosmonauts to proceeding, except very slowly.[45]

Done primarily in response to the polio pandemics of the mid-20th century, early bedrest studies in the United States had already alerted NASA to the possibility of cardiovascular and muscle deconditioning after long flights. The Agency had begun sponsoring university studies of the issue in the mid-1960s, and it included exercise countermeasures in early mission planning. In the early 1970s, Ames Research Center began a multidecadal series of prolonged bedrest studies using volunteers recruited from surrounding communities, attempting to duplicate the physiological changes expected for Shuttle and Station astronauts, and to experiment with various techniques, such as recumbent exercise bikes, that might prevent or counteract physical decline.[46] Later, Ames Research Center pharmacologist Dr. Emily Morey-Holton began "hindlimb unloading" studies using rats to simulate zero-g. This became the standard protocol for an animal model of bone loss in microgravity. By taking the weight off the animal's back legs while it moved about, she could simulate what happened to human bones in microgravity. (Typically weight-bearing bones in astronauts lost mass, while wrists and hands gained.)[47] Johnson Space Center had conducted its own bedrest studies but began another program, or "campaign," in 2004 with the Flight Analogs Project. Carried out jointly with the University of Texas Medical Branch in Galveston, Texas, it included studies of artificial gravity as a countermeasure.[48]

On very rare occasions an astronaut candidate would make it through the entire screening process only to be ruled out for psychiatric reasons.[49] On orbit,

---

45. Abram Genin, interview by Cathleen S. Lewis, IMBP, 29 November 1989, transcript, pp. 8–11, Smithsonian Archives, Washington, DC.
46. The literature on NASA bedrest studies is enormous and influenced research overseas as well. The journal of the Aerospace Medical Association, *Aviation, Space, and Environmental Medicine*, is a good source of published results.
47. Emily Morey-Holton, Ruth K. Globus, Alexander Kaplansky, and Galina Durnova, "The Hindlimb Unloading Rat Model: Literature Overview, Technique Update and Comparison with Space Flight Data," in *Experimentation with Animal Models in Space*, ed. G. Sonnenfeld (Amsterdam, Netherlands: Elsevier, 2005), pp. 7–40. This article confirmed its usefulness, with various modifications, in a survey of 1,064 scientific journal articles written over a 27-year period.
48. "Past and Present Campaigns," Human Adaptation and Countermeasures Division, JSC, last updated 26 February 2008, available at *http://hacd.jsc.nasa.gov/projects/flight_analogs_pastpresent.cfm.*
49. "Class of 1989 Astronaut Candidate Recommendations," presentation to William Lenoir, folder "Astronaut Selections," box 31, Richard H. Truly U.S. Space Program Collection, Regis University, Denver, CO. Former JSC flight surgeon Patricia Santy argued for more rigorous psychological screening in her book *Choosing the Right Stuff: The Psychological Selection of Astronauts and Cosmonauts* (Westport, CT: Praeger Scientific, 1994).

ISS crews began keeping journals and taking self-evaluation surveys for NASA-funded studies of crewmember and crew-ground interactions and the effects of isolation.[50] Much more psychological research happened on the ground, however, and it had been under way for decades. Harkening back to the U.S. Air Force School of Aviation Medicine closed-cabin space simulations in the 1950s, JSC and its contractors carried out many such studies in the 1970s and 1980s, preparing for Shuttle and planetary missions.[51] Story Musgrave, Charles Sawin, Dennis Morrison, and R. S. Clark stood in for a Spacelab crew in 1975 and 1976 simulations, while others played the part of Principal Investigators on the ground.[52] Ames Research Center conducted flight crew research with the FAA, universities, or airlines, in the air and in simulators, to better understand communications, leadership style, environmental health, and other human factors issues in the close confines of the cockpit. An example was the Fatigue Countermeasures Program in the 1980s and 1990s.[53] Members of JSC's lunar and Mars habitat study group made parabolic flights in the KC-135 and "lived" as a group in a mock lunar base, studying air circulation, group dynamics, mobility, noise, vibration, and the overall ability to live and work in tight quarters.[54] NASA also supported research on psychological adaptation to extreme environments, such as the Antarctic, with its severe weather, isolation, and issues with organizational leadership and integration of foreign personnel.[55] The National Science Foundation's Office of Polar Programs, which is responsible for the U.S. Antarctic Program, cosponsored the Antarctic Space

---

50. "Experiment List," Station Science, Current Missions, available at *http://www.nasa.gov/mission_pages/station/science/experiments/List.html* (accessed 28 February 2008).
51. Mackowski, *Testing the Limits*, pp. 176–177.
52. "Spacelab Mission Simulation, Life Sciences Payload Test I, General Summary," JSC 09928, August 1975; "Life Sciences Spacelab Mission Simulation II, SMS Crew Debriefing Transcripts," DE-SMS-II-050; and "Engineering and Operations Report," DE-SMS-II-055, JSC, April and September 1976, folder 50.2, "Spacelab Mission Simulation Test (SMS-1) Oct. 1974," and folder 50.3, "Spacelab Mission Simulation Test II (SMS-2) 1976," shelf 2, cabinet 2, Chambers files, NASA Headquarters.
53. Publications from these studies can be found on the NASA Technical Reports Server. *Aviation, Space, and Environmental Medicine* published a series of articles in the late 1990s headed "Flight Crew Fatigue."
54. Martha E. Evert to Nathan R. Moore and David Gutierrez, 8 July 1991, folder "Action Item 91-12. July 2, 1991," and "Initial Lunar Habitat Simulation Questions" and "Initial Lunar Habitat Simulation 1st Crew Shift Results," both in folder "Initial Lunar Habitat—Simulation Schedule and Questions. June 1991," all in box 8, Center Series-Habitability Studies, JSC History Collection, UH-CL.
55. A useful comparison with the Antarctic and ISS mission support is Lawrence A. Palinkas, "Psychosocial Issues in Long-Term Space Flight: Overview," *Gravitational and Space Biology Bulletin* 14, no. 2 (June 2001): 25–33. A helpful introduction to the history of behavioral studies at the South Pole is Peter Suedfeld, "Polar Psychology: An Overview," *Environment and Behavior* 23, no. 6 (November 1991): 653–665.

Analog Program. The JSC and Antarctic analog studies both began in 1991 as part of President George H. W. Bush's SEI, and they were to aid in designing layout and environmental systems for Mars habitats, as well as to study behavior and performance.[56]

To maximize participation by the university community, lure top-rated medical researchers away from the better-bankrolled federal science programs, piggyback on existing research, and expand industrial participation, NASA set up many consortiums in the 1990s. One was the Center for Macromolecular Crystallography at the University of Alabama-Birmingham. National Specialized Centers of Research and Training (NSCORTs) were dedicated to life sciences, and the overall project lasted from 1990 to 2002. Purdue University had a Center for Research on Controlled Ecological Life Support Systems from 1990 to 1995, and Kansas State University opened its Center for Gravitational Studies in Cellular and Developmental Biology the same year. Lawrence Berkeley National Laboratory and Colorado State University formed an NSCORT in Space Radiation Health in 1992, the same year that Scripps Institution of Oceanography and the University of California-San Diego teamed on an exobiology NSCORT. North Carolina State University and Wake Forest University began a collaborative NSCORT in Plant Gravitational Biology and Genomics in 1996. Planning for what eventually became the National Space Biomedical Research Institute (NSBRI) began in 1992, under JSC's Director of Space and Life Sciences Carolyn Huntoon.[57] The NSBRI idea also aligned with new Administrator Daniel Goldin's idea of the space agency as being an R&D, rather than operational, organization.[58] In March 1997, a consortium led by Baylor College of Medicine won the contract to create essentially a virtual institute, but one with a physical base and presence in Houston, Texas.[59] By 2005, it had expanded to include 12 institutions, an Industry Forum, and col-

56. "The Next Frontier," *National Science Foundation Annual Report 1991*, NSF 92-1 (Washington, DC: NSF, 5 October 1992), updated 31 October 1995, available at *http://www.nsf.gov/pubs/stis1992/nsf921/nsf921.txt*.

57. Archived records of these NSCORTS are available at *http://www-cyanosite.bio.purdue.edu/nscort/NSCORT.html*; Carolyn Huntoon, interview by Rebecca Wright, 5 June 2002, Barrington, RI, transcript, pp. 29, 32–34, "Administrators," JSC Oral History Project, available at *http://www.jsc.nasa.gov/history/oral_histories/oral_histories.htm*.

58. W. Henry Lambright, "Transforming Government: Dan Goldin and the Remaking of NASA" (Arlington, VA: The PricewaterhouseCoopers Endowment for The Business of Government, March 2001), p. 21.

59. Michael Braukus, "NASA Names a New National Biomedical Research Institute," PR 97-43, NASA Headquarters, 14 March 1997.

laboration with the Department of Energy's Brookhaven National Laboratory, where NASA had helped build the new Space Radiation Lab two years earlier.[60]

Space Stations

Crewed, Earth-orbiting space stations had existed in the mind of engineer Wernher von Braun since the 1940s but were a long time coming in real life. Plans were on the drawing board during the 1950s at LaRC and contractor firms.[61] In the 1960s, Boeing and Douglas Aircraft teamed on LaRC's Manned Orbiting Research Lab idea, and the McDonnell Douglas Corporation tried to persuade NASA and the Air Force that one of their Gemini capsules, supersized, could serve as a shuttle to a station that was a cluster of their "Big Gs."[62] The U.S. and USSR briefly considered a joint space laboratory in 1971, fabricated from anticipated spare *Skylab* and ASTP hardware and extra Salyut science modules.[63] In 1973, *Skylab* became NASA's first space station; as a demonstration project under lead Center MSFC, it was occupied for 171 days by three crews of three.[64] By that time, three cosmonauts of space archrival the Soviet Union had lived aboard their station, *Salyut 1*, for three days. Tragically, they died on reentry when their Soyuz craft depressurized. None of them was wearing a pressure suit.

After *Skylab*, NASA did grant General Electric a study contract for a human-tended orbiting vivarium (animal habitat) called the Biomedical Experiments Scientific Satellite (BESS), and both JSC and MSFC let study contracts on stopgap

---

60. "Annual Scientific and Technical Report, October 1, 2004–September 30, 2005, Cooperative Agreement NC 9-58 with the National Aeronautics and Space Administration" (Houston, TX: NSBRI, 30 September 2005); "Brookhaven National Laboratory's Marcelo Vazquez Selected as Space Radiation Liaison for the National Space Biomedical Research Institute," *News Releases*, NSBRI, 14 July 2004, available at *http://www.nsbri.org/NewsPublicOut/* (accessed 19 August 2008); "NASA Space Radiobiology Research Takes Off at New Brookhaven Facility," *Discover Brookhaven* 1, no. 3 (fall 2003), available at *http://www.bnl.gov/discover/Fall_03/NSRL_1.asp* (accessed 17 August 2008); "Space Radiobiology," Brookhaven National Laboratory, last modified 1 February 2008, available at *http://www.bnl.gov/medical/NASA/LTSF.asp* (accessed 17 August 2008).
61. For more details, see James R. Hansen, *Spaceflight Revolution: NASA Langley Research Center from Sputnik to Apollo* (Washington, DC: NASA SP-4308, 1995).
62. McDonnell Douglas Astronautics Company, "Background and Viewpoints on Space Station" (November 1981); McDonnell Douglas Corporation, "Big G" (internal publication, December 1967), pp. 2, 4–5, 7, 26, 32.
63. McDonnell Douglas Astronautics Company, Eastern Division, "U.S./USSR Cooperative Space Laboratory (Skylab/Salyut)" (internal publication, 23 June 1972); McDonnell Douglas Astronautics Company-West, "International Skylab Space Station Technical Considerations" (April 1973).
64. Hansen, *Spaceflight Revolution*, chap. 9.

biology stations and platforms for materials processing.[65] Space Station *Freedom* famously got under way at President Reagan's 1984 State of the Union address, but it took until 1998 and the demise of the USSR for NASA to put its first (and markedly downsized) "permanent" Station component into low-Earth orbit. Russia, Japan, and ESA furnished follow-on nodes and labs; Canada contributed the robotic "hand" Dextre, and the Italians, logistics modules. As of 2009, the ISS is still unfinished.

Cost overruns and errant accounting methods, waxing and waning support (both political and public), economic ups and downs, and even just having six NASA Administrators and four Presidents since Reagan announced Space Station *Freedom* all contributed to the delay.[66] "Big science" was cast as a villain in the partisan budget battles of the 1990s, and the price of keeping the Space Station from cancellation in Congress in 1993 was the death of the Superconducting Supercollider project in Texas, just as the lesser-known Lifesat had been deleted in 1991. The latter was intended to gather radiation and immunological data on living biospecimens. It came under suspicion when questioning during congressional subcommittee hearings revealed that the figure NASA had requested for the program was a tiny fraction of what would actually be needed to develop, deploy, and operate the four-satellite series. The Senate denied funding on the grounds that NASA was cannibalizing its hoped-for Station.[67] NASA repeatedly "scrubbed" the Space Station *Freedom* design, reconfiguring and downsizing it, and ultimately giving the Station's diverse constituents, mainly scientists, a much-diminished return on investment.[68]

Then, in 2004, less than four years after Station assembly began, George W. Bush's Vision for Space Exploration radically changed the present and future of ISS life sciences research in three key ways.

---

65. General Electric Space Division, "Biomedical Experiments Scientific Satellite, Second Program Review" (8 December 1975) and "Biomedical Experiments Scientific Satellite Preliminary Design Study," vol. 2, "System Design" (n.d.); W. E. Berry, J. W. Tremor, and T. C. Aepli, "Biomedical Experiments Scientific Satellite (BESS)" (contributed by the Aerospace Division of the American Society of Mechanical Engineers for presentation at the Intersociety Conference on Environmental Systems, San Diego, CA, 12–15 July 1976); McDonnell Douglas Astronautics Company, "Background and Viewpoints on Space Station."
66. An interesting example is in an audit report by NASA's Inspector General, *Barters on the International Space Station Program*, IG-02-024, 6 September 2002, available via FOIA request to NASA Headquarters. The Inspector General found that the Agency undervalued and improperly recorded some of the $1.5 billion worth of Station components provided by international partners, making NASA likely to not deliver agreed-upon services, including flight time for foreign astronauts to experiment aboard ISS labs. One Center accountant "did not know NASA was required to provide services to the partner in exchange" and recorded an Italian module as a "donation."
67. William Gilbreath, "LifeSat, a New Satellite for Biological Space Research," *ASGSB Newsletter* 3, no. 1 (March 1987): 11; *Congressional Record*, 27 September 1991, p. S13917.
68. "Scrubbed" is used here in the sense of a redesign in search of cost and labor savings, not "canceled."

1) *Emphasis*—The Station had been sold (and strongly affirmed in the 1990 Report of the Advisory Committee on the Future of the U.S. Space Program) as, first and foremost, a life sciences lab.[69] Instead it became "an operational medicine stopover" on the road back to where the space agency had been in 1969—the Moon—then on to Mars, maybe. Agreeing to implement the Vision for Space Exploration would mean that NASA could only carry out research aboard the ISS that supported Bush's specific goal.[70] Evolutionary biology did not, and some fundamental biological research was questionable. Operational medicine and radiation biology looked to be the sole survivors.

2) *Money*—Funds to support lab functions were redirected to development of the follow-on Orion crew vehicle. The replacement of Administrator Sean O'Keefe with Mike Griffin speeded up this process.[71] Congress, in its fiscal year 2005 budget deliberations, directed NASA to designate the ISS as the nation's newest National Laboratory.[72]

3) *Science Capability*—The "centerpiece" of the ISS was to have been a centrifuge to study artificial gravity as a preventive for bone loss and muscle wasting in microgravity and in the one-sixth or one-third gravity of the Moon and Mars. An orbiting centrifuge would have produced better science with true controls, as one specimen remained in microgravity while an identical biospecimen, in the identical noise, vibration, and radiation environment, spun nearby in 1 g, much like the Frog Otolith experiment of 1970. As it trimmed costs, NASA downgraded the centrifuge from human-rated to animal- and plant-rated. Much later it assigned the device, the animal holding racks, the Life Sciences Glovebox, and the module they would all fit inside to Japan as a way to pay for their Kibo lab launches.[73] Post-Vision for Space Exploration,

---

69. U.S. Advisory Committee on the Future of the U.S. Space Program, *Report of the Advisory Committee on the Future of the U.S. Space Program* (Washington, DC: GPO, 1990), pp. 7, 47.
70. Marcia S. Smith and Daniel Morgan, *The National Aeronautics and Space Administration's FY2006 Budget Request: Description, Analysis, and Issues for Congress*, RL32988, 17 November 2005, Congressional Research Service, Library of Congress, pp. CRS-18, CRS-27.
71. Smith and Morgan, *The National Aeronautics and Space Administration's FY2006 Budget Request*, summary and pp. CRS-21, CRS-23, CRS-27, CRS-29, CRS-48.
72. See "Overview: International Space Station," National Lab, last updated 24 December 2008, available at *http://www.nasa.gov/mission_pages/station/science/nlab/*.
73. "Article 6: Respective Responsibilities, Section 6.3 Additional Responsibilities," Memorandum of Understanding between the National Aeronautics and Space Administration of the United States of America and the Government of Japan Concerning Cooperation on the Civil International Space Station, signed 24 February 1998, NASA-Japan Agreement, Space Station Assembly, ISS, last updated 23 November 2007, available at *http://www.nasa.gov/mission_pages/station/structure/elements/nasa_japan.html*.

NASA canceled it for being "nonessential."[74] Congress ordered NASA to reassess that decision, realizing taxpayers were not going to get what they paid for. NASA did not reinstate the centrifuge.

In the course of its 50 years, life science at NASA has made a number of contributions in areas not entirely expected. It has been an agent of social change, as thousands of K–12 students and teachers worldwide have taken part in missions via televised lessons from space, through ham radio programs, and by designing and flying experiments via Shuttle, balloon, and sounding rockets. NASA demanded that contractors and academic affiliates provide equal access to programs and contracts and actively recruit underrepresented peoples. Groups like People for the Ethical Treatment of Animals (PETA) criticized military and NASA life sciences researchers—often justifiably—for using animal test subjects.[75] In response, NASA redesigned and refined its animal safeguards and made agreeing to these a prerequisite for foreign participation. University researchers tried to develop closed-loop environmental systems that would clean air and water in space and provide edible plants. The researchers also created new methods of recycling water on Earth and new food crop varieties that took less space to grow and would provide greater nutrition.[76] Telemedicine of the Mercury program grew into a means for medical care providers to help patients worldwide, and portable defibrillators for the space program have wound up in airports and other public venues for timely response to cardiac emergencies. Ames in 1995 became lead Center for astrobiology and opened the virtual Astrobiology Institute in 1999.[77] By that time, astrobiology was becoming a recognized field of university study. NASA trained and funded numerous college students, postdoctoral academics, and

---

74. Marcia S. Smith and Daniel Morgan, *The National Aeronautics and Space Administration's FY2006 Budget Request: Description, Analysis, and Issues for Congress*, RL32988, 24 January 2006, Congressional Research Service, Library of Congress, pp. CRS-29, CRS-34.
75. N. G. Khruschov to V. E. Sokolov and Mr. N. Saenko, and E. A. Ilyin, D. O. Meshkov, and V. I. Korolkov to The Chancery of the Russian Federation President, Moscow, 15 April 1996, folder 13.2.1.19.13, "Protocol #24, IMBP Biomedical Ethics Committee Reports," shelf 4, cabinet 1, Chambers files, NASA Headquarters. Animal rights protestors carried their arguments to Red Square with letters to Boris Yeltsin, protesting planned Bion 11 primate experiments on the grounds of cruelty.
76. "CELSS: Supplying Humans in Space," *NASA Life Sciences Report 1987* (Washington, DC: NASA-OSSA Life Sciences Division, 1987) outlines some of this work; Utah State University, ARC, and KSC led the Closed Environmental Life Support Systems (CELSS) research. See also B. C. Wolverton and John D. Wolverton, *Growing Clean Water: Nature's Solution to Water Pollution* (Picayune, MS: WES, Inc., 2001). Bill Wolverton designed experimental plant-based wastewater treatment facilities under NASA contract, used at SSC in Mississippi since 1974.
77. A virtual institute required little capital outlay because its members remained physically at their home institutions and communicated primarily via the Internet, telephone, and video link.

professors, some of whom pondered the likelihood of life in unexpected places, fostering Antarctic and ocean-floor research. Observatories sought and found over 100 planets where life might exist in the Milky Way alone. More landers followed the Vikings of 1976 and, in the early years of the second millennium, were still hoping to find signs of life, or even water so that *human* life could be possible, on the Red Planet.

In spite of its many incredible successes, its failures and frustrations have several times been the cause of calls for the Agency's dismantling or demise. This schizophrenic atmosphere makes space life sciences an especially patchy element of NASA history to cover. Searching for the elusive life sciences "spinoff" has lured some authors to look for touted benefits that didn't really happen, an example being the impressive crystals that have come back from orbit but produced a cure for nothing after millions of dollars being spent.[78] In the life sciences, "spinoffs" have more often been a new application for, or repackaging of, existing technology (and someone else's at that). The stop-start nature of NASA programs greatly delayed or aborted ideas with real potential, such as the ISF free flyer, the many Crew Return Vehicle designs since the 1960s, and the canceled ISS centrifuge. Records of those efforts, especially electronic files, essentially go away, as if they had never existed. International politics always had an outsized seat at the table, and consequently diplomacy goals were not always weighed against the cost of knowingly giving away flight time and payload space for patently little or no scientific return. Transparency has also been an issue. University scientists "publish or perish," but there is/was no consistent rule that *everyone*, even Agency researchers, do so.[79] Shuttle press kits don't confirm that a manifested experiment actually flew or that it even worked. Space Shuttle Mission Reports did contain that information, but they exist for only a fraction of the missions. Astronaut debriefs in the Apollo days were "right between the eyes," while Shuttle crew debriefs read suspiciously like a canned document prepared ahead of time, with one astronaut, likely the commander, speaking for everyone else. Overconcern with public image—"what would the neighbors think"—such as the near panic at Headquarters,

---

78. The 2009 Nobel Prize in Chemistry was awarded for ribosome research done, in part, as Shuttle protein crystallization experiments.
79. One such charge was made concerning the Bevelac particle accelerator, which closed in 1993. According to an article in *Nature*, "most of the facility's unique heavy-ion data had never been published in any form." See Stephen M. Maurer et al., "Science's Neglected Legacy," *Nature* 405 (11 May 2000): 117–120.

which canceled participation in Bion 12 after a Russian monkey died, drove other decisions about using animals for research.[80] Semantics often redefined success as having learned something in the process of failing, even when the failure might have been prevented in the first place. In summary, NASA's life sciences personnel and programs have done well considering the many naysayers, doubters, critics, and outright enemies they have faced both internally and externally. Describing their last 50 years, though, makes the Agency seem like a home for the sanity challenged. Maybe it has been, but maybe the next half century will see it evolve into something more sane or reach some breaking point that will produce positive change.

Most of the research for this paper was done under contract NNH05CC40C with the NASA History Division for a study of the Agency's life sciences research from 1980 to 2005.

80. The archival files of Bion Program Manager Lawrence Chambers contain a great deal of information on the monkey controversy, including a shelf of folders on Bion 11 and the canceled Bion 12 program.

Chapter 15

Voyages to Mars

Laurence Bergreen

NASA has been undertaking voyages to Mars since the Mariner missions of the 1960s, frequently updating and revolutionizing our knowledge of the Red Planet. I had occasion in my nonfiction book, *Voyage to Mars*, published in 2000, to witness and to evoke for a general audience some of the varied and unexpected inspiration for the constantly evolving Mars program.[1] I should emphasize how unusual this project was for me, as a biographer and historian, to undertake. During the eventful years (1997 to 2000) that I worked on the book, I was fortunate to receive almost unlimited access to a cross section of NASA's managers and scientists; I also had access to the Agency's ever-increasing stores of data about the universe in general and Mars in particular. Dr. Claire Parkinson of NASA's GSFC in Greenbelt, Maryland, and Dr. James Garvin, currently Chief Scientist at GSFC, acted as guides and reference points during my extended sojourns among the planetary science community. As they studied Mars, I studied them, their colleagues, and their professional quests.

Mainstream newspaper coverage of the era (with John Noble Wilford's perceptive dispatches in the pages and on the Web site of the *New York Times* serving as one prominent exception) seemed to focus on two themes: NASA's failures in execution and NASA's seemingly unbounded spending. As I came to realize, "failure" in space does not equate to failure in business or politics; it is the flip side of exploration, the yardstick to measure success. "Success"

1. Laurence Bergreen, *Voyage to Mars: NASA's Search for Life Beyond Earth* (New York, NY: Riverhead Books [Penguin Putnam], 2000).

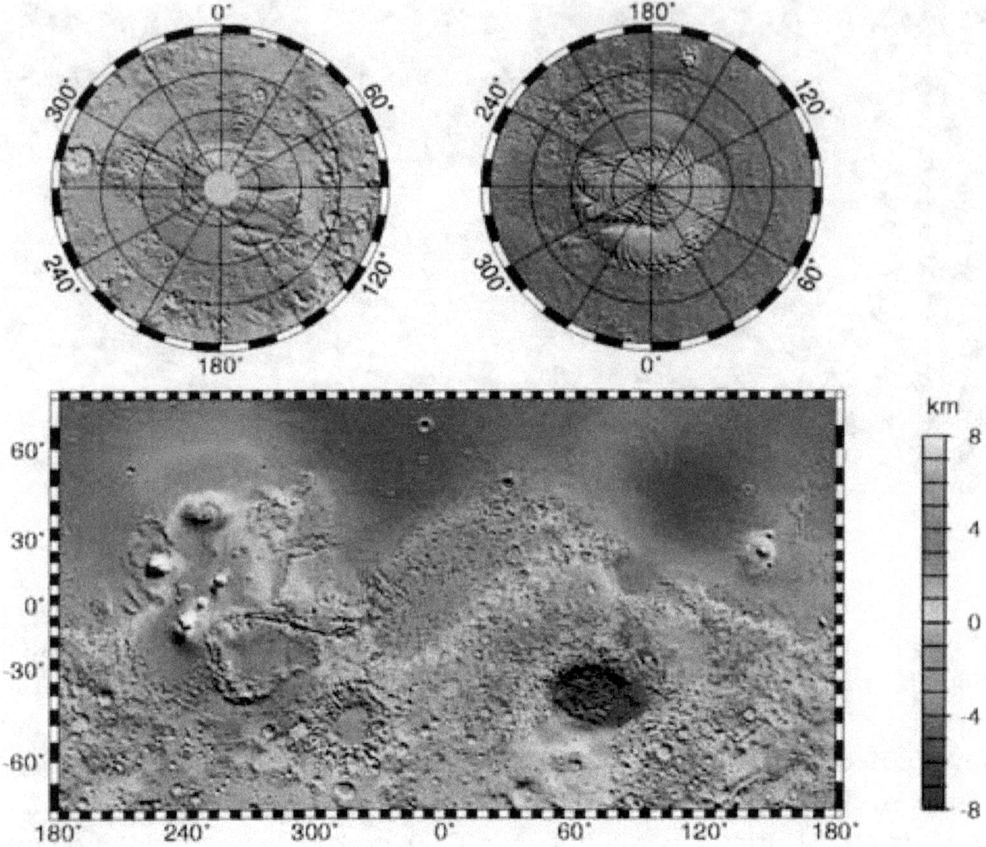

Figure 1: Topographic contour map of Mars from the MOLA. *NASA/JPL–Caltech/GSFC and the MOLA Science Team*

carries its own burden, because it can mean the end of a program (as with NASA's exploration of the Moon), just as "failure" sometimes implies the need for further investigation. And NASA's budget, as is well known, amounts to just 0.4 percent of the U.S. discretionary budget, yet it is much more visible than other, more mundane or politically expedient components of federal spending.

The more time I spent with NASA's scientists, the more I came to appreciate how limited the Agency's resources actually are and how skillfully the managers and scientists leverage the resources of other institutions. I also came to appreciate how a significant component of the mainstream popular press promoted unrealistic expectations for NASA's Mars program, which by its nature contained a significant element of risk. I vividly recall one distinguished planetary scientist expressing frustration with being asked by science reporters what, exactly, a science team planned to discover about Mars. "If we knew what we were going to discover," she said, "they wouldn't be discoveries." Another scientist found that the press had been so selective in its reporting of NASA's varied findings concerning global warming, or climate change, that it became increasingly difficult to give interviews because they might be edited in a way to distort the evidence. If one were to ask members of the public if NASA has points of view about various scientific matters, many respondents would be tempted to answer in the affirmative, but another scientist had occasion to tell me that NASA science does not proceed from fixed beliefs; rather, NASA's experiments "test hypotheses." That was my experience while observing dozens of science meetings devoted to the topographic mapping of Mars, among other subjects, as carried out by the Mars Orbiter Laser Altimeter (MOLA), aboard Mars Global Surveyor. In the interest of accuracy, there was often sharp debate among the participants, until those present "came to clarity," in their words, forming a consensus based on the best available data. These hard-won moments of clarity were all subject to further review, new data, and more testing.

During the course of scores of extended interviews that I conducted over a three-year period, beginning in 1997 with NASA-funded scientists, it became apparent that many drew analogies from those distant eras to NASA's current search for precise scientific understanding of the geologic history of Mars and indirect signs, if any, of life, ancient or otherwise, on the Red Planet. This last item was often referred to as the "Holy Grail" of Mars discovery because of its implications for science, philosophy, and cosmology.

Mars has long exerted a singular allure for scientists and all manner of observers since time immemorial. Few, if any, other planets exert the same level of visceral fascination. For many of the scientists—my particular area of study—working on missions beginning in the late 1990s and continuing up to the present, interest in Mars began early in life, seeded in large part by science fiction and other imaginative literature that lent the Red Planet a mystique appealing especially to younger minds. In particular, the writings of Ray Bradbury cropped up frequently in my conversations with scientists whom I asked about their initial inspirations, especially his best-known work, the beguiling rhapsody known as *The Martian Chronicles*.[2]

In the years after the Second World War, Bradbury, then a young writer living in Southern California, began publishing stories about Mars. In 1950, they were collected and issued as *The Martian Chronicles*, which eventually became a cult favorite. Carl Sagan, among others, cited it as the most captivat-

2. Ray Bradbury, *The Martian Chronicles* (New York, NY: Bantam, 1979 [originally published in 1950]). See also Carl Sagan, *Mars and the Mind of Man* (New York, NY: Harper and Row, 1973).

ing work of fiction about the Red Planet. Bradbury skillfully blended colorful, poetic descriptions. "They had a house of crystal pillars on the Planet Mars by the edge of an empty sea," he writes of a Martian family, "and every morning you could see Mrs. K eating the golden fruits that grew from the crystal walls, or cleaning the house with handfuls of magnetic dust" with visionary scientific predictions. For instance, his denizens of Mars have talking laptop computers, as envisioned by Bradbury decades before they became a reality: "You could see Mr. K himself in his room, reading from a metal book with raised hieroglyphics over which he brushed his hand, as one might play a harp. And from the book, as his fingers stroked, a voice sang, a soft ancient voice, which told tales of when the sea was red steam on the shore and ancient men had carried clouds of insects and electric spiders into battle."

Bradbury's intuitive prescience has its limits. He tends to speed up the exploration and colonization of Mars to what now seems like an unrealistic degree. In his book, the *third* human expedition to Mars was supposed occur in April 2000. In fact, NASA's Mars exploration developed in ways that no one could have imagined, with a number of initiatives and setbacks that no writer or reporter could have predicted. For example, the success of Viking missions in the late 1970s generated the expectation of even more ambitious Mars exploration, but it never materialized, at least not in the immediate aftermath of Viking.

Other inspirations concerning the Red Planet are more elusive, yet no less significant. None of the scientists, engineers, or NASA managers whom I interviewed at the time mentioned Wernher von Braun's blueprint for travel to the Red Planet, *The Mars Project*, yet it has had a lasting effect on the Agency's thinking about Mars. The work was published in this country in 1953, three years after Bradbury's work, but written some years earlier, during von Braun's transition from German rocket scientist during World War II to the linchpin of NASA's peacetime rocketry program. *The Mars Project* is also a classic of its kind, but it was not intended as a visionary or fanciful work; instead von Braun offered a how-to manual of Mars exploration, a practical plan for getting from Earth to Mars and back again.[3] He did his best to minimize the complexities and risks and expense of such a mammoth undertaking, claiming at one point, "the logistic requirements for a large elaborate expedition to Mars are no greater than those for a minor military operation extending over a limited theater of war." Others of the era tended to portray travel through space to

---

3. Wernher von Braun, *The Mars Project* (Urbana, IL: University of Illinois Press, 1991 [originally published in 1953]).

Mars as a poignantly lonely undertaking, but von Braun took exception. He outlined a large flotilla or fleet of spacecraft, orbiting Mars before sending smaller craft to set down on the surface of the Red Planet in orderly fashion. He prescribed the use of three-stage rocket "ferries" to transport people and items to and from the fleet. To the extent that NASA eventually developed embryonic plans for reaching Mars, von Braun provided much of the practical and strategic approach, while Bradbury and other science fiction writers provided the less predictable, and more individualistic, inspiration to those men and women who would actually devote their careers to carrying it out.

Yet even that was not sufficient for a complete Mars program. For all his gifts, von Braun did not make plans for the scientific study of Mars, nor did he—or anyone else at the time—fully appreciate the difficulties confronting humans in deep space, especially harmful radiation.

Although the romance of finding some sort of life on Mars—intelligent, primitive, or otherwise—informed the mystique of exploration, NASA's first attempt to explore the Red Planet robotically with the Mariner 4 flyby mission in 1965 nearly brought the project to a jarring conclusion. On 14 July, this spacecraft's black-and-white camera sent back stark, low-resolution images of the surface of Mars that appeared to depict landscape features devoid of life-favorable environments. Where Bradbury had suggested seductive mental constructs and some scientists had hoped to find at least some signs of primitive vegetation, or ideally even liquid water or some other liquid, those early Mariner images showed unrelieved desert. There were no rivers, no forests, and no water-related channels, as some had thought once flowed across the surface of Mars. The Red Planet revealed itself as the Dead Planet, that is, desiccated, and from certain perspectives, it did indeed present an environment that was inhospitable to life. There were no oceans, obvious vegetation, or signs of life (intelligent or otherwise). But Mariner 4's smudged images, 22 in all, recorded only a tiny part of Mars's story; they covered merely 1 percent of the planet's surface, and, it would emerge, they concealed as much as they revealed. Furthermore, the images were unable to resolve human-scale features considered necessary to understanding the history of water and environments.

By this time, Mars communities were flourishing. In 1996, Robert Zubrin, a former engineer at Lockheed Martin, published *The Case for Mars: The Plan to Settle the Red Planet and Why We Must*. An organization known as the Mars Society took root and lent its support and commentary to NASA's robotic exploration of the Red Planet. Zubrin's book is in its way a thorough updating of von Braun's *The Mars Project*, a rigorous effort to adapt Mars exploration to current political, scientific, and cultural realities. Zubrin makes reference to the ways in which various influential politicians might best advocate Mars exploration, emphasizing practical benefits, the kind that elected leaders could sell to their constituencies. Zubrin also considered problems that had occurred to few others, for instance, the precise length of the Martian day, which is about 40 minutes longer than an Earth day. There were serious implications in this disparity for computer clocks and geographical coordinates—all of which could make managing communications with, and navigation of, Mars extremely complex and prone to accidents. "The practical answer is simple," Zubrin proposed, "just divide up the Martian day into 24 Martian hours, each composed of 60 Martian minutes, each of which is composed of sixty Martian seconds. The conversion factor between Martian days, hours, minutes, and seconds and their terrestrial equivalents would this be 1.0275 across the board . . . . Such a clock solves all the practical problems associated with daily timekeeping on Mars."[4]

This solution, as critical as it was, did not address the issues raised by maintaining a healthy environment for people spending days, weeks, months, or longer on Mars. Despite his considerable care and thought, Zubrin occasionally overlooked the limits of our understanding of Mars and underplayed the hostility of the Martian environment for humans. It may be true that, as Zubrin states, "among extraterrestrial bodies in our solar system, Mars is singular in that it possesses all the raw materials required to not only support life, but a new branch of human civilization," but obtaining, refining, storing, and deploying those materials requires new levels of understanding and technological sophistication. It is not clear that a "new branch of human civilization" is a widely endorsed or understood goal at present.

So much for the theorists and visionaries. The reality of Mars exploration proceeded in unpredictable fits and starts. In September 1993, NASA lost its Mars Observer mission, which had cost nearly a billion dollars, including the launch vehicle. The ill-fated orbiter had carried no less than a dozen experiments designed to map Mars and had the potential to greatly enhance or even revolutionize our knowledge of the Red Planet.

This discouraging event gave rise to public outcry and ridicule, including a satirical monologue delivered on television by the popular comedian David Letterman. To reinvigorate robotic exploration, Dan Goldin, on becoming NASA's Administrator in 1992, instituted a so-called "faster, better, cheaper" approach,

4. Robert Zubrin, *The Case for Mars: The Plan to Settle the Red Planet and Why We Must* (New York, NY: Touchstone [Simon & Schuster], 1996), p. 163.

which, for planetary exploration, meant that individual missions would make the journey from idea to launch in less time and at much less cost. The goal was to keep the cost of each mission to under 200 million dollars, less than one-fifth what Mars Observer had cost to execute, and thus to have more of them. "Faster, better, cheaper" went along with Goldin's style of management, aimed to shake up NASA, which, he liked to say, had become "too male, stale, and pale." This approach widened NASA's doors to capable scientists, engineers, and managers from a wider pool, to the Agency's credit. At the height of the "faster, better, cheaper" era, Goldin explained to me its reason for being: "There were so many experiments on spacecraft that if you lost it, you lost the whole system. Those spacecraft cost too much. They took too long to build. Instead of using next-generation technology, which is what NASA is supposed to do, to drive the technical base of the country, it went back in time and used old, proven technology, which made the spacecraft bigger and more expensive. It was very inefficient."

Goldin functioned as an agent provocateur. He had come out the intelligence arena, and he was accustomed to a certain amount of confidentiality, even as NASA entered an era of transparency at the end of the Cold War. Russian rocket scientists and engineers were now working for NASA, an unthinkable development only a few years earlier. Goldin seemed to some to run the space agency as a Skunk Works in plain sight. It was said that he would pit rival teams against one another to accomplish tasks, an effective, if unsettling, management technique. Few doubted his brilliance, his forceful personality, and his commitment to achievement in space, but he achieved his ends not without a certain amount of stress and perhaps bruised egos. Nevertheless, he remained in the job throughout the decade and publicly declared himself as a proponent of the Agency's Mars program.

Over time, "faster, better, cheaper" gave rise to one closely spaced robotic Mars mission after another: Mars Global Surveyor (a success), Mars Pathfinder (a success), Mars Polar Lander (a failure), and Mars Climate Orbiter (another failure, and painful one at that). Each of these failures, while less costly than Mars Observer, took its toll on morale at NASA and tarnished the early gleam of the "faster, better, cheaper" approach. The Agency had not anticipated the extent to which the American public would fasten on to the failures and overlook the successful missions, which completed their goals without making headlines or causing crises. The idea that NASA was exploring Mars and the rest of the solar system on the cheap came to annoy Congress and the public, yet it was Congress that, year by year, at the urging of the public, imposed steadily increasing budget cuts that led to the institution of "faster, better, cheaper" missions in the first place. For instance, NASA's budget came to $13.8 billion in 1998, declining to $13.7 billion the next year and $13.6 billion the following year, despite a robust, expanding economy, increased tax revenues, and a federal budget surplus. (NASA's budget for 2009 has increased to $17.6 billion, an apparent gain diminished somewhat by inflation.) By way of comparison, DOD was receiving funding that was 20 times greater, and the Department of Housing and Urban Development received twice the amount received by NASA. Private industry and advocacy urged NASA to contract missions out to the corporate sector, but Goldin remained unpersuaded, having come from the private sector. He was deeply skeptical that private enterprise would be willing to risk the capital, the time, and the resources on a scale necessary for space exploration to succeed. Ultimately, the operative word of his approach became "cheaper." Even the successes seemed compromised by their cutting of costs, which meant lost opportunities to gather data. As one Mars scientist, James Garvin, was given to lament, the Viking missions of the 1970s had better television cameras than the missions of the 1990s.

Yet the 1990s had its successes, as well. In 1997, a relatively small NASA robotic spacecraft named Mars Pathfinder landed on the surface of Mars and revived interest in the Red Planet, both scientific and popular. For a time, the press suspended accounts of failures and cost overruns to report on data gathered by Pathfinder and posted on the Internet in something close to real time. Pathfinder's miniature remote weather stations recorded temperatures ranging from 60°F at noon to -100°F at night. Its tiny rover, named Sojourner, captivated the public imagination. Mars Pathfinder and its rover perceived smudges across the Martian sky caused by the shadows of the Red Planet's two small moons, Phobos ("fear") and Deimos ("terror"). Later, Mars Global Surveyor captured high-resolution images of these objects. As a result, Mars became more than a scientific construct; it became a place, an address, and a destination.

On 11 December 1998, I was on hand at Cape Canaveral, Florida, to observe the liftoff of Mars Climate Orbiter, a low-budget spacecraft perched atop the usually reliable Delta II launch vehicle. The sight and sensation of so much power harnessed toward a single objective was awe-inspiring and admirable, but also deceptive. Mars Climate Orbiter was scheduled to arrive at the Red Planet 10 months later, after a journey of 416 million miles. "A bullet has been fired at the planet Mars," I noted at launch time. Expectations ran high and overlooked the likelihood of failure. Mars Climate Orbiter came out of the era's "faster, better, cheaper" approach for robotic spacecraft, which emphasized, for very good reasons, getting more spacecraft into space as soon as possible to lower costs and to distribute risk.

The fate of Mars Climate Orbiter became a test case for "faster, better, cheaper" and, to a certain extent, for the entire Mars Surveyor program of its era, and even for the Goldin-inspired approach to space exploration. It should be noted that Mars Climate Orbiter was an orbiter, and orbiters were prone to failure. Orbital insertion around Mars (or any other planet) requires extreme precision and planning because the spacecraft is too far from Earth to respond to navigation signals in real time. If the spacecraft's navigational computers were programmed correctly with appropriate data, and if the rocketry functioned precisely as calculated, and if nothing unexpected occurred in the vicinity of Mars, all would go as planned.

Those familiar with the risks inherent in the exploration of Mars occasionally invoked the fanciful "Great Galactic Ghoul," the demon of misfortune that destroyed vulnerable missions such as the Mars Climate Orbiter mission, and, on this occasion, the Great Galactic Ghoul ate Mars Climate Orbiter.

The cause of the failure was soon discovered and, when revealed, proved embarrassing to NASA. One group designing the mission habitually calculated thrust in pounds, the traditional British unit, while another relied on the metric unit known as the newton. A newton is not a pound, and a pound is not a newton, especially in space. (One pound of thrust accelerates 1 pound of mass 1 foot per second, squared; a newton accelerates 1 kilogram of mass 1 meter per second.) The discrepancy was not insignificant: a pound of thrust is nearly four times greater than a newton. The resulting error sent Mars Climate Orbiter into orbit around Mars at an altitude of 37 miles rather than about 70 miles, as planned. How tiny that distance seemed coming after a mission that had operated flawlessly for hundreds of millions of miles, yet it was sufficient to send the fragile spacecraft low enough to burn or break up in the thin Martian atmosphere. Although easily rectified, the error had a sobering, not to say unnerving, effect on NASA because it could have been easily avoided. It slowly began to dawn on engineers that too many failures, too much urgency, and excessive frugality posed significant hazards.

Despite these problems, the Viking missions of the 1970s, Mars Pathfinder (1997), and Mars Global Surveyor (1997 to 2007), all of them flying laboratories with multiple objectives and research strategies, have sustained the promise of exploring our planetary neighbor with ample rewards in a sequence of discovery paralleling the explorers of the Renaissance. As a result, a vibrant new world, rich in water and in water history, as well as potential habitats for life, has continuously unfolded, as Mars slowly yields its secrets.

Throughout this period, the impulse to indulge in occasionally perverse fantasy regarding the Red Planet flourished, abetted by the masses of data that NASA and allied organizations, universities, and companies made available. One example of stubborn insistence on belief unsupported by data was the "Face on Mars," first observed, so the theory ran, by Viking's cameras back in the 1970s. In the way that clouds can resemble familiar objects, a mesa-like landform on Mars in the Cydonia region came to be seen as a monument to a human face, offering tantalizing evidence of an ancient civilization and clues to the destiny of our own. In his book *The Mars Mystery: The Secret Connection Between Earth and the Red Planet* (1998), Graham Hancock advanced the theory that "Cydonia is indeed some sort of signal—not a radio broadcast intended for an entire universe, but a specific directional beacon transmitting a message that was intended exclusively for mankind."[5] Hancock's book was published by a mainstream publisher despite a widely known article by Carl Sagan in 1985 explaining that the Face was an optical illusion. It seemed as if Ray Bradbury's vision had returned, minus the whimsy and spiked with an urge to distort and mislead. In time, some of the intensity surrounding the Face had dissipated, in part because newer, high-resolution images demonstrated that it is, in fact, just what NASA and Carl Sagan had been stating all along: a mesa-like landform; but the controversy persists as example of irrationality despite the ample evidence to the contrary.

Researching my book, I noticed the frustration of NASA scientists who felt obligated to address these unrealistic and distracting theories. Some refused; others pointed out obvious scientific realities; and still others revealed a sense of ironic whimsy, decorating their desks and blackboards with replicas of Martian gremlins and other imaginative creatures. No effort has been made to suppress the spurious theories. During the years I worked on my book, I was struck again and again by the transparency of NASA's science. Although NASA scientists, managers, and engineers occasionally disagreed with my conclusions or assumptions, they were generally glad to have the opportunity to explain their position; at no time did my research suffer from censorship or secrecy—quite the opposite.

In this relatively transparent and tumultuous era, NASA contributed to the revolution in Mars science with the 1996 announcement of the discovery of possible ancient microbial life on Mars. Today, the controversy about the

5. Graham Hancock, *The Mars Mystery: The Secret Connection Between Earth and the Red Planet* (New York, NY: Crown, 1998).

evidence persists, but at the time, a Martian meteorite recovered in 1984—known as ALH84001—suggested that primitive, microbial life once existed on Mars and might still be there today, concealed beneath its parched, toxic surface, hidden deep beneath subsurface rocks. When *Science* magazine posted an early version of the article, titled "Search for Past Life on Mars: Possible Relic Biogenic Activity in Martian Meteorite ALH84001," on its Web site in July 1996, it received a million hits, one gauge of interest.[6] If accurate, the article offered the first scientific evidence of life on Mars, not a theory, whim, or conspiracy. The implications, both scientific and philosophical, were immense, and even scientists, in fact especially scientists, were apt to become emotional on the subject. It might be expected that Carl Sagan endorsed the article and its implications, but an earlier comment of his—"Extraordinary claims require extraordinary evidence"—was as far as he went. The remark implied that the article's findings did not quite meet the test. Nevertheless, the possibility that some form of life, no matter how simple, had once existed gave NASA's Mars program a new impetus, focus, and newly enhanced area of research: astrobiology.

The implications of life in some form on Mars held out the promise of a new paradigm concerning life throughout the universe. If it was established that life, even in the form of simple microbial activity, existed on Earth and Mars, it became plausible that life, in one form or another, primitive or advanced, was widely distributed throughout the solar system and the universe. If true, astrobiologists were in a position to pose questions as to how and where life first arose. Had it spread from Mars to Earth via meteors, or vice versa, or from a common source?

At the same time, many scientists remained deeply skeptical, even hostile, to the finding of life on Mars, arguing that the meteorite was contaminated with terrestrial life, that it offered nothing more, and possibly even less, than evidence of life on Earth. What had seemed to rank as one of the seminal scientific discoveries of the era gradually lost support and credibility in the scientific community, and NASA, after endorsing the article at the highest levels, gradually moved on. Even if the ALH84001 contained Martian nanofossils, which came to appear somewhat unlikely, the discovery (if that is what it

6. David S. McKay, Everett K. Gibson, Jr., Kathie L. Thomas-Keprta, Hojatollah Vali, Christopher S. Romanek, Simon J. Clemett, Xavier D. F. Chillier, Claude R. Maechling, and Richard N. Zare, "Search for Past Life on Mars: Possible Relic Biogenic Activity in Martian Meteorite ALH84001," *Science* 273, no. 5277 (August 1996): 924–930.

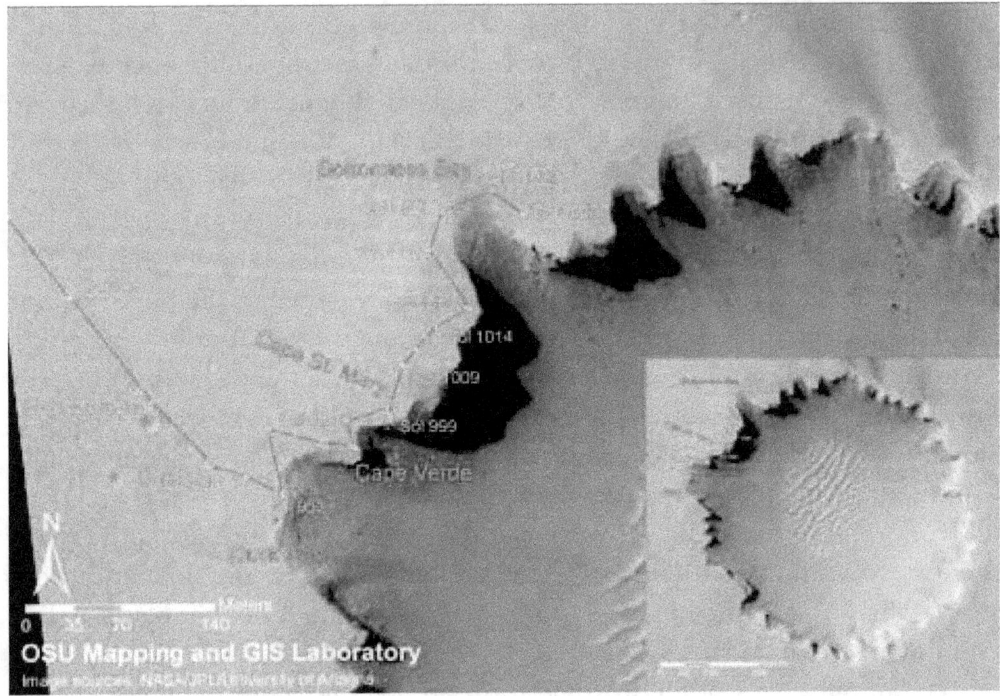

Figure 2: A traverse map of Victoria Crater illustrating the path taken by the MER Opportunity rover. *USGS/University of Arizona/NASA/JPL*

was) required confirmation, which could best be found on Mars, rather than in meteorites presumed to have originated on the Red Planet, and that evidence (while entirely plausible and consistent with what is known about Mars) has yet to be found and confirmed. As a result, the Holy Grail of NASA's Mars program remains a tantalizing possibility.

Despite this flurry of Mars-related activity in the recent past, NASA has continued its painstaking robotic exploration of Mars, with emphasis on a pair of MERs, as they have come to be known, Spirit and Opportunity.

Recently, the rover Opportunity, enjoying an extended mission lasting nearly five years longer than originally planned, or budgeted, has been exploring Victoria Crater, and that, in turn, brought my experience with NASA's Mars robotic exploration program full circle. To help underscore analogies between the exploration of planets and the Age of Discovery and our own era, Steven Squyres, the Principal Investigator of the scientific payload on the MER mission, asked me to contribute place-names for those features around Victoria Crater after those discovered by Ferdinand Magellan during his first-ever circumnavigation (1518 to 1521). Magellan served as the protagonist of my book, *Over the Edge of the World: Magellan's Terrifying Circumnavigation of the Globe*

(2003).[7] And the subject of this historical account had been suggested by my earlier book about the Agency's Mars program, where scientists often cited Magellan as a source of inspiration. James Garvin characterized Magellan's voyage as an example of "intelligent exploration," that is, knowing where you are going, and why; and Magellan's voyage was certainly more sophisticated and ambitious than any recorded prior journey of its type. But by contemporary standards, it was woefully simplistic and wishful.

Although Magellan was convinced that he was divinely chosen to succeed, most people of his time believed he was attempting the impossible; in those days, it was believed that ships would never make it as far south as the equator, let alone more exotic destinations, that boiling seas would scald sailors to death, or magnetic islands pull the very nails from the planks of their ships, sending them to the bottom of the sea. The size and shape of the world was misunderstood; even the most advanced minds in Europe did not realize that the Pacific Ocean was the largest body water on the planet. Had Magellan known its true extent, he—or his cautious backers—would not have undertaken the voyage. In addition, the art of navigation was still in its infancy. It was still impossible for mariners to determine longitude, and even the length of a degree of latitude was subject to debate. There were maps, but, as might be expected, the further from home, the more inaccurate they became, until by the time Magellan reached South America, they contained more geographical fantasy than fact. Eventually Magellan became so exasperated with his useless maps that he threw them overboard, declaring they were not to be trusted.

Given these multiple hazards and difficulties, why did Magellan go? And why did his backers risk their capital and prestige on his expedition? The answer can be summed up in two words: greed and glory. If Magellan accomplished his goal, Spain hoped to seize control of the spice trade and, by extension, the emerging global economy. Magellan himself hoped to claim lands and titles and unimaginable wealth to pass on to his heirs. We tend to forget that the first Age of Discovery was often driven by some very disagreeable goals. The idea of scientific exploration did not become prominent until the 18th century with the voyages of Captain Cook and, still later, Charles Darwin.

One of the momentous events of the Age of Discovery occurred on 6 September 1522, when tiny *Victoria*, the sole survivor of Ferdinand Magellan's first-ever circumnavigation of the globe, returned to her home port of Sanlúcar

7. Laurence Bergreen, *Over the Edge of the World: Magellan's Terrifying Circumnavigation of the Globe* (New York, NY: William Morrow, 2003).

de Barrameda, Spain. By any conventional reckoning, the expedition was a disaster; four of the fleet's five ships were lost, and out of the 260 men who had set out from Spain three years earlier, only 18 made it all the way around the world. Magellan himself, the Captain General, was not among them, having been killed in battle with tribal warriors in the Philippines, where he paused en route to the Spice Islands in Indonesia. Magellan, an abrasive Portuguese nobleman sailing for Spain, sacrificed his life in the course of disproving centuries of accumulated superstition and outright ignorance concerning the nature of our world, and today we prize this tragic expedition for its many contributions to our knowledge of the world: geography, peoples, cultures, and climates, to name a few.

To memorialize Magellan's exploits, I selected names used by Magellan that seemed appropriate to the locations explored by Opportunity and submitted them to Steve Squyres, who assigned them to various Martian features. In October 2008, when I asked how the mapping of this particular region of Mars with names inspired by Magellan was proceeding, he wrote back, "About a year ago, we"—meaning the rover and, by extension, its handlers back on Earth—"entered Victoria Crater at the place we named Duck Bay (*Bahia de los Patos*) and we spent nearly a year inside the crater, working our way slowly and methodically through all of the stratigraphy presented in the wall there. Several weeks ago we exited the crater. We are now driving counterclockwise around the crater for some final imaging before we head off toward our next exploration goal. A week or so ago we drove out onto the northern end of the promontory we have named Cape Victory, imaging northward toward the south-facing wall of Cape Pillar. We are now on the promontory we have named Cape Agulhas, imaging the south-facing wall of Cape Victory. Cape Agulhas will be our last stop at Victoria before we turn southward."[8]

If Magellan explored Earth to a greater extent than anyone before him, it is safe to assume that he never imagined that he was also going to explore Mars, or any other planet, by analogy. The annals of NASA's robotic exploration of Mars demonstrate that sometimes history can be made by looking backward, with informed reference to the distant past, as well as forward.[9]

8. Steven Squyres, *Roving Mars: Spirit, Opportunity, and the Exploration of the Red Planet* (New York, NY: Hyperion, 2005).
9. For another, more recent popular view of NASA's exploration of Mars, lavishly illustrated, see Andrew Chaikin, *A Passion for Mars: Intrepid Explorers of the Red Planet* (New York, NY: Harry N. Abrams, 2008).

Chapter 16

The Space Age and Disciplinary Change in Astronomy

David DeVorkin

How has the Space Age changed the astronomical profession? Historians such as Robert Smith, Robert Seidel, Michael Dennis, and of course Paul Forman have all discussed what can happen when physical scientists of all stripes shift to new and very big technologies, altering their methods of inquiry, defining their competitiveness in terms of these machines and systems, and, moreover, relying on sources of funding whose motivations lie outside the realm of science. Smith, in particular, shows poignantly from his case study of the Hubble Space Telescope that

> Space astronomy placed new demands on astronomers, not only in terms of the reliability of their instruments and their methods of work, but also in how that work was to be directed and controlled. Thus it changed what it means to be an astronomer.[1]

Indeed, a full assessment of the impact of the Space Age, and of NASA (in particular) for the American component of the discipline, on the practice

1. Robert W. Smith, "The Biggest Kind of Big Science: Astronomers and the Space Telescope," in *Big Science: The Growth of Large-Scale Research*, ed. P. Galison and B. Hevly (Stanford, CA: Stanford University Press, 1992), p. 194; Paul Forman, "Behind Quantum Electronics: National Security as Basis for Physical Research in the United States, 1940–1960," *Historical Studies in the Physical and Biological Sciences* 18, pt. 1 (1987): 149–229; R. W. Seidel, "Accelerating Science: The Postwar Transformation of the Lawrence Radiation Laboratory," *Historical Studies in the Physical Sciences* 13 (1983): 375–400.

of astronomy requires, as Smith implies, looking not only at the relationship of the instruments of observation to their builders and users, or even to the problems attempted and discoveries made, but at the relationship of these practitioners to those who paid the bills and, to a large part, controlled what would fly and what would not fly. These include the individuals, institutions, and governments that supported astronomy for reasons ranging from national identity and security, to economic and intellectual competitiveness, to merit-based peer review, and even to idealist forms of inclusiveness and curiosity.

There are many ways to explore how astronomy as a discipline changed in the latter half of the 20th century and how what it means to be an astronomer changed as well. Complementing the views of historians like Smith, at least one prominent late 20th-century astronomer, Leo Goldberg, observed that "astronomy has always been inseparable from developments in experimental and theoretical physics." In his experience, "discoveries in physics have been applied to astronomy by physicists, some of whom, like Prof. [V. L.] Ginzburg, Bengt Edlén, Bruno Rossi and Hans Bethe, have retained their identity as physicists, while others, for example S. Chandrasekhar, E. E. Salpeter, M. Ryle and R. Giacconi have chosen to become affiliated with astronomy."[2] Indeed, assessing disciplinary change requires that one consider all relevant views on how a discipline is described or defined. Do those who affiliate through problem choice alone have an impact any different from those who also affiliate through professional association? Here we will look at only two characteristics that describe disciplinary change. We will consider first the influence of straight growth in size on the discipline, and then we will consider how that growth was stimulated by the nature of the projects that were attempted by astronomers, both indigenous and migrants from physics.

We will start by looking at what astronomers were like and how astronomers behaved before the Space Age, examining aspects of how their institutions were structured and how they changed. We will look at growth, at specialization, at problem choice, and finally, through two brief case studies, at how some astronomers responded to the opportunities of the Space Age and what that response did in turn to change their institutions and profession. This discussion

2. Leo Goldberg, "Quantum Mechanics at the Harvard Observatory in the 1930s," in *Problems in Theoretical Physics and Astrophysics: A Collection of Essays dedicated to V. L. Ginzburg on his 70th Birthday*, ed. L. V. Keldysh and V. Ia. Fainberg (Izdatel'stvo, Russia: Nauka, 1989), pp. 21–22. Karl Hufbauer also makes this point in part 1 of his *Exploring the Sun: Solar Science Since Galileo* (Baltimore, MD: Johns Hopkins University Press, 1991).

is both stimulated and informed by one of the most persistent questions asked by historians of post-World War II science: if "the desire of scientists to retain as much authority as possible over the direction and management of research" has in fact been met or has been lost.[3] We offer no definitive answers here, though the suggestion will be made that no matter where authority resided, if it was retained or lost, the fundamental nature of the discipline and its members profoundly changed. This essay is a contribution to appreciating the nature of that change.

Astronomy's First Contact with the Promise of Space Research

In late 1945, learning of what captured German V2 missiles could do, some astronomers were initially very excited about the possibility of sending instruments on rockets and seeing the universe from space. Leo Goldberg, then at Michigan, told Harvard's Donald Menzel, in September 1945, that he'd be willing to shave his head and live in a cell for the next 10 or 15 years for an opportunity to view the Sun's spectrum from space.[4]

Through the spring and summer of 1946, U.S. Army Ordnance initiated the first flights of reconstructed missiles at White Sands with warhead payloads filled with cosmic-ray counters, solar spectrographs, and other devices prepared by military laboratories at the Applied Physics Laboratory (APL) and the Naval Research Laboratory (NRL), but the results were very disappointing to astronomers like Lyman Spitzer and Leo Goldberg, who had formed an Office of Naval Research (ONR)-supported "Astrophysical Consulting Bureau" to assist the physicists at APL and NRL in the analysis of their hoped-for ultraviolet solar spectra. Spitzer and others worried that making these devices work would require technologies they were not familiar with, as well as investments in time and energy far beyond reason. Goldberg could not imagine spending $5,000 on an instrument that would be destroyed each time it was used.[5] After their first brush with doing science on a rocket, astronomers like Spitzer, Goldberg, and Yerkes Observatory astronomer Jesse Greenstein all shied away, preferring the traditional mode of using reliable instruments that could be incrementally

3. As posed by Robert W. Smith, *The Space Telescope: A Study of NASA, Science, Technology, and Politics* (Cambridge, U.K.: Cambridge University Press, 1989), p. 187.
4. Goldberg to Menzel, 28 September 1945, Menzel Papers, Harvard University Archives (HUA), quoted in David DeVorkin, *Science with a Vengeance: How the Military Created the US Space Sciences after World War II* (New York, NY: Springer-Verlag, 1992 [reprinted in 1993, paperback study edition]), p. 207.
5. Ibid., p. 209.

improved. In contrast, NRL staff, especially experimental physicists like Ernst Krause and Richard Tousey, were interested in instrument development and the creation of a capability. Krause's recollections of his excitement in 1946 remained indelible in his mind 30 years later:

> Now, this is a good way to do some experimentation. We're going to get away from this business of having a complicated, costly set of apparatus in a physics laboratory in a basement in some university, and because it is complicated and costly, it lasts for 50 years and generation after generation grinds out theses on that same equipment because it's expensive and new equipment is more expensive. We've got a set-up here which by its very definition is going to get destroyed each time. How good can you have it?[6]

Those who did dedicate themselves to doing science from rockets were not astronomers, and they did not engage in astronomical practice or ask questions astronomers would ask. Most of these workers were based in military laboratories or in physics departments and spent much of the decade of the 1950s primarily refining their instruments and techniques, asking questions that would help to improve the rocketry itself, knowledge of the medium through which rockets traveled, and knowledge of the solar influences upon global radio communications networks. To some extent, a few looked for a discipline to associate with. Herbert Friedman, who explored high-energy solar phenomena using his modified x-ray-sensitive proportional (Geiger) counters, recalls having one of his early papers refused by the editors of the *Astrophysical Journal*, and so he directed most of his effort to physicists and geophysicists. There was no question, looking at his publishing history, that Friedman was looking for a receptive audience. After 1958, he had no trouble finding one.

Friedman and Tousey at NRL, as well as J. J. Hopfield at the Johns Hopkins University (working with APL staff) and William Rense at the University of Colorado, were typical of those few physicists who applied their craft to astronomy without prior interest or activity in the field. They stood apart from mainstream astronomy, based on campuses and at a few highly prominent observatories. As John Lankford and others have shown, astronomy as a

6. Ibid., 1993, p. 214.

TABLE 17.7 Friedman's Publication History, with Collaborators, 1946–57

Journal	1946	1947	1948	1949	1950	1951	1952	1953	1954	1955	1956	1957
Phys. Soc. (U.K.)										2		
Jet Prop.												1
Elec. Eng.												1
Yale Sci.												1
Astronautics												1
Phys. Rev.						1		2	1	2		2
Op. Soc. Am. J.							1		2		2	
Ann. Geophys.										1		
Rev. Sci. Instr.										1		
JGR											1	1
Ap. J.											1	
Science											1	
ICSU Rpt.									1			1
IAGA Bull.												1
Mil. Elect.												1
A.J.												1
Nature												2
AFCRC Chem. Aeronomy												2

Note: Herbert Friedman was engaged in classified research prior to 1950, and then for some time in instrument and program development. Primary focus at first is the *Physical Review*, and then an explosive search for a larger audience begins in 1956–57.

Figure 1: Herbert Friedman's efforts to get the results of his team's efforts out to various audiences are portrayed here. At first his attention was focused on physicists and instrumentalists, but in the peak IGY of 1957, he clearly was trying to reach a much wider audience. His first astrophysical paper was published in 1956. *Table 17.7 in DeVorkin 1993, p. 333*

discipline was still very much defined by a small circle of observatory directors who enjoyed considerable independence and autonomy of action.[7] Problems were defined by a set of centrally shared goals in stellar astronomy identified through vast empirical photometric and spectroscopic surveys in the first half of the century. Cosmology was limited to the largest observatories, and, indeed, problem choice was most commonly defined by available instrumentation and institutional history. National facilities did not yet exist, and access to the largest and most powerful telescopes was through elite channels: membership on a faculty or staff at an institution possessing that equipment.

This highly centralized and selective system, employing purely optical techniques and largely photographic recording, enjoyed acknowledged world leadership in observational astronomy. As a result, astronomers reacted cautiously to the prospect of federal funding first from the military in the 1940s and then from NSF in the 1950s, displaying considerable conservatism, and even resistance, to the rapid growth potential made possible by federal funding after World War II.[8] American observatory directors had been extremely

7. John Lankford and Ricky L. Slavings, *American Astronomy: Community, Careers, and Power, 1859–1940* (Chicago, IL: University of Chicago Press, 1997).
8. David DeVorkin, "Who Speaks for Astronomy? How astronomers responded to government funding after World War II," *Historical Studies in the Physical and Biological Sciences* 31, pt. 1 (2000): 55–92.

successful with traditional philanthropic support, as well as with local support from state governments, and wanted to keep it that way. In 1947, the leaders of the community, such as Washburn Observatory Director Joel Stebbins, speaking on the occasion of the American Astronomical Society's 50th year, envisioned that the profession would not grow by more than 10 members per year, the pace already established by the growth of the present infrastructure. In that year, about one-half of the some 625 members were at observatories; one-quarter were at colleges and scientific institutions; and the remainder had no formal connection to astronomy.[9]

Stebbins's predictions held reasonably accurate for another decade. In 1948, astronomers advised ONR (and NSF after 1951) on astronomers' needs (averaging all major construction costs and operating costs known in astronomy, from 1923 to 1948) and determined that the discipline could utilize not more than $400,000 per year in individual research contracts and grants. But at the same time, NRL alone was spending millions of dollars on upper atmosphere research.[10] Such expenditures seemed staggering to mainstream optical astronomers.

In another decade, on the eve of Sputnik, even after significant developments in electronics and digital computing and, most of all, a slow but accelerating warming to federal funding as rank-and-file fears of loss of autonomy under both ONR and NSF funding faded largely through experience in their peer-reviewed research grant programs, events surrounding the IGY, and the emergence of an egalitarian national observatory movement fostered by NSF, no astronomical institution overtly embraced scientific rocketry. Instrument stabilization, rocket reliability, and payload retrieval had improved to the point where astrophysically useful spectroscopic data on solar phenomena in ultraviolet and x-ray regions were being obtained by three groups, one in a university physics department and two at military laboratories. And although the leaders of these groups became known and respected in astronomical circles, no astronomers, and no known departments of astronomy, were willing to consider the investment required to use rockets for science.[11] Then, virtually

9. Joel Stebbins, "The American Astronomical Society, 1897–1917," *Popular Astronomy* 55 (October 1947): 412, reprinted in David DeVorkin, ed., *The American Astronomical Society's First Century* (New York, NY: AIP Press, 1999), pp. 53–57, on p. 53.
10. DeVorkin, "Who Speaks?", pp. 70–72.
11. Ibid., and DeVorkin, *Science with a Vengeance*, pp. 213–215; chaps. 17 and 18.

overnight, popular and political reaction to Sputnik led to a series of appeals that changed astronomers' attitudes toward space research.

Early Appeals

The IGY itself certainly played a strong role in heightening interest in upper atmosphere research, solar research, and plans for satellite-based research. At the 10th meeting of the Upper Atmosphere Rocket Research Panel in January 1956, convened by James Van Allen, Panel Chair, and his colleagues, most of them "seasoned veterans of physical research at high altitudes, using rockets as vehicles," the hope was that they could, by virtue of their collective experience, "record the heart-beat of this field of research" and predict for a wider audience what could be done with orbiting vehicles.[12] Based upon this foundation, others were able to react quickly when, in the wake of Sputnik, demands came in for assessments of what would be possible, given the present state of launch vehicle development.

The first to report was W. W. Kellogg of the RAND Corporation, who in November 1957 brought to conclusion a report he had been developing for some time, at the request of the IGY Working Group on Internal Instrumentation of the Earth Satellite Program. Thanks to Sputnik, Kellogg's analysis, entitled "Basic Objectives of a Continuing Program of Scientific Research in Outer Space," was the first topic for the National Academy IGY committee agenda, but his conclusions reflected pre-Sputnik values. He plotted out a scientific program assuming that development would be gradual and not revolutionary; each stage of the program would help to design the next stage, and manned spaceflight would occur eventually, but not immediately:

1. Immediate: continue IGY type sounding rocket and balloon programs: capability 10–30 lbs;
2. Within the year: "Lightweight satellite experiments" (50 to 75 lbs.);
3. Within 5 years: "Advanced Satellite Experiments." (100–500 lbs) 2 and 3 axis stabilization.[13]

---

12. James A. Van Allen, "Preface," in *Scientific Uses of Earth Satellites*, ed. Van Allen (Ann Arbor, MI: University of Michigan Press, 1956), pp. v–vi.
13. W. W. Kellogg, "Basic Objectives of a Continuing Program of Scientific Research in Outer Space," 9 December 1957, Dow Papers, box 8.4, University of Michigan Archives, Bentley Historical Library. Kellogg's analysis, and those of subsequent panels examined here, are addressed in John E. Naugle and John M. Logsdon, "Space Science: Origins, Evolution, and Organization," in *Exploring the Unknown*, ed. John M. Logsdon, vol. 5, *Space and Earth Science* (Washington, DC: NASA SP-2001-4407, 2001), pp. 1–15, and somewhat more fully in David DeVorkin, "Solar Physics from Space," in *Exploring the Unknown*, ed. John M. Logsdon, vol. 6, *Space and Earth Science* (Washington, DC: NASA SP-2004-4407, 2004), pp. 1–36.

In February and March of 1958, fueled by Sputnik fever, the NACA convened a study by a Special Committee on Space Technology to consider space research objectives. This body called for a more rapid payload growth:

1. Immediate: launching 30 lbs class once per month into earth orbit.
2. By sometime in 1959: 300 lbs class supplements class 1, one every 2 months including stabilization and capability of UV imaging.
3. By sometime in 1961: 3000 lbs, supplements 2 and 3: once every 4 to 5 months.[14]

Then, in June 1958, the new Space Science Board (SSB) of the National Academy of Sciences, created in part to complement elements of existing IGY technical panels, called at first for "an orderly extension and continuation of the rocket and satellite work of the USNC/IGY."[15] But at its first meeting, it listened to Herb York, of the newly created DOD ARPA, predict that

1. By 1960: 3000 pounds into orbit, with the possibility of lunar and planetary exploration
2. By 1962, double that
3. By the mid 60s, 30 times that
4. By late 1960s, 50 tons "by multiplexing of rockets"[16]

By implication, in the heat of competition with the Soviets, York challenged scientists to plan accordingly. Faced with this opportunity, in a state of rush, by night letter on 3 July, Lloyd V. Berkner, Chairman of the SSB, sent out an appeal to scientists to suggest ways to exploit these vehicles for research in, or from, space. He asked for "possible experiments" that could be flown within two years, weighing as much as 100 pounds, but compatible with other "smaller non-conflicting experiments." These packages would have to be ready for environmental testing earlier than mid-1959. He wanted answers *within the week* to include a short description, "its scientific value," the instruments involved, weights, all costs required to build "four hardware units," manpower requirements, and time needed. He concluded, "Regret need to ask for such information on so short notice but cannot avoid."[17]

---

14. "Minutes of Meeting, Working Group on Space Research Objectives," 30 April 1958, Lyman Spitzer Papers, NACA file, Princeton University Library Manuscripts Division.
15. "Minutes of the First Meeting of the Space Science Board, 27 June 1958," p. 2, SSB files, NASA Historical Reference Collection, NASA History Division, NASA Headquarters, Washington, DC.
16. Ibid., p. 6.
17. L. V. Berkner to United States National Committee (USNC)/IGY/European Southern Observatory (ESO), "Space Science Board Requests for Support: Proposals 1958," 3 July 1958, folder 1.3, SSB Papers, National Academy of Sciences Archives (SSB/National Academy of Sciences).

The General Response 1: Immediate

There was widespread, but not totally positive, response to Berkner's breathless appeal. Many senior astronomers were just then packing their bags for the triennial General Assembly of the International Astronomical Union meeting that August in Moscow, and, as one astronomer responded, Berkner's telegram "caught us pretty flat footed."[18] Kitt Peak National Observatory Director Aden Meinel, one of the most creative instrument builders in the discipline, felt his staff was too committed to building their new observatory to propose active work, though they were soon to take steps in this direction. Meinel knew that other groups were proposing "simple early experiments in optical astronomy" and concluded that "it would appear to me that several years of laboratory work lie ahead for any more sophisticated experiments before they approach the stage of reliability to warrant vehicle flights."[19] Others were unwilling to consider employing new methods of data retrieval, saying they would propose when physical recovery would be possible.[20] At least one prominent radiation biologist recognized the importance of assessing "the very serious environmental hazards to be overcome if man is to survive during extended flights" but felt that Berkner's appeal was premature for his expertise: "Perhaps I am being a pessimist about the value of radiation space experiments so far as the biologist is concerned, but without the physical evidence to go on (and I am not aware of any detailed data available) such experiments are in my opinion a waste of time and good money."[21] One respondent felt strongly that nothing new should be considered until the packages that were prepared for the failed Vanguards were flown.[22]

Berkner sent his appeal very broadly, and about 200 scientists and institutions responded one way or another, most with advice and counsel, some with specific suggestions for problems to attack, and more than half with concrete

---

18. D. W. R. McKinley (National Research Council of Canada Radio and Electrical Engineering Division) to Berkner, 4 July 1958, responding for Peter Millman, then in Paris bound for Moscow, folder 1.3, SSB/National Academy of Sciences.
19. A. B. Meinel to Berkner, 7 July 1958, folder 1.3, SSB/National Academy of Sciences. Meinel would soon create a division at Kitt Peak National Observatory for space-based research led by his own proposal for a 50-inch telescope in space. See Frank K. Edmondson, *AURA and its U.S. National Observatories* (Cambridge, U.K.: Cambridge University Press, 1997), pp. 80–81, 109–111.
20. Bertram Stiller to Berkner, 10 July 1958. Stiller worked with M. M. Shapiro at NRL. J. J. Lod to Berkner, 10 July 1959, folder 1.3, SSB/National Academy of Sciences.
21. C. P. Swanson to Berkner, 8 July 1958, folder 1.3, "Johns Hopkins biologist—radiation biology," SSB/National Academy of Sciences.
22. Hans Ziegler, Assistant Director of Research, U.S. Army Signal Research and Development Laboratory, to Berkner, 10 July 1958, folder 1.3, SSB/National Academy of Sciences.

proposals. Some 70 proposals came in from physical scientists; there were 30 responses for some 60 projects from life scientists; and there were 10 separate suggestions for some 17 distinct engineering studies. Of these, about 30 were clearly astronomical, proposed by a wide array of institutions and individuals.

Berkner and his staff spent the summer sorting through the responses, and Berkner responded to some inquiries personally. To Donald Menzel, Director of the Harvard College Observatory, he rejoiced that the "response was voluminous." This was, of course, his primary goal because the SSB wanted to encourage as many people as possible to get involved: "One of the great problems in getting our country into the space science field more actively is to find laboratories and groups where suitable hardware can be developed to carry on the more sophisticated experiments," Berkner observed, hoping that many universities and civilian laboratories would respond "so that the scientific aspects of the space program are not forced into a single government bureau for lack of competent facilities and men outside the government." But, on the other hand, there were clearly too many proposals to handle in the first few years; so the SSB was taking steps to group together similar interests, specifically, "there appear to be a number of proposals to get a telescope into space, and it would appear most sound to bring these proposals together so that the best of each of them could be actually projected into space and they could cover the whole of the spectrum."[23]

Indeed, the SSB encouraged many disparate groups to work together to submit revised joint institutional proposals, in all fields. They asked Yale University to join forces with the National Radio Astronomy Observatory, and they asked other institutions to develop ultralong- and shortwave radio observations from satellites, "In the belief that a broad-based participation of the scientific community is necessary for a successful program of space research." They asked the Geophysical Research Directorate of the Air Force Cambridge Research Center (AFCRC) to join with NRL and the University of Colorado to coordinate extreme ultraviolet and x-ray solar monitoring programs.[24] There were at least 11 proposals to perform various observations to study relativistic effects through performing precision orbit measurements

---

23. Berkner to Menzel, 2 September 1958, folder 1.3, SSB/National Academy of Sciences.
24. Lieutenant Commander L. M. Cormier, Secretary to the SSB Committee on Optical and Radio Astronomy, to D. S. Heeschen, 17 December 1958; Cormier to Lilley, 17 December 1958; draft, Cormier to James Gallagher GRD, n.d., folder 1.3, SSB/National Academy of Sciences.

and distinguishing the effects of geodetic and atmospheric drag, and these people were encouraged to collaborate.

At the end of the summer, two committees continued to deliberate over priorities. Leo Goldberg at Michigan, an SSB member and a proposer of several projects, was asked to chair an ad hoc Committee for Astronomy and Radio Astronomy, and it fell to Goldberg to coordinate the major astronomy proposals. Informing J. Allen Hynek at the Smithsonian Astrophysical Observatory, among other astronomers, of this decision in early September, Berkner naturally hoped they would indeed work together. There would be plenty of room to work, no doubt: "you, Menzel and Goldberg will have an interesting time in the study of the telescope project. You will, of course, have the whole spectrum range before you involving certainly more than 16 octaves."[25] Berkner may not have appreciated that, among optical astronomers, the playing field was barely one or two new octaves, but there was still plenty of room. The problem was not so much between astronomers vying for some version of a satellite telescope; it lay at higher levels in the formation of the nation's emerging space program. In sum, five distinct proposals came back quickly from astronomers Leo Goldberg, Lawrence Aller, Fred Whipple, Lyman Spitzer, and Arthur Code, and one from a collection of NRL physicists, Talbot Chubb, Herbert Friedman, and James Kupperian, for variations of what Berkner viewed as the "telescope project"—ultraviolet studies of the Sun, the stars, and the spaces between the stars. It would be NASA's job to manage this response, and much of this fell eventually to Nancy Grace Roman.[26]

The Emergence of NASA and the Creation of the Orbiting Astronomical Observatory (OAO) Mission Concept

Berkner's initial reaction to the responses he generated by his 3 July 1958 telegram was to promote cooperative proposals between competing institutions that would somehow be vetted by the SSB. The extent to which this actually happened remains to be determined, but, for the purpose of the present study, it suffices to say that the SSB's goals as a deliberative and coordinating agency became moot by the end of the year, when the SSB submitted its findings to

25. L. Berkner to J. A. Hynek, 2 September 1958, Smithsonian Astrophysical Observatory, Smithsonian Institution Archives (SAO/SIA).
26. Nancy Grace Roman, "Exploring the Universe: Space-Based Astronomy and Astrophysics," in *Exploring the Unknown*, ed. John M. Logsdon, vol. 5, *Exploring the Cosmos* (Washington, DC: NASA SP-2001-4407, 2001), pp. 501–543.

NASA. In March 1959, Hugh Odishaw retroactively acknowledged this fact when he finally got around to thanking all the initial respondents, saying that the SSB was now an advisory body and did not have "executive responsibility for the support of the U.S. program in science and space." The funding for these activities now resided with the agencies that did: NASA, NSF, and ARPA.[27]

By the time of Berkner's notice, of course, NASA had already opted to control executive responsibilities for the nation's civilian space program, a detail left ambiguous by the National Space Act.[28] Administrator T. Keith Glennan appointed the NACA engineer Abe Silverstein to direct the Office of Space Flight Programs, and Silverstein appointed Homer E. Newell, a charter member of NRL's space science program, as Assistant Director for Space Sciences. Newell in turn formed a Space Science Working Group and various subunits to establish an overall plan. The details of this effort, in the case of optical astronomy at least, have been touched upon elsewhere but still require considerable elaboration.[29] The result was the formation of a small set of large programs that would lead to a series of space missions throughout the 1960s and into the early 1970s, in fact the largest NASA would mount in the strictly speaking pure (that is, unmanned) space sciences. These missions served groups at both academic and NASA facilities, and in every case, they required infrastructures at each university, as well as within NASA, that grew in size to levels unknown in astronomy but quite familiar to those in physics and in the space sciences at national laboratories such as NRL.

Since the focus of this paper is disciplinary change, I will leave out the rather extensive deliberations over the structure of what became the OAO and Orbiting Solar Observatory (OSO) programs and the debates that occurred over the general mission concept between NASA, the SSB, and ultimately the PSAC. Suffice it to say, the framework that emerged within NASA promoted the creation of teams of unprecedented size within the astronomical community. Here are two of the institutional responses that led to the first successful OAO. Both stimulated institutional growth; in the first, it was moderated by a strong agenda with clear goals, whereas in the second, it was unbridled, reflecting an opportunistic agenda with initially mixed goals.

---

27. H. Odishaw to mailing list of names who responded to original Berkner night letter, 30 March 1959, SSB/National Academy of Sciences.
28. Naugle and Logsdon, "Space Science," pp. 8–9.
29. Ibid.; David H. DeVorkin, "SAO During the Whipple Years: The Origins of Project Celescope," in *The New Astronomy: Opening the Electromagnetic Window and Expanding our View of Planet Earth*, ed. W. Orchiston (New York, NY: Springer, 2005), pp. 229–250.

The Wisconsin Experiment Package (WEP)

In their historical assessment of WEP, a battery of seven small telescopes feeding ultraviolet nebular and stellar photometers and spectrometers devised by A. D. Code and his associates at the Space Astronomy Laboratory (SAL) of the University of Wisconsin, Marché and Walsh identify the scale of the enterprise required to manage its development for the second OAO.[30] Arthur Code created the laboratory as a unit of the Astronomy Department when he returned to the University of Wisconsin as director of the Washburn Observatory. He could have stayed in a tenured position at Caltech, with direct access to Palomar, but, as he recalls,

> The thing that changed was Sputnik and the possibility of making observations from above the earth's atmosphere. And lots of people were using ground-based telescopes, and nobody seemed to be interested in doing space astronomy.[31]

Code was less interested in the directorship and more in the freedom it would provide him to pursue space research:

> I would not have accepted it on just that basis. I wasn't looking for being a director of an observatory. But at the same time, there was this letter that Lloyd Berkner circulated in academic circles, that if you had a 100-pound satellite, what would you do with it? I thought about that and responded to that letter.[32]

His SAL was housed in rented space, outgrowing the department itself, and from there, over the next decade, his staff of astronomers and technicians produced a successful series of developmental sounding rocket payloads, and eventually the highly successful WEP.

As Marché and Walsh demonstrate,[33] Code followed clear lines of personal research interest and established technical expertise to accomplish his goals.

---

30. Jordan D. Marché and Adam J. Walsh, "The Wisconsin Experiment Package (WEP) Aboard the Orbiting Astronomical Observatory (OAO-2)," *Journal of Astronomical History and Heritage* 9, no. 2 (2006): 185–199.
31. A. D. Code oral history, 30 September 1982, p. 24, Space Astronomy Oral History Project (SAOHP)/ NASM Archives.
32. Ibid.
33. Marché and Walsh, "The Wisconsin Experiment Package."

He was a creative and adept instrumentalist who was intimately familiar with photoelectric and spectrophotometric techniques, and he had performed observational research at Caltech on the spectral energy distributions of stars, exactly what he was proposing for WEP. He also enjoyed a healthy and dedicated instrument-based expertise in the University of Wisconsin department, honed from two former generations of directors, Joel Stebbins and Albert Whitford, who were pioneers in photoelectric techniques and associated instrumentation. One can easily rationalize how WEP and SAL were organic extensions of the University of Wisconsin program. This is not to say the path was easy and without frustration, because there was at least one vehicle failure.

The battery of seven telescopes comprising WEP, however, did not emerge full-blown from Code's mind or reflect his initial ambitions. As he recalls, based upon the initial SSB appeal, he and his staff developed a single 100-pound telescope, within the limit set by Berkner. It was a 10-inch, off-axis reflecting telescope feeding a single photomultiplier. But when NASA had secured SSB's responses and deliberations by early 1959 and decided to combine as many instruments as possible on a single stabilized platform, Code and his colleagues were faced with developing a 2,000-pound payload. "Well, that looked like a pretty big spacecraft," he recalled, "especially after you throw away all the satellite [housekeeping] part[s] that we had in our proposal. So what we did basically was cluster a whole bunch of these and fill up a lot of this space."[34]

The Response from Harvard-Smithsonian

The deliberations of the Space Science Working Group led to multiple institutions preparing for the same space platform. By the spring of 1959, Code's WEP would share a single platform with a similar proposal emerging from SAO. By the time Berkner sent out his appeal, Fred Whipple was already deeply involved preparing for satellite astronomy. As the director of SAO, housed at Harvard since 1955 and already growing into a very large enterprise through space-related service programs such as a worldwide optical tracking network to determine high-precision artificial satellite orbital behavior for a wide range of pure and applied goals, Whipple and his staff by then had already taken a visible and leading role in the IGY satellite program. This so-called Baker-Nunn tracking network and its popular counterpart, Project Moonwatch, were high-visibility efforts that caught public attention, and its fascination, symbolizing,

34. Code Oral History, SAOHP/NASM, p. 35.

as they did, a new worldwide infrastructure required by the Space Age and a means of engaging public involvement, approval, and support.[35] Whipple had been a pioneer member of the V-2 Panel since the 1940s, and he was the only astronomer on von Braun's "Hoover Panel," which promoted the Army's bid for an IGY payload, Project Orbiter.[36] He also had been intimately involved, as Deborah Warner has shown us, in plans for utilizing meteors and artificial satellites for geodesy, combined with his ongoing studies of the upper atmosphere and his hyperballistic studies—all technical prerequisites for a reliable ICBM system.[37] He had been deeply involved on the V-2 Panel and its successors with Van Allen, Friedman, Tousey, and the other rocket scientists, and he lent scientific legitimacy and access to a wide range of interconnected military and civilian advisory panels in postwar Washington, DC, but never once seriously proposed an instrument for a flight on a rocket.

Whipple had, of course, all along been thinking deeply about satellites and space stations, his "dear dream," as Harvard's McGeorge Bundy said of him in 1955.[38] In 1955 and 1956, he and SAO staff had been considering a geodetic satellite, either a highly reflective polyhedron or a rotating visible beacon.[39] On 4 February 1958, however, Whipple, along with Harvard College Observatory Director Donald Menzel, called together a large group of Harvard and Smithsonian astronomers to ponder SAO's involvement with rockets and satellites. The main purpose of the meeting was to get ideas for space vehicles, to set up committees, and to report ideas. "We are looking forward to a real space platform" was the sense of the meeting, as described in notes kept by the youngest member of the group, a graduate student, Robert Davis, who would soon be tapped to manage what came to be known as the Celescope program. Whipple's burgeoning SAO staff had expertise in geodesy, as well as in meteoritics, cosmic rays, the outer atmosphere, and meteorology. But the Harvard astronomers were also interested in the Sun and stars as well as

35. Patrick McCray, *Keep Watching the Skies! The Story of Operation Moonwatch and the Dawn of the Space Age* (Princeton, NJ: Princeton University Press, 2008); Teasel Muir-Harmony, "Tracking Diplomacy: The IGY and American Scientific and Technical Exchange with East Asia, 1955–1973," in *Globalizing Polar Science: The Legacies of the International Polar and Geophysical Years*, ed. James Fleming, Roger D. Launius, and David DeVorkin (New York, NY: Palgrave–MacMillan, 2010).
36. Michael J. Neufeld, *Von Braun: Dreamer of Space, Engineer of War* (New York, NY: Vintage, 2007).
37. Deborah Jean Warner, "From Tallahassee to Timbuktu: Cold War Efforts to Measure Intercontinental Distances," *Historical Studies in the Physical and Biological Sciences* 30, pt. 2 (2000): 393–415.
38. McGeorge Bundy to "Dear Nate," 26 January 1955, HUA, UA III 5.55.26, folder "HCO," box 12.
39. Charles A. Whitney and George Vis, "A Flashing Satellite for Geodetic Studies," *SAO Special Report* 19, pt. 3 (1958): 9–20.

galactic structure, and Harvard itself had a long tradition in mapping the sky by traditional photographic techniques.[40]

Whipped up by Whipple, the group felt that their combined staffs had "a certain degree of proficiency" and, moreover, "an operational advantage as of the present moment." They sensed that their only competitors were at NRL, Michigan, the Geophysics Research Directorate at Air Force Cambridge Research Laboratory, and of course "Dr. Van Allen," Whipple's compatriots on the Rocket Sonde Research Panel. The group's first impressions, recorded by Davis, were that "We want to aim as high as possible." Indeed, this was Whipple's strategy. In April, well aware of the fact that a National Space Act was circulating in Congress, Whipple testified that the role this new agency might take in the nation's space program was as a coordinator of expanded university-based scientific centers, not as a controlling agent.[41]

Although Davis's notes from February were enthusiastic and ambitious, by June he and Whipple had made a conservative proposal to the ABMA to launch a small reflecting telescope to obtain ultraviolet brightnesses for stars that would extend knowledge of the energy distributions in stars and would help to calibrate bolometric data acquired from ground-based telescopes. Their idea, very similar to Code's initial design for a single telescope, was to develop something quickly for a first glimpse of the ultraviolet sky. Unlike Code, for Whipple, the satellite telescope itself was only part of a much larger system he had in mind.

Whipple's efforts to track satellites and analyze their data for a wide range of Space Age interests required a virtual army: teams of amateurs and volunteers in military and technical organizations performing visual reconnaissance of satellite orbits and the development of large and sophisticated optical tracking stations across the globe. Managed by a considerable staff of computers and administrators in Cambridge, Massachusetts, SAO possessed sufficient expertise in orbit analysis and computational expertise to perform a wide range of tracking, data reduction, and geodetic analysis. In the spring of 1958, Whipple and the Smithsonian had proposals in to the ABMA to manage optical tracking for the Explorers. Whipple's plan for a Harvard-Smithsonian-based space operation, where liaison with NASA was a relatively minor component, could be gleaned from a timeline for Project Celescope, as it was called by 1960. The Smithsonian Astrophysical Observatory was to be the center not only for the definition and design of the payload, as well as its fabrication, but after

---

40. DeVorkin, "SAO During the Whipple Years," pp. 234–237.
41. Ibid., p. 237.

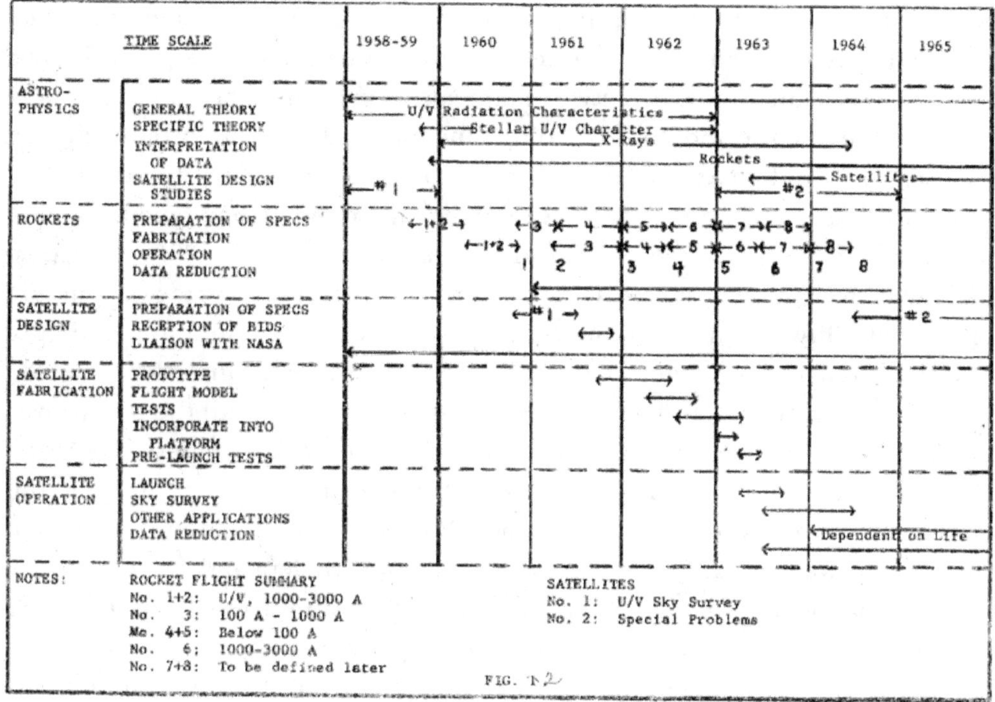

Figure 2: Celescope flowchart as envisioned by Whipple's staff in May 1960. The project was seen as the nucleus of a general capability to design, build, test, and fly packages on satellites, with sounding rocket flights as developmental steps but also as part of data generation, reduction, and analysis. The SAO sounding rocket flights were not supported by NASA and so never happened. *Smithsonian Institution Astrophysical Observatory, Quarterly Progress Report No. 3 Project Celescope, 29 April 1960, SIA RG 522, folder "Progress 1960," box 10*

launch for its operation, the collection of data, and subsequent analysis and publication as a contribution to astrophysics.

By late 1958, it was clear to Whipple that his full plan would not be approved. NASA would manage the infrastructure and would invite scientists to propose payload instruments, referred to by NASA in the same terms Berkner had used: "experiments," a term rarely, if ever, used by astronomers.[42] And at first

42. Astronomers prior to this period rarely, if ever, referred to an operational instrument on a telescope as an experiment, unless it was a speculative technical venture to try out some new technique. As the venerable astronomer Joel Stebbins succinctly put it in a banquet speech in 1950, recounting his pioneering work with photoelectric cells since 1910: "The electrical photometry of stars involves the technique of experimental physics at the end of a telescope," Joel Stebbins George Darwin Lecture, Royal Astronomical Society (RAS), 13 October 1950, *Monthly Notices of the Royal Astronomical Society* 110 (1950): 416. But, of course, that's exactly what NASA program managers and Administrators wished that astronomers would be doing. Instead of a telescope, NASA would provide berths and invite astronomers to get aboard.

NASA had in mind a universal platform that would accommodate both solar and stellar experiments. These large, multifunctional platforms, attributed by Homer Newell and others to a style of large-scale experimental management in shared infrastructure and housekeeping functions, a design philosophy familiar to NACA leaders like Abe Silverstein, envisioned an economy of scale that, in some respects, paralleled the SSB's urging that like-minded experimenters band together to propose single unified programs. But unlike the SSB, engineers like Silverstein, now at NASA, did not think as critically in terms of function or compatibility. When astronomers hotly objected to the lumping together of solar and stellar experiments, each requiring very different infrastructure and capability from their standpoint, NASA refused to budge until the matter was taken to the White House via members of the PSAC.[43] In the end, a compromise was reached separating the solar and stellar programs, but in each, the missions remained multifunctional, which Silverstein and his subordinates, like Homer E. Newell, ardently believed would be more efficient and cost effective.[44]

The goal of SAO's Celescope grew just as SAL's WEP, from a simple, quick-and-dirty peek with a single telescope to a battery of instruments. Ultimately, SAO's Celescope became four separate telescopes, physically bundled and working in parallel, each tuned to a different ultraviolet band for wide-field photometry. The goal was to produce a catalog of ultraviolet stellar characteristics for every bright star in the sky: not hundreds, but thousands of stars. In order to do this, Whipple and his staff decided to take a technological risk: utilize area detectors based upon television technology to image star fields, rather than conventional means of photometric measurement, star by star, with photoelectric sensors.

A full discussion of this decision lies outside the boundaries of the present paper. Here we limit attention to the significance of this decision insofar as it came from a desire to offer up a competitive proposal to NASA. "By the time NASA existed, we'd pretty much settled on television," Robert Davis recalls, but the team did not appreciate at first how large the technological challenge would be until Davis started talking with Westinghouse and realized that commercial broadcast devices would not suffice; and when asked the question another way, it was clear that, although using enhanced image-intensified television

43. Martin Schwarzschild oral history, 20 April 1983, pp. 16–19; 19 July 1979, pp. 232–233, Sources for History of Modern Astronomy, American Institute of Physics (SHMA/AIP).

44. Homer E. Newell, *Beyond the Atmosphere: Early Years of Space Science* (Washington, DC: NASA SP-4211, 1980), p. 207.

systems for low light level image detection by remote control was far from a casual decision for Whipple and Davis, making that decision required, as Davis recalls, a "'think big' philosophy. Don't do anything small and primitive. Go for something big and spectacular."[45]

Whipple certainly encouraged such thinking, but so did NASA. And indeed there were practical aspects of the decision. At the time, one could collect far more photometric data on stellar brightnesses with an area detector in a given amount of time than with a point source device. But choosing this technology also required contracting for expertise on a scale far larger than was required for photoelectric systems. As Marché and Walsh point out, the University of Wisconsin group was proud that they possessed an internal capability to design and test their space telescope systems and that what contracting for construction they needed was local.[46] In contrast, SAO had to search far and wide for highly specialized and somewhat independently minded contractors who could manage the design and construction of electronic detectors considered at the time by leading authorities to be beyond the state of the art for prolonged reliable operation on a satellite in 1959; chancy at best, and, even by 1961, it was considered by its own leading practitioners as a "business for serious professionals who have chosen a difficult field."[47]

At this point in our story, it should be clear that two of the four scientific institutions engaged in the OAO program made very different choices and developed very different programs in order to achieve what were, in effect, different goals.[48] The University of Wisconsin created a distinct laboratory within a traditional campus-based observatory to pursue traditional problems in stellar astronomy. At first, SAO pushed to establish an overall capability in space research that would assume a large part of the infrastructure that would be required. It also entered problem areas in which it did not possess prior

45. R. D. Davis oral history, 15 October 2005, p. 84, author's working files, Department of Space History/ NASM.
46. Marché and Walsh, "The Wisconsin Experiment Package," pp. 188, 191.
47. Ray V. Hembree, "Summary of British Image Tube Symposium," in *Proceedings from the Image Intensifier Symposium* (Washington, DC: NASA SP-2, 1961), p. 3; J. D. McGee, "Image Detection by Television Signal Generation" in *Astronomical Techniques*, ed. G. Kuiper and B. Middlehurst (Chicago, IL: University of Chicago Press, 1962), pp. 302–329; McGee to Baum, 14 March 1962, quoted in Samantha Thompson, "The Best is the Enemy of the Good: The story of James Dwyer McGee and the forgotten technology that helped shape modern astronomy" (unpublished master's thesis, Imperial College, 2007).
48. The payloads from GSFC and Princeton, along with the full history of the OAO program, await adequate attention by historians. See Robert S. Rudney, "A Preliminary History of the OAO Program (1966–1968)" (Washington, DC: NASA History Division, HHN-115, September 1971).

expressed interest or expertise. The two institutions were also distinct in achieving stated scientific goals: The University of Wisconsin eventually enjoyed a full measure of successful data flow, whereas the ultimate SAO ultraviolet star catalog that was issued in the 1970s was compromised by calibration problems in the detector systems that could not be fully rectified.[49] The two institutions shared in institutional growth, of course, but of the two, SAO grew far more rapidly to extreme levels. I will now examine the SAO growth as an extreme case of disciplinary change.

Institutional Growth at SAO

Whipple had restarted SAO at Harvard, hiring 7 people with $50,000 support from the Smithsonian in 1955. When he turned over the directorship in 1973, "there were 307 people on the staff, and a basic Congressional budget of three million a year" and another $10 million in competitive grants.[50] One can sense the growth from annual reports he prepared for the Smithsonian Secretary and for the American Astronomical Society. In 1956, Whipple identified three separate units within SAO: solar astrophysics, meteoritic studies, and the "Satellite Program" for the IGY. The first was a legacy program from Washington, DC, which had no staffing that year; the second was led by John Rinehart and included hyperballistic studies of reentry ablation effects; and the third was led by Hynek and Armand Spitz. In 1957, these three units retained their identities, but their subunits, essentially identifiable problem areas with specific funding from the Smithsonian, the Air Force, the National Academy of Sciences, and NSF, grew from 6 to 15. Theodore Sterne had been hired to head "solar astrophysics." By the end of 1958, the three units had become four divisions, adding the upper atmosphere explicitly, and within that section was a program identified as "uv and x-ray space telescope" and another as "stellar scintillation," among 17 distinct problem areas funded as well by the ABMA and the Aero Medical Command at Holloman Air Force Base, in collaboration with Winzen Laboratories and the MIT Instrumentation Laboratory under Charles Stark Draper. By the early 1960s, the Upper Atmosphere division had become "Space Studies" and by 1964, there were six primary divisions, now described in

49. Davis oral history, pp. 85–87; Gene Avrett oral history, 14 October 2005, working files, Department of Space History/NASM.
50. Fred Whipple oral history, 29 April 1977, SHMA/AIP, p. 132. On the establishment of SAO at Harvard, see Ron Doel, "Redefining a Mission: The Smithsonian Astrophysical Observatory on the Move," *Journal for the History of Astronomy* 21 (1990): 137–153, and David DeVorkin, "Defending a Dream—Charles Greeley Abbot's Years at the Smithsonian," *Journal for the History of Astronomy* 21 (1990): 121–136.

terms of scientific problem areas, not space activity areas. Satellite Tracking, as a major division from the start, no longer appeared explicitly that year, among divisions called Planetary Sciences, Meteoritic Science, Cometary Science, Solar Observations, Stellar Observations, and Stellar Theory. Within the latter two were distinct problem areas identified as "Project Celescope" and associated groups devoted to developing the ultraviolet television systems, techniques for machine reduction of data, and the production of the SAO star catalog. On the theoretical side, a multifaceted team of astrophysicists was identified as devoted to model stellar atmospheres in support of Celescope. Overall staff growth increased by over 50 positions per year until 1960 and by over 100 each from 1963 to 1964 and 1964 to 1965, reaching a full staff complement in 1965 (scientific and support, both in Cambridge, Massachusetts, and at stations around the world) of over 500 positions. Celescope-related programs alone had over 100 positions in 1961.[51]

As I have shown in another study, Whipple achieved this explosive growth employing what he called a "brinksmanship principle" where he secured large government contracts for Space Age services and scientific programs, and then he appealed to private donors to support bricks and mortar so that he could expand to honor those contracts. To one donor, he might claim he had the best manpower to solve a specific problem that required a huge investment in a computer or a laboratory, whereas to another he would say that to solve that problem he had to staff that laboratory with more physicists. He could attract good workers by offering secure employment with federal civil service status. So he would finance administration and support through federal grants and contracts, and he would hire scientists and obtain their support directly from the Smithsonian's congressional appropriation. And in parallel, he campaigned with private benefactors, pleading for a place to house all these people.[52]

Whereas Code at the University of Wisconsin already had a staff and students capable of taking advantage of the data that would be forthcoming from their instruments, Whipple's SAO initially had no staff expertise in stellar spectrophotometry or in the stellar atmospheres theory that would

51. Fred Whipple, "Reports to the Secretary 1956–1964," "Observatory Reports, BAAS 1956–1964," and "Yearly Reports compiled by SI Office of Human Resources," examined by Louise Thorn, S2006, Office of Human Resources/SIA.
52. Fred Whipple oral history, June 1976, RU 9520 SIA, pp. 32–35; DeVorkin, "SAO During the Whipple Years."

GROWTH OF HARVARD COLLEGE OBSERVATORY AND
SMITHSONIAN ASTROPHYSICAL OBSERVATORY

	Personnel in Cambridge		General Budget *		Contracts and Grants *	
	HCO	SAO	HCO	SAO	HCO	SAO
1956	78	12	227	46	48	85
1960	90	214	216	250	735	3,393
1965	164	325	821	921	2,844	6,100
1966	193	375	870	1,100	3,818	10,200

* (in thousands of dollars)

Figure 3: Growth of Harvard College Observatory and SAO from 1956 to 1966. In 1955, SAO was a suite of offices within the Harvard College Observatory. By the 1960s, it had far outstripped its host in funding, manpower, external contracts, and grants. *"Harvard College Observatory 1965–1972," folder 104.5, Jesse Greenstein Papers, Caltech Archives, n.d., circa 1966*

form the basis for a full analysis of their data. Naturally, he anticipated this deficit and began to hire staff to rectify it using the dual lines of federal funding available to him and the patronage Harvard had long enjoyed. In the early 1960s, SAO staff grew rapidly, so much so that Whipple's administrative staff created a weekly newsletter just to keep track of all the new faces. Some of them, like Eugene Avrett and Charles Whitney, knew that they had been hired for their expertise and that, sooner or later, they would be called upon for one of Whipple's needs.[53] In this way, Whipple established the largest astronomical empire on the planet by the mid-1960s, one that dwarfed even its host, the Harvard College Observatory. Most of Whipple's hires were not astronomers, but experimental physicists, engineering specialists, optical and electronics experts, machinists, data processors, and applied mathematicians.

53. Avrett oral history; Charles Whitney oral history, 13 October 2005, working files, Department of Space History/NASM, pp. 40–46.

Of course, by 1966, Harvard College Observatory was also experiencing significant growth. In 1966, Leo Goldberg, who had moved to Harvard in 1960 from Michigan and was developing a very active program of solar research contributing instruments to the OSO satellite series, had replaced Donald Menzel as Chairman of the Harvard College Observatory. Unlike Menzel, Goldberg worried about SAO's apparently unbridled growth. More problematic, SAO staff members were eager to teach undergraduate and graduate courses and were mentoring a disproportionate number of Harvard Ph.D. theses. Goldberg felt that Whipple was building something that, though astronomical in name, would not be perceived to be astronomical in nature by influential astronomers and by the Harvard Corporation. Ironically, Whipple's federal and contract hires, even though they rarely if ever got Harvard appointments, were far more enthusiastic about teaching than were Harvard's own tenured professors. By the late 1960s, with massive NASA and NSF cutbacks already straining resources, Goldberg and others like Bart Bok, now in Arizona, started wondering: where were all these new Ph.D. graduates going to get positions, and would Harvard's graduates be competitive if it were known that they were not trained by tenured Harvard faculty? Harvard was not the only burgeoning Ph.D. mill: the University of Texas, the University of California–Los Angeles (UCLA), and the University of Michigan had all increased Ph.D. production due in no small part to major NASA graduate fellowships and traineeships.[54] In a 1971 report, according to Bok, Harvard ranked only sixth among graduate departments in astronomy.[55] By the mid-1970s, he was far from alone in worrying about the oversupply of Ph.D.'s due not only to overproduction at elite institutions like Harvard and the University of Texas, but due to the rapidly increasing number of Ph.D.-granting institutions.

Plotting Out the General Response: The First 15 Years

It must be appreciated that at both the University of Wisconsin and at Harvard, the influence of NASA funding on the growth of the discipline, and at like institutions, was far broader than the creation of the infrastructure needed to design,

54. B. T. Lynds, "Employment Problems in Astronomy: Report of the Astronomy Manpower Committee on Science and Public Policy" (Washington, DC: ED 112 724 HE 006 653, National Academy of Sciences, March 1975). See also anon., "Measuring the Impact of NASA on the Nation's Economy" (Washington, DC: NASA TM-109753, 1990); Bart Bok oral history, 15 May 1978, SHMA/AIP, pp. 145–150; Alex Dalgarno oral history, 6 December 2007, working files, Division of Space History/NASM, pp. 56–59; Leo Goldberg oral history, 10 October 1983, NASM/AIP; and Goldberg oral history, 17 May 1978, SHMA/AIP, pp. 124–126.
55. Bart Bok oral history, p. 99.

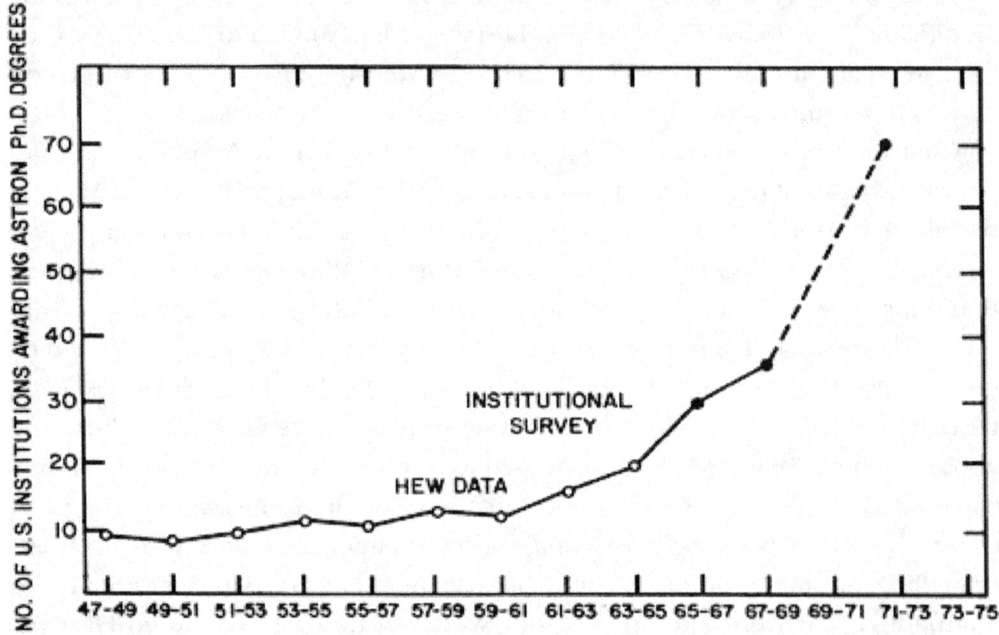

Figure 4: Ph.D. production in astronomy. Growth of number of institutions awarding Ph.D.'s in astronomy in the United States. *Figure 4 in Lynds 1975*

build, test, and then operate these instrument payloads. Well into the mid-1960s, the rapidly expanding horizons enjoyed by many of the larger departments of astronomy, driven largely by NASA funding, did not seem to raise any flags. In fiscal year 1968, federal support was still on the rise, but inflation made it effectively less than the year before. To make matters worse, at the end of 1969, Senator Mike Mansfield introduced an amendment to the Military Authorization Act that, if passed, would severely limit DOD funding "to carry out any research project or study unless such project or study has a direct and apparent relationship to a specific military function." The potential impact on funding for civilian programs like NSF or NASA would be huge; overall for science, over $300 million per year had been supported by the military services, and a sizable fraction would have to be moved to the civilian sector. And to be sure, the new Nixon administration after 1968 did not help; the continued effective drop in funding for science overall portended a disaster brewing.[56]

56. National Science Board (NSB), "A History of Highlights 1950–2000," available at *http://www.nsf.gov/nsb/documents/2000/nsb00215/nsb50/1970/mansfield.html* (accessed 1 February 2009).

Joe Tatarewicz has well described how NASA relations with the scientific community were at a low ebb in the late 1960s. Nixon had directed Agnew to lead a study to determine a course for NASA in the post-Apollo era. This interagency Space Task Group deliberated, seeking politically neutral territory in an atmosphere charged with fears of overspending as the war in Vietnam expanded. Still and all, the focus once again became manned space adventures, specifically manned planetary exploration with the next stop Mars, possibly as soon as 1981. NASA's response to this report was understandably giddy, pushing for very large space programs: the Space Shuttle, a space station, lunar bases, a nuclear Shuttle, a Grand Tour of the planets, and more than one manned mission to Mars. As Tatarewicz relates, the scientific community howled, and at least a few leaders in NASA's space sciences areas listened, appealing quietly to appease the sciences by paying more attention to improving the infrastructure of ground-based planetary astronomy.[57] Places like the University of Texas were delighted to get their 107-inch telescopes and continued support for graduate training.

One only has to look briefly at the growth in membership of the American Astronomical Society during the latter half of the 20th century to sense how the reaction to Sputnik and the emergence of NASA as a major funder of astronomy changed the discipline overall. As a consequence of this growth, meetings of the American Astronomical Society became larger and more complex. The use of parallel sessions was resisted throughout the 1950s, and only by late 1956 were simultaneous sessions allowed. These were on demand for some years until, by the mid-1960s, they became standard; not only double, but triple simultaneous sessions soon followed. These parallel sessions changed the flavor of the meetings and naturally stimulated subdisciplinary concentrations that inevitably grew apart from one another.[58]

By the end of the 1960s, another stratification was taking place in the discipline, the creation and growth of specialist divisions. Knowing that solar astronomers and physicists, as well as high-energy astronomers and physicists, were advocating for identity in the rapidly growing society, the American Astronomical Society responded quickly by establishing formal division structures in August 1968. The Division of Planetary Sciences formed soon after, in December 1968, followed by Solar Physics, High Energy Astrophysics, and

57. Joseph N. Tatarewicz, *Space Technology and Planetary Astronomy* (Bloomington, IN: Indiana University Press, 1990), pp. 103–104. See also J. Tatarewicz, "Federal Funding and Planetary Astronomy, 1950–1975: A Case Study," *Social Studies of Science* 16, no. 1 (1986): 79–103.
58. DeVorkin, "The American Astronomical Society."

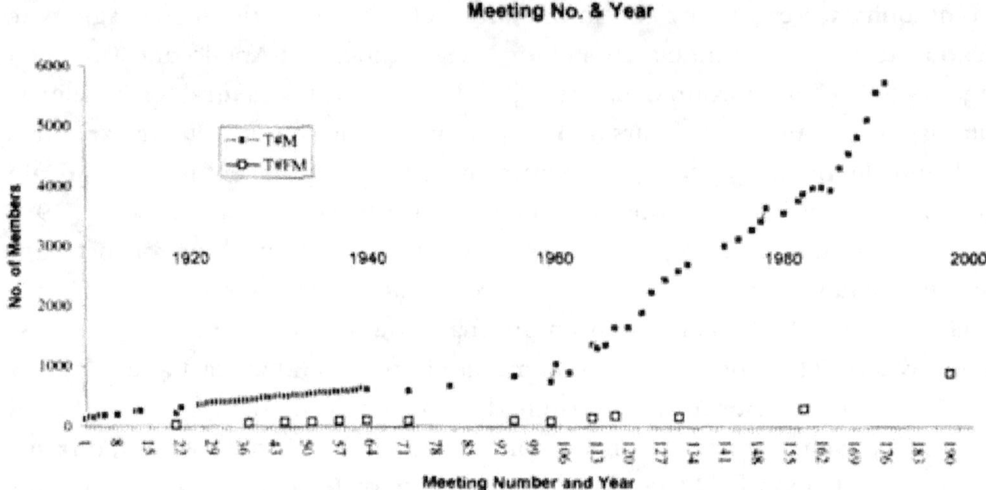

Figure 5: Growth of the American Astronomical Society. One can see a rapidly increasing population due not only to vastly increased federal funding, but also to changes in the society's own definitions for membership. The society had been making quiet moves to increase affiliations with radio engineering groups in the late 1950s and, by 1960, agreed to add member categories of "junior" and "associate" to accommodate a rapidly increasing population of undergraduate students enrolled as astronomy majors and the rapidly increasing numbers of nonastronomers interested in the discipline. *DeVorkin and Routly, "The Modern Society, Changes in Demographics," in* The American Astronomical Society's First Century, *ed. DeVorkin (AIP: American Astronomical Society, 1999), p. 128*

Dynamical Astronomy. In each case, the "demands of the burgeoning Space Age for trained personnel" fueled growth.[59]

Stimulated by suggestions from Henry J. Smith in 1965, the leading solar program manager at NASA Headquarters, solar astronomers began talking about the need for disciplinary identity and focused activities. By 1969, a Solar Physics Division was formed within the American Astronomical Society, but not without expressed concerns that it would separate solar astronomers from the rest of the discipline. Smith, in fact, had initially suggested wholly separate annual meetings of solar physicists funded on NASA grants. In response, the president of the American Astronomical Society, the solar astronomer Leo Goldberg, suggested that this meeting take place with the American Astronomical Society by recognizing the group as a division of the society.[60]

59. R. L. Duncombe, "The Founding of the Division on Dynamical Astronomy—A Few Recollections," in *The American Astronomical Society*, pp. 269–276.
60. John H. Thomas, "The Solar Physics Division," in *The American Astronomical Society*, pp. 238–251.

Planetary astronomers did much the same thing, starting a bit after the solar astronomers but converging on a decision earlier. And again their newly recognized identity was stimulated largely, but not solely, by NASA-related programming. As Joe Tatarewicz has related, in the early 1960s, NASA program managers actively searched many disciplines looking for expertise to satisfy its needs in determining the suitability of the Moon as a base of operations, the atmospheric contents of the terrestrial planets, and other questions relating to the nature of bodies in the solar system.[61] Both NASA and NSF funding created explosive growth, drawing in talent from many and diverse disciplines, leading to the creation of new specialist journals like *Icarus* and *Planetary and Space Science* and creating broader linkages between astronomers and the geophysical communities. The Planetary Sciences Division of the Kitt Peak National Observatory, combined with Gerard Kuiper's Tucson-based Lunar and Planetary Laboratory, created the largest concentration of such activities in the United States and led to a series of annual conferences, stimulating the need for professional identity.[62] Calls for a separate society were debated within this informal group in the late 1960s, but reticence by Kuiper, along with the advice of Carl Sagan and others, caused Tobias Owen to ask the American Astronomical Society to consider forming a branch that would "be concerned primarily with Solar System problems."[63]

Among all of these divisions, the High Energy Astrophysics Division was dominated (but not exclusively) by specialists from areas outside mainstream optical astronomy and those areas most active in space research. The leaders in the field in the 1960s were all physicists, as Richard Hirsh has shown, trained in physics and identified with physics.[64] But those who actually formed the High Energy Astrophysics Division included mainstream astronomers who were interested in the astronomical phenomena that indicated "high energy per photon" processes at play. As Virginia Trimble has observed, the first three divisions arose "almost entirely out of dissatisfaction" with mainstream astronomy and its reward system. But the high-energy astrophysicists also found themselves feeling disaffected "within the space astronomy community"

---

61. Tatarewicz, *Space Technology*.
62. Cruikshank and Chamberlain, "The Beginnings of the Division for Planetary Sciences of the AAS," in *The American Astronomical Society*, pp. 252–268.
63. Ibid., p. 256.
64. Ibid., p. 259. A thorough and convincing early study is Richard Hirsh, *Glimpsing an Invisible Universe: The Emergence of X-Ray Astronomy* (Cambridge, U.K.: Cambridge University Press, 1983), pp. 62–65.

itself, which seemed to be less interested in their work than were mainstream astronomers.[65]

The fourth division to form consisted of mathematical orbit specialists or celestial mechanicians who, in consequence of the need for their expertise by both national security interests and space travel ventures, also grew in number and splintered in the 1960s. The person who in fact first suggested a division for celestial mechanics was Samuel Herrick of the UCLA School of Engineering, where he headed the astrodynamics program. Credited by some with actually inventing the term,[66] Herrick's brand of astrodynamics was specifically directed to orbit calculations for artificial satellites and probes. The petition, prepared by J. Derral Mulholland, who was at that time a research scientist in astronautics at JPL, was signed by only 15 people but was given full consideration by the American Astronomical Society, which advised that the petitioners consider defining their boundaries more carefully.[67] After prolonged deliberation over whether the Dynamical Astronomy Division would be concentrated on applied problems or would include mainstream issues like astrometry and galactic structure and dynamics, the petitioners decided to change the name of the division to Dynamical Astronomy to favor the broader agenda and a blend of pure and applied science. Nevertheless, among the some 45 names on the revised petition, 19 came from NASA-related Centers and 5 from the U.S. Naval Observatory.[68]

By the end of the 1960s, the discipline of astronomy in the United States had experienced profound change. Beyond growth in sheer numbers and the formation of strong subdisciplinary identities, and, most significantly, beyond the rapid expansion of the technologies and specialties required to design and build devices that could take advantage of the new accessibility of the many octaves of information available in nonoptical regimes, there was also a significant change in organizational and political action among astronomers. National planning became essential, first stimulated in the early 1960s by fears that funding priorities for astronomy were not being determined by

---

65. Virginia Trimble, "The Origins of the Divisions of the American Astronomical Society and the History of the High-Energy Astrophysics Division," in *The American Astronomical Society*, p. 228.
66. Samuel Herrick, *Astrodynamics: Orbit Correction, Perturbation Theory, Integration*, vol. 2, available at *http://www.amazon.com/Astrodynamics-Correction-Perturbation-Theory-Integration/dp/0442033710* (accessed 19 January 2009).
67. R. L. Duncombe, "The Founding of the Division on Dynamical Astronomy—A Few Recollections," in *The American Astronomical Society*, pp. 269–276.
68. Ibid., p. 273, table 1.

astronomers themselves, but by military interests; second, by the end of the decade, that NASA's manned spaceflight priorities would channel funds into astronomical programs that were, again, not of high priority; and third, most definitely, there was increased pressure within the subcommunities and subdisciplines of the astronomical sciences that found they were now in direct competition with each other for support. Astronomers as a community found themselves debating their priorities under the aegis of the National Academy of Sciences. The first of the "Decadal Surveys," as they were called, chaired by Lick Observatory's Albert Whitford in the mid-1960s, provided mainly a list of desired projects and programs, but they did not include space research or NASA astronomy missions. In February 1963, Whitford told the astronomical community that the first Decadal Survey panel "will direct its [attention to] ground-based facilities in optical and radio astronomy and to related auxiliary instruments and data-handling equipment." Even though it claimed that it would plan to examine the "relative roles of ground-based astronomy and space astronomy and the relationship between independent research observatories and university-connected observatories," these latter issues were claimed to not involve planning or the setting of priorities, when in fact both were deeply embedded in the value systems of differing groups, characteristic of American science, mainly the "haves" (the institutions on the West and East Coasts with the largest observatories or the greatest endowments) and the "have-nots" (those institutions mainly in the middle of the country that are state and campus based).[69]

The second survey, chaired by Jesse Greenstein from Caltech, did include space missions like the Large Space Telescope but, more significantly, also set clear priorities, listing the Large Space Telescope as the top priority among space projects but ninth overall behind radio facilities, such as the Very Large Array, and the support of a national infrastructure for ground-based optical astronomy.[70] In addition to these surveys fostered by astronomers, NASA and

69. A. E. Whitford to members of the American Astronomical Society, 15 February 1963; American Astronomical Society Records, AIP. The "East-West Split in American Astronomy" has been touched upon by various historians and astronomers. See, for instance: D. H. DeVorkin, "Where to Put it? The East-West Split over the site for the 200-inch telescope," American Astronomical Society, 192nd American Astronomical Society meeting, no. 20.03; *Bulletin of the American Astronomical Society* 30 (1998): 847.

70. W. Patrick McCray, *Giant Telescopes: Astronomical Ambition and the Promise of Technology* (Cambridge, MA: Harvard University Press, 2004), pp. 80–81; DeVorkin, "Who Speaks"; Smith, *Space Telescope*, pp. 131–134; Michael A. G. Michaud, *Reaching for the High Frontier: The American Pro-Space Movement 1972–84* (Westport, CT: Praeger Publishers, Greenwood Publishing Group, Inc., 1986), chap. 10.

NSF maintained advisory bodies that included astronomers. The planetary community engaged in similar advisory functions. Overall, by 1970, the field experienced significant pressures between competing interests and found it harder and harder to plan in the face of uncertain budgets in an increasingly unstable political climate.

Beyond Complexity—Limits to Growth and Retrenchment in a Changing Political Climate

Relentless growth of the astronomical community continued throughout the 1960s, even though overall federal funding for astronomy, space astronomy in particular, peaked in 1968 and went through serious decline through 1972.[71] NASA funding, very large compared to all others, flowed for training and missions, but it was deemed inadequate by astronomers for full data analysis, for maintaining the infrastructure, or for increasing the number of real jobs. It was also erratic; many of the fears astronomers expressed in the 1940s and early 1950s when faced with dependencies on federal funding started to come home to roost.

Some astronomers started pointing fingers in various directions. The astronomer turned educator, historian, and ultimately university administrator Richard Berendzen was one of the first to sound the alarm. In a paper given at the New York City annual meetings of the National Science Teachers Association in April 1972, Berendzen concluded that "a major cause of the very limited employment opportunities in astronomy has been the migration into this field by scientists from other disciplines, especially physics, where the job opportunities are worse."[72] Even though the discipline definitely benefited from this inflow, sustained graduate training was producing a serious oversupply of new Ph.D.'s that was leading, by the early 1980s, to suggestions for rather draconian measures, like requiring that foreign students leave the United States after graduation.[73] On the heels of Berendzen's observations, American Astronomical Society President Bart Bok agreed, warning that employment problems in physics would spill over to astronomy: "Many such physicists will

71. B. T. Lynds, "Employment Problems in Astronomy: Report of the Astronomy Manpower Committee on Science and Public Policy" (Washington, DC: ED 112 724 HE 006 653, National Academy of Sciences, March 1975), table XI, p. 22.
72. Richard Berendzen, "Manpower and Employment in American Astronomy" (paper presented at the National Science Teachers Association Annual Meeting, New York City, New York, April 1972, sess. B-8, ERIC #ED064101).
73. I. Peterson, "Homeward Bound for Graduate Students?" *Science News* 121, no. 22 (29 May 1982): 359.

be seeking employment in astronomy, even though their doctoral research may have been in an unrelated area." Bok, however, called for serious curtailment in graduate enrollment, as well as significant increases in federal funding from NSF for ground-based astronomy. Overall, however, he pointed fingers at NASA: "Because government funding for space research and astronomy has failed to keep pace with the rising number of astronomers . . . ," it is clear that jobs need to be created elsewhere.[74]

Both Bok and Berendzen knew that there was a forum to face challenges like this: the National Academy Committee on Science and Public Policy had, by then, created an Astronomy Manpower Committee at Leo Goldberg's urging. By 1970, frustrated with what Whipple was building at SAO, Goldberg had quit his position as chair and observatory director at Harvard College Observatory in protest, a move that led to the creation of a combined Harvard and Smithsonian "Center for Astrophysics" under a single director, George B. Field, in 1972. Goldberg was now director at Kitt Peak National Observatory, and his concerns for what was happening at Harvard extended to the astronomical community overall.

The committee met at Woods Hole in June of 1974 to review preliminary statistics and get briefings on forecasts for federal funding. They called for more detailed studies, extending a 1973 AIP survey, and then met again in September after collecting information from 45 scientists representing some 70 observatories and universities. Their primary conclusion, simply put, was that

> The ratio of astronomers under forty years of age to those over 50 years is nearly 6 to 1 and if all of the younger cohort seek permanent employment in the traditional modes, only one in 6 will be successful. The possibility of an unemployment rate of near 600 percent is indeed staggering![75]

Clearly what was happening at Harvard-Smithsonian was not unique; it was just an extreme example of what was happening nationally. One of the primary concerns was that NASA had concentrated on missions and training, but not enough attention was paid to data analysis or maintaining the growing

---

74. Bart J. Bok and Donald W. Goldsmith, "Present Employment Trends in Astronomy," *Mercury* 2 (July/August 1973): 2–3; Bok's remarks on "The State of Astronomy," reported in R. Weyman "Now is the Time for all Good Men," *Mercury* 2, no. 2 (March/April 1973): 2–3, 19.
75. Lynds, "Employment Problems," p. 43.

infrastructure. Somehow, a better balance had to be attained, partly through curtailing Ph.D. production and partly through opening up new employment opportunities for Ph.D.'s in nontraditional areas, like industry and education.[76] Stating the obvious, the committee recognized that government support was very critical to increasing astronomical facilities and the pace of research. Government support had been critical to increasing the means to offer graduate training and had contributed to the increase in Ph.D. production. What government funding had not provided or had even maintained, however, and this was especially evident for NASA funding priorities, had been continuing support for the analysis of data from space missions, in effect, for the science. NASA had supported a host of missions, to be sure, that had returned spectacular results and opened vast new fields for investigation, but, the committee concluded:

> It is imperative to achieve the optimum balance between the support of major space missions on the one hand, and the support of the specialists who obtain, analyze and interpret the wealth of new astronomical data produced by such programs on the other. The young space scientists who must be counted on to design the experiments to be flown in the shuttle are actually being forced out of the field because they cannot find employment.[77]

Despite the 1975 survey and Bart Bok's continuing campaign to reduce the size of the Ph.D. population as well as find new sources of employment for astronomers in education and industry, the discipline continued to grow as before. Fortunately, other events minimized the damage. When Ford replaced Nixon in 1975, followed by Carter, new policies ushered in an era of recovering levels of funding for science. The precipitous decline in the early 1970s that had caused so much concern and calls for triage was slowly reversed enough that prospects for new big starts like the Space Telescope became possible, bubbling up slowly to the point where, as Robert Smith has shown, sympathetic members of the executive branch were receptive to the prospect of a "Large" Space Telescope as a "fine example of the kind of science program that was worthy of support."[78] But once again the choice was for one huge mission, to

76. Ibid., p. 35.
77. Ibid., p. 48.
78. Robert W. Smith, *The Space Telescope*, p. 387.

be replenished by human visits, instead of a series of stepwise, ever improving smaller missions, tailored to special interests and needs.

So, in sum, what had changed? Within the warming political climate just noted, NASA did respond to the 1975 Lynds Committee call for data analysis and infrastructure support. Between 1975 and 1976, the NASA request for data analysis for physics and astronomy grew from $6.5 million to $13.8 million, out of which $10.8 million was actually programmed, representing a doubling in two years. Just less than half of this was for *Skylab*, but even in 1977 the request was considerably above previous levels.[79] Similarly, requests to support research and technology funding for physics and astronomy grew from $13.8 million in 1974 to $25 million in 1975, even though once again substantial portions were for specific NASA missions like Spacelab, the Large Space Telescope, and the Solar Maximum Mission.[80] Overall, however, physics and astronomy experienced a serious dip in 1974 programmed funding, from $126.2 million in 1973 to $94 million in 1974; but physics and astronomy had a larger increase in the following years, to $136.3 million in 1975 and up to $224.2 million in 1978.[81]

The report of the Lynds Committee also stimulated NASA scientists and science managers to think more about the critical importance of data analysis not only for knowledge production but for the health of the field. The old question of who was in charge of space science, scientists or politicians, led to the creation of the semi-independent Space Telescope Science Institute, championed by Noel Hinners, in particular, once he was convinced of its need in the mid- to late 1970s by astronomers including John Bahcall, C. R. O'Dell, A. D. Code, and the Science Working Group.[82] Hinners, then Associate Administrator for Space Science, and with him other science managers like Charles Pellerin, believed that "NASA had," in the words of Peter Boyce, then Executive Director of the American Astronomical Society, "a responsibility to support the field if they were to be able to have groups available to propose in the future. This was a big change and very important in funding the field."[83]

---

79. Linda N. Ezell, *NASA Historical Data Book*, vol. 3 (Washington, DC: NASA SP-4012, 1988), table 3-4, p. 133, and table 3-19, p. 140.
80. Ibid., table 3-18, p. 139. Partly offsetting these rises were substantial drops in mission funding for OAO and OSO, tables 3-7 and 3-8, p. 135.
81. Ibid., table 3-5, p. 134.
82. Smith, *Space Telescope*, pp. 197–201.
83. Peter Boyce to author, 6 February 2009, working files, Department of Space History, NASM.

Although this response was heartening, it was not the solution. Certainly the scale of the profession had changed drastically, and with this change came increased pressures to gain support through innovation and novelty. These pressures were also accompanied by deeper changes in the complexion of the astronomical community regarding changes in the source of innovation. At the midpoint of the 20th century, American astronomy was almost exclusively optical, and the means of recording data was still almost completely photographic. When mainstream astronomy thought of expansion of capability, it was along these two traditional lines. Large investments in fewer and fewer instruments, those requiring elite involvement, and in turn increasing a "stratification pyramid" in the discipline, concentrated the largest and most powerful telescopes into the fewest hands, typically the elite observatory directors.[84]

Through the 1950s, the trend continued but began to be countered by the national observatory movement made possible by federal funding and by the maturation and adoption of electronic means of recording data. The national observatory system fostered by NSF, for both radio and optical centers making world-class instrumentation available to all astronomers by peer review, resulted in profound cultural shifts in astronomers' behavior, including increased attention to collective planning beyond their particular institutions and, as a result, an expanded vision of new possibilities and scales of operation made possible by the wealth of a great nation. Both counteracting trends rapidly expanded in the Space Age of the 1960s and beyond, with NASA's engineering enthusiasms fueling novelty and innovation, drawing in expertise from wider and wider circles beyond astronomy and adopting the national observatory model in its "Great Observatories" third-generation satellite programs. Larger new technology programs continued to trump smaller ones, of course, because the larger ones garnered wider political support and, to be sure, broader appeal within the scientific community as well. But now, many of those proposing these new programs were not mainstream astronomers. By the end of the century, physicists dominated training, technique, and the professional culture of the astronomical community. The Lynds Committee showed that between 1970 and 1973 the number of Ph.D. physicists with stated specialization in astronomy jumped from 623 to 1,313, and the number of general workers who trained in physics but were doing astronomy went from 1,074 to 1,906. Among physicists,

---

84. Michael John Halliwell, "Prestige Allocation in Astronomical Research: A study of Dysfunctional Aspects," *Pacific Sociological Review* 25, no. 2 (April 1982): 233–249. See also Hirsh, *Emergence*.

astronomy ranked ninth as a physics specialty in 1970 and third in 1973.[85] Further analysis showed that the flow of physicists into astronomy was not primarily a flow of new Ph.D.'s in physics, but it was due to a "flow of more mature Ph.D. physicists in the field." Subfields of physics that provided the most manpower flowing into astronomy came from elementary particle physics, nuclear physics, atomic and molecular physics, and theoretical physics—with no reverse flow. And they were flowing not to optical astronomy, but to the new astronomies—the high-energy regimes understood best by x-ray and gamma-ray physicists. And they were attracted to the biggest projects, like the Large Space Telescope and the other Great Observatories. To be sure, plenty of excellent science came with it, but with it came a very different culture open to structural as well as technical innovation, including a broadened field for awarding large institutional and mission contracts and rapid public access to data produced by these extremely expensive federally funded missions.[86]

When Peter Boyce surveyed the membership of the American Astronomical Society in 1997, looking back upon his years as American Astronomical Society Executive Officer, he observed that "Fuelled by increasing support from NASA for the astronomical research community, astronomy in the United States grew even more rapidly in the 1980s than ever before." Boyce also singled out the fact that NASA also fostered many changes in the profession—in the structure and nature of the American Astronomical Society itself—in the way it serves and mediates the profession and in the way it presents itself to the world of Washington, DC, patronage. In 1979, with some 3,500 members, with national meetings attracting upward of 600 people, and with a yearly budget of $2 million, the responsibilities of managing the American Astronomical Society required a full-time professional office staff "assuring compliance with increasingly complex governmental regulations" and "developing effective relations with Congress and Governmental agencies." Thus the executive office moved from the sylvan landscape of Princeton's meadows and fields to Dupont Circle, Washington, DC, and is unlikely ever to return.[87]

Both SAO and the Harvard College Observatory experienced profound change, though it cannot be said that in its hybrid form since 1955 it ever

---

85. Lynds, "Employment Problems," p. 4.
86. The author is indebted to Peter Boyce and others for helpful discussions leading to these conclusions. They are suggestive of further work to be done.
87. Peter Boyce, "Moving the AAS Executive Office to Washington," in *The American Astronomical Society's First Century*, p. 153.

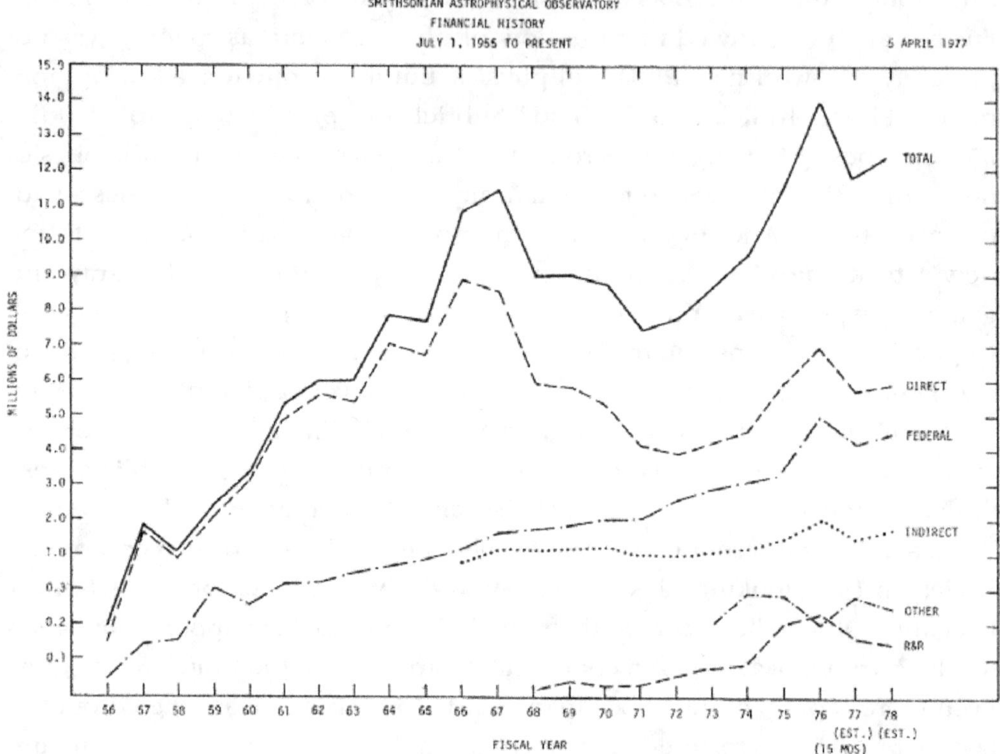

Figure 6: SAO funding showing multiple federal sources that moderated sudden drops in NASA funding. *SAO Financial History, "Report by the Director to the SAO Visiting Committee," April 1975–April 1977, p. 8; Exhibit 3, "Report by the Director to the Smithsonian Astrophysical Observatory Visiting Committee," in "For the Record" by M. Malec, 6 October 1977, p. 8, RU 468, box 22, "Congressional Investigation of SAO"*

represented a traditional form of astronomical institution. The crisis in management and oversight that led to Goldberg's departure and Whipple's retirement in the early 1970s and the combination of the two bodies under a single title, the Center for Astrophysics (CfA), and a single director were not caused by NASA funding; though one can argue that the availability of the opportunities NASA or a NASA-like agency (in another world, it could certainly have been military support) afforded facilitated growth and, hence, the problem.

But look closer. During the 1970s, the CfA, like all institutions, suffered from initial cutbacks and unstable science policies. However, with its multiplicity of federal funding sources, SAO was less affected than most. Looking at its broad combination of funding, unique among American astronomical institutions, one can see that as direct federal funding through competitive grants and contracts fell, the losses were partially offset by a steady increase in congressional appropriations through the Smithsonian itself. The Smithsonian

Astrophysical Observatory did substantially trim its staffing in programs that were no longer critical in the post-Whipple era, mainly through attrition due to the waning of satellite tracking and Moonwatch programs. Celescope expenditures and manpower requirements also peaked at its launch in 1968, leaving mainly the astrophysical groups intact. Some of the technical staff were picked up by the growth of ground-based activities in Arizona or by a rapidly expanding non-optical, high-energy astrophysics division led by the migration of Riccardo Giacconi's x-ray group to SAO in the mid-1970s.[88]

Like the CfA, the American Astronomical Society has outgrown many perceived traditional boundaries. At his 1947 banquet address, on the occasion of the society's 50th year, Joel Stebbins remarked with satisfaction how visitors from other disciplines like mathematics and physics often remarked that society meetings were "like that of a club while that of larger societies . . . much like a market." Stebbins added poignantly that "We should hate to lose the intimate contacts which we get in the small Society."[89] Some 30 years hence, the fears Stebbins expressed definitely came true, with parallel sessions, town meetings highlighting national programs and priorities, job centers, balkanized and competing specialties, and vastly expanded news media attention. Even so, the loss of intimacy, which could as well be regarded as the weakening of the traditional observatory system led by directors like Stebbins, was not on the minds of the leaders of the discipline in following years. Responding to a critical editorial on astronomy funding in the *Christian Science Monitor*'s "Research Notebook" in April 1982, the CfA director succeeding Goldberg and Whipple, George Field, who was then chair of the third Decadal Survey in the early 1980s, pointed to the fact that governmental funding agencies had indeed listened to the recommendations of astronomers. Almost all of the examples he gave of new facilities for astronomy that had been envisioned and placed at high priority were NASA missions ranging from the Apollo Telescope Mount, to Solar Max, to the IUE, with the exception of the SAO-Arizona Multiple Mirror Telescope, a radical departure for ground-based optical astronomy, and a millimeter-wave radio telescope. Field cautioned that astronomers needed to remain wary about "prospective budgets for U.S. science," adding that "Serious

88. SAO financial history in "Report by the Director to the SAO Visiting Committee," April 1975–April 1977, pp. 8–9; Exhibit 3, "Report by the Director to the Smithsonian Astrophysical Observatory Visiting Committee," in M. Malec, "For the Record," 6 October 1977, RU 468, box 22; "Congressional Investigation of SAO," SIA.
89. Joel Stebbins, "The American Astronomical Society, 1897–1917," *Popular Astronomy* 55 (October 1947): 412, reprinted in *The American Astronomical Society's First Century*, pp. 53–57.

cuts have been made, and further cuts can be anticipated in the future." But overall he was adamant that they had exhibited fiscal restraint in their priority listings and were hardly "tilting at windmills . . . ."[90] Field and his contemporaries well appreciated the fact that with careful planning and constant vigilance, the bargain they and their cohorts had made with federal patronage did lead to a vastly increased opportunity for exploring the universe. What autonomy they retained, moreover, was directly related to how well they could remain united as astronomers on the national level.

Acknowledgments

Over the past few years a number of summer students and research assistants contributed to this effort. Chief among them, in chronological order, are Louise Thorn, Teasel Muir-Harmony, Caroline Tung, Samantha Thompson, Chris Hearsey, Fernanda Luppani, and Joan Mathys. Patrick McCray's collaboration in an NSF-sponsored grant project was the launch point of this research, which was funded by the NASA History Division and the Smithsonian Institution. I am particularly indebted to the archivists at the Smithsonian Institution Archives, mainly Ellen Alers; to the NASA archives and Jane Odom; to the Caltech Archives; and to the Archives Division at NASM and at the American Institute of Physics's Center for History of Physics for access to oral histories. Readers providing helpful commentary included Peter Boyce, B. T. Lynds, Robert Smith, Michael Dennis, and Marc Rothenberg.

90. George B. Field, "Letter to the Natural Science Editor on Astronomy," *Christian Science Monitor* (2 June 1982): 10.

Chapter 17

Planetary Exploration in the Inner Solar System

Joseph N. Tatarewicz

Introduction

Exploring the inner solar system animated the visions of the pioneers of rocketry and spaceflight. Very soon after Sputnik, as the Soviet Union launched probes to the Moon, Venus, and Mars, NASA organized itself for a broad program of space exploration. The state of propulsion, communications, and on-board computer technology made the small, rocky planets of the inner solar system an obvious, if challenging, choice, with exploration of the gas giant outer planets deferred until considerably longer into the future. The first 50 years of planetary exploration saw a spirited contest between the United States and the Soviet Union, sending a wide variety of flyby probes, orbiters, and landers to the so-called terrestrial planets, Mercury, Venus, Earth, and Mars, as well as human explorers and an automated sample return mission to the Moon. Reconnaissance missions visited each of these bodies in the first two decades, before any missions were launched to the outer planets, while more focused and ambitious successor missions returned repeatedly over the next three decades. While Mercury is still at the reconnaissance stage, with the first return since the early 1970s now en route, the other planets and satellites have been subject to extensive exploration by probes, orbiters, and landers.[1]

1. The author is grateful to the NASA History Division for generous and patient support over many years. Themes and interpretations in this chapter are drawn in part from Joseph N. Tatarewicz, *Exploring the Solar System: The Planetary Sciences Since Galileo*, manuscript in preparation. For the latter, the author is especially grateful to Roger Launius and Robert J. Brugger. Due to the vast scope of this essay, references below are to representative literature and reviews.

Scientific understanding of the state, history, and origins of these bodies was transformed by in situ observations, but not before an intensive NASA program of remedial planetary astronomy, institution building, and social engineering bolstered the fund of basic knowledge available from Earth-based study and created a community of planetary scientists. Synergistic interplay between planetary astronomy, laboratory studies, and probe-based study overturned expectations about Venus and Mars, provided the first real information on tiny Mercury, and settled major controversies about the state and history of the Moon. Comparative studies of impact cratering helped establish a basic chronology for the evolution of the planetary system and prepared scientists to understand the data returned from probes to the outer planets. Laboratory analysis of returned lunar samples and remote chemical analysis of Martian samples even allowed identification of lunar and Martian meteorites recovered from Antarctica. Planetary science itself, as an integrative interdisciplinary approach, cut its teeth in the inner solar system and developed into an established and productive enterprise that transformed our understanding of the solar system.[2]

The Inner Solar System and the Terrestrial Planets Viewed from Afar

For the first two millennia of our preoccupation with the cosmos, Earth was our spherical home, at rest in the center of the universe and surrounded by

2. The NASA efforts to develop planetary astronomy and socially engineer a planetary science community are recounted in Joseph N. Tatarewicz, *Space Technology and Planetary Astronomy* (Bloomington, IN: Indiana University Press, 1990). The broader context and alternate interpretations can be found in Ronald E. Doel, *Solar System Astronomy in America: Communities, Patronage, and Interdisciplinary Science, 1920–1960* (New York, NY: Cambridge University Press, 1995). The increasing and changing knowledge of the solar system in response to spacecraft visits and Earth-based study can be seen easily in successive editions of J. Kelly Beatty et al., ed., *The New Solar System* (New York, NY: Cambridge University Press, 1981, 1982, 1990, 1999). Anticipating the impact of spacecraft exploration on planetary science and the need for a contemporary summary of current knowledge, astronomer Gerard P. Kuiper, with Barbara M. Middlehurst, edited four volumes of *The Solar System* (Chicago, IL: University of Chicago Press, 1958–1963). Individual volumes appeared: *The Sun*; *The Moon, Meteorites, and Comets*; *The Planets*; and *The Earth as a Planet.* A fifth, unpublished volume exists in manuscript fragments in Kuiper's papers at the University of Arizona–Tucson's university archives. The University of Arizona, where Kuiper established his Lunar and Planetary Laboratory in the early 1960s, continued the publishing tradition with a series of volumes from the University of Arizona Press, Space Science Series. Many NASA solar system exploration programs have been the subjects of individual volumes in the NASA History Series. An extensive listing of the latter, many available for download in full, may be found at *http://history.nasa.gov/series95.html.* The three volumes of Stephen G. Brush, *A History of Modern Planetary Physics* (New York, NY: Cambridge University Press, 1996), cover the 19th and 20th centuries comprehensively. Ronald A. Schorn covers Earth-based studies in *Planetary Astronomy: From Ancient Times to the Third Millennium* (College Station, TX: Texas A&M University Press, 1998).

planets (including the Sun and Moon) and stars that were, by definition, completely inaccessible and entirely "other." There was simply no prospect at all of exploring or knowing them. They were even made of an exotic, separate, "fifth element," totally unlike earth, water, air, and fire here below. During the so-called Copernican and Scientific Revolutions, the move from a cozy, tightly constructed, Earth-centered universe to one in which the Sun (a star) is surrounded by a collection of planets (all worlds in their own right) was positively transforming. The Sun stopped being just a planet and became a star; Earth stopped being the unique home of humanity and became just another planet; the Moon ceased being a planet and became one planetary satellite among many; and the other planets became worlds in themselves—accessible, understandable, and familiar—rather than totally alien. The planets—Earth included—now formed a generic group, capable of comparative study using common techniques, the results of which would be applicable to all of the others, to some degree.[3]

It is no accident that the ancient and venerable Aristotelian and scholastic assertion that a man could never journey to pierce the heavenly spheres, because there would be neither space nor place to receive him, gave way almost immediately to Johann Kepler's *Dream*, in which he journeyed to the Moon to find it Earth-like and gazed back upon Earth; the assertion also gave way to Galileo's telescopic observations that taught him that the sunrise over the lunar mountains viewed from Earth was the same as the sunrise over Bohemia viewed from the Moon.[4]

There followed three and a half centuries of telescopic observations. Telescopes got bigger and ever more capable. The Moon swelled to reveal a breathtaking topography of plains, basins, mountains, and craters. Jupiter showed complex and dynamic equatorial bands and an enigmatic feature creatively named the Great Red Spot. Saturn revealed fixating striated rings. Newly discovered planets, Uranus and Neptune, showed pale blue-green discs. Other moons—visible as only tiny specks of light—multiplied around most of the planets. The puzzling

---

3. In the Aristotelian-Ptolemaic system, the "planets," or "wandering stars," were generally arranged as follows: Moon, Sun, Mercury, Venus, Mars, Jupiter, and Saturn, although there were ancient and medieval variations concerning the relative order of the Sun, Mercury, and Venus. See Edward Grant, *Planets, Stars, and Orbs: The Medieval Cosmos, 1200–1687* (Cambridge, U.K.: Cambridge University Press, 1996), pp. 310–312. Once Copernicus put the Sun at the center, the modern order for the (now) five planets was followed.
4. Galileo Galilei, *Sidereus Nuncius; or, the Sidereal Messenger*, translated with introduction, conclusion, and notes by Albert Van Helden (Chicago, IL: University of Chicago Press, 1989).

Figure 1: The solar system. While produced with the benefit of information from spacecraft exploration, this modern schematic is nonetheless typical of depictions of the solar system over the past two centuries. The solar system features eight planets, seen in this artist's diagram. Although there is some debate within the science community as to whether Pluto should be classified as a planet or a dwarf planet, the International Astronomical Union has decided on the term "Plutoid" as a name for dwarf planets like Pluto. This representation of the solar system is intentionally fanciful, as the planets are depicted far closer together than they really are. Similarly, the relative sizes of the bodies are inaccurate. Drawn to scale, the Sun would be a mere speck, and the planets—even the majestic Jupiter—would be far too small to be seen. *NASA/JPL*

space between Mars and Jupiter filled up with a panoply of so-called "minor planets." (Much later, planetary scientists at a professional conference would adopt [tongue-in-cheek] a more politically correct nomenclature: "dimensionally challenged objects" for these so-called "asteroids".)[5]

Astronomers adopted this extended band of minor planets and debris as a demarcation between an "inner" and an "outer" solar system. Combining bootstrapped gravitational analyses of orbits with telescopic measurements of diameters, they realized that the inner planets were small, dense, and rocky, just like Earth, and therefore called "terrestrial." The outer planets were exotic

5. Telescopic study of the planets is treated in Michael Hoskin, ed., *The Cambridge Concise History of Astronomy* (Cambridge, U.K.: Cambridge University Press, 2008) and the much more extensive multiple volumes of Hoskin, ed., *The Cambridge General History of Astronomy*, particularly vol. 2, *Planetary Astronomy from the Renaissance to the Rise of Astrophysics* (in two parts, 1995 and 2003).

gas giants, their misleading visible surfaces merely cloud decks, deep within which might be small, metallic cores; therefore, they were called "Jovian" after Jupiter, the exemplar of their class. Comets, loosely bound collections of sand and perhaps ice, commuted from the exurbs. But, except for a bunch of well-defined planets and moons; an asteroidal band of debris; and the few, tiny commuters, there was nothing but empty space, solar radiation, and the invisible glue of gravity in between. This is the canonical Copernican-Keplerian Solar System, successor to the Aristotelian-Ptolemaic-Scholastic Cosmos, which would stand until spacecraft made their journeys.[6]

Ironically, the gas giants of the outer solar system revealed more telescopic detail and dynamism than the more familiar bodies of the inner solar system. While the Moon was a field day for mappers and catalogers on Earth, its consistently gray, unchanging, and airless surface was disappointing. Earth-based spectroscopy found its surface to resemble Earth-like volcanic basalts, and while the morphology of the craters suggested an impact origin, there were some features suggestive of ancient volcanic calderas. Nonetheless, it was an obvious first destination for exploration, and some held hope for active volcanism among the craters. Just after Sputnik, the U.S. Army's classified Project Horizon proposed establishing a fully staffed base for nuclear missiles on it, the better to defend the United States from Soviet attack from the ultimate high ground.[7]

Mercury, tiny and always in solar glare, barely showed a disc, and the few features a handful of astronomers thought they saw were mostly discounted by their colleagues. In any case, it was too hot (any atmosphere it might have had was long since burned off) and so close to the Sun that even getting there was problematic. Not until 1974 would the first probe arrive, a triumph of innovative spacecraft engineering and navigation.

Venus, visibly larger, more easily observed, and showing a full suite of phases, was shrouded in dense, white clouds. Optimistic thermodynamic calculations and the assumption that the clouds were water vapor led some astronomers to conclude a balmy surface beneath, with oceans, continents, and probably a carboniferous swamp. Even the first radio observations of the late 1950s, which produced puzzling results, did not dissuade the majority opinion

---

6. Steven J. Dick, *Plurality of Worlds: The Origins of the Extraterrestrial Life Debate From Democritus to Kant* (New York, NY: Cambridge University Press, 1982).
7. Ralph Baldwin, *The Face of the Moon* (Chicago, IL: University of Chicago Press, 1949); Dwayne A. Day, "Take off and nuke the site from orbit (it's the only way to be sure . . .)," *Space Review* (4 June 2007), available at *http://www.thespacereview.com/article/882/1* (accessed 4 April 2009).

that Venus was at least capable of sustaining life and that it might well harbor indigenous beings. A young Carl Sagan, already then cocky and annoying most of his fellows, concluded in his Ph.D. dissertation that Venus was the victim of a runaway greenhouse effect and an inferno. The first successful U.S. planetary probe, Mariner 2 in 1961, added to the weight of ground-based evidence; and later probes measuring the atmosphere in situ finally confirmed that Venus was inhospitable in the extreme.

Unique among the inner and outer planets was Mars—a positive obsession. While only the Moon showed more telescopic detail, Mars showed every plausible indication of domesticity. It was near to the size of Earth and closer than any other planet; its year was just under twice that of Earth; its atmosphere was detectable; its white polar caps expanded and shrank with the seasons; dust blew across the surface; and a wide variety of surface features and greenish colors came and went. Venus and Mars were always the preferred destinations of the rocketry pioneers and the science fiction writers. But Mars was first among equals until the first Mariner spacecraft visited Venus, and it was the only hope thereafter. Not dissuaded by Sagan's gloomy analysis, the U.S. Air Force included both Venus and Mars in its "Strategic Interplanetary Systems" proposal, whereby it asserted the mission to protect peaceful U.S. exploratory bases on both planets from Soviet incursion.[8]

So, on the eve of the Space Age, this was the state of the solar system. While planetary astronomy on Earth included much more than just visual observation, the best telescopic images of the planets serve as a reminder of just how little was evident, and contemporary charts of the solar system, consisting mostly of lines and little circles drawn against vast blank space, remind us of just how empty and tidy it was thought to be.[9]

First Plans for Voyages

Very soon after Sputnik, as the Soviet Union launched probes to the Moon and prepared to journey to Venus and Mars, NASA organized itself for a broad program of space exploration. At a time when successfully getting Earth orbital

---

8. Dwayne A. Day, "Take off and nuke the site from orbit (it's the only way to be sure . . .)." The Army efforts were instigated by Wernher von Braun, then at the ABMA, and all but ceased once von Braun and his team were transferred to NASA; the Air Force interest in the Moon and nearer planets continued through the 1960s, with a dedicated basic research institute for lunar and planetary study at the Air Force Cambridge Research Laboratory under John W. Salisbury.
9. A good example is Henry Norris Russell, *The Solar System and Its Origin* (New York, NY: Macmillan, 1935).

satellites only a few hundred miles into orbit and functioning was itself a challenge, missions to even the Moon and nearer planets were the highest challenge and evoked the most skepticism. In 1959, as Caltech's JPL negotiated transfer of its contract to NASA from the U.S. Army, JPL settled on seeking the lunar and planetary mission for its own. The state of propulsion, communications, and on-board computer technology made the small, rocky planets of the inner solar system an obvious, if challenging, choice, with exploration of the gas giant outer planets deferred until considerably longer into the future. In JPL's eyes, the Moon might serve as an interim test bed for such exploration, but it did not seem intrinsically interesting in its own right. The paradigm that eventually followed was evident in these early studies. It called for 1) one or more swift flyby probes for reconnaissance, 2) one or more orbiters for mapping and characterizing the environment, 3) one or more automated soft landers, and 4) human exploration.[10]

But, we should not take this tidy exploration paradigm too far—choices of which bodies when, which instruments to fly, and which questions to ask were a hoary suboptimization of scientific and engineering agendas, political directives, budgetary feast or famine, and responses to discoveries as well as failures. And, while JPL became most identified with this exploration, other NASA Centers were involved as well. Goddard Space Flight Center, for example, took near-Earth orbit as its purview, and its instruments and expertise found application around other bodies as well. Ames Research Center, with its special expertise in hypervelocity aerodynamics, worked on entry probes and also adapted its successful interplanetary Pioneer series to outer solar system exploration. Johnson Space Center and Marshall Space Flight Center, of course, were most deeply involved in Apollo, including lunar science, but contributed to various planetary missions as well. If we recall that Earth is itself a planet, it is interesting to note that the exploration paradigm was followed in this case, more or less, in reverse. Centuries of human exploration of our own planet had amassed great detail, but with varying coverage and a synoptic view assembled only with great difficulty. Sounding rockets and later satellites moved ever farther from Earth, providing the kinds of observations and global perspective that formed the starting point for studying other planets.[11]

---

10. Clayton R. Koppes, *JPL and the American Space Program* (New Haven, CT: Yale University Press, 1982).
11. Noel W. Hinners, "The Golden Age of Solar System Exploration," in *The New Solar System*, ed. J. Kelly Beatty and Andrew Chaikin (New York, NY: Cambridge University Press, 1990), pp. 3–14.

Overview

The cleanest example of the astronomy-flyby-orbiter-lander-humans paradigm was the Apollo program. The Jet Propulsion Laboratory's initial plan to test its probes, orbiters, and landers at the Moon was quickly overtaken by President Kennedy's 1961 directive and NASA's subsequent transformation. Ranger impact probes, Surveyor Landers, and Lunar Orbiters preceded the Apollo missions and gathered advance data for development and operation of the Apollo system, and they secondarily performed a scientific reconnaissance to prepare for astronaut surface operations. Before all this, JPL and NASA had formed a relationship with astronomer Gerard P. Kuiper and his Lunar and Planetary Laboratory in Tucson, Arizona, to provide remedial astronomical reconnaissance and analysis. It was Kuiper and his sometime colleague and ofttimes nemesis, Harold Urey, who provided the early scientific justification and guidance for Apollo: that the Moon was a "Rosetta Stone" for the solar system, a reliquary that had frozen in time the conditions and chemistry of its formation. That proved only partially correct, but the detailed study of the surface and interior of the Moon from all platforms, however skewed and compromised by the mandates of engineering and astronaut safety, was a continuing bonanza.[12]

Not only did Apollo succeed in characterizing the Moon (and Earth's) formation and conditions, but the stratigraphic analysis of cratering brought to a high art by the U.S. Geological Survey and its Astrogeology Division in Flagstaff, Arizona, gave the first detailed chronology of the chaotic conditions that accompanied the accretion, melting, and surface molding of the Moon and Earth out of planetesimals. The Astrogeology Division, led by geologists Eugene Shoemaker, Harold Masursky, and others, trained the Apollo astronauts to be field geologists, and the division even sent one of their own, Harrison "Jack" Schmitt, to be the first (and last) scientist-astronaut to the Moon on Apollo 17. So unlikely a location as Houston, Texas, became the repository of the returned lunar samples, the later identified lunar and Martian meteorites recovered from Antarctica, and other extraterrestrial materials that happened to pass close enough to Earth to be collected. From shortly after Apollo 11 until today, lunar and planetary scientists hovered around Houston, Texas, the lunar receiving laboratory and curatorial facility, and the Lunar and Planetary Institute. The

---

12. Dale P. Cruikshank, "Gerard Peter Kuiper (December 7, 1905–December 24, 1973)," *Biographical Memoirs of the National Academy of Sciences* 62 (1993): 259–295; James R. Arnold, Jacob Bigeleisen, and Clyde A. Hutchison, Jr., "Harold Clayton Urey (April 29, 1893–January 5, 1981)," *Biographical Memoirs of the National Academy of Sciences* 68 (1995): 363–411.

astronauts themselves, much to their annoyance, were initially quarantined along with their samples for a time—something for which they held Carl Sagan and another portion of his Ph.D. dissertation personally responsible.[13]

Apollo was the full paradigm, calibrated with ground truth from six sampled landing sites. For entirely separate reasons, outbound planetary spacecraft did not return equivalently detailed imagery and sophisticated other observations of Mercury, Venus, and Mars until after Apollo—when planetary geologists were primed and ready to apply the Apollo knowledge and technique. Similarly, observations of the icy satellites of the outer planets began arriving only in 1979, when the full complement of technique and knowledge gained in the inner solar system could be applied to them. From the standpoint of internal structure, surface morphology, and other characteristics, the satellites of the outer solar system are closer kin to the planets of the inner solar system, while the gas giant outer planets are vastly different from the inner planets. The only two known planetary satellites of the inner solar system (besides the Moon), Mars's tiny Phobos and Deimos, were eventually recognized as closer kin to most of the asteroids than other planetary satellites, inner or outer. The Moon, Earth's satellite, is unique in the inner solar system but resembles the larger asteroids—differentiated, evolved bodies with a complex, stable surface and little or no ice.

The tidy distinction between inner and outer solar system still stands for the major planets (Pluto's recent reappraisal being an example, leaving only gas giants in that region), but 50 years of spacecraft exploration has reshuffled the earlier characterizations of the other bodies. When the other bodies were all but points of light in telescopes with masses of various reliabilities and conditions inferred from the general environment, their definitions and names were driven by location and gravity. The accumulated knowledge of this first half century of exploration has revealed a very different solar system, a diversity of individual bodies that now seem, at times, to defy earlier classifications and call for new groupings. There are two major stories from this phase of exploration and study. First, the major terrestrial planets have been confirmed to be more similar to one another than different, even if each has revealed its own stunning surprises. Formed from roughly equivalent materials, Mercury lost almost all its atmosphere in the

13. Don E. Wilhelms, *To a Rocky Moon: a Geologist's History of Lunar Exploration* (Tucson, AZ: University of Arizona Press, 1993); Elbert A. King, *Moon Trip: A Personal Account of the Apollo Program and Its Science* (Houston, TX: University of Houston, 1989).

intense heat of its location, while Earth, Venus, and Mars took vastly different evolutionary paths. Similarly, Jupiter, Saturn, Uranus, and Neptune reflect the frigid conditions of their locations in the original solar nebula, retaining most of their atmospheres. Second, and perhaps even more important, what had been that vast, empty space between the planets is now known to be filled—with electromagnetic fields and particles and bodies of all sizes dynamically interacting and extending throughout the solar system. While the admittedly fascinating details of each of the inner planets are captivating, we are only beginning to appreciate the significance of all the other smaller bodies and other phenomena in between.[14]

The many missions sent to the inner solar system, with their long planning cycles and episodic bursts of information, make for a confusing and interlocking chronology. Because each of the planets and the Moon revealed such individuality, it is best to discuss them by body rather than by year. Because it was exploration of the Moon that unfolded rapidly in the early decades, that will come first. Missions to Mars and Venus alternated at almost every celestial mechanics opportunity, with Mars garnering more attention. Mercury did not receive its first visitor until 1974. The "asteroid belt" between Mars and Jupiter was not crossed by spacecraft until about that same time, and individual asteroids did not receive attention until much later.

The Moon

The early plans for robotic exploration of the Moon drove the designs of what became the Ranger impactors, Lunar Orbiters, and Surveyor Landers. The 1961 presidential decision to land humans on the Moon before the end of the decade changed the emphasis of these missions more toward engineering support than pure science, but many of the scientific and engineering observations and studies desired were common. After a difficult start, from 1961 to 1964, during which engineers and managers had to learn the severe requirements of getting spacecraft to survive and operate to lunar distances, Rangers 7 though 9 succeeded, from 1964 to 1965, in returning the first high-resolution glimpses of the cratered surface of the Moon, over a variety of terrain, and

14. Optical and spectroscopic studies of the smaller bodies were extremely difficult, if not impossible, even for the nearest members, but the nearer bodies were within reach of various powerful radar instruments: Andrew J. Butrica, *To See the Unseen: A History of American Planetary Radar Astronomy, 1946–1991* (Washington, DC: NASA SP-4218, 1996).

they returned some hints of the mechanical properties of the surface through their own destruction and the ensuing artificial craters.[15]

From 1966 to 1968, the Surveyor series sent seven automated landers, of which five were fully successful, to a large variety of sites. In addition to providing early (the Soviets had landed Luna 9 four months before Surveyor 1) in situ images of the terrain, the Surveyors assessed the mechanical characteristics of the regolith (soil) just by landing, as well as through digging and manipulating rocks with a pantograph arm and scoop. This settled a scientific controversy over the depth and particle size of the lunar dust. Simple magnets, as well as an alpha scattering instrument on some of the Surveyors' scoops, allowed the earliest and simplest analysis of the composition of the lunar material, already known from Earth-based spectroscopy to be similar to volcanic basalts. Apollo 12, in 1972, was able to land close enough to Surveyor 3 that the astronauts could remove a portion of its camera system for return to Earth and subsequent analysis to characterize the meteorite environment over time.[16]

At the same time (from 1966 to 1967) five Lunar Orbiters, carrying an innovative imaging system that used high-resolution film processed on board and then scanned for transmission (as opposed to low-resolution television) covered nearly all of the lunar surface with 60-meter resolution and selected areas at 2-meter resolution. The first three orbiters were dedicated to selecting Apollo landing sites, and so they concentrated on equatorial near side regions. The final two orbiters provided scientists with synoptic information on the entire lunar surface as well as detailed information on areas of interest. While the Apollo Command and Service Modules did continue this orbital study somewhat, expanding it to other wavelengths and using other techniques to characterize the mineralogy, the Lunar Orbiter data stood as the only comprehensive lunar coverage until 1994, when the Clementine orbiter performed extensive studies in a joint Ballistic Missile Defense Organization-NASA effort. In addition to using more robust imaging sensors, Clementine also added unprecedented altimetric data through an infrared laser device. Heights of lunar features had to be inferred from shadow analysis in Lunar Orbiter images. In 2008, nearly forgotten Lunar Orbiter tapes recorded at ground stations were resurrected and modern data recovery and processing techniques applied to extract even more information. The Soviet Luna and Zond series sent a variety of flyby,

---

15. R. Cargill Hall, *Lunar Impact: A History of Project Ranger* (Washington, DC: NASA SP-4210, 1977).
16. Donald A. Beattie, *Taking Science to the Moon: Lunar Experiments and the Apollo Program* (Baltimore, MD: Johns Hopkins University Press, 2001).

orbiter, lander, rover, and even sample return missions from 1959 to 1976, of which 20 were successful.[17]

The Apollo program provided the most extensive collection of lunar samples, from six diverse sites, the last three of which included lunar rovers and allowed a wider range of coverage, as well as some orbital studies on two additional missions. The landings, from 1969 to 1972, were initially in safe, bland terrain near the equator with short stays, but they grew increasingly ambitious. In addition to returning more lunar samples, of greater diversity and from a wider range (22 kilograms on the first landing, reaching 111 kilograms on the last), the Apollo missions also conducted other experiments and left automated instruments that returned data for years. The orbiting Command and Service Modules were outfitted with remote sensing instruments of increasing sophistication as well. The samples, dated with precision in Earth laboratories, showed ages between 3.2 and 4.6 billion years old—representing the early formation of the Moon and the later melting and resolidifying of the smooth maria after relatively recent large impacts. This complements the analysis of the oldest Earth rocks and recovered meteorites. Chemical composition of the lunar samples was not that expected of a body that would have formed directly out of the solar nebula but more resembled Earth's crust, itself depleted of metals that are presumed to have sunk to the core. This and other data helped move to the fore the "giant impactor" hypothesis that the Moon had formed after Earth had begun to differentiate, when a glancing blow from a passing Mars-sized body removed mostly crustal material. Laboratory analysis of returned samples, combined with synoptic orbital remote sensing capable of mineral analysis, and with similar analysis of Earth and meteorites, demonstrated the power of these mutually reinforcing approaches. In addition, conventional stratigraphy was extended to use statistical studies of superposed cratering, providing knowledge of the more recent history of the lunar surface, after the presumed magma ocean that had once covered its entire body solidified and began to suffer impacts.[18]

While the Moon did not turn out to be precisely the "Rosetta Stone" of the solar system (a primordial object retaining evidence from the earliest accretion

---

17. Bruce K. Byers, *Destination Moon: A History of the Lunar Orbiter Program* (Washington, DC: NASA TM X-3478, 1977).
18. Wilhelms, *To a Rocky Moon*; Beattie, *Taking Science to the Moon*; William David Compton, *Where No Man Has Gone Before: A History of Apollo Lunar Exploration Missions* (Washington, DC: NASA SP-4212, 1989).

of the terrestrial planets), as Harold Urey had suggested, it did nonetheless provide crucial evidence of the conditions and events relatively soon thereafter. These would be applied to the other terrestrial planets and even to the icy satellites of the Jovian planets.

Mars

The subject of an intense and systematic campaign of observation and study, the Moon revealed information steadily through the early space program, and the information gained fueled studies long thereafter. Mars, however, was not so easily accessible, generally only every two years, and even then, the long mission times and difficulty of returning information meant that understanding this planet would be an episodic enterprise. Initially, fleeting glimpses of small portions of the planet meant that Mars would appear differently each time a mission succeeded, which in the early period was only about one-third of the time for both the United States and the Soviets. The accumulated lore of Mars, its polar caps, seasonal changes, and even the spurious detection of chlorophyll in certain Earth-based observations, not long before spaceflight, made it a special place. When, at last, Mariner 4 in 1965 and then Mariners 6 and 7 in 1969 finally flew by and returned images of a lunar-looking cratered surface, lacking anything resembling the "canals" that had been glimpsed in Earth-based telescopic study, many scientists were crushed. Still, even in the "dead Mars" period, the tiny percentage of the surface that had been seen buoyed hopes for a more systematic reconnaissance. Mariner 9 entered orbit in 1971 (as did a Soviet orbiter) and found quite a different place. It was not the Mars of Edgar Rice Burroughs, to be sure, but once the planetwide dust storm cleared, scientists found two new and stunning types of features: volcanic calderas and huge valleys and canyons. The volcanic features suggested that, at least geologically, Mars was not so dead as the Moon; the canyons suggested that water might have once flowed in abundance. Mariner 9 mapped more than 80 percent of the planet during its yearlong mission, documenting an active, if thin, atmosphere and providing the first images of the two Martian satellites. It is safe to say that Mariner 9 reinvigorated the study of Mars, providing the justification and the will to continue with the Viking missions five years later.[19]

---

19. Edward Clinton Ezell and Linda Neuman Ezell, *On Mars: Exploration of the Red Planet, 1958–1978* (Washington, DC: NASA SP-4212, 1984).

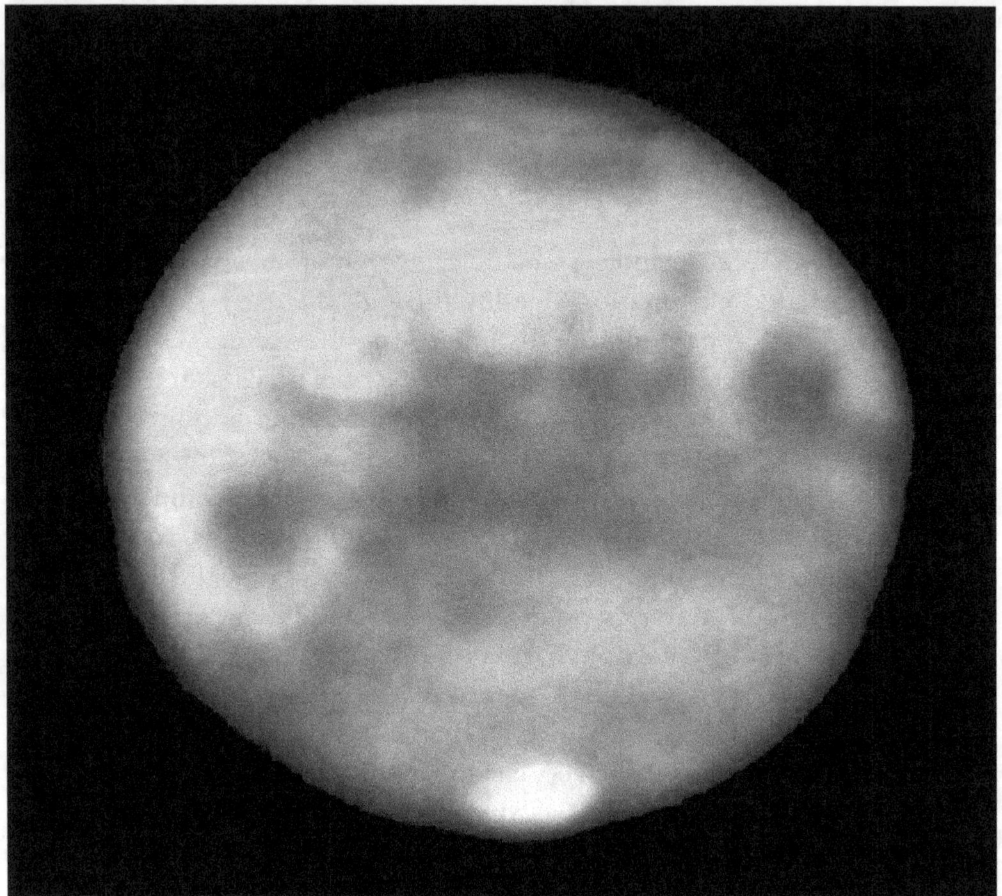

Figures 2 to 5: These four images indicate the progression in visual knowledge from a typical, best Earth-based telescopic view of Mars (19th century through the late 20th), to an Earth orbital-based Hubble Space Telescope image, to a synthesis of Mars orbital mapping, and, finally, to a lander on the surface (Pathfinder). To be complete, this sequence would also include a photomicrograph of exposed Mars rock obtained from the microscopic imagers on the MERs.
Figure 2: Earth Telescope View of Mars (0°N, 43°W). This 1988 image from the Lowell Observatory was obtained at the start of spring in the Southern Hemisphere, so the southern polar cap is prominent. Valles Marineris is the narrow feature protruding to the left of the dark region in the center of the image. Because of the obscuring effects of Earth's atmosphere, even the best ground-based telescopes usually can resolve features no smaller than about 300 kilometers across when Earth and Mars are closest to one another. *Courtesy of Leonard Martin*

Figure 3: Space Telescope View of Mars (0°N, 270°W) This image was obtained by the Hubble Space Telescope in 1995 as part of an observing program to monitor seasonal changes in the atmosphere and surface of Mars. The prominent dark feature in the center of the image is called Syrtis Major. The differences in color, which are exaggerated in this computer-enhanced image, are thought to be caused by differences in the deposits of dust and sand covering different regions. Features as small as 50 kilometers are seen in this image. The white region at the top of the image is the north polar cap. On the right, white clouds cover the Elysium volcanoes. *Hubble Space Telescope Image STScI-PR95-17*

Figure 4: Viking Orbiter Mosaic (5°S, 80°W). This mosaic shows a global view of Mars as seen by NASA's Viking spacecraft in the late 1970s. The linear structure stretching east-west across the center of the image is Valles Marineris, a very large trough system. The two brown, circular objects on the left side of the image are Pavonis Mons and Ascraeus Mons, two of the large shield volcanoes in the Tharsis region. Most of the volcanic and tectonic activity on Mars in the last 3 billion years has been concentrated in the Tharsis region. *Image processing by the U.S. Geological Survey*

Figure 5: Mars Pathfinder: Twin Peaks (19°N, 34°W). This image shows the view to the west of the Pathfinder landing site. At the bottom, portions of the spacecraft's solar panels and airbags are visible. Pathfinder landed in the outflow region of the Ares Vallis, an outflow channel similar to Maja Valles. Many rocks are visible in this image and may have been transported to this region by massive floods early in the history of Mars. A pair of hills known as "Twin Peaks," located about 1 kilometer from the Pathfinder lander, can be seen on the horizon. This slide is actually a mosaic of many individual images, and there are some misalignments visible in the mosaic. *Mars Pathfinder Image 81957*

The twin Viking orbiter landers had their origins in an ambitious and ill-fated program during the Apollo era called Voyager (not the outer planet mission by the same name). Although they had Mariner and Surveyor inheritance, these were extremely sophisticated spacecraft, so heavily laden with instruments that they each required the largest launch vehicle in the fleet, the Titan III, which itself was sufficient to send an interplanetary Voyager spacecraft to all the outer planets. The twin pairs arrived in orbit around Mars in the summer of 1976, the U.S. bicentennial, waiting for the orbiters to return imagery of potential landing sites at high enough resolution to select the most appropriate. Each lander was targeted to a site of somewhat different characteristics. Viking I landed 20 July 1976, and Viking II landed on 3 September 1976. Viking I was set down near the equator, and Viking II was set well into the Northern Hemisphere. They operated for six and four years on the surface, respectively, while their orbiters had similar longevity. Their data would feed Mars science for nearly two decades before another attempt at a Mars landing.

In light of the numerous other investigations carried and the voluminous high-quality data returned by the orbiters and landers, which marked the high point of an extended, systematic, and multifaceted study of another planet from orbit and surface, it is perhaps surprising that the results were so disappointing. The search for life on Mars had been designated as the primary goal, and perhaps it had been oversold. Given the public and scientific obsession with Mars through the ages, however, it was perhaps inevitable that anything less than confirmation would be perceived as failure. Four separate technologies analyzed soil samples. The so-called Labeled Release Experiment showed a positive result, but the others were enigmatic or disputable. The Gas Chromatograph-Mass Spectrometer, an instrument considered agnostic with respect to life and extremely reliable, identified no organic molecules down to parts per billion. The controversy continues to this day, although the majority opinion is that Viking detected an exotic chemical reaction, not metabolism of Martian organisms.[20]

In the 1990s, a series of new missions heralded a return to Mars, but with very mixed results. This time orbiters and landers were sent separately, though they were coordinated. To the present, more than 15 such missions, most with more limited goals and innovative technologies, have been attempted. The majority

---

20. Ezell, *On Mars*. For an alternative interpretation of the biology results, see Norman H. Horowitz, *To Utopia and Back: The Search for Life in the Solar System* (New York, NY: W. H. Freeman, 1986); Steven J. Dick, *The Biological Universe: The Twentieth Century Extraterrestrial Life Debate and the Limits of Science* (New York, NY: Cambridge University Press, 1996).

of these missions have succeeded, but those that have failed have done so in a troublesome or traditionally mysterious way. Those missions that have succeeded have increased steadily, and almost inconspicuously, the store of data of all kinds, showing Mars to be a dynamic and fascinating place at present, and with a likely past that bolsters the prospects for life at some time or another. More than ever, the consensus is that the right ingredients and conditions for life are, or have been, there. This most recent suite of orbiters, landers, and even rovers has captured the public imagination, even with the search for life having been kept rather quiet. Mars seems now capable of being appreciated on its own. In the most recent comprehensive planning, the Vision for Space Exploration that grew out of the Shuttle *Columbia* accident and led to the Constellation program, Mars was selected as the ultimate destination, as it has been so many times before.[21]

Venus

Only slightly easier to get to than Mars from the standpoint of celestial mechanics and, like Mars, similar enough to Earth in its mass, size, and space environment, cloud-shrouded Venus was just as inviting and even more mysterious. As United States and Soviet spacecraft set out to explore in 1961, it was still plausible, if not likely, that a fairly benign environment might be found beneath the clouds. The Soviet attempts did not fare well until 1965, when Venera 3 became the first spacecraft to impact another planet, followed by Venera 4 in 1967, which relayed the first data from a suite of instruments within Venus's atmosphere. The U.S. Mariner 2 (Mariner 1 did not survive launch) relayed the first data from the vicinity of Venus in late 1962. It seemed to confirm the more sober interpretation of previous Earth-based radiometry measurements, suggesting hot surface temperatures, high pressure, and a carbon dioxide atmosphere. Mariner 5, in 1967, refined these measurements, and without the ability to probe beneath the clouds it concentrated on the interaction of the planet's atmosphere with the interplanetary environment. The combined United States and Soviet results of 1967 provided confidence in the overall interpretation that Venus was very inhospitable beneath the clouds. While the United States did not return until 1973 (and then only as part of a combined Venus-Mercury mission using the first gravity-assist trajectory), the Soviets made Venus a primary target of their program. For the 10 years after 1967, the Soviets made

---

21. Steve Squyres, *Roving Mars: Spirit, Opportunity, and the Exploration of the Red Planet* (New York, NY: Hyperion, 2006); Tatarewicz, "The 'Vision for Space Exploration' of President George W. Bush, Space Science, and U.S. Space Policy," *Futures* (in press).

eight attempts, several of which survived to increasingly greater depths in the atmosphere, until Venera 7, late in 1970, landed on the surface and survived for 23 minutes, and Venera 8 in mid-1972 survived for 50 minutes. In October 1975, twin orbiter probes Veneras 9 and 10 obtained the first images from the surface and made the first radar observations from orbit.[22]

In 1978, two orbiter-lander combinations, Veneras 11 and 12, were joined by the first comprehensive United States attempt, the Pioneer Venus orbiter and a separate suite of multiple probes targeted to various locations. The Soviets sent their last orbiter-lander combinations at the 1981 opportunity and their final orbiters at the 1983 opportunity. In 1990, the United States' Magellan orbiter began providing detailed radar mapping of the entire planet. Aside from the 2005 ESA Venus Express mission, which engaged in a variety of observations, Venus has since only been visited by spacecraft en route to other destinations and performing observations while getting a gravity-assist.

Initially, with the inability to see beneath the clouds, Venus provided a unique opportunity to observe somewhat exotic interactions with the interplanetary medium, magnetic fields, and solar wind. Through the 1970s and 1980s, as scientists began to understand the exotic atmospheric and surface conditions, Venus provided a counterpart to Mars and Earth. On Mars, with a thin atmosphere, rugged surface, blowing sand, and polar caps of water ice and carbon dioxide, mechanical and chemical interactions and transports could be observed that were analogous to those on Earth. Venus was more extreme. The high pressures and temperatures and the atmospheric chemistry of a runaway greenhouse effect showed another evolutionary path from that followed by Earth or Mars, otherwise, from their locations and presumed conditions of formation not too different. The geology of Venus also displayed extremes not seen anywhere else. On planets and satellites without appreciable atmospheres like the Moon and Mercury, the surfaces retained crisp impact craters and a clear record of bombardment, with some local melting and flow from the impacts. On Mars, this record was more subtle, due to erosion from familiar processes like wind and, in the past, water. On Venus, however, the intense pressure and heat made the surface plastic, so craters and other relief slumped and, slowly, almost flowed. While the rocks glimpsed by landers before they failed were recognizable, the entire environment was far different in the extreme from anywhere else.

---

22. David Harry Grinspoon, *Venus Revealed: A New Look Below the Clouds of Our Mysterious Twin Planet* (New York, NY: Basic Books, 1998).

Mercury

Mercury was visited by Mariner 10 three times from 1974 to 1975, after a gravity-assist from Venus. Because of the interlocking orbital mechanics, the extremely slow rotation of the planet, and illumination of the Sun, Mariner 10 obtained images of less than half the surface. The crisp, cratered topography, apparently unmodified since the earlier bombardment, seemed at odds with the surprising discovery of reasonably strong magnetic fields. Since magnetic fields are thought to require a molten core, which would also drive various tectonic processes, and Mercury should also be subject to intense tidal pumping effects and the heat of the near-solar region, the apparently stable surface was an enigma. However, for the next 30 years, proponents of further study of Mercury could not win approval for another mission. The MErcury Surface, Space ENvironment, GEochemistry, and Ranging (MESSENGER) probe was launched in 2004, and a complex set of Earth-Venus-Mercury gravity-assists has allowed it to return flyby images of Mercury from perspectives missed by Mariner 10. It is expected to achieve Mercury orbit in 2011, making Mercury the least-visited and least-studied of the inner planets.[23]

Asteroids, Comets, and Other Visitors

The so-called asteroid belt was traditionally the demarcation between the inner and outer solar system. From the first discovery of a body between Mars and Jupiter in 1801 through the ever-increasing population of this region with seemingly innumerable bodies of all sizes, the primary question has been, was this material the remains of a disrupted planet or the material from which a putative planet never coalesced? Studies of the orbits of the known bodies in the 19th century suggested the region had orbital structure, consisting of families of bodies. Spectroscopic studies in the 20th century revealed classes of bodies. But only with increasingly sophisticated theories of planetary formation and differentiation, in which the heavier, metallic materials sink to the core while the lighter, rocky materials rise to form the crust, did the significance of these spectral signatures become clear. The knowledge and theories that issued from spacecraft missions to the inner planets, combined with basic ground-based and theoretical studies, suggested a diversity of bodies in this zone long before any spacecraft—until very recently bound for somewhere else—made close observations. "Asteroids" now seems far

---

23. Robert G. Strom and Anne L. Sprague, *Exploring Mercury: The Iron Planet* (New York, NY: Springer, 2003).

too broad a category, since it includes the larger, spherical, differentiated bodies such as Ceres, as well as presumed broken fragments from the crusts of these kinds of bodies, exposed fragments of metallic cores, and a whole variety of materials in between. Jupiter perturbs these objects gravitationally, sometimes sending them deeper into the inner solar system, just as it does other objects from the outer solar system, including cometary nuclei. None of these objects shows any telescopic detail, so their morphology was a mystery until spacecraft visits and encounters. As a result of ground-based optical and radar study, theoretical analysis, the knowledge gained from planetary exploration proper, and a few visits beginning late in the last century, the proper terminology for all these objects is now in flux.[24]

Mariner 9's 1971 images of the two satellites of Mars showed them to be irregular and small, probably captured asteroids. In 1991, the Galileo spacecraft on its way to Jupiter was the first to acquire images of asteroids Gaspra, Ida, and the latter's own satellite, Dactyl. Only in 1997 did the first dedicated mission to an asteroid, the Near Earth Asteroid Rendezvous-Shoemaker spacecraft, visit Mathilde and then rendezvous with and land on the surface of Eros in 2001. In 2007, almost exactly 50 years after Sputnik, NASA launched the Dawn mission, headed toward Ceres, Vesta, and possibly Pallas—three of the four asteroids first discovered in the first decade of the 19th century. Earth-crossing bodies of all sizes, cometary nuclei, meteoritic materials, and a wide variety of other such tiny bodies, not to mention the vast variety of particles and fields, have joined the (now renamed) minor planets and asteroids as subjects of serious study and inclusion into the definition of the solar system—inner and outer.[25]

Perhaps nothing symbolizes this reappraisal better than the controversy over the status of Pluto. Discovered by Clyde Tombaugh at the Lowell Observatory in 1930 and quickly judged to be the "Planet X" thought to be perturbing the orbit of Neptune, its planetary status was challenged shortly thereafter. A complex interplay of ground-based, planetary spacecraft, and Earth orbital

24. Curtis Peebles, *Asteroids: A History* (Washington, DC: Smithsonian Institution Press, 2001).
25. David W. Hughes and Brian G. Marsden, "Planet, asteroid, minor planet: A case study in astronomical nomenclature," *Journal of Astronomical History and Heritage* 10, no. 1 (2007): 21–30; National Academy of Sciences, SSB, *New Frontiers in the Solar System: An Integrated Exploration Strategy*, chap. 1, "Primitive Bodies: Building Blocks of the Solar System" (Washington, DC: National Academies Press, 2003). The extensive series of studies and reviews of solar system exploration, which provide a rich chronicle of scientific results and changing scientific agendas, is available at *http://www.nationalacademies.org/ssb/*.

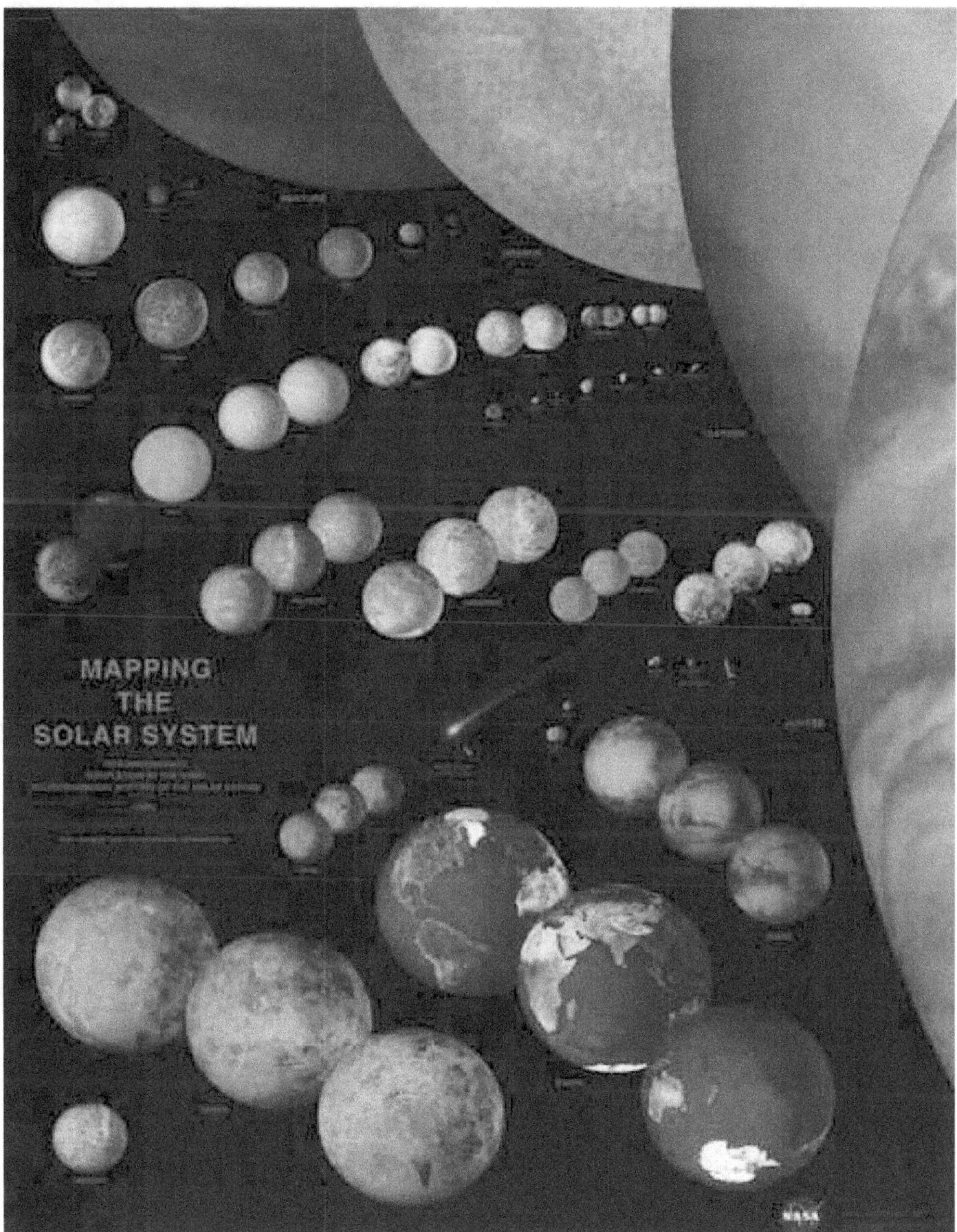

Figure 6: The wealth of detail, inclusion of many planetary satellites, and addition of numerous smaller bodies in this montage suggests the degree to which solar system objects, not just the major planets, have become appreciated as worlds in themselves. The information presented in the "Browse the Solar System" Web pages was originally published as a poster, "Atlas of the Solar System—Topographic Series—Mapping the Solar System," I-2447, 1995. The poster was prepared on behalf of the Solar System Exploration Division, Office of Space Science, by the U.S. Geological Survey Astrogeology Research Program "to commemorate completion of the first reconnaissance mapping of the solar system."

astronomical observations fueled observational and theoretical reappraisals that culminated in a 2006 International Astronomical Union resolution attempting to enact a new and more precise definition for "planet," which Pluto promptly failed. It has been reappraised as a minor planet, a Kuiper Belt object, a trans-Neptunian object, a dwarf planet, and also as the first of a new class, Plutoids. The New Horizons probe, launched in 2006, has taken its first image of its target and is expected to arrive in 2015, though Pluto's status at that time cannot be predicted.[26]

Conclusion

In the past 50 years of spaceflight, what started as planetary exploration has truly become solar system exploration. The larger planetary bodies have become for us truly worlds in themselves, with kinship to our home planet and yet a staggering diversity. With contributions from many of the nations of our home planet, what began as a national effort amid international competition has become, in just 50 years, an enterprise of vast extent, incorporating a diversity of individuals in a common bond—much like the solar system itself.

The distinction between the inner terrestrial planets and the outer gas giants has stood, although the latter are now subdivided between gas and ice giants. Mercury, in the most extreme environment, has shown familiar planetary processes; and with the recent, first return glimpses since the initial flybys of the early 1970s, Mercury is garnering new respect and attention. While early expectations about Venus and Mars were largely overturned, they have remained the closest analogs to Earth and have contributed significantly to an evolutionary perspective that considers all three together. Earth, of course, has been studied most extensively and in a full planetary perspective. These four large, evolved, and differentiated bodies dominated early interest, and extensive study has answered first-order questions and bequeathed an extensive agenda for sophisticated and detailed further study.[27]

While these major bodies now appear conceptually bound to a degree exceeding expectations of the explorer and scientist visionaries before spacecraft voyages, the surprising result of this first half century of exploration is

---

26. Neil deGrasse-Tyson, *The Pluto Files: The Rise and Fall of America's Favorite Planet* (New York, NY: W. W. Norton, 2009). For the Lowell Observatory search and earlier controversies, see William Graves Hoyt, *Planets X and Pluto* (Tucson, AZ: University of Arizona Press, 1980).
27. National Academy of Sciences, SSB, *New Frontiers in the Solar System: An Integrated Exploration Strategy*, chap 1.

new respect and appreciation for a panoply of smaller bodies, from atoms and molecules up to near-planetary-sized, fully differentiated and evolved members, not to mention charged particles and magnetic fields. Discoveries and developing understanding in this area have challenged early and comfortable definitions, and they have also contributed a satisfying, integrative understanding across space and time. As Carl Sagan often said, we are a privileged generation to have witnessed the first steps in the exploration of the solar system.[28]

---

28. Carl Sagan, *Pale Blue Dot: A Vision of the Human Future in Space* (New York, NY: Random House, 1994), preface.

Chapter 18

NASA's Voyages to the Outer Solar System

Michael Meltzer

NASA has become the world leader in space exploration, journeying further and learning more about our surrounding universe than any other group of scientists and engineers on our planet. Beginning in the 1970s with Pioneer, NASA has conducted many voyages to our outer solar system, where planets very unlike our own reside. Employing flybys (Pioneer and Voyager) and orbiters and probes (Galileo and Cassini-Huygens), NASA missions have analyzed these strange bodies and their systems of moons, particles, and fields. These trips generated a variety of benefits, including engineering advances, scientific discoveries, political influence, defense advantages, and less concrete benefits related to ethical questions and national vitality. This paper examines some of the benefits that emerged from two notable journeys beyond the asteroid belt—Galileo and Cassini-Huygens.

Outer Solar System Exploration

In the late 17th century, it became quite the thing for young aristocrats from northern Europe to visit Paris, Venice, Florence, and especially Rome. The more adventurous pushed on further, perhaps to Turkey. Such journeys became rites of passage for those aspiring to positions of influence and national leadership, and they were the culmination of years of classical education. In those days, travel was arduous, costly, and more than a little risky. Travelers who participated in such trips were expected to return home with knowledge and understanding from their exposure to momentous sights as well as souvenirs from the exotic places. These would allow the folks back home to experience their adventures vicariously.[1]

---

1. Jean Sorabella, "The Grand Tour," Heilbrunn Timeline of Art History (New York, NY: The Metropolitan Museum of Art, October 2003), available at *http://www.metmuseum.org/toah/hd/grtr/hd_grtr.htm.*

Such journeys gave concrete form to ideas regarding other places and the different ways that people lived, helping to foster new, more worldly views and ideals. These voyages represented travel for the sake of curiosity and learning. They helped those fortunate enough to partake to understand better the realities of the world. The cross-border friendships and alliances made on these wanderings helped the travelers when they assumed positions of influence. These life-changing, rite-of-passage explorations flourished for centuries, until the advent of mass railroad transit in the 1840s made such trips no longer the exclusive domain of men of means, and thus not nearly so special or important.

Ambitious voyages of discovery are still conducted, albeit with greatly extended boundaries of travel. Some of these journeys explore other celestial bodies. Soon after NASA was established in 1958, its explorations into space gave its engineers and scientists positions of prominence in the world. They became pioneers in a new age of exploration. One area of particular interest for many space scientists was the outer solar system. NASA has conducted a multifaceted exploration of this region, where planets and satellites markedly different from our own reside. The Agency began its investigations in the 1970s with the Pioneer and Voyager flyby spacecraft, followed by the orbiters and probes of Galileo and Cassini-Huygens, the milestone missions that are the main subjects of this paper.

NASA's Early Voyages to the Outer Solar System

During the mid-1960s, NASA moved from conceptual studies of journeys to the outer planets of our solar system to actual mission planning activities. President Lyndon B. Johnson strongly supported ambitious space exploration, viewing "the quest to capture outer space from the evil hands of the Soviets"[2] as a 20th-century equivalent to the Roman Empire's road system or the British Empire's mighty navy in the 19th century.

NASA took a key step toward outer planet exploration in January 1968, when it kicked off planning activities for two missions using Pioneer-series spacecraft that would eventually fly by Jupiter and points beyond. These missions functioned in some sense as scouts for future, more intensive outer

2. Joan Hoff, "The Presidency, Congress, and the Deceleration of the U.S. Space Program in the 1970s," in *Spaceflight and the Myth of Presidential Leadership*, ed. Roger D. Launius and Howard McCurdy (Urbana and Chicago, IL: University of Illinois Press, 1977), p. 93; Michael Meltzer, *Mission to Jupiter: A History of the Galileo Project* (Washington, DC: NASA SP-2007-4231, 2007), p. 16.

planet efforts such as Galileo and Cassini-Huygens, which would orbit the target planets many times and send probes down to study some of the bodies in greater detail.[3]

Pioneer 10, the first outer solar system endeavor, launched in March 1972 and flew within 120,000 miles of Jupiter in December 1973, studying the planet's satellites as well as its magnetic fields, atmosphere, hydrogen abundance, radiation belts, aurorae, and radio waves. In December 1974, Pioneer 11 passed much closer—only 21,000 miles—by Jupiter, and then it went on to Saturn. One of the most important sets of observations for future voyages that was made by the Pioneer missions determined that Jovian radiation levels were not as severe as had been feared, presenting less danger to spacecraft operating systems and scientific experiments than was previously thought.[4] These findings helped in developing an appropriate design for the Galileo orbiter and atmospheric probe that would eventually visit Jupiter.[5] Pioneer findings at Jupiter and Saturn also helped set the scientific agendas for Galileo and Cassini-Huygens. For example, Pioneer discovered that Saturn's magnetic axis almost exactly lines up with its rotational axis, dramatically different from both Earth's and Jupiter's, "where magnetic and geographic poles lie far apart,"[6] and the reasons for this difference are currently being researched by the Cassini orbiter team.[7]

Following the Pioneer launches were those of two Mariner-class vessels, Voyagers 1 and 2, both lifting off in 1977 and flying by Jupiter in 1979. The mission team designed Voyager 1's trajectory so that when the craft reached Saturn the following year, it flew by the planet's largest moon, Titan, an unusual satellite that possessed a thick, opaque atmosphere. Voyager 2 flew

---

3. Meltzer, *Mission to Jupiter*, p. 21.
4. "Pioneer 10," *National Space Science Data Center National Catalog*, Spacecraft ID: 72-012A, available at *http://nssdc.gsfc.nasa.gov/nmc/spacecraftDisplay.do?id=1972-012A* (accessed 24 January 2008); "Pioneer 11," *National Space Science Data Center National Catalog*, Spacecraft ID: 73-019A, available at *http://nssdc.gsfc.nasa.gov/nmc/spacecraftDisplay.do?id=1973-019A* (accessed 29 January 1998); Meltzer, *Mission to Jupiter*, pp. 28–29.
5. Craig B. Waff, "Jupiter Orbiter Probe: The Marketing of a NASA Planetary Spacecraft Mission," (paper presented at American Astronomical Society meeting, session "National Observatories: Origins and Functions (The American Setting)," Washington, DC: American Astronomical Society, 14 January 1990), pp. 3–4.
6. Henry C. Dethloff and Ronald A. Schorn, Voyager's Grand Tour (Washington, DC: Smithsonian Books, 2003), p. 179.
7. Andy Ingersoll, interview by author, Rome, Italy, 12 June 2008.

by Saturn during the next year and then went on to Uranus and Neptune, reaching them in 1986 and 1989, respectively.[8]

As with Pioneer, the Voyager mission "opened up a whole new set of [research] questions"[9] for future outer solar system voyages. For instance, when Voyager flew by Saturn's moon Titan, "it saw an orange ball."[10] In other words, it saw an atmosphere opaque to visible light that hid characteristics of the moon's surface. So mission planners included among Cassini-Huygens's instruments a sophisticated radar whose long wavelengths would be able to pass through Titan's atmosphere and image its surface. And like Voyager, Cassini-Huygens will teach the next mission to Saturn some of the questions it needs to ask and some of the instruments that will be needed on its journey.

The Galileo Mission to Jupiter

The two Voyager spacecraft carried out flyby visits to all of the outer planets known to ancient astronomers. The Galileo mission began a different phase in the study of the outer solar system: a deeper, more thorough exploration featuring extended orbital visits; repeated, close flybys of satellites; and use of atmospheric probes, resulting in in-depth analyses of planetary system characteristics. Galileo was not a planetary mission in the traditional sense of the term, which typically referred to efforts focusing on specific targets. Galileo's objectives were much broader, encompassing a holistic analysis of the entire Jovian system of satellites, primary planet, magnetic field, and particle distributions.[11]

The Cassini-Huygens Mission to Saturn

In a March 2007 talk to the TED Corporation, Carolyn Porco, Cassini-Huygens Imaging Science Team leader, asserted that the mission to Saturn, as well as other robotic space journeys such as Galileo, were "part of a bigger human journey: a voyage . . . to get a sense of our cosmic place, to understand something of our origins and how we living on Earth came to be." Cassini-Huygens

---

8. Craig B. Waff, former NASA contract historian, telephone conversation with author, 22 May 2000; Craig B. Waff, "The Struggle for the Outer Planets," *Astronomy* 17, no. 9 (1989): 49–51; Jet Propulsion Laboratory, "Planetary Voyage," *Voyager—Celebrating 25 Years of Discovery*, available at *http://www.jpl.nasa.gov/multimedia/voyager_record/* (accessed 24 January 2008); National Space Science Data Center, "Voyager Project Information," available at *http://nssdc.gsfc.nasa.gov/planetary/voyager.html*, updated 11 March 2008 (accessed 3 September 2009); Dethloff and Schorn, *Voyager's Grand Tour*.
9. Trina Ray, interview by author, JPL, 22 October 2008.
10. Ibid.
11. Meltzer, *Mission to Jupiter*, p. 3.

has made a noble attempt to do this. Built on the shoulders of the Pioneer, Voyager, and Galileo endeavors, Cassini-Huygens has to date traveled well over 2 billion miles, carrying 18 sophisticated scientific experiments and a probe that it sent through the atmosphere and to the surface of Titan, a Saturnian moon larger than the planet Mercury. Moreover, Cassini-Huygens took, as did Galileo, important steps in answering a question central to our weltanschauung, our perception of the universe: are we, the human race and all that is alive on our planet, alone in the cosmos? Or are we joined by other forms of life on other worlds?[12]

Scientific Achievements of the Galileo and Cassini-Huygens Expeditions

The scientific discoveries that emerged from the Galileo and Cassini-Huygens missions helped to rewrite textbooks on the outer planets. These expeditions were products of a new template for mission organization that involved increased international cooperation and shared investments in space exploration.[13] A few examples from the long list of notable discoveries made on these missions included finding the following:

- a probable warm-water, salty ocean under the ice of Jupiter's Europa;
- incredible volcanism on Jupiter's Io;
- an Earth-like landscape on Saturn's Titan; and
- water jets on Saturn's Enceladus.

Europa, a Jovian moon, was one of those discovered by the Renaissance scientist Galileo. The spacecraft named Galileo sent back various types of data from Europa that, over time, strengthened the case for a salty, warm-water ocean under the moon's icy surface (see figure 1). Imaging data revealed ice floes on the surface that appeared to have moved and rotated, like they would if they were floating on soft ice or water. Images also showed that Europa has relatively few craters, suggesting a young surface, possibly renewed by upwellings of ice or water from below. Mass spectrometer data showed the presence of salt deposits on the moon's most recently disrupted areas. A brine ocean under these areas was a likely source for such salts. Gravity measurements indicated that the moon's surface layers had a low density similar to that of either liquid

12. Carolyn Porco, "Fly Me to the Moons of Saturn" (speech to the TED Conference, Monterey, CA, 7 March 2007), available at *http://www.ted.com/talks/view/id/178*.
13. Michael Meltzer, *Meeting the Lord of Rings: A History of the Cassini-Huygens Mission to Saturn* (Washington, DC: NASA, under development).

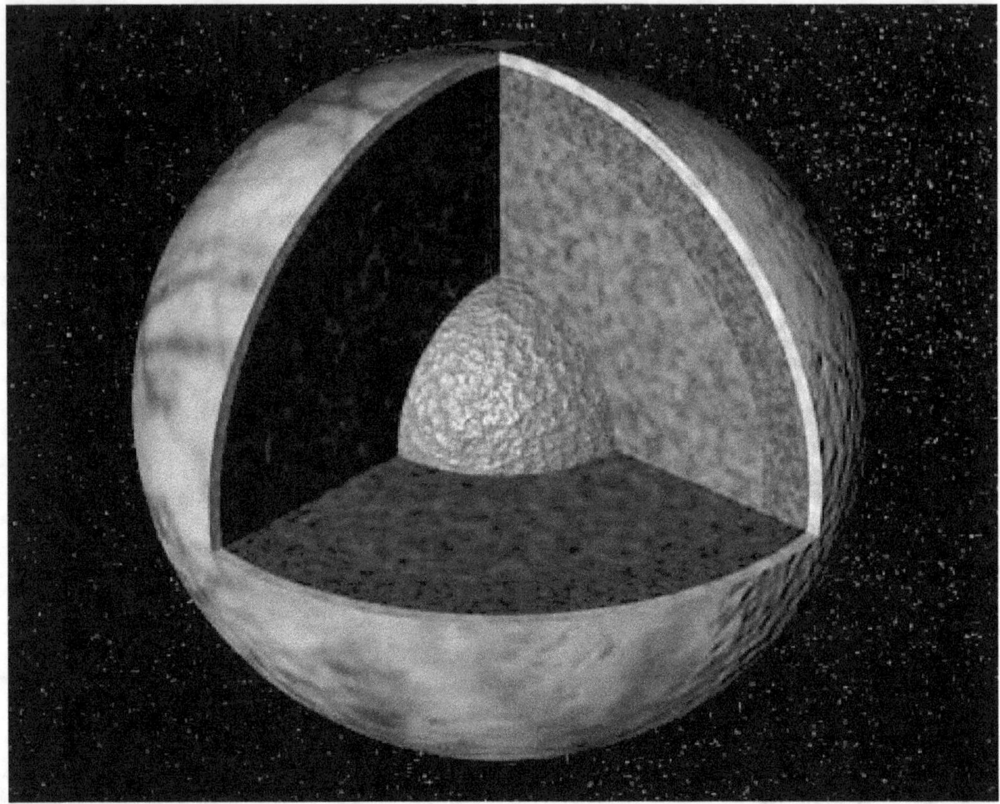

Figure 1: An ocean under the icy surface of the Jovian moon Europa? *NASA Image PIA01130*

or frozen water. But the strongest evidence came from magnetometer readings showing that the direction of Europa's magnetic field depended on the local direction of Jupiter's field, which indicated that the Europan field was likely induced by the larger field from the mother planet. Such an occurrence could best be explained by the presence of electrically conductive material, most likely salt water, just under the moon's surface.[14] What makes the existence of a salty ocean especially interesting is that the three main factors exobiologists seek when searching for extraterrestrial life would be present on Europa: liquid water, organic chemicals, and energy.

Another moon of Jupiter discovered by the scientist Galileo was Io (see figure 2). It generates such wonders as "gigantic lava flows and lava lakes, and towering, collapsing mountains. Io makes Dante's Inferno seem like another

---

14. Michael Meltzer, *Mission to Jupiter: A History of the Galileo Project* (Washington, DC: NASA SP-2007-4231, 2007), pp. 259–264.

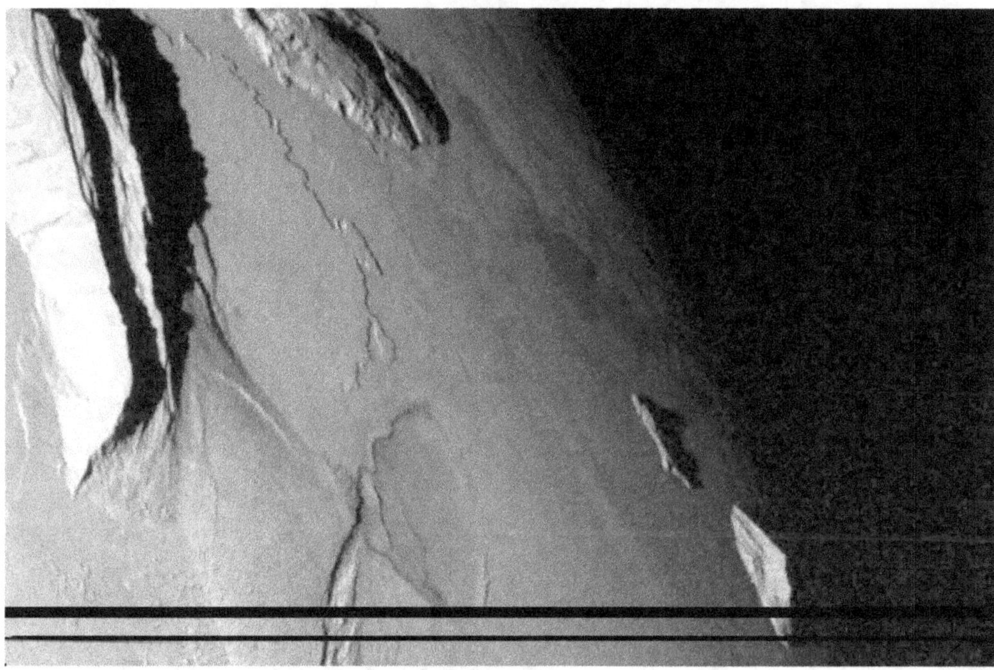

Figure 2: Io mountains at sunset (February 2000). Mongibello Mons is the jagged ridge at the left of the image, rising a mere 23,000 feet above the plains of Io, far lower than the satellite's tallest peaks. *NASA/JPL/University of Arizona/Arizona State University Image PIA03886*

day in paradise."[15] Earth has not seen volcanism like this for 15 million years. Io thus provides "the next best thing to traveling back in time to Earth's earlier years. It gives us an opportunity to watch, in action, phenomena long dead in the rest of the solar system."[16] The scale of these occurrences is enormous, especially considering the relatively small size of the moon Io. For instance, a lava lake near the moon's Pele volcano is estimated to be *100 times* larger than those found in Hawaii.[17] And the heat generated by Io's Loki, an even larger volcano than Pele, exceeds that from *all* of Earth's active volcanoes combined.

Although Io is tiny compared to Earth, it thrusts parts of itself up far higher. On Thanksgiving night, 1999, the Galileo spacecraft was incredibly lucky to observe a fountain of sparkling molten rock shooting more than a mile above the satellite's surface. Such fountains on Earth rarely exceed a

15. Jane Platt, "Jupiter's Moon Io: A Flashback to Earth's Volcanic Past," JPL news release no. 99-097, 19 November 1999.
16. Ibid.
17. Ibid.; Ron Cowen, "Close Encounter: Galileo Eyes Io," *Science News* (11 December 1999): 382.

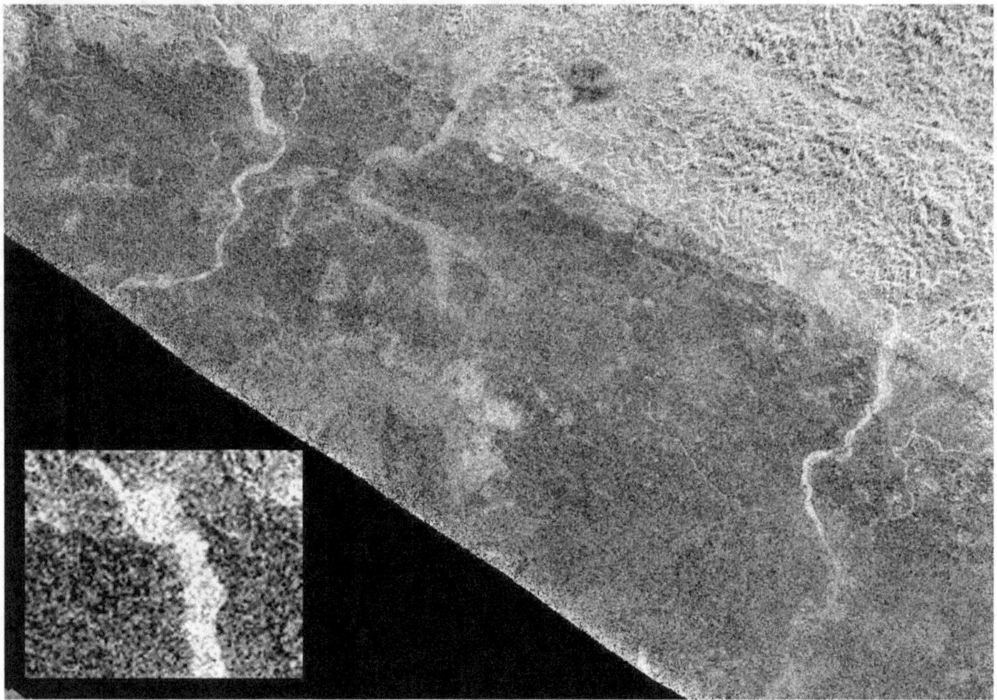

Figure 3: "Riverbeds—this was for me so striking . . . a network of channels. I'd never seen this before on a planet or moon," said former Cassini-Huygens Project Scientist Dennis Matson (interview by author, Rome, Italy, 9 June 2008). *NASA Image PIA10956*

few hundred yards.[18] And Io's mountains soar up to 52,000 feet, *over 4 miles higher* than Mount Everest.[19]

When Cassini-Huygens peered through the mists of Saturn satellite Titan's atmosphere, it found more surprises: riverbeds, alluvial plains, lakes, and coastlines strangely like those on Earth (see figures 3 and 4), except with the water replaced by hydrocarbons. Former Cassini-Huygens Project Scientist Dennis Matson had never seen such features before on another moon or a planet other than Earth,[20] while Huygens probe Project Scientist Jean-Pierre Lebreton found sections of Titan's coasts reminiscent of the Cote D'Azur along the French Riviera.[21]

---

18. JPL, "Galileo Sees Dazzling Lava Fountain on Io," JPL Media Relations news release, 17 December 1999.
19. Meltzer, *Mission to Jupiter*, p. 242.
20. Dennis Matson, interview by author, Rome, Italy, 9 June 2008.
21. Jean-Pierre Lebreton, interview by author, Rome, Italy, 9 June 2008.

Figure 4: Titan coastlines reminiscent of those on Earth. *NASA Image PIA09211*

Some of Titan's lakes had the round shapes of those on Earth formed from collapsed calderas (see figure 5), such as Crater Lake in Oregon. The largest liquid hydrocarbon body observed on Titan was the size of the Caspian Sea, the biggest enclosed body of water on Earth; the next largest Titan body of liquid was about the size of Lake Superior. But Titan is six times smaller than Earth. In terms of percentage of the surface covered, the largest hydrocarbon body on Titan is equivalent to Earth's Bay of Bengal.[22]

Since the Voyager flybys of the early 1980s, planetary scientists knew there was something unusual about Saturn's ice-covered satellite Enceladus, a body only 300 miles across whose width is smaller than that of Arizona.[23] Then the Cassini orbiter conducted a series of close flybys starting in 2005, spying places where the satellite surface had recently cracked and contorted, and

22. NASA/JPL/Space Science Institute, "Exploring the Wetlands of Titan," available at *http://ciclops.org/view/2631/Exploring_the_Wetlands_of_Titan?js=1, 15 March 2007* (accessed 17 January 2009).
23. Joanne Baker, "Tiger, Tiger Burning Bright," *Science* 311 (10 March 2006).

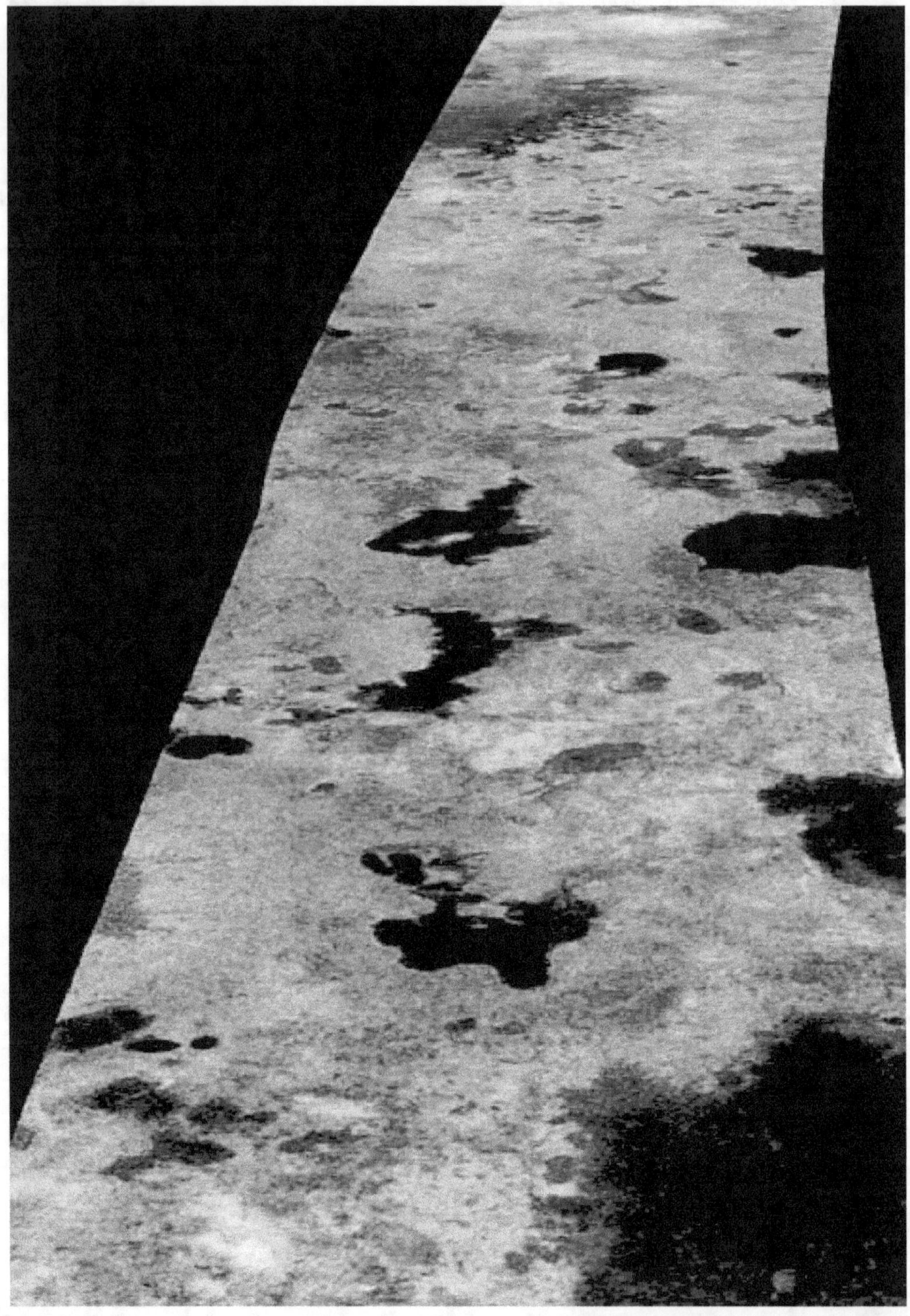

Figure 5: Circular lakes on Titan resembling those on Earth formed from collapsed calderas. *NASA Image PIA09112*

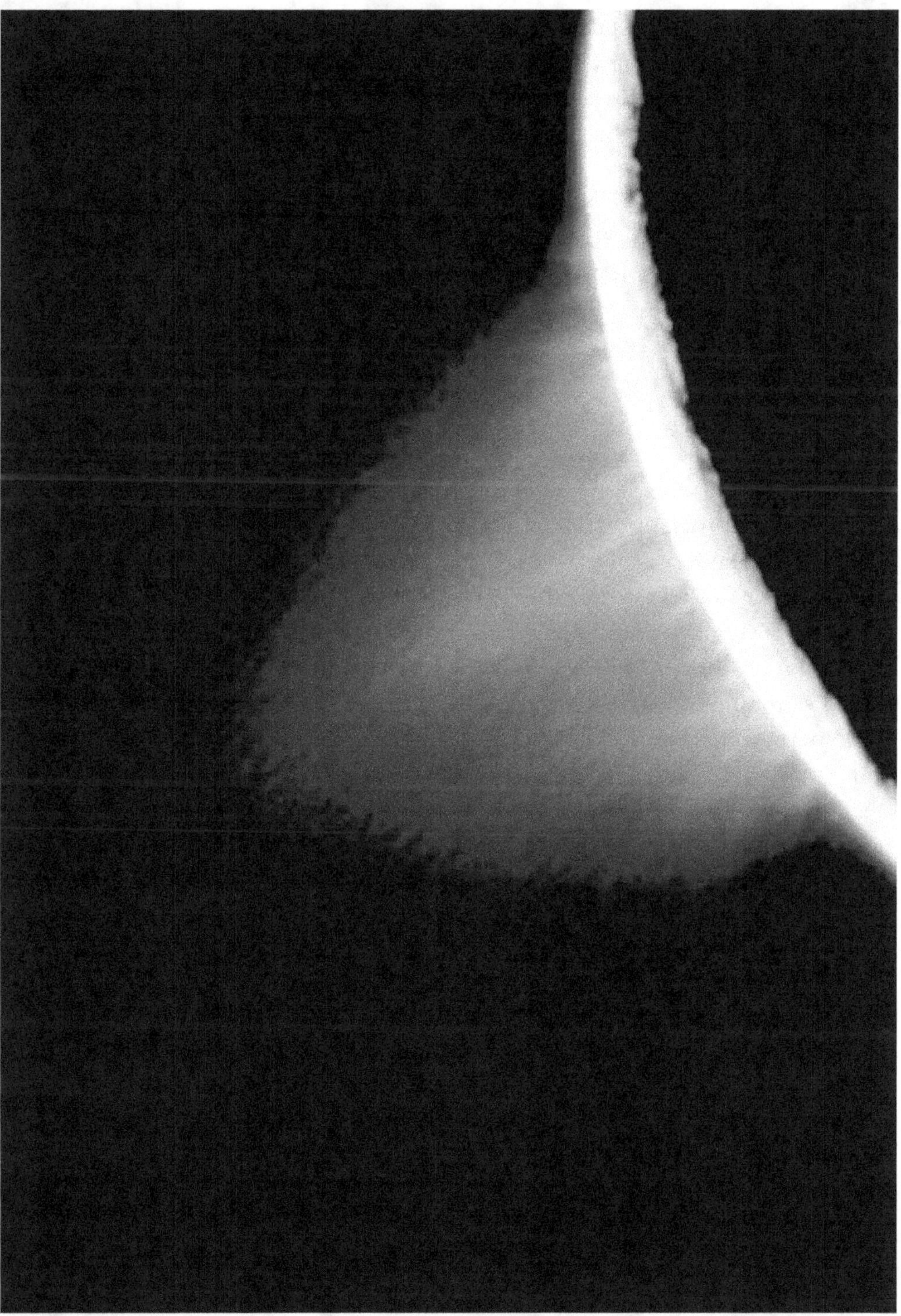

Figure 6: Water vapor and ice jets erupting from giant fissures in Enceladus's southern polar region. *NASA Image PIA08386*

water vapor and ice crystals were spewing out (see figure 6). How could such a small body, so far from the Sun, generate enough energy for such geologic activity? Neither of the usual sources of planetary heat—decay of radioactive elements in deep rock or deformations caused by orbital interactions with Saturn and other moons—appeared to be great enough, according to Caltech planetary physicist David Stevenson, who added, "Enceladus is somehow special."[24] While no one is saying there is life on Enceladus, some basic conditions of life seem to be present inside the moon: liquid water, sources of energy, and organic materials.

Erosion of the Boundary Between Robotic and Human Missions

Different factions of NASA have debated, since the first years of the Agency, the relative merits of human versus robotic exploration. Sending people into space and then to the Moon gave the United States enormous cachet in terms of political and social prestige and showed the USSR how technically capable we were. As Galileo Project Manager John Casani said, "Having heroes is an important part of any enterprise like this."[25] But human exploration requires a considerable part of the spacecraft filled with life-support equipment, cutting down on the amount of scientific instrumentation in the craft and thus severely limiting the observations and experiments able to be performed. Many in the planetary science community have thus favored more robotic and fewer human missions.

But with the advent of technology that allows humans sitting on Earth to see, hear, and nearly feel what the spacecraft does, as well as finely controlling its movements, the line between robotic and human exploration is becoming somewhat blurred. One aspect of the Galileo and the Cassini-Huygens missions was that, in a way, *we* were up there. The crews sat in a control room on Earth rather than in the space vessel, but they still saw and measured, and to some extent experienced, what the vessel did. During the Cassini-Huygens mission, thousands of people around the world shared, in near-real time through their Internet connections, the highs and lows of the space voyage. The connections between us and our spacecraft are now such that, in the words of Galileo Project Scientist Torrence Johnson, "We are *all* standing on the bridge of the

---

24. Richard A. Kerr, "Cassini Catches Mysterious Hot Spot on Icy-Cold Enceladus," *Science* 309 (5 August 2005): 859–860.
25. John Casani, telephone interview by author, 29 May 2001.

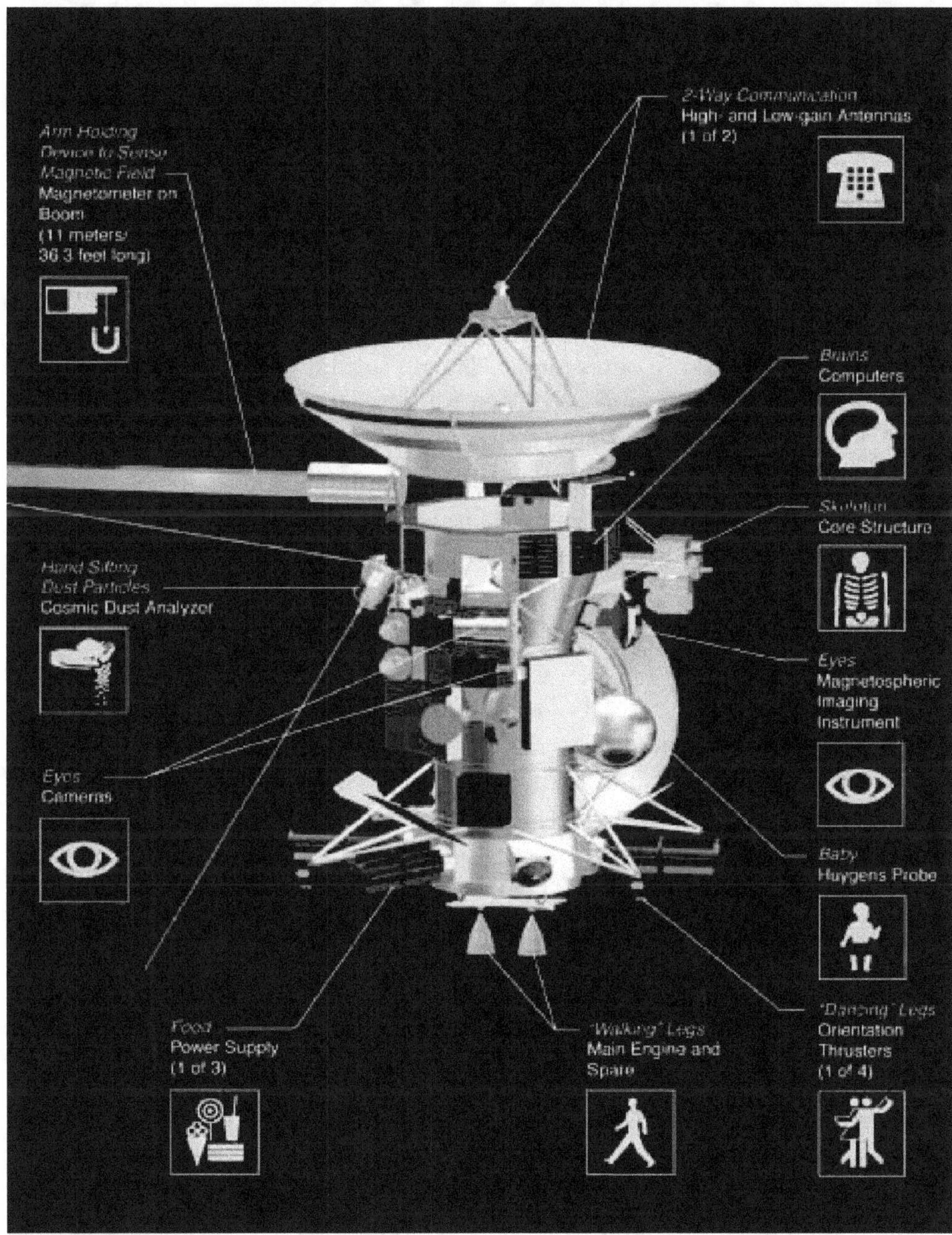

Figure 7: This figure identifies the spacecraft's cameras and magnetospheric imagers as its *eyes*, the cosmic dust analyzer as its *hands*, the ship's computer as its *brain*, the main engine and thrusters as its "walking" and "dancing" *legs*, the radioisotope thermoelectric generator (RTG) electrical output as its *food*, the Huygens probe as the *baby* it will birth, and so on. But envisioning the craft in this way is more than simply entertaining. It's a recognition that the Cassini space vehicle really has become our eyes and hands at Saturn, the means by which we humans now visit distant locations that our corporeal bodies cannot easily reach. The line between human and robotic space exploration has indeed become blurred. *NASA/JPL*

Starship Enterprise."[26] Our space vessels have become *our* hands, ears, and eyes examining distant worlds (see figure 7).

Galileo and Cassini-Huygens Engineering Achievements

The scientific success of NASA's outer solar system missions was only possible because of the resounding technical success that mission engineers achieved in the development of their spacecraft. In designing the Galileo space vehicle for comprehensive exploration of Jupiter's atmosphere, physical environment, and satellites, NASA engineers and technicians learned how to better equip robotic vessels for journeys to the outer planets lasting several years, so that the vessels would be more reliable and perform at a higher level. Key developments in this regard included more durable technologies with greatly extended lifetimes and the ability to operate independently of mission control. Deep space vessels have become models of resilience and tough construction, able to withstand the impacts of vacuum, far higher *and* lower temperatures than those found on Earth, intense radiation belts, and interplanetary storms of tiny particles.

As successful as the Galileo craft was, it experienced many problems with some of its mechanical systems, notably the moving parts in its antenna and tape recorder. Among the most impressive engineering achievements of the mission were those that occurred *after* liftoff, when operational problems had to be solved from millions of miles away. The failure of Galileo's main antenna to open, for instance, required NASA staff to develop a way of using the spacecraft's small backup antenna, which had a very modest transmitting capability. NASA personnel had to develop new software to compress data and thus enhance transmission rates; the on-board tape recorder needed to assume a more important data-storage role than originally planned; and the Deep Space Network of antennas on Earth required improvements that would augment the amount of useful data it could discern from the faint signal the spacecraft was able to send. Even with the Galileo orbiter's extremely impaired transmission capability, mission engineers still found ways to work around the problems so that the scientists could meet most of their goals.[27]

Years of long space missions have taught NASA engineers the myriad ways in which complex subsystems can fail. NASA built the Cassini-Huygens vehicle

---

26. Torrence V. Johnson, telephone interview by author, 12 January 2006.
27. Michael R. Johnson, "The Galileo High Gain Antenna Deployment Anomaly" (NASA TR N94-33319 in the proceedings of the 28th Aerospace Mechanisms Symposium, Lewis Research Center, May 1994), pp. 360, 364; Meltzer, *Mission to Jupiter*, chap. 7.

with far fewer moving parts than Galileo, greatly reducing the mechanical problems and component failures that plagued the mission to Jupiter. NASA's expeditions have enabled an evolution from the *mechanical* to the *nonmechanical* that included replacing the following:[28]

- traditional toggle switches with solid state devices;
- dish antennas that used to open like umbrellas with fixed-in-place antennas;
- reel-to-reel data storage tape recorders with solid-state digital devices capable of recording *and* playing back simultaneously; and
- spinning gyros, which maintained a vessel's orientation, with "hemispherical resonator gyros" made of radiation-resistant pieces of machined quartz,[29] whose lifetimes can *exceed 10 million* hours.

Advances in space vessel *autonomy* have enabled the craft to operate longer and perform more complex tasks while out of touch with Earth. By the time Cassini-Huygens reached Saturn, for instance, the vessel had traveled so far from Earth that radio communication between it and mission control, limited by the speed of light, had grown quite slow. The spacecraft had to perform its sensitive Saturn orbit insertion maneuver by itself, guided only by its sensors, computers, and memory. Developing such independent space vessels that can "operate for decades, continue to send back data without failing, and be smart enough to take care of themselves out of communication with the Earth"[30] was a major step forward for expeditions far from the mother planet.

Societal Benefits Gained from Outer Solar System Tours

Societal benefits gained from trips to the outer planets included a variety of new technologies for U.S. industries as well as improved defense capabilities for fulfilling important military goals. Some of the novel technologies developed for the Cassini-Huygens outer solar system mission that have already been spun off to U.S. industries include the following:[31]

---

28. "Scientific Benefits of the Cassini Mission," Insert A in "Integrated JPL General Comments—Cassini Draft Environmental Impact Statement (10/1 9193 SAIL Draft)," in Richard J. Spehalski memos, November 1993, JPL Cassini Technical Library (CASTL).
29. Don Barteld, "Exclusive Northrop Grumman Resonating Gyro Achieves 10 Million Operating Hours in Space," Northrop Grumman Electronic Systems news release, 19 May 2008, available at *http://www.irconnect.com/noc/press/pages/news_releases.html?d=142997* (accessed 7 January 2009).
30. Torrence Johnson, telephone interview by author, 31 July 2001.
31. "Scientific Benefits of the Cassini Mission," Insert A in "Integrated JPL General Comments—Cassini Draft Environmental Impact Statement (10/1 9193 SAIL Draft)," in Richard J. Spehalski memos, November 1993, JPL CASTL.

- *Computerized resource trading system* to resolve conflicting cost, data rate, and electrical power needs for the spacecraft's science instruments and other subsystems. This tool has been utilized by an environmental agency for market-based regulation of air pollution.
- *Integrated circuit advances* such as new application-specific integrated circuit (ASIC) parts that reduce mass and replace one hundred or more traditional chips.
- *Solid-state power switch* for eliminating transient current surges and extending part lifetimes and efficiencies.
- *Solid-state data recorder* with no moving parts (mentioned above) that has seen use in a variety of fields, from aerospace to the entertainment industry, and has found applications in consumer electronics.
- *Inertial reference unit "hemispherical resonator gyros"* (also mentioned above) with greater reliability and less vulnerability to mechanical failure because they contain no moving parts.

Other societal benefits emerging from NASA's expeditions to the outer planets include better knowledge of how to design and build spacecraft with augmented *defense* capabilities. Of major importance for defense as well as scientific applications is space vessel *autonomy*—having the ability to conduct operations independently for long periods of time, even if radio contact with Earth is broken. Autonomy could be critical if, for instance, ground control facilities are damaged or shielded from the craft during a war. *Durability* of the craft also has high value. Of particular use is *radiation hardening,* enabling a vessel to survive high-energy particle hits, intense gamma rays, and so on. Radiation hardening helps protect defense-oriented spacecraft in a military situation, as well as science-oriented space vessels in an environment such as Jupiter's radiation belts.

Political Influence

Successful U.S. missions to far away targets have conferred upon us an influential voice in the international political arena. Galileo Project Manager John Casani expressed it this way—our noteworthy civilian space exploration advances send a strong message to other nations, particularly potentially hostile ones, of what the U.S. could do if we turned our peaceful mission capabilities to militaristic purposes. This gives us considerable international political authority.[32] Having such technological capabilities in space is vital in the following areas:

32. John Casani, telephone interview with author, 29 May 2001.

- Helping to ensure that control of space is not used by other nations to threaten us.
- Preparing us to use our access to space to defend ourselves if necessary.

The 17th-century voyages to far-off places discussed earlier in this paper enabled well-heeled travelers to create friendships, alliances, and business relationships across various national borders that could be used to their advantage at a later time, in both the political and business arenas. And so it is with NASA's outer solar system expeditions. The Galileo project and, to an even greater extent, the Cassini-Huygens mission were especially good examples of this, for their development led to a range of international alliances that grew from multinational partnerships and working relations. Sometimes, it was beneficial to the mission that international partners were on different sides of the world. This was often the case on Galileo. According to Torrence Johnson, Galileo "was one of the first missions in which the analysis of data more or less continued around the clock, around the globe. We'd wake up in the morning and hear that our colleagues in Berlin had processed some images overnight and brought new data to the table—and we could look at it immediately, while they had a chance to sleep."[33]

The cross-border relationships on Cassini-Huygens were much more complex, and they have strongly bound together colleagues working on both sides of the Atlantic who have depended on each other in a variety of different ways. NASA furnished the mission's launch vehicle and had the main responsibility for developing the Cassini orbiter, while ESA oversaw development of the Huygens probe. But many sections of various science instruments and engineering systems on the orbiter were developed in Europe, while the United States furnished some systems for the probe. As examples, the complex orbiter antenna system was designed and constructed by an Italian company, and two of the major Huygens probe science instruments were built in the United States. In addition, key Huygens probe subsystems, such as its batteries, were provided by the United States, and some of the Huygens probe's parachute testing was performed in a U.S. wind tunnel. These intimate connections and dependencies between the United States and many European nations are viewed by participating governments as important mechanisms for maintaining friendly, cooperative international relations.

---

33. Torrence Johnson, foreword to *Mission to Jupiter*, by Michael Meltzer.

Maintaining the Vitality of a Nation

James Michener once said when addressing the U.S. Senate Subcommittee on Science, Technology, and Space that "it is extremely difficult to keep a human life or the life of a nation moving forward with enough energy and commitment to lift it into the next cycle of experience . . . . There are moments in history when challenges occur of such a compelling nature that to miss them is to miss the whole meaning of an epoch. Space is such a challenge."[34]

Voyages of discovery, such as Galileo and Cassini-Huygens, constituted strong responses to the challenge offered by space. Both of these missions altered the way we view our surroundings and breathed new vitality into our quest to understand ourselves and our universe.[35] NASA's outer planet tours have been a source of pride and inspiration for millions of Americans and have demonstrated our country's position of preeminence in space. NASA's planetary exploration efforts have also served as magnets attracting talented students into research-oriented, technological careers, thus renewing our nation's vigor and ability to push the envelope of scientific endeavor.

Ethical Benefits Emerging from Space Exploration

Questions that a 17th-century traveler visiting other lands might have asked include the following: how do people in those lands live, and what relationship shall I seek with them? NASA's *robotic* travelers to other heavenly bodies seek answers to questions such as these: *do* other beings live on those bodies? And if so, then how shall spacefaring nations treat them?

Missions such as Galileo and Cassini-Huygens confronted fundamental ethical issues regarding these questions, including the following:

- What procedures should we include on expeditions to bodies that might harbor life?
- Shall we protect any possible lifeforms at all cost, and if so, for how long?

By confronting such issues, our society comes face to face with its core ethics, including the value we place on life and on different *types* of life. Outer solar system voyages elicit some unique ethical issues, but they also resemble issues that people of conscience and humanity have wrestled with for the last several millennia. Human beings have always longed to understand our place

---

34. James A. Michener, "Space Exploration: Military and Non-Military Advantages" (speech delivered before the U.S. Senate Subcommittee on Science, Technology, and Space, Washington, DC, 1 February 1979), published in *Vital Speeches of the Day* (Southold, NY: City News Publishing Co., 15 July 1979).
35. Meltzer, *Mission to Jupiter*, p. xxii.

in the cosmos. This inquiry has been reexamined and reformulated through the centuries as our understanding of space science progressed and our technology for observing celestial bodies and making measurements improved.

A basic question we ask when we send a vessel deep into space to bodies never before explored is, are we on Earth the lone forms of life in the universe? This is not a new question. In 300 BC, for instance, the Greek philosopher Epicurus wrote to Herodotus regarding the infinite number of worlds that existed, arguing that he saw no reason why these bodies "could not contain germs of plants and animals and all the rest of what can be seen on Earth."[36] Epicurus was expressing his human curiosity to understand whether forms of life exist beyond the confines of our planet. The Roman philosopher Lucretius conveyed similar musings when he said, "Confess you must that other worlds exist in other regions of the sky, and different tribes of men, kinds of wild beasts."[37]

The debate on how to conduct interplanetary exploration tends to polarize into three different camps: *preserving* extraterrestrial environments in unchanged states; *stewarding* other bodies in a way that will maximize the benefits to all parties concerned; and *exploiting* these bodies, treating them as resources that can greatly aid our species.

The *preservation ethic* suggests that human action in nature should be minimized, and this translates to the imperative to leave an extraterrestrial body unaltered—"to neither enhance its environment for the indigenous biology, if any, nor to introduce life from Earth."[38] The perceived need to preserve a body's biosphere indefinitely, beyond the period of biologic exploration, arises from a belief in the inherent worth of any life present, no matter how humble. This is a view that was eloquently expressed by Albert Schweitzer.

The great theologian and physician Albert Schweitzer wrote about topics very relevant to space exploration in his Reverence for Life ethic, which held that "It is good to maintain and to encourage life; it is bad to destroy life or to obstruct it."[39] Schweitzer's German term for this ethic was "Ehrfurcht vor dem Leben," which is also translated as "to be in awe of the mystery of life."

---

36. Paul Clancy, Andre Brack, and Gerda Horneck, *Looking for Life, Searching the Solar System* (New York, NY: Cambridge University Press, 2005), pp. 180–181.
37. Ibid., p. 181.
38. Richard O. Randolph, Margaret S. Race, and Christopher P. McKay, "Reconsidering the Theological and Ethical Implications of Extraterrestrial Life," *Center for Theology and Natural Sciences (CTNS) Bulletin* 17, no. 3 (Berkeley, CA: summer 1997): 1–8.
39. Albert Schweitzer, "The Ethics of Reverence for Life," in *The Philosophy of Civilization*, trans. C. T. Campion (Buffalo, NY: Prometheus, 1987), chap. 26, available at *http://www1.chapman.edu/schweitzer/sch.reading1.html.*

Schweitzer strongly defended his belief that we should respect all "wills to live" as we do our own, and he thought that we are truly ethical only when we help all life that we are able to and shrink from harming anything that lives.[40]

What is relevant to the search for life on other planets is that Schweitzer refuted *sentiency* (consciousness) as the discriminator for protecting life, favoring instead *conativity*—having only the minimal characteristics of life. These bare essentials were, in Schweitzer's mind, sufficient for the organism to be a thing of value and thus nurtured.[41] According to Schweitzer's ethic, if our space vessels find even the crudest one-celled organisms struggling to survive on another heavenly body, we need to protect them as best we can.

Schweitzer was hardly the first to articulate the value of all forms of life. The Bible stated this belief as well. The first chapter of Genesis affirmed grass, herbs, trees bearing fruit, ocean life, and every living creature that crept across Earth as "kee-tov"[42] (good), and even "tov m'od"[43] (very good). A biblical view of space exploration ethics might thus recognize the inherent value of any living creature we find on another world.

The conservationist Aldo Leopold took a step beyond simply identifying all lifeforms as good things to be preserved. He recognized that no living creature exists unconnected to other such creatures. In his essay "The Land Ethic," he presented reasons, both ethical and aesthetic, for *why* an organism had value: "A thing is right when it tends to preserve the integrity, stability, and beauty of the biotic community. It is wrong when it tends otherwise."[44] In this statement, he envisioned the connected life skeins and interdependence necessary to maintain a bionetwork's health. In other words, he argued that an organism's importance to the web of community that surrounded it was a reason to protect it.

A key question that arises when preservation of a solar system body is discussed is, what exactly has intrinsic worth on the body? Is it only biological life, or should all the natural attributes of the body be preserved, including its rocks and its dirt? I. Almar, in a paper presented at the 34th COSPAR Scientific Assembly, expressed the concern that damage caused by

---

40. A. Schweitzer, *Out of My Life and Thought* (Baltimore, MD: Johns Hopkins University Press, 1998).
41. Lucy Goodwin (reviewer), "J. Baird Callicott, 'Moral Considerability and Extraterrestrial Life,'" *Reviews of Ethics and Animals Literature* I (fall 1997), available at *http://core.ecu.edu/phil/mccartyr/Animals/Real97/goodwin.htm* (accessed 3 September 2009).
42. J. H. Hertz, ed., *The Pentateuch and Haftorahs* (London, U.K.: Soncino Press, 1972), pp. 3–4.
43. Ibid., p. 5.
44. Aldo Leopold, "The Land Ethic," in *A Sand County Almanac and Sketches Here and There*, by Aldo Leopold (Oxford, U.K.: Oxford University Press, 1949), pp. 201–226, available at *http://www.luminary.us/leopold/land_ethic.html.*

any human intervention on a lifeless world would be irreversible. One possible reason for protecting the lifeless space environment was its scientific aspect—areas and objects could exist of the highest scientific priority on different celestial bodies. As an example of this, much can be learned about volcanism and the impacts of tidal forces by studying the Jovian moon Io, which is almost certainly lifeless.[45]

A logical extension of the *preservation ethic* is to reduce the risk of contaminating an extraterrestrial body by visiting it solely with *robot spacecraft*. Sending human explorers to Jupiter's moon Europa would be, for some observers, more exciting. But in the view of much of the space science community, robot missions are the way to accomplish the maximum amount of scientific inquiry, since valuable fuel and shipboard power do not have to be expended transporting and operating the equipment to keep a human crew alive and healthy. And very important to preserving extraterrestrial ecosystems is that robot craft can be thoroughly sterilized, while humans cannot. Such a difference could be critical in protecting a sensitive planetary ecosystem.

The concept of *stewardship* is articulated by the mission of the nonprofit Association of Forest Service Employees for Environmental Ethics: "to forge a socially responsible value system for the Forest Service based on a land ethic which ensures *ecologically and economically sustainable* [author's italics] resource management."[46] The key is resource management rather than preservation of the forest in a pristine, unaltered state. Applied to extraterrestrial bodies, stewardship "would imply that the broad scientific and economic benefits from having a second planetary-scale biosphere [in addition to Earth] would justify planetary alteration."[47]

An interesting combination of the preservationist and stewardship approaches arises from the belief that only *biotic life* has intrinsic value, not a body's geology. Thus, the stewardship perspective, which desires that we humans use nature wisely for our own benefit, would consider terraforming a celestial body to be ethical, even obligatory, if it promoted the growth of indigenous life on that

---

45. I. Almar, "Protection of the Lifeless Environment in the Solar System" (presented at the 34th COSPAR Scientific Assembly, Second World Space Congress, 10–19 October 2002, Houston, TX, 2002); Meltzer, *Mission to Jupiter.*
46. Lawrence M. Hinman, "Environmental Ethics," University of San Diego Ethics Update—Environmental Ethics Resources, 17 August 2006, available at *http://ethics.sandiego.edu/Applied/Environment/index.asp* (accessed 13 October 2006).
47. Richard O. Randolph, Margaret S. Race, and Christopher P. McKay, "Reconsidering the Theological and Ethical Implications of Extraterrestrial Life," *Center for Theology and Natural Sciences (CTNS) Bulletin* 17, no. 3 (Berkeley, CA, summer 1997): 1–8.

body, even as-yet-undiscovered life, and even if such terraforming destroyed the beauty of the body's mountains, valleys, or other geological features.

Exploring and colonizing other bodies might offer a huge social benefit: providing a long-term, unifying project on which humans, cooperating around the globe, could focus. Such an effort might prove to be one important step toward world peace. Constructing active biospheres on other bodies could also become critical for our own survival, serving as refuges for terrestrial life in the event of nuclear war or some other global catastrophe.[48] Some voices in the space science community, in fact, called for colonization of other worlds in very strong terms. Michael J. Rycroft of the International Space University has argued that "the overarching goal of space exploration for the twenty-first century should be to send humans to Mars, with the primary objective of having them remain there,"[49] so that our human species might have a second home in the event that a disaster on Earth rendered it uninhabitable.

Rycroft believed that many factors could cause such a catastrophe, including overpopulation; global terrorism; nuclear or biological war or accidents; occurrence of a supervirus; natural disaster (e.g., from an asteroid collision, flood, volcano, and so on); depletion of vital resources such as oil or natural gas reserves; climate change, global warming, and sea level rise; and stratospheric ozone depletion. Rycroft held that the colonization of a habitable world was thus an *imperative* human endeavor of this century and an insurance policy, and he emphasized his point by quoting M. Rees's opinion that "the odds are no better than 50–50 that our present civilization on Earth will survive to the end of the present century."[50]

But can colonization of a world containing indigenous life be performed while at the same time following the strictures of planetary protection? It could be argued that if planetary protection measures seriously delayed colonizing another world, they would be *unethical* to perform, since they would endanger the safety and future of our own species. Compulsory colonization as soon as it is feasible, on the other hand, will likely contaminate a body with Earth organisms and may well extinguish any indigenous lifeforms. The human race has arguably done a terrible job of protecting its own planet's environment, so can we even imagine that we will appropriately protect other bodies we visit?[51]

---

48. Clancy et al., *Looking for Life*, pp. 187–189; Christopher P. McKay, Owen B. Toon, and James F. Kasting, "Making Mars Habitable," *Nature* 352 (8 August 1991): 489–496.
49. Michael J. Rycroft, "Space Exploration Goals for the 21st Century," *Space Policy* 22 (2006): 158–161.
50. M. Rees, *Our Final Century* (London, U.K.: William Heinemann, 2003), p. 228, as reported in Rycroft, p. 159.
51. Clancy et al., *Looking for Life*, p. 188.

Deciding which course should be followed—colonization as soon as it is feasible, or waiting until a thorough search for life has been performed—depends on the intrinsic value we decide to give to extraterrestrial lifeforms, even nonsentient forms, and to biotic communities of those organisms. Arguing for the planetary protection approach, however, is not always easy. How does one make a convincing case for protecting the possible existence of some nonsentient microbes when their ecological niche may be required by the human race for its own survival?

The United Nations Educational, Scientific and Cultural Organization (UNESCO) has compared the ethics of outer space exploration with those of terrestrial environmental ethics, believing that respect for Earth's environment also applies to respect of other celestial bodies. Then when is it all right to *exploit* extraterrestrial bodies by mining and farming them? Should undeveloped lands of great natural beauty or high scientific interest instead be carefully preserved, even if we are sure they don't harbor life? Our country's history has demonstrated the devastating speed at which natural resources can be exploited and destroyed; but throughout U.S. history, we have also taken pride in, and placed high value on, the extraordinary beauty of our country.[52] I. Almar argued for intelligence and restraint in our utilization of extraterrestrial bodies when he identified the need "not to prevent any commercial utilization of Solar System resources, but to make space exploration and exploitation of resources a controlled and well-planned endeavor."[53] I. Almar also included bodies without life in his idea of responsible exploitation, recommending a large-scale discussion on the ethical values of the lifeless environment.

One rather arrogant view that emerged during discussions of planetary exploitation was that "the destiny of humanity is to occupy space, a destiny written in our genes."[54] This is a position reminiscent of the political philosophy of manifest destiny, held by many U.S. statesmen and business leaders in the 19th century, that our country *deserved* to conquer the heart of North America from the Atlantic Ocean to the Pacific Ocean and use its resources.[55] The United States

---

52. Adam Rome, "Conservation, Preservation, and Environmental Activism: A Survey of the Historical Literature," National Park Service History, available at *http://www.cr.nps.gov/history/hisnps/NPSThinking/nps-oah.htm*, last modified 16 January 2003 (accessed 13 October 2006).
53. I. Almar, "Protection of the Lifeless Environment in the Solar System" (presented at the 34th COSPAR Scientific Assembly, Second World Space Congress, held 10–19 October 2002, Houston, TX, 2002).
54. Ibid., p. 185.
55. Hermon Dunlap Smith Center for the History of Cartography, glossary in "Historic Maps in K–12 Classrooms," Newberry Library Web site, 2003, available at *http://www3.newberry.org/k12maps/glossary/* (accessed 4 October 2006).

would do this no matter the price paid by indigenous people or the environment. As Democratic leader and editor John L. O'Sullivan insisted in 1845, "our manifest destiny [is] to over spread and to possess the whole of the continent which Providence has given us for the development of the great experiment of liberty . . . ."[56] In the not too distant future, we will have to decide if we have this same right as we explore other bodies of our solar system. What impact will we have on other worlds if we operate according to a manifest destiny ethic?

While the concept of manifest destiny was once, and still may be, popular among many U.S. citizens, we were never unanimous in our support, either in the past or in the present. Some statesmen have recognized, and warned against, our country's tendency to run roughshod over delicate nature. In 1837, for instance, William E. Channing wrote to Henry Clay that "We are a restless people, prone to encroachment, impatient of the ordinary laws of progress . . . forgetting that, throughout nature, noble growths are slow . . . . It is full time that we should lay on ourselves serious, resolute restraint."[57] This belief in restraint and care in our expansion through the universe is expressed in many forms today, and it is part of the ongoing debate on how to explore space. One particularly elegant opinion regarding the wisdom of restraint is expressed in the book *Looking for Life, Searching the Solar System*: "The Earth can be seen as a spaceship driven by humanity acting as a crew, and it is the destiny of a crew to stay onboard the ship . . . ."[58]

Exploration for the Improvement of the World

Dave Scott, commander of Apollo 15, believed that "there's a fundamental truth to our nature—Man must explore,"[59] and it is this urge that has propelled adventurers toward both new lands and new worlds. John Young, a veteran of Gemini, Apollo, and Space Shuttle flights, realized how important this urge is

---

56. John Louis O'Sullivan, from an editorial supporting the annexation of Texas in the July-August 1845 edition of the *United States Magazine and Democratic Review*, as reported in Michael T. Lubragge, "Manifest Destiny: The Philosophy That Created A Nation," in *From Revolution to Reconstruction*, 2003, from the site "A Hypertext on American History from the Colonial Period until Modern Times, Department of Humanities Computing, University of Groningen, The Netherlands," available at *http://www.let.rug.nl/usa/E/manifest/manif1.htm* (accessed 4 October 2006).
57. John M. Blum, William S. McFeely, Edmund S. Morgan, Arthur M. Schlesinger, Jr., and Kenneth M. Stampp, *The National Experience: A History of the United States*, 6th ed. (New York, NY: Harcourt Brace Jovanovich, 1985), p. 276.
58. Clancy et al., *Looking for Life*, pp. 185–186.
59. NASA, "Man Must Explore," available at *http://www.history.nasa.gov/alsj/UL15MustExplore.html* (accessed 18 January 2009).

to our nation when he noted that "the things we learn out there will be making life better for a lot of people who won't be able to go."[60]

Journeys to the outer solar system are of necessity robotic journeys, but the technologies now available enable us to hitchhike on those faraway voyages, experience the new worlds encountered by the spacecraft, and reap various rewards: a sense of wonder at what lies beyond our planet; the vitality that comes from glimpsing what is possible and achievable; and the intellectual, social, political, and ethical benefits that can help us maintain an influential place in the world.

60. "Great Quotes," *WordPress.com*, available at *http://todayinspacehistory.wordpress.com/great-quotes/* (accessed 18 January 2009).

Chapter 19

Deep Space Navigation, Planetary Science, and Astronomy

A Synergetic Relationship

Andrew J. Butrica

The many technological and scientific accomplishments of NASA's half century of solar system exploration succeeded in no small part because of the efforts and expertise of deep space navigators. Their crucial contribution to the Agency's successes (and failures) is the ongoing determination of a space probe's position. Spacecraft navigation assures the collection of vital scientific data, which Earth-bound scientists in turn interpret within a paradigmatic framework indigenous to their discipline. In short, behind the scientific successes of NASA's missions of solar system exploration is deep space navigation.

Far less apparent, but no less far-reaching, have been the contributions of deep space navigation to astronomy, a scientific discipline on which navigation is especially dependent, and the planetary sciences. This paper partially addresses this lacuna by summarizing deep space navigation's general evolution over the past five decades and by pointing out some examples of navigation's role in advancing both astronomy and the planetary sciences. First, however, we need to understand what deep space navigation is.

Navigators and Navigation

All solar system missions include an approach to at least one celestial body. The simplest missions consist of a flyby or a hard impact, while more complicated missions require placing a probe in orbit, trimming an orbit to enable television or radar imaging, or descending through an atmosphere to execute a soft landing. Navigation entails determining the position and motion of a spacecraft at any given time from data furnished by the Deep Space Network and comparing them with the position and motion predicted by preflight computations.

Navigators compute probe flightpaths using a repetitive orbit determination process. Trajectory-correcting maneuvers performed during the mission require special, highly accurate measurements of the craft's velocity and position as well as highly precise orbit determinations, not to mention the calculation of the velocity and directional changes required to adjust the probe's flightpath. Navigators perform these vital calculations by means of a complex collection of computer software called the Orbit Determination Program that operated exclusively on large mainframe computers well into the 1980s. The institutional home of NASA's deep space navigators is the Systems Division of JPL.

Through their efforts, the navigators at JPL enabled NASA's successful exploration of the solar system beginning with the Moon-bound Pioneers, Rangers, Surveyors, and Lunar Orbiters of the 1960s; subsequently, they enabled the Mariner and other missions to Mars, Venus, and Mercury, as well as the Pioneer, Voyager, Galileo, and Cassini voyages to the outer planets and the latest exploration of asteroids and comets.

Doppler Era

For the purposes of this contribution, I divide the history of deep space navigation into three broad, overlapping eras. The first is that of the initial decade of solar system exploration, when navigators relied solely on Doppler data. The era witnessed the first use of a parking orbit, the first Hohmann transfer orbit, the first trajectory correction maneuver, and the first hard landing on another celestial body—all accomplished by the Ranger program—as well as the first soft landing (the Surveyors) and the first planetary flyby (Mariner 2). These were all firsts of necessity, as these were NASA's initial forays into deep space.

One of the great navigational milestones of this era laid the foundation for the ongoing quest for greater and greater navigational accuracy that has marked the history of navigation. That quest had its formal beginnings in July 1965 with the establishment of the Inherent Accuracy Project, renamed and greatly expanded as the JPL Navigation Program in December 1968.[1] The experiments and studies initiated under the aegis of this ongoing project attempted to understand and correct for the many sources of inaccuracies that held navigators back from deriving the full benefit of the accuracy intrinsic

1. William R. Corliss, *A History of the Deep Space Network* (Washington, DC: NASA CR-151915, 1 May 1976), pp. 124–125; Thomas W. Hamilton, "Introduction," in *Space Programs Summary No. 37-38, Volume III, for the period January 1, 1966, to February 28, 1966* (Pasadena, CA: JPL, 31 March 1966), p. 8 (hereafter SPS 37-38).

in upgrading the Deep Space Network tracking stations from the L-band to the higher frequencies of the S-band. They formulated better models of the effects of the stratosphere and ionosphere over each tracking station, strove to improve timekeeping and frequency standard precision, and computed highly accurate tracking station locations using the Orbit Determination Program and the Doppler data collected during NASA missions.[2] In the process of improving Deep Space Network tracking station locations, navigators contributed to Earth geodesy in a variety of ways worthy of study on their own.

Navigators established a synergistic relationship with those studying planetary atmospheres on NASA missions beginning with the flight of Mariner 4. After NASA selected all the science teams and instruments for that mission, a proposal came forth to perform an occultation experiment—the first of its kind—in which researchers studied the properties of the Martian atmosphere by examining the changes experienced by radio waves passing through it. Later missions included occultation investigations of planetary atmospheres, ionospheres, rings, and magnetic fields. They were a feature of Mariner and Viking flights to Mercury, Venus, and Mars, as well as the Pioneer 10 and Pioneer 11 (Jupiter and Saturn), Voyager (Jupiter, Saturn, Uranus, Neptune, and Triton), Ulysses (Jupiter), Magellan (Venus), Galileo (Jupiter, Callisto, Io, Ganymede, and Europa), and Cassini (Saturn and Titan) missions.[3]

The ideas for the occultation experiment arose independently at Stanford University and JPL, with Stanford researchers coming up with the first proposal in 1962 and JPL contributing later in the spring of 1964.[4] The Stanford proposal grew out of the research at the Center for Radar Astronomy on bistatic radar and its uses for exploring the Moon and planets.[5] The idea for the occultation

---

2. Jordan Ellis, "Large Scale State Estimation Algorithms for DSN Tracking Station Location Determination," *Journal of the Astronautical Sciences* 28 (January–March 1980): 15–30.
3. *Mariner-Mars 1964 Final Project Report* (Washington, DC: NASA SP-139, 1967), p. 6; Douglas J. Mudgway, *Uplink-Downlink: A History of the Deep Space Network, 1957–1997* (Washington, DC: NASA SP-2001-4227, 2001), pp. 514–516; Sami W. Asmar and Nicholas A. Renzetti, *The Deep Space Network as an Instrument for Radio Science Research*, rev. 1 (Pasadena, CA: JPL Publication 80-93, 15 April 1993), p. 13.
4. Arvydas J. Kliore, Dan L. Cain, Gerald S. Levy, Von R. Eshleman, Frank D. Drake, and Gunnar Fjeldbo, "The Mariner 4 Occultation Experiment," *Astronautics & Aeronautics* 3 (July 1965): 73; Asmar and Renzetti, *The Deep Space Network*, p. 13; Mudgway, *Uplink-Downlink*, p. 515; Von R. Eshleman, interview by author, Stanford University, 9 May 1994, transcript and tape, pp. 23–25, NASA Historical Reference Collection, NASA History Division, NASA Headquarters, Washington, DC.
5. Asmar and Renzetti, *The Deep Space Network*, p. 13; Gunnar Fjeldbo and Von R. Eshleman, "The Bistatic Radar-Occultation Method for the Study of Planetary Atmospheres," *Journal of Geophysical Research* 70, no. 13 (July 1, 1965): 3217–3225; Gunnar Fjeldbo, Von R. Eshleman, Owen K. Garriott, and F. L. Smith III, "The Two-Frequency, Bistatic Radar-Occultation Method for the Study of Planetary

experiment at JPL had rather different roots, not in radar but in deep space navigation. There, navigator Dan L. Cain proposed conducting the occultation experiment on Mariner 4. As a result, Cain and Arvydas "Art" Kliore, another JPL navigator, joined Stanford on the occultation science team. They and other navigators also were part of subsequent occultation experiments performed by Stanford on NASA missions.[6]

Opening up the Mariner 4 mission had larger repercussions for both science and navigation. Navigators now participated on NASA mission science teams and contributed to new discoveries. For example, during the Lunar Orbiter flights, JPL navigators William L. "Bill" Sjogren and Paul M. Muller discovered mass concentrations ("mascons"), regions of the Moon's crust exhibiting large gravitational anomalies, while analyzing navigational data from the mission. Later, while examining Viking Orbiter 2 navigation data, Bill Sjogren announced the discovery of mascons at several locations on Mars.[7]

By instigating the institution of rules for outside access to navigation data,[8] the Mariner 4 occultation experiment also opened the Deep Space Network to radio astronomers beginning in 1967, following coincidentally the discovery earlier that year by S. Jocelyn Bell, a Cambridge University graduate student working under Sir Anthony Hewish, of a scintillating radio source dubbed

---

Ionospheres," *Journal of Geophysical Research* 70, no. 15 (1 August 1965): 3701–3710; Andrew J. Butrica, *To See the Unseen: A History of Planetary Radar Astronomy* (Washington, DC: NASA SP-4218, 1996), pp. 57, 155. Fjeldbo's dissertation was "Bistatic-Radar Methods for Studying Planetary Ionospheres and Surfaces" (Ph.D. diss., Stanford University, 1964), later published as Fjeldbo, *Bistatic-Radar Methods for Studying Planetary Ionospheres and Surfaces* (Stanford, CA: Radioscience Laboratory, Stanford Electronics Laboratory, SR 2, 1964).

6. Eshleman interview, pp. 25, 28; Kliore, Cain, Levy, Eshleman, Drake, and Fjeldbo, "The Mariner 4 Occultation Experiment," p. 73; NASA, "Mariner IV Pre-Encounter Press Conference," pp. 30–31, folder 5193, NASA Historical Reference Collection, NASA History Division, NASA Headquarters, Washington, DC; Asmar and Renzetti, *The Deep Space Network*, p. 13; Thomas W. Hamilton, interview by José Alonso, n.d., JPL, tape and transcript, p. 6, JPL Archives, JPL, Pasadena, CA; Arvydas J. Kliore, interview by José Alonso, 13 July 1992, JPL, tape and transcript, JPL Archives, Jet Propulsion Laboratory, Pasadena, CA; Oran W. Nicks, *A Review of the Mariner IV Results* (Washington, DC: NASA SP-130, 1967), p. 32.
7. Paul M. Muller and William L. Sjogren, "Mascons: Lunar Mass Concentrations," *Science* 161, no. 3842 (16 August 1968): 680–684; Roger J. Phillips, James E. Conel, Elsa A. Abbott, William L. Sjogren, and John B. Morton, "Mascons: Progress Toward a Unique Solution for Mass Distribution," *Journal of Geophysical Research* 77 (10 December 1972): 7106–7114; William L. Sjogren and Wilber R. Wollenhaupt, "Gravity: Mare Humorum," *Moon* 8, no. 1-2 (1973): 25–32; William L. Sjogren, "Mars Gravity: High-Resolution Results from Viking Orbiter 2," *Science* 203, no. 4384 (March 9, 1979): 1006–1010; Paul D. Lowman, Jr., *Exploring Space, Exploring Earth: New Understanding of the Earth from Space Research* (New York, NY: Cambridge University Press, 2002), p. 72.
8. See Homer E. Newell to William H. Pickering, 8 November 1964, and attachment, "Policy on the Utilization for Scientific Purposes of Tracking Data from the Deep Space Network," 13 August 1964, Historian's Files, record no. 07 00024 BF, JPL Archives.

Little Green Men 1 (LGM 1), but known today as a pulsar. The announcement spurred a worldwide hunt for more of these pulsating radio sources, and radio astronomers now could take advantage of the Deep Space Network in that hunt.[9]

VLBI and ΔDOR

The second era of deep space navigation began around 1970 as NASA was planning its first voyages to the outer planets. The accuracy of navigation improved with the gradual shift to X-band frequencies. Navigators added range to their repertoire of data types and experimented with several new varieties of data, such as Differenced Range Versus Integrated Doppler (DRVID). The purpose of DRVID was to correct for the effects of charged particles in the ionosphere via computational methods rather than by making measurements using satellites in geostationary orbit, as had been the case. The results of the DRVID experiments on the Mariner Mars 1969 mission were promising, despite many difficulties, and both Mariner 9 and Pioneer Venus used DRVID, while Viking used a variant called "pseudo-DRVID."[10] More importantly in the long run, however, navigators added Very Long Baseline Interferometry (VLBI) and optical navigation to their repertoire of techniques.

Very Long Baseline Interferometry is a technique borrowed from radio astronomy, specifically a variety of VLBI equipment and software developed by the National Radio Astronomy Observatory (NRAO) as well as by the Caltech radio astronomy community with funding from NSF.[11] The VLBI technique

---

9. Nicholas A. Renzetti, Gerald S. Levy, Thomas B. H. Kuiper, Pamela R. Wolken, and R. C. Chandlee, *The Deep Space Network: An Instrument for Radio Astronomy Research*, rev. 1 (Pasadena, CA: JPL 82-68, 1 September 1988), p. 2-1; Anthony Hewish, S. Jocelyn Bell, John D. Pilkington, P. F. Scott, and R. A. Collins, "Observation of a Rapidly Pulsating Radio Source," *Nature* 217 (February 1968): 709–713; Sir Bernard Lovell, *Astronomer by Chance* (New York, NY: Basic Books, 1990), pp. 294–295; Benjamin K. Malphrus, *The History of Radio Astronomy and the National Radio Astronomy Observatory: Evolution Toward Big Science* (Malabar, FL: Krieger, 1996), pp. 137–138.
10. Nicolas A. Renzetti, James F. Jordan, Allen L. Berman, Joseph A. Wackley, and Thomas P. Yunck, *The Deep Space Network: An Instrument for Radio Navigation of Deep Space Probes* (Pasadena, CA: JPL TR 82-102, 15 December 1982); G. A. Madrid, "Charged Particles," in *Tracking System Analytic Calibration Activities for the Mariner Mars 1971 Mission*, ed. G. A. Madrid, C. C. Chao, H. F. Fliegel, R. K. Leavitt, N. A. Mottinger, F. B. Winn, R. N. Wimberly, K. B. Yip, and J. W. Zielenbach (Pasadena, CA: JPL TR 32-1587, 1 March 1974), pp. 43–60; P. S. Callahan, "A Preliminary Analysis of Viking S-X Doppler Data and Comparison to Results of Mariner 6, 7, and 9 DRVID Measurements of the Solar Wind Turbulence," in *The Deep Space Network Progress Report 42-39, March and April 1977* (Pasadena, CA: JPL, 15 June 1977), pp 23–29.
11. Kurt M. Liewer, "DSN Very Long Baseline Interferometry System Mark IV-88," in *The Telecommunications and Data Acquisition Progress Report 42-93, January–March 1988*, ed. E. C. Posner (Pasadena, CA: JPL, 15 May 1988), pp. 239–246; J. W. Layland and L. L. Rauch, *The Evolution of Technology in the Deep Space Network: A History of the Advanced Systems Program*, TDA Progress Report 42-130

Angular Tracking Using Station-Differenced Observables

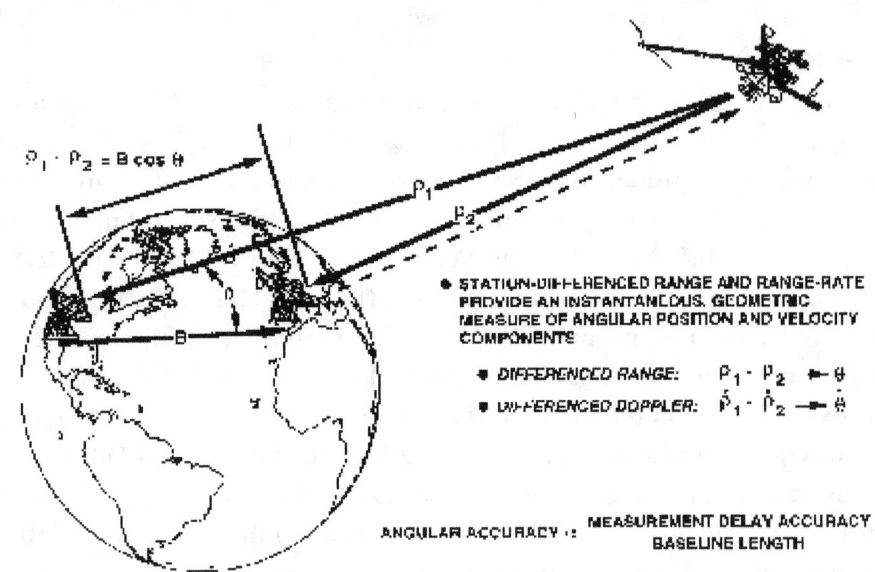

Figure 1: Satellite tracking using differenced observations from stations located across a long baseline (VLBI). *Lincoln J. Wood, "The Evolution of Deep Space Navigation: 1962–1989," paper 08-051, read at the 31st Annual AAS Guidance and Control Conference, Breckenridge, Colorado, 1–6 February 2008*

uses two widely separated antennas (for example, one in California and the other in Australia) receiving radio signals from the same distant source, the spacecraft's transponder. By studying the differences between the two signals, navigators can calculate spacecraft positions far more accurately than with traditional Doppler and range techniques.[12]

From the standpoint of deep space navigation, a far more preferable approach is to take readings from a spacecraft, then from a distant extragalactic radio source (a quasar), at two separated antennas. Navigators then analyze the differences in readings obtained at the two antennas, which provides an

(Pasadena, CA: JPL, 15 August 1997), p. 18; Martin A. Slade, Robert A. Preston, Alan W. Harris, Lyle J. Skjerve, and Donovan J. Spitzmesser, "ALSEP-Quasar Differential VLBI," *Earth, Moon, and Planets* 17, no. 2 (October 1977): 133–147.

12. Lincoln J. Wood, "The Evolution of Deep Space Navigation: 1962–1989" (AAS paper 08-051, read at the 31st Annual AAS Guidance and Control Conference, 1–6 February 2008, Breckenridge, CO), published in Michael E. Drews and Robert D. Culp, eds., *Advances in the Astronautical Sciences*, vol. 131 (San Diego, CA: Univelt, 2008), pp. 299–301.

even more accurate navigational result. The generic term for this version of VLBI is delta-VLBI.[13]

The first test of VLBI for navigation actually took place between Earth and the Moon. Between 1969 and 1972, astronauts left the Apollo Lunar Surface Experiments Package (ALSEP), a laser reflector array, on the lunar surface. Carried out by Martin A. "Marty" Slade and others at JPL, along with Charles C. Counselman III and Irwin I Shapiro at MIT, the experiment involved having four antennas make simultaneous observations (two at each end of an intercontinental baseline) of a quasar and a number of ALSEP transmitters and obtaining the differential interferometric phase between the two sources. The researchers also hoped that these highly precise observations would help to test gravitational theories and to measure the Earth-Moon tidal friction interaction.[14] Thus VLBI became an instrument of both deep space navigation and Earth geodesy.

The first mission to benefit from this important navigational tool was Voyager.[15] In order to have a map of quasars useful for navigation, JPL began to participate in the Quasar Patrol, an informal group of scientists and engineers from MIT (Irwin Shapiro) and its Haystack Observatory (Alan Rogers), GSFC (R. J. Coates), and the University of Maryland (Thomas A. Clark) that made VLBI observations of quasars during the early 1970s.[16]

---

13. Wood, "The Evolution of Deep Space Navigation: 1962–1989," p. 301; Liewer, "DSN Very Long Baseline Interferometry System Mark IV-88," pp. 239–240.
14. Martin A. Slade, Robert A. Preston, Alan W. Harris, Lyle J. Skjerve, and Donovan J. Spitzmesser, "ALSEP-Quasar Differential VLBI," *Earth, Moon, and Planets* 17, no. 2 (October 1977): 133–147; Robert W. King, Charles C. Counselman III, Irwin I. Shapiro, and Hans F. Hinteregger, "Study of Lunar Librations using Differential Very-Long-Baseline Observations of ALSEPs," *Recent Advances in Engineering Science* 8 (1977): 431–438; Martin A. Slade, interview by author, JPL, 24 May 1996, transcript and tape, pp. 5–6, NASA Historical Reference Collection, NASA History Division, NASA Headquarters, Washington, DC; Irwin I. Shapiro, interview by author, Harvard-Smithsonian Center for Astrophysics, Cambridge, MA, 1 October 1993, tape and transcript, pp. 4–6, NASA Historical Reference Collection, NASA History Division, NASA Headquarters, Washington, DC.
15. Layland and Rauch, p. 27; D. Lee Brunn, Robert A. Preston, Sien C. Wu, Herbert L. Siegel, David S. Brown, Carl S. Christensen, and David E. Hilt, "Δ VLBI Spacecraft Tracking System Demonstration: Part 1: Design and Planning," in *The Deep Space Network Progress Report 42-45, March and April 1978* (Pasadena, CA: JPL, 15 June 1978), pp. 111–132; Christensen, B. Moultrie, Philip S. Callahan, F. F. Donivan, and S. C. Wu, "Differential Very Long Baseline Interferometry (Delta VLBI) Spacecraft Tracking System Demonstration: Part 2: Data Acquisition and Processing," in *The Telecommunications and Data Acquisition Progress Report 42-60, September and October 1980*, ed. N. A. Renzetti (Pasadena, CA: JPL, 15 December 1980), pp. 60–67.
16. Karl W. Linnes, "Radio Science Support," in *The Deep Space Network Progress Report for November and December 1972*, vol. 13 (Pasadena, CA: JPL TR 32-1526, 15 February 1973), pp. 37–40; O. J. Sovers, C. D. Edwards, C. S. Jacobs, G. E. Lanyi, K. M. Liewer, and R. N. Treuhaft, "Astrometric Results of 1978–1985 Deep Space Network Radio Interferometry: The JPL 1987-1 Extragalactic Source Catalog," *Astronomical Journal* 95 (June 1988): 1647–1658.

Very Long Baseline Interferometry

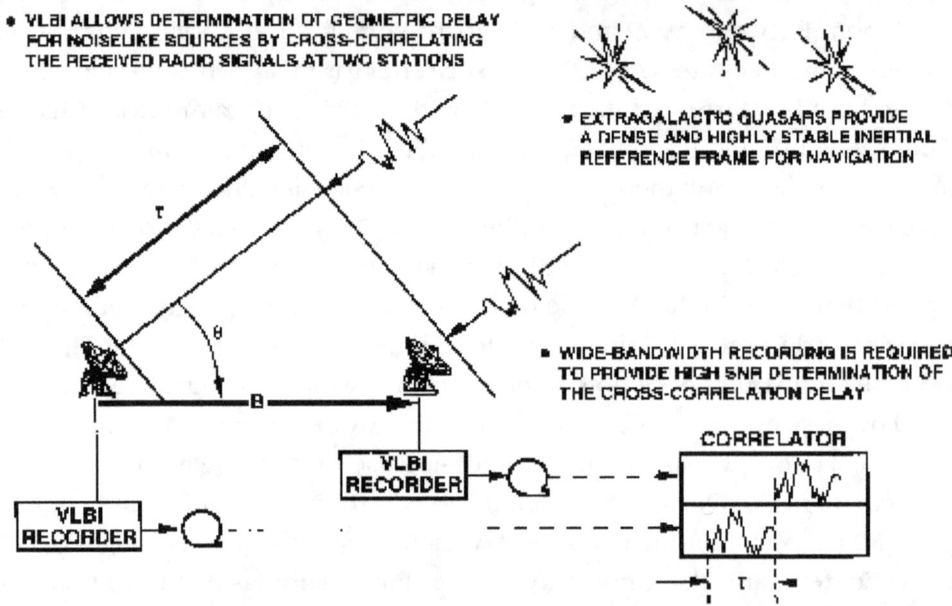
• VLBI ALLOWS DETERMINATION OF GEOMETRIC DELAY FOR NOISELIKE SOURCES BY CROSS-CORRELATING THE RECEIVED RADIO SIGNALS AT TWO STATIONS
• EXTRAGALACTIC QUASARS PROVIDE A DENSE AND HIGHLY STABLE INERTIAL REFERENCE FRAME FOR NAVIGATION
τ
θ
B
• WIDE-BANDWIDTH RECORDING IS REQUIRED TO PROVIDE HIGH SNR DETERMINATION OF THE CROSS-CORRELATION DELAY
CORRELATOR
VLBI RECORDER
VLBI RECORDER
τ

Figure 2: Very Long Baseline Interferometry using quasars as a reference frame for deep space navigation. *Lincoln J. Wood, "The Evolution of Deep Space Navigation: 1962–1989," paper 08-051, read at the 31st Annual AAS Guidance and Control Conference, Breckenridge, Colorado, 1–6 February 2008*

The VLBI technique, known as ΔDOR, also became an important, though not primary, tool for improving navigational accuracy beginning with the Voyager flights.[17] Its use soon came to an end, however, when navigators on the Galileo mission showed mathematically that high-precision, two-way ranging and Doppler were "somewhat better" than ΔDOR alone and noted that two-way ranging was simpler to schedule and process.[18] NASA Headquarters subsequently withdrew ΔDOR funding.[19] But the space agency reinstated ΔDOR in response to the conclusions and recommendations of the Mars Climate

---

17. Jordan Ellis, "Deep Space Navigation with Noncoherent Tracking Data," in *The Telecommunications and Data Acquisition Progress Report 42-74, for April–June 1983*, ed. E. C. Posner (Pasadena, CA: JPL, 15 August 1983), pp. 1–12.
18. Vincent M. Pollmeier and S. W. Thurman, "Application of High-Precision Two-Way Ranging to Galileo Earth-1 Encounter Navigation," in *The Telecommunications and Data Acquisition Progress Report 42-110, April–June 1992*, ed. Renzetti (Pasadena, CA: JPL, 15 August 1992), pp. 21–32.
19. Laureano A. "Al" Cangahuala, interview by author, JPL, 20 April 2007, transcript, pp. 24–25, NASA Historical Reference Collection, NASA History Division, NASA Headquarters, Washington, DC; Lincoln J. Wood, interview by author, JPL, 13 November 2008, tape, NASA Historical Reference Collection,

Navigation Measurements -- Optical Data

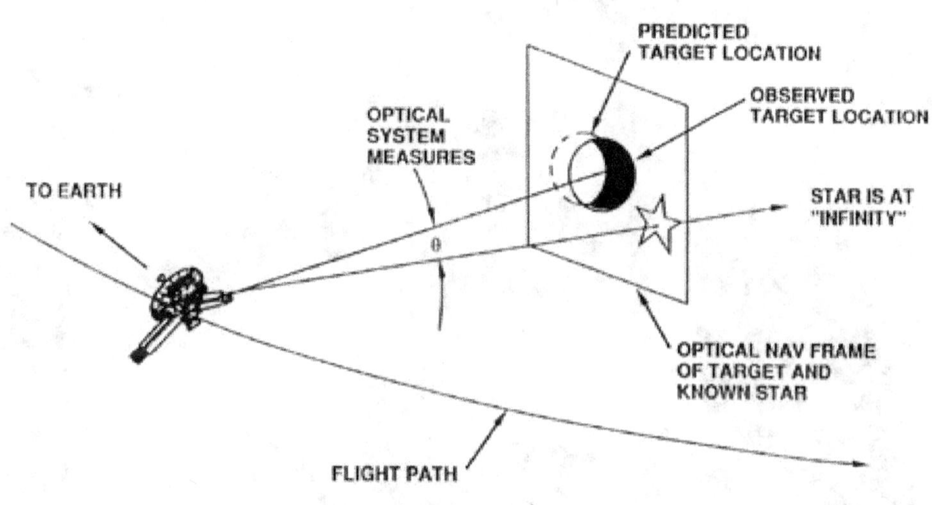

Figure 3: Optical navigation images a target body along with a known nearby star in order to determine a spacecraft's position relative to the target. *Lincoln J. Wood, "The Evolution of Deep Space Navigation: 1962–1989," paper 08-051, read at the 31st Annual AAS Guidance and Control Conference, Breckenridge, Colorado, 1–6 February 2008*

Observer mishap board. The board concluded that the mission navigation team had been unable to observe the deleterious changes to the probe's flightpath because those effects were perpendicular to the line of sight. Standard Doppler measurements provided only a line-of-sight evaluation of the craft's trajectory. The mishap board recommended therefore that "several other navigation methods should be compared to the prime navigation method to help uncover any mismodeled small forces" on future missions, namely the Mars Polar Lander.[20]

The board's language—"other navigation methods"—was understood to be a call for the reinstatement of ΔDOR,[21] and consequently the Mars Observer of

NASA History Division, NASA Headquarters, Washington, DC; Shyam Bhaskaran, interview by author, JPL, 14 November 2008, tape, NASA Historical Reference Collection, NASA History Division, NASA Headquarters, Washington, DC.

20. Mars Climate Observer Mishap Investigation Board, *Phase I Report* (Washington, DC: NASA, 10 November 1999), p. 18.
21. Bhaskaran interview.

Figure 4: Mariner 9 marked a number of milestones in the history of deep space navigation. Gathered here are members of the navigation team during Mars orbit insertion (13 November 1971). *William M. Owen, Jr., Thomas C. Duxbury, Charles H. Acton, Jr., Stephen P. Synnott, Joseph E. Riedel, and Shyam Bhaskaran, "A Brief History of Optical Navigation at JPL," paper AAS 08-053, read at the 31st Annual AAS Guidance and Control Conference, Breckenridge, Colorado, 1–6 February 2008*

2001 became only the second spacecraft to use ΔDOR routinely for navigation, though only as a supplement to range and Doppler data.[22]

Optical Navigation

The other powerful navigational tool that emerged during this second era was optical navigation. Optical navigation requires an on-board camera that one can point toward selected targets and image them against a background of stars. The first attempts at optical navigation used the camera reserved for

22. Peter M. Kroger, James S. Border, and Sumita Nandi, "The Mars Observer Differential One-Way Range Demonstration," in *The Telecommunications and Data Acquisition Progress Report 42-117, January–March 1994*, ed. Joseph H. Yuen (Pasadena, CA: JPL, 15 May 1994), pp. 1–15.

science experiments, and data processing took place on the ground.[23] Optical navigation is most useful when the motion of the target body is not understood with sufficient precision, as in the case of the moons of Jupiter or other outer planet bodies or asteroids and comets. Navigators use it in conjunction with such standard data types as Doppler, range, and ΔDOR during the approach, rendezvous, and orbit phases of a mission.

The first experimental optical navigation attempt took place during the Mariner 6 and 7 (Mariner Mars 1969) flights using the "Far Encounter Planet Sensor" (the on-board science camera).[24] Navigators followed that success with a second demonstration on Mariner 9, which took off for Mars on 30 May 1971, again using the science camera.[25] One of the major benefits of the experiment was the creation of two sets of programs that became standard and continued in use in modified form. The Optical Navigation Image Processing System (ONIPS) displayed images and located their center, while the Optical Navigation Program (ONP) set made predictions, generated residuals and partials, and filtered data.[26] For their development of the optical navigation method, JPL navigators Thomas C. "Tom" Duxbury and Charles H. "Chuck" Acton, Jr., received NASA's Exceptional Scientific Achievement Medal as well as the Institute of Navigation's Samuel M. Burka prize for their paper,[27] along with a modest financial award of $175 each.[28]

---

23. James Frank Jordan, interview by author, JPL, 19 April 2007, transcript and tape, pp. 20–21, 33–34, 36, 38–39, 41–42, NASA Historical Reference Collection, NASA History Division, NASA Headquarters, Washington, DC.
24. Jordan interview, pp. 31, 64; Thomas C. Duxbury, "Navigation Data from Mariner Mars 1969 TV Pictures," *Navigation* 17, no. 3 (1970): 219–225; Thomas C. Duxbury and William G. Breckenridge, "Mariner Mars 1969 Optical Approach Navigation" (AIAA paper 70-70, AIAA 8th Aerospace Sciences Meeting, New York, NY, 1970); William M. Owen, Jr., Thomas C. Duxbury, Charles H. Acton, Jr., Stephen P. Synnott, Joseph E. Riedel, and Shyam Bhaskaran, "A Brief History of Optical Navigation at JPL" (paper AAS 08-053, read at the 31st Annual AAS Guidance and Control Conference, Breckenridge, CO, 1–6 February 2008, copy provided by Lincoln Wood and Shyam Bhaskaran), pp. 2, 3 (hereafter "A Brief History of Optical Navigation at JPL").
25. George H. Born, Thomas C. Duxbury, William G. Breckenridge, Charles H. Acton, Srinivas N. Mohan, Navin Jerath, and Hiroshi Ohtakay, *Mariner Mars 1971 Optical Navigation Demonstration* (Pasadena, CA: JPL TM 33-683, 15 April 1974); William G. Breckenridge and Charles H. Acton, Jr., "A Detailed Analysis of Mariner Nine TV Navigation Data" (paper 72-866, AIAA Guidance and Control Conference, Stanford, CA, 1972); Acton, "Processing Onboard Optical Data for Planetary Approach Navigation," *Journal of Spacecraft and Rockets* 9, no. 10 (1972): 746–750.
26. J. F. Jordan and A. J. Fuchs, "Autonomy in Space Navigation," *Astronautics and Aeronautics* 17 (May 1979): 46–49; "A Brief History of Optical Navigation at JPL," p. 3.
27. Thomas C. Duxbury and Charles H. Acton, "On-Board Optical Navigation Data from Mariner 71," *Navigation* 19, no. 4 (1972): 295–307.
28. "A Brief History of Optical Navigation at JPL," p. 4.

Figure 5: Thomas C. Duxbury (left) and Charles H. Acton, Jr., at the Institute of Navigation's 1974 meeting in San Diego, California, receiving the Samuel M. Burka prize for their pioneering paper on optical navigation. *William M. Owen, Jr., Thomas C. Duxbury, Charles H. Acton, Jr., Stephen P. Synnott, Joseph E. Riedel, and Shyam Bhaskaran, "A Brief History of Optical Navigation at JPL," paper AAS 08-053, read at the 31st Annual AAS Guidance and Control Conference, Breckenridge, Colorado, 1–6 February 2008*

In 1976, the Viking mission also used optical navigation, targeting again the planet's moons, Phobos and Deimos, to guide the spacecraft not only on approach to Mars, but also in orbit.[29] However, the most dramatic and certainly the most scientifically rewarding application of optical navigation was the Voyager mission to the outer planets starting in 1977. Voyager marked the "coming of age" of optical navigation.[30]

Equipped to operate in both the S-band and the X-band, the Voyager spacecraft also were the first to utilize the X-band as the primary telemetry link. Optical data enhanced navigational accuracy at Jupiter; ground-based Doppler and range alone were sufficiently accurate to fulfill scientific objectives.

---

29. Navin Jerath, *Interplanetary Approach Optical Navigation with Applications* (Pasadena, CA: JPL TR 78-40, 1 June 1978); "A Brief History of Optical Navigation at JPL," pp. 4, 5.
30. "A Brief History of Optical Navigation at JPL," p. 5.

For the Saturn encounters, however, radio waves alone lacked the accuracy needed to accomplish scientific goals, so optical data for the first time were the primary method for meeting mission objectives.[31] Later, for the encounters with Uranus (1986), Neptune (1989), and their moons, optical navigation was a prime requisite for accomplishing the mission's scientific goals, even with the assistance of a concerted effort to expand astronomer knowledge of the ephemerides of those planets and their moons.[32]

The use of optical navigation involved navigators in updating the astronomers' star catalog. The Mariner and Viking missions, for example, had relied on the best whole-sky star catalog available, namely, the one that SAO compiled and published in 1966.[33] However, most of the material in that catalog came from the Yale Zone Catalogs of the 1930s, meaning that the star positions used on the Mariner and Viking missions already were 30 years out of date.[34]

Beginning with Voyager, JPL contracted with Lick Observatory for the creation of ad hoc star catalogs. Navigators from JPL provided the coordinates, and Lick staff pointed their 20-inch dual astrograph (one lens corrected for yellow light, one for blue) and exposed glass photographic plates. Navigators from JPL traveled to Santa Cruz, California, to use Lick's survey machine to select the stars to be measured, a rather tedious process. The observatory reduced the data and sent JPL a catalog on magnetic tape. Navigators used Lick's services for all the Voyager encounters, as well as for Galileo's flybys of asteroids Gaspra and Ida.[35]

The release of ESA's Hipparcos and Tycho catalogs[36] in 1997 changed the world of stellar astrometry overnight and benefited the accuracy of optical

---

31. Wood, "The Evolution of Deep Space Navigation: 1962–1989," pp. 302–303; Jordan, "Navigation Systems," *Journal of the British Interplanetary Society*, vol. 38 (1 October 1985): 444–449; J. K. Campbell, Stephen P. Synnott, and G. J. Bierman, "Voyager Orbit Determination at Jupiter," *IEEE Transactions on Automatic Control*, vol. AC-28 (March 1983): 256–268.
32. Robert A. Jacobson and E. Myles Standish, "Satellite Ephemerides for the Voyager Uranus Encounter" (AIAA paper 84-2024, AIAA/AAS Astrodynamics Conference, Seattle, WA, August 1984); Jacobson, "Satellite Ephemerides for the Voyager Neptune Encounter," in *Advances in the Astronautical Sciences: Astrodynamics 1987*, vol. 65, part 1, ed. John K. Soldner, Arun K. MIsra, Robert E. Lindberg, and Walton Williamson (San Diego, CA: Univelt, 1988), pp. 657–680; Wood, "The Evolution of Deep Space Navigation: 1962–1989," p. 303.
33. Smithsonian Astrophysical Observatory, *Star Catalog: Positions and Proper Motions of 258,997 Stars for the Epoch and Equinox of 1950.0* (Washington, DC: Smithsonian Institution, 1966).
34. "A Brief History of Optical Navigation at JPL," p. 17.
35. "A Brief History of Optical Navigation at JPL," p. 17.
36. European Space Agency, *The Hipparcos and Tycho Catalogues: Astrometric and Photometric Star Catalogues Derived from the ESA Hipparcos Space Astrometry Mission*, 17 volumes (Noordwijk, Netherlands: ESA Publications Division, 1997).

Figure 6: Arthur J. "Joe" Donegan, Edwin S. Travers, Linda A. Morabito, and Stephen P. Synnott, members of the Voyager optical navigation team, gathered at the Modcomp IV minicomputer (1979). *Courtesy of Frank Jordan*

navigation. The catalog gave the positions and motions of stars to an unprecedented accuracy. Norbert Zacharias and his colleagues at the U.S. Naval Observatory also created a new Astrograph Catalog based on images of the whole sky made with charge-coupled device (CCD) cameras. Version two of the observatory's CCD Astrograph Catalog (UCAC) catalog,[37] which covers 86

37. Norbert Zacharias, Sean E. Urban, Marion I. Zacharias, Gary L. Wycoff, David M. Hall, David G. Monet, and Theodore J. Rafferty, "The Second U.S. Naval Observatory CCD Astrograph Catalog (UCAC2)," *Astronomical Journal* 127, no. 5 (2004): 3043–3059.

percent of the sky (omitting the north polar regions), was used by the Deep Impact asteroid mission and currently is in use by Cassini.[38]

The optical navigation team on Voyager also continued the tradition of contributing to the advancement of solar system science. Well known is the discovery of volcanic activity on Jupiter's moon Io by Linda A. Morabito (Kelly), a member of the Voyager optical navigation team, in March 1979. Hers was the first-ever sighting of active volcanism outside of Earth.[39] Less known to the public, however, are the discoveries made by another navigator, Stephen P. Synnott, during the same mission. Synnott discovered two new satellites of Jupiter in the Voyager 1 images that scientists later dubbed Thebe and Metis, respectively. He, too, made these discoveries while using the optical navigation software. Synnott later found more "rocks" at Saturn, Uranus, and Neptune, including 10 small satellites alone at Uranus.[40]

Current Era

The current navigation era is one of missions to the outer planets, Mars, asteroids, and comets on spacecraft operating in the X-band with initial incursions into the Ka-band. Navigators continue to use ΔDOR and optical navigation, which have become customary data types for supplementing Doppler and range. With the Magellan radar-mapping mission to Venus, launched in May 1989, deep space navigation began to use the Global Positioning System (GPS), mainly to improve

38. "A Brief History of Optical Navigation at JPL," p. 17.
39. Linda A. Morabito, Stephen P. Synnott, P. N. Kupferman, and Stewart A. Collins, "Discovery of Currently Active Extraterrestrial Volcanism," *Science* 204, no. 4396 (June 1979): 972; The Planetary Society, "Space Topics: Voyager: The Stories Behind the Mission: Linda Morabito Kelly, As Told to A. J. S. Rayl in 2002 on the Occasion of Voyager's 25th Anniversary," available at *http://www.planetary.org/explore/topics/space_missions/voyager/stories_kelly.html* (accessed 10 October 2008). The asteroid 3106 Morabito was found by Edward L. G. Bowell on 9 March 1981 and named after the navigator. The discovery tale is told succinctly by Eric Burgess in *By Jupiter: Odysseys to a Giant* (New York, NY: Columbia University Press, 1982), p. 89; and Richard O. Fimmel, William Swindell, and Eric Burgess, *Pioneer Odyssey* (Washington, DC: NASA SP-349, 1977), p. 74.
40. Stephen P. Synnott, "1979J2: Discovery of a Previously Unknown Jovian Satellite," *Science* 210, no. 4471 (14 November 1980): 786–788; Stephen P. Synnott, "1979J3: Discovery of a Previously Unknown Satellite of Jupiter," *Science* 212, no. 4501 (19 June 1981): 1392; Stephen P. Synnott, "Evidence for the Existence of Additional Small Satellites of Saturn," *Icarus* 67 (August 1986): 189–204; Stephen P. Synnott, C. F. Peters, B. A. Smith, and Morabito, "Orbits of the Small Satellites of Saturn," *Science* 212, no. 4491 (10 April 1981): 191–192; William M. Owen, Jr., and Stephen P. Synnott, "Orbits of the Ten Small Satellites of Uranus," *Astronomical Journal* 93 (May 1987): 1268–1271; William M. Owen, Jr., R. M. Vaughan, and Stephen P. Synnott, "Orbits of the Six New Satellites of Neptune," *Astronomical Journal* 101 (April 1991): 1511–1515.

Deep Space Network tracking (and navigation) accuracy.[41] Both optical navigation and ΔDOR helped to navigate Magellan, which operated in the S-band and X-band.[42] That mission also saw navigation software migrate from mainframes to minicomputers and, in the early 1990s, to high-performance workstations.[43]

The Galileo mission additionally marked some transitions in navigation history. It encountered two asteroids on its way to Jupiter, and asteroid research would be an integral part of the current era of space navigation. Also, Galileo optical navigation did not rely on a traditional on-board television camera, but on a CCD.[44]

A number of significant Agency-wide changes also impacted deep space navigation, such as the advent of the "faster, better, cheaper" management philosophy.[45] As a result of some of these changes, navigators found themselves on multimission teams, which meant that individual navigators concurrently worked on multiple projects, including Earth-orbiting satellites, the Space

---

41. Lincoln J. Wood, "The Evolution of Deep Space Navigation: 1989–1999" (AAS paper 08-311, read at the AAS F. Landis Markley Astronautics Symposium, Cambridge, MD, 30 June 2008), p. 3, available at *http://trs-new.jpl.nasa.gov/dspace/bitstream/2014/40880/1/08-2094.pdf* (accessed 15 October 2008). Prior to Magellan, the first navigational use of GPS by the Deep Space Network was on the Topography Experiment (TOPEX)/Poseidon project for precise orbit determination. Layland and Rauch, p. 19.
42. Jon D. Giorgini, Eric J. Graat, Tung-Han You, Mark S. Ryne, S. K. Wong, and John B. McNamee, "Magellan Navigation Using X-Band Differenced Doppler During Venus Mapping Phase," in *Proceedings of the AIAA/AAS Astrodynamics Conference, Hilton Head, SC, August 1992* (Washington, DC: AIAA, 1992), pp. 351–360; E. J. Graat, M. S. Ryne, James S. Border, and D. B. Engelhardt, "Contribution of Doppler and Interferometric Tracking During the Magellan Approach to Venus," in *Advances in the Astronautical Sciences: Astrodynamics 1991*, vol. 76, part II, ed. Bernard Kaufman, Kyle T. Alfriend, and Robert R. Dasenbrock (San Diego, CA: Univelt, 1992), pp. 919–939; Jordan interview, p. 52; Wood, "The Evolution of Deep Space Navigation: 1989–1999," p. 2; Wood, "The Evolution of Deep Space Navigation: 1962–1989," pp. 286, 294, 295.
43. Wood, "The Evolution of Deep Space Navigation: 1989–1999," p. 8.
44. R. M. Vaughan, Joseph E. Riedel, Robert P. Davis, William M. Owen, Jr., and Stephen P. Synnott, "Optical Navigation for the Galileo Gaspra Encounter," AIAA paper 92-4522 in *1992 AIAA/AAS Astrodynamics Conference, Hilton Head Island, SC, August 10–12, 1992, Technical Papers* (Washington, DC: AIAA, 1992), pp. 361–369; Pieter H. Kallemeyn, Robert J. Haw, Vincent M. Pollmeier, Francis T. Nicholson, and D. W. Murrow, "Galileo Orbit Determination for the Venus and Earth-1 Flybys," in *Advances in the Astronautical Sciences: Astrodynamics 1991*, vol. 76, part II, ed. Bernard Kaufman, Kyle T. Alfriend, and Robert R. Dasenbrock (San Diego, CA: Univelt, 1992), pp. 1013–1026; Shyam Bhaskaran, Joseph E. Riedel, and Stephen P. Synnott, "Demonstration of Autonomous Orbit Determination Around Small Bodies" (paper AAS 95-387, AAS/AIAA Astrodynamics Specialist Conference, Halifax, NS, 1995); "A Brief History of Optical Navigation at JPL," pp. 8, 10.
45. On "faster, better, cheaper," see Howard E. McCurdy, *Faster, Better, Cheaper: Low-Cost Innovation in the U.S. Space Program* (Baltimore, MD: Johns Hopkins University Press, 2001). Andrew J. Butrica, *Single Stage to Orbit: Politics, Space Technology, and the Quest for Reusable Rocketry* (Baltimore, MD: Johns Hopkins University Press, 2003), pp. 134–140, puts this management approach in historical context.

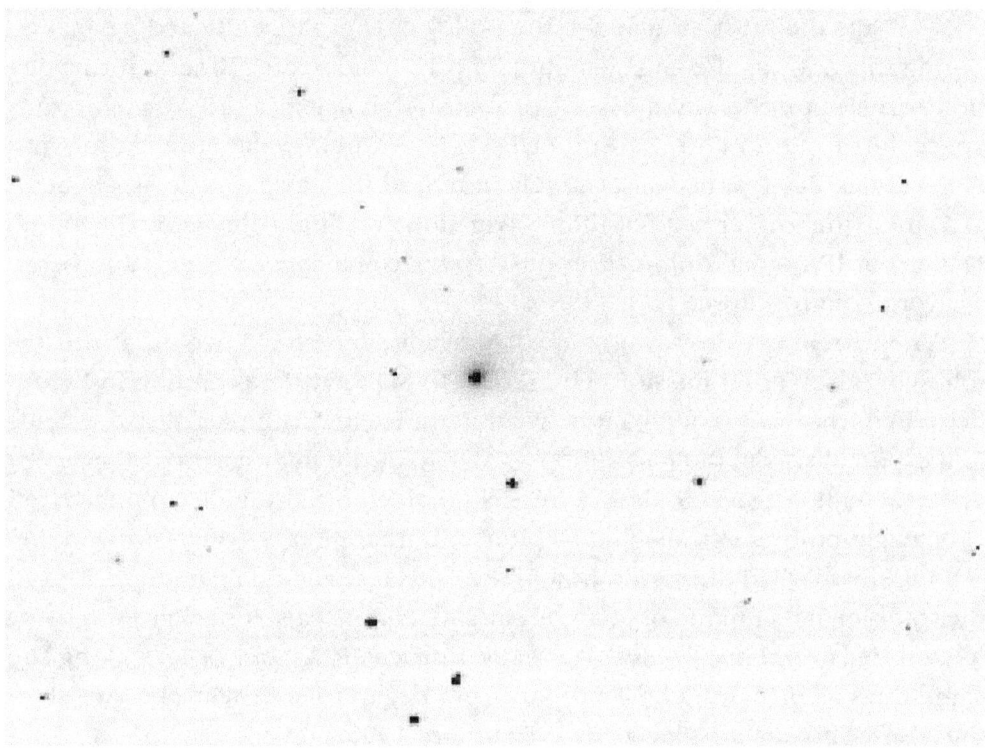

Figure 7: The first optical navigation image of Comet Tempel taken by the Deep Impact spacecraft (25 April 2005). *NASA Image PIA0788*

Shuttle, and launches by foreign governments. Navigation on the Cassini mission, launched in 1997, typified the current era in that a Multimission Navigation Team performed the navigation and the spacecraft operated in both the new experimental Ka-band and the X-band.[46]

46. See, for example, Duane C. Roth, Mark D. Guman, Rodica Ionasescu, and Anthony H. Taylor, "Cassini Orbit Determination from Launch to the First Venus Flyby" (AIAA paper 98-4563, AIAA/AAS Astrodynamics Specialist Conference, Boston, MA, 10–12 August 1998), available at JPL, "Beacon eSpace," *http://hdl.handle.net/2014/19073* (accessed 14 October 2008); Mark D. Guman, Duane C. Roth, Rodica Ionasescu, T. D. Goodson, Anthony H. Taylor, and J. B. Jones, "Cassini Orbit Determination from First Venus Flyby to Earth Flyby" (AAS paper 00-168, read at AAS/AIAA Space Flight Mechanics Meeting, Clearwater, FL, 23–26 January 2000), available at JPL, "Beacon eSpace," *http://hdl.handle.net/2014/13709* (accessed 14 October 2008); and Sami Asmar, Randy Herrera, John Armstrong, Elias Barbinis, Don Fleischman, Mark Gatti, Gene Goltz, and Luciano Iess, "First Deep Space Operational Experience with Simultaneous X- and Ka-bands Coherent Tracking" (paper read at SpaceOps 2002 Conference, Houston, TX, 9–12 October 2002), available at JPL, "Beacon eSpace," *http://hdl.handle.net/2014/10518* (accessed 14 October 2008).

Perhaps the most dramatic of the policy changes that affected navigation during this era was the engagement of a private company, KinetX, for navigation services on the MESSENGER probe to Mercury and the New Horizons spacecraft to Pluto and the Kuiper Belt. The KinetX Space Navigation and Flight Dynamics Team consists largely of retired JPL navigators led by James P. McDanell, the former head of JPL's Navigation and Flight Mechanics Section.[47] The era of JPL's total domination of deep space navigation seemed to be on the verge of an eclipse.

The current era also saw a spate of Mars missions after a two-decade hiatus following the Viking mission. The return to Mars, at times ending in failure, benefited from increasingly more accurate range and Doppler measurements and eventually from the restoration of ΔDOR. Meanwhile, a series of missions aimed at exploring asteroids and comets provided opportunities to polish optical navigation dramatically and to pioneer autonomous navigation.

Cassini marked the maturation of the optical navigation paradigm. It used the traditional technique of imaging satellites against a star background using its two CCD cameras. Optical navigation proceeded more or less along the same lines as on Voyager.[48] That paradigm began to shift with the launch of the NEAR-Shoemaker spacecraft in February 1996.

Operating in the X-band, NEAR-Shoemaker's first encounter was with the main belt asteroid 253 Mathilde on 27 June 1997. Hindering navigation by traditional Doppler and range was the probe's passage through a solar conjunction (15 February to 23 February 1997) and the need to have the Deep Space Network track it at zero declination (22 April 1997) on approach. Range and Doppler also were unusable for about a month around conjunction because of solar corona effects. Assisting these Earth-bound navigational

---

47. William L. Sjogren, interview by author, JPL, 4 November 2008, tape, NASA Historical Reference Collection, NASA History Division, NASA Headquarters, Washington, DC; Wood interview; Jordan interview, p. 61; KinetX, "Team Experience," available at *http://www.kinetx.com/snafd_experience.shtml*; KinetX, "SNAFD Programs," available at *http://www.kinetx.com/snafd_programs.shtml*; and KinetX, "SNAFD," available at *http://www.kinetx.com/snafd_space_nav.shtml* (all accessed 8 October 2008).

48. Duane Roth, Vijay Alwar, John Bordi, Troy Goodson, Yungsun Hahn, Rodica Ionasescu, Jeremy Jones, William Owen, Joan Pojman, Ian Roundhill, Shawna Santos, Nathan Strange, Sean Wagner, and Mau Wong, "Cassini Tour Navigation Strategy" (AAS paper 03-546, read at the 2003 AAS/AIAA Astrodynamics Specialist Conference, Big Sky, MT, 3–7 August 2004), available at JPL, "Beacon eSpace," *http://hdl.handle.net/2014/7042*; Stephen D. Gillam, William M. Owen, Jr., A. T. Vaughan, T. C. Wang, Vijay Alwar, J. D. Costello, R. Jacobson, D. Bluhm, and Joan L. Pojman, "Optical Navigation for the Cassini/Huygens Mission" (paper AAS 07-252, AAS/AIAA Astrodynamics Specialists Conference, Mackinac Island, MI, 19–23 August 2007), published in *Advances in the Astronautical Sciences* 129, part 1 (2008): 3–20; Jordan interview, p. 52; "A Brief History of Optical Navigation at JPL," p. 12.

efforts were numerous observations of Mathilde made at observatories around the world during the asteroid's 1995 and 1996 oppositions. The availability of the Hipparcos and Tycho star catalogs also helped to make those observations particularly accurate.[49]

For orbital operations at Eros in 2000 and 2001, the NEAR-Shoemaker mission relied on optical navigation. This was the first use of landmark-based (versus star-based) optical navigation at small bodies. Locating the target against a background of stars would not have worked well, because the high surface brightness of Eros would have required overexposing the camera in order to see the stars. Also, the asteroid's irregular shape would have complicated the task of finding its center, a requisite for optical navigation. The chosen solution was to use identifiable surface features ("landmarks") and rely on the telemetered spacecraft attitude for camera orientation.[50]

Subsequently, the Stardust mission used optical navigation in 2004 for its approach to comet Wild 2, and in 2005, the Deep Impact mission used the technique for its approach to comet Tempel 1. Although not an explorer of comets or asteroids, the Mars Reconnaissance Orbiter marked a milestone in optical navigation in 2001. Instead of taking advantage of whatever science camera already was on the spacecraft, the Mars Reconnaissance Orbiter carried its own "navcam," the Mars Optical Navigation Camera, designed and built specifically for optical navigation. The probe acquired images of Phobos and Deimos against star backgrounds, with Mars outside of the field of view

---

49. Daniel J. Scheeres, David W. Dunham, Robert W. Farquhar, Clifford E. Helfrich, James V. McAdams, William M. Owen, Jr., Stephen P. Synnott, B. G. Williams, P. J. Wolff, and Donald K. Yeomans, "Mission Design and Navigation of NEAR's Encounter with Asteroid 253 Mathilde," in *Advances in the Astronautical Sciences: Spaceflight Mechanics 1998*, vol. 99, part II, ed. Jay W. Middour, Lester L. Sacket, Louis A. D'Amario, and Dennis V. Byrnes (San Diego, CA: Univelt, 1998), pp. 1157–1173; Donald K. Yeomans, J. P. Barriot, David W. Dunham, Robert W. Farquhar, Jon D. Giorgini, C. E. Helfrich, A. S. Konopliv, James V. McAdams, J. K. Miller, William M. Owen, Jr., Daniel J. Scheeres, Stephen P. Synnott, and B. G. Williams, "Estimating the Mass of Asteroid 253 Mathilde from Tracking Data During the NEAR Flyby," *Science* 278, no. 5346 (19 December 1997): 2106–2109; Wood, "The Evolution of Deep Space Navigation: 1989–1999," pp. 15, 16.
50. William M. Owen, Jr., T. C. Wang, A. Harch, M. Bell, and C. Peterson, "NEAR Optical Navigation at Eros" (paper read at 2001 AAS/AIAA Astrodynamics Specialist Conference, Quebec City, QC, 30 July–2 August 2001), available at JPL, "Beacon eSpace," *http://hdl.handle.net/2014/12992*; William M. Owen, Jr., Thomas C. Duxbury, Charles H. Acton, Jr., Stephen P. Synnott, Joseph E. Riedel, and Shyam Bhaskaran, "A Brief History of Optical Navigation at JPL" (paper AAS 08-053, read at the 31st Annual AAS Guidance and Control Conference, Breckenridge, CO, 1–6 February 2008), p. 11.

between 10 February and 7 March 2006. Navigators at JPL processed the images on the ground in the usual fashion.[51]

Dawn, launched in September 2007, also will use landmark-based optical navigation for its orbital operations around main belt asteroid 4 Vesta and the dwarf planet Ceres. However, the image processing will use a new mapping technique, known as stereophotoclinometry, to model the terrain surrounding the landmarks. The technique creates maps from images taken under varying lighting and viewing angles and promises to be far more accurate than manual image processing.[52]

The replacement of ground-based processing of optical navigation images with spacecraft-based navigation is the objective of autonomous navigation. Autonomous navigation is a software system that transfers certain navigational computations to the spacecraft, so that it can calculate its own orbit and maneuvers from the optical navigation data to achieve mission goals. To date, only three missions have used autonomous navigation: Deep Space 1, Stardust, and Deep Impact.

The initial impetus for developing autonomous navigation came from a broader drive to develop autonomous spacecraft as a cost-saving measure during the 1970s and led to a modest research and development effort at JPL. As navigator Frank Jordan mused about that first step: "We had R&D in autonomous navigation ahead of its time. Evidently a hundred years ahead of its time."[53]

The opportunity to demonstrate autonomous navigation finally arrived with the establishment of NASA's New Millennium Program and the experimental Deep Space 1 mission, the purpose of which was to demonstrate a panoply of experimental technologies, including autonomous navigation. Development of the prototype "AutoNav" system began in 1995 as a research endeavor under-

51. Tung-Han You, Allen Halsell, Dolan Highsmith, Jah Moriba, Stuart Demcak, Earl Higa, Stacia Long, and Shyam Bhaskaran, "Mars Reconnaissance Orbiter Navigation" (read at the AIAA/AAS Astrodynamics Specialist Conference, Providence, RI, 16–19 August 2004), available at JPL, "Beacon eSpace," *http://hdl.handle.net/2014/40425* (accessed 3 October 2008); Jordan interview, p. 67; "A Brief History of Optical Navigation at JPL," p. 15.

52. Robert W. Gaskell, Oliver S. Barnouin-Jha, and Daniel J. Scheeres, "Modeling Eros with Stereophotoclinometry" (paper read at 38th Lunar and Planetary Science Conference, League City, TX, 12–16 March 2007), abstract 1333, available at *http://www.lpi.usra.edu/meetings/lpsc2007/pdf/1333.pdf* (accessed 17 October 2008); Robert W. Gaskell, "Landmark Navigation and Target Characterization in a Simulated Itokawa Encounter" (paper read at AAS/AIAA Astrodynamics Specialists Conference, Lake Tahoe, CA, 7–11 August 2005), available at JPL, "Beacon eSpace," *http://hdl.handle.net/2014/37465*; "A Brief History of Optical Navigation at JPL," p. 17.

53. Jordan interview, p. 61. See also JPL Research and Technology Objectives and Plans (RTOP), "Autonomous Navigation for Unmanned Spacecraft," 15 May 1978, copy provided to author by Frank Jordan.

written with JPL internal funds; it then evolved into the system accepted for inclusion on Deep Space 1, which launched in October 1998.[54]

During the probe's cruise phase, AutoNav obtained weekly images of several "beacon asteroids," brighter main belt asteroids whose ephemerides were well known thanks in part to an intensive ground-based observing campaign at JPL's Table Mountain Observatory.[55] Three hours before its closest approach to asteroid 9969 Braille, the Deep Space 1 AutoNav system was supposed to calculate and execute a maneuver to control the aiming point to an accuracy of 3 kilometers and provide information for timing the science images. However, a failure in the flight version of AutoNav prevented it from doing the maneuver calculation (the ground version worked fine, and its results were used), and inadequacies in the experimental camera made it impossible for AutoNav to detect the asteroid. Deep Space 1, on an extended mission, next flew by comet 19P/Borrelly. The problems experienced during the Braille encounter were resolved, and AutoNav performed flawlessly.[56]

Subsequently, the Stardust mission, launched in February 1999, flew with an autonomous navigation system. During a "dress rehearsal" of the system on 2 November 2002, it successfully tracked asteroid 5535 Annefrank, and on 2 January 2004, it successfully tracked comet 81P/Wild 2. Initiated 20 or 30 minutes prior to encounter, the autonomous navigation system controlled the camera mirror as well as the spacecraft's attitude. The system tracked the

---

54. Shyam Bhaskaran, Joseph E. Riedel, Shailen D. Desai, Philip J. Dumont, George W. Null, William M. Owen, Jr., Stephen P. Synnott, and R. A. Werner, "Orbit Determination Performance Evaluation of the Deep Space 1 Autonomous Navigation System" (paper AAS 98-193, AAS/AIAA Space Flight Mechanics Meeting, Monterey, CA, 9–11 February 1998), available at *http://trs-new.jpl.nasa.gov/dspace/bitstream/2014/22668/1/97-1182.pdf* (accessed 14 October 2008); Shyam Bhaskaran, Joseph E. Riedel, Stephen P. Synnott, and T. Wang, "The Deep Space 1 Autonomous Navigation System: A Post-Flight Analysis" (paper AIAA 2000-3935, AIAA/AAS Astrodynamics Specialist Conference, Denver, CO, 14–17 August 2000), available at *http://trs-new.jpl.nasa.gov/dspace/bitstream/2014/15617/1/00-1323.pdf* (accessed 14 October 2008); Jordan interview, p. 62; Bhaskaran interview.

55. William M. Owen, Jr., Synnott and George W. Null, "High-Accuracy Asteroid Astrometry from Table Mountain Observatory," in *Modern Astrometry and Astrodynamics*, ed. Rudolf Dvorak, H. F. Haupt, and K. Wodnar (Vienna, Austria: Verlag der Österreichischen Akademie der Wissenschaften, 1998), pp. 89–102; Marc D. Rayman, Philip L. Varghese, David H. Lehman, and Leslie L. Livesay, "Results from the Deep Space 1 Technology Validation Mission," *Acta Astronautica* 47 (2000): 475–487, esp. p. 478; and "A Brief History of Optical Navigation at JPL," p. 14.

56. Shyam Bhaskaran, Joseph E. Riedel, Brian Kennedy, and T. C. Wang, "Navigation of the Deep Space 1 Spacecraft at Borrelly" (paper AIAA 2002-4815, AAS/AIAA Astrodynamics Specialist Conference, Monterey, CA, 5–8 August 2002), available at *http://trs-new.jpl.nasa.gov/dspace/bitstream/2014/8839/1/02-1389.pdf* (accessed 14 October 2008); Bhaskaran interview; "A Brief History of Optical Navigation at JPL," p. 14.

target only during the time around closest approach; calculation of the final maneuver took place on the ground.[57]

The most demanding use of autonomous navigation to date has been the Deep Impact mission. Launched in January 2005, its mission was to send an Impactor spacecraft into the nucleus of comet 9P/Tempel 1 while the main Flyby spacecraft, the Hubble Space Telescope, and various ground observatories watched the event on 4 July 2005. During the two months prior to encounter, navigators used traditional ground-based optical navigation. Two hours before impact, the autonomous navigation system on both vehicles started up. On the Impactor, the system commanded pictures, performed orbit determination, and computed and executed three targeting maneuvers in the last 90 minutes. The same software ran in parallel on the Flyby spacecraft, but it operated in a different mode, concentrating on determining the time of closest approach to the comet, so that the Flyby's cameras could take pictures of the nucleus before, during, and after impact. The spectacular images that Deep Impact acquired attested to the success of the autonomous navigation.[58]

Celestial Mechanics

At the heart of all navigation success is the accuracy achieved in knowing the location and motion of the target body. Over the centuries, astronomers have created tables to describe such motions. These ephemerides, as they are known, have served to improve navigation over the deep seas and oceans of our own planet. Creating those tables was—and remains—the job of national almanac

---

57. Shyam Bhaskaran, Joseph E. Riedel, and Stephen P. Synnott, "Autonomous Nucleus Tracking for Comet/Asteroid Encounters: The STARDUST Example" (paper AAS 97-628, AAS/AIAA Astrodynamics Specialist Conference, Sun Valley, ID, 4–7 August 1997), available at *http://trs-new.jpl.nasa.gov/dspace/bitstream/2014/22498/1/97-0995.pdf* (accessed 14 October 2008); Shyam Bhaskaran, Nickolaos Mastrodemos, Joseph E. Riedel, and Stephen P. Synnott, "Optical Navigation for the STARDUST Wild 2 Encounter," in *Proceedings of the 18th International Symposium on Space Flight Dynamics, Held October 11–15, 2004, Munich, Germany*, ed. Oliver Montenbruck and Bruce Battrick (Noordwijk, Netherlands: ESA SP-548, 2004), p. 455; Shyam interview; "A Brief History of Optical Navigation at JPL," p. 14.
58. William M. Owen, Jr., Nickolaos Mastrodemos, Brian P. Rush, Tseng-Chang M. Wang, Stephen D. Gillam, and Shyam Bhaskaran, "Optical Navigation for Deep Impact," in *Spaceflight Mechanics 2006—Part II*, ed. Srinivas Rao Vadali, L. Alberto Cangahuala, Paul W. Schumaker, Jr., and Jose J. Guzman (San Diego, CA: Univelt, 2006), pp. 1231–1250; Mark Ryne, David Jefferson, Diane Craig, Earl Higa, George Lewis, and Prem Menon, "Ground-Based Orbit Determination for Deep Impact," in *Spaceflight Mechanics*, ed. Vadali et al., pp. 1179–1202; Nicholas Mastrodemos, Daniel G. Kubitschek, Robert A. Werner, Brian M. Kennedy, Stephen P. Synnott, George W. Null, Joseph E. Riedel, Shyam Bhaskaran, and Andrew T. Vaughan, "Autonomous Navigation for Deep Impact," in *Spaceflight Mechanics*, ed. Vadali et al., pp. 1251–1270; "A Brief History of Optical Navigation at JPL," p. 15; Bhaskaran interview.

offices, such as the one that the U.S. Naval Observatory established in the 19th century.[59] The core of the astronomy that went into those ephemerides was a set of physical constants, such as the astronomical unit (the average distance between Earth and the Sun).

With the advent of deep space exploration, JPL navigators needed extremely precise ephemerides of the Moon, planets, and other solar system bodies in order to compute the trajectories of NASA spacecraft with the necessary degree of accuracy. Initially, they digitized the ephemerides tables published by the U.S. Naval Observatory, which were in the form of books, not computer-compatible cards or tape.[60] The almanac office's tables, however, quickly proved to lack the precision that deep space navigation needed. Instead, they had to rely on the calculations of physical constants made by their orbit determination program following each mission as well as the highly precise measurements furnished by radar astronomers. The need for more precise values for physical constants also drove JPL to establish its own set of constants in cooperation with engineers and scientists from other NASA Centers involved in Earth satellite and human spaceflight navigation. As a result, those values became the ad hoc values applied across the board in the NASA space program, entirely without the approval of NASA Headquarters.[61]

At the same time, astronomers were attempting to improve their own system of constants. In 1964, the International Astronomical Union adopted what came to be called the so-called 1968 system of constants, meaning they went into effect in 1968, which incorporated the radar-derived value for the astronomical unit.[62] The growing influence of values computed by navigators was visible in the 1974 International Astronomical Union system of constants and peaked during the organization's 1979 meeting, when it approved, effective 1 January 1984, the JPL Development Ephemeris 200 as the source of values for constants as well as the basis for the theories and tables used by participating

---

59. See, for example, Steven J. Dick, "Foundations of the American Nautical Almanac Office," *Sky and Ocean Joined: The U.S. Naval Observatory, 1830–2000* (New York, NY: Cambridge University Press, 2002), chap. 3, pp. 118–139.
60. R. Henry Hudson, *Subtabulated Lunar and Planetary Ephemerides* (Pasadena, CA: JPL TR 34-239, 2 November 1960), pp. 1–2; Douglas B. Holdridge, *Space Trajectories Program for the IBM 7090 Computer* (Pasadena, CA: JPL TR 32-223, 1 September 1962), p. 2; and P. R. Peabody, James F. Scott, and Everett G. Orozco, *Users' Description of JPL Ephemeris Tapes* (Pasadena, CA: JPL TR 32-580, 2 March 1964), pp. 9, 10.
61. Victor C. Clarke, Jr., *Constants and Related Data for Use in Trajectory Calculations as Adopted by the Ad Hoc NASA Standard Constants Committee* (Pasadena, CA: JPL TR 32-604, 6 March 1964), pp. 1, 2, 10.
62. Butrica, *To See the Unseen*, pp. 46–49.

countries' almanac offices.[63] In effect, navigators at JPL were now in charge of the ephemerides within the United States as well as overseas. The solutions of the practical world of deep space navigation had become the theory and practice of astronomy.

Conclusion

Clearly, in order for NASA's missions of exploration to achieve their scientific goals, they need accurate navigation, and navigational accuracy in turn depends critically on astronomy. Navigation has borrowed techniques from radio astronomy, and radar astronomy has lent its precise measurements. Dynamic astronomy furnished the initial ephemerides and constants for navigation, and throughout the sustained period of exploration of the outer planets, asteroids, and comets, ground-based astronomers assisted in refining the ephemerides of those bodies, and their revised star catalogs were crucial in the development of optical navigation.

Meanwhile, the additional accuracy of deep space navigation techniques facilitated the contributions of navigation to astronomy by, for instance, refining physical constants and ephemerides. That accuracy also contributed to the planetary sciences with the discovery of mascons and the study of planetary atmospheres with radio waves, to cite two examples. Optical navigation, in particular, led to the discovery of more than a few new moons around the outer planets as well as the first sighting of volcanism on another celestial body. In turn, optical navigation relies on astronomers' star catalogs, ephemerides, and observations. Navigation, astronomy, and the planetary sciences thus are intertwined in a relationship that is both mutually beneficial and mutually necessary.

---

63. Dick, pp. 532–533, 537–542.

Chapter 20

NASA's Earth Science Program

The Space Agency's Mission to Our Home Planet

Edward S. Goldstein

> Let us remember as we chase our dreams into the stars that our first responsibility is to our Earth, to our children, to ourselves. Yes, let us dream, and let us pursue those dreams, but let us also preserve the fragile world we inhabit.
>
> —President George H. W. Bush, 1989[1]

A Storm and Its Aftermath

A good point to start understanding NASA's Earth science program and its place in an outward-focused space agency is to examine the events surrounding Hurricane Katrina, as well as two different ways in which the storm affected NASA.

Hurricanes are often considered episodic acts of God, which only briefly command the public's attention. Katrina was a dramatic exception to this rule. Four years after the storm's 29 August 2005 landfall on the southeast Louisiana coast at Category 3 strength on the Saffir-Simpson hurricane scale,[2] people are still passionately talking about Katrina's economic, political, scientific, and social implications. NASA is not immune from those discussions.

The first NASA-Katrina connection involves the heroic mythology of the space agency's human space exploration program. During the early morning

1. J. A. Angelo and I. W. Ginsberg, eds., *Earth Observations and Global Change Decision Making, 1989: A National Partnership*, vol. 1 (Malabar, FL: Krieger Publishing, 1990), p. i.
2. Chris Mooney, *Storm World: Hurricanes, Politics, and the Battle Over Global Warming* (Orlando, FL: Harcourt Publishing Co., 2007), p. 2.

hours of 29 August, when the storm hammered greater New Orleans, Louisiana, a ride-out volunteer crew, known as the "Marshworks Team," hunkered down at the Michoud Assembly Facility in east New Orleans, where external tanks are assembled for Space Shuttle flights. In their work to protect the facility and its space hardware from the storm's fury, the Marshworks Team performed with incredible bravery. Two employees, Joe Barrett and Dan Doell, fought through the night at Michoud's pump house to keep flooding from occurring. And when they were forced to abandon their post as the hurricane's winds grew in intensity, Barrett and Doell made a "gut call" on where to place the pump settings in order to keep the waters at bay. As NASA Michoud Chief Operating Officer Patrick Scheuermann later wrote, "In hindsight, they made exactly the right call, because any other setting would have either run the pumps dry or they would have not kept up with the incoming flood. This single action . . . may possibly have saved the Space Shuttle Program. If not for this team there would have been at least eight feet of standing water in the factory."[3]

From the standpoint of this history, the role of NASA's Earth observing satellites in monitoring Katrina's formation, path, and devastation represents the second major relationship between Katrina and the space program. Viewers of the NASA Web site, *http://www.nasa.gov*, could see near-real-time details of Katrina's cloud-top heights and cloud-tracked wind velocities imaged by the Multi-angle Imaging SpectroRadiometer (MISR) on the Terra satellite; information about Katrina's precipitation, energy, and winds obtained by the Tropical Rainfall Measuring Mission (TRMM), the Advanced Microwave Scanning Radiometer-EOS (AMSR-E) instrument on the Aqua satellite, and the Quick Scatterometer (QuikSCAT) satellite; and images of New Orleans flooding documented by the Landsat 7 Earth resources monitoring satellite and Moderate Resolution Imaging Spectroradiometer (MODIS) instrument on the Terra satellite.[4]

What we learned about Katrina from these satellites showcased the results of a multidecade, multi-billion-dollar commitment NASA undertook in 1991 to develop a comprehensive Earth Observing System (EOS) of satellites and instruments to better understand the dynamics of our changing planet. The system is aligned with the space agency's strategic goal 3A: "Study Earth from

---

3. Patrick Scheuermann, "Not on My Watch: The Saga of the Michoud Marshworks Heroes," in *NASA: 50 Years of Exploration and Discovery* (Tampa, FL: Faircount Media Group, 2008), p. 302.
4. NASA Web site, "Hurricane Season 2005: Katrina," available at *http://www.nasa.gov/vision/earth/lookingatearth/h2005_katrina_prt.htm* (accessed 18 December 2008).

space to advance scientific understanding and meet societal needs,"[5] and EOS is part of a larger Earth science program at NASA. This program, with the aid of a comprehensive Earth Observing System Data and Information System (EOSDIS), has supported the work of a generation of Earth science researchers, and it has led to significant improvements in the comprehensiveness of computer climate models. It has helped to revolutionize the way we understand our dynamic, ever-changing planet. Today, there is a constellation of 15 Earth observing satellites in orbit, 6 satellites under development,[6] and 2 satellites recommended by the Decadal Survey for Earth Science and Applications of the National Research Council under formulation.

In its comprehensiveness, NASA's multisatellite Earth science program represents a departure from NASA's early history of pioneering new satellite technologies for Earth observations, with weather satellites being the classic example, and turning them over to an operational agency such as NOAA. In the case of EOS, NASA not only developed new technologies, but also operated them in space for several years and invested heavily in scientific research related to the observations of its satellite missions. Also, the information produced by the suite of EOS satellite instruments is integrated, so rather than just measuring a few distinct variables such as temperature, precipitation, vegetation, transport of air pollution, ocean wind speed, and sea ice, we are viewing them in an integrated fashion, allowing scientists to understand the relationship between these variables.

While many observers view the space program as being primarily about human spaceflight—a perception that is understandable when one looks at NASA's budget, the statements of its senior leaders, and media coverage—and secondarily about planetary exploration and astrophysics (for example, the Hubble Space Telescope), with Earth science relegated to a lesser role, the relevance to society of NASA's Earth science efforts after Katrina prompted new discussion. People in the scientific and public policy community asserted that Katrina represented an example of how tropical storms can intensify as

5. NASA, *2006 NASA Strategic Plan* (Washington, DC: NASA NP-2006-02-423-HQ, 2006), p. 8.

6. The NASA satellites under development will help measure ocean topography (Ocean Surface Topography Mission, 2008); global aerosol and cloud properties, along with total solar irradiance (Glory, 2009); sea surface salinity (Aquarius, 2009); global land cover change (Landsat Data Continuity Mission, 2011); and global rainfall from tropical to mid-latitudes (Global Precipitation Measurement, 2012). The two missions recommended by the decadal survey for Earth Science and Applications from Space will observe ice cover (ICESat-II) and soil moisture (Soil Moisture Active and Passive). The OCO mission failed on launch in February 2009. Congress has directed NASA to begin work on a replacement OCO satellite mission.

a result of ocean temperature warming related to increasing greenhouse gas emissions,[7] thus underscoring the scientific relevance of NASA's EOS in helping expand knowledge about climatic change.

In commenting on the larger issue of climate change and NASA's contributions to its study, the *Washington Post* editorially contended, on 31 July 2006, that NASA more than NOAA was "uniquely qualified to launch and maintain weather satellites" and stated that "[President George W.] Bush needs to get his head out of the stars. Even though scientists agree that Earth is warming, they still need to investigate how, and how fast, the phenomenon is proceeding—a much more pressing task than landing on the Red Planet. The White House has to either pay responsibly for its exploration programs or cancel them."[8] Also, the *Boston Globe* argued on 15 June 2006:

> Trips to the moon or Mars, which the president also favors, fit the better-known part of NASA's mission to explore space. But at a time when climate change, in particular, is threatening the well-being of the planet, NASA should be increasing, not decreasing, funding for projects like the canceled satellite mission to measure global soil moisture. A climate observatory in deep space [Triana], which would monitor this planet's solar radiation, ozone, clouds, and water vapor, has also been dropped . . . . But whether the cuts in earth science are punitive or simply a case of misguided priorities, they are a mistake that Congress should rectify. The potential long-term benefits of moon and Mars flights should not be allowed to squeeze out the worthwhile work NASA is doing in its deep-space probes, its attempt to repair the Hubble Space Telescope, and its earth-science research.[9]

---

7. The scientific debate about hurricanes largely concerns the contention that hurricanes become more intense when they pass over bodies of water that are warmer as a result of greenhouse gas emissions. According to NASA, "Research shows a link between the intensity of hurricanes in the region and oceanic heat content. In late August 2005, when Katrina passed over the Loop Current and large warm eddies called the core ocean ring, it evolved over the Gulf of Mexico from a Category 3 to Category 5 hurricane in only nine hours. The warm waters of the Loop Current appeared to have rapidly fueled the storm while the warm core rings seemed to have sustained the storm's intensity" (NASA, *Fiscal Year 2005 Performance and Accountability Report* [Washington, DC: NASA NP-2005-11-417-HQ, 2005], p. 16).
8. "Meanwhile, Back on Earth," *Washington Post* (31 July 2006): A14.
9. "Earth to NASA: Help!" *Boston Globe* (15 June 2006): A20.

Even voices not associated with liberal newspaper editorial pages chimed in on the subject. In one of his last public statements, retired "Mercury 7" astronaut Wally Schirra objected to the decision by NASA's leadership in 2006 to delete the goal of "understanding and protecting our home planet" from its mission statement. In the *New York Times*, he wrote: "My theme when I ran an environmental company after leaving the space program was, and still is: 'I left Earth three times, and found no other place to go. Please take care of Spaceship Earth.'"[10]

For my 2007 doctoral dissertation at the George Washington University, "NASA's Earth Science Program: The Bureaucratic Struggles of the Space Agency's Mission to Planet Earth," I documented how Earth science at NASA, currently representing one-tenth of the Agency's budget and two-thirds of the U.S. government's climate change research effort, developed over time into a comprehensive program that continues to make important contributions to scientific understanding of climate change and other Earth processes, underwent several major structural changes as it was buffeted by technical and political challenges, and became a maelstrom of contention as NASA's priorities were under debate during the George W. Bush administration. This paper revisits many of the issues I wrote about and anticipates how Earth science at NASA may fare in the Agency's second half century.

The Four Stages of NASA's Earth Science Program

To provide some perspective, NASA's Earth science efforts grew from initial attempts to launch weather and Earth resources monitoring satellites into an activity that received more focused attention beginning in about 1972. From that point on, NASA's Earth science program largely unfolded in four distinct stages related to planning and public policy agenda access, decision-making, and policy implementation:

Stage One: *Program Planning* (1972 to 1986). During this stage, NASA became interested in the subject of Earth science as a potential Agency mission focus, as a means to broaden the Agency's portfolio and to address specific issues of concern to NASA from an institutional standpoint as well as to society. As techniques for Earth remote sensing from space matured, NASA policy entrepreneurs plotted to move beyond a sporadic satellite mission or two focused on Earth.

---

10. Walter Schirra, "An Astronaut's View," *New York Times* (1 August 2006): A18.

Stage Two: *Agenda Access and Program Adoption* (1987 to 1990). In this stage, the Reagan and George H. W. Bush White Houses, under the auspices of an interagency Committee on Earth Sciences (CES), became receptive to a proposed course of action involving coordinated interagency scientific research of climate and other change occurring on Earth. Under the rubric of this interagency effort, NASA proposed to pursue a Mission to Planet Earth (MTPE) that would provide an EOS of coordinated satellite and instrument measurements of Earth's atmosphere, land, and oceans, to be obtained over a period of 15 years. Contributing to government leaders' willingness to support this new "mission" for NASA was the desire to give the Agency new purpose and focus following the morale-bursting Space Shuttle *Challenger* disaster.

In 1989, the George H. W. Bush administration formally proposed NASA's pursuit of EOS, along with a more ambitious plan for human exploration of the Moon and Mars, the so-called "Space Exploration Initiative." Congress would only approve EOS, with the initial funding promise of $17 billion over a 10-year period, while rejecting the projected $400 billion SEI.

Stage Three: *Program Restructuring* (1990 to 1996). The adoption of the EOS program was rapidly followed by an early implementation stage in which NASA found its initial approach of launching large Earth observing platforms wanting, as it began to face the fiscal realities of the 1990s. During this stage, the design of EOS was changed several times to reduce its cost and complexity, with medium-sized satellites replacing large platforms.[11] After the 1994 congressional elections, members of the new Republican majority—especially conservative House Republicans skeptical about the global warming issue—questioned the purposes and costs of mounting a large-scale Earth science program. In response to this challenge, NASA fought congressional attempts to slash EOS funding and focused some of its efforts on obtaining practical applications that would expand the constituency for the program beyond scientists and those interested in the subject of global warming.

Anticipating slim future budgets, NASA also decided during this stage to drop plans to produce replacements for the three main EOS satellites, Terra, Aqua, and Aura. Instead, NASA would transfer technology to the National Polar-orbiting Operational Environmental Satellite System (NPOESS) being jointly planned by NOAA and DOD so that NPOESS sensors could make climate-quality measurements and extend the data record begun by EOS.

---

11. For example, Aqua and Aura, two of the three main EOS satellites currently orbiting Earth, are smaller versions of the large platform originally designated EOS PM-1.

Stage Four: *Program Implementation* (1997 to 2008). In the fourth stage, NASA launched and operated the flagship EOS satellites and instruments, obtaining an impressive scientific return. However, NASA struggled to maintain long-term momentum for Earth science as its science budget was squeezed to pay for the Vision for Space Exploration, President George W. Bush's plan to orient the Agency's future work toward the exploration of the Moon, Mars, and beyond. A potential guiding star for the program is the National Research Council's Decadal Survey for Earth Science and Applications from Space, released in 2007, which has provided recommendations for a coherent program of Earth science missions at the Agency through 2020, based on an annual budget of $2 billion, representing an annual $500 increase over current NASA Earth science funding. In his first months in office, President Barack Obama targeted $325 million in American Recovery and Reinvestment Act stimulus funding to accelerate the development of NASA's Earth science climate research missions. President Obama also stated in April 2009 remarks to the annual meeting of the National Academy of Sciences that strengthening Earth observations from space would be a priority of his Council of Advisers on Science and Technology.

Stage One: Program Planning (1972 to 1986)

New Approaches to Studying Our Home Planet

Following initial, but sporadic, launches of weather and Earth resources monitoring satellites from 1960 to 1972, Earth science actually began to draw the attention of NASA leadership for reasons largely to do with NASA's top mission priority, human spaceflight. NASA's concern that ozone-depleting emissions from the proposed Space Shuttle's engine exhaust might pose a barrier to congressional approval for the program led the Agency to organize a research effort to address this question. The people conducting this research, including physical chemist Shelby G. Tilford and climate modeler James E. Hansen, prompted NASA to conduct other Earth science research activities.

In the late 1970s, an Upper Atmosphere Research Office was set up in the Agency's Office of Space Science and Applications (OSSA) to manage satellite development and research activities. NASA's ozone research program eventually consisted of Nimbus 7 (1978), which studied the upper atmosphere, and the Upper Atmosphere Research Satellite (1991), a satellite dedicated to ozone layer

Figure 1: Shelby Tilford. *Courtesy of Shelby Tilford*

depletion. In 1985, NASA's work with other agencies in confirming ground-based observations of an ozone hole forming over Antarctica dramatized the significance of environmental change and underscored the relevance of NASA satellite technology to understanding the risk. This and other findings contributed to the 1988 Montreal Protocols, an international agreement requiring signatory nations to develop nondestructive alternatives to chlorofluorocarbons.

Program Manager Shelby Tilford said the input of science to this agreement marked an important turning point:

> For all practical purposes, to . . . decide that man-made activities were having an impact, and to do something about it on an international basis was really . . . the first time that science ever made an impact, one way or another, with respect to international regulation of anything.[12]

From NASA's internal standpoint, added Tilford, the ozone work also affected the Agency's ambitions to conduct Earth science on a more sustained basis:

> When I came to NASA in 1976 essentially there was no real research program in Earth science. There were a number of instrument demonstration programs or instrument development programs. But most of them were related to the technology and handing it off to someone else to interpret . . . . In general NASA flew the instruments and someone else for the most part did the analysis. For the ozone issue for the first time NASA really got a charter to do something in Earth science from a science point of view . . . . After we had some success with that and consolidating a lot of scientists from a lot of universities, a lot of research centers, other agencies, we all became a fairly cohesive unit with respect to the upper atmosphere. But nothing much was happening otherwise. A few ocean instrument development programs were being done

12. Shelby Tilford, interview by author, 2 June 2006.

> . . . . But a number of us thought perhaps its time for us to put together a program for Earth science similar to what we did with the ozone issue.[13]

From that point on, Tilford, aided by atmospheric scientist Francis Bretherton, the head of the NASA Advisory Committee's Earth Systems Science Committee, took the lead inside NASA and within the scientific and public policy community in championing the idea of NASA taking a more aggressive role in conducting Earth science.

James Hansen and the Goddard Institute for Space Studies

Today, James Hansen is well known due to his strong statements about the need for global society to reduce greenhouse emissions and because of the clumsy attempts of NASA Public Affairs personnel to control his public interviews from 2005 to 2006.[14] Yet his actual scientific output and the origins of his work on climate also deserve attention. Hansen, current director of NASA's Goddard Institute of Space Studies (GISS) in New York City, New York, explained how GISS's pioneering work in climate modeling with high-capacity general processing computers began with the ozone question:

> The way the greenhouse effect came up was in the 1970s there was a realization, or a hypothesis that humans might affect the ozone layer . . . . I was interested in that problem because . . . if you change the ozone you could change the climate . . . . I decided to propose that we would take a [computer] weather model and convert it into a climate model and use it for that problem, to see what is the effect of ozone changes and what is the effect of other [changes] . . . and NASA got identified as the agency to address that stratospheric problem . . . . So that gave us an opportunity to propose to NASA that we would make this climate model . . . . I started this little group and got some support for it. By the time a few years later when I

---

13. Ibid.
14. See, for example, Mark Bowen, *Censoring Science: Inside the Political Attack Against James Hansen and the Truth of Global Warming* (New York, NY: Dutton/Penguin, 2007).

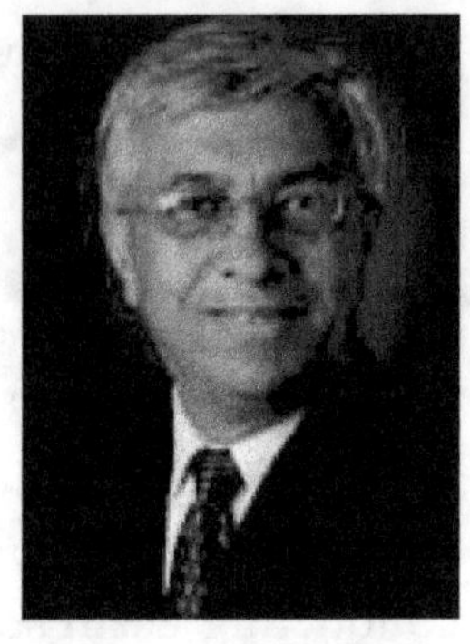

Figure 2: Ghassem Asrar. *Courtesy of Ghassem Asrar*

> had the opportunity to be the director here in 1981, I decided to make our focus on global change.[15]

The work of Hansen's GISS team in developing general atmospheric circulation models tangibly led to improved climate modeling. Author Spencer Weart wrote that "by simplifying some features while adding depth to others, the Hansen team managed to get a quite realistic-looking simulation that ran an order of magnitude faster than rival general circulation models. That permitted the group to experiment with multiple runs, varying one factor or another to see what changed."[16] By the late 1970s, noted journalist Bill McKibben, Hansen wrote a paper for *Science* magazine based on

> this first climate model, of the planet's climate that allows its user to, say, add a layer of carbon dioxide to the atmosphere and see what happens. What happens . . . is that the temperature goes up, a lot: He predicted that "the continuing increase in fossil fuel use would lead to about 4.5-degree Fahrenheit global warming by the end of the twenty-first century." The incoming Reagan administration . . . did not want to hear that and so they cut his funding to the bone, forcing him to lay off most of his NASA staff.[17]

A Systematic Approach to Understanding Climate Change

Along with climate modelers, other prominent scientists began thinking that studying Earth processes on a systematic basis might hold the key to understanding climate change phenomena. This was a change broadly occurring in the scientific disciplines related to Earth such as atmospheric physics, geology, and oceanography. As Ghassem Asrar, who led NASA's Earth Science program from 1998 to 2005, observed,

15. James Hansen, interview by author, 19 August 2005.
16. Spencer Weart, *The Discovery of Global Warming* (Cambridge, MA: Harvard University Press, 2003), pp. 110–111.
17. Bill McKibben, "Too Hot to Handle," *Boston Globe* (5 February 2006), available at *http://boston.com/news/science/articles/2006/02/05/too_hot_to_handle?mode=PF* (accessed 10 February 2006).

> What began as a series of space-technology demonstrations led the broader scientific community to think in new ways about Earth science. Just as satellites view the Earth without regard to national boundaries, they see the Earth without regard to traditional boundaries of scientific disciplines. From space, the interactions among continents, oceans, atmosphere, ice caps, and life itself are most striking. The transitions from circulation ocean to sea ice to ice caps across seasons and years; the global respiration of terrestrial and marine vegetation; the transport of Saharan dust across the Atlantic; the regional differentiation as global average temperatures rise; the human influences on global change and their consequences—these emerge as questions of interest that no one Earth science discipline can adequately address. In the 1980s the science community, under NASA sponsorship, developed the interdisciplinary concept of *Earth System Science* as the framework within which to pursue research on questions of global change and regional changes in their global context.[18]

Beyond the germination of new concepts, tangible technological developments were occurring that enabled greater leaps in Earth observations. With the TIROS-1 weather satellite, NASA had pioneered the use of passive remote sensing, or the use of instruments to receive either reflected or radiated energy from the object being observed. By the 1980s, NASA was beginning to develop active remote sensing sensors that "actually transmit energy from the instrument to the object and then observes the characteristics of the 'echo' coming back"[19] These instruments, placed on orbiting satellites, would form the core of EOS.

Earth Science Becomes a NASA Strategic Priority

While the work of committed government scientists was a moving force behind the rise of Earth science at NASA, also essential was leadership from the top. Prior to this period, the strong insistence of NASA's second Administrator, James Webb, that NASA be much more than just a human space exploration agency provided the foundation for NASA's later commitment to Earth science and applications. In

18. Linda Glover, ed., *The National Geographic Encyclopedia of Space* (Washington, DC: The National Geographic Society, 2004), p. 296.
19. Ibid., p. 265.

a recorded meeting with President Kennedy in the White House Cabinet Room on 21 November 1962, Webb told the President, "If I go out and say that this [Project Apollo] is the number one priority and that everything else must give way to it, I'm going to lose an important element of support for your program and for your administration."[20] Webb's Associate Administrator, Robert Seamans, added, "When you say something has a top priority, in my view it doesn't mean that you completely emasculate everything else if you run into budget problems on the Apollo and the Gemini. Because you could very rapidly completely eliminate your meteorological program, your communications program . . . ."[21]

A decade later, President Nixon had canceled the final three planned Apollo lunar landings, and the Agency was striving to demonstrate its relevance to a public that had become increasingly blasé about human space missions. Historian Roger Launius noted in a profile of James C. Fletcher's two stints as NASA Administrator (1971 to 1977 and 1986 to 1989) that Fletcher's decision to engage the Agency in ozone research also represented a calculated decision to present NASA to policy-makers as a multimission agency that could help provide knowledge to solve social problems. Launius wrote that Fletcher, the son of pioneering Mormons, drew on the "peculiarly Mormon stewardship principle, conservation ethic, and zionic/utopian ideal" to "emphasize space exploration as a means of helping to preserve the Earth and to make it something better than what it already was."[22] Fletcher told a Senate Committee on Aeronautical and Space Sciences the following in 1973:

> As you know, NASA is called the space agency, but in a broader sense, we could be called an environmental agency. It is not just that space is our environment, but it is rather that, as you have seen, virtually everything we do, manned or unmanned, science or applications, helps in some practical way to improve the environment of our planet and helps us understand the forces that affect it. Perhaps that is our essential task, to study and understand the Earth and its environment.[23]

---

20. Transcript of presidential meeting in the Cabinet Room of the White House, "Supplemental appropriations for the National Aeronautics and Space Administration (NASA)," 21 November 1962, p. 18.
21. Ibid., p. 19.
22. Roger Launius, "A Western Mormon in Washington, DC: James C. Fletcher, NASA, and the Final Frontier," *Pacific Historical Review* 64 (May 1995): 220.
23. James Fletcher, *Spaceship Earth: A Look Ahead to a Better Life*, NASA Congressional Testimony (Washington, DC: GPO, 1973), p. 28.

Robert Frosch, the research scientist who was President Carter's NASA Administrator from 1977 to 1981, brought a scientist's intuition into how Earth science should be viewed at the Agency. Frosch recalled observing a change of thinking in internal NASA deliberations about potential future missions:

> The bigger battles in fact tended to be over the balance between Earth looking applications, which were just sort of arbitrarily defined as applications, and space [science]. It was sort of a three way science thing—earth looking applications, space science, which tended to come into two major divisions, solar system and deep space, astrophysics roughly speaking, with the moon as an interesting subset of the solar system. And, in fact, one of the things that was going on was the beginnings of the generalization of the Earth to, "Well the Earth is another planet, and we can learn things about the Earth by looking at the other planets, and about the other planets by looking at the Earth and so on," which is yet another area of connecting disciplines and interests.[24]

NASA's next Administrator, James Beggs (1981 to 1985), took concrete action to elevate the role of Earth science at the Agency. In addition to seeking President Reagan's approval for a permanently crewed space station, Beggs encouraged Burton "Burt" Edelson, NASA's Associate Administrator for Space Science and Applications, to pursue a larger role for Earth science. Historian Henry Lambright wrote that Edelson decided

> to fasten on global monitoring by satellites as an idea whose time had come, one that would do OSSA *and* NASA the most good. Edelson, an engineer, envisioned the launch of large platforms in space carrying a range of new sensors. Such a program had the scale and growth potential to help keep NASA going if Space Station ran up against delays in approval. Beggs agreed.[25]

---

24. Robert Frosch, interview by author, 12 January 2006.
25. Henry Lambright, "Administrative Entrepreneurship and Space Technology: The Ups and Downs of 'Mission to Planet Earth,'" *Public Administration* 54, no. 2 (March/April 1994): 99.

Ideas for a new satellite initiative to collect data on the oceans, atmosphere, ice sheets, and land took form at a 1982 NASA workshop, chaired by Harvard University atmospheric physicist Richard Goody, entitled "Global Change: Impacts on Habitability." Work began in earnest on the project that year, under the direction of Shelby Tilford and a group of scientists who were tasked to come up with new ideas for space activities in advance of the United Nations Conference on the Peaceful Uses of Outer Space in Vienna, Austria.

In Vienna, wrote Lambright, Beggs announced that NASA would launch a new program of Earth observations called "Project Habitat," and he called for "an international cooperative project to use space technology to address natural and manmade changes affecting habitability of Earth."[26] Contrary to NASA's expectations, the international audience did not greet "Project Habitat" with open arms. "The developing countries didn't want any part of it because they thought this was another concept of the rich countries to take over their resources and limit what they were able to do," recalled Tilford.[27] "It was a humiliating experience for all of us," added Beggs.[28]

The domestic response to the proposal also was hostile. "This was a preemptive strategy," wrote Lambright, "and NASA did not have the power in 1982 to be preemptive. Many scientists, other federal agencies and even some governments were angry that they had not been informed. Some members of the Reagan White House thought Habitat was too much like something President Jimmy Carter and his environmentalists would propose. The idea of a comprehensive Earth-monitoring system might have been technically ripe for development, but it was not politically ready for adoption."[29] Hans Mark, Beggs's Deputy Administrator, who played a behind-the-scenes role in supporting the initiative, said in retrospect, "Basically the administration killed it. They took the view that the private sector ought to be doing it."[30]

Proposing a Mission to Planet Earth

Beggs was undaunted by the negative reaction to Project Habitat. He allowed Edelson's group to quietly develop research ideas and build up support within the scientific community for an Earth monitoring system, should the opportunity

26. Ibid.
27. Shelby Tilford, interview by author, 2 June 2006.
28. James Beggs, interview by author, 24 August 2004.
29. Lambright, "Administrative Entrepreneurship": 99.
30. Hans Mark, interview by author, 19 January 2006.

arise to restart the project. Shelby Tilford described this period as one in which the NASA planners adopted a new strategy:

> We thought we can't do this again ["Project Habitat"] because it simply would not work because of the international politics. We thought the first thing we would do is come back home and try to get a group of scientists from all the different disciplines who had an interest and who needed to play a part. If we could get them together and do an integrated look at what do we know, what don't we know in various areas, and what do we need to know, and sort of lay out an integrated approach to what measurements were required and how long they should be required, then we might be able to involve the international community up front rather presenting them with something as a kind of fait accompli. So we set out to put together this group headed by Francis Bretherton (Director, National Atmospheric Research Center, Boulder, Colorado, 1974–1981), and we struggled a long time for a name as it might play a significant role. And that's when the Mission to Planet Earth concept came out.[31]

While Bretherton's group quietly conducted its work, NASA used a low-key announcement to publicly resurrect the idea of conducting a broader Earth science program. In a 25 January 1985 editorial in *Science*, Burt Edelson expressed NASA's intention to use satellites to focus on natural and humanmade changes on a planetary scale in cooperation with other domestic and international organizations. "In some ways we know more about our neighboring planets than we do about the Earth," he wrote.[32] Edelson called for an interdisciplinary research program involving "many organizations and countries."

This time, the proposal was received more positively. The name Mission to Planet Earth resonated with the public, evoking the idea of a NASA focused on issues important to ordinary people. The proposed expansion of NASA's Earth science efforts also complemented a larger ongoing effort in the scientific community to focus more comprehensively on global climate processes. The National Academy of Sciences was then organizing physicists, chemists, and

---

31. Shelby Tilford, interview by author, 2 June 2006.
32. Burt Edelson, "Mission to Planet Earth," *Science* 227, no. 4685 (25 January 1985): 367.

biologists to contribute to an International Geosphere-Biosphere Program, known generally as Global Change. Many of these scientists recognized that their research could benefit from satellite data, and they welcomed Edelson's ideas.

The Bretherton Report and Earth Systems Science

A groundbreaking report authored by NASA Advisory Council member Francis Bretherton, officially known as *Earth Systems Science: A Program For Global Change* (1986), is deemed crucial by observers to the development of the interdisciplinary concept of Earth systems science. It partly owed its existence to the struggles of NASA's planetary science and astronomy groups to justify their programs in the early days of the Reagan administration. In 1981, David Stockman, President Reagan's budget director, sought to make deep cuts in NASA's planetary program, arguing that we'd already visited the planets. NASA responded to this challenge by developing an in-house solar system science report that provided a justification for continued planetary exploration. Edelson and Tilford then decided to do the same for Earth sciences. They charged the Bretherton Committee to 1) review the science of Earth as an integrated system of interacting components, 2) recommend an implementation strategy for global Earth studies, and 3) define NASA's role in such a program of Earth system science.[33, 34]

The Bretherton Report's impact was wide ranging. It would lead to new ways of thinking and acting throughout the scientific community regarding the study of Earth as a complex system and help establish the framework for the multibillion interagency U.S. Global Change Research Program (USGCRP).

Bretherton said he initially was worried that scientists from different disciplines wouldn't talk to each other and work cooperatively to formulate a coherent Earth science program. But he said all the participants at the first meeting of the larger NASA Advisory Council realized

> that if we, the Earth-looking community didn't get our act together quick, in five years there would be no Earth-looking program because the astronomers and the planetary exploration people were all highly organized. They had a case that they

---

33. NASA Earth System Science Committee, NASA Advisory Committee, *Earth Systems Science Overview: A Program for Global Change* (Washington, DC: NASA, 1986), p. i.
34. While the NASA Earth System Science Committee conducted its work, planning proceeded at NASA on a series of Earth-orbiting platforms, first known as System Z and later as EOS.

> could make. They had got their community all lined up. And they were speaking outwardly at least with one voice. And they were all prepared with good documents that made their case. We had nothing.[35]

Once the Committee got rolling, Bretherton felt he had an ace up his sleeve, the simple fact that the "really important problem was the human footprint on this planet," and that the needed observations could best, but not totally, "be done from space."[36] The Committee defined the key human footprint issues as being humanity's influence on the climate and the impact of deforestation on plant and animal diversity. The Committee concluded:

> All these human-induced changes are difficult to assess and measure accurately, but it is already evident that they are playing a role in shaping present and future global conditions. Now is the time to document these processes on a global scale and to identify the causal relationships among them while there is still time to respond effectively.[37]

The Bretherton Report also stated that the anticipated achievements of Earth system science include the following:

- *Global Measurements:* Establishment of the worldwide observations necessary to understand the physical, chemical, and biological processes responsible for Earth evolution on all timescales.
- *Documentation of Global Change:* Recording of those changes that will occur in the Earth system over the coming decades.
- *Predictions:* Use of quantitative models of the Earth system to anticipate future global trends.
- *Information Base:* Assembly of the information essential for effective decision-making to respond to the consequences of global change.[38]

To achieve these goals, the Bretherton Committee recommended that NASA's EOS rely on "polar-orbiting platforms now planned as part of the U.S. Space

35. Francis Bretherton, interview by author, 6 July 2006.
36. Ibid.
37. NASA Earth System Science Committee, NASA Advisory Committee, *Earth Systems Science Overview*, p. 12.
38. Ibid., p. 6.

Station Complex."[39] This recommendation, consistent with NASA planning, showed, said Berrien Moore, a member of the Earth System Science Committee, that

> We were kind of willing to walk into the future holding our hands with the manned program, as we got a free co-orbiting platform out of it, which happened to go into polar orbit. So I think we found ourselves uncomfortably joined at the hip. But at least initially it gave the Earth scientists a way to compete with the astrophysicists . . . . In some ways we used the manned program to jump start this big program.[40]

But Dixon Butler, who helped develop EOS's science plan, observed that NASA officials soon recognized the impracticality of tying EOS to polar orbiting platforms and making them serviceable by the Space Shuttle:

> We initially wanted to be serviceable too but robotic servicing in polar orbit was nuts. And, by the way, EOS had figured out that we could not do servicing missions from the shuttle launched from the west coast, before *Challenger* and Slick Six [planned shuttle launch site at Vandenberg AFB, California] being mothballed. We just didn't get a chance to announce it. So in the history it may look like it was a consequence of the *Challenger* accident, but we had already come to that conclusion because by the time you carried EVA equipment in the shuttle to polar orbit you would have had 400 kilograms with which to service a 5,000 pound kilogram spacecraft. No way. That just wasn't happening . . . .[41]

By the end of the decade, the whole idea of using an Earth observing platform alongside the Space Station was abandoned. The salvation for the Earth science program, now based on the concept of co-orbiting satellites, said Berrien Moore, was its overall "scientific credibility."[42]

---

39. Ibid., p. 5.
40. Berrien Moore, interview by author, 2 April 2006.
41. Dixon Butler, interview by author, 9 June 2006.
42. Berrien Moore, interview by author, 2 April 2006.

Stage Two: Agenda Access and Program Adoption (1987 to 1990)

Crisis—The Challenger Disaster and Its Aftermath

In the second stage of NASA's Earth science program, the loss of the Space Shuttle *Challenger* and its seven-member crew on 28 January 1986 forced the space agency to address criticism about its lack of coherent priorities. James Fletcher, brought back as Administrator to lead NASA's organizational recovery, tasked astronaut Sally Ride, America's first woman in space, with developing a report on NASA's future goals. The Ride Report listed Mission to Planet Earth, "a program that would use the perspective afforded from space to study and characterize our home planet on a global scale,"[43] first among four recommended "leadership initiatives" for the future.[44] The Ride Report concluded:

> Mission to Planet Earth is not the sort of major program the public normally associates with an agency famous for *Apollo, Viking,* and *Voyager.* But this initiative is a great one, not because it offers tremendous excitement and adventure, but because of its fundamental importance to humanity's future on this planet. This initiative directly addresses the problems that will be facing humanity in the coming decades and its continuous scientific return will produce results which are of major significance to all the residents of the planet. The benefits are clear to a public that is increasingly concerned about global environmental problems like ozone depletion, buildup of greenhouse gases, and acidification of lakes and forests. And as the environment and its preservation become more pressing issues, the initiative retains its importance for many generations to come. For this reason it should enjoy sustained public and Congressional support and interest. The U.S. is the only country currently capable of leading a Mission

---

43. NASA, *Leadership and America's Future in Space* (Washington, DC, 1987), p. 21.
44. The other recommended initiatives in the Ride Report were programs to expand solar system exploration, building permanent lunar scientific outposts, and to send astronauts to Mars within two decades.

> to Planet Earth, but the program is designed around, and requires, international cooperation.[45]

The Ride Report was never formally adopted as Reagan administration policy, but it did bolster the case for elevating the role of Earth science at NASA.

While the Ride Report's stated motivation for MTPE was solely for the purpose of demonstrating value to society, panel member James Garvin said that "behind closed doors" there was a different motivation at work related to NASA's dominant space exploration culture. He noted that those members of the panel enamored with human solar system exploration viewed MTPE as providing a testing bed for developing systematic studies of other planets. Garvin said, "The discussion was: We want to go out. We can't go out without understanding within . . . . So a lot of us realized that the key to exploring Mars, and even the Moon, ultimately Venus and beyond was really right here at home."[46]

Interagency Linkage and the U.S. Global Change Research Program

Concurrent with the post-*Challenger* review of NASA's goals, the Reagan administration recognized that more attention needed to be paid to the looming issue of global warming and considered funding and coordinating an expanded research program on the subject. The research coordination function, which was carried on in the George H. W. Bush and subsequent administrations, helped to link NASA's proposed EOS to a larger governmental initiative, the USGCRP. As such, interagency coordination became an important bureaucratic strategy to bolster the prospects for EOS, as Shelby Tilford aide Peter Backlund observed:

> If you can nationalize your program, then it's more than just an agency priority. I think that many of the most successful NASA programs have had that attribute. Even going back to human spaceflight, clearly Apollo was a national presidential program. The [space] station was a presidential program. So everybody in program development at NASA wanted to get their programs designated as a national priority. I think where Shelby [Tilford] was particularly insightful in that regard was

45. Ibid., pp. 51–52.
46. James Garvin, interview by author, 19 October 2004.

> not just succeeding in making it a national presidential-level priority, but in making it also an interagency priority.[47]

In March 1987, presidential Science Adviser William Graham established an interagency CES, later renamed Committee on Earth and Environmental Sciences (CEES), to coordinate the administration's climate research programs. On the committee, NASA's Shelby Tilford, NOAA's Mike Hall, and NSF's Robert Corell developed a unique strategy to lobby OMB to consider their global change research budgets together, outside of OMB's normal give and take with individual agencies. Office of Management and Budget official Jack Fellows supported this strategy and later wrote that a confluence of unique circumstances aided the USGCRP's development:

> In 1988 our nation had (1) a very hot summer, (2) a research community ready to engage in research on global change (e.g., the Bretherton Report, technologies to make global satellite observations, etc.), (3) a policy community beginning to awaken to the issues of global warming, and (4) a group of dedicated and creative researchers that happened to be in key Federal positions in Washington.[48]

Indeed, 1988 was an eventful year in which various indicators (for example, rising global temperature readings, warming ocean temperatures, major hurricanes, droughts, and the Yellowstone National Park wildfires) lent credence to the idea that Earth's changing climate could be defined as a public policy problem demanding the attention of political leaders. In June, at the invitation of Senator Tim Wirth (D-CO), NASA's James Hansen testified before a congressional committee about the link between greenhouse gases and global warming. Spencer Weart writes that Hansen said he could "state with '99 percent confidence' that there was a long-term warming trend under way, and he strongly suspected that the greenhouse effect was to blame." He and like-minded scientists testified that global warming could bring more frequent storms and floods as well as life-threatening heat waves. Talking with reporters afterward, Hansen said it was time to "stop waffling, and say that the evidence is pretty

---

47. Peter Backlund, interview by author, 2 February 2006.
48. "Fellows Receives the Flinn Award (1997)," American Geophysical Union, available at *http://www.agu.org/inside/awards/bios/fellows_jack.html* (accessed on 28 May 2006).

strong that the greenhouse effect is here."[49] That December, *TIME* designated endangered Earth as the "Planet of the Year." For leaders on both sides of the political fence, it seemed to make sense to invest in an increased NASA Earth science effort, in the context of the USGCRP, to help inform decision-making on the developing issue of global climate change.

A Kickoff for Mission to Planet Earth

As momentum built for the USGCRP, NASA proposed implementing the centerpiece of its MTPE, a $30 billion EOS over a 15-year development and demonstration period.[50] The program's other components would be small environmental probes and the EOS data system.[51] The EOS plan also called for NASA to coordinate its efforts with remote sensing activities conducted by Japan, Europe, the Soviet Union, and Canada.

Shortly after his inauguration, President George H. W. Bush, in January of 1989, reported to Congress his intent to make EOS the primary program of the USGCRP. That summer, on the 20th anniversary of the first Moon landing, Bush also proposed a new Space Exploration Initiative (SEI) aimed at sending astronauts back to the Moon and on to Mars. The SEI was essentially pronounced dead on arrival by Congress, largely because of NASA's massive internal $400 billion budget estimate. Fifteen years later, Bush's son George W.

---

49. Spencer Weart, *The Discovery of Global Warming* (Cambridge, MA: Harvard University Press, 2003), pp. 154–155.
50. The plan called for NASA to begin shortly after project initiation to launch a series of satellites that would collect data on ozone depletion, tropical rainfall, and ocean wind patterns. The plan called for a 1998 launch of the first EOS platform, which would orbit in tandem with the Space Station at a height of 437 miles, with sensors simultaneously monitoring surface temperatures, winds, clouds, rain, lightning, and radiation. The second EOS platform would launch in 2001. It was hoped the data gathered from these platforms and other satellites would allow scientists to obtain a precise understanding of how radiation, rainfall, ocean currents, and clouds interact and how humanmade pollutants alter and accelerate climate change. Each platform was expected to be replaced three times over its 15-year life. The ESA and Japanese National Space Development Agency also had committed, at the time, to launching three observation platforms within 10 years, two by the Europeans and one by Japan.
51. NASA was tasked, beginning in 1991, with developing a data management system using the world's fastest computers that would allow the information from EOS to be integrated into models predicting future climate change. As envisioned, this would be the largest and most complex unclassified computer data system in the world. This was a stern challenge, as the computing infrastructure and software needed for the project was not available when MTPE began. In addition, what became known as the EOSDIS, or Earth Observing System Data and Information System, was designed prior to the World Wide Web coming into widespread use. As EOSDIS finally came into being, its designers scrambled to link the data management system into this new capability.

Bush essentially embraced the goals of the SEI with his January 2004 "Vision for Space Exploration" proposal.

In 1990, without significant controversy, Congress did approve EOS for a new start as part of NASA's fiscal year 1991 budget, although some members were reported to have sticker shock about the project's $17 billion price tag over 10 years. The projected costs of EOS also drew the attention of OMB Director Richard Darman, who quipped, "I didn't know we needed a $30 billion [the program's long term budget estimate] thermometer."[52]

Stage Three: Program Restructuring (1990 to 1996)

Budgetary Politics Takes Hold

By the summer of 1990, budget cutting was the order of the day in Washington, DC, with the agreement between the Bush administration and Congress to raise taxes and limit discretionary spending. "All of a sudden, 'normal' budget politics geared to incremental increases were replaced by a new politics of constraint on overall, government-wide expenditures," wrote Henry Lambright. "The change in the environment of NASA and its allies occurred just at the point when EOS needed increased funds for development."[53] This budgetary politics required the Agency's Earth science planners to undergo several painful exercises aimed at simplifying and reducing the scope and duration of the ultimate EOS and to rely on medium-sized satellites flying in close formation rather than on large platforms. Indeed, shortly after the EOS program was enacted, its planned 10-year budget was slashed from $18 billion to $7.25 billion. These changes eliminated the promise that the program would provide the twin benefits of comprehensiveness and simultaneity of measurements. These latter measurements were prized by scientists who were seeking to narrow down the cause and effect of global warming mechanisms.

NASA's revised approach to EOS during this period was in sync with new NASA Administrator Dan Goldin's "faster, better, cheaper" strategy for NASA

52. Jerome Cramer, "A Mission Close to Home," *TIME*, available at *http://www.time.com/time/magazine/article/0,9171,973799,00.html* (accessed on 4 August 2006).
53. Henry Lambright, "Administrative Entrepreneurship and Space Technology: The Ups and Downs of 'Mission to Planet Earth,'" *Public Administration* 54, no. 2 (March/April 1994): 101.

missions. Under "faster, better, cheaper," NASA sought to tap developments in microelectronics and other technological fields to redesign its unpiloted spaceflight and to reform its big-program-oriented culture. Goldin, unlike former astronaut Richard Truly, whom he was hired to replace as Administrator in April 1992, saw future Agency budget cutbacks as inevitable and decided to lead the retrenchment process rather than have change forced on NASA from the President or Congress. In addressing his approach to the redesign of EOS, Goldin said the following:

> The program needed further restructuring [when I arrived] because it was still much too unwieldy . . . . It didn't have an ability to exert more time responsive science, and allow the flexibility to the Earth science community to really be able to in a rapid fashion get the data that they needed . . . . The dilemma with that was if you just launched those spacecraft on five year sensors, and given how long it would take to develop those platforms over a fifteen year period, you wouldn't have time to get data back and restructure the payloads for each series of spacecraft. So what we set out to do was see how we could decouple the payloads so . . . instead of having multiple payloads on a spacecraft—and we kept just a few of them—what we wanted to do if we have single or double payload spacecraft because there were twenty-four measurements that had to be made, and we would fly them in formation . . . . That's where faster, better, cheaper came from.
>
> We went to fly them in formation, in orbit, because one of the big arguments of the original EOS program when they had two spacecraft, I mean Battlestar Galacticas, an AM crossing spacecraft and a PM crossing spacecraft, each with twenty-four instruments, the argument was you had to put all the Earth science instruments on one spacecraft because of the necessity for simultaneity of measurements. That is, if I wanted to measure twenty-four different phenomena, I wanted to be sure all the instruments were on the same spacecraft because of the time coherence that I would need to make those measurements. And that was a completely false assumption. And it was felt that if I had independent spacecraft flying in orbit—co-orbital—because of the distance separation and the variability in their orbital velocity you could never get the

simultaneity of measurements, you couldn't do co-registration when you'd want to overlay one image over another. And that is one of the reasons why they (wanted to) build those big platforms. But the dilemma in building those big platforms was that if I put twenty-four instruments on one platform, I might not ever be able to get that thing ready to launch, it was so big, so complex . . . complexity goes up as the number of elements squared because of all the interconnectivity.

So the thing that we tried to do when I got to NASA, and by the way in cooperation with Ed Frieman,[54] was to take the next step and start sending up single and double instrument platforms and put them into co-orbital formation and prove that we could get simultaneity of measurements. And by the time that I had left we had twenty-four spacecraft in development, and I think that they are all up there. The co-orbital functions works, and if you could put a spacecraft together in three years you can have quite a bit of data and you could get the information back and every five years, every seven years when you replenish, you could bring in new technology. So instead of being retrograded technology, because of the necessity of replicating things all the time and technology changes, I could now go very pro-technology, and I could keep adapting my system as technology moves along. So that was the concept and my recollection is that we got that number down to about $8 billion for the first series of spacecraft and every one of them worked. So it really gave me great pleasure to see that happen.[55]

---

54. Edward Frieman, a physicist at the Scripps Institution of Oceanography, led an outside panel (Engineering Review Board) appointed by the National Research Council that reviewed NASA's plans for downsizing EOS. The Engineering Review Board's marching orders were to examine options for restructuring EOS and to determine if its key values—simultaneity and comprehensiveness—could be gained sooner through simpler and cheaper technology. The Frieman Board determined that NASA could get enough simultaneity from a group of satellites flying in formation. The Frieman Board's influential report called for the twin EOS platforms to be replaced by a fleet of smaller satellites and recommended that NASA rely on expertise in operating small satellite systems held in DOD and the Department of Energy.
55. Dan Goldin, interview by author, 18 February 2006.

Earth Observing System Science Measurements

Earth Science Category	EOS 24 Measurements
Atmosphere	Cloud properties Radiative energy fluxes Precipitation Tropospheric chemistry Stratospheric chemistry Aerosol properties Atmospheric temperature Atmospheric humidity Lightning
Solar Radiation	Total solar irradiance Solar spectral irradiance
Land	Land cover and land use change Vegetation dynamics Fire occurrence Volcanic effects Surface wetness
Ocean	Surface temperature Phytoplankton and dissolved organic matter Surface wind fields Ocean surface topography
Cryosphere	Land ice Sea ice Snow cover

Shelby Tilford left NASA in October 1993, unconvinced that Goldin's "faster, better, cheaper" approach was necessarily "better." He was replaced by UCLA space scientist Charles Kennel, whom Goldin ordered to reinvigorate the program under the Administrator's preferred approach. Goldin also sought to transition Earth observing technologies developed by NASA to NOAA, in accord with his philosophy that "NASA was formed as a development organization, research and development organization, to always push back boundaries of the unknown, not to be an operational agency, not to do things that are comfortable, but to always go to the edge, and to take risk."[56] While the

Figure 3: Space scientist Charles Kennel of UCLA. *Courtesy of Charles Kennel*

56. Dan Goldin, *The MacNeil/Lehrer News Hour*, New York, NY, and Washington, DC, Public Broadcasting Service, 29 November 1996.

Tilford era was characterized by coalition building, agenda setting, and planning, the Kennel era would be more devoted to the building of spacecraft and the refinement of EOS's research aims, with scientists outside the Agency helping him juggle the EOS launching schedule in order to avoid further budget cuts.

Kennel says the challenge he inherited was reducing the high projected costs of EOS and making the ensuing program sound on a scientific basis:

> When I came to NASA in 1994 there was a serious job ahead of us at NASA. And that was the initial conception of the EOS as three launches every five years of massive . . . experimental satellites launched by the Shuttle. There was a run-out cost close to $20 billion dollars. That initial conception was clearly not viable . . . primarily financially. However, there was a secondary aspect to it. The critics of that system said with some justice, that as planned it was the largest and most expensive scientific experiment ever put on the books at that time and that the Principal Investigators and so forth for the instruments were being locked in. There was going to be a fifteen year free ride for those people however competent they were and there was going to be little or no opportunity for a change of goals and the addition of younger and newer people. That was the criticism. I didn't think it would actually unfold that way but it was quite clear to me and it was quite clear to Goldin that politically that level of commitment couldn't be sustained . . . .
>
> What happened during my three years was that we restructured and reshaped it [EOS] and got to many more spacecraft with smaller payloads and more focused payloads . . . . The question I had at the end of that whole process was: At what stage do you know when you've cut enough? At what point have you cut so much out of the system that you've reduced the technical reliability? And the other question was: How do I know where we've hit a stable budget that the congressional and administration budget process can tolerate this project? My job was to stabilize the system and stabilize the budget and keep as much of the science as possible.[57]

---

57. Charles Kennel, interview by author, 22 July 2004.

The Republican takeover of both houses of Congress following the 1994 elections also had important implications during this third stage of the Earth science program. Because of the program's association with global warming, an unpopular issue in conservative circles, NASA was forced to fight back several partisan attempts by the new Republican majority in the House of Representatives to impose EOS budget cuts larger than NASA was willing to accept. Members of the new Republican majority in the Senate, however, were more supportive of EOS because of its perceived relevance and its impact on local economies. One strategy utilized by NASA during this period was to stress Earth science applications, or the "down to Earth" elements of EOS, such as assisting with agricultural forecasting, transportation planning, and disaster relief.

Stage Four: Program Implementation (1997 to 2008)

Progress in Space, Turmoil on the Ground

In the fourth stage of NASA's Earth science program, program managers focused on implementing the existing program and planning for future missions. In 1999, NASA launched Terra,[58] the first of the main EOS satellites, and continued to obtain useful scientific results from its entire fleet of Earth observing satellites. On 14 July 2004, with the launch of the Aura satellite designed to provide global scale information about Earth's atmospheric composition, NASA fulfilled its commitment to develop and launch the first phase of EOS. Still, scientists continued to complain that the program was not sufficiently focused on its main goal of understanding Earth as a system, due to NASA's drift away from its original EOS plans.

The period between 2004 and 2008 was dominated by concerns about the program's long-term prospects. President George W. Bush's 2004 proposal to renew NASA with a new program to send humans and robots to the Moon, Mars, and beyond has squeezed the EOS budget to the point that missions in

---

58. Carrying five instruments that operate simultaneously on a single platform, Terra is examining such things as the impact of clouds and sulfate particles on atmospheric cooling; the way climatic processes are affected by changes in land use and by volcanoes and fire; and the impact on the atmosphere from changes on land and at sea, such as ice cover and phytoplankton growth. Terra instruments also are designed to provide scientists with data on methane and carbon dioxide growth rates in the atmosphere and on the impact of solar irradiance on Earth's surface temperatures.

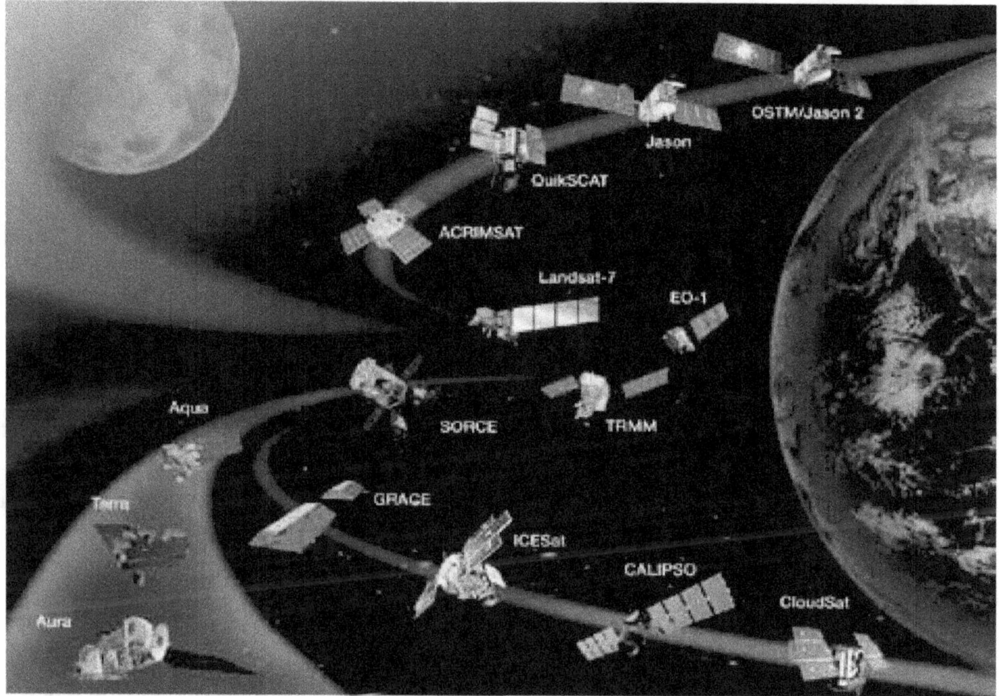

Figure 4: Nearly two decades after the initiation of NASA's MPTE, the space agency has successfully launched and operated a flotilla of EOS satellites, including the flagship Terra, Aqua, and Aura missions. *NASA*

development were delayed and efforts to follow up on the existing EOS satellites and instruments were dropped.

The George W. Bush NASA plan also spurred a lively debate about whether NASA's Earth science program should be administratively transferred to NOAA. In the face of this uncertainty, the National Research Council of the National Academies undertook, at the request of NASA, NOAA, and the U.S. Geological Survey, the first-ever decadal survey aimed at recommending a priority-ranked list of Earth science missions for NASA. The decadal survey recommended that NASA launch 15 missions from 2010 to 2020 to address the following themes: Earth science applications and societal benefits; land use change, ecosystem dynamics, and biodiversity; weather science and applications; climate viability and change; water resources and the global hydrological cycle; human health and security; and solid-Earth hazards, natural resources, and dynamics.

NASA's Earth Science Program Under the George W. Bush Administration

In 2001, the incoming George W. Bush administration quickly postponed, and later canceled, the Deep Space Climate Observatory. This satellite was intended to tally Earth's energy budget—how much energy comes in from the Sun, how much immediately reflects, and how much is absorbed to be reradiated at longer infrared wavelengths—an important element of understanding climate change. The satellite was actually a scientifically improved version of Triana, the satellite proposed by former Vice President Al Gore, derisively labeled "Goresat" by its critics, to constantly observe Earth from a high orbit.[59] The Bush administration's action was consistent with an attitude expressed in the NASA Transition Policy Book prepared by the incoming Bush-Cheney team, which asserted, "Heavily politicized missions imposed on NASA by the Clinton Administration over the last eight years (global warming and Russian space station participation are prime examples) have poisoned the policy well with several key Members of Congress, particularly in the House."[60]

More broadly, the Transition Policy Book observed that due to the budget constraints that NASA faced throughout the 1990s (an estimated budget reduction of $40 billion in the 1993 to 2001 time period from what the Agency had expected entering this timeframe), and with the Agency itself understanding "that there is little room for growing their budget in a dramatic way,"[61] "a return to the era of large-, multi-billion dollar spacecraft is not desirable or likely to be feasible."[62] To provide prudent reserves to address expected cost growth, the Transition Policy Book pointed out that "in Earth Science, the solution has been to defer development of the next generation of spacecraft."[63] The bottom line, the report stated, is that "NASA has too much on its plate for the total budgets it receives."[64] This refrain would be repeated almost word for word by NASA Administrator Michael Griffin five years later when he announced that, under a constrained budget, progress on NASA's science programs would have to be delayed more than previously anticipated.

Bush's first budget proposed cutting NASA's Earth science funding by $200 million and leaving future spending flat for five years. The Bush administration did request funding for five new satellites to mark the second phase of NASA's EOS effort. The new missions were proposed to monitor global precipitation (in tropical and temperate zones), ocean topography

---

59. Following Barack Obama's victory in the 2008 presidential election, a sign was observed at GSFC that stated, "Free Triana."
60. "NASA 2001 Transition Policy Book" (Washington, DC: unpublished, 4 January 2001), p. 31.
61. Ibid., p. 23.
62. Ibid., p. 21.
63. Ibid., p. 27.
64. Ibid.

(measuring phenomena such as El Niño and La Niña), ocean surface winds (useful in hurricane monitoring), atmospheric ozone and aerosols, and the energy input to Earth's atmosphere from the Sun.

In June 2001, saying he wanted to create "an effective and science-based response to the issue of global warming," President Bush announced several new programs to study greenhouse gases and climate change and vowed to fully finance climate change science over the next five years. Bush proposed a new U.S. Climate Change Research Initiative (USCCRI) to take up the work previously done by the USGCRP, with an emphasis on short-term uncertainties in climate change. In July 2003, the Bush administration hosted an international summit on global climate change led by NOAA Administrator Vice Admiral Conrad C. Lautenbacher. More than 40 nations and 20 international organizations agreed at the State Department to form an ad hoc international Group on Earth Observations, with the goal of forming a Global System of Systems producing a comprehensive, coordinated, and sustained system of Earth observations. This agreement did not come with any promises of more funding for NASA, yet it did underscore the continuing relevance of the EOS program.

The *Columbia* Accident and Its Aftermath: A New Mission for NASA

The loss of the Space Shuttle *Columbia* and its crew of seven on 1 February 2003, like that of the *Challenger*, prompted a major reassessment of NASA's priorities. Encouraged by NASA's 10th Administrator, Sean O'Keefe (2001–05), the Bush White House undertook an internal review of space program goals. The review culminated in President Bush's 14 January 2004 announcement at NASA Headquarters tasking NASA to undertake a long-term human and robotic program to renew the exploration of the Moon and eventually to land astronauts on Mars.

Initially, O'Keefe promised that NASA's science and aeronautics programs would not suffer in order to fund Bush's "Vision for Space Exploration." NASA briefing charts provided to the White House prior to the Vision's announcement, however, suggested that funding would come from "reallocating resources and internal restructuring," with the biggest hits coming to the "Space Science and Earth Science components of the existing NASA program."[65]

65. NASA, "Briefing for the President: Future U.S. Space Exploration—Alternative Visions, Key Elements, and Issues for Decision" (Washington, DC: unpublished document, 19 December 2003), p. 11.

Indeed, the next administration budget request (for fiscal year 2005) proposed taking $1.2 billion out of the NASA Earth science budget over a five-year period. Later in 2004, NASA Associate Administrator for Science Al Diaz reportedly ordered his managers to find at least $400 million in cuts to space and Earth science efforts to address the rising costs NASA was facing in getting the Space Shuttle ready to fly again.

Prior to the Moon-Mars program's announcement, Bush administration officials discussed transferring NASA's EOS program to NOAA, or perhaps some of NOAA's satellite programs to NASA. While no decisions were made on these complex issues, Bush's Commission on Implementation of United States Space Exploration Policy, chaired by former Air Force Secretary Edward "Pete" Aldridge, Jr., suggested the possibility of moving NASA's Earth science activities to another agency.

Ghassem Asrar, NASA's Earth Science Program Director, addressed the Aldridge Commission to defend the place of EOS within the space agency. In making his presentation, Asrar said:

> My main objective was to present what the Earth science program is all about and what are its unique contributions to the study of Earth and Earth Systems Science and answer the questions about the role of the private sector, commercial remote sensing, the role of NASA and other agencies. [We tried] to convince them that what we did was really unique and didn't infringe on any commercial sector or business, because we never built anything that was finer than 10 meters. With all the high resolution stuff that the private sector is building we are the users; we are not competing with them. And as a matter of fact we use those observations for our scientific research. We buy that data; we are the user of those data sets.
>
> And we also tried to convince them if it weren't for NASA, NOAA would not be there today. The fact is that if you look at today's weather satellites, they are [using] 1970s technologies. There was a hiatus, there was a period when NASA and NOAA parted their ways and NASA focused on climate and NOAA focused on weather . . . .
>
> NOAA by charter is an operational agency. They cannot introduce new capabilities into their mainstream of operational activities unless the risks of those new technologies are well understood, demonstrated, retired . . . . The models that we

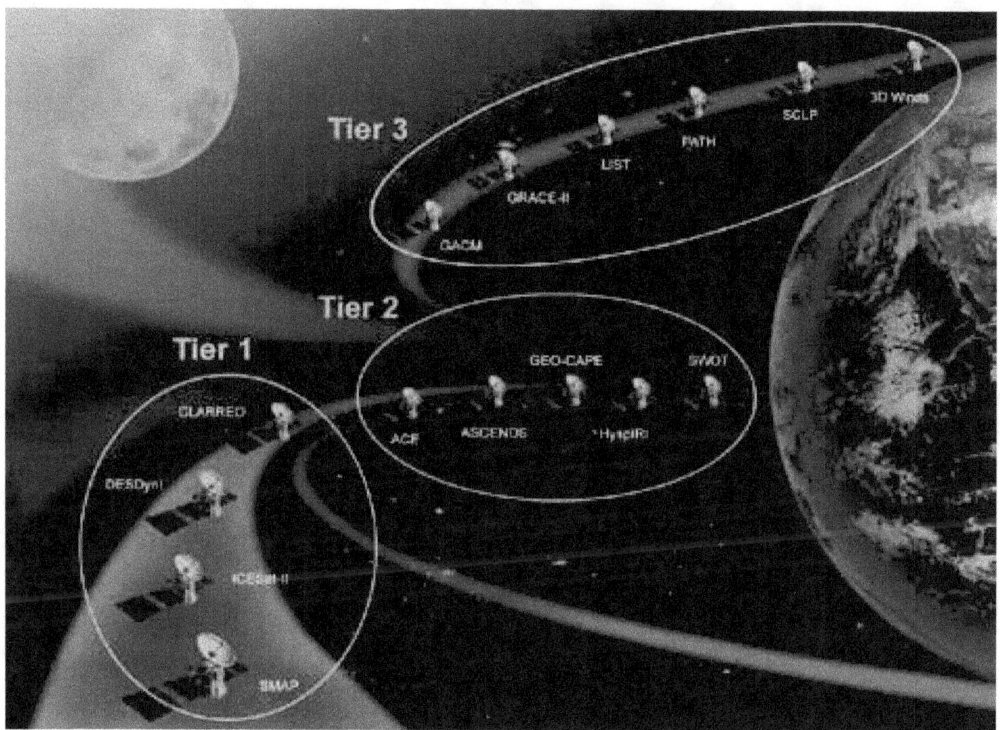

Figure 5: The National Research Council's Decadal Survey for Earth Science and Applications recommended that NASA launch 15 missions from 2010 to 2020 to address the following themes: Earth science applications and societal benefits, land use change, ecosystem dynamics and biodiversity, weather science and applications, climate viability and change, water resources and the global hydrological cycle, human health and security, solid-Earth hazards, and natural resources and dynamics. In the National Research Council plan, the missions would be sequenced by time in three tiers. *NASA*

> have developed to use these data sets improve the weather prediction models. So all these capabilities that NASA has developed over the last 10–15 years are becoming part of the next generation operational weather satellites. I told them [Aldridge Commission] that unless . . . NASA had developed these, NOAA would not be in a position to do that. We would still be flying the old technology . . . .[66]

The Aldridge Commission also recommended that the NASA Headquarters organization be realigned to support the Moon-Mars program and that the

66. Ghassem Asrar, interview by author, 19 July 2005.

Agency's science enterprises be consolidated. In response, O'Keefe merged the Earth Science Enterprise and Space Science Enterprise into a single Science Mission Directorate, a move seen as a downgrading of the importance of Earth Science. O'Keefe argued, however, that the move made sense because NASA could apply what the EOS program was learning about Earth's climate to the scientific investigation of Mars and other planets, the motivation that had previously inspired members of the Sally Ride Task Force in 1987.

A Decadal Survey for Earth Science and Applications from Space

In a move to ensure that NASA's Earth science program will have a future as well as a present, the National Research Council undertook to conduct an Earth Science and Applications from Space decadal survey, an effort like those conducted in the astronomy and planetary science fields, to provide authoritative recommendations from the science community on program priorities. The decadal survey, requested and supported by NASA, NOAA, and the U.S. Geological Survey, urged the U.S. government to spend $2 billion a year ($500 above NASA's current Earth science budget) to fund a series of 15 NASA Earth science missions between 2010 and 2020, with two other missions proposed for NOAA.[67]

---

67. The recommended NASA missions from 2010 to 2013 and their mission descriptions are as follows: Climate Absolute Radiance and Refractivity Observatory (CLARREO): Solar and Earth radiation; spectrally resolved forcing and response of the climate system; Soil Moisture Active and Passive (SMAP): Soil moisture and freeze-thaw for weather and water cycle processes; Ice, Cloud, and land Elevation Satellite (ICESat)-II: Ice sheet height changes for climate change diagnoses; and Deformation, Ecosystem Structure, and Dynamics of Ice (DESDynI): Surface and ice sheet deformation for understanding natural hazards and climate; vegetation structure for ecosystem health. The recommended NASA missions from 2013 to 2016 and their mission descriptions are as follows: Hyperspectral Infrared Imager (HyspIRI): Land surface composition for agriculture and mineral characterization; vegetation types for ecosystem health; Active Sensing of CO2 Emissions over Nights, Days, and Seasons (ASCENDS): Day/night, all-latitude, all-season carbon dioxide column integrals for climate emissions; Surface Water Ocean Topography (SWOT): Ocean, lake, and river water levels for ocean and inland water dynamics; Geostationary Coastal and Air Pollution Events (GEO-CAPE): Atmospheric gas columns for air quality forecasts; ocean color for coastal ecosystem health and climate emissions; and Advanced Composition Explorer (ACE): Aerosol and cloud profiles for climate and water cycles; ocean color for open ocean biogeochemistry. The recommended NASA missions from 2016 to 2020 and their mission descriptions are as follows: Lidar Surface Topography (LIST): Land surface topography for landslide hazards and water runoff; Precipitation and All-weather Temperature and Humidity (PATH): High-frequency, all-weather temperature and humidity soundings for weather forecasting and sea surface temperature; Gravity Recovery and Climate Experiment (GRACE)-II: High-temporal-resolution gravity fields for tracking large-scale water movement; Snow and Cold Land Processes (SCLP): Snow accumulation for freshwater availability; Global Atmospheric Composition Mission (GACM): Ozone and related gases for intercontinental air quality and stratospheric ozone layer prediction; and 3D-Winds: Tropospheric

While calling for NASA to reenergize the major commitment to Earth science undertaken in the 1990s, the report's authors warned that due to attrition and funding issues, by 2010 the number of operating Earth observing instruments on NASA satellites is likely to drop by 40 percent, leaving gaps in data used to forecast severe weather events such as hurricanes, tornadoes, and tsunamis, and also leaving gaps in data used to understand the dynamics of global climate change. Decadal Survey Cochairman Richard Anthes, president of the American Meteorological Society, argued that the reductions in NASA resources devoted to Earth science had led to a decline in our country's capabilities to monitor environmental change. While NASA welcomed the decadal survey as a useful tool for guiding the development of future missions, Administrator Michael Griffin at the Goddard Space Symposium in 2007 criticized the survey's proposed budget increases as being unrealistic.

The larger issue of what NASA's future priorities should be was featured more prominently than some people expected during the 2008 presidential campaign. The tight race for Florida's 25 electoral votes contributed to Barack Obama's backing off of an earlier campaign statement that he would transfer funding from NASA to educational needs and asserting down the campaign stretch that he would fully fund NASA's programs. The Obama campaign also detailed its views on the importance of monitoring Earth from space in a Web forum sponsored by several scientific organizations:

> Barack Obama has proposed bold initiatives to put America on the path to stop global climate change. His administration will set standards based on rigorous scientific inquiry that, in turn, cannot take place without a capable space program. The task of researching and understanding the forces that affect our home planet will require a constellation of climate monitoring space platforms. As president, Obama will ensure that NASA has the funding necessary to play its part in the fight against global climate change.
>
> As president, I will establish a robust and balanced civilian space program. Under my administration, NASA not only will inspire the world with both human and robotic space exploration,

---

winds for weather forecasting and pollution transport ("Satellite Observations to Benefit Science and Society: Recommended Missions for the Next Decade," *National Research Council* [Washington, DC: National Academies Press, 2008], p. 9).

> but also will again lead in confronting the challenges we face here on Earth, including global climate change, energy independence, and aeronautics research.[68]

Looking Back, Looking Forward

Has NASA's Earth science program been successful? And what are its likely prospects for the future? One way to judge the successes of EOS and related activities is in terms of its original criteria for mission accomplishment. Specifically, when EOS was first initiated, NASA described the program's objectives as follows:

> The Earth Observing System (EOS) will be a science and observation program that will provide long-term (fifteen year) data sets for Earth system science in order to gain an understanding of the interactions between Earth's land, atmosphere, oceans, and life. Areas of study will include the global hydrological cycle, global biogeochemical cycle, and global climate processes with a focus on greenhouse gases and the role of clouds. EOS will provide for the interdisciplinary evaluation of EOS data. This includes the funding of interdisciplinary science team grants, Principal Investigator research grants, and post-graduate fellowships. To process the data from EOS, the program will include development of a comprehensive Earth Observing System Data and Information System (EOSDIS), designed to maximize the Earth science research community's access to, and processing of, the necessary measurements through an open data policy.[69]

By these and other broad objectives associated with the program, EOS has been successful on a number of fronts, including the following:

- Pioneering new Earth monitoring technologies.
- Successfully launching and operating a current constellation of 15 EOS satellites.
- Providing to the scientific community 24 key climate measurements.

68. "ScienceDebate2008.com presents Presidential answers to the top 14 science questions facing America," Science Debate 2008 Web site, available at *http:/www.sciencedebate2008.com/www/index.php?id+42* (accessed on 23 October 2008).
69. NASA, *Earth Observing System (EOS)* (Washington, DC: Author, 1992).

Figure 6: NASA has five new Earth science missions already under way or under development. They will help measure ocean topography (Ocean Surface Topography Mission), global aerosol and cloud properties (Glory), sea surface salinity (Aquarius), global land cover change (Landsat Data Continuity Mission), and global rainfall from tropical to mid-latitudes (Global Precipitation Measurement Mission). The planned OCO mission to measure atmospheric carbon dioxide had a launch failure in February 2009 but may be reflown. *NASA*

- Utilizing Earth measurements to improve climate models and to enhance monitoring and understanding of severe climate events such as hurricanes and tsunamis.
- Applying Earth observations to applications benefiting a number of user communities.
- Developing productive new partnerships with other nations' space and science agencies on Earth science research.
- Helping to build up the Earth science research community through research grants and fellowships for graduate students.

Yet any evaluation of the record of NASA's Earth science program must honestly address failures to meet expectations, including the following:

- The inability to deliver on the EOS program's promise of providing 15 years of continuous and comprehensive data on Earth processes.
- The abandonment of specific science goals due to budget priorities.

- Cost growth and delays in mission developments that constrained NASA's ability to increase technology investments.[70]
- A lack of effective strategic planning to develop a coherent and achievable followup plan for the initial EOS.

These failures can be largely explained by NASA's inability to reconcile its intentions at various points in its history to be a leading multimission agency with the reality that, when push comes to shove, NASA has always tended to anchor its hopes on its human spaceflight mission and make budgetary decisions accordingly.

As we turn to the future, however, the new Decadal Survey for Earth Science and Applications and the advent of an Obama administration that is determined to do more about climate change, which by implication will require more knowledge about climate change processes, have led to increased hopes that Earth science may well have a stronger future at NASA.

NASA does have plans for the long term that might lead to a renaissance for Earth science. Current NASA Earth science director Dr. Michael Frelich said his primary objective "is to expand the leading role of NASA measurements and NASA-supported analyses in advancing Earth system science—improving our quantitative understanding of the Earth as an integrated system.[71]" He said that such measurements and analysis will address the need to develop "long, multi-decadal, global, consistently processed measurements of particular processes . . . especially of the oceans because the ocean is the giant flywheel of the [Earth] system . . . about half the heat that's put in near the equator and goes off toward the poles is carried by the atmosphere and half is carried in the ocean."[72]

NASA's long-term strategic plan imagines a scenario for beyond 2016 that includes a technological leap in Earth observations to detect changes in the Earth system as they happen. The Agency envisions using constellations of smart satellites placed in various orbits to augment airborne sensors and surface-based sensors to form an integrated, interactive "sensorweb" observing system.

As we anticipate such a future, perhaps the greatest testimony to the lasting relevance of NASA's Earth science program comes from those select few

70. Greg Williams, "The Difficult Journey of a Great Idea: An Inside, Informal History of Earth System Science at NASA" (Washington, DC: unpublished manuscript, 2005), p. 16.
71. Edward Goldstein and Tabatha Thompson, "Earth Science: NASA's Mission to Our Home Planet," in *NASA: 50 Years of Exploration and Discovery* (Tampa, FL: Faircount Media Group, 2008), p. 181.
72. Ibid.

humans who have seen our fragile Earth from space. Among them, astronaut Piers Sellers said the following in 2004:

> Our technical ability to view the Earth from space is coincident with our ability to change our planetary environment. So at the very time that we are able to see our planetary home in its entirety, we are powerfully motivated to do so—to understand how the Earth system works, to help us assess the kind and degree of changes, both manmade and natural, that are ongoing, and ultimately to help us predict the future consequences of these changes . . . . The public and their representatives in government need better information on which to base all kinds of decisions involving the planetary environment; from targeting famine relief in Africa during droughts, to the continuing discussion on global climate change. Satellite observations provide the sole means to observe the whole planet almost every day using the same instrument.[73]

73. Piers Sellers, "Presentation on Earth Observations from Space at the Smithsonian's National Air and Space Museum" (Washington, DC, September 2004).

Chapter 21

Earth Observations from Space

Achievements, Challenges, and Realities

James R. Fleming

> Space technology affords new opportunities for scientific observation and experiment, which will add to our knowledge and understanding of the earth.
>
> —President's Science Advisory Committee, 1958

For the past 50 years, Earth observations from space have provided absolutely unique perspectives and have given a tremendous boost to the interdisciplinary geophysical sciences and to public awareness of the planet's beauty and fragility. Ironically, however, our ability to continue such unprecedented observations and provide critical information about Earth processes may be severely challenged over the course of the next several years. The Space Age officially began on 4 October 1957, with the dramatic and historic launch of Sputnik I by the Soviet Union; the United States joined the club less than four months later with its launch of Explorer 1 on 31 January 1958. These and subsequent accomplishments represent the fulfillment of an ancient quest for altitude and for a panoptic perspective; but Earth observations have much deeper historical roots.

Ancient History

When asked recently when humans got interested in and began to get serious about climate change, I responded immediately, "in the Pleistocene." In other words, the emergence of *Homo sapiens sapiens* was contemporaneous with the cycles of ice ages and interglacials. Our distant ancestors, with cranial capacity equal to ours, were acute observers of their environment and would have certainly engaged in serious deliberations in response to rapid climate change events, for example, the Younger Dryas event some 12,000 years ago, which cooled large areas of the globe from 5 to 15°C and may have contributed to the Neolithic revolution. In this and similar ways, Earth observations have a history as old as our species.

Seeing conditions were great in the past, with no light pollution and only occasional smoke from biomass burning. All that was needed to view the sky was a wide horizon, a hillside, or a mountaintop, with perhaps a stick or gnomon to make consistent measurements over time. Ancient observers viewed the changing conditions below their feet and identified patterns in the sky, for example, the Great Bear (Ursa Major), the Eagle (Aquila), and the Dragon (Draco). Our ancestors soon discovered that if they just ascended a little, by climbing a hill, a pyramid, or even a mountain, they could see farther and have a bigger horizon, gaining both added perspective on the world and perhaps added authority. The monumental structures at Stonehenge, England, provided such a perspective on the heavens, as does the ancient solar observatory at Chankillo, Peru, the oldest in the New World. In the modern era, Piazzi Smyth installed a high-altitude telescope in 1856 on the Teide Volcano, Tenerife, while the first permanent mountain observatory was Lick Observatory in California, which saw first light in 1888.

The Quest for Altitude

Concerning Earth observations at altitude, in 1648, Blaise Pascal convinced his brother-in-law, Florin Perier, to transport a barometer to the top of the Puy de Dôme (1,465 meters). The level of the mercury decreased during ascent of the mountain and returned to its original level at the base, indicating that the air had weight. In 1749, Dr. Alexander Wilson, a Scottish physician, tied detachable thermometers to a string of kites and initiated the tradition of sounding the vertical structure of the free air. The key attached to Benjamin Franklin's kite string represented a specialized static electricity sensor. In the 1830s, James Espy used kites flown by the Franklin Kite Club of Philadelphia, Pennsylvania, to determine the height of the cloud base. He related this height to the surface temperature and dew point, and he was able to estimate the temperature lapse rate. Kites equipped with meteorographs were in widespread use as atmospheric probes in the late 1890s. They were typically flown from ships and mountain stations and could routinely reach altitudes of 2.5 kilometers, with record-setting flights of over 7 kilometers. Theorists such as Vilhelm Bjerknes credited the kite soundings of Abbot Lawrence Rotch at the Blue Hill Observatory with providing the observational data necessary for the development of Bjerknes's cyclone model.[1]

Early surface-based coordinated observing programs were aimed at extending international observations. The International Polar Year from 1882 to 1883 involved 11 nations in a coordinated effort to study atmospheric changes and "electrical weather" as evidenced by magnetic disturbances and the polar lights. The International Cloud Year from 1896 to 1897 measured the altitude and motion of clouds with the goal of producing a global view of atmospheric circulation. Such efforts had many predecessors, but all suffered from coordination problems and lack of aereal coverage; most were confined to surface observations.[2]

About 1809, Thomas Foster in England began the practice of releasing small free balloons (called pilot balloons) and tracking them to obtain information about the winds aloft. A small light was attached to the balloon at night to make it visible. Balloons could be tracked with an optical theodolite to about 5 kilometers in good weather conditions, but their flightpaths were often obscured by clouds.

Writers also fantasized about flights of discovery beyond Earth, as in Francis Godwin's 1638 celestial chariot ride to the Moon powered by geese and gunpowder. In 1783, the French physician Jean François Pilâtre de Rozier and the Marquis François-Laurent d'Arlandes ascended in a hot air balloon designed by the Montgolfier brothers; later that same year, the scientist Jacques-Alexandre-César Charles (1746–1823) and an assistant, Nicholas Louis Robert, carried a barometer aloft in a balloon filled with inflammable air (hydrogen). French balloonist Jean-Pierre Blanchard and American John Jeffries flew across the English Channel in 1785, producing temperature and pressure measurements that extended above 2.5 kilometers.[3]

In the 19th and early 20th centuries, aeronauts reached unprecedented heights in balloons: In 1804, the French scientists Joseph Louis Gay-Lussac and Jean-Baptiste Biot made balloon ascents of over 7 kilometers to study

1. James Rodger Fleming, "Meteorology," in *A History of Modern Science and Mathematics*, vol. 3, ed. Brian S. Biagre (New York, NY: Scribner's, 2002), pp. 184–217.
2. James Rodger Fleming and Cara Seitchek, "Advancing Polar Research and Communicating Its Wonders," in *Smithsonian at the Poles: Contributions to International Polar Year Science*, ed. Igor Krupnik, Michael Lang, and Scott Miller (Washington, DC: Smithsonian Institution, 2009).
3. Francis Godwin, *The Man in the Moone; or, A Discourse on a Voyage Thither* (London, U.K.: John Norton, 1638); Charles Coulston Gillispie, *The Montgolfier Brothers and the Invention of Aviation, 1783–1784* (Princeton, NJ: Princeton University Press, 1983).

the composition of the air and the effect of altitude on terrestrial magnetism. James Glaisher reached 11 kilometers in 1865, a feat that almost killed him and subjected his copilot, Henry Coxwel, to severe frostbite. Clearly, human endurance limits had been reached. In 1931, Auguste Piccard and an assistant ascended to over 15 kilometers in a balloon with a pressurized cabin. They spent their time aloft studying cosmic rays. Malcolm Ross and Victor Prather exceeded 34 kilometers in 1961 in a balloon called "Lee Lewis Memorial."

Self-Recording and Transmitting Instruments

Increasingly, self-recording instruments were sent aloft using kite and balloon sondes. By the 1890s, balloon-borne meteorographs were reaching altitudes as high as 20 kilometers, and a coherent picture of the structure of the atmosphere was beginning to emerge. In 1902, Léon Teisserenc de Bort noted explicitly the existence of an isothermal zone beginning at about 10 kilometers with a warmer layer above that. Almost simultaneously, Richard Assmann in Germany, using different equipment and techniques, announced similar results. In 1908, de Bort coined the terms "troposphere" and "stratosphere" to denote the lower and upper layers of the atmosphere, with the isothermal zone between coming to be called the "tropopause." Additional results showed that the tropopause was higher in the summer, higher in the tropics, and higher above anticyclonic circulations.

By the mid-1920s, experimental radio transmitters had been fitted to free balloons, and by 1936, radiosondes were in widespread use by meteorologists interested in vertical profiles of temperature, pressure, and humidity. Radiosondes were renamed "rawindondes" when advances in tracking allowed the recovery of information about upper-level winds. More recently, radiosondes have been fitted with miniaturized instruments, computers, and telemetry, and they are tracked using GPS.

During World War I, meteorographs were attached to airplanes that flew over and around a military theater of operations. Networks of aircraft sounding stations emerged in the 1920s and 1930s in support of the growing needs of commercial aviation. Still, the data collected by airplanes were less than ideal: the soundings were horizontally oriented, limited in altitude, and collected only when the weather was good enough for safe flying! Moreover, the data could not be analyzed until the airplane landed. Still, airplanes were extremely useful for specialized missions and research. In 1943, Colonel Joseph P. Duckworth and Lieutenant Ralph O'Hair completed the first intentional flight into the eye of a hurricane. The following year, Colonel Floyd B. Wood, Major Harry Wexler, and Lieutenant Frank Reckord flew into a hurricane in a Douglas A-20 to gather scientific data.

To give an indication of the excitement being generated by newly available technologies, consider the letters received by Willis R. Gregg, Chief of the U.S. Weather Bureau, who, in 1938, had asked his colleagues to speculate on what the meteorological profession might look like in 50 years. W. C. Devereaux published a summary of the responses. Some emphasized the growing importance of upper air measurements using radiosondes and broadcasts that would allow "records to be flashed to all parts of the world." Charles Franklin Brooks foresaw remote sensing of the atmosphere using ultra-high-frequency radio transmissions. J. Cecil Alter suggested that "sky-sweeping robots of electric eyes will explore the upper atmosphere for air mass demarcations, depths, direction and velocity movement, moisture content, and other factors. Zig-zag tracings or photographic replicas, automatically registered, will be made of the shape of the course of the refracted ray from the electric eye, as it passes through different air masses." William Jackson Humphreys wrote of "robot reporters—instruments that not only keep a continuous record of the weather elements, but which, at the touch of a button, or automatically at regular intervals, also tell all about the weather there at the time."[4] These predictions were largely realized through the development of weather radar and other forms of remote sensing. Also in 1939, George W. Mindling foretold, in doggerel, of the "coming perpetual visiontone show" of perfect surveillance and perfect prediction using television and infrared sensors, a technology instituted in the Television Infrared Observation Satellite (TIROS) meteorological satellite program in 1960.

> In the coming perpetual visiontone show
> We shall see the full action of storms as they go.
> We shall watch them develop on far away seas,
> And we'll plot out their courses with much greater ease.
>
> Then a new day will come in electrical lore
> When the pictures will register very much more . . .
> Then a day there will be when predictions won't fail,
> Though describing the weather in every detail,
> Just what minute 'twill rain, even when it will hail.

---

4. W. C. Devereaux, "A Meteorological Service of the Future," *Bulletin of the American Meteorological Society* 29 (May 1939): 212–221. See also Robert C. Landis, "Future of International Cooperation in Meteorological and Related Services," available at *http://ams.confex.com/ams/pdfpapers/139163.pdf* (accessed 8 September 2009).

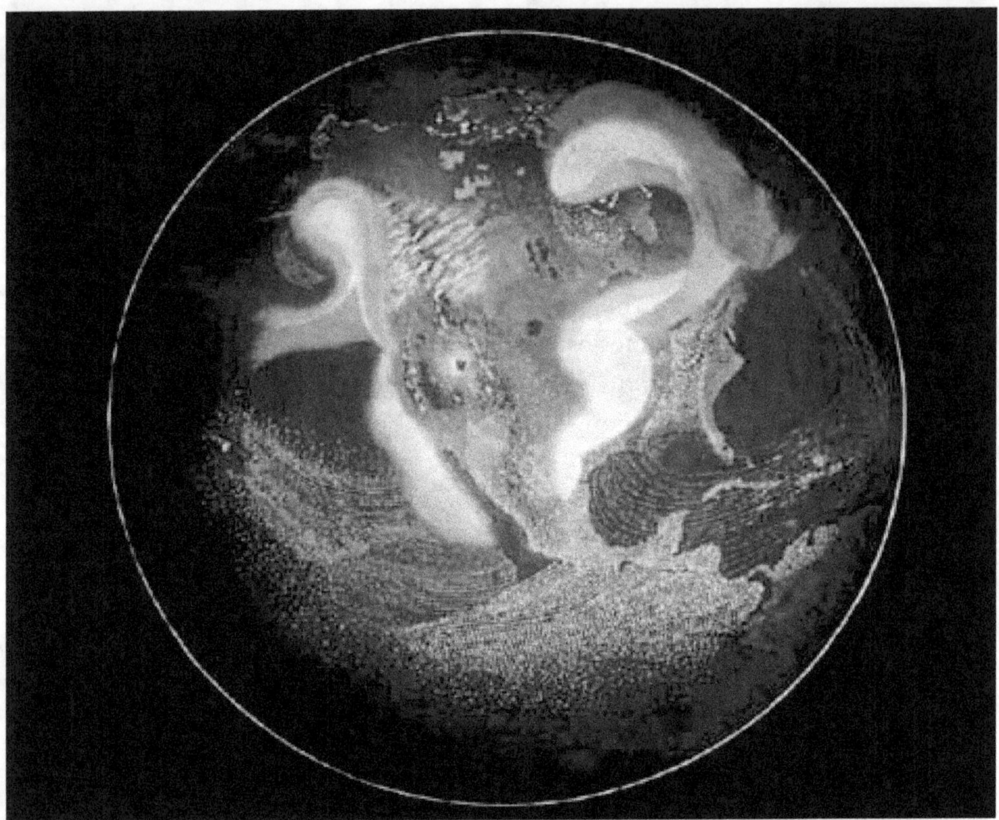

Figure 1: Weather systems over North America as they would appear from a satellite 4,000 miles above Amarillo, Texas, on 21 June. The painting was commissioned in 1954 by Dr. Harry Wexler, Director of Meteorological Research at the U.S. Weather Bureau. Surface features are drawn taking into account Earth's normal colors, reflectivity of sunlight, and scattering and depleting effects of light passing through the atmosphere, with calculated brightness of various cloud types. Weather features include a family of three cyclonic storms extending southwest from Hudson Bay to Texas; a similar system over the Bay of Alaska; a small hurricane developing near Puerto Rico; a line squall in the eastern United States; scattered cumulus clouds over heated land areas; lenticular clouds usually found where the jet stream crosses mountains, as over the northern Canadian Rockies; and low stratus and fog off the California coast, over the Great Lakes, and in the Newfoundland area. *Harry Wexler Papers, Library of Congress; the original painting hangs in the conference room of the National Environmental Satellite, Data, and Information Service, Silver Spring, Maryland*

Mindling's verses, reproduced here, are preceded by seven stanzas praising the radiosonde, and they are followed by two stanzas anticipating that weather forecasting might someday attain the accuracy of astronomical predictions.[5]

5. George W. Mindling, "The Raymete and the Future," 29 March 1939, available at *http://www.history.noaa.gov* (accessed 31 December 2008).

Precursors to the Space Age

Immediately following World War II, the United States began to launch rockets into near space. In 1947, a V-2 photographed clouds from an altitude of 160 kilometers; three years later, a two-stage Bumper V-2, launched from Cape Canaveral, Florida, reached an altitude of 400 kilometers carrying temperature and cosmic-ray sensors. In 1954, a photograph of a previously undetected tropical storm in the Gulf of Mexico was taken by an Aerobee sounding rocket launched over Texas.[6] Anticipation of the capabilities of a future Earth satellite ran high. In a 1954 symposium on space travel at New York's Hayden Planetarium, Harry Wexler, Chief of Scientific Services at the U.S. Weather Bureau, lectured on the possibilities of observing Earth's weather from a satellite vehicle.[7]

Wexler used the image in figure 1 in his public lectures to make a strong claim for the utility of the meteorological satellite, not only as a "storm patrol," but also as a potentially revolutionary new tool with global capabilities:

> Since the satellite will be the first vehicle contrived by man which will be entirely out of the influence of weather it may at first glance appear rather startling that this same vehicle will introduce a revolutionary chapter in meteorological science—not only by improving global weather observing and forecasting, but by providing a better understanding of the atmosphere and its ways. There are many things that meteorologists do now know about the atmosphere, but one thing they are sure of is this—that the atmosphere is indivisible—that meteorological events occurring far away will ultimately affect local weather. This global aspect of meteorology lends itself admirably to an observation platform of truly global capability—the Earth satellite.[8]

---

6. L. F. Hubert and Otto Berg, "A Rocket Portrait of a Tropical Storm," *Monthly Weather Review* 83 (June 1955): 121.
7. James Rodger Fleming, "A 1954 Color Painting of Weather Systems as Viewed from a Future Satellite," *Bulletin of the American Meteorological Society* 88 (October 2007): 1525–1527; Harry Wexler Papers, Library of Congress.
8. Harry Wexler, "Observing the Weather from a Satellite Vehicle," *Journal of the British Interplanetary Society* 13 (1954): 269–276; Wexler, "The Satellite and Meteorology," *Journal of Astronautics* 4 (spring 1957): 1–6; "Meteorological Satellites," *Exploring the Unknown: Selected Documents in the History of the U.S. Civil Space Program*, ed. John M. Logsdon, vol. 3, *Using Space* (Washington, DC: NASA SP-4407, 1998), p. 156 passim, available at *http://history.nasa.gov/SP-4407/vol3/cover.pdf* (accessed 31 December 2008). See also *New Dictionary of Scientific Biography*, vol. 25 (Detroit, MI: Charles Scribner's Sons/Thomson Gale, 2008), pp. 273–276.

The Blue and Brown Marble

Thus, as of the middle of the 20th century, we could say that Earth observations are very old; historically, they were made mainly by the unaided eye, but also by photography and other sensors; and there was a growing anticipation of what might be possible in the near future. We could also say that although rockets had recently ventured into near space, the Space Age was just about to begin. Adventurers had been flying since 1784, and commercial flight had recently made it possible for most everyone to fly in an airplane and look down at Earth and its clouds, but space travel has been for the privileged few—and oh, what a difference space travel has made! As Richard Somerville pointed out:

> Astronauts, who are technical people—pilots and engineers who say things like "10-4" and "affirmative" when they mean "yes"—waxed absolutely poetic when they were in space. The word you heard from all of them was *beautiful.* Alan Shepard, the first of the American astronauts, said, "What a beautiful view!" The Soviet cosmonauts said the same thing: "Our planet is uncommonly beautiful and looks wonderful from cosmic heights." There probably aren't any astronauts who haven't said words to that effect.[9]

The impressions of astronauts were reinforced by images of the "Blue Marble" from satellites and popularizations. For example, on 10 November 1967, the Applications Technology Satellite ATS-3 took the first color pictures of Earth from geosynchronous orbit using Verner Suomi's spin-scan camera. Parked above the equator just east of Brazil, the satellite image clearly showed Earth suspended in the blackness of space with three tawny continents, two blue oceans, and numerous white swirling cloud forms clearly visible. This extremely powerful image was "motivating for a lot of people," according to Stewart Brand, creator of the *Whole Earth Catalog*, "because it gave the sense that Earth is an island, surrounded by a lot of inhospitable space. And it's so graphic, this little blue, white, green, and brown jewel-like icon amongst a quite featureless black vacuum."[10] This is the "ooh" factor of Earth observations from space.

---

9. Richard C. J. Somerville, *The Forgiving Air: Understanding Environmental Change* (Berkeley, CA: University of California Press, 1996), p. 153.
10. "First photo of the whole Earth," available at *http://sciencetrack.blogspot.com/2007/07/first-photo-of-whole-earth.html* (accessed 31 December 2008).

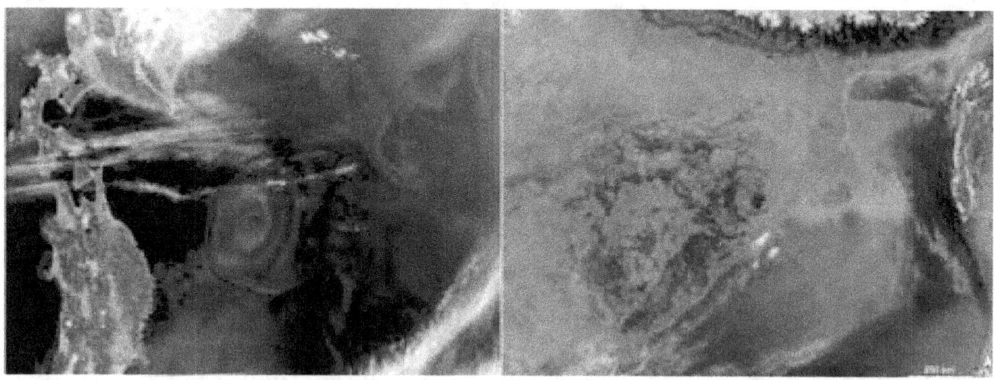

Figure 2: (Left) Ocean gyre of junk in the Western Pacific Garbage Patch east of Japan and west of Hawaii. The swirling mass of trash is currently twice as large as Texas. And there are two of them! (Right) Southeast Asian pollution cloud as captured by MODIS on the Terra satellite in 2006. Rivers of grayish haze follow the course of the Ganges River and its tributaries and flow out over the Bay of Bengal. *"An Island of Garbage Twice the Size of Texas,"* Buffalo Readings *(30 October 2007), available at* http://visibleearth.nasa.gov/view_rec.php?id=20461 *(accessed 31 December 2008)*

But it is also clear from space that there is a "Brown Marble" and that Earth has some major environmental problems. Figure 2 presents two visual examples of this "yuck" factor.

Earth Observations from Space

The balance of this chapter revisits some of the major Earth science accomplishments of the Space Age, as documented in the National Research Council 2007 publication *Earth Observations from Space: The First 50 Years of Scientific Achievements*, and then points to the ongoing challenges.[11] In 1958, the PSAC pointed out that a satellite in orbit could be used for three scientific purposes: 1) it can sample the strange new environment through which it moves; 2) it can look down and see Earth as it has never been seen before; and 3) it can look out into the universe and record information that can never reach Earth's surface because of the intervening atmosphere.[12]

The very first launches made important new discoveries. The Soviet Sputniks "observed" two things: the orbital drag of Earth's exosphere and the absence

11. National Research Council, *Earth Observations from Space: The First 50 Years of Scientific Achievements* (Washington, DC: National Academies Press, 2007). Erik Conway, *Atmospheric Science at NASA: A History* (Baltimore, MD: Johns Hopkins University Press, 2008), is a reliable source for further details on programs and sensors.
12. President's Science Advisory Committee, *Introduction to Outer Space* (Washington, DC: GPO, 1958).

of micrometeorite punctures on their pressurized capsules. On 1 May 1958, University of Iowa scientist James Van Allen announced that Geiger-Müller counters aboard JPL's Explorer 1 and Explorer 3 satellites had been swamped by high radiation levels at certain points in their orbits, indicating that powerful radiation belts, later known as the Van Allen Belts, surround Earth. These first launches revealed a new feature of the planet: the magnetosphere.[13] Vanguard 1, the fourth artificial satellite ever launched and the oldest piece of space junk still in orbit, provided important geodetic information about the shape of Earth, specifically its "pear-shaped" north-south asymmetry.

NASA launched the world's first weather satellite, TIROS 1, on 1 April 1960, with Harry Wexler in charge of the meteorology program. The TIROS satellites took television and (on later flights) infrared photos of weather patterns from space, serving as a "storm patrol" for early warnings, an aid to weather analysis and forecasting, and a research tool for atmospheric scientists. The TIROS satellites were also used in support of the Mercury launches. In a posthumous article published in 1965, Wexler wrote, "the TIROS satellites disclosed the existence of storms in areas where few or no observations previously existed, revealed unsuspected structures of storms even in areas of extensive observational coverage, depicted snow fields over land, ice floes over water, and temperature patterns on land and ocean as well as temperatures of tops of cloud layers." It also opened up the possibility of global weather coverage and the measuring of Earth's heat budget.[14]

Satellites serve to supplement and extend ground-based and radiosonde "Truth Sites" with orbital "Truth Trajectories" that provide panoptic and synoptic views. There have always been major gaps in observing networks over the oceans and sparsely inhabited areas. Such networks are labor intensive, expensive to maintain and supply, and difficult to standardize and calibrate. But in one day, the orbits of a Sun-synchronous satellite can cover the globe and a single instrument can view the entire Earth. An example of this is the first complete view of the world's weather, laboriously reconstructed from 450 different images taken by TIROS 9 in 1965. Another example is the dense

---

13. *New Dictionary of Scientific Biography*, vol. 25 (Detroit, MI: Charles Scribner's Sons/Thomson Gale, 2008), pp. 118–126.
14. Harry Wexler, "Future Forecast and Weather Control," in *From Atoms to Infinity: Readings in Modern Science*, ed. Clifford D. Simak (New York, NY: Harper & Row, 1965), pp. 96–103; National Research Council, *Earth Observations from Space*, figures 2.4 and 2.5.

satellite coverage of remote areas such as Antarctica, compared with the sparse data available during the IGY from 1957 to 1958.[15]

As an example of proof of concept, in 1953 Verner Suomi used radiometers to measure the heat budget of a Wisconsin cornfield for his doctoral thesis. This could have led him further into micrometeorology, but instead, Suomi began to think about the future possibility of measuring the heat budget of the entire planet using satellites. He developed the spin-scan camera and used it successfully on a number of satellites, including ATS-1 and ATS-3, to view an entire hemisphere; take full Earth disk, high-quality cloud-cover pictures from equatorial geostationary orbit; and provide data for the Earth Radiation Budget Experiment.[16] Additional satellite-based discoveries, proofs of concept, transformational science, and environmental monitoring services are provided below, proceeding in order from studies of the solid Earth to higher altitudes.

Solid Earth

Recently, the precise figure of a rather "lumpy" Earth was reconstructed using departures of the geoid as low as 106 meters below the ellipsoid and as high as 85 meters above it, as measured by the Gravity Recovery and Climate Experiment (GRACE) mission using precise orbital tracking and radar altimetry. The same satellite revealed much smaller scale structures from gravity anomalies, allowing for the construction of maps of seafloor topography and Earth deformations that reveal earthquake stress patterns. Such images provide real-time monitoring of fault movements and volcanic deformations prior to eruptions, and they form the basis of early warning systems. Land subsidence from ground water extraction can be monitored from space, and sunken rivers (for example, under the sands of the Sahara) can be revealed by ground-penetrating radar.[17]

---

15. The concepts of truth sites and truth trajectories are illustrated in National Research Council, *Earth Observations from Space*, figures 3.1, 3.2, and 7.1. The TIROS 9 image of world weather systems is reproduced in figure 3.3.
16. A discussion of Suomi's 1971 heat budget experiment is found in James Rodger Fleming, *Climate Change and Anthropogenic Greenhouse Warming: A Selection of Key Articles, 1824–1995, with Interpretive Essays* (Arlington, VA: National Science Foundation, National Science Digital Library, 2008), available at *http://wiki.nsdl.org/index.php/PALE:ClassicArticles/GlobalWarming* (accessed 31 December 2008). See also *New Dictionary of Scientific Biography*, vol. 24 (Detroit, MI: Charles Scribner's Sons/Thomson Gale, 2008), pp. 553–558.
17. National Research Council, *Earth Observations from Space*, figures 6.4, 11.1, 11.2, 11.3, 11.6, and 11.7.

Oceans

Thermal differences and turbulent eddies in the ocean are clearly revealed by satellite remote sensing. Maps of the Gulf Stream, for example, which date to the time of Benjamin Franklin, were reconstructed by measurements made from ships, a practice that continued into the 1960s. However, in the mid-1970s, the synoptic view provided by satellite thermal infrared imagery showed that the Gulf Stream was a single filament, albeit following a tortuous and time-changing path. Satellite measurements also have replaced widely scattered truth sites and limited measurements along ship trajectories with continuous monitoring. This has revealed the new phenomena of internal tides, leading scientists to conclude that the deep oceans are much more dynamic and well mixed than previously imagined. This is important in many related fields, such as global carbon cycling.

Some of the more iconic images of the Pacific Ocean show sea surface temperatures, heights, and biological productivity during El Niño and La Niña conditions, as revealed by the U.S.-French TOPEX/Poseidon satellite and by the Sea-viewing Wide Field-of-view Sensor (SeaWiFS). During the 1997 El Niño, for example, warm surface waters in the eastern Pacific Ocean were 14 to 32 centimeters higher than normal, and they were abnormally low in chlorophyll because the supply of nutrients was greatly reduced due to suppressed upwelling. The following year, under La Niña conditions, the same waters were cold; sea levels were low; and chlorophyll concentrations were higher than average due to enhanced upwelling and an extensive phytoplankton bloom at the equator. This was the first, but certainly not the last, El Niño Southern Oscillation (ENSO) to be closely monitored from space. Biological oceanographers are now able to look in great detail at formerly inaccessible regions. The Amazon River plume stretching thousands of kilometers into the Atlantic Ocean is an example of a new discovery resulting from the first ocean color observations from space.[18]

Land Surface

In 1972, NASA launched the Landsat program (previously called Earth Resources Technology Satellite) to study the features of Earth's landscapes and monitor its natural resources. Landsat data demonstrated early success in monitoring Earth's croplands, forests, and other natural resources. It has

18. National Research Council, *Earth Observations from Space*, figures 8.1, 8.2, 8.4, and 12.1. Only two of the cards in the NASA 50th anniversary card deck refer directly to Earth observations; in this case, it is the Jack of Hearts.

since become the workhorse for mapping land-use and land-cover change across the world and now provides the longest continuous record of Earth's changing land cover. Landsat 7, launched in 1999, has functioned beyond its expected five-year lifespan, but an instrument malfunction in 2003 has compromised some of the data it collects. Notable Earth land surface studies (there have been many) include quantification of deforestation in the Amazon; photographs of urban sprawl that reveal different patterns and dynamics in different nations; and the World Fire Atlas, presenting both real-time data and seasonal and annual patterns. The sheer beauty of landform photography from space is also worthy of note, as in the delicate fractal geometry of Iceland's fjords.[19]

Cryosphere

Satellite observations have revolutionized cryosphere research. The example of Antarctic traverses then and now was mentioned earlier. Access to remote, frozen regions has always involved often-great risks and logistical difficulties. The synoptic view from satellites increases the data coverage by multiple orders of magnitude, and access is no longer restricted by seasons. The discovery of variability in the flow velocity of ice sheets is an example of how the dynamics of a major system went undetected until reliable and repeated satellite observations were available. The rapid and dramatic collapse of the Larsen B Ice Shelf in 2002 was documented by satellite imagery and became a popular example of the sensitivity of ice sheets in a changing climate. NASA animations were in the news as well, showing dramatic loss of Arctic sea ice since satellite coverage began in 1979.[20]

Atmosphere

Satellites have contributed so much to observing and visualizing atmospheric processes that only a few examples can be given here. The most widely recognized are the polar stratospheric ozone measurements from the Total Ozone Mapping Spectrometer (TOMS), updated by the Microwave Limb Sounder on the Upper Atmosphere Research Satellite (UARS). Chlorine monoxide concentra-

---

19. National Research Council, *Earth Observations from Space*, figures 10.3, 10.4, 10.5, and 10.7; "RST Section 17," available at *http://www.fas.org/irp/imint/docs/rst/Sect17/Sect17_4.html* (accessed 31 December 2008).
20. National Research Council, *Earth Observations from Space*, figures 7.2 and 7.3; "Arctic Sea Ice Continues to Decline," available at *http://www.nasa.gov/centers/goddard/news/topstory/2005/arcticice_decline_prt.htm* (accessed 31 December 2008).

tions are correlated with low ozone concentrations, confirming ground-based measurements and the proposed mechanisms for ozone depletion.[21]

Closer to the ground, massive and intense African dust storms can be imaged and monitored by MODIS aboard NASA's Terra satellite. Dust storms in Asia and elsewhere often generate public health and safety warnings. Hurricanes were first photographed and routinely followed by TIROS. Before that, tropical storms would sometimes get "lost" between encounters with ships, airplanes, or island observers. Notable storms, such as Hurricane Katrina in 2005 and many others, have been photographed from space in visual and infrared wavelengths and probed by radar. The TRMM uses infrared and microwave images to depict the horizontal distribution of rain intensity inside a storm. This makes isolated, and otherwise hidden, hot towers visible and may indicate storm intensification.[22]

Satellite measurements in both the Northern and Southern Hemispheres can be linked to computer models of numerical weather prediction to improve data initialization and overall forecasting skill. Figure 3 shows the overall improvement and convergence of skill in the Northern and Southern Hemispheres since 1980, with the critical three-day forecast skill improving from 85 to 95 percent in the Northern Hemisphere and from 70 to 95 percent in the Southern Hemisphere, where land observations are scarce.

Tracking pollution from space is also possible. For example, data from NASA's TOMS satellite instrument were used to monitor smoke, smog, and tropospheric ozone from fires in Indonesia and Africa in 1997 while the satellite moved across the Indian Ocean. The Terra Measurement of Pollution in the Troposphere (MOPITT) satellite observed the seasonally changing global distribution of carbon monoxide (CO) pollution. Northern Hemisphere pollution sources are predominantly urban and industrial, while high CO in the tropics and Southern Hemisphere often results from biomass burning.[23] The 2009 failure of the Orbiting Carbon Observatory (OCO) was a serious, but hopefully temporary, setback for precise monitoring of carbon dioxide emissions and concentrations. Such a satellite will allow this critical trace gas to be studied both in its global background concentration and in its temporal and spatial variations.[24]

---

21. National Research Council, *Earth Observations from Space*, figure 5.3; NASA 50th anniversary card deck, Five of Clubs.
22. National Research Council, *Earth Observations from Space*, figures 3.5, 4.10, and 6.1.
23. National Research Council, *Earth Observations from Space*, figures 3.8, 5.8, and 5.9.
24. "Project Vulcan," available at *http://www.purdue.edu/eas/carbon/vulcan/index.php* (accessed 31 December 2008).

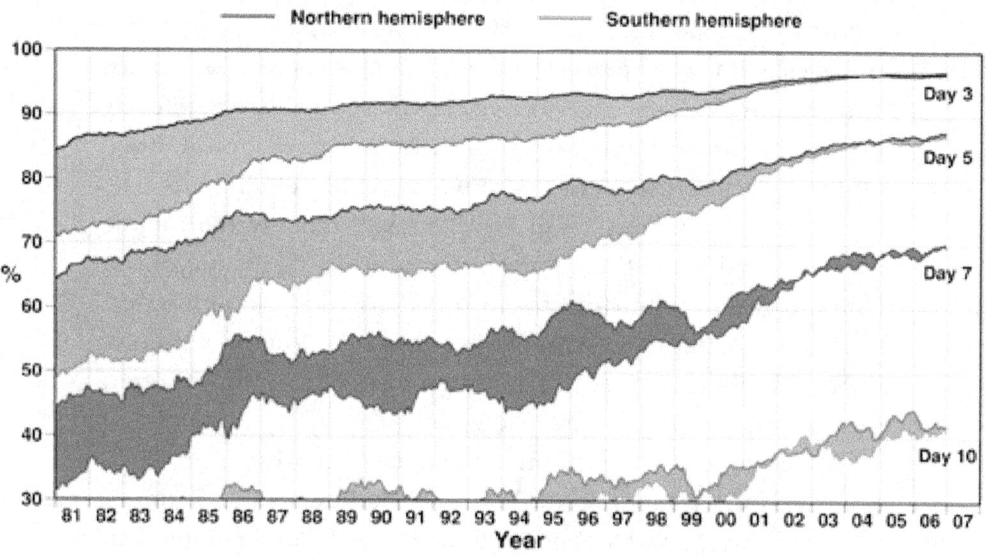

Figure 3: Anomaly correlation of 500 hectopascal height forecasts by the European Centre for Medium Range Forecasting showing overall improvement since 1980 for day 3, 5, and 7 forecasts and convergence of skill for forecasting in the Northern and Southern Hemispheres. *Courtesy of the Royal Meteorological Society*

Space

When the IGY got under way in 1957, Earth had one Moon. Now, our increasingly electronic and space-based infrastructure has given rise to the new field of "space weather," defined as "conditions in the Sun and in the solar wind, magnetosphere, ionosphere and thermosphere that can influence the performance and reliability of space-borne and ground-based technological systems and can endanger human life or health."[25] These ionized and electronically active realms eventually merge with the extended solar atmosphere, raising the question of where the atmosphere ends and where "space" begins.

The 2007 National Research Council-EOS report focused mainly on the scientific accomplishments of Earth observations from space and reached the following conclusions:

---

25. Solar and Heliospheric Observatory (SOHO), available at *http://sohowww.nascom.nasa.gov/spaceweather/* (accessed 31 December 2008).

Conclusion 1: The daily synoptic global view of Earth, uniquely available from satellite observations, has revolutionized Earth studies and ushered in a new era of multidisciplinary Earth sciences, with an emphasis on dynamics at all accessible spatial and temporal scales, even in remote areas. This new capability plays a critically important role in helping society manage planetary scale resources and environmental challenges.

Conclusion 2: To assess global change quantitatively, synoptic data sets with long time series are required. The value of the data increases significantly with seamless and inter-calibrated time series, which highlight the benefits of follow-on missions. Further, as these time series lengthen, historical data sets often increase in scientific and societal value.

Conclusion 3: The scientific advances resulting from Earth observations from space illustrate the successful synergy between science and technology. The scientific and commercial value of satellite observations from space and their potential to benefit society often increase dramatically as instruments become more accurate.

Conclusion 4: Satellite observations often reveal known phenomena and processes to be more complex than previously understood. This brings to the fore the indisputable benefits of multiple synergistic observations, including orbital, suborbital, and in situ measurements, linked with the best models available.

Conclusion 5: The full benefits of satellite observations of Earth are realized only when the essential infrastructure, such as models, computing facilities, ground networks, and trained personnel, is in place.

Conclusion 6: Providing full and open access to global data to an international audience more fully capitalizes on the investment in satellite technology and creates a more interdisciplinary and integrated Earth science community. International data sharing and collaborations on satellite missions lessen the burden on individual nations to maintain Earth observational capacities.

Conclusion 7: Over the past 50 years, space observations of the Earth have accelerated the cross-disciplinary integration of analysis, interpretation, and, ultimately, our understanding of the dynamic processes that govern the planet. Given this momentum, the next decades will bring more remarkable

> discoveries and the capability to predict Earth processes, critical to protect human lives and property. However, the nation's commitment to Earth satellite missions must be renewed to realize the potential of this fertile area of science.[26]

Challenges

Several important challenges confront us at this juncture: 1) the challenges of discovery vs. monitoring, 2) the challenges of international cooperation and data sharing, and 3) the looming "Satellite Gap." Charles David Keeling faced the first challenge directly when he measured rising levels of carbon dioxide in Earth's atmosphere, between 1958 and 1960, and then had to scramble to find support to continue his monitoring program to generate the Keeling curve, which has become an environmental icon. To assess global change quantitatively, synoptic data sets with long time series are required. The value of the data increases significantly with seamless and intercalibrated time series, which highlight the benefits of follow-on missions. This need for continuity and intercalibration is exacerbated in the expensive world of satellite launches. Two examples come to mind: global sea level rise, which is more of a monitoring issue than a new discovery, and polar ice trends that are quite different in the Arctic and the Antarctic. For example, deviations in monthly sea ice extent for the Northern and Southern Hemispheres from November 1978 through December 2004 were derived from satellite passive-microwave observations. The Arctic sea-ice *decreases* are significant, with a loss of about 38,200 ± 2,000 square kilometers per year, and they have led to much concern about Arctic and global warming. Antarctic sea ice has been *increasing*, although at a much lower rate of 13,600 ± 2,900 square kilometers per year. Notably, both trends are statistically significant.[27]

Providing full and open access to global data to an international audience more fully capitalizes on the investment in satellite technology and creates a more interdisciplinary and integrated Earth science community. The Global Weather Experiment (GWE) of 1984 employed five geostationary satellites from the United States, the USSR, Japan, and ESA to make global observations of cloud-tracked winds.[28] At the time it was the largest coordinated experiment ever undertaken, but today, more data arrive every instant than during the entire

---

26. National Research Council, "Conclusions," *Earth Observations from Space*, pp. 99–106.
27. National Research Council, *Earth Observations from Space*, figure 7.4.
28. Ibid., figure 2.7.

GWE. International data sharing and collaborations on satellite missions lessen the burden on individual nations to maintain Earth observational capacities and also allow for better weather prediction and disaster warning systems for all nations. However, space is not used only for scientific discovery, monitoring, and services. On a clear night, we may look at the ancient constellations Ursa Major, Aquila, and Draco and recall that Russia, the United States, and China are not the only three spacefaring nations up there looking down at us and that not all satellites are scientific or commercial. Space is a busy place both militarily and politically.

A final important challenge is what I call "The Satellite Gap." According to the current *NASA Strategic Management Handbook*, the mission statement "summarizes the accomplishments of the organization in fulfilling its vision, its main purpose for existing, and the basic social or political needs that are to be met. The mission statement addresses the unique products and services that the organization delivers to its customers."[29] In the National Aeronautics and Space Act, which established the Agency in 1958, the first objective was listed as "the expansion of human knowledge of the Earth and of phenomena in the atmosphere and space." Traditionally, it has always been the case that NASA's mission statement highlighted the "advancement and communication of scientific knowledge and understanding of the Earth." In 2002, it was even more explicit: "To understand and protect our home planet . . . ." In early February 2006, this statement was suddenly and summarily deleted, which, combined with the cancellation or delay of a number of Earth science missions, raised eyebrows and led to talk of a pending satellite gap. NASA's current mission statement calls on the Agency "to pioneer the future in space exploration, scientific discovery and aeronautics research." It is the first time since NASA's founding in 1958 that the mission statement does not explicitly include mention of Earth.[30] Perhaps this omission can be remedied.

The National Research Council's 2007 Decadal Survey warned that a number of measurements providing critical information about Earth processes would cease to be made over the next few years. By 2010, it says, the number of Earth-observing missions will drop dramatically, and the number

29. *NASA Strategic Management Handbook*, available at *http://www.hq.nasa.gov/office/codez/strahand/planning.htm* (accessed 31 December 2008).
30. Andrew Revkin, "NASA's Goals Delete Mention of Home Planet," *New York Times* (22 July 2006): A1, A10; "About NASA," available at *http://www.nasa.gov/about/highlights/what_does_nasa_do.html* (accessed 31 December 2008).

of operating sensors and instruments on NASA spacecraft will decrease by 40 percent, putting our extraordinary foundation of global observations at great risk. It strongly advised that "the U.S. government, working in concert with the private sector, academe, the public, and its international partners, should renew its investment in Earth-observing systems and restore its leadership in Earth science and applications."[31]

Conclusion

Over the past 50 years, it has been very good viewing from space. As unique observing platforms, satellites have fundamentally transformed the practice of Earth science. We have fulfilled an ancient quest for altitude and provided scientists and the general public with a panoptic viewpoint. The launches of Sputnik and Explorer were "Galilean moments," with many other such moments following. Historians argue that ever since 1543, Earth has been systematically and progressively demoted from its ancient status as the center of the universe, yet many would also agree that the Space Age launched a new "Copernican type revolution" in returning our values to an Earth-centric focus, both through Blue Marble inspiration and Brown Marble pollution studies.[32]

Observing Earth from space over the past 50 years has fundamentally transformed the way people view our home planet. The daily synoptic global view of Earth, uniquely available from satellite observations, has revolutionized Earth studies and ushered in a new dynamic era of multidisciplinary Earth sciences, with an emphasis on dynamics at all accessible spatial and temporal scales, even in remote areas. This new capability plays a critically important role in helping society manage planetary-scale resources and understand environmental challenges.

Satellites have contributed in unique ways to scientific understanding of the solid Earth, the oceans, the land, the cryosphere, the atmosphere, and the near-space environment. They have enhanced our ability to predict variations in the Earth system and promise new opportunities to improve Earth science research—and there still is a tremendous amount of knowledge to be gained and discoveries to be made. Satellites and their instrumentation exist in synergy with deep space probes to other planets. They constitute a wonderful context

31. National Research Council, *Earth Science and Applications from Space: National Imperatives for the Next Decade and Beyond* (Washington, DC: National Academies Press, 2007).
32. National Research Council, *Earth Observations from Space*, figure S.1.

for discovery, but they are also essential monitoring devices. Our world simply could not function as it does without satellites.

Recall the 1958 statement by the PSAC that serves as the epigraph to this chapter. While the cost of Earth observations from space over the past five decades is measured in billions of dollars, the value of understanding our home planet is *priceless!*

Chapter 22

Earth Science and Planetary Science

A Symbiotic Relationship?

Erik Conway[1]

As NASA celebrates its first 50 years, JPL has two robots on Mars, carrying out what can best be described as late-19th-century field geology with 21st-century instruments. NASA has a flotilla of spacecraft studying Earth "as a planet," to borrow one nearly ubiquitous catchphrase. It has Earth scientists of nearly every discipline at its Centers—glaciologists at GSFC, atmospheric scientists at LaRC, and oceanographers at JPL. It dispenses tens of millions of dollars in grants and contracts to university-based scientists.

All of this activity raises the important question, what has been the impact on the practice of Earth science? In his 1980 memoir, former Agency Chief Scientist Homer Newell argued that space science had been an "integrative force." It had broken the geosciences "loose from a preoccupation with a single planet."[2] But it has also provided a planetary perspective, enabling local and regional phenomena to be placed within a still larger context. "Planetary methods," initially dismissed by American Earth scientists, gradually became a routine part of their endeavor.

In the *History of Atmospheric Science at NASA*, I go into great detail about the intersection of planetary and Earth science. In particular, I examine the way that the striking discovery, in the late 1950s and 1960s, that Mars and Venus

1. Jet Propulsion Laboratory, California Institute of Technology, Pasadena, CA. © 2008. California Institute of Technology. Government sponsorship acknowledged.
2. Homer Newell, *Beyond the Atmosphere: Early Years of Space Science* (Washington, DC: NASA SP-4211, 1980), p. 328.

were climatically far different than expected forced scientists to reconsider the way Earth's climate has evolved over the last 4.5 billion years.[3] Here I focus on how the *practice* of science changed as a result of planetary science.

After I recently reread Ron Doel's *Solar System Astronomy in America*, it occurred to me there was an obvious way to approach the subject of scientific practice.[4] Doel uses a number of cases of individual research activity to build his argument for the fundamentally interdisciplinary nature of planetary astronomy in the 1940s and 1950s. I'm going to take a similar approach. First, I will explore the career of a single JPL infrared spectroscopist who made important contributions both to the question of whether there is water on Mars and to the science of stratospheric ozone depletion on Earth. Broadening the view, I will then look at the intersection of planetary modeling and an old Earth science question, whether volcanic eruptions cause short-term cooling. Finally, I will examine the role of planetary exploration in forming the new discipline of Earth system science, NASA's rather ambitious attempt to remake the geosciences.

Spectroscopy in Earth and Planetary Sciences

Spectroscopy is not new to the Space Age. It is a product of the 19th century, based on the discovery that all objects radiate electromagnetic energy. In fact, atoms and molecules radiate at very specific, and characteristic, wavelengths. This means that the radiated energy can be used to determine what an unknown object, or light source, is composed of. Historian David DeVorkin has written about the astronomers' quest to measure the spectral signature of hydrogen in the Sun, for example.[5] Calculations based on the Sun's gravity and size had already suggested it was mostly hydrogen, but confirmation required measuring a set of spectral lines known as "Lyman Alpha," which happen to be in the ultraviolet portion of the spectrum. They are blocked by Earth's ozone layer. Lyman Alpha was finally measured 11 December 1952, using an Aerobee sounding rocket.

The race for Lyman Alpha was one use of spectroscopy in astronomy. Another historical use was for study of the composition of planetary atmospheres using ground-based telescopes. Carbon dioxide, which has spectral

3. Erik M. Conway, *Atmospheric Science at NASA: A History* (Baltimore, MD: Johns Hopkins University Press, 2008).
4. Ronald E. Doel, *Solar System Astronomy in America* (New York, NY: Cambridge University Press, 1996).
5. David H. DeVorkin, *Science with a Vengeance: How the Military Created the US Space Sciences After World War II* (New York, NY: Springer-Verlag, 1992), pp. 221–230.

characteristics in the thermal infrared, was detected in the Venus atmosphere by astronomers during the opposition of 1932.[6] It was also detected in the Martian atmosphere in 1947 by Gerard Kuiper at the McDonald Observatory.[7] Water vapor, radiatively active throughout the infrared spectrum, was detected in the Martian atmosphere in 1963 using the 100-inch reflector on Mount Wilson.[8]

Earth's atmosphere contains both carbon dioxide and lots of water vapor, so a professional hazard of astrophysicists in the 1950s was obscuration of the weak planetary spectral lines by the strong absorption of these same gases in Earth's atmosphere. The water vapor and carbon dioxide also happen to absorb interesting spectral features, like Lyman Alpha, so a great deal simply couldn't be measured from the ground. There were some means of correcting for these distortions. Kuiper, for example, had removed the effects of terrestrial carbon dioxide from his Mars spectra by first determining the spectrum of Earth's atmosphere via reflected moonlight. But in turn, these corrections raised new questions about the adequacy of the corrections.

Crofton B. Farmer, born in Wales, U.K., in 1931, was originally interested in measuring the solar spectrum. He completed a Ph.D. in physics at King's College, London, on measuring the solar spectrum from visible wavelengths out to the far infrared at 300 microns. Atmospheric water vapor absorbs throughout that entire wavelength region, and his approach had been to haul a spectrometer up to 18,000 feet to get above most of it. He did his Ph.D. work during several expeditions to Mount Chacaltaya in the Bolivian Andes, finishing in 1966.

Water on Mars?

At about the same time, he started consulting at JPL on the establishment of an infrared spectroscopy lab. He recalled many years later, "the goals at that lab were to look at the infrared spectrum of simulated planetary atmospheres. We would look at all sorts of gases and mixtures of gases for Mars and Venus and Jupiter and Saturn and so on."[9] This fundamental spectroscopic work was necessary to understand what spectrometers sent off to other planets would actually be seeing. Mixtures of gases produce more complicated spectra than pure gases do. Gases also produce slightly different spectra depending on

---

6. Mikhail Ya. Marov and David H. Grinspoon, *The Planet Venus* (New Haven, CT: Yale University Press, 1998), p. 28.
7. William Sheehan, *The Planet Mars: A History of Observation and Discovery* (Tuscon, AZ: University of Arizona Press, 1996), p. 150.
8. Ibid., p. 160.
9. Crofton B. Farmer, OHI by Erik Conway, 20 May 2004, JPL Archives.

temperature and pressure. So measurements had to be made of many different combinations at many different pressures and temperatures. After consulting for a year, Farmer realized that JPL was probably the only place that could support the technical development necessary to build the very high spectral resolution instruments he wanted to pursue, and he moved his family to Pasadena, California, permanently.

His first real planetary effort was aimed toward the water on Mars question. Photographs sent back in 1965 by Mariner 4 showed an unexpected, cratered, Moon-like surface. But they also seemed to show features that looked like runoff features, suggesting Mars once had liquid water. Water vapor, of course, absorbs in the infrared, so Farmer got involved in an effort to distinguish water vapor lines in spectra of the Mars atmosphere. These had first been detected in 1963, but the detection had been weak; it was questioned for several years. Three different groups sought to confirm the detection in the winter conjunction of 1968 to 1969, including Farmer's, and with Ronald Schorn and Stephen Little, Farmer was able to publish a paper in 1969 sketching out the latitudinal distribution of water vapor.[10]

In turn, this work got Farmer the opportunity to build an instrument for the two Viking orbiters planned for 1976. This was the Mars Atmospheric Water Detector (MAWD), a grating spectrometer.[11] These both operated successfully, as it turned out, and Farmer got one and a half Mars years of data. The data allowed him to examine the seasonal distribution of water vapor with both latitude and altitude. Latitudinal distribution varied widely throughout the year, revealing exchange between the two hemispheres as well as making clear that there was a reservoir of water—almost certainly water ice, not liquid—in communication with the atmosphere. The polar ice caps were the obvious source. These had been thought to be carbon dioxide, as the surface temperature was often below the condensation point for that gas. But MAWD clearly showed the atmosphere above the north polar ice to saturate with water vapor quickly in late spring.[12] So the permanent, or "residual," ice cap was water.

---

10. Ronald A. Schorn, C. B. Farmer, and Stephen J. Little, "High Dispersion Spectroscopic Studies of Mars III," *Icarus* 11 (1969): 283–288.
11. While operating in the 1.4-micron water vapor band, MAWD used reflectance from the surface. See Crofton B. Farmer and Daniel D. LaPorte, "The Detection and Mapping of Water Vapor in the Martian Atmosphere," *Icarus* 16 (1972): 34–46. On the Viking project, see Edward Clinton Ezell and Linda Neuman Ezell, *On Mars: Exploration of the Red Planet, 1958–1978* (Washington, DC: NASA SP-4212, 1984).
12. C. B. Farmer and P. E. Doms, "Global Seasonal Variation of Water Vapor on Mars and the Implications for Permafrost," *Journal of Geophysical Research* 84, no. B10 (10 June 1979): 2881–2888.

Working with several others, Farmer also ran laboratory experiments designed to establish how water ice bound in dirt ("regolith," geologists call it) might interact with the thin Martian atmosphere. This combination of the remote sensing data and the laboratory experiments allowed Farmer to argue in a 1979 article that, in addition to the water ice in the residual polar caps, water ice was also likely to be found within a meter of the surface nearly everywhere poleward of about 45 degrees latitude.

Stratospheric Chemistry on Earth

There had been a long delay between delivering MAWD and the arrival at Mars, and in the interim, Farmer got involved in two more initiatives. He proposed an instrument for the Voyager spacecraft and was not selected. And, more importantly, he proposed a balloon instrument to measure oxides of nitrogen in Earth's stratosphere. These had been fingered as possible ozone destroyers in 1970. Internal combustion engines release the chemicals; and at the time, there appeared to be a future for supersonic, stratospheric airliners.[13] No one knew what the "natural" level of nitrogen oxidcs were at the time, so it wasn't possible to determine whether fleets of stratospheric aircraft would significantly increase their amounts or not.

Oxides of nitrogen happen to have absorption lines in the infrared, so Farmer knew that at least in theory, he could measure them. He built a series of Fourier transform spectrometers to be hoisted up into the stratosphere on balloons, where they would produce spectra via solar occultation. He flew one of these aboard a prototype Concorde six times in 1973. He also started hiring more spectroscopists to help analyze all the data, building a research team in the infrared that's still active at JPL.

In 1974, Farmer also got one of his spectrometers flown aboard NASA Ames Research Center's U-2 stratospheric research aircraft, where he was able to make the first measurements of hydrogen chloride absorption lines in Earth's atmosphere.[14] This was a challenging experience, involving more than a little risk. To ensure the spectra were absolutely clean, the U-2 pilot shut down the plane's engine during the solar occultation, risking a very long glide back to Earth if the engine refused to relight. But it did. The risk had been considered worth it,

---

13. Erik M. Conway, *High Speed Dreams: The Technopolitics of Supersonic Transportation* (Baltimore, MD: Johns Hopkins University Press, 2005), chap. 6.
14. C. B. Farmer, O. F. Raper, and R. H. Norton, "Spectroscopic Detection and Vertical Distribution of HCl in the Troposphere and Stratosphere," *Geophysical Research Letters* 3, no. 1 (1976): 13–16.

though, because that same year saw F. Sherwood Rowland and Mario Molina proposing that chlorofluorocarbons could destroy the ozone layer. They argued this would happen because ultraviolet radiation in the stratosphere would break up the large chlorofluorocarbon molecules into fluorine and chlorine compounds, and the chlorine compounds would then deplete the stratospheric ozone layer.

Farmer's interferometer could measure both hydrogen chloride lines and hydrogen fluoride lines, enabling him to confirm that the principal source of chlorine and fluorine in the stratosphere was chlorofluorocarbons that were released at the surface and were broken down by ultraviolet radiation when they reached the stratosphere. Hydrogen fluoride has no natural sources that reach the stratosphere, so with the Mark III's interferometer ability to measure both hydrogen fluoride and hydrogen chloride simultaneously, Farmer's team could also produce a ratio that would permit partitioning of the human and natural stratospheric chlorine inventories.[15]

So Farmer's group had demonstrated the ability to measure not only the nitrogen oxides they had set out to get, but also several other key ozone-related trace molecules. This put them in a position to contribute observations relevant to what became the "ozone war" of the late 1970s.[16]

The final form of Farmer's balloon instrument series was only ever known as the Mark IV spectrometer. It was built after a disastrous flight of the Mark III out of Palestine, Texas, in 1982. NASA Upper Atmosphere Research Program Manager Robert T. Watson had established a set of "Balloon Intercomparison Campaigns" designed to boost the credibility of a whole host of stratospheric chemistry instruments.[17] Remote sensing techniques, while familiar to astrophysicists, were not widely accepted by Earth scientists, and given the politically controversial nature of ozone science, disagreements over measurement technique often appeared in the political realm.[18] Several *very* large stratospheric

---

15. C. B. Farmer, O. F. Raper, and R. H. Norton, "Spectroscopic Detection and Vertical Distribution of HCl in the Troposphere and Stratosphere," *Geophysical Research Letters* 3, no. 1 (January 1976): 13–16; C. B. Farmer and O. F. Raper, "The HF:HCl Ratio in the 14–38 km Region of the Stratosphere," *Geophysical Research Letters* 4, no. 11 (November 1977): 527–529; C. B. Farmer et al., "Simultaneous Spectroscopic Measurements of Stratospheric Species: O3, CH4, CO, CO2, N2O, H2O, HCl, and HF at Northern and Southern Mid-Latitudes," *Journal of Geophysical Research* 85, no. C3 (20 March 1980): 1621–1632.
16. Lydia Dotto and Harold Schiff, *The Ozone War* (Garden City, NY: Doubleday, 1978).
17. C. B. Farmer et al., "Balloon Intercomparison Campaigns—Results of Remote-Sensing Measurements of HCl," *Journal of Atmospheric Chemistry* 10, no. 2 (February 1990): 237–272.
18. NASA had also embarked on an effort to improve the scientific standing of its Earth science efforts more generally. See Conway, *Atmospheric Science at NASA: A History*, chaps. 5 and 6.

balloons were used to hoist the instrumented gondolas. The first attempt for Farmer's gondola failed when the balloon burst, but the instrument survived the fall. On the second attempt, it reached the stratosphere safely and got data, but the gondola was cut down at too high an altitude, and the parachute didn't open.[19] The gondola free-fell to the ground, and the instrument and gondola were crushed. The accident got Farmer $2 million to build the new, and much improved, Mark IV.

Farmer's Mark I through III interferometers used the solar occultation observational method, which has the advantage of a high signal-to-noise ratio so that the very high spectral resolution needed to detect and measure trace gases could be achieved. But the technique can only produce vertical profiles for altitudes up to the observing platform (about 40 kilometers for balloons) together with the total column above that. To overcome this limitation, the next step was to move to spaceborne observations. So after demonstrating the Mark III, Farmer also received funds to build a Shuttle-based version, which became known as the Atmospheric Trace Molecules Spectroscopy (ATMOS) experiment. Shelby Tilford, who was head of NASA's Earth science and observations programs, thought this would be a revolutionary instrument, and he supported it through many significant developmental problems.[20]

From 30 April to 6 May 1985, ATMOS flew as part of the Spacelab 3 payload on Space Shuttle *Challenger*. This first flight was troubled; a manufacturing error caused the instrument to lose pressure fairly rapidly, and its internal laser calibration system failed after only a few days. However, frantic rearrangement of the Shuttle schedule enabled it to collect about 500 spectra from Earth's atmosphere during the first 19 sunrise and sunset occultations from the Shuttle. This bounty included 30 different molecular constituents of the atmosphere, including a number of "first detections" of trace species predicted by ozone depletion theory but never actually seen.[21] Adrian Tuck, a meteorologist at the NOAA Aeronomy Laboratory in Colorado, later reflected that these results provided a big boost in credibility for ozone chemistry overall.[22] It had

---

19. C. B. Farmer, OHI by Erik Conway, 20 May 2004; Joe Waters, OHI by Erik Conway, 31 October 2005, both in JPL Archives.
20. Larry Simmons, OHI by Erik Conway, 31 January 2005, JPL Archives.
21. See, for example, C. B. Farmer and O. F. Raper, *High Resolution Infrared Spectroscopy from Space: A Preliminary Report on the Results of the Atmospheric Trace Molecule Spectroscopy Experiment on Spacelab 3* (Washington, DC: NASA CP-2429, May 1986); R. Zander et al., "Infrared Spectroscopic Measurements of Halogenated Source Gases in the Stratosphere with the ATMOS Instrument," *Journal of Geophysical Research* 92, no. D8 (20 August 1987): 9836–9850.
22. Adrian Tuck, OHI by Erik Conway, 3 November 2003, NASA History Division.

been difficult to accept chemical models that depended on the existence of molecules never seen outside a laboratory.

That same year happened to be the year that the British Antarctic Survey announced the existence of the ozone "hole" over Antarctica.[23] In fact, the announcement occurred within a few weeks of the ATMOS flight. The spectrometer had flown at the wrong time of year to investigate the "hole," and due to the grounding of the Shuttles after the *Challenger* accident, ATMOS didn't fly again until 1992. So ATMOS was of no further assistance in understanding the phenomenon.

Instead, Farmer, his brand new Mark IV, and JPL colleague Geoffrey Toon were shipped to Antarctica in August 1986 to spend the winter. This first Antarctic expedition was known as the National Ozone Experiment (NOZE), and the expedition was led by Susan Solomon of the NOAA Aeronomy Laboratory. The large stratospheric balloons that Mark IV had been designed for couldn't be flown from McMurdo Station, so the instrument operated in a modified shipping container for the season. It also had to be modified to look upward instead of downward, but this was not a big technical challenge. Toon recalled later that his biggest trouble was in keeping the Mark IV operating without being able to see the resulting data. The spectra it recorded could only be processed back at JPL, on JPL's mainframe computer. So he couldn't know whether it was producing good results during their months of Antarctic seclusion.

But it did produce good results, and the next year a larger Mark IV team went back to Antarctica for the Airborne Antarctic Ozone Experiment (AAOE), which was flown out of Punta Arenas, Chile. This was carried out from August to September 1987. The Mark IV flew aboard the NASA DC-8, again using an upward-scanning mode as the primary region of ozone depletion was at around 22 kilometers, far above the DC-8's cruising altitude. A NASA ER-2 carried in situ instruments at that altitude, providing corroboration for some (though not all) of the Mark IV's data.[24]

There were many other investigators from other NASA Centers, NOAA, the British Antarctic Survey, and universities involved in this effort. So my purpose in following Farmer and his team is not to claim pride of place for Farmer or

---

23. J. C. Farman, B. G. Gardiner, and J. D. Shanklin, "Large Losses of Total Ozone in Antarctica reveal seasonal ClOx/NOx interaction," *Nature* 315 (16 May 1985): 207–210.
24. A. F. Tuck, R. T. Watson, E. P. Condon, J. J. Margitan, and O. B. Toon, "The Planning and Execution of ER-2 and DC-8 Aircraft Flights Over Antarctica, August and September 1987," *Journal of Geophysical Research* 94, no. D9 (30 August 1989): 11181–11222.

JPL, as it were, but only to illuminate the intersection of planetary and Earth sciences through specific example. Much more complete discussions of the complex AAOE can be found elsewhere.[25]

These expeditions led to the determination that chlorofluorocarbon breakdown products, combined with catalyzing reactions on the surfaces of aerosols and polar stratospheric clouds, result in the formation of the ozone hole each Antarctic spring.[26] Similar, though far less intense, reactions also occur in the Arctic spring. NASA's support for these expeditions provided the observational data that allowed atmospheric scientists to unambiguously identify the chemical and physical processes responsible for the Antarctic ozone hole, which in turn led to the Montreal Protocol and the international agreement to phase out the use of those industrial chemicals responsible for this threat to the ozone layer.[27]

While I've focused on Farmer's infrared spectroscopy team in this paper, global monitoring of ozone and the active ozone destroyer chlorine monoxide is provided by a microwave instrument, the Microwave Limb Sounder. Joe Waters developed this instrument beginning in the mid-1970s, after coming to JPL from David Staelin's microwave laboratory at MIT. While there, Waters had used a radio telescope to investigate upper atmosphere water vapor, another case of astrophysics meeting Earth science. Farmer's infrared spectrometers were not able to measure chlorine monoxide, a chemical active in the microwave portion of the spectrum, yet chlorine monoxide is the trace species that is directly responsible for ozone destruction. So different techniques had to be deployed to fully understand the phenomenon. In September 1991, Space Shuttle *Atlantis* deployed the Microwave Limb Sounder aboard UARS. By simultaneous measures of stratospheric temperature and ozone, the Microwave Limb Sounder was able to provide key context for chlorine monoxide, too.[28] The Microwave Limb Sounder and its successor on the Aura satellite constitute the basis of long-term monitoring capacity of stratospheric ozone conditions.

---

25. Conway, *Atmospheric Science at NASA: A History*, chap. 6; Edward A. Parson, *Protecting the Ozone Layer: Science and Strategy* (New York, NY: Oxford University Press, 2003); Richard Elliot Benedick, *Ozone Diplomacy: New Directions in Safeguarding the Planet* (Cambridge, MA: Harvard University Press, 1991); Maureen Christie, *The Ozone Layer: A Philosophy of Science Perspective* (New York, NY: Cambridge University Press, 2000).
26. Susan Solomon, "Stratospheric Ozone Depletion: A Review of Concepts and History," *Reviews of Geophysics* (August 1999): 275–316.
27. Parson, *Protecting the Ozone Layer*, pp. 3–4, 62–109.
28. J. W. Waters et al., "Stratospheric ClO and Ozone from the Microwave Limb Sounder on the Upper-Atmosphere Research Satellite," *Nature* 362 (15 April 1993): 597–602; Waters, OHI by Conway, 31 October 2005, JPL Archives.

Chlorofluorocarbons were almost completely banned by 1990 and 1992 revisions to the Montreal Protocol of 1987. They represent the first class of chemicals to be essentially eliminated by international fiat due to environmental destructiveness. And as I have pointed out elsewhere, NASA's role in proving the case scientifically has made it a lightning rod for environmental controversy.[29]

To link recursively back to planetary science, in 2002, William Boynton of the Lunar and Planetary Laboratory at the University of Arizona finally tested Farmer's hypothesis about the probable location of Mars's water ice. His gamma-ray spectrometer aboard Mars Odyssey could detect hydrogen emission of gamma rays and neutrons liberated by cosmic-ray bombardment. From a year's worth of data, Boynton produced a map of near-subsurface hydrogen on Mars that was very similar to Farmer's hypothesized ice belt of 20 years earlier.[30] There is subsurface hydrogen, and thus probably water ice, nearly everywhere poleward of 50 degrees. It was reconfirmed by the 2008 Mars Phoenix lander, which touched down at about 68 degrees north latitude, near the edge of the permanent northern ice cap, and identified water ice chemically.[31]

Models, Volcanoes, and Climate Change

Earlier in this chapter, we examined the career of a spectroscopist and used him to posit the applicability of spectroscopy to Earth and planetary science questions. But remote sensing is only one part of what I see as a transformation in Earth science methodology triggered by NASA's planetary explorations. A second major part is in the deployment of modeling. There are many kinds of models in modern science, and modeling didn't originate in NASA. Kristine Harper has just published a history of the development of "numerical weather prediction" in the 1950s, which is simply the use of atmospheric models to forecast the weather. NASA was formed just as numerical atmosphere models were a "hot new thing" in science, and their role in Earth and planetary science is what I'm interested in illuminating.

---

29. Conway, *Atmospheric Science at NASA: A History*, pp. 1–10; Erik M. Conway, "Satellites and Security: Space in Service to Humanity" in *The Societal Impact of Spaceflight*, ed. Steven J. Dick and Roger D. Launius (Washington, DC: NASA SP-2007-4801, 2007), pp. 282–283.
30. W. V. Boynton et al., "Distribution of Hydrogen in the Near-Surface of Mars: Evidence for Subsurface Ice Deposits," *Science Express* (30 May 2002), doi:10.1126/science.1073722.
31. "NASA Spacecraft Confirms Martian Water, Mission Extended," 31 July 2008, available at *http://phoenix.lpl.arizona.edu/07_31_pr.php* (accessed 9 September 2009).

One of the most famous early atmosphere models was the Mintz/Arakawa model. Its creator was Yale Mintz of UCLA, who shared it widely. Mintz, according to modeler Conway Leovy of the University of Washington, "painted this immense canvas of applying numerical models to every atmosphere in the solar system" in lectures he gave during the early 1960s.[32] Atmospheres are physical entities bound to behave in accordance with known physical laws, primarily fluid mechanics and radiative transfer. These are difficult to compute, and because they are nonlinear, they often can only be approximated numerically. But Mintz and his followers believed they could be used as useful tools to help understand the behavior of atmospheres universally.

The Mintz/Arakawa model became the basis of at least two modeling centers at Ames and GISS. Ames had been founded as an aeronautical center in 1939, and during the 1950s, it had built a computing infrastructure to foster the development of CFD for use in aircraft design.[33] While the goal of numerical atmosphere modeling is different from that of CFD, in principle they are very similar. Both derive from the equations of fluid motion, and both require powerful computers. So Ames's adoption of planetary atmosphere modeling was an obvious extension of existing expertise and resources.

Founded in 1960, GISS was to be NASA's theoretical studies center. Computational models are theories expressed as equations and embedded in computer code, and as such are fundamental to GISS. They serve modern scientists as machine-assisted "gedankenexperiments."[34] The idea to form GISS came from astrophysicist Robert Jastrow, but the basic concept undergirding GISS was the same as Yale Mintz's: if you could model one atmosphere numerically, you could model them all. During the 1960s, GISS built a new weather forecast model around the Mintz-Arakawa model's dynamics and used it to help refine the need for weather satellites. In the 1970s, Jule Charney of MIT used it to help understand regional climate processes.

One of the key challenges in modeling is validation. How does one determine that a numerical simulation has validity, a connection to the real Earth (or Mars or Jupiter?) This had been an important issue to Charney, who had to build a set of forecast reconstructions against which to test his initial weather

32. Conway Leovy, OHI by Erik Conway, 22 February 2006, JPL Archives.
33. See Glenn Bugos, *Atmosphere of Freedom: Sixty Years at the NASA Ames Research Center* (Washington, DC: NASA SP-4134, 2000) for an overview of Ames; CFD is not yet the subject of its own history.
34. N. Oreskes, K. Shrader-Frechette, and K. Belitz, "Verification, Validation, and Confirmation of Numerical Models in the Earth Sciences," *Science* 263, no. 5147 (4 February 1994): 641–646; Conway, *Atmospheric Science at NASA: A History*, pp. 318–319.

forecast models in the 1950s. Observational data gleaned from the atmospheres of other planets was a bit harder to come by than forecast analyses of Earthly weather. While the National Weather Service collects and archives the daily weather data in the United States, there was no such data for other planets. Instead, planetary atmosphere modelers used various aspects of Earth's atmosphere to examine model physics.

The effort to validate climate model physics in the 1970s caused the NASA Ames and GISS researchers to reopen an argument from the 19th century about whether large volcanic eruptions produce climate changes on Earth. In 1815, the titanic explosion of Tambora in Indonesia had seemed to produce a "year without a summer," leading to crop failure and famine. But scientists of the 19th and early 20th centuries could not even agree on whether a measurable planetary cooling had resulted; the very large eruption of Krakatau in 1883 did not clarify the issue. Scientists lacked both adequate measurements and, as Matthias Dörries has recently argued, an explanatory framework.[35]

Exploration of Mars and Venus during the 1960s and early 1970s by American and Soviet spacecraft revived the stalled argument over volcanic climate change. The Mariner 9 mission to Mars in 1971 enabled scientists to watch a planetary-scale climate change occur before their very eyes. A dust cloud that started in the Southern Hemisphere spread across the entire planet as the spacecraft approached; it, and its science teams on Earth, waited for weeks for the dust to settle. An infrared interferometer aboard Mariner 9 revealed that the dust cloud created a strong temperature inversion that helped sustain it. The dust reflected sunlight back into space, causing the surface to cool very rapidly. But it also absorbed some of the sunlight, heating itself and the atmospheric layer around it. One of Pollack's colleagues at Ames, O. Brian Toon, commented in 1975 that it was "the only global climate change whose cause is known that man has ever scientifically observed."[36] While the dust storm was not volcanic in nature, on the apparently waterless Mars the dust was probably chemically similar to volcanic dust on Earth. So the data could be used to help model the effects of volcanic dust in Earth's atmosphere, at least as a first approximation.

---

35. Matthias Dörries, "In the public eye: Volcanology and climate changes studies in the 20th century," *Historical Studies in the Physical and Biological Sciences* 37, no. 1 (2006): 87–124.
36. Quoted in Spencer Weart, *The Discovery of Global Warming* (Cambridge, MA: Harvard University Press, 2003), p. 88.

Venus-bound spacecraft encountered a very different set of phenomena. Astronomers had long known that the planet was permanently veiled in clouds; radiotelescope data suggested that the surface was extraordinarily hot by 1960, which was confirmed by Mariner 2 in 1962 and then repeatedly by Soviet landers. Carl Sagan had argued, in 1960, that the high surface temperature was probably the result of a super-greenhouse effect from huge amounts of carbon dioxide in the atmosphere; this, too, seemed confirmed by the mid-1960s. But the composition of the cloud sheet remained unknown until 1972, when examination of the clouds' index of refraction by ground-based observation suggested they were sulfuric acid. James Pollack, using the Ames's airborne infrared observatory, confirmed the finding via infrared spectra in 1974.[37] There was an obvious mechanism to explain this cloud composition: volcanoes (on Earth, at least) release sulfate, which is transformed chemically into sulfuric acid in the atmosphere. Sulfate is highly reflective in the visible spectrum, while also having some infrared opacity. Which would dominate in the atmosphere(s)?

In 1976, Pollack and his research group published an article laying out an argument that these sulfate aerosols were probably the dominant cause of volcano-induced climate change on Earth. Using their Mars and Venus observations, laboratory studies of the radiative characteristics of dust and sulfate aerosols, and data collected from volcanic eruptions on Earth and atmospheric atomic bomb tests in the early 1960s, they argued that volcanic dust fell out of the stratosphere over a few weeks' time, and the dust would not produce a long-term cooling. But the sulfates could remain for a few years, producing an average cooling of up to 1°C globally.[38]

At the same time, James Hansen and Andrew Lacis at GISS were starting to build their own climate models. They were interested in anthropogenic climate change induced by greenhouse gases and aerosols, which, they thought, were

37. Ronald Schorn, *Planetary Astronomy: From Ancient Times to the Third Millennium* (College Station, TX: Texas A&M University Press, 1999), p. 259; James Pollack et al., "Aircraft observations of Venus' near-infrared reflection spectrum: implications for cloud composition," *Icarus* 23 (1974): 8–26; James Pollack et al., "A determination of the composition of the Venus clouds from aircraft observations in the near infrared," *Journal of the Atmospheric Sciences* 32 (1975): 376–390.

38. James B. Pollack, Owen B. Toon, Carl Sagan, Audrey Summers, Betty Baldwin, and Warren Van Camp, "Volcanic Explosions and Climatic Change: A Theoretical Assessment," *Journal of Geophysical Research* 81, no. 6 (20 February 1976): 1071–1083; O. Brian Toon, interview by Erik Conway, 13 February 2004, JPL Archives. The group detailed their sulfate aerosols work in R. C. Whitten, O. B. Toon, and R. P. Turco, "The Stratospheric Sulfate Aerosol Layer: Processes, Models, Observations, and Simulations," *Pure and Applied Geophysics* 118 (1980): 86–127.

likely to be of the same magnitude as volcanic emissions. They explored the Mount Agung eruption of 1963 with an early one-dimensional model to help illuminate that question.[39]

Interest in examining volcanic eruptions on Earth for model validation and Earth understanding purposes was not confined to the two modeling centers. NASA's LaRC had researchers interested in measuring atmospheric aerosols via both laser remote sensing and visible and infrared remote sensing from aircraft and from space. M. P. McCormick, James Russell, David Winker, and others developed instruments for ground, airborne, and spaceborne measurements of aerosols. With Pollack, they also promoted a joint NASA/NOAA program called Research on Volcanic Eruptions (RAVE), which was approved and just put in place when Mount Soufriere erupted in 1979, followed almost immediately by Mount Saint Helens in 1980. As a result, these two explosions were extraordinarily well documented scientifically. But these also turned out to be very low in sulfate, and (as Pollack's group had surmised earlier) the volcanic dust and ash fell out of the atmosphere fairly quickly. Without much sulfate, there also wasn't a measurable cooling, so the eruptions didn't resolve the question of temperature response.[40] Nor did El Chichón in 1983, which actually produced a very substantial increase in stratospheric sulfate

39. James E. Hansen, "Mount Agung eruption provides test of a global climatic perturbation," *Science* 199 (1978): 1065–1068.
40. "Researchers Track Volcanic Plume," *Langley Researcher* (30 May 1980): 4–5; M. P. McCormick, interview by Erik Conway, 7 April 2004, JPL Archives; W. I. Rose, Jr., and M. F. Hoffman, "The 18 May 1980 Eruption of Mount St. Helens: The Nature of the Eruption, with an Atmospheric Perspective," in *Atmospheric Effects and Potential Climatic Impact of the 1980 Eruptions of Mount St. Helens*, ed. Adarsh Deepak (Washington, DC: NASA CP-2240), pp. 1–14; Owen B. Toon, "Volcanoes and Climate," in *Atmospheric Effects and Potential Climatic Impact of the 1980 Eruptions of Mount St. Helens*, ed. Adarsh Deepak (Washington, DC: NASA CP-2240), pp. 15–36; E. C. Y. Inn, J. F. Vedder, E. P. Condon, and D. O'Hara, "Precursor Gases of Aerosols in the Mount St. Helens Eruption Plumes at Stratospheric Altitudes," in *Atmospheric Effects and Potential Climatic Impact of the 1980 Eruptions of Mount St. Helens*, ed. Adarsh Deepak (Washington, DC: NASA CP-2240), pp. 47–54; M. P. McCormick, "Ground-Based and Airborne Measurements of Mount St. Helens Stratospheric Effluents," in *Atmospheric Effects and Potential Climatic Impact of the 1980 Eruptions of Mount St. Helens*, ed. Adarsh Deepak (Washington, DC: NASA CP-2240), pp. 125–130; R. P. Turco, O. B. Toon, R. C. Whitten, R. G. Keese, and P. Hamill, "Simulation Studies of the Physical and Chemical Processes Occurring in the Stratospheric Clouds of the Mount St. Helens Eruptions of May and June 1980," in *Atmospheric Effects and Potential Climatic Impact of the 1980 Eruptions of Mount St. Helens*, ed. Adarsh Deepak (Washington, DC: NASA CP-2240), pp. 161–190.

aerosols, but after which whatever cooling might have happened was masked by a strong El Niño.[41]

The eruption of Mount Pinatubo in 1991 finally provided the scientific community with an adequate test of the new models and measurement capabilities. Pinatubo's eruption caused the largest aerosol injection into the stratosphere of the 20th century, and the eruption was estimated to be the third largest perturbation of the industrial era (behind Tambora in 1815 and Krakatau in 1883). Of the estimated 30 teragrams of mass shot into the stratosphere, the old TOMS instrument's data suggested about two-thirds was sulfur dioxide. This transformed into radiatively active sulfate as it aged.

Satellite instruments (SAGE II and TOMS) showed the plume moving around the world in 22 days, spreading relatively quickly southward to 10 degrees south latitude, then more slowly dispersing to higher latitudes. The Upper Atmosphere Research Satellite measured the rapid warming of the stratosphere that followed, and the heating effect also lofted the aerosols, moving them higher in the stratosphere. But despite the large increase in sulfate, there was no measurable increase in active chlorine immediately after the eruption. Instead, as the plume spread to the poles, the aerosols appeared to increase the surface area available for the wintertime production of chlorine dioxide (the inactive precursor species involved in ozone depletion), leading to significantly larger ozone loss than in prior years. The Upper Atmosphere Research Satellite's scientists were able to characterize the stratosphere's chemical response to the volcano and its recovery. Two and a half years later, the aerosol loading had diminished to about one-sixth of the original amount.[42]

Almost immediately after the eruption, NASA Headquarters called an interagency meeting in Washington, DC, to discuss preliminary information gathered about the eruption; after this, using data gleaned from some of his colleagues, Hansen used his three-dimensional climate model, known as the Model II, to make a forecast of the volcano's climate impact. It predicted an immediate

41. M. P. McCormick, G. S. Kent, G. K. Yue, and D. M. Cunnold, "Stratospheric Aerosol Effects from Soufriere Volcano as Measured by the SAGE Satellite System," *Science* (4 June 1982): 1115–1118; Michael R. Rampino and Stephen Self, "The Atmospheric Effects of El Chichon," *Scientific American* (January 1984): 48–57; David J. Hofmann, "Perturbations to the Global Atmosphere Associated with the El Chichon Volcanic Eruption of 1982," *Reviews of Geophysics* 25, no. 4 (May 1987): 743–759. See also Matthias Dörries, "In the Public Eye: Volcanology and Climate Change Studies in the 20th Century," *Historical Studies in the Biological and Physical Sciences* 37, no. 1 (2006): 117.
42. M. P. McCormick, Larry W. Thomason, and Charles R. Trepte, "Atmospheric Effects of the Mt. Pinatubo Eruption," *Nature* 373 (2 February 1995): 399–404.

low-latitude cooling, becoming essentially global by mid-1992 and peaking at about −0.5°C (globally averaged, of course) late in 1992. In the resulting 1992 paper, he argued the volcano provided an "acid test for global climate models." The expected cooling was about three times the standard deviation of global mean temperature; this should, he thought, be measurable despite the apparent onset of an El Niño (which tends to warm the troposphere).[43]

The surviving Earth Radiation Budget Experiment, aboard the Earth Radiation Budget Satellite (ERBS), showed that the eruption increased Earth's albedo, as expected, by a significant amount. Sulfate aerosols injected into the stratosphere were the culprit; as both Ames and GISS groups expected, in addition to their reflective characteristics, they also warmed the stratosphere. Reduced insolation caused cooling of the troposphere, which, despite an El Niño that year, resulted in the year's being measurably cooler than the 26-year mean. In fact, weather satellite data showed almost exactly the amount of cooling predicted by GISS Model II, and in a spatial and temporal pattern that was highly consistent with the model's, too. (Hansen was quick to point out in a 2006 interview that the forecast hadn't been perfect). A 1999 reviewer called the consistency between the prediction and the independent analyses "highly significant and very striking": they had led, he argued, to increased confidence in the models' representation of climate processes.[44] Mount Pinatubo confirmed that the relevant model physics were reasonable facsimiles of the real atmosphere, and it ended the argument over whether volcanic explosions cause climatic effects: some do, some don't, depending (as often happens in Earth sciences) on the details.

The volcano/climate nexus is just one thread of the larger story of planetary climate evolution. It drew so much attention from scientists because eruptions occur on human timescales; unlike greenhouse gas-induced warming, which no one in the 1970s expected to become measurable in less than decades, an entire cycle of eruption, cooling, and recovery could happen over a couple of years. Eruptions could be subjected to familiar investigation techniques, while greenhouse-gas warming was far more difficult to test with traditional techniques on reasonable timescales. I have argued elsewhere that global warming

43. James Hansen, Andrew Lacis, Reto Reudy, and Makiko Sato, "Potential Climate Impact of Mount Pinatubo Aerosol," *Geophysical Research Letters* 19, no. 2 (24 January 1992): 215–218.

44. D. J. Carson, "Climate Modeling: Achievements and Prospects," *Quarterly Journal of the Royal Meteorological Society* 125, part A (January 1999): 10; M. P. McCormick, Larry W. Thomason, and Charles R. Trepte, "Atmospheric effects of the Mt. Pinatubo Eruption," *Nature* 373 (2 February 1995): 399–404; James Hansen, interview by Erik Conway, 16 January 2006, JPL Archives.

has forced scientists to address timescale problems through new approaches to measurement and calibration, as well as confronting the institutional bias of American science in favor of short-term results ("publish or perish!")[45] But volcanoes were accessible, time-limited phenomena, and so they could be used to examine the parts of model physics that concerned short-lived phenomena.

Toward an Interplanetary Perspective

By the early 1980s, the leaders of NASA's scientific community—scientists that worked at its research centers as well as university scientists that collaborated in its efforts—believed that the new techniques that they had pioneered justified a change in the very disciplinary structure of American science. One result of their efforts was Earth system science, a new discipline focused on the dynamics of planetary-scale processes.

The realization that scientists needed to look again at integrated planetary-scale processes to understand the evolution of planets came, again, from the early Mariner missions. It was clear from the Mariners that Mars had once been warm enough to have liquid water on its surface (if not for long), yet some early one-dimensional model calculations by Carl Sagan showed that *both* Earth and Mars should have started their planetary careers as frozen balls of ice. That was because astronomers believed the Sun had been 30 to 40 percent dimmer when it first formed than it is now. But the geological evidence available on Earth during the 1970s did not support an "iceball Earth," and Sagan called this problem at the intersection of astronomy and Earth science the Faint Early Sun Paradox.[46]

Thinking about the Faint Early Sun Paradox helped lead James Lovelock to his controversial proposal that biological activity regulated Earth's climate (familiarly known as the Gaia hypothesis).[47] Lovelock had started arguing back in the 1960s that the place to look for life on other planets was in their atmospheres. All organisms produce waste products; those waste products, like ammonia and methane, would be found most easily in the planet's atmosphere.

---

45. Conway, *Atmospheric Science in NASA: A History*, pp. 8–10, 318. For another view of trying to sustain long-term research, see Charles D. Keeling, "Rewards and Penalties of Monitoring the Earth," *Annual Review of Energy and Environment* 23 (1998): 25–82.
46. Carl Sagan and George Mullen, "Earth and Mars: Evolution of Atmospheres and Surface Temperatures," *Science* 177 (7 July 1972): 52–56.
47. Conway, *Atmospheric Science in NASA: A History*, pp. 114–116; Lynn Margulis and J. E. Lovelock, "Biological Regulation of the Earth's Atmosphere," *Icarus* 21 (1974): 471–489; J. E. Lovelock, *Gaia: A New Look at Life On Earth* (New York, NY: Oxford University Press, 1974), pp. 13–32.

These waste products are chemically reactive, so a living planet, in Lovelock's view, would have an atmosphere in a state of chemical disequilibrium.[48]

When Lovelock was thinking about the subject, Earth's atmosphere consisted of 21 percent oxygen, by volume, and 78 percent nitrogen, with the remaining 1 percent made up by various trace gases. Carbon dioxide, for example, was 0.03 percent of the atmosphere's volume. Yet in terms of chemical equilibrium, this was highly improbable. Over Earth's billions of years of existence, oxygen, a highly reactive gas, should have been extracted by chemical weathering of surface material and be, as it was on Venus and Mars, undetectable. Similarly, the most chemically stable form of nitrogen was in the form of nitrate ions in the oceans, not as a noble gas in the atmosphere. Hence in a chemically stable version of Earth, the atmosphere would be mostly carbon dioxide, as were the atmospheres of Mars and Venus, and contain neither nitrogen nor oxygen.[49]

Lovelock was not the first to recognize the unstable nature of Earth's atmosphere. Rather, he was building on a minority view in geochemistry.[50] In the majority view, Earth's unlikely atmosphere was explained as a product of planetary outgassing, with the oxygen provided by the photodissociation of water vapor in the upper atmosphere. The resulting hydrogen, as it had on Venus, would then escape into space, leaving oxygen free in the atmosphere. Yet this view did not comport with evidence available by the late 1960s regarding the dissociation rate of water in the upper atmosphere, or with the rates of consumption of oxygen in the weathering processes. It also did not square with the interplanetary view. Lacking both a magnetic field and an ozone layer, Venus experienced much larger high-energy fluxes at the top of its atmosphere than Earth did, which would lead to a higher dissociation rate and more rapid hydrogen escape and oxygen production. And, of course, whatever water Venus had was gone. But there was no measurable residual oxygen. Hence the majority view no longer explained the available evidence.

Lovelock argued that life itself maintained the relative abundances of these gases. Photosynthetic plants consumed carbon dioxide and released oxygen, while animal life consumed oxygen and released carbon dioxide. The trace amounts of methane in the atmosphere, about a billion tons, were already well

48. Lovelock, *Gaia*, pp. 6–7.
49. Ibid., pp. 35–36.
50. Ibid., p. 35.

known to be a mostly biological byproduct.[51] The presence of these gases was, for Lovelock, the ultimate proof of life, and in a 1965 article for *Nature*, he set out his argument that the atmosphere was the place to search for Martian life.[52]

But in the process of thinking about how to find life, he began to reconceive Earth as a single, self-regulating organism. After a 1969 presentation in Boston, he began working with Lynn Margulis, then Sagan's wife, to refine and flesh out the idea. They eventually published two important articles in 1973 and 1974 in which they described their hypothesis. They used the metaphor of a planetary engineer, whose employer had assigned him a planet and directed him to maintain a specific set of temperature and acidity specifications for several billion years. Then they reviewed the tools available to the engineer for temperature control: control of the planet's radiation balance, its surface emissivity, the composition of its atmosphere, and the distribution of dust and aerosols. As seemed to be the case with Mars, small changes in planetary albedo could effect sizeable changes in temperature. The engineer could change this by, for example, darkening the polar regions. Similarly, organisms could impact albedo by changing their colors, by changing the color of the sediments they trapped and fixed, and even by altering the color of snow and ice.

Organisms also altered the chemical composition of Earth's atmosphere, impacting its radiative qualities. Nearly all organisms either consumed or produced carbon dioxide. Ammonia, the gas Sagan and George Mullen had proposed as maintaining Earth's warmth under the faint early Sun, was also a "very active product of microbial metabolism."[53] It was a waste product of many organisms, and it also was consumed by nearly all bacteria and fungi. Hence while the amount of it in the current atmosphere was vanishingly small, this was because virtually all of the billion or so tons produced each year by biologic processes were also being consumed.

To Margulis and Lovelock, microbial consumption explained the near disappearance of ammonia from the early atmosphere; ammonia-fixing microbes would have thrived on the young Earth, and as they drew down the atmospheric reservoir of ammonia, these microbes would have been increasingly pressured into environments where they would be in contact with ammonia-producing

---

51. Ibid., p. 72.
52. J. E. Lovelock, "A Physical Basis for Life Detection Experiments," *Nature* 207 (1965): 568–570. See also D. R. Hitchcock and J. E. Lovelock, "Life Detection by Atmospheric Analysis," *Icarus* 7, no. 2 (1967): 149–159.
53. Lynn Margulis and J. E. Lovelock, "Biological Regulation of the Earth's Atmosphere," *Icarus* 21 (1974): 481.

microbes. This would have had a large radiative impact on Earth's atmosphere, as under the early faint Sun, removal of ammonia at too high a rate would have sent Earth into an "iceball" mode from which it could not recover. Indeed, their reading of Earth's chemical history suggested a crisis for its thermal equilibrium in the late Precambrian, but the geologic record did not seem to contain evidence of one.[54] They took this as evidence of active control of the climate by biologic actors, postulating that selection pressures on local populations produced a response to the cooling Earth that eventually counteracted it. The need for an active control agent led the two to conceive of Earth as a single organism they named Gaia, for the Earth goddess of the ancient Greeks (also known as Ge, from which derived the names for geology and geography).[55]

In their seminal 1974 *Icarus* article, Margulis and Lovelock commented " . . . probably a planet is either lifeless or it teems with life. We suspect that on a planetary scale sparse life is an unstable state implying recent birth or imminent death."[56] The combination of living processes and evolutionary ones was so powerful, in their view, that organisms could remake a planetary environment to facilitate their own spread. Hence over the eons of deep time, life would take over a planet, make it more suitable, and eventually be found everywhere. In this view of life, there were no marginal environments. Life would be found in any local environment of an inhabited planet—or nowhere. This did not bode well for NASA dreams of finding life on Mars, or anywhere else in the solar system. If life existed at all off Gaia, it would be readily apparent from its impact on the composition of planetary atmospheres. Telescopes and telescope-aided infrared spectroscopy were all one needed.

But stripping away the mysticism inherent in the Gaia label, the two were presenting a view of Earth that could be grasped by systems engineers, a profession that specialized in (nonliving) feedback control systems. In his 1979 popular exegesis of the Gaia hypothesis, Lovelock even devoted a chapter to cybernetic theory, the mathematical basis for feedback systems. For Earth scientists, Margulis and Lovelock were presenting a view of the world that required

---

54. Recent evidence suggests that the climate crisis that J. E. Lovelock thought should have happened actually did, and Earth became an "iceball" for between 35 and 100 million years in the late pre-Cambrian. Its recovery seems to have been a product of volcanic outgassing over eons, raising the greenhouse effect sufficiently to break the "iceball" climate. See Robert E. Kopp et al., "The Paleoproterozoic snowball Earth: A climate disaster triggered by the evolution of oxygenic photosynthesis," *Proceedings of the National Academy of Sciences* 102, no. 32 (9 August 2005): 11131–11136.
55. Lovelock, *Gaia*, p. 10; Margulis and Lovelock, "Biological Regulation of the Earth's Atmosphere," p. 471.
56. Margulis and Lovelock, "Biological Regulation of the Earth's Atmosphere," pp. 478–479.

examination of complex, interlocking feedback loops. Some of these feedback loops, such as the hydrologic cycle that was of great interest to meteorologists, were primarily physical. At least in the 1970s, evapotranspiration was perceived as only a minor participant in the water cycle. Other obvious cycles, such as the carbon cycle, were both physical and biological. Understanding them required the very interdisciplinary research that the American scientific community did not consider "serious science" and was not set up to foster.[57]

Shelby Tilford, at NASA Headquarters, set out to change that. In 1982, he organized a committee to formulate a discipline around the idea of studying Earth as a dynamic system. Tilford thought this was "simply the next logical step. We'd been trying to do things piecemeal, some atmospheric satellites, some ocean satellites, some of the solar observation satellites. But no one had sat down to figure out how all this fits together," he said in 2004.[58] Wesley T. Huntress, who moved from JPL to NASA Headquarters to become Tilford's deputy, reflected later that NASA "was the right place to do it because we were the ones who were going to be studying the planet from a global perspective, looking at it at large scales and trying to understand how to put the smaller scales together."[59]

The committee chairman Tilford chose was Francis Bretherton, who had just stepped down as the director of the National Center for Atmospheric Research in Boulder, Colorado. It took him several years to wrangle an agreement out of his group. At its root, this was the product of classic disciplinary disputes over standards of evidence, measurement methodologies, and even timescales. The atmosphere changes far faster than does solid Earth, so an observing system suitable for atmospheric measurement was not necessarily suitable for solid Earth sensing. This made it very difficult to produce agreement within the committee. As a result, the committee's report, "Earth System Science: A Program for Global Change," was not released until 1986.

Earth system science was immediately controversial. In addition to the disciplinary challenges it suggested, it looked to many Earth scientists as a NASA grab for their money. NASA was using the idea of a new science to help promote an EOS program costing $30 billion, which, if approved, would be by

57. Lovelock, *Gaia*, pp. 48–63; Margulis and Lovelock, "Biological Regulation of the Earth's Atmosphere," p. 487.
58. Shelby Tilford, OHI by Erik Conway, 11 February 2004, JPL Archives.
59. Wesley T. Huntress, OHI by Erik Conway, 17 March 2008, JPL Archives.

far the most expensive science program in American history.[60] Many scientists believed it would vacuum up all the nation's Earth science funds for the next couple of decades, depriving more traditionally trained and motivated investigators of research funds.[61] So it didn't have smooth sailing.[62] The Earth Observing System took until 1990 to gain approval, ultimately as NASA's contribution to the U.S. Global Change Research Program. But as an organizing theme in Earth and planetary science, Earth system science has grown steadily.[63] There is now an undergraduate textbook for Earth system science, and a number of universities have formed interdisciplinary centers of Earth system science.[64]

Conclusion

NASA's role in the geosciences has altered both the disciplinary structure of American science as well as the methodologies employed by scientists. One major change has been the addition of remote sensing to a scientist's "toolkit." This, too, was not without controversy. Remote sensing depends on the interpretation of spectra, a complex, computer-assisted process that seems arcane to those who are not physicists. Remote sensing was far less controversial among astronomers (who had nothing else) than among Earth scientists, and part of Robert Watson's motivation in holding the Balloon Intercomparison Campaign flights was to help make remote sensing of the atmosphere more credible among proponents of more traditional methodologies. His larger idea had been side-by-side comparison of the results of many kinds of instruments, improving credibility throughout the atmospheric measurements community.[65]

Just as there was controversy surrounding remote sensing, there was, and still is, controversy surrounding the use of modeling. Some of this controversy is manufactured by the public relations effort to deny the existence of anthropogenic climate change. Some of it is legitimate—models are routinely used

---

60. M. Mitchell Waldrop, "Washington Embraces Global Earth Sciences," *Science* (5 September 1986): 1040–1042.
61. This was not a "real" possibility, because at the time NASA was funded by a different congressional appropriations subcommittee than the other science agencies of the government. So while EOS's funds jeopardized other NASA-funded efforts, such as space astronomy, it did not threaten the funds of other agencies that supported geosciences.
62. Conway, *Atmospheric Science at NASA: A History*, chap. 8.
63. John Lawton, "Earth System Science," *Science* 292, no. 5524 (15 June 2001), doi:10.1126/science.292.5524.1965; Conway, *Atmospheric Science at NASA: A History*.
64. Michael C. Jacobson, *Earth System Science: From Biogeochemical Cycles to Global Change*, 2nd ed. (New York, NY: Academic Press, 2000).
65. Robert T. Watson, OHI by Erik Conway, 14 April 2004, JPL Archives.

and misused in public policy.[66] Indeed, the use of scientific models for policy purposes is a subject in need of examination by science historians. Yet modeling has become an essential part of many sciences, even those in which NASA has no role, such as genetics. Models serve as digital thought experiments, allowing Earth and planetary scientists to pose questions about complex processes and assess means of testing hypotheses through observation and experiment.[67]

In his 1980 memoir, former Agency Chief Scientist Homer Newell argued that space science had been an "integrative force." It had broken the geosciences "loose from a preoccupation with a single planet."[68] The interdisciplinary research that Doel found in planetary astronomy of the 1940s and 1950s has resulted in a gradual merging of once very different fields of study. Discipline boundary erasures are starting to occur, revealed perhaps most clearly by the increasing number of university departments named with variations of "Earth and planetary sciences." Early solar system exploration produced a substantial change in the scientific worldview. It has also produced change in the structure of American science.

---

66. Orrin H. Pilkey and Linda Pilkey-Jarvis, *Useless Arithmetic: Why Environmental Scientists Can't Predict the Future* (New York, NY: Columbia University Press, 2006); Erik M. Conway, "Review of *Useless Arithmetic*," *Quarterly Review of Biology* (December 2007), doi:10.1086/527637.
67. Conway, *Atmospheric Science at NASA: A History*, pp. 196, 319; N. Oreskes et al., "Verification, Validation, and Confirmation of Numerical Models in the Earth Sciences": 641–646.
68. Homer Newell, *Beyond the Atmosphere: Early Years of Space Science*, p. 328.

Chapter 23

Exploration, Discovery, and Culture

NASA's Role in History

Steven J. Dick

Introduction: Space Exploration in Context

Like the facets of a jewel, the overall importance of NASA and the Space Age over the last 50 years may be considered from many viewpoints, ranging from the geopolitical and technological to the educational and scientific. But no facet is more central than exploration, a concept that encompasses most of the other possibilities and arguably constitutes one of the main engines of human culture, spanning millennia. In its simplest and purest form, the Space Age may be seen as the latest episode in a long tradition of human exploration. Surveying the vast panoply of history, historians have often found "symmetry in the narrative arc of the Great Ages of Discovery" or traced that tradition back even to the Paleolithic Era in an attempt to find a "global historical context" for the Space Age.[1]

---

1. Stephen J. Pyne, "The Third Great Age of Discovery," in *Space: Discovery and Exploration*, ed. Martin Collins and Sylvia Fries (New York, NY: Beaux Arts Editions, 1993); Stephen J. Pyne, "Seeking Newer Worlds: An Historical Context for Space Exploration," in *Critical Issues in the History of Spaceflight*, ed. Steven J. Dick and Roger Launius (Washington, DC: NASA SP-2006-4702, 2006), pp. 7–35, available at *http://history.nasa.gov/SP-2006-4702/frontmatter.pdf*; J. R. McNeill, "Gigantic Follies? Human Exploration and the Space Age in Long-term Historical Perspective," in *Remembering the Space Age*, ed. Steven J. Dick (Washington, DC: NASA SP-2008-4703, 2008), pp. 3–16. An interesting exemplar of the continuous exploration theme is Richard S. Lewis, *From Vinland to Mars: A Thousand Years of Exploration* (New York, NY: New York Times Book Company, 1976).

The Paleolithic Era aside, prior to the Space Age, historians often distinguished two modern Ages of Exploration, the Age of Discovery in the 15th and 16th centuries associated with Prince Henry the Navigator, Columbus, Magellan, and other European explorers, and the Second Age in the 18th and 19th centuries characterized by further geographic exploration such as the voyages of Captain Cook, underpinned and driven by the scientific revolution.[2] Some now distinguish a Third Age, beginning with the IGY and Sputnik, primarily associated with space exploration, but also with the Antarctic and the oceans.[3] If one accepts this framework, it makes sense to compare one age of exploration with another, constantly keeping in mind the differences as well as the similarities and with full realization of the unlikelihood of any predictive ability. Here we choose to compare the Age of Space with the European Age of Discovery, in the hope of revealing symmetries and differences and casting in a new light some of the chief characteristics of the last 50 years in space.

The overarching theme and structure of our argument for the primacy of exploration as a key to understanding the Space Age is inspired by the distinguished Harvard maritime historian J. H. Parry, who 30 years ago published his classic volume *The Age of Reconnaissance: Discovery, Exploration and Settlement, 1450 to 1650*.[4] NASA's first 50 years may also be characterized as "The Age of Reconnaissance," or to put it more broadly, as the first stages of "The Age of Discovery." There have been discovery and exploration, but not yet settlement—unsurprisingly, since we are only 50 years into the Age of Reconnaissance for space. Parry tackled his theme by discussing the conditions for discovery, then the story of the discoveries themselves, and finally the "fruits of discovery." A parallel tripartite structure provides a framework for examining the importance of NASA and the Space Age: what were the conditions for the Space Age, the story of its voyages, and their impact? Much of the meaning of NASA and the Space Age may be found in the context of those three questions.

By drawing such comparisons we are engaging in the time-worn method of analogy, and we need to ask whether analogy is a valid framework for analysis, a proper method of reasoning? In making use of analogy, I am following

---

2. William H. Goetzmann, *New Lands, New Men: America and the Second Great Age of Discovery* (New York, NY: Penguin, 1986).
3. Pyne, "Seeking Newer Worlds," pp. 7–35.
4. J. H. Parry, *The Age of Reconnaissance: Discovery, Exploration and Settlement, 1450 to 1650* (Berkeley, CA: University of California Press, 1981; 1st ed., London, U.K.: Weidenfeld and Nicolson, 1963).

a methodology pioneered almost 50 years ago in another classic book, *The Railroad and the Space Program*, whose subtitle is *An Exploration of Historical Analogy*. This volume, edited by MIT Professor Bruce Mazlish and populated with well-known scholars, addressed the problem of analogy in considerable detail. Mazlish himself spoke of "attempting to set up a new branch of comparative history: the study of comparative or analogous social inventions and their impact on society." The authors went on to give what is, almost 50 years later, perhaps still the best treatment of the general use of historical analogy. Although originally suspicious of parallels with the past, present, and future, the contributors to this volume found it a useful tool; historian Thomas P. Hughes saw "the possibility of moving up onto a level of abstraction where the terrain of the past is suggestive of the topography of the present and its future projection."[5] The authors cautioned that as much empirical detail should be used as possible and that analogies drawn from vague generalities should be avoided. Confident in the use of historical analogy as suggestive but not predictive of the future, Mazlish and his coauthors went on to elaborate their analogy with the railroad and the space program with such a degree of success that their work is still discussed today.

The utility of analogy is suggested by its frequent use: throughout the Space Age, and indeed the history of science in general, scientists have been drawn to this mode of reasoning.[6] The Antarctic dry valleys have been studied as analogs to conditions for life on Mars, the subglacial Antarctic Lake Vostok as an analog to the ocean of Jupiter's satellite Europa, and extremophiles on Earth as analogs to possible alien life. More similar in kind to the railroad and the space program analogy, NASA Administrator Michael Griffin has invoked the "highway to space" to emphasize the sustaining effort required

---

5. Bruce Mazlish, "Historical Analogy: The Railroad and the Space Program and Their Impact on Society," in *The Railroad and the Space Program: An Exploration of Historical Analogy*, ed. Bruce Mazlish (Cambridge, MA: MIT Press, 1965), p. 12; Thomas Parke Hughes, "The Technological Frontier: The Railroad," in *The Railroad and the Space Program*, p. 53, note 1. The circumstances of this volume are discussed by Jonathan Coopersmith, "Great (Unfulfilled) Expectations: To Boldly Go Where No Social Scientist or Historian Has Gone Before," in *Remembering the Space Age*, ed. Steven J. Dick (Washington, DC: NASA SP-2008-4703, 2008), pp. 135–154.
6. For example, in his book *At Home in the Universe* (New York, NY: Springer-Verlag, 1996), pp. 13–16, pioneering physicist John A. Wheeler speaks of analogy as a stimulus to creativity. For another use of analogy in the history of physics, see Daniel Kennefick, *Traveling at the Speed of Thought: Einstein and the Quest for Gravitational Waves* (Princeton, NJ: Princeton University Press, 2007). The "method of analogy" is an important subject in the philosophy of science, for example, R. Harré, *The Philosophies of Science: An Introductory Survey* (London, U.K.: Oxford University Press, 1972), pp. 172–176, and Mary B. Hesse, *Models and Analogies in Science* (Notre Dame, IN: University of Notre Dame Press, 1966).

in space exploration. "Space exploration by its very nature requires the planning and implementation of missions and projects over decades, not years," he wrote. "Decades of commitment were required to build up our network of transcontinental railroads and highways, as well as our systems for maritime and aeronautical commerce. It will be no quicker or easier to build our highways to space, and the commitment to do it must be clear and sustaining."[7] Speaking of the new systems being built for the current space exploration vision, Griffin wrote that "NASA will build the 'interstate highway' that will allow us to return to the moon, and to go to Mars." Similarly, he has compared polar exploration to lunar exploration, arguing that the Apollo program was like the singular forays of Scott or Byrd, while the current plans to establish a base on the Moon are more like the permanent presence that several countries have had in the Antarctic since the 1950s, requiring international collaboration.[8]

Analogies are never perfect, but they can be useful and illuminating as guides for thought. They can also be overstated and misleading, as in the case of the "frontier analogy" so prominent in American space exploration. There is no doubt that exploration is part of the American character and that federally funded exploration has been a significant part of American history.[9] But the very idea of the American frontier and its meaning have been questioned, especially as popularized at the end of the 19th century by historian Frederick Jackson Turner. Turner saw many of the distinctive characteristics of American society, including inventiveness, inquisitiveness, and individualism, as deriving from the existence of a frontier, and he therefore saw the closing of the Western frontier about 1890 as cause for worry.[10] It was natural for Americans

---

7. Michael Griffin, "Leadership in Space," lecture to California Space Authority, 2 December 2005, in *Leadership in Space: Selected Speeches of NASA Administrator Michael Griffin, May 2005–July 2008* (Washington, DC: NASA SP-2008-564, 2008), pp. 1–8, esp. p. 4.
8. Michael Griffin, "NASA and the Business of Space," American Astronautical Society 52nd Annual Conference, 15 November 2005, in *Leadership in Space*, pp. 175–186, esp. p. 181, where the point of the discussion was the role the commercial market could play in the infrastructure that comes along with the highway. On the Antarctic, see *NASA Update*, NASA TV, 12 September 2008.
9. William H. Goetzmann, *Exploration and Empire* (New York, NY: Knopf, 1966). Any highlights of 19th-century American exploration would include the Lewis and Clark expedition from 1803 to 1806, the U.S. Exploring Expedition headed by Charles Wilkes from 1838 to 1842, and the exploration of the American West by the likes of John Wesley Powell. The Lewis and Clark literature is voluminous, but on the Wilkes expedition, see Nathaniel Philbrick, *Sea of Glory: America's Voyage of Discovery: The U.S. Exploring Expedition, 1838–1842* (New York, NY: Viking, 2003).
10. Frederick Jackson Turner, "The Significance of the Frontier in American History," in *Rereading Frederick Jackson Turner: The Significance of the Frontier in American History and Other Essays*, ed. John M. Faragher (New Haven, CT: Yale University Press, 1994).

to find a new frontier in space as an analog to their Western frontier and to argue that conquering the new frontier would perpetuate those characteristics described by Turner. The problem is that many historians do not agree with Jackson's "frontier thesis" as the sole, or even the primary, source of these characteristics in the United States. And by extension, they are skeptical of the benefits of the new frontier. Historians notwithstanding, space as a new frontier has always been a driver of the U.S. space program and remains very much in NASA's lexicon. Nevertheless, it is an analogy that needs to be used with qualification and caution.[11]

If we accept analogical reasoning as a useful tool applied with caution, are exploration and discovery the right analogies? Certainly exploration was not the only, or even the chief, motivation for the space program. But, abstract and even metaphysical as it may seem, it was surely one of the motivations, and a major one at that—the philosophical apex of a pyramid that, of necessity, included more practical motivations. The concepts of "discovery" and "exploration" are frequently found throughout space literature, most recently in the Vision for Space Exploration, billed as "a new spirit of discovery," enunciated by President George W. Bush in January 2004. The same concepts are emphasized in the Aldridge Commission's Report on the Implementation of United States Space Exploration Policy, titled *A Journey to Inspire, Innovate, and Discover*, and yet again in NASA's subsequent new strategic objectives released in a report titled "The New Age of Exploration."[12] One can easily trace the concept back to the dawn

---

11. See Howard McCurdy, *Space and the American Imagination* (Washington, DC, and London, U.K.: Smithsonian Institution Press, 1997), pp. 144–145 and references therein. Roger Launius discusses the controversy over the space frontier analogy in Dick and Launius, *Critical Issues*, pp. 44–45, as do Howard McCurdy and Asif Siddiqi on pages 84–85 and 437–438, respectively, of the same volume. The noted historian of the American West, Patricia Nelson Limerick, has argued especially vigorously that the American frontier, with its history of exploitation and conquest, should not be used as an analogy for space exploration; see Limerick, "Imagined Frontiers: Westward Expansion and the Future of the Space Program," in *Space Policy Alternatives*, ed. Radford Byerly (Boulder, CO: Westview Press, 1992), pp. 249–261, and "Space Exploration and the Frontier," in *What Is the Value of Space Exploration?* 18–19 July 1994, NASA Historical Reference Collection, NASA History Division, NASA Headquarters, Washington, DC.
12. The White House, "A New Spirit of Discovery: The President's Vision for U. S. Space Exploration" (January 2004), available at *http://www.ostp.gov/pdf/renewedspiritofdiscovery.pdf*; E. C. "Pete" Aldridge, Jr., et al., *A Journey to Inspire, Innovate, and Discover: Report of the President's Commission on the Implementation of the United States Space Exploration Policy*, transmitted from E. C. "Pete" Aldridge to President George W. Bush (Washington, DC: GPO, June 2004); NASA, *The New Age of Exploration: NASA's Direction for 2005 and Beyond* (Washington, DC: NASA NP-2005-01-397-HQ, February 2005), available at *http://www1.nasa.gov/pdf/107490main_FY06_Direction.pdf*. The

of the Space Age, an omnipresent, if insufficient, driver of the new age that anchored it in history.

As space science practitioners and supporters like to emphasize, exploration and discovery apply not only to human spaceflight, but also (they would say especially) to space science. That, indeed, is the broad definition encompassed in NASA's documentary history series over the last two decades, *Exploring the Unknown*.[13] Moreover, "The New Age of Exploration" speaks of a human and robotic partnership for exploration—robotic reconnaissance, followed by human voyages that satisfy that desire to explore in person and up close. In 2005, A National Research Council study also concluded that "the expansion of the frontiers of human spaceflight and the robotic study of the broader universe can be complementary approaches to a larger goal." This is easy to say and difficult to implement. To achieve that balanced partnership with the limited resources at hand, in the midst of turbulent events and ever-changing economic and political conditions on Earth, has been one of NASA's great challenges over the last 50 years.[14]

Exploration parallels have, of course, been drawn before. Wernher von Braun was fond of comparing his proposed voyages to Mars to the voyages of Magellan. When Laurence Bergreen researched his book *Voyage to Mars: NASA's Search for Life Beyond Earth*, about the Pathfinder, the Mars Global Surveyor, and the unsuccessful 1999 voyages to Mars, he found references to the Age of Discovery and Magellan rampant within NASA. "After the tenth or maybe the twentieth time the name Ferdinand Magellan was mentioned to

---

renewed emphasis on exploration at NASA raises the question of the relation between exploration, discovery, and science—and not just for academic reasons. One formulation holds that exploration and science are one and the same and that when it comes to spaceflight, exploration equals science. A National Research Council study, *Science in NASA's Vision for Space Exploration* (2005), asserted that "Exploration is a key step in the search for fundamental and systematic understanding of the universe around us. Exploration done properly is a form of science." Yet, while it is clear that there is a synergy between exploration and science, they are not one and the same. After all, Magellan was an explorer, not a scientist or a natural philosopher. And many scientists undertake routine science that can hardly be called exploration; though even routine science can lead to discovery, often it does not. Exploration can also lead to discovery, but not necessarily. In either case, exploration and science are not the same.

13. John M. Logsdon, ed., *Exploring the Unknown: Selected Documents in the History of the U.S. Civil Space Program*, 7 vols. (Washington, DC: NASA SP-4407, 1995 to 2008), available at *http://history.nasa.gov/series95.html*. Volume 8, the final volume, is in preparation.
14. National Research Council, *Science in NASA's Vision for Space Exploration* (Washington, DC: National Academies Press, 2005), available at *http://books.nap.edu/openbook.php?record_id=11225&page=R1*.

me," he recalled, "a dim light bulb eventually illuminated in my mind."[15] The experience led him to write his gripping account, *Over the Edge of the World: Magellan's Terrifying Circumnavigation of the World.* Moreover, references to exploration in the American context are even more common and reached their height in 2003 with the bicentennial of the Lewis and Clark expedition. Such analogies were used to sell the space program and, more recently, the Vision for Space Exploration.[16]

Finally, the imagery of the oceans of Earth and the ocean of space has often been employed in space rhetoric, evoking past exploration. It is one thing when the President of the United States proclaims, as he did a few months after setting the course for the Moon in 1961, that "We set sail on this new sea because there is new knowledge to be gained, and new rights to be won, and they must be won and used for the progress of all people. For space science, like nuclear science and all technology, has no conscience of its own. Whether it will become a force for good or ill depends on man, and only if the United States occupies a position of preeminence can we help decide whether this new ocean will be a sea of peace or a new, terrifying theater of war." And it is significant when historians and journalists build on the analogy, as in the official history of project Mercury, entitled *This New Ocean*, or William Burrows's classic history of the Space Age with the same title.[17] But it is even more significant when NASA workers see themselves in the tradition of the Age of Discovery, for that idea, once individually and institutionally internalized, becomes a part of NASA culture and a powerful force in itself.[18]

With analogy as our guide, exploration as our theme, and Parry's work as our framework, let us examine NASA and the Space Age with all the caution and boldness due such a complex and all-encompassing theme.

---

15. Laurence Bergreen, "Over the Edge of the World," in *Risk and Exploration: Earth, Sea and the Stars*, ed. Steven J. Dick and Keith Cowing (Washington, DC: NASA SP-2005-4701), p. 131. The book referred to is Laurence Bergreen, *Voyage to Mars: NASA's Search for Life Beyond Earth* (New York, NY: Riverhead Books, 2000).
16. Historians Glen Asner and Stephen Garber have pointed this out in their history of events leading up to the Vision for Space Exploration (forthcoming).
17. Loyd S. Swenson, Jr., James M. Grimwood, and Charles C. Alexander, *This New Ocean: A History of Project Mercury* (Washington, DC: NASA SP-4201, 1966, repr. 1998); William E. Burrows, *This New Ocean: The Story of the First Space Age* (New York, NY: The Modern Library, 1998). John F. Kennedy's words were spoken at Rice University Stadium in Houston, TX, 12 September 1962.
18. On NASA culture, see Howard McCurdy, *Inside NASA: High Technology and Organizational Change in the U.S. Space Program* (Baltimore, MD: Johns Hopkins University Press, 1993), as well as Dick and Launius, *Critical Issues*, section V, "NASA Cultures," pp. 345–428.

The Conditions for the Space Age

Analysis of a sampling of the many major factors in common between the Age of Discovery and the Age of Space will suffice to demonstrate the utility of making comparisons: motivations, infrastructure, voyagers, funding, and risk were clearly important considerations in both eras. It is no surprise that similar narrative arcs should generate similar general categories. But the interest lies in the details, the analogies and the dis-analogies, all placed in the proper context of their time, and allowing us to see the Space Age in the light of long historical perspective.

Motivations

As a necessary condition of existence, both ages had their motivations, but they were very different. In the 15th century, exploring nations were in search of empire, and their motivations were twofold: economic gain, through trading or land acquisition, and religious conversion. As Parry put it in his classic study, "Among the many and complex motives which impelled Europeans, and especially the peoples of the Iberian peninsula, to venture oversea in the fifteenth and sixteenth centuries, two were obvious, universal, and admitted: acquisitiveness [wanting to acquire land for empire] and religious zeal. Many of the great explorers and conquerors proclaimed these two purposes in unequivocal terms."[19]

The motivation for the Space Age was neither of these. In the wake of Sputnik, under the Eisenhower administration, the newly formed PSAC, chaired by James R. Killian, identified four factors that gave "importance, urgency and inevitability" to entering space. The first of these was exploration. Foreshadowing the theme of *Star Trek* 10 years later, the report spoke of "the compelling urge of man to explore and to discover, the thrust of curiosity that leads men to try to go where no one has gone before." With an explicit nod to past exploration, the authors of the report noted that "Most of the surface of the earth has now been explored and men now turn to the exploration of outer space as their next objective."[20]

---

19. Parry, *Age of Reconnaissance*, p. 19.
20. President's Science Advisory Committee, "Introduction to Outer Space," 26 March 1958, in *Exploring the Unknown*, ed. Logsdon, vol. 1, *Organizing for Exploration*, pp. 332–333.

The second rationale posed in 1958 for entering space was national defense. "We wish to be sure that space is not used to endanger our security. If space is to be used for military purposes, we must be prepared to use space to defend ourselves." Third was national prestige. "To be strong and bold in space technology will enhance the prestige of the United States among the peoples of the world and create added confidence in our scientific, technological, industrial, and military strength." Science was the fourth factor, for space "affords new opportunities for scientific observation and experiment which will add to our knowledge and understanding of the earth, the solar system, and the universe.[21] In the Soviet Union, the only other space power at the time, the motivations were much the same.

Among these motivations for spaceflight, national prestige was paramount for the first decades of the Space Age, as historical analyses, such as Walter McDougall's . . . *The Heavens and the Earth*, have shown.[22] The motivations are much the same today, although economic competitiveness and survival of the species are now at least part of the discussion.[23] Since the end of the Cold War in the early 1990s, and arguably since the end of the Apollo era, we have entered a period that will determine whether international cooperation, exploration, and commercial gain can provide the same impetus to space that international competition once did. The ISS is a prime example of the cooperation, albeit sometimes difficult, of 16 countries over the last decade. Still, the utility and cost of the ISS have often been called into question, and analysts such as Woody Kay have asked with more than just rhetoric, "Can Democracies Fly in Space?" Without the impetus of outside competition, under always-difficult economic conditions, and in the midst of so many other priorities in a democratic society, this remains an important question of public policy.[24]

---

21. Ibid., p. 333.
22. Walter McDougall, . . . *The Heavens and the Earth: A Political History of the Space Age* (New York, NY: Basic Books, 1985).
23. Roger Launius, "Compelling Rationales for Spaceflight? History and the Search for Relevance," in *Critical Issues*, ed. Dick and Launius, pp. 37–70.
24. On international cooperation, see John Krige, "Technology, Foreign Policy, and International Cooperation in Space," in *Critical Issues*, ed. Dick and Launius, pp. 239–268, and John Logsdon, "The Development of International Cooperation in Space," in *Exploring the Unknown*, ed. Logsdon, vol. 2, *External Relationships*, pp. 1–15 and associated documents. A full history of NASA's international relations is being written by John Krige and his colleagues. Woody Kay, *Can Democracies Fly in Space? The Challenge of Revitalizing the U.S. Space Program* (Westport, CT: Praeger Publishers, 1995).

Infrastructure

Both ages of discovery required a certain infrastructure, none more important than the means of conveyance—ships for the Age of Discovery and rockets for the Age of Space. Beginning with Prince Henry the Navigator in the 15th century, the vessel of choice for ocean exploration was the small, maneuverable, and relatively fast caravel with its "lateen" triangular sail, in contrast to the galley or other vessels with fixed sails or oarsmen (see figure 1).[25] Caravels were used for everyday trade routes in Western Europe, and typically new types of vessels were not constructed for the early long, transoceanic voyages. But caravels were small, crowded, and uncomfortable, and as the Age of Reconnaissance continued, mixed types of ship designs were developed, and fleets sailed with a balanced mix of ships when possible: "one or two caravels, which they employed for dispatch-carrying, inshore reconnaissance, and other odd jobs which later admirals would entrust to frigates. Such ships and such fleets first became available, through a strenuous process of experiment and change, to Europeans in the late fifteenth century. This was the development which made the Reconnaissance physically possible." Caravels could also carry cannons, and some historians argue that "Caravels and cannon were the technological developments that made European expansion overseas possible, not astrolabes and improved maps."[26]

By contrast, because nothing had ever entered the ocean of space, designers had to invent motive power and spaceships through a combination of old and new technologies and sometimes from scratch (see figure 2). It is true that both the Soviet Union and the United States adapted older military missiles as the motive power to enter space, but both also independently designed new rockets.[27] Unlike ships, the motive power was no longer natural wind

---

25. Peter Russell, "The Caravels of Christ," chap. 9 in *Prince Henry the Navigator: A Life* (New Haven, CT, and London, U.K.: Yale University Press, 2001), pp. 225–238.
26. Parry, "Ships and Shipbuilders," chap. 3 in *Age of Reconnaissance*, and on caravels, pp. 65–66; Ronald H. Fritze, *New Worlds: The Great Voyages of Discovery 1400–1600* (Phoenix Mill, U.K.: Sutton Publishing, 2002), pp. 70–73.
27. The best overall history of rocketry in the United States is J. D. Hunley, *Preludes to U.S. Space-Launch Vehicle Technology: Goddard Rockets to Minuteman III* (Gainesville, FL: University Press of Florida, 2008), and J. D. Hunley, *U.S. Space-Launch Vehicle Technology: Viking to Space Shuttle* (Gainesville, FL: University Press of Florida, 2008), and their one-volume companion *The Development of Propulsion Technology for U.S. Space-Launch Vehicles, 1926–1991* (College Station, TX: Texas A&M Press, 2007). See also Roger Launius and Dennis R. Jenkins, eds., *To Reach the High Frontier: A History of U.S. Launch Vehicles* (Lexington, KY: University Press of Kentucky, 2002), and histories of particular rockets such as Bilstein's history of Saturn and Virginia P. Dawson and Mark D. Bowles, *Taming Liquid Hydrogen: The Centaur Upper Stage Rocket, 1958–2002* (Washington, DC: NASA SP-2004-4230, 2004).

power. The core of the new rockets was their engines, and the history of engine development is fraught with uncertainty and contingency. At every stage, from the V-2s and their successors, to the Apollo first-stage F-1 engines with their famous early "combustion instability" problems, and to the SSMEs, it was never assured that access to space would be possible, and it is still not cost-effective.[28] Another of the perennial debates of the Space Age was whether reusable or expendable launch vehicles were best; history records that despite its utility and magnificent engineering, even the reusable Space Shuttle was never cost-effective.[29] With the projected return to expendable rockets after 2010, human winged spaceflight may prove to have been only an ephemeral 30-year phenomenon, at least for the 20th and 21st centuries.

Human spaceflight also required the design of capsules and later the reusable Shuttle to carry humans on their epic early piloted programs. Spacecraft design pioneers like Max Faget (who played a role in the design of every American piloted spacecraft), as well as a variety of unsung heroes, were no less essential to the Space Age than were rocket engineers, and both were as indispensable as the shipbuilders of 500 years before.[30] The design of robotic

---

28. On the F-1 engines, see Roger E. Bilstein, *Stages to Saturn: A Technological History of the Apollo/Saturn Launch Vehicles* (Washington, DC: NASA SP-4206, 1980). Charles Murray and Catherine Bly Cox, *Apollo: The Race to the Moon* (New York, NY: Simon & Schuster, 1989), chap. 10, tell the story based on interviews with participants. For developments placed in their institutional context, see Andrew J. Dunar and Stephen P. Waring, "Crafting Rockets and Rovers: Apollo Engineering Achievements," chap. 3 in *Power to Explore: A History of the Marshall Space Flight Center, 1960–1990* (Washington, DC: NASA SP-4313, 1999). On the SSMEs, see Robert E. Biggs, ed., *Space Shuttle Main Engines: The First Twenty Years and Beyond*, vol. 29 (San Diego, CA: American Astronautical Society, 2008). On the transition from aircraft to spacecraft engines, see Virginia P. Dawson, *Engines and Innovation: Lewis Laboratory and American Propulsion Technology* (Washington, DC: NASA SP-4306, 1991), especially chaps. 8 and following.
29. See Andrew J. Butrica, "Reusable Launch Vehicles or Expendable Launch Vehicles? A Perennial Debate," in *Critical Issues*, ed. Dick and Launius, pp. 301–341, and John Logsdon, "'A Failure of National Leadership': Why No Replacement for the Space Shuttle?," in *Critical Issues*, ed. Dick and Launius, pp. 269–300.
30. For an excellent overview of the types of questions that can be asked about spaceflight infrastructure and design, including spacecraft, see Philip Scranton, "NASA and the Aerospace Industry: Critical Issues and Research Prospects," in *Critical Issues*, ed. Dick and Launius, pp. 169–198. On robotic spacecraft design, see Michael Gruntman, *Blazing the Trail: The Early History of Spacecraft and Rocketry* (Reston, VA: AIAA, 2004). While no general history of spacecraft design exists, histories of individual programs generally cover design. See Swenson et al., *This New Ocean*, esp. chaps. 6–8 on Project Mercury; Courtney Brooks, James Grimwood, and Loyd S. Swenson, Jr., *Chariots for Apollo: A History of Manned Lunar Spacecraft* (Washington, DC: NASA SP-4205, 1979), available at *http://www.hq.nasa.gov/office/pao/History/SP-4205/cover.html*; Tom Heppenheimer, *Development of the Space Shuttle, 1972–1981* (Washington, DC: Smithsonian Institution Press, 1984); and Edward C. Ezell and Linda N. Ezell, *On Mars: Exploration of the Red Planet, 1958–1978* (Washington, DC: NASA SP-4212, 1984) for Viking spacecraft design, as exemplars. On the debate over expendable vs. reusable launch vehicles, see

spacecraft and the perennial debate over human versus robotic spacecraft, on the other hand, find no parallel with the Age of Discovery.[31] Robotic spacecraft design, with its communications, thermal, and electronic subsystems, is especially part of the histories of JPL, GSFC, and their aerospace partners. Indeed, an entire industry sprang up on the foundations of the aviation industry to cater to both the human and the robotic rocket and spacecraft needs of the Age of Space.[32]

The engineering challenges inherent in the design of rockets and spacecraft were legion.[33] Design decisions were sometimes brilliant, often modified, and occasionally second-guessed after accidents and failures, whether human or robotic, and the agonizing but detailed accident reports of those failures make for compelling reading about the importance and far-reaching consequences of engineering decisions.[34] As far as we know, no such ex post facto analysis was undertaken in the Age of Discovery, where the whims of nature at sea were

---

John M. Logsdon, "A Failure of National Leadership: Why No Replacement for the Space Shuttle," and Andrew J. Butrica, "Reusable Launch Vehicles or Expendable Launch Vehicles? A Perennial Debate," in *Critical Issues*, ed. Dick and Launius, pp. 263–344.

31. On the human and robotic debate in spaceflight, see Howard E. McCurdy, "Observations on the Robotic Versus Human Issue in Spaceflight," Slava Gerovitch, "Human-Machine Issues in the Soviet Space Program," and David A. Mindell, "Human and Machine in the History of Spaceflight," all in *Critical Issues*, ed. Dick and Launius, pp. 73–164. See also Roger D. Launius and Howard E. McCurdy, *Robots in Space: Technology, Evolution, and Interplanetary Travel* (Baltimore, MD: Johns Hopkins University Press, 2008).
32. Roger Bilstein, *The American Aerospace Industry: From Workshop to Global Enterprise* (New York, NY: Twayne, 1996); Joan Lisa Bromberg, *NASA and the Aerospace Industry* (Baltimore, MD: Johns Hopkins University Press, 1999).
33. On engineering at NASA, see Sylvia D. Fries, *NASA Engineers and the Age of Apollo* (Washington, DC: NASA SP-4104, 1992). See also Stephen Johnson, *The Secret of Apollo* (Baltimore, MD: Johns Hopkins University Press, 2002). For a specific example of engineering, see David Mindell, *Digital Apollo: Human and Machine in Spaceflight* (Cambridge, MA: MIT Press, 2008).
34. Alexander Brown, "Accidents, Engineering, and History at NASA, 1967–2003," in *Critical Issues*, ed. Dick and Launius, pp. 377–402; Diane Vaughan, *The Challenger Launch Decision: Risky Technology, Culture, and Deviance at NASA* (Chicago, IL: University of Chicago Press, 1996). Among the seminal original accident reports are Apollo 204 Review Board, *Report of Apollo 204 Review Board to the Administrator, National Aeronautics and Space Administration* (Washington, DC: GPO, 1967), available at *http://history.nasa.gov/Apollo204/as204report.html*; William P. Rogers, Chair, *Report of the Presidential Commission on the Space Shuttle Challenger Accident* (Washington, DC, June 1986), available at *http://history.nasa.gov/rogersrep/genindex.htm*; and Columbia Accident Investigation Board, *Report* (Washington, DC: GPO, August 2003), available at *http://history.nasa.gov/columbia/CAIB_reportindex.html*. The Mars Climate Orbiter failure investigation found that the root cause of failure was the failure to translate English units into metric units in a segment of ground-based, navigation-related mission software; see *The Mars Climate Orbiter Mishap Investigation Board Phase I Report* (10 November 1999), available at *ftp://ftp.hq.nasa.gov/pub/pao/reports/1999/MCO_report.pdf*. The Mars Polar Lander accident report and others are available at *http://sunnyday.mit.edu/accidents/*.

most often at fault (though one might question some of Magellan's decisions, including the final one leading to his death).

Ships and rockets alike required specialized points of departure, where they could prepare for the journey (see figure 3). Unlike the ancient ports from which the ships of the 15th and 16th centuries departed, spaceports were built from scratch or on sites of military missile launches. Their locations were determined not so much by water (though an uninhabited overflight path was a factor), but by the latitudes at which Earth's rotation could impart additional motive power, among other considerations. Those spaceports, with now-legendary names like Cape Canaveral, Vandenberg, Kourou, Pletetsk, and Baikonur, were the equivalents of Palos, Lisbon, and Sanlúcar de Barrameda. Except for the ever-popular KSC, the launch sites so essential to spaceflight are often unappreciated by the public, as is other necessary infrastructure such as ground tracking stations, navigation, and mission control. Scientists, engineers, and historians, however, are fully aware that the Space Age could not exist without them.[35]

Both the Age of Discovery and the Age of Space had their navigators, their users and producers of maps that increased in accuracy as a result of the voyages of discovery. The Age of Discovery had its world cosmographic maps and its portolan maps, the latter to actually help in navigating. The Age of Space, too, had its general cosmography, as backdrop, and its practical star maps for celestial navigation, though its methods of navigation—gravitational assists from planetary flybys, for example—were strikingly novel. As in the 16th century, Space Age voyages of discovery produce ever more accurate maps of their routes and their destinations, and the astrogeology branch of the U.S. Geological Survey, funded largely by NASA, carries out

---

35. On spaceports, see John T. Sheahan and Francis T. Hoban, "Spaceports," in *Defining Aerospace Policy: Essays in Honor of Francis T. Hoban*, ed. Kenneth Button, Julianne Lammersen-Baum, and Roger Stough (Burlington, VT: Ashgate, 2004), and for one of the most important spaceports, see Ken Lipartito and Orville Butler, *A History of the Kennedy Space Center* (Gainesville, FL: University Press of Florida, 2007), and Charles D. Benson and William B. Faherty, *Gateway to the Moon, Building the Kennedy Space Center Launch Complex* (Gainesville, FL: University Press of Florida, 2001) and *Moon Launch! A History of the Saturn-Apollo Launch Operations* (Gainesville, FL: University Press of Florida, 2001). On tracking stations, see Sunny Tsiao, "Read You Loud and Clear!" The Story of NASA's Spaceflight Tracking and Data Network (Washington, DC: NASA SP-2007-4233, 2008); Douglas Mudgway, *Uplink-Downlink: A History of the NASA Deep Space Network, 1957–1997* (Washington, DC: NASA SP-2001-4227, 2001); and Douglas Mudgway, *Big Dish: Building America's Deep Space Connection to the Planets* (Gainesville, FL: University Press of Florida, 2005). On the evolution of mission control, see Gene Kranz, *Failure Is Not an Option: Mission Control from Mercury to Apollo 13 and Beyond* (New York, NY: Berkley Books, 2000); and Chris Kraft, *Flight: My Life in Mission Control* (New York, NY: Plume, 2002).

the same role for mapping new worlds as 16th-century cartographers did for the New World (see figure 4).[36]

Voyagers

Voyagers, whether human or robotic, are also essential to the exploration enterprise. Both ages had their heroes, leaders of the voyages of discovery. Columbus and Magellan were men of daring and adventure who personally argued for government funding of their voyages. Cosmonauts, astronauts, and taikonauts were also daring, but unlike explorers from the Age of Discovery, it was not they who argued for government funding for the space program; it was scientists and managers like Wernher von Braun and a sequence of NASA Administrators, now enmeshed in a growing technocratic complex.

At another level, crews in the Age of Discovery, as in the case of Magellan's circumnavigation, were often hard to come by. There is no parallel to this situation among myriad astronaut applicants, who outnumbered successful candidates by more than 1,000 to one. While many ship captains were men of some learning, their crews varied greatly, from people off the streets to religious seekers, profiteers, and pirates. By contrast, the nearly 500 astronauts, cosmonauts, and taikonauts who have ventured into Earth orbit or beyond over the last 50 years were the products of refined technical training, as were the eight X-15 pilots who flew high enough to be qualified as astronauts, and even the two pilots who flew on *SpaceShipOne* in 2004. Beginning with the Mercury 7, they all had what writer Tom Wolfe immortalized as "the right stuff" (see figure 5).[37]

In the United States in 1962, JSC in Houston, Texas, became the home of the astronauts, where they underwent (and still undergo) rigorous training.

---

36. The history of deep space navigation is being written under contract to NASA by Andrew Butrica. On the astrogeology research program, see *http://astrogeology.usgs.gov/*, and on its history, see David Levy, *Shoemaker by Levy: The Man Who Made an Impact* (Princeton, NJ: Princeton University Press, 2000), and Gerald G. Schaber, *The U.S. Geological Survey, Branch of Astrogeology—A Chronology of Activities from Conception through the End of Project Apollo (1960–1973)*, available at *http://pubs.usgs.gov/of/2005/1190/*.
37. Tom Wolfe, *The Right Stuff* (New York, NY: Farrar, Strauss and Giroux, 1979; illustrated edition, New York, NY: Black Dog & Leventhal, 2004). In both the Soviet and American cases, the first astronauts and cosmonauts had military backgrounds. When in April 1959 NASA selected its first astronauts, all seven had aviation experience in the military. As Asif Siddiqi has shown in *Challenge to Apollo: The Soviet Union and the Space Race, 1945–1974* (Washington, DC: NASA SP-2000-4408, 2000), although the Soviets considered individuals from aviation, the Soviet navy, rocketry, and car-racing backgrounds, their Air Force physicians insisted that the initial pool be limited to qualified Air Force pilots. By the end of 1959, they had chosen 20 cosmonauts, formally approved on 7 March 1960.

The Soviet/Russian counterpart is the legendary Cosmonaut Training Center in Star City, near Moscow, where training began in 1965. At these two locations, the vast majority of space explorers have prepared for their journeys prior to launch from their countries' respective spaceports into the "new ocean."[38]

Institutions and Funding

The space programs of the world required massive efforts in institution building, management, and funding. The Age of Discovery explorers were funded in part by nation states such as Spain and Portugal, often without the intermediary of an organizing institution. By the time of the Age of Space, the infrastructure had grown so complicated and expensive that national governments had to form new agencies dedicated to the task.[39] Paramount among these was NASA, and its story of "organizing for exploration" is well known.[40] Along with the

---

38. Numerous firsthand accounts have been written by the astronauts themselves, ranging from the Mercury astronauts' collective book *We Seven* (New York, NY: Simon & Schuster, 1962) recalling Charles Lindbergh's book *We* (New York, NY: G. P. Putnam & Sons, 1927); Scott Carpenter and Kris Stoever's *For Spacious Skies: The Uncommon Journey of a Mercury Astronaut* (New York, NY: New American Library, 2004); Wally Schirra's *Schirra's Space* (Annapolis, MD: U.S. Naval Institute Press, 1995); John Glenn's *John Glenn: A Memoir* (New York, NY: Bantam Books, 1999); Deke Slayton and Michael Cassut's *Deke! An Autobiography* (New York, NY: Forge Books, 1995); and Gordon Cooper's wild *Leap of Faith: An Astronaut's Journey into the Unknown* (New York, NY: HarperCollins, 2000), complete with space aliens; to Apollo astronauts Jim Lovell and Jeffrey Kluger's *Apollo 13: Lost Moon: The Perilous Voyage of Apollo 13* (Boston, MA, and New York, NY: Houghton Mifflin Co., 1994) and Gene Cernan's *The Last Man on the Moon* (New York, NY: St. Martin's Press, 1999) and Space Shuttle astronauts Mike Mullane's irreverent *Riding Rockets: The Outrageous Tales of a Space Shuttle Astronaut* (New York, NY: Scribner, 2006) and Tom Jones's *Sky Walking: An Astronaut's Memoir* (New York, NY: HarperCollins, 2006). A unique dual autobiography is David Scott and Alexei Leonov's *Two Sides of the Moon: Our Story of the Cold War Space Race* (New York, NY: St. Martin's Press, 2004). A few astronauts have been the subject of full-scale biographies, including Neal Thompson's *Light This Candle: The Life & Times of Alan Shepard—America's First Spaceman* (New York, NY: Crown Publishers, 2004); Ray Boomhower's *Gus Grissom: The Lost Astronaut* (Indianapolis, IN: Indiana Historical Society, 2004); and James Hansen's definitive *First Man: The Life of Neil A. Armstrong* (New York, NY: Simon & Schuster, 2005). Others have been the subjects of biographies by relatives, as in Nancy Conrad's *Rocketman: Astronaut Pete Conrad's Incredible Ride to the Moon and Beyond* (New York, NY: NAL Trade, 2006).
39. The best single volume covering the history of the world's space agencies is P. V. Manoranjan Rao, ed., *50 Years of Space: A Global Perspective* (Hyderabad, India: Universities Press, 2007). This volume, produced for the 50th anniversary of the Space Age, is composed of histories of each agency written by a high-level representative from each agency.
40. NASA's founding documents and relevant materials for its first 35 years are found in *Exploring the Unknown*, vol. 1, *Organizing for Exploration* (1995), available at *http://history.nasa.gov/series95.html*. For NASA in the broader context of history, see Robert R. MacGregor, "Imagining an Aerospace Agency in the Atomic Age," in *Remembering the Space Age*, ed. Steven J. Dick (Washington, DC: NASA SP-2008-4703, 2008), and Walter McDougall, . . . *The Heavens and the Earth: A Political History of the Space Age* (Baltimore, MD: Johns Hopkins University Press, 1985). The story from the point of view of NASA's first Administrator is in J. D. Hunley, ed., *The Birth of NASA: The Diary of T. Keith Glennan*

technical aspects, the development of management techniques appropriate to a high-technology, high-reliability organization has been essential to its success, and Apollo management techniques have been especially studied.[41]

No less crucial has been funding. Over the last 50 years, aside from the anomalous Apollo era, NASA's budget has remained relatively stable at below 1 percent of the federal budget (see figure 6). Still, NASA leads the world in its space budget as a percentage of government spending.[42] As with other government agencies, and especially because NASA's reach exceeds its grasp, the search for more funding is a never-ending enterprise. Yet, in the case of the United States, polls show most of the public is content with this level (see figure 7).[43] Whether in the next 50 years harsh economic realities drive the budget percentage down, or whether international competitive pressures from Europe, China, and India drive it up, the budget must remain, for now, one of the great unanswered questions of the future Space Age.

Risk

Finally, it is important to emphasize that both the Age of Discovery and the Age of Space had, and will continue to have, their risks and their tragedies. Out of five ships and 260 men who departed Spain with Magellan on 20 September 1519, only one ship and 18 bedraggled men returned in 1522—and Magellan

---

(Washington, DC: NASA SP-4105, 1993), available at *http://history.nasa.gov/SP-4105/sp4105.htm.* See also John M. Logsdon, moderator, *Legislative Origins of the National Aeronautics and Space Act of 1958: Proceedings of an Oral History Workshop*, Monographs in Aerospace History, No. 8 (Washington, DC: NASA History Division, 1998), available at *http://www.hq.nasa.gov/office/pao/History/40thann/legorgns.pdf*; David S. Portree, *NASA's Origins and the Dawn of the Space Age*, Monographs in Aerospace History, No. 10 (Washington, DC: NASA History Division, September 1998), available at *http://www.hq.nasa.gov/office/pao/History/monograph10/*; and Robert L. Rosholt, *An Administrative History of NASA, 1958–1963* (Washington, DC: NASA SP-4101, 1966). The National Aeronautics and Space Act of 1958, as amended, with legislative history showing changes over time, is available at *http://history.nasa.gov/spaceact-legishistory.pdf.*

41. See, for example, Johnson, *Secret of Apollo.*
42. For the world context, see Organization for Economic Cooperation and Development (OECD), *The Space Economy at a Glance 2007* (Paris, France: OECD, 2007); also *Understanding the Space Economy: Competition, Cooperation and Commerce* (Oxford, U.K.: Oxford Analytica, June 2008). It is important to note that during the earlier decades of the space program, the U.S. federal budget was almost all discretionary, whereas now most of it goes to entitlements and interest on the debt. So a more proper comparison is with today's discretionary budget, totaling some $900 billion, of which NASA's budget of $17 billion is about 2 percent.
43. On public opinion, see William Sims Bainbridge, "The Impact of Space Exploration on Public Opinions, Attitudes and Beliefs," in *Historical Studies in the Societal Impact of Spaceflight*, ed. Steven J. Dick (NASA History Series, forthcoming); Deborah D. Stine, *U.S. Civilian Space Policy and Priorities: Reflections 50 Years after Sputnik* (Washington, DC: Congressional Research Service, 3 December 2007), prepared for members and committees of Congress, available at *http://fas.org/sgp/crs/space/RL34263.pdf.*

was not one of them. In a sense, there is a huge difference between the two ages in this regard; while both ages recognized risk, little was done to manage risk in the Age of Discovery. By contrast, in the Age of Space, risk is managed to the extent that agencies such as NASA, and by association the entire nation, are sometimes accused of being risk averse. One of the greatest policy challenges is to find the proper balance between risk and exploration, and this, too, should be informed by history.

One of the greatest lessons of history, emphasized by studies from the Augustine Report of 1990 to the Columbia Accident Investigation Board Report of 2003, is that risk will always be associated with exploration. The Augustine Report conjoined both themes of risk and exploration and the Age of Discovery when it opined that "In a very real sense, the space program is analogous to the exploration and settlement of the new world. In this view, risk and sacrifice are seen to be constant features of the American experience. There is a national heritage of risk taking handed down from early explorers, immigrants, settlers, and adventurers. It is this element of our National character that is the wellspring of the U. S. space program."[44] At times during the last 50 years, that element of willingness to take risk in the space program has hung by a thread in the aftermath of searing accidents in both human and robotic spaceflight. The easy course after losing both the Mars Climate Orbiter and the Mars Polar Lander in 1999, and after losing the second Space Shuttle in 2003, would have been to cancel the programs. But despite deep personal losses to families, careers, and the American sense of exceptionalism, the programs moved ahead. Just as the original Age of Discovery faded, and the preeminence of their nation states along with it, there is no guarantee the Space Age will not suffer the same fate, despite its literally infinite possibilities.

In summary, both symmetries and asymmetries exist in the general narrative arc of the Age of Space and the Age of Discovery, whether in terms of motivation, infrastructure, funding, people, risk, or many other factors not mentioned here. The particular conditions were very different, and both ages can only be understood in the context of their times. Nevertheless, both ages indisputably produced great voyages of discovery, and it is to those voyages we now turn.

---

44. Norm Augustine, Chair, *Report of the Advisory Committee on the Future of the U.S. Space Program* (Washington, DC, December 1990); Columbia Accident Investigation Board, *Report.* That risk is the constant companion of exploration, and that the public needs to understand this, is one of the main conclusions of the essays in Steven J. Dick and Keith Cowing, ed. *Risk and Exploration: Earth, Sea and the Stars* (NASA SP-2005-4701, 2005), available via the NASA History Division or at *http://history.nasa.gov/SP-4701/frontmatter.pdf.*

The Story of the Space Age

Even with its multifaceted and fascinating policy, infrastructure, and engineering aspects, the Age of Space is best characterized not by its conditions, but by its results. Space exploration has generated many narratives, but its central narrative is simple, straightforward, and profound: a continuous story of voyages further and further from the home planet. The Age of Discovery began in the 15th century with Portuguese sailors hugging the west coastline of Africa, then sailing outward to increasingly distant islands—Madeira in the 1420s, the Azores in the 1430s and 1440s, and the Cape Verde islands in the 1450s (and a long unsuccessful attempt at the Canary Islands controlled by Castile).[45] At the end of the century, the Portuguese denied Columbus the funding he requested, and it was the Spanish who funded the first plunge across the ocean in a remarkable story we all learn in school.[46] By about 1650, in Parry's estimation, the Age of Reconnaissance was over, as Africa, Asia, and the Americas had become routine destinations.

Unlike the Age of Discovery, which ran its course in about two centuries, the Age of Space is a process that has only begun and that potentially has no end, but that is nonetheless fundamentally a story of exploration and discovery now played out on an unimaginably vaster scale. Not by accident have spacecraft been named Mariner, Voyager, Viking, Ulysses, *Challenger*, *Endeavor*, and Magellan, hearkening back to that long exploring tradition. The 50-year narrative trajectory of spacecraft, ranging from Earth's atmosphere and Earth orbit to the solar system and the universe at large, is full of remarkable discoveries that will echo down the ages and that will someday also be part of the standard school curriculum.

The Realm of Earth

The journey begins with atmospheric flight, which takes place within a thin skin surrounding Earth to an altitude of a few tens of miles (see figure 8). Like

45. Russell, "Lord of the Isles," chap. 4 in *Prince Henry the Navigator*; Parry, *Age of Reconnaissance*; Fritze, *New Worlds*.
46. Among the many histories of this event, see most recently Hugh Thomas, *Rivers of Gold: The Rise of the Spanish Empire, from Columbus to Magellan* (New York, NY: Random House, 2003).

15th-century coastal navigation in relation to oceanic navigation, aeronautics was a preparation, but in this case, for leaving Earth. Aside from its own intrinsic practical value, flight has been essential to spaceflight in numerous ways, ranging from supersonics to hypersonics and the Space Shuttle.[47] It is therefore no surprise that NASA's technical history has close connections to the history of flight (see the aeronautics section of this volume); indeed, the institution was built on the foundations of the NACA, dating back to 1915.[48] The X-15 research of the 1960s is legendary, but aeronautics continues to be important for spaceflight in ways not usually appreciated by the public (see figure 9).

Voyages to Earth orbit have a special meaning of their own. Climbing Earth's gravitational well put one "half way to anywhere in the solar system," as science fiction writer Robert Heinlein once put it, a necessary step toward more distant explorations. But even from Earth orbit, humans and robots saw the planet anew and viewed it in unprecedented fashion, whether for reconnaissance, for environmental remote sensing, or as "high ground" for providing a means of navigation and communication. Each of these programs has its own history of technical problems and achievements, though some of the history is better known than others, and some is classified.[49] Reconnaissance for reasons of national security was one of the earliest drivers of the space program, featuring satellites from CORONA to the KH series, among others.[50] Communications satellites also enjoyed early successes with the likes of Telstar, followed by Intelsat and a variety of domestic satellite systems.[51] And after early successes with weather satellites such as TIROS 1, Earth science observation and research from space began to

47. See Erik M. Conway, *High-Speed Dreams: NASA and the Technopolitics of Supersonic Transportation, 1945–1999* (Baltimore, MD: Johns Hopkins University Press, 2005), and T. A. Heppenheimer, "Hypersonics and the Space Shuttle," in *Facing the Heat Barrier: A History of Hypersonics* (Washington, DC: NASA SP-2007-4232, 2007).
48. The classic volumes of NACA history are Alex Roland, *Model Research: The National Advisory Committee for Aeronautics, 1915–1958* (Washington, DC: NASA SP-4103, 1985), two volumes, and James R. Hansen, *Engineer in Charge: A History of the Langley Aeronautical Laboratory, 1917–1958* (Washington, DC: NASA SP-4305, 1987). On the relation of the NACA to the founding of NASA, see McDougall, "The Birth of NASA," chap. 7 in . . . *The Heavens and the Earth*, pp. 157–176.
49. For an overview, see David J. Whalen, "For All Mankind: Societal Impact of Application Satellites," in *Societal Impact of Spaceflight*, ed. Steven J. Dick and Roger Launius (Washington, DC: NASA SP-2007-4801, 2007), pp. 288–312.
50. For an entrée into this field, see Stephen B. Johnson, "The History and Historiography of National Security Space," in *Critical Issues*, ed. Dick and Launius, pp. 481–548, and Glenn Hastedt, "Reconnaissance Satellites, Intelligence and National Security," in *Societal Impact of Spaceflight*, ed. Dick and Launius, pp. 369–383.
51. Joseph N. Pelton, "The History of Satellite Communications," in *Exploring the Unknown*, vol. 3, *Using Space*, pp. 1–154; Andrew J. Butrica, ed., *Beyond the Ionosphere: Fifty Years of Satellite Communication* (Washington, DC: NASA SP-4217, 1997), available at *http://history.nasa.gov/SP-4217/sp4217.htm.*

find global coherence in NASA's flagship Earth Observing System, a system of satellites that monitors Earth at many wavelengths (see figures 10 and 11).[52]

For the human spaceflight program, Earth orbit is where humans first learned that the human body could function under the harsh conditions of space, including the new experience of weightlessness, as long as they could carry the necessities of life with them in their hermetically sealed spacecraft. It is also where they learned to "fly in space," with Vostok, Voskhod, and Soyuz on the Soviet/Russian side and Mercury and Gemini on the U.S. side—the indispensable prelude to Apollo (see figures 12 and 13).[53] Earth orbit also provided a microgravity environment for experiments, both on the Space Shuttle and on space stations (see figures 14 and 15). Taken together, and perhaps most importantly, all these endeavors provided a new perspective on the home planet. Although still "hugging the coastline" in terms of the analogous maritime history, these endeavors were nonetheless voyages of discovery, yielding data on huge issues of great practical import, such as global climate change, land use, and meteorology, and providing the essential infrastructure for global navigation and communication.

Important as low-Earth orbit and geosynchronous orbit are for utilitarian applications and way station status, it is the voyages beyond Earth that captured the public imagination. It is not surprising that we turned first to the nearest celestial body, our own Moon—a nearby "island" less than two light-seconds away (where light travels at 186,000 miles per second), still gravitationally in the realm of Earth. The Luna, Ranger, Surveyor, and Lunar Orbiter spacecraft served as the prelude to the piloted Moon landings and gave us the first iconic images of the Space Age (see figures 16 and 17).[54] Above all are the epic piloted

---

52. For original documents and accompanying essays related to Earth science research from space to 1997, see Pamela E. Mack and Ray A. Williamson, "Observing the Earth from Space," in *Exploring the Unknown*, vol. 3, *Using Space*, pp. 155–384, and John H. McElroy and Ray A. Williamson, "The Evolution of Earth Science Research from Space: NASA's Earth Observing System," in *Exploring the Unknown*, vol. 6, *Space and Earth Science*, pp. 441–690. A succinct and up-to-date overview is Andrew J. Tatem, Scott J. Goetz, and Simon I. Hay, "Fifty Years of Earth-observation Satellites," *American Scientist* 96 (September–October 2008): 390–398, and a beautiful volume on environmental remote sensing from space of Earth's land, oceans, atmosphere, and the effects of humans is Claire Parkinson, Kim C. Partington, and Robin G. Williams, ed., *Our Changing Planet: The View from Space* (Cambridge, U.K.: Cambridge University Press, 2007). On the development of one particular system, see Pamela Mack, *Viewing the Earth: The Social Construction of the Landsat System* (Cambridge, MA: MIT Press, 1990).
53. On project Mercury, see Swenson et al., *This New Ocean*, and on project Gemini, Barton C. Hacker and James M. Grimwood, *On the Shoulders of Titans: A History of Project Gemini* (Washington, DC: NASA SP-4203, 1977).
54. Cargill Hall, *Lunar Impact: A History of Project Ranger* (Washington, DC: GPO, 1977), available at *http://history.nasa.gov/SP-4210/pages/Cover.htm*; Bruce K. Byers, *Destination Moon: A History of the Lunar Orbiter Program* (Washington, DC: NASA TM 3487, 1977), available at *http://www.hq.nasa.gov/office/*

voyages of the United States that resulted in 12 humans walking on the Moon, a feat that many think 500 years from now will be viewed in the same way as we now look back on the Age of Discovery. The stories of Neil Armstrong and Buzz Aldrin touching down on the Moon in July 1969, followed by 10 others by 1972; the harrowing experiences of the ill-fated Apollo 13; the astronauts roving over the surface of another world; are seared in memory and will remain monuments to ingenuity, the force of geopolitics, and exploration (see figures 18 and 19).[55]

The achievements of Apollo culminated in 1972, and since then only our robotic surrogates have left the vicinity of Earth. A single voyage, or set of voyages, does not make an age, and the jury is still out on whether our descendants 20 generations from now will view Apollo as a unique set of bold achievements or the beginnings of an era of human space exploration. Historian Arthur M. Schlesinger, Jr., special assistant to President Kennedy, ventured one opinion when he wrote in support of the new Vision for Space Exploration in January 2004, "It has been almost a third of a century since human beings took a step on the Moon—rather as if no intrepid mariner had bothered after 1492 to follow up on Christopher Columbus. Yet 500 years from now (if humans have not blown up the planet), the 20th century will be remembered, if at all, as the century in which man began the exploration of space." On the other hand, there are some, historians among them, who think the Apollo program was time and money misspent and that analogies to Columbus are misplaced. In reviewing Andrew Chaikin's book *A Man on the Moon: The Voyages of the Apollo Astronauts* in the *New York Times Review of Books*, historian of technology Alex Roland called Chaikin's retelling of the Apollo story "the great American legend of the late 20th century," replete with heroic astronauts and epic tales. Eschewing Apollo's role in exploration, and pointing to the lack of science on the missions, he downplayed the significance of the voyages of Apollo.[56] More perspective is needed; in part, the course of the next 50 years will determine

---

pao/History/TM-3487/top.htm; NASA Office of Space Science and Applications, *Surveyor Program Results* (Washington, DC: NASA SP-184, 1969), available at *http://ntrs.nasa.gov/archive/nasa/casi.ntrs.nasa.gov/19690027073_1969027073.pdf.*

55. Andrew Chaikin, *A Man on the Moon: The Voyages of the Apollo Astronauts* (New York, NY: Viking Penguin, 1994).

56. Arthur M. Schlesinger, Jr., "State of the 'Vision Thing,'" *Los Angeles Times* (21 January 2004): B13, available at *http://www.commondreams.org/views04/0121-06.htm*; Alex Roland, "How We Won the Moon," *New York Times*, sec. 7 (17 July 1994): 1. The importance of the Apollo missions is one of the themes in Dick and Launius, eds., *Societal Impact of Spaceflight* (Washington, DC: NASA SP-2007-4801, 2007), esp. Andrew Chaikin, "Live from the Moon: The Societal Impact of Apollo," pp. 53–66.

whether Apollo was a beginning or an ending. In any event, it was only a tiny first step into the immensity of space.

The Realm of the Planets

Even as we began lunar exploration, scientists and engineers were looking beyond to the realm of the planets (now light-hours away rather than light-seconds for the Moon) and the preserve of robotic, rather than piloted, spacecraft.[57] In the 1960s, the Mariner spacecraft took us to the nearest planets, first to Venus in 1962, revealing an extremely hot planet with a runaway greenhouse effect and a dense and weird atmosphere dominated by carbon dioxide and sulfuric acid rain. By 1965, it was on to Mars, where Mariner IV imagery revealed a cratered surface, a shocking discovery at the time, indicating a dead planet, like the Moon, rather than the canalled Mars of Percival Lowell. But by 1972, Mariner 9 revealed ancient riverbeds and a much more active geological history, reviving interest in Mars as an abode of life (see figures 20 and 21). The exploration of Mars has been continued by the likes of Viking, Mars Global Surveyor, Mars Odyssey, ESA's Mars Express, the MERs, and Pathfinder and Phoenix. After four years, the MERs Spirit and Opportunity still roamed the surface of the Red Planet during NASA's 50th anniversary (see figures 22 through 25).[58] Mariner 10 reached the inner planet Mercury only in 1974, a planet not to be visited again until 2008, when the MESSENGER spacecraft produced stunning imagery and scientific data from the planet closest to our Sun. Meanwhile, the exploration of the other inner planet, Venus, was continued by the Soviet Venera spacecraft, Pioneer Venus, and the ingenious radar mapper aboard Magellan, which pierced the thick clouds (see figures 26 and 27).

In what Carl Sagan and others have called the Golden Age of Exploration, in the 1970s and 1980s, the Pioneer and Voyager spacecraft took us to Jupiter, Saturn,

57. For general histories of robotic exploration, see Ronald A. Schorn, *Planetary Astronomy From Ancient Times to the Third Millennium* (College Station, TX: Texas A&M University Press, 1998); William Burrows, *Exploring Space: Voyages in the Solar System and Beyond* (New York, NY: Random House, 1990); Robert S. Kraemer, *Beyond the Moon: A Golden Age of Planetary Exploration, 1971–1978* (Washington, DC: Smithsonian Institution Press, 2000); and Bruce Murray, *Journey into Space: The First Three Decades of Space Exploration* (New York, NY: W. W. Norton, 1989).
58. On planetary probes, see Asif Siddiqi, *Deep Space Chronicle: A Chronology of Deep Space and Planetary Probes*, Monographs in Aerospace History, No. 24 (Washington, DC: NASA SP-2002-4524, 2002). On the individual programs, see for example, Edward C. Ezell and Linda N. Ezell, *On Mars: Exploration of the Red Planet, 1958–1978* (Washington, DC: NASA SP-4212, 1984). On the MERs, see Steve Squyres, *Roving Mars: Spirit, Opportunity, and the Exploration of the Red Planet* (New York, NY: Hyperion, 2005). On the voyages of the 1990s, including Mars Global Surveyor, see Laurence Bergreen, *Voyage to Mars.*

and, in the case of Voyager 2, all the way to Uranus and Neptune at the edge of the solar system (see figures 28 to 31).[59] Galileo revisited Jupiter and its retinue of moons in the 1990s, and Cassini is now exploring Saturn, with its Huygens companion having landed on the huge Saturnian moon Titan.[60] New Horizons is on the way to Pluto, classified in 2006 by the International Astronomical Union as a dwarf planet, much to the chagrin of some planetary scientists. Other spacecraft have visited comets (Giotto, Deep Impact) and orbited and even landed on an asteroid (NEAR Shoemaker) (see figures 32 and 33).[61] Moreover, the Voyager spacecraft (renamed the Voyager Interstellar Mission), with their engraved greetings from Earth, are traveling beyond the solar system on their way to the stars.[62]

In the process of exploring our solar system, fundamental discoveries were made. We learned the geological and atmospheric histories of new worlds. We found planetary rings to be more common than once thought, though still not surpassing those of Saturn. And we discovered an entire retinue of new and unique worlds—the planetary satellites (see figures 34 to 37). Whereas when the Space Age began, about 30 natural satellites were known, now more than 145 are known and named, many of them imaged up close by spacecraft. The new worlds do not end there. Since 1995, we have discovered hundreds of new planets beyond the solar system, and the Kepler spacecraft, launched in 2009, will doubtless carry that number into the thousands; some, perhaps, are worlds like our Earth.

Finally, spacecraft such as Ulysses and SOHO have observed the nearest star, our life-giving Sun, returning spectacular images of solar activity and inaugurating the new field of heliophysics (see figure 38). The Sun is our entrée into another, more far-reaching realm, the realm of the stars.

---

59. Stephen J. Pyne, "A Third Great Age of Discovery," in Carl Sagan and Stephen J. Pyne, *The Scientific and Historical Rationales for Solar System Exploration* (Washington, DC: Space Policy Institute [SPI], SPI 88-1, George Washington University, 1988), pp. 13–77; Henry C. Dethloff and Ronald A. Schorn, *To the Outer Planets and Beyond: Voyager's Grand Tour* (Washington, DC: Smithsonian Institution Press, 2003).
60. Michael Meltzer, *Mission to Jupiter: A History of the Galileo Project* (Washington, DC: NASA SP-2007-4231, 2003), available at *http://history.nasa.gov/sp4231.pdf*. Michael Meltzer is also working on a book-length Cassini history.
61. The Web site for SOHO is available at *http://sohowww.nascom.nasa.gov*; the Web site for Ulysses is available at *http://ulysses.jpl.nasa.gov/*; Howard McCurdy, *Low-Cost Innovation in Spaceflight: The Near Earth Asteroid Rendezvous (NEAR) Shoemaker Mission*, Monographs in Aerospace History, No. 36 (Washington, DC: NASA SP-2005-4536, 2005).
62. NASA, "Voyager Interstellar Mission," available at *http://voyager.jpl.nasa.gov/*; Carl Sagan, Frank Drake, Ann Druyan, Timothy Feerris, Jon Lomerg, and Linda Salzman Sagan, *Murmurs of Earth: The Voyager Interstellar Record* (New York, NY: Random House, 1978). *Science* magazine cover story and special section, Brooks Hanson, "Voyager 1 Passes the Termination Shock," *Science* 309 (23 September 2005): 2015–2029.

Figures 1 to 4: Infrastructure for the Age of Discovery and the Age of Space.

Figure 1: (Right) The already-existing caravel was often the vessel of choice in the Age of Discovery, while rockets had to be built *de novo* or based on military rockets. (Above) Symbolizing the Age of Discovery and the Age of Space, replicas of Christopher Columbus's sailing ships *Santa Maria*, *Niña*, and *Pinta* sail by the Space Shuttle *Endeavour* at KSC's Launch Complex 39B awaiting liftoff on its maiden voyage in 1992. The *Niña* and *Pinta* were caravels, whereas the *Santa Maria* was a merchant ship known as a carrack. Next to the launchpad are the sound suppression water system tower and the liquid hydrogen (LH_2) storage tank, all part of the complex infrastructure of the Space Age. *NASA JSC Image S92-39077*

Figure 2: The Apollo 11 Saturn V space vehicle lifts off with astronauts Neil A. Armstrong, Michael Collins, and Edwin E. "Buzz" Aldrin, Jr., at 9:32 a.m. EDT, 16 July 1969, from KSC's Launch Complex 39A. The Saturn V-Apollo system was the only system capable of a Moon voyage. *NASA Image 69PC-0447*

Figure 3: Spaceports were also essential infrastructure for the Age of Space. This aerial view of Missile Row, Cape Canaveral Air Force Station, was taken in November 1964. The view is looking north, with the VAB under construction in the upper left-hand corner. *NASA Image 64PC-0082*

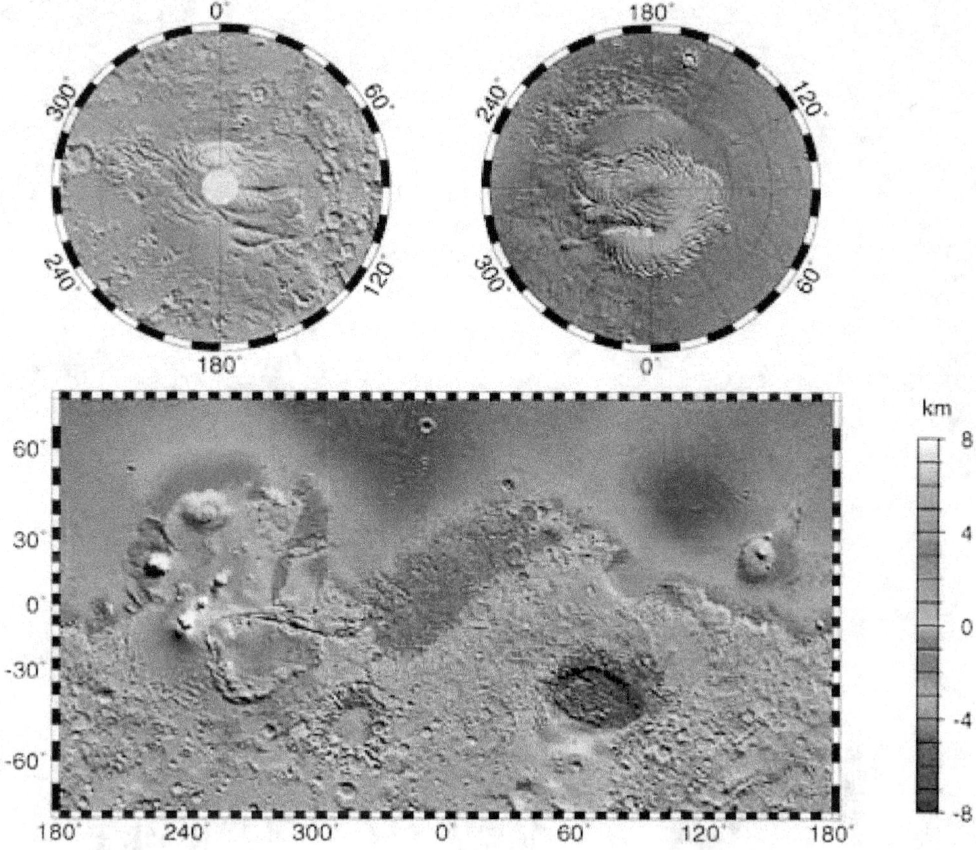

Figure 4: Just as the Age of Discovery charted new territory and produced more accurate maps, the Space Age has charted entirely new worlds. Released on 27 May 1999, this topographic map of Mars, produced with data from the Mars Global Surveyor laser altimeter, clearly shows the Tharsis volcanoes in the west (including Olympus Mons), Valles Marineris to the east of Tharsis, and Hellas Basin in the southern hemisphere. North is at top. *NASA/JPL-Caltech/GSFC*

Figure 5: As with the Age of Discovery, voyagers were essential for the Age of Space. Project Mercury astronaut selection was announced on 9 April 1959, only six months after NASA was formally established on 1 October 1958. All were military test pilots. This iconic image, taken at LaRC, shows the pilots. In the front row, from left to right, are Walter H. Schirra, Jr., Donald K. Slayton, John H. Glenn, Jr., and Scott Carpenter; in the back row are Alan B. Shepard, Jr., Virgil I. "Gus" Grissom, and L. Gordon Cooper. Despite the iconic status of the image, its precise date is unknown, but it was taken by Ralph Morse for *LIFE* magazine prior to Alan Shepard's suborbital flight in May 1961. *NASA Image 84PC-0022*

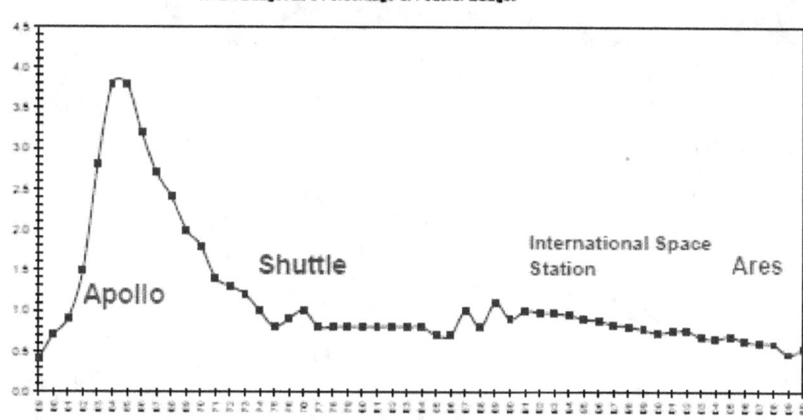

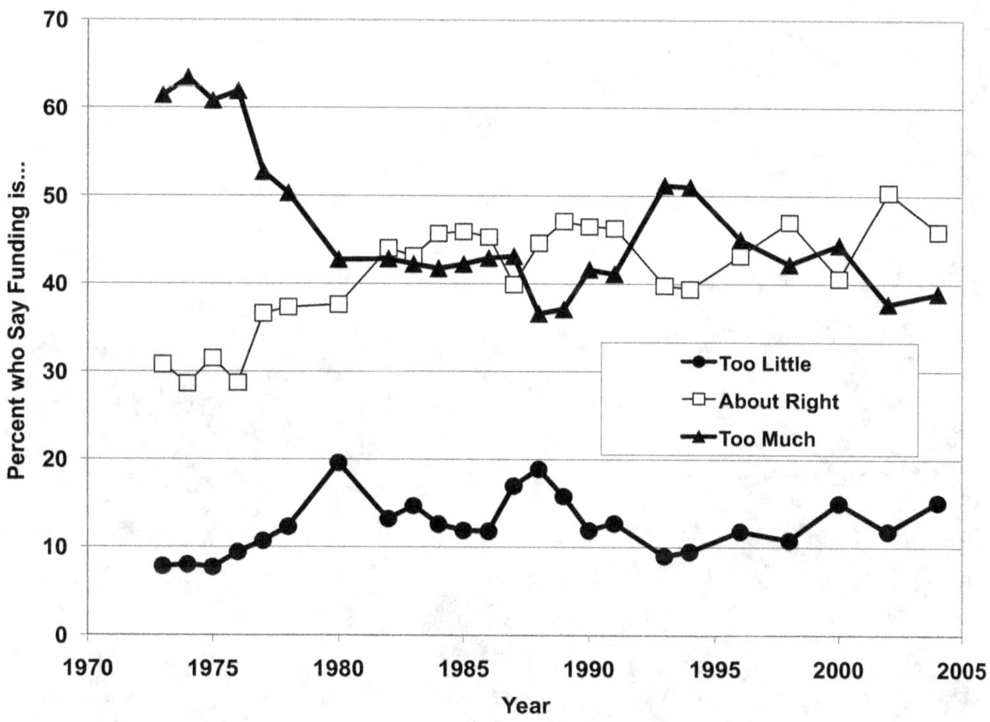

Figures 6 and 7: Funding as a condition for the Space Age. (Upper) NASA's budget over its first 50 years as a percentage of federal discretionary spending. Except for the Apollo bump at the far left, the budget has been relatively stable below 1 percent. (Lower) Attitudes toward space program funding, 1973 to 2004. Public opinion polls show most of the public is content with the level of NASA funding. *Courtesy of William Simms Bainbridge*

Figures 8 to 19: The realm of Earth.
Figure 8: The thin skin of Earth's atmosphere is clearly visible in this image taken from the Space Shuttle *Discovery* during the STS-96 mission. All of aeronautics in the 20th century developed within this thin layer surrounding Earth. *NASA Image STS096-705-066*

Figure 9: Dryden pilot Neil Armstrong is seen here next to the X-15 after a research flight. The X-15 was a rocket-powered aircraft 50 feet long with a wingspan of 22 feet, flown over a period of nearly 10 years, from June 1959 to October 1968. On 22 August 1963, Joe Walker flew the X-15 to 354,200 feet (67 miles or 107 kilometers) to set the world altitude record for winged vehicles. Research from the X-15 project was vital to the design and construction of the Space Shuttle and also contributed to the development of the Mercury, Gemini, and Apollo piloted spaceflight programs. The X-15s made a total of 199 flights. *NASA Image E60-6286*

FIRST TELEVISION PICTURE FROM SPACE
TIROS I SATELLITE APRIL 1, 1960

Figure 10: Images from Earth orbit have dramatically improved since TIROS 1 took this first TV picture from space on 1 April 1960. In 1962, TIROS satellites began continuous coverage and enabled accurate worldwide weather forecasts. *NASA*

Figure 11: This spectacular "Blue Marble" image, taken by the MODIS sensor on the Terra satellite in February 2008, is the most detailed true-color image of the entire Earth to date. It was taken from about 700 kilometers above Earth. Using a collection of satellite-based observations, scientists and visualizers stitched together months of observations of the land surface, oceans, sea ice, and clouds into a seamless, true-color mosaic of every square kilometer (.386 square mile) of our planet. *NASA GSFC image by Reto Stöckli (land surface, shallow water, and clouds). Enhancements by Robert Simmon (ocean color, compositing, three-dimensional globes, and animation). Data and technical support by MODIS Land Group, MODIS Science Data Support Team. Many similar Earth images are available at* http://visibleearth.nasa.gov/

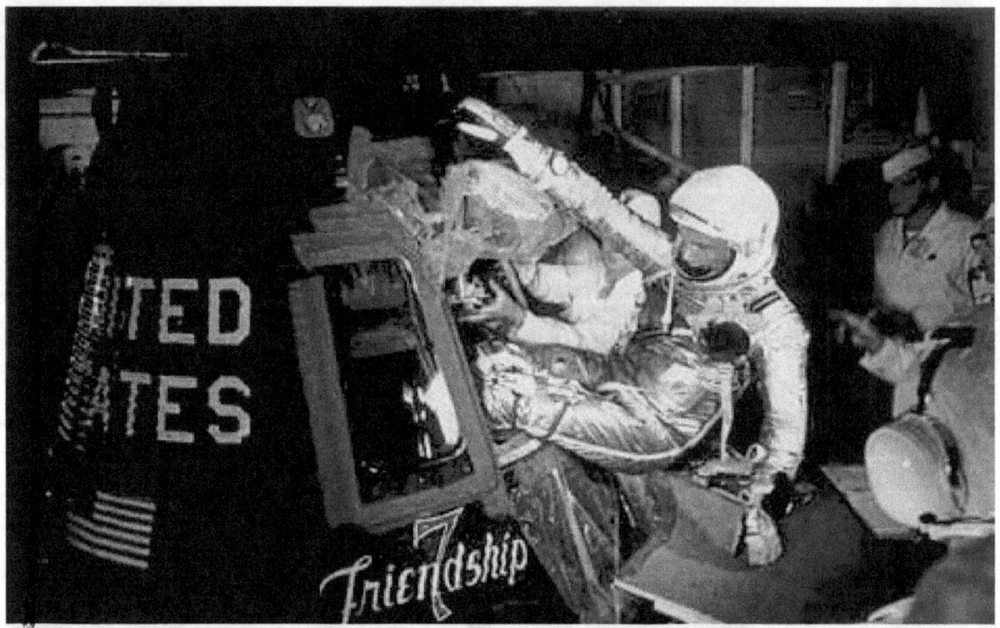

Figures 12 and 13: Americans first learned to fly in space in the Mercury and Gemini spacecraft. (Top) Astronaut John Glenn, Jr., enters *Friendship 7* prior to launch on 20 February 1962. (Bottom) NASA successfully completed its first rendezvous mission with two spacecraft—Gemini VII and Gemini VI—in December 1965. This photograph, taken by Gemini VII crewmembers Frank Lovell and Frank Borman, shows Gemini VI in orbit 160 miles (257 kilometers) above Earth. Although the principal objectives of both missions differed, they were both carried out so that NASA could master the technical challenges of getting into and working in space. *NASA Images 87PC-0069 (top) and S65-63221 (bottom)*

Figure 14: The Space Shuttle and International Space Station inaugurated advanced human activity in Earth orbit in the wake of Russian success with their space stations. A new era in human spaceflight began on 12 April 1981 with the launch of the first Space Shuttle, mission STS-1. This timed exposure of the Shuttle at Launch Pad A, Complex 39, turns the space vehicle and support facilities into an evening fantasy of light. The structures to the left of the Shuttle are the fixed and the rotating service structures. *NASA Image 81PC-0136*

Figure 15: Backdropped by Earth's horizon and the blackness of space, the Space Station is seen from Space Shuttle *Discovery* as the two spacecraft separated during the STS-119 mission in March 2009. During this, the 28th Shuttle mission to the Space Station, the fourth set of solar arrays was deployed. *NASA*

Figure 16: Charles Conrad, Jr., Apollo 12 commander, examines the robotic Surveyor III spacecraft during the second extravehicular activity. The Lunar Module *Intrepid* is in the right-hand background. This picture was taken by astronaut Alan L. Bean, Lunar Module pilot. The *Intrepid* landed on the Moon's Ocean of Storms only 600 feet from Surveyor III. The television camera and several other components were taken from Surveyor III and brought back to Earth for scientific analysis. Surveyor III soft-landed on the Moon on 19 April 1967. From 1966 to 1968, five Surveyor spacecraft successfully landed on the Moon and two crash-landed. *NASA Image AS12-48-7136*

Figure 17: The first view of Earth taken by a spacecraft from the vicinity of the Moon. The photo was transmitted to Earth by the United States Lunar Orbiter I and received at the NASA tracking station at Robledo De Chavela near Madrid, Spain. This crescent of Earth was photographed 23 August 1966 at 16:35 GMT, when the spacecraft was on its 16th orbit and just about to pass behind the Moon. In 1966 and 1967, five Lunar Orbiters returned imagery of 99 percent of the Moon, with a resolution of 60 meters or better. Lunar Orbiter images have recently been reprocessed at a much higher resolution. *NASA Image 67-H-218*

Figure 18: Astronaut Edwin E. "Buzz" Aldrin, Jr., Lunar Module pilot, is photographed during the Apollo 11 EVA on the lunar surface Sea of Tranquillity. In the right-hand background is the Lunar Module *Eagle*. On Aldrin's right is the deployed Solar Wind Composition experiment. This photograph was taken by Neil A. Armstrong with a 70-millimeter lunar surface camera. *NASA Image AS11-40-5873*

Figure 19: Apollo 15 Lunar Module pilot James B. Irwin loads the rover with tools and equipment in preparation for the first lunar EVA at the Hadley-Apennine landing site. A portion of the Lunar Module *Falcon* is on the left. The undeployed Laser Ranging Retro-Reflector lies near the Lunar Module. Hadley Delta and the Apennine Front are in the background to the left. Saint George crater is approximately 5 kilometers (about 3 statute miles) in the distance behind Irwin's head. *NASA Image AS15-86-11602*

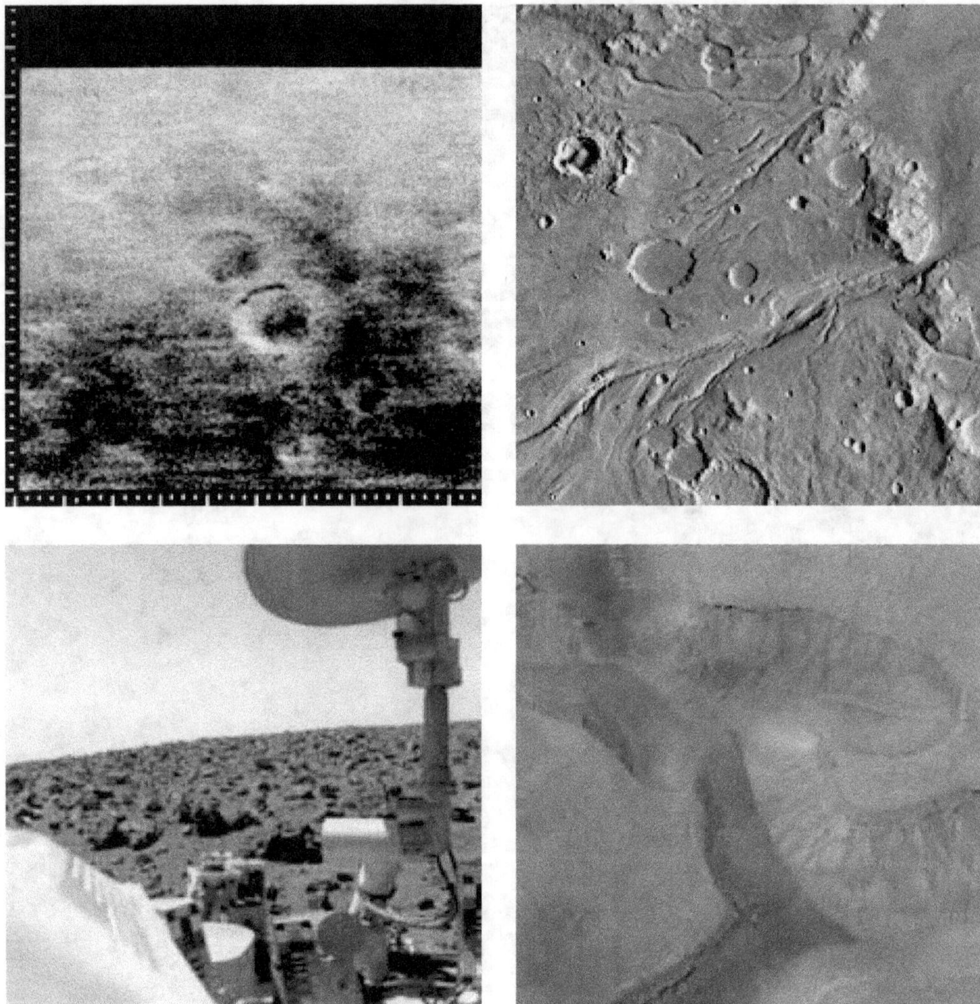

Figures 20 to 38: The realm of the planets. Mars, an object of fascination because of its possibilities for life, was the first planet from which images were returned by spacecraft.

Figure 20: (Upper left) This Mariner IV image, taken on 15 July 1965 from a range of 13,400 kilometers, was the second picture showing unambiguous craters on the surface of Mars, taken as an indication of a dead planet. *National Space Science Data Center (NSSDC)*

Figure 21: (Upper right) The Mariner 9 mission showed evidence of past water flow and resulted in a global mapping of the surface of Mars, including the first detailed views of the Martian volcanoes, Valles Marineris, the polar caps, and the satellites Phobos and Deimos. *NASA*

Figure 22: (Lower left) In this Viking 2 lander image taken on 25 September 1977, dark boulders are prominent against the reddish soil. The landing site, Utopia Planitia, is a region of fractured plains. *NASA Image 22A158*

Figure 23: (Lower right) This image, acquired by the Mars Global Surveyor Mars Orbiter Camera (MOC) in May 2000, shows numerous examples of Martian gullies. These features are located on the south-facing wall of a trough in the Gorgonum Chaos region, an area found to have many examples of gullies proposed to have formed by seepage and runoff of liquid water in recent Martian times. *NSSDC*

Figure 24: The realm of the planets (cont.). Mars Exploration Rover view from the Spirit spacecraft at the top of Husband Hill, 23 August 2005. *NASA/JPL/Caltech/Cornell*

Figure 25: This color image was acquired by NASA's Phoenix Mars Lander's Surface Stereo Imager on the 20th day of the mission, 13 June 2008. White material, possibly ice, is located at the upper portion of the trench. The Phoenix Mission is led by the University of Arizona–Tucson, on behalf of NASA. *NASA/JPL/Caltech/University of Arizona/Texas A&M University*

Figures 26 and 27: The realm of the planets (cont.). The inner planets Mercury and Venus.

Figure 26: Mariner 10 first revealed the cratered surface of Mercury in 1974, as shown in this mosaic of Mercury taken by the spacecraft during its approach on 29 March. It was three decades later before the MESSENGER spacecraft returned even more detailed images. *NSSDC*

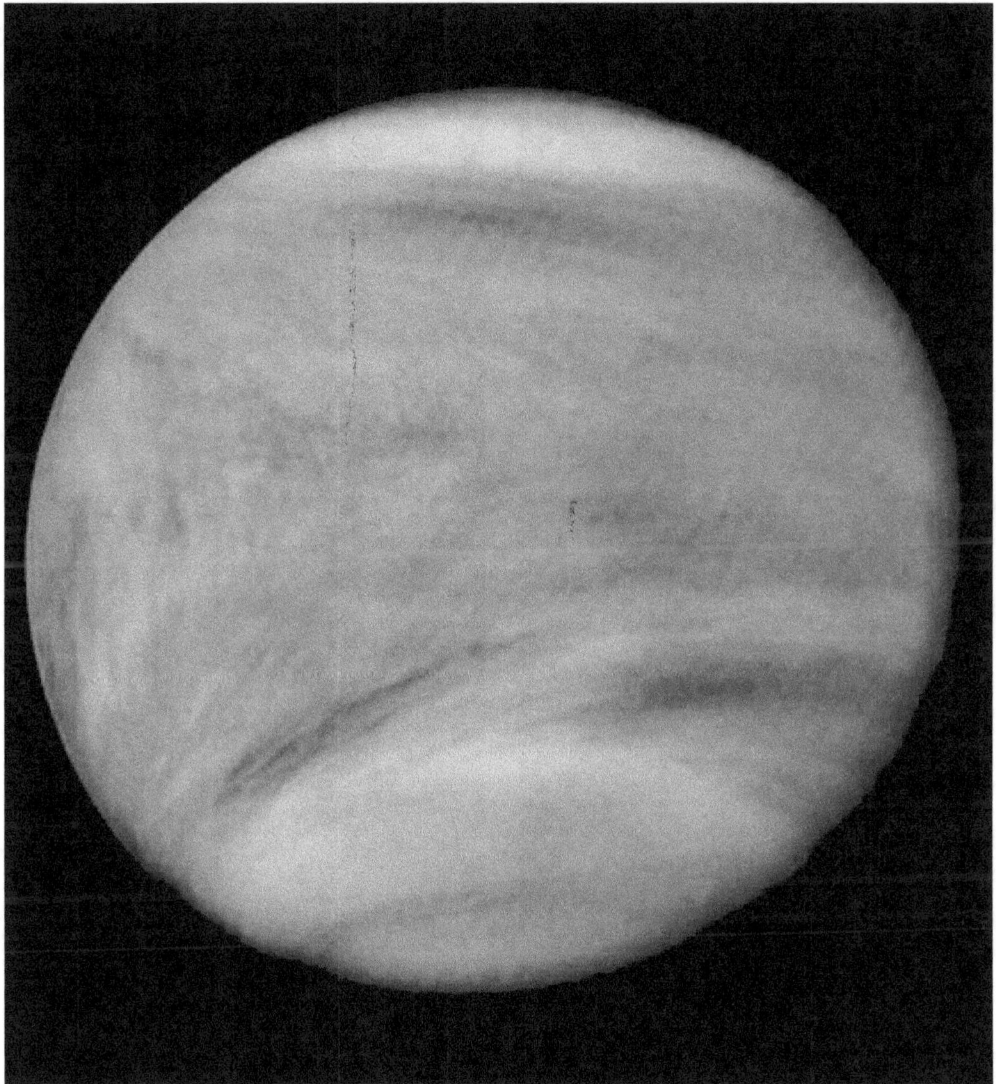

Figure 27: Following the Soviet Venera Lander photos from the surface on 22 October 1975, ultraviolet views of cloud-enshrouded Venus were imaged by Pioneer Venus Orbiter on 5 February 1979. The Magellan spacecraft later penetrated the clouds using radar, producing spectacular images of the Venusian surface from orbit. *NSSDC*

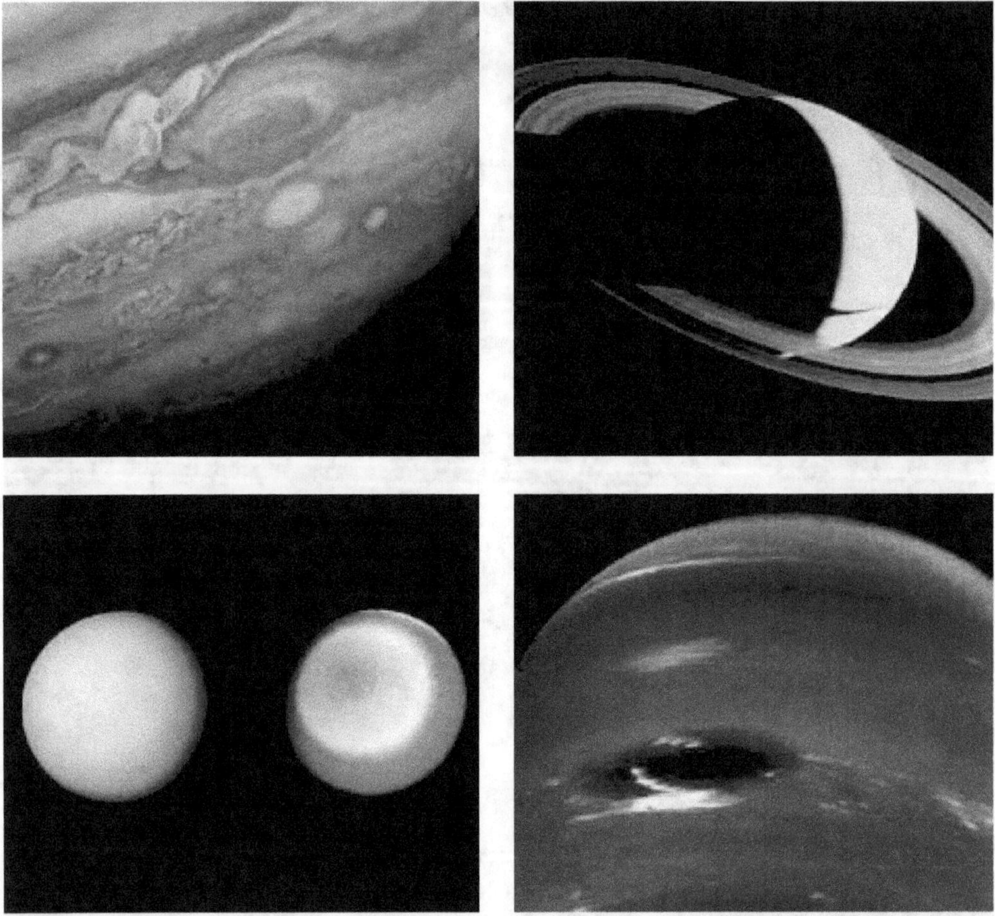

Figures 28 to 31: The realm of the planets (cont.). Exploration of the outer solar system began first with the Pioneer spacecraft, followed by Voyager 1.
Figure 28: (Upper left) Jupiter's Great Red Spot and its surroundings were imaged by Voyager 1 from a distance of 5.7 million kilometers, just over a week before its 5 March 1979 closest approach. Note the complex wave motion in the clouds to the left of the Great Red Spot, which is roughly 12,000 kilometers from top to bottom (Voyager 1, P-21151). *NSSDC*
Figure 29: (Upper right) Voyager 1 image of Saturn from 5.3 million kilometers, taken 6 November 1980, four days after its closest approach. This perspective allows a view of Saturn looking back toward the Sun. The shadow of Saturn can be seen on the rings, and Saturn can be seen through the rings as well (Voyager 1, P-23254). *NSSDC*
Figure 30: (Lower left) Moving ever further away from Earth, these two images of Uranus were taken on 17 January 1986 by Voyager 2 at a distance of 9.1 million kilometers. The picture on the left is a composite using images from the blue, green, and orange filters processed to approximate Uranus as the human eye would see it. The image on the right was produced using ultraviolet, violet, and orange filters to exaggerate the contrast (Voyager 2, P-29478). *NSSDC*
Figure 31: (Lower right) Voyager 2 image of Neptune, taken in August 1989 from 6.1 million kilometers, showing the Great Dark Spot and associated bright clouds and a bright "Scooter" cloud to the lower left. All the features are moving to the east at different speeds with the strong global winds. The Great Dark Spot is about 6,000 kilometers from top to bottom (Voyager 2, P-34632). *NSSDC*

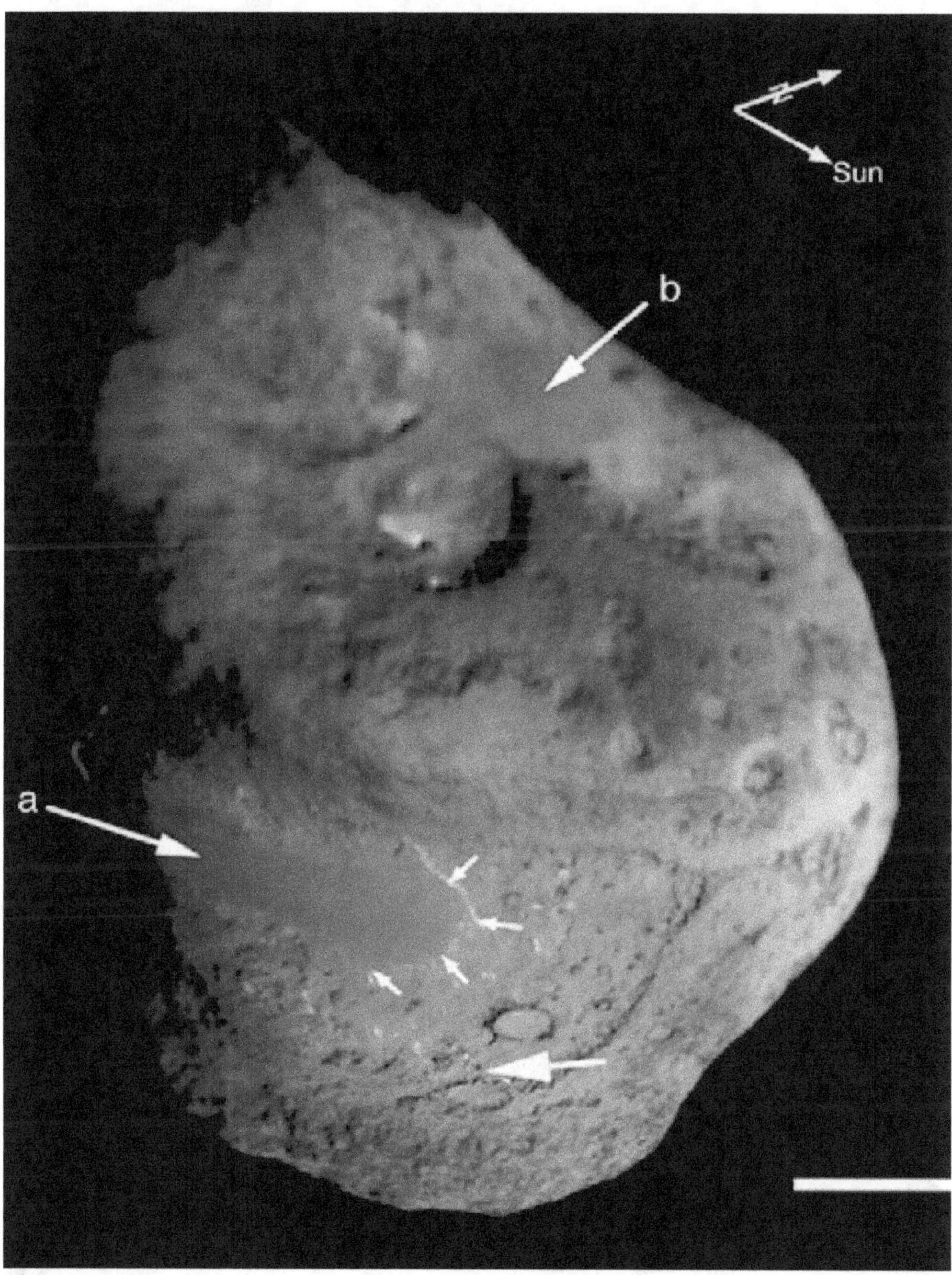

Figures 32 and 33: The realm of the planets (cont.).
Figure 32: Comet Tempel 1 from the Deep Impact spacecraft. Arrows "a" and "b" point to large, smooth regions. The impact site is indicated by the third large arrow. Small grouped arrows highlight a scarp that is bright due to illumination angle. They show a smooth area to be elevated above the extremely rough terrain. The white scale bar in the lower right represents 1 kilometer across the surface of the comet nucleus. *NASA/UM M. F. A'Hearn et al.,* Science *310, no. 258 (2005)*

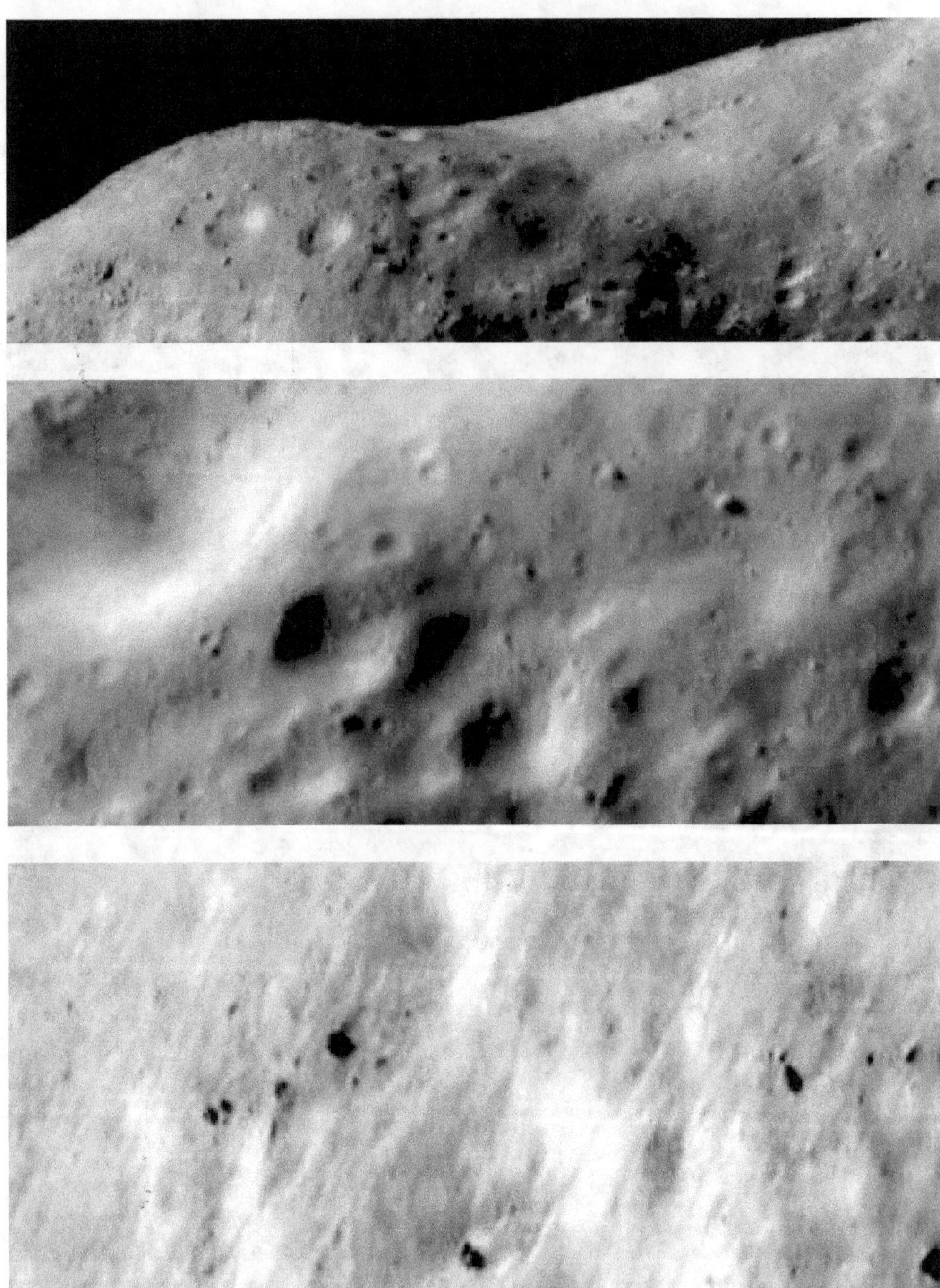

Figure 33: NEAR Shoemaker took these images of the asteroid Eros on 16 October 2000, while orbiting 54 kilometers (34 miles) above the asteroid. *NASA/Johns Hopkins University/Applied Physics Laboratory Images 0147090361-0147090659 (top), 0147067621-0147067625 (middle), and 0147089675-0147089679 (bottom)*

Figures 34 to 37: The realm of the planets (cont.). Rings and satellites.
Figure 34: Panoramic view of Saturn's rings created by combining a total of 165 images taken by the Cassini wide-angle camera over nearly 3 hours on 15 September 2006. The mosaic images were acquired as the spacecraft drifted in the darkness of Saturn's shadow for about 12 hours, allowing a multitude of unique observations of the microscopic particles that compose Saturn's faint rings. *NASA/JPL/Space Science Institute*
Figure 35: Voyager 2 captured a continuous distribution of small particles throughout the Uranus ring system. Voyager took this image while in the shadow of Uranus, at a distance of 236,000 kilometers (142,000 miles) and a resolution of about 33 kilometers (20 miles). *NASA/JPL*
Figure 36: Jupiter and its four planet-size moons, known as the Galilean satellites, were photographed in early March 1979 by Voyager 1 and assembled into this collage. They are not to scale but are in their relative positions. Reddish Io (upper left) is nearest Jupiter, then Europa (center), then Ganymede and Callisto (lower right). Not visible is Jupiter's faint ring of particles, seen for the first time by Voyager. *NASA/JPL/Caltech Image 1-149 P-21631 C*
Figure 37: False color enhances the visibility of features in this composite of three images of the Minos Linea region on Jupiter's moon Europa taken on 28 June 1996 by the solid state imaging camera on NASA's Galileo spacecraft. The linear features are believed to be cracks in the ice, beneath which is an ocean, possibly with life. *NASA*

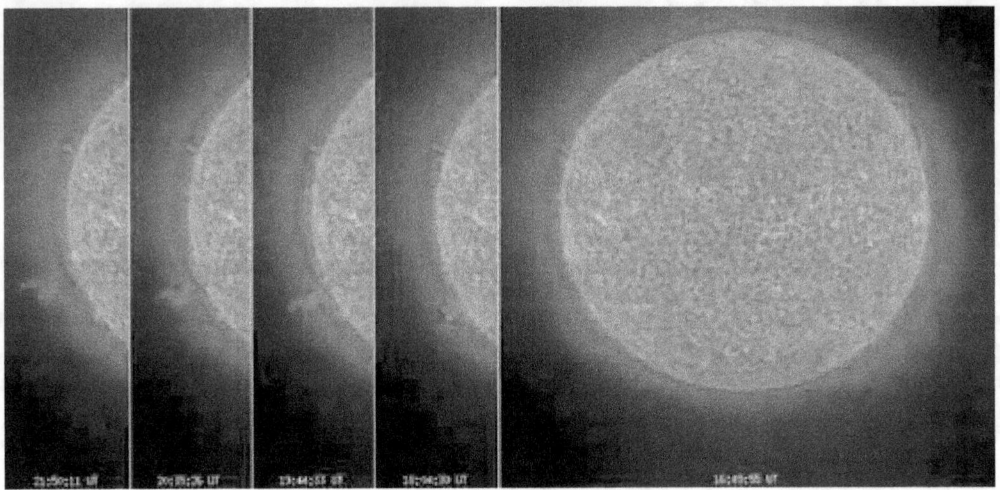

Figure 38: A joint project of ESA and NASA, SOHO took this sequence of images with its Extreme Ultraviolet Imaging Telescope, one of the observatory's 12 instruments. Easily visible on the lower left side is an "eruptive prominence" or blob of 60,000°F (33,315°C) gas measuring more than 80,000 miles (128,747 kilometers) long. When the observatory took the image on 11 February 1996, the blob was traveling at more than 15,000 mph (24,140 kph). SOHO observed these events during the minimum phase of the Sun's 11-year activity cycle. *NASA Image 091*

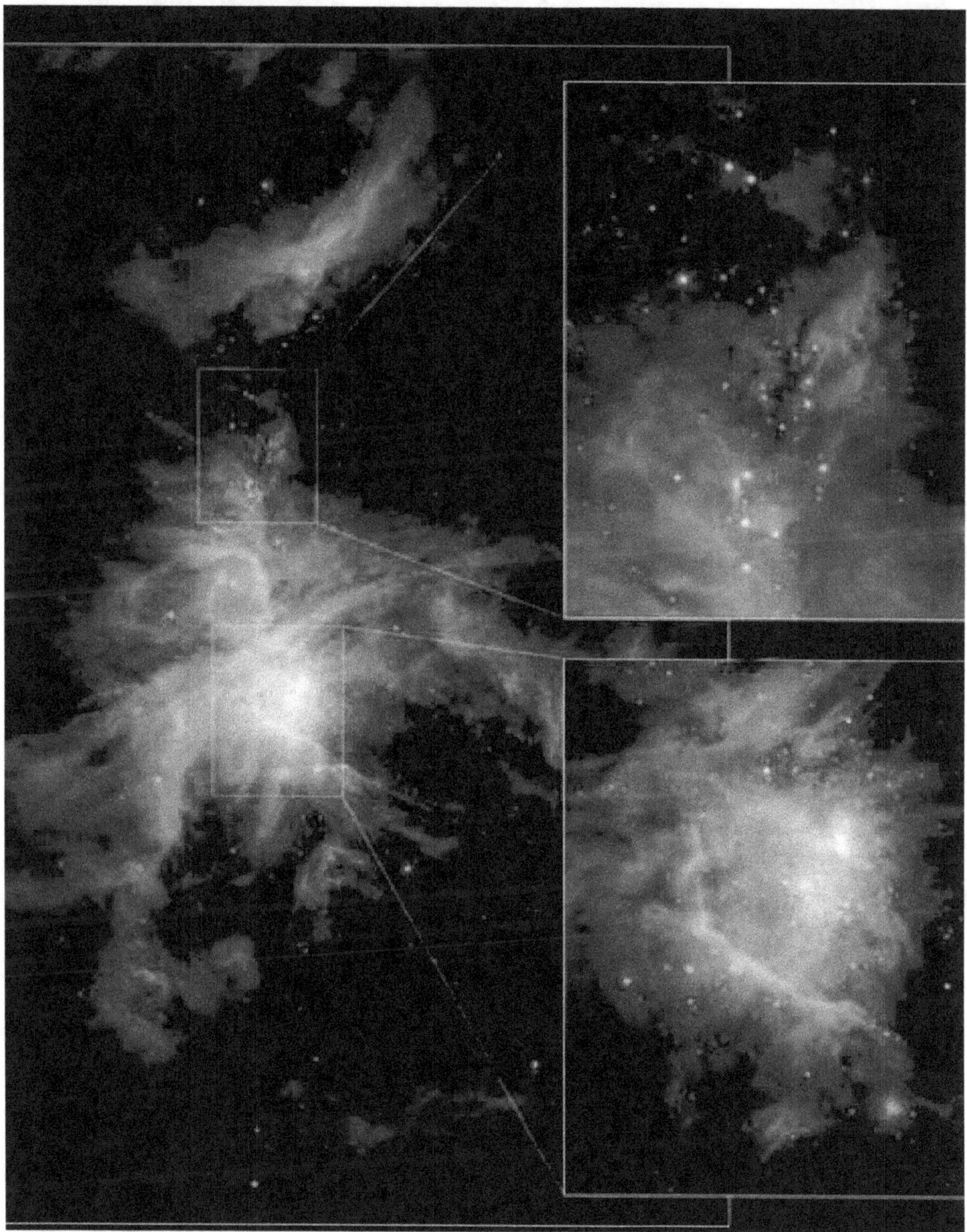

Figures 39 to 41: Realm of the stars.
Figure 39: Star birth. This infrared image from NASA's Spitzer Space Telescope shows the Orion Nebula, our closest massive star-making factory, 1,450 light-years from Earth. The nebula itself is located on the lower half of the image, surrounded by a ring of dust. It formed in a cold cloud of gas and dust and contains about 1,000 young stars. These stars illuminate the cloud, creating the beautiful nebulosity, or swirls of material, seen here in infrared. Stellar disks made of gas and dust whirl around young suns. Each disk has the potential to form planets and its own solar system. Released 14 August 2006. *NASA/JPL-Caltech/T. Megeath (University of Toledo)*

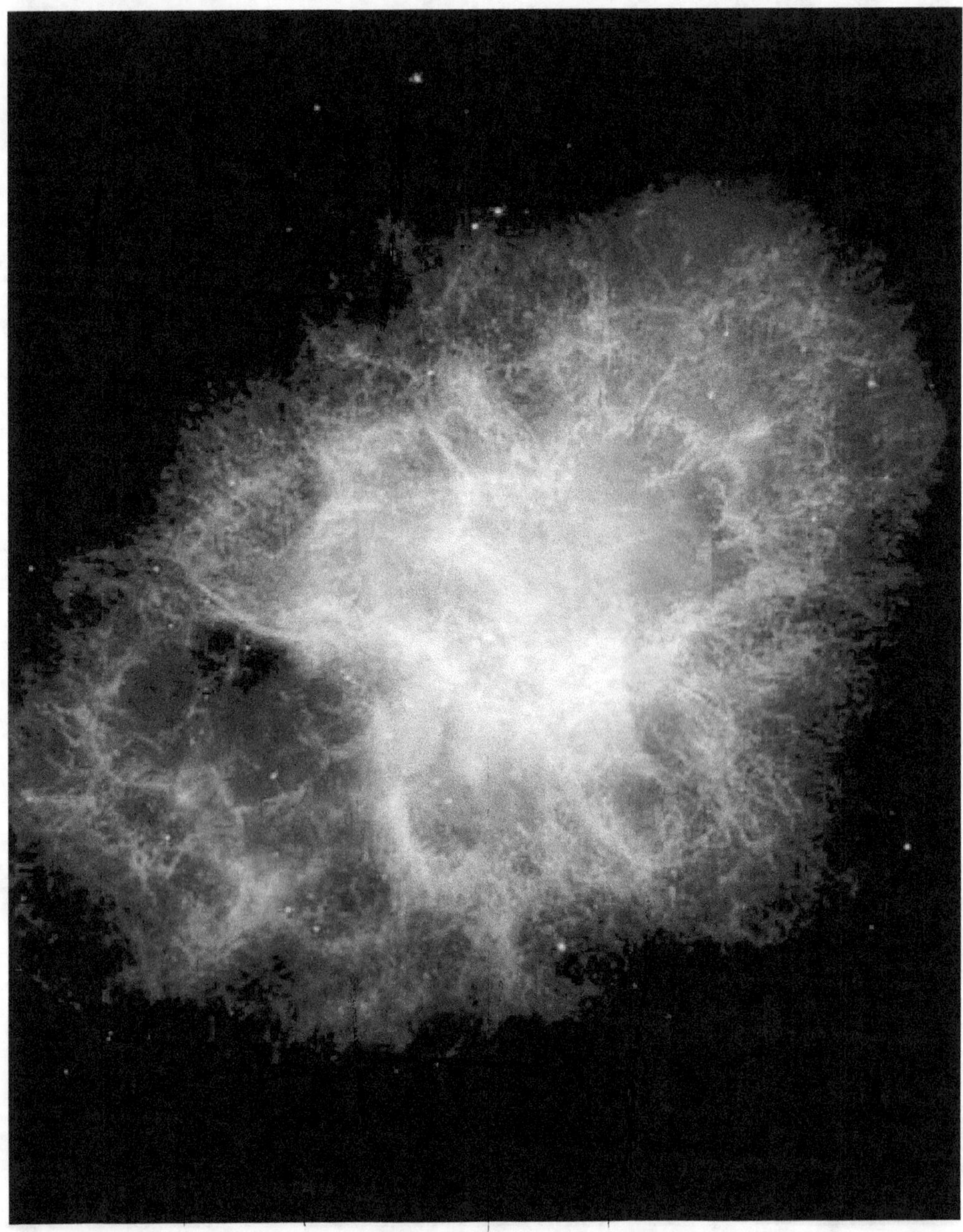

Figure 40: Stellar explosion. The Crab Nebula is an expanding remnant of a star's supernova explosion, recorded by Japanese and Chinese astronomers nearly 1,000 years ago in 1054. This composite image uses data from three of NASA's Great Observatories. The Chandra X-ray image is shown in light blue; the Hubble Space Telescope optical images are in green and dark blue; and the Spitzer Space Telescope's infrared image is in red. The neutron star, which has the mass equivalent to the Sun crammed into a rapidly spinning ball of neutrons 12 miles across, is the bright white dot in the center of the image. *NASA, ESA, Chandra X-ray Observatory, JPL-Caltech, J. Hester and A. Loll (Arizona State University), R. Gehrz (University of Minnesota), and Space Telescope Science Institute (STScI)*

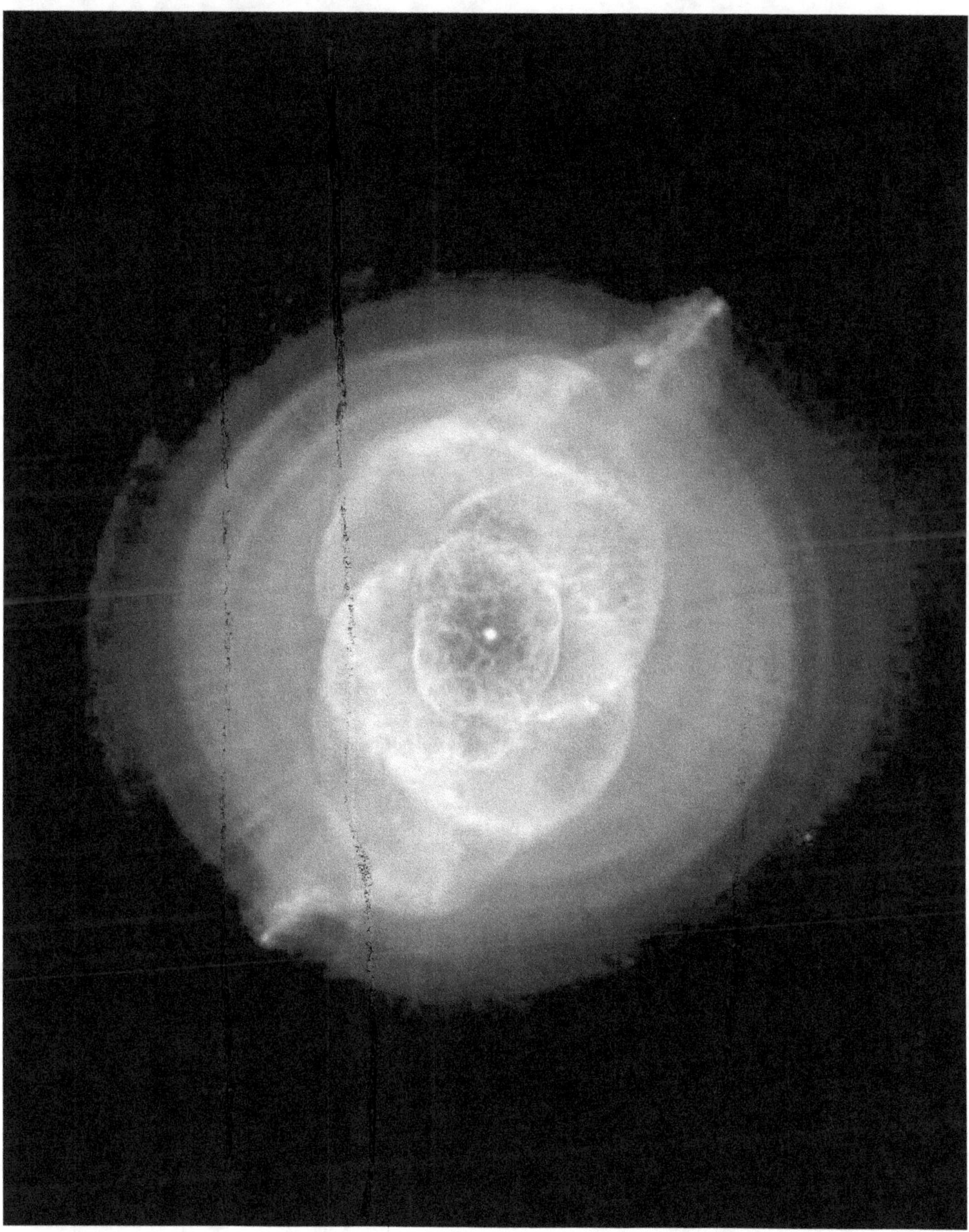

Figure 41: Star death. The so-called Cat's Eye Nebula is one of the most complex such nebulae seen in space. A planetary nebula forms when Sun-like stars gently eject their outer gaseous layers that form bright nebulae with amazing and confounding shapes. This image, taken with the Hubble Space Telescope's Advanced Camera for Surveys (ACS), reveals the full beauty of a bull's eye pattern of 11 or even more concentric rings, or shells, around the Cat's Eye. Each "ring" is actually the edge of a spherical bubble. These concentric shells make a layered, onionskin structure around the dying star. *NASA, ESA, Hubble European Space Agency Information Centre (HEIC), and the Hubble Heritage Team (STScI/AURA), as well as R. Corradi (Isaac Newton Group of Telescopes, Spain) and Z. Tsvetanov (NASA)*

Figure 42: Realm of the galaxies. When this image was released on 15 January 1996, it was the "deepest-ever" view of the universe, called the Hubble Deep Field because it was made with NASA's Hubble Space Telescope. Almost every image on this photograph, which covers a speck of sky only 1/30 the diameter of the full Moon, is a galaxy. Besides the classical spiral- and elliptical-shaped galaxies, a variety of other galaxy shapes and colors provide important clues to understanding the evolution of the universe. Some of the galaxies may have formed less than one billion years after the Big Bang. The image was assembled from many separate exposures with the Wide Field Planetary Camera 2, for 10 consecutive days from 18 to 28 December 1995. Other Hubble Deep Field images have been released since this time. *Robert Williams and the Hubble Deep Field Team (STScI) and NASA Image STScI-PRC96-01a*

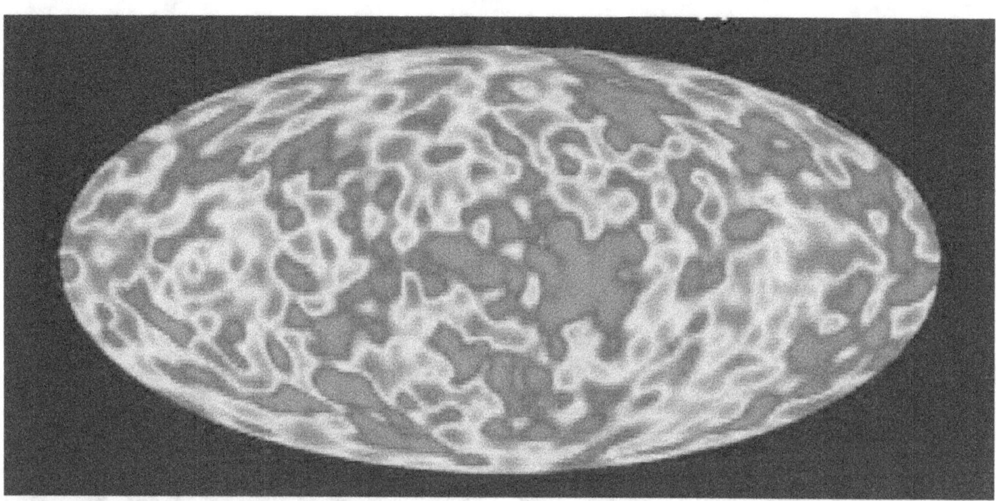

Figures 43 and 44: Realm of the galaxies (cont.).
Figure 43: Cosmic background radiation data from the Cosmic Background Explorer (COBE) during its first two years of operation. The cosmic microwave background fluctuations are extremely faint (red is hotter), only 1 part in 100,000 in the 2.73 K average temperature of the radiation field. The cosmic microwave background radiation is a remnant of the Big Bang, and the fluctuations are the imprint of density contrast in the early universe. The density ripples are believed to have given rise to the structures that populate the universe today: clusters of galaxies and vast regions devoid of galaxies. *NASA/DMR/COBE Science Team*

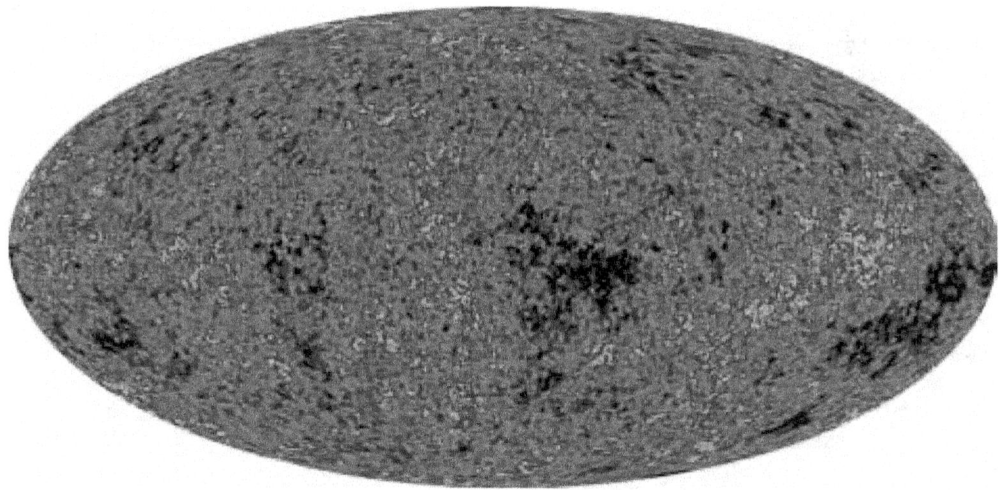

Figure 44: Where COBE measured temperature variations to 1 part in 100,000, 15 years later, the Wilkinson Microwave Anisotropy Probe (WMAP) spacecraft measured those variations to less than 1 part in 1,000,000. This image shows a temperature range of ± 200 micro-K. *NASA/WMAP Science Team*

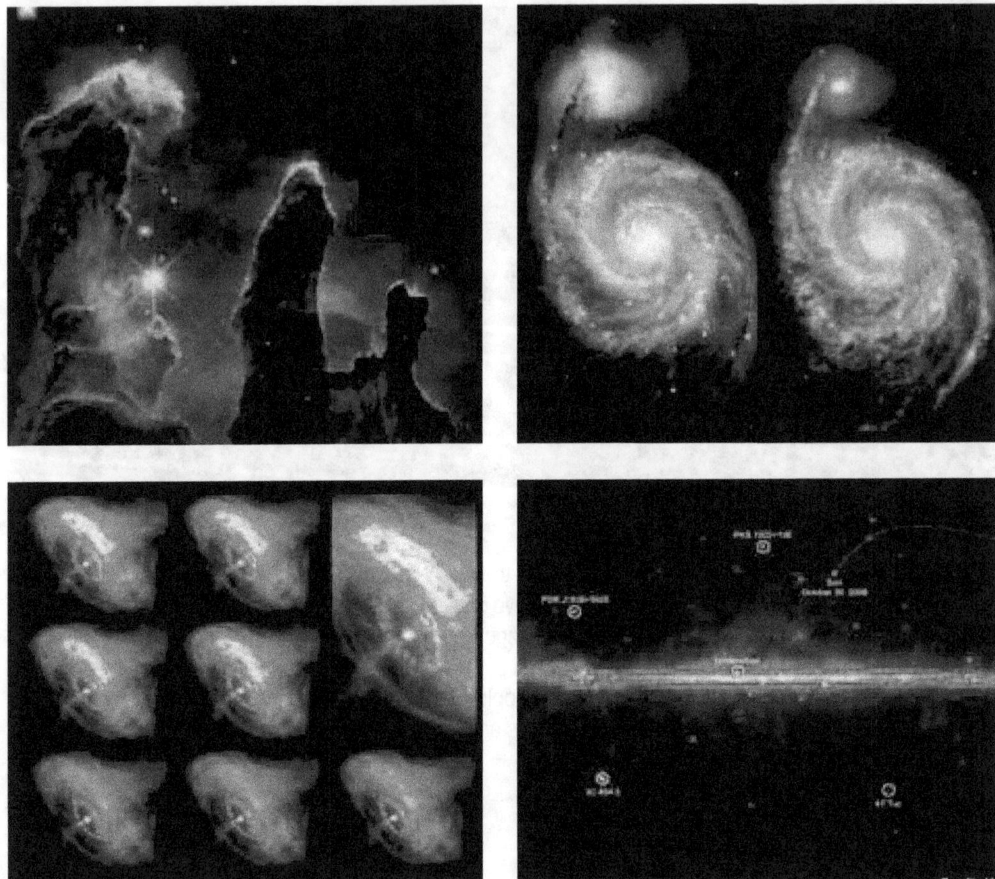

Figures 45 to 48: The opening of the electromagnetic spectrum. Space telescopes have opened the electromagnetic spectrum for space observations, ranging from the visible and infrared wavelengths to the ultraviolet, x-ray, and gamma-ray regime.

Figure 45: (Upper left) The Eagle Nebula is a Hubble Space Telescope image in the visible portion of the spectrum, showing "pillars of creation" in a star-forming region. These eerie, dark, pillar-like structures are columns of cool interstellar hydrogen gas and dust that are also incubators for new stars. The image was taken on 1 April 1995. *Jeff Hester and Paul Scowen (Arizona State University) and NASA Image STScI-PRC95-44a*

Figure 46: (Upper right) Comparison of images of the galaxy M 51, taken in visible light by the Kitt Peak National Observatory 2.1-meter telescope on the left, and in the infrared by NASA's Spitzer Space Telescope on the right. M 51 is 37 million light-years away. *NASA/JPL-Caltech/R. Kennicutt (University of Arizona)*

Figure 47: (Lower left) The Chandra X-ray Observatory images in this collage were made over a span of several months (ordered left to right, except for the close-up). They provide a stunning view of the activity in the inner region around the Crab Nebula pulsar, a rapidly rotating neutron star seen as a bright white dot near the center of the images. Compare to the image in figure 40. *NASA/CXC/ASU/J. Hester et al.*

Figure 48: (Lower right) This view from NASA's Fermi Gamma-ray Space Telescope is the deepest and best-resolved portrait of the gamma-ray sky to date. The image shows how the sky appears at energies more than 150 million times greater than that of visible light. Among the bright pulsars and active galaxies labeled here is a faint path traced by the Sun. *NASA/DOE/Fermi LAT Collaboration*

Figure 49: Cosmic evolution is depicted in this image from the exobiology program at NASA Ames Research Center, 1986. (Upper left) The formation of stars, the production of heavy elements, and the formation of planetary systems, including our own. (Left) Prebiotic molecules, RNA, and DNA are formed within the first billion years on primitive Earth. (Center) The origin and evolution of life leads to increasing complexity, culminating with intelligence, technology, and astronomers (upper right) contemplating the universe. The image was created by David DesMarais, Thomas Scattergood, and Linda Jahnke at NASA Ames Research Center in 1986 and reissued in 1997. *NASA*

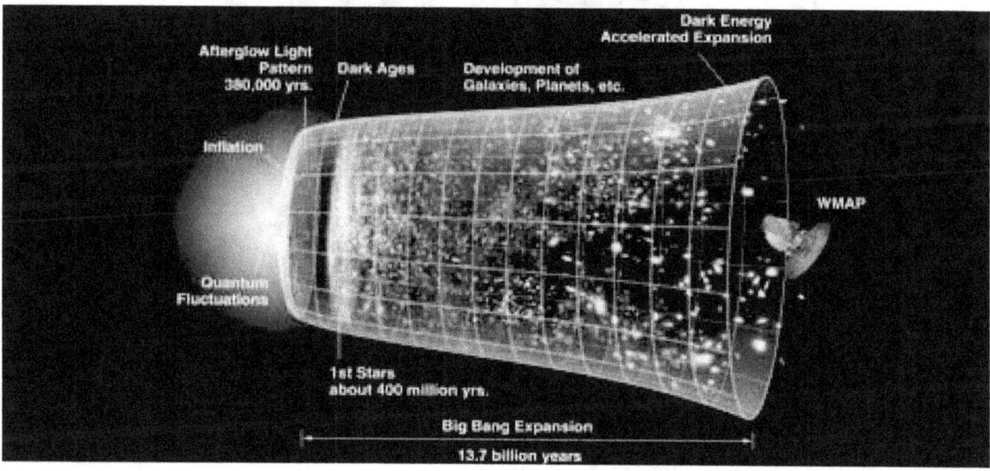

Figure 50: A representation of the evolution of the universe over 13.7 billion years. The far left depicts the earliest moment we can now probe, when a period of "inflation" produced a burst of exponential growth in the universe. (Size is depicted by the vertical extent of the grid in this graphic.) For the next several billion years, the expansion of the universe gradually slowed down as the matter in the universe pulled on itself via gravity. More recently, the expansion has begun to speed up again as the repulsive effects of dark energy have come to dominate the expansion of the universe. The afterglow light seen by WMAP was emitted about 380,000 years after inflation and has traversed the universe largely unimpeded since then. The conditions of earlier times are imprinted on this light; it also forms a backlight for later developments of the universe. *NASA/WMAP Science Team*

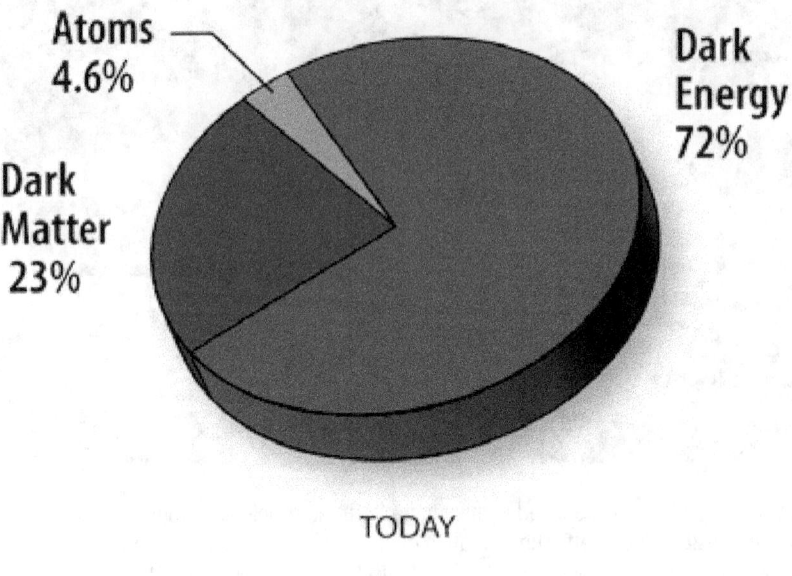

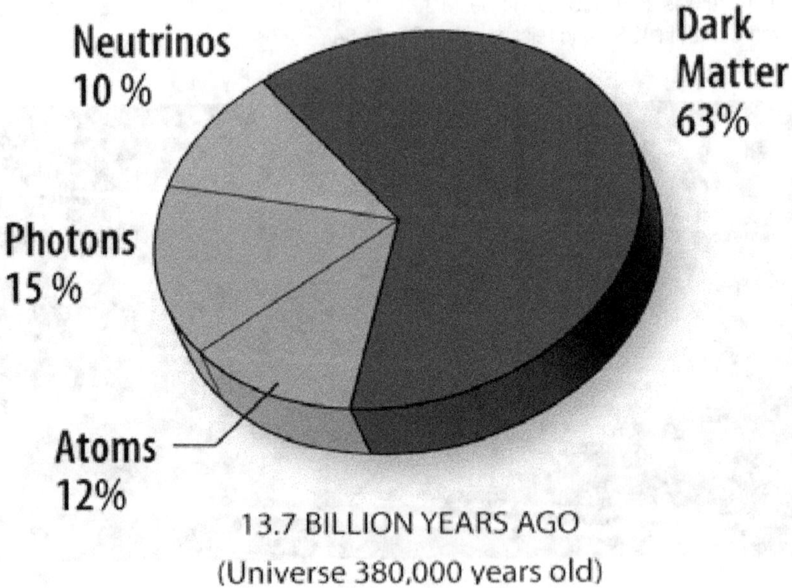

Figure 51: Content of the universe. Data from WMAP reveal that the content of the universe includes only 4.6 percent atoms, the building blocks of stars and planets. Dark matter comprises 23 percent of the universe. This matter, different from atoms, does not emit or absorb light. It has only been detected indirectly by its gravity. "Dark energy," which acts as a sort of antigravity, composes 72 percent of the universe. This energy, distinct from dark matter, is responsible for the present-day acceleration of the universal expansion. Data from WMAP are accurate to two digits, so the total of these numbers is not 100 percent. This reflects the current limits of WMAP's ability to define dark matter and dark energy. *NASA/ WMAP Science Team*

Figures 52 and 53: Societal impact of spaceflight: seeing Earth from space.
Figure 52: This view of the rising Earth greeted the Apollo 8 astronauts as they came from behind the Moon after the lunar orbit insertion burn in December 1968. The photo is displayed here in its original orientation, though it is more commonly viewed with the lunar surface at the bottom of the photo. Earth is about five degrees left of the horizon in the photo. The surface features on the left are near the eastern limb of the Moon as viewed from Earth. The lunar horizon is approximately 780 kilometers from the spacecraft. *NASA Image 68-HC-870*

Figure 53: The classic "Blue Marble" view of Earth was captured by the Apollo 17 crew traveling toward the Moon on 7 December 1972. The photograph extends from the Mediterranean Sea area to the Antarctica south polar ice cap. Heavy cloud covers the Southern Hemisphere. Almost the entire coastline of Africa is clearly visible. The Arabian Peninsula can be seen at the northeastern edge of Africa. The large island off the coast of Africa is the Malagasy Republic. The Asian mainland is on the horizon toward the northeast. *NASA/JSC Image AS17-148-22727*

The Realm of the Stars and Galaxies

Beyond the realm of the planets, we pass from the regime of light-minutes and light-hours to the realm of the stars—light-years to tens of thousands of light-years distant in our own Milky Way Galaxy, and then to the realm of the galaxies millions or billions of light-years distant. Space telescopes in Earth orbit, or its vicinity, have taken us only vicariously on voyages beyond the solar system. Those sensors that have pointed upward rather than downward—after a prelude of pioneering observatories, such as the OAOs and the Infrared Astronomical Satellite (IRAS); the "Great Observatories" including the Hubble Space Telescope, Spitzer, Compton, and Chandra; as well as the Fermi Gamma Ray Telescope—have probed the depths of the universe and produced stunning images and pioneering data of star birth such as the Orion Nebula, stellar explosions like the Crab Nebula, and star death, visible in a stunning array of planetary nebulae (see figures 39 to 41). Their images and data gave a sense of reality to the various phases of cosmic evolution, proving that robotic spacecraft results can also capture the public imagination.[63]

In the realm of the galaxies, the Hubble Space Telescope played a key role in discovering "dark energy" and the apparent acceleration of the expansion rate of the universe. It narrowed the age of the universe to 13 to 14 billion years, an accuracy of about 10 percent. The Hubble Deep Fields provided snapshots of the early universe within a few hundred million years of the Big Bang. Two spacecraft, COBE and WMAP, studied the details of the background radiation remaining from the Big Bang, pinpointed the age of the universe to 13.7 billion years (plus or minus 100 million years), and detected the seeds from which galaxies grew, a result that yielded NASA's only Nobel Prize winner.[64] As we once mapped Earth in the wake of the Age of Discovery, we are now mapping the heavens, both in space and time and in the entire range of the spectrum (see figures 42 to 44).

---

63. On these programs, see especially Robert Smith et al., *The Space Telescope: A Study of NASA, Science, Technology, and Politics* (Cambridge, U.K., and New York, NY: Cambridge University Press, 1989, 2nd edition with new introduction, 1993); Robert Zimmerman, *The Universe in a Mirror: The Saga of the Hubble Space Telescope and the Visionaries Who Built It* (Princeton, NJ: Princeton University Press, 2008); David DeVorkin and Robert Smith, *Hubble: Imaging Space and Time* (Washington, DC: National Geographic, 2008); Wallace Tucker and Karen Tucker, *Revealing the Universe: The Making of the Chandra X-Ray Observatory* (Cambridge, MA: Harvard University Press, 2001); and George Rieke, *The Last of the Great Observatories: Spitzer and the Era of Faster, Better, Cheaper at NASA* (Tucson, AZ: University of Arizona Press, 2006).
64. John Mather, *The Very First Light: The True Inside Story of the Scientific Journey Back to the Dawn of the Universe* (New York, NY: Basic Books, 1998); George Smoot and Keay Davidson, *Wrinkles in Time* (New York, NY: William Morrow and Company, 1993).

Three main themes emerge from this master narrative of Space Age voyaging. First, science has benefited tremendously from the journey into space. The Earth Observing System and its predecessors have brought unprecedented knowledge of our home planet. The lunar probes and the Apollo program (though often maligned for its scientific return) have returned data not only important for its science, but also crucial to human settlements that will undoubtedly come in the future.[65] In the realms of the planets, stars, and galaxies, we have added infinite detail to a story previously grasped only through ground-based telescopes, which, fantastic as they have become with adaptive optics and other stunning innovations, must still peer through the Earth's atmosphere, as through a glass darkly. By making the universe a real place filled with a bestiary of fantastic but scientifically comprehensible objects, space exploration has provided almost infinite space for free reign of the human imagination.

Secondly, although prior to the Space Age we learned much from 350 years of ground-based telescopic observations, in carrying out their missions, space telescopes during the last 50 years have opened the electromagnetic spectrum for astronomy in a way that could, by definition, not have been done from Earth, revealing the relatively calm sights of the infrared to the extreme violence of the x- and gamma-ray universe. The discoveries of Spitzer and its predecessors (especially IRAS) in the infrared; of International Ultraviolet Explorer (IUE), Far Ultraviolet Spectroscopic Explorer (FUSE), and Galaxy Evolution Explorer (GALEX) in the ultraviolet; of Chandra and its predecessors (the High-Energy Astronomical Observatory [HEAO] series, X-ray Multi-Mirror Mission [XMM]-Newton, Rossi, and Röntgen Satellite [ROSAT]) in the x ray; and of Compton, Swift, and Fermi in the gamma ray, reveal a universe totally unknown when the Space Age began. Along with ground-based optical, infrared, and radio wavelength observations, the next

---

65. Aside from its geopolitical goals, and despite the clear backseat status of science, a considerable amount of science was, in fact, returned from the Moon. As Donald Beattie has described in his book *Taking Science to the Moon: Lunar Experiments and the Apollo Program* (Baltimore, MD, and London, U.K.: Johns Hopkins University Press, 2001), almost 5,000 pounds of experimental equipment was landed on the Moon, including ALSEP on each of the last five Apollo missions. Eight hundred forty pounds of lunar material was returned and analyzed. On foot or in the lunar rover, 65 miles were traversed in support of field geology and geophysical studies. And during the last three missions, detailed data were collected from the orbiting Command and Service Modules. The overall result is a much better understanding of the nature and origin of the Moon and its relation to Earth. The top 10 science discoveries from the Apollo missions, as ranked by the office of the curator for planetary materials at NASA's JSC, are available at *http://www.lpi.usra.edu/expmoon/science/lunar10.html*.

50 years will see new discoveries with a range of new spacecraft spanning the entire electromagnetic spectrum (see figures 45 to 48).

Thirdly, along with stunning advances in ground-based astronomy, the Space Age has, for the first time, revealed our place in cosmic evolution in general through its spacecraft and more particularly through NASA's Origins and Astrobiology programs and similar programs in other space agencies around the world. Though cosmic evolution is an idea that dates back at least a century, it has been taken seriously only in the last 50 years—not coincidentally, the same length of time as the Space Age.[66] To a large extent, space science since that time has filled in the epic of cosmic evolution in increasing detail, revealing for the first time in detail our real place in the universe. And it has revealed that the visible universe represents less than 5 percent of the content of the universe, the remainder constituted by dark matter and dark energy. That 95 percent of the universe remains to be explored (see figures 49 to 51).

The Impact of the Space Age

It was recognized early in the Space Age that access to outer space would affect society. NASA's founding document, the National Aeronautics and Space Act of 1958, specifically charged the new agency with eight objectives, including "the establishment of long-range studies of the potential benefits to be gained from, the opportunities for, and the problems involved in the utilization of aeronautical and space activities for peaceful and scientific purposes."[67] Despite a few early studies,[68] the mandate to study the societal impact of spaceflight went largely

66. On scientific aspects of cosmic evolution, see Eric Chaisson, *Cosmic Evolution: The Rise of Complexity in Nature* (Cambridge, MA: Harvard University Press, 2001). In 1958, Harlow Shapley, the Harvard College Observatory director, wrote his popular book *Of Stars and Men* (Boston, MA: Beacon Press, 1958), beginning the modern career of the idea of cosmic evolution. For the further history of the idea, see Steven J. Dick, "Cosmic Evolution: History, Culture and Human Destiny," in *Cosmos and Culture: Cultural Evolution in a Cosmic Context*, ed. Steven J. Dick and Mark Lupisella (Washington, DC: NASA SP-2009-4802, 2009).
67. The National Aeronautics and Space Act and its complete legislative history are available at *http://www.hq.nasa.gov/office/pao/History/spaceact-legishistory.pdf.* The passage quoted here is on page 6. Although the Space Act has been often amended, this provision has never changed.
68. In addition to *The Railroad and the Space Program*, there have been sporadic studies of the societal impact of spaceflight. On the occasion of the 60th anniversary of the British Interplanetary Society, NASA was heavily involved in a special issue of its journal devoted to the impact of space on culture:

unfulfilled as NASA concentrated on the many opportunities and technical problems of spaceflight itself. Only recently has NASA made a serious attempt to examine, with historical objectivity, the broad impact of the Space Age.[69]

Once again, because of the symmetry of the narrative arc, studies of the impact of the Age of Discovery offer a framework for analysis. The impacts of the Age of Discovery were complex and bidirectional, encompassing sometimes disastrous effects on the New World and not always positive effects on the Old World. This suggests that not all impacts of spaceflight may be good, though we must at the outset take into account that the often insidious effects of culture contact are unlikely to be a factor in space exploration in the near term. Moreover, the eminent historian J. H. Elliott has delineated three components in the impact of the New World on the Old: intellectual (challenging European assumptions about geography, theology, history, and the nature of man), economic (as an extension of European business and a source of produce), and political (affecting the balance of power).[70] These broad categories also apply to the Space Age, some in the short term and others in the long term.

Intellectual Impact

Perhaps the most profound, and as yet largely unrealized, effect of the Space Age is the intellectual impact. As the story of the Space Age demonstrates, the science returned from spaceborne instruments over the last 50 years has been truly transformational, most immediately for scientists, but also for our general

---

British Interplanetary Society, "The Impact of Space on Culture," *Journal of the British Interplanetary Society* 46, no. 11 (1993). In 1994, the Mission from Planet Earth program in the Office of Space Science at NASA sponsored a symposium entitled "What Is the Value of Space Exploration?" 18–19 July 1994, NASA Historical Reference Collection, NASA History Division, NASA Headquarters, Washington, DC. More recently, in 2005, the International Academy of Astronautics (IAA), which has a commission devoted to space and society, sponsored the first international conference on space and society in Budapest, Hungary (IAA, 2005). The meeting agenda is available at *http://www.iaaweb.org/iaa/Publications/budapest2005fp.pdf.* The IAA and ESA jointly sponsored a study published as *The Impact of Space Activities upon Society*, in which well-known players on the world scene briefly discussed their ideas of societal impact, ranging from the practical to the inspirational (ESA BR-237, 2005).

69. Dick and Launius, eds., *Societal Impact of Spaceflight* (Washington, DC: NASA SP-2007-4801, 2007). The NASA History Division also has commissioned a series of specific studies on the societal impact of spaceflight, Historical Studies in the Societal Impact of Spaceflight, to appear in the NASA History series. *Cosmos and Culture: Cultural Evolution in a Cosmic Context* (Washington, DC: NASA SP-2009-4802, 2009) has also appeared in NASA's Societal Impact series.
70. For the impact of the European discovery of America on Europe, see J. H. Elliott, *The Old World and the New, 1492–1650* (Cambridge, U.K.: Cambridge University Press, 1970). For the longer-term impact on the Americas, see J. H. Elliott, *Empires of the Atlantic World: Britain and Spain in America, 1492–1830* (New Haven, CT: Yale University Press, 2006).

worldview. Although not everyone has yet absorbed the impact, that worldview has been altered or completely transformed by the images of "Earthrise" and the "Blue Marble" from space, with consequences that have affected, or will eventually affect, philosophy, theology, and the view of our place in nature (see figures 52 and 53). In *Rocket Dreams: How the Space Age Shaped Our Vision of a World Beyond*, Marina Benjamin argues that "The impact of seeing the Earth from space focused our energies on the home planet in unprecedented ways, dramatically affecting our relationship to the natural world and our appreciation of the greater community of mankind, and prompting a revolution in our understanding of the Earth as a living system."[71] She finds it no coincidence that the first Earth Day on 20 April 1970 occurred in the midst of the Apollo program, or that one of the astronauts developed a new school of spiritualism.[72]

More broadly, the same master narrative of cosmic evolution that over the last 50 years has shown us our true place in the universe has also spread to many areas of society, from history and education to religion and theology (see figures 51 and 52). Some historians have begun a movement toward "Big History," in which the usual political, social, and economic factors of human history are fully integrated and analyzed in the context of the billions of years of cosmic evolution it took to arrive at *Homo sapiens*.[73] Some educators have integrated cosmic evolution into the standard school curriculum with the same goal of perspective.[74] And some theologians have even called cosmic evolution "Genesis for the Third Millennium."[75] Cosmic evolution is the ultimate master

---

71. Marina Benjamin, *Rocket Dreams: How the Space Age Shaped Our Vision of a World Beyond* (New York, NY: Free Press, 2003). For the full story of the first Earth images from space, see Robert Poole, *Earthrise: How Man First Saw the Earth* (New Haven, CT, and London, U.K.: Yale University Press, 2008).
72. On NASA and the environmental movement, see William H. Lambright, "NASA and the Environment: Science in a Political Context," in *Societal Impact*, ed. Dick and Launius, pp. 313–330; also William H. Lambright, *NASA and the Environment: The Case of Ozone Depletion* (Washington, DC: NASA SP-2005-4538, 2005).
73. David Christian, "Maps of Time": An Introduction to "Big History" (Berkeley, CA: University of California Press, 2004); Fred Spier, *The Structure of Big History: From the Big Bang Until Today* (Amsterdam, Netherlands: Amsterdam University Press, 1996); and most recently, Cynthia Stokes Brown, *Big History: From the Big Bang to the Present* (New York, NY: New Press, 2007).
74. One curriculum, developed by the SETI Institute, the California Academy of Sciences, NASA Ames Research Center, and San Francisco State University, is available on CD-ROM. This and other educational curricula are described and available at *http://www.seti.org/epo/litu-curriculum/*. The Wright Center program on cosmic evolution, directed by Eric Chaisson, is available at *http://www.tufts.edu/as/wright_center/cosmic_evolution/docs/splash.html*.
75. Arthur Peacocke, "The Challenge and Stimulus of the Epic of Evolution to Theology," in *Many Worlds: The New Universe, Extraterrestrial Life and the Theological Implications*, ed. Steven J. Dick (Philadelphia, PA, and London, U.K.: Templeton Foundation Press, 2000), pp. 89–117. Theological aspects of cosmic evolution are also discussed in Ursula Goodenough, *The Sacred Depths of Nature* (Oxford, U.K., and

narrative within which the future of humanity will be played out. The discovery of our place in the universe made possible by studies of cosmic evolution and the search for extraterrestrial life, and the embodiment of these and other themes in literature and the arts, is surely an important effect of space exploration not yet fully realized. Exploration shapes world views and changes cultures in unexpected ways, and so does lack of exploration. The full extent of the intellectual impact of the Space Age remains to be seen.

Economic Impact

The economic impact of spaceflight has been considerable, but it has only begun to be felt. That impact ranges from a far-reaching aerospace industry at one end of the spectrum to the famous (and sometimes literally legendary) "spinoffs" at the other end; it is a part of national and international political economy; and it has sometimes measurable, but often elusive, effects on daily life and commerce. Recent rigorous historical studies suggest the scope of the impact of the Space Age, while emphasizing the complexity and richness of this topic.[76]

Economic impact is also closely related to applications satellites. We now take for granted photographs of weather and Earth resources data from space, as well as navigation and worldwide communications made possible by satellite. Along with human and robotic missions, the late 20th century will be remembered collectively as the time when humans not only saw Earth as a fragile planet against the backdrop of space, but also utilized near-Earth space to study the planet's resources, to provide essential information about weather, and to provide means for navigation that both were life-saving and had enormous economic implications. Worldwide satellite communications brought the world closer together, a factor difficult to estimate from a cost-

New York, NY: Oxford University Press, 1998) and Michael Dowd, *Thank God for Evolution: How the Marriage of Science and Religion Will Transform Your Life and Our World* (New York, NY: Viking, 2007). On the more general societal impact of cosmic evolution, see Dick and Lupisella, eds., *Cosmos and Culture: Cultural Evolution in a Cosmic Context* (Washington, DC: NASA SP-2009-4802, 2009). The modern career of the concept of cosmic evolution began with Harlow Shapley, *Of Stars and Men: The Human Response to an Expanding Universe* (Boston, MA: Beacon Press, 1958); Shapley's role in spreading that idea into broader society is discussed in JoAnn Palmeri, "Bringing Cosmos to Culture: Harlow Shapley and the Uses of Cosmic Evolution," in *Cosmos and Culture*, ed. Dick and Lupisella.

76. On the economic impact of spaceflight, see the "Commercial and Economic Impact" section of Dick and Launius, eds., *Societal Impact*, pp. 212–266. Measuring economic impact is never easy, but see Henry R. Hertzfeld, "Space as an Investment in Economic Growth," in *Exploring the Unknown*, ed. Logsdon, vol. 3, *Using Space*, pp. 385–402 and the following documents.

benefit analysis. Names like Landsat, Geostationary Operational Environmental Satellites (GOES), Intelsat, and GPS may not be household words (though the latter is now becoming one), but they affect humanity in significant ways not always appreciated.[77]

Applications satellites are, in turn, inseparable from environmental issues and national security. Imaging Earth from space and global space surveillance have played an arguably central role in the increasingly heated debate over global climate change and altered the manner in which national security issues are understood and interpreted. Despite political and economic hurdles, monitoring our home planet is likely to be an important and sustained space activity over the next 50 years, with concomitant impact on society.[78]

The greatest economic potential will come after space travel becomes cheaper, opening up new resources on the Moon and in the solar system. There has been no lack of specific proposals for exploiting such resources, especially with regard to the Moon. Senator Harrison Schmitt, the only scientist to fly in the Apollo program (Apollo 17), has argued that the Moon is a resource for the clean generation of fusion energy and for the mining and processing of materials; he also has argued that the Moon is a logical outpost from which more cost-effective exploration of the solar system can take place. For decades, some visionaries have proposed schemes for harnessing solar power, mining asteroids, and exploiting other resources of the solar system. In the far future, some have even proposed large-scale astroengineering projects, such as the Dyson spheres that astronomers have searched for as evidence of advanced extraterrestrial civilizations. While such proposals have been criticized as being impractical, pie-in-the-sky, and in the long-term future, history shows that it is likely only a matter of time before some of them become realities.[79]

The economic impact of the Space Age has been real and significant in certain segments of society over the last 50 years, but it is only a taste of things to come. In a democratic free-market society, once outer space becomes eco-

---

77. Henry Hertzfeld and Ray A. Williamson, "The Social and Economic Impact of Earth Observing Satellites," in *Societal Impact of Spaceflight*, pp. 237–266, and section IV of the same volume, "Applications Satellites, the Environment, and National Security."
78. Dick and Launius, *Societal Impact of Spaceflight*, section IV. For just how politically sensitive the study of global climate change became within NASA in the early 21st century, see Mark Bowen, *Censoring Science: Inside the Political Attack on Dr. James Hansen and the Truth of Global Warming* (New York, NY: Dutton, 2008).
79. Harrison Schmitt, *Return to the Moon: Exploration, Enterprise, and Energy in the Human Settlement of Space* (New York, NY: Praxis Publishing, 2006); Dennis Wingo, *Moonrush: Improving Life on Earth with the Moon's Resources* (Burlington, ON: Apogee Books, 2004).

nomically viable in the marketplace, commercial ventures will find a way into that market. Space tourism is likely to be one of the earliest such ventures.

Geopolitical Impact

The third area of societal impact of spaceflight is geopolitical, and as our discussion of motivations indicated, there is no denying that this aspect has played a central role over the last 50 years. The Moon race between the United States and the Soviet Union was totally driven by geopolitical considerations. Satellite reconnaissance has been an important part, at times even a driver, of national space activities, certainly in the United States, where the space budgets of DOD and NRO far exceed those of NASA. The weaponization and militarization of space are huge issues with immense consequences for the future of both Earth and activities in outer space.[80] Space has become both an instrument of foreign policy and a strategic asset, and the interactions of Russia, China, India, Europe, and the United States in the space arena are likely to be a dominant theme for the next 50 years.[81]

Social Impact

To the intellectual, economic, and political, we may add a fourth domain, that of social impact. Space activities have affected science, math, and engineering education; embodied questions of status, civil rights, and gender among other social issues; and led to the creation of "space states" such as California, Florida, and Texas. Others have demonstrated the complex relation of such space goals to social, racial, and political themes. One such study, De Witt Kilgore's *Astrofuturism: Science, Race, and Visions of Utopia in Space*, examines the work of Wernher von Braun, Willy Ley, Robert Heinlein, Arthur C. Clarke, Gentry Lee, Gerard O'Neill, and Ben Bova, among others, in what he calls the tradition of American astrofuturism.[82] Such studies remind us that, like it or not, the idea of space exploration has been woven into the fabric of society over the last 50 years, even as exploration has raised our cosmic conscious-

---

80. For an overview of national security space and an entrée into its literature, see Stephen Johnson, "The History and Historiography of National Security Space," in *Critical Issues*, ed. Dick and Launius, pp. 481–548. For some of the policy issues, see Glenn Hastedt, "Reconnaissance Satellites, Intelligence and National Security," in *Societal Impact*, ed. Dick and Launius, pp. 369–383. NASA has a long history of interaction with DOD; see Peter Hays, "NASA and the Department of Defense: Enduring Themes in Three Key Areas," in *Critical Issues*, ed. Dick and Launius, pp. 199–238.
81. Joan Johnson-Freese, *Space as a Strategic Asset* (New York, NY: Columbia University Press, 2007).
82. De Witt Douglas Kilgore, *Astrofuturism: Science, Race, and Visions of Utopia in Space* (Philadelphia, PA: University of Pennsylvania Press, 2003).

ness. The historical analysis of that transformation, in ways large and small, should help justify space exploration as an integral part of society rather than a burden on it as sometimes perceived by the public.

Important as they are, the social effects thus far pale in significance compared to what space may represent for the future of humanity. While some argue that robotic spacecraft are cheaper and less risky than human spaceflight, it is most likely that humans will follow robotic reconnaissance as night follows day—perhaps not immediately, but in the long-term future of humanity. Humans will not be content with a space odyssey carried out by robotic surrogates any more than the other great voyages of human history. Robots extend the human senses but will not replace the human mind in the foreseeable future, even with advances in artificial intelligence. HAL in Arthur C. Clarke's famous novel and movie was not as smart as he thought, and he will not be for a long time. As President Bush said in announcing his new initiative in January 2004, humans will spread through the solar system, fulfilling the vision of what British philosopher Olaf Stapledon 55 years ago called "interplanetary man."[83] Eventually humans will spread into the cosmos at large. Space enthusiasts tend to argue that is the nature of humans, with their inbuilt curiosity and penchant for exploration; one might say that it is the very definition of what it is to be human. Not all historians and social scientists agree, however, that the utopian ideal of spreading humanity to outer space is a valid reason for going or that utopia is what we will build when we get there.

There are also more practical reasons for going into space: the survival of the species may depend on the human space program. Specifically, it would seem prudent to remove some of our species from the planet in case of natural or human-induced catastrophe, whether an asteroid impact or nuclear war. In that context, space exploration would seem a small price to pay for survival of the species, as opposed to having to start over from 3.8 billion years of evolution after, for example, a Near-Earth Object impact.

This theme treads dangerously close to "manifest destiny," the belief that spreading a culture, or a species, is part of its destiny, to be attained by any means. Although the concept has been a red flag for historians, who like to recall that Manifest Destiny led to slaughter as Americans spread westward and pushed out Native Americans, the analogy is not a good one. Though *Star*

---

83. Steven J. Dick, "Interstellar Humanity," *Futures: The Journal of Forecasting Planning and Policy* 32 (2000): 555–567; Olaf Stapledon, "Interplanetary Man?," in *An Olaf Stapledon Reader*, ed. Robert Crossley (Syracuse, NY: Syracuse University Press, 1997), pp. 218–241.

Wars makes good entertainment, it is unlikely to become reality as humans spread throughout the solar system. Nor should we a priori shrink from the idea of destiny, though no destiny will be achieved without proper funding.

Indeed, one feature unlikely to be paralleled with the Age of Discovery, or the Second Age of Discovery in the 18th and 19th centuries, is contact with other cultures. Ship crews often included naturalists to study exotic new flora and fauna, and the ultimate experience in the Age of Discovery was contact with exotic human cultures. In the Age of Space, the search for microbial life has been a main driver of space exploration, in particular with regard to Mars, but also now extended to more exotic environments like the Jovian moon Europa. This activity has generated its share of ethical conundrums.[84] And with the search for life on new worlds, planetary protection protocols—sometimes controversial—have been put in place, both for our own planet and for others.[85] Contact with intelligent extraterrestrials beyond the solar system will remain a more remote possibility, and when and if it happens, it is more likely to be radio rather than physical contact. Difficult as they are, such impacts have been studied in some detail at NASA and elsewhere.[86]

As NASA's *Societal Impact of Spaceflight* study shows, unpacking the nature and extent of societal impact is no simple task. "Society" is not monolithic, and "impact" can be an elusive concept. Determining the impact of anything is problematic, especially in the short term, and especially in the hands of academics. If we succeed in the near future in going back to the Moon on a permanent basis, perhaps Columbus may be a good analogy for the Apollo program, and the Age of Discovery a good analogy for the Age of Space; if not, it will have been an abortive attempt more akin to Leif Erickson and the Vikings.

---

84. Steven J. Dick and James E. Strick, *The Living Universe: NASA and the Development of Astrobiology* (New Brunswick, NJ, and London, U.K.: Rutgers University Press, 2004); Constance M. Bertka, ed., *Exploring the Origin, Extent, and Future of Life: Philosophical, Ethical, and Theological Perspectives* (Cambridge, U.K.: Cambridge University Press, 2009).
85. Michael Meltzer, *When Biospheres Collide: A History of NASA's Planetary Protection Programs* (NASA History Series, in press).
86. Just prior to launching its SETI observations in 1993, NASA conducted a series of three Cultural Aspects of SETI (CASETI) workshops, later published as J. Billingham, R. Heyns, D. Milne, et al., *Social Implications of the Detection of an Extraterrestrial Civilization* (Mountain View, CA: SETI Press, 1999). See also Allen Tough, ed., *When SETI Succeeds: The Impact of High-Information Contact* (Bellevue, WA: Foundation for the Future, 2000), and Steven J. Dick, "Consequences of Success in SETI: Lessons from the History of Science," in *Progress in the Search for Extraterrestrial Life*, ed. Seth Shostak (proceedings of Santa Cruz, CA, meeting on SETI, August 1993; San Francisco, CA: ASP Conference series, 1995), pp. 521–532.

Despite the difficulty, analysis of the societal impact of spaceflight is not just an academic exercise. NASA's plans for the next 50 years—multidecade programs to explore the planets, build and operate large space telescopes and space stations, or take humans to the Moon and Mars—require that the public have a vested interest. Whether or not those ambitious space visions of the United States and other countries are fulfilled, the question of societal impact over the past 50 years remains urgent, and it may in fact help fulfill current visions or, at least, raise the level of debate.

In the end, it is difficult to determine how much society has really been affected by spaceflight during its first 50 years because society is composed of individuals, and each individual has been affected in different ways, even when witnessing a transformational event such as the first Moon landing. "The horror of the Twentieth Century," Norman Mailer declared in his account of the first Moon landing, *Of a Fire on the Moon*, "was the size of each new event, and the paucity of its reverberation."[87] The "paucity of reverberation" may reflect a lack of appreciation in the minds of the average citizen about the role space has played, rather than the absolute role itself, which in fact has arguably been very significant. Whether a boon or a burden to society, the impact of space activities is likely to increase over the next 50 years.

Conclusions—Ad Astra?

I do not wish to imply that exploration is the only interpretive framework for the Age of Space. There are real-life, more immediately compelling, and strategic considerations that impel the United States and other countries into space. But in my view, far from being the metaphysical, esoteric, or empty conceit of its critics, exploration is an unchanging, long-term, stimulating, and useful framework for understanding why any country with a claim to greatness must go into space. Moreover, while the analogies discussed here are only suggestive, placing space exploration within the deep history of exploration gives a context to space history that it otherwise might not have, integrating space history into the broader history of humanity and going

87. Norman Mailer, *Of a Fire on the Moon* (New York, NY: Little, Brown and Co., 1969), p. 29.

some way toward eliminating the isolation of space history from other historical subdisciplines.

I do wish to claim that by conquering the third dimension of space—as maritime explorers did in two geographical dimensions during the Age of Discovery, as the 18th- and 19th-century explorers did on both land and sea with improved transportation methods, and as aviation has in the thin skin of our atmosphere during a century of flight—in the long run, the space program has the potential to have an impact that far exceeds any of these advances. Despite historians' qualms about the negative effects of these developments, especially the conquest mode of the Age of Discovery, the Space Age opens a vast new future to humanity, most likely not utopian, but one already imagined in science fiction and, for the first time in history, contemplated in science fact. In contemplating that future, it is well to remember that history need not repeat itself, either in its positive or negative aspects.

The experience of the railroad with which we opened this essay illuminates the Space Age from a different angle and scale. The railroad was, the authors of *The Railroad and the Space Program* concluded, an engine of social revolution that had its greatest impact only 50 years after the start of the railways in America. As a transportation system, the railway had to be competitive with canals and turnpikes, and 20 years after the start of railways in America, more miles of canals were being built than railroads. It was not clear how they could be economically feasible. And though many technological, economic, and managerial hurdles needed to be overcome, railroads are still with us. In the course of the 19th century, they represented human conquest of natural obstacles, with consequences for the human view of nature and our place in it. Secondary consequences often turned out to have greater societal impact than the supposed primary purposes for which they were built. The space program has had, and still has, its technological challenges, and the economic benefits may be even longer term than those of the railroad. But by conquering the third dimension of space, it has the potential to have an exceedingly large impact on the human story, as we expand into the solar system and find our place in the scheme of cosmic evolution.

For its part, the United States has much at stake in the debate over the importance of space exploration. Pulitzer Prize-winning historian William Goetzmann saw the history of the United States as inextricably linked with exploration. "America has indeed been 'exploration's nation,'" he wrote, "a culture of endless possibilities that, in the spirit of both science and its component, exploration, continually looks forward in the direction of the new."[88] The direction of the new is now outer space, and the space exploration debate should accordingly be seen in that context. At the same time, we need to be fully aware that pro-space ideology is often driven by the problematic idea of "progress," an idea with a long history in which Americans are deeply invested. As one scholar concluded, "Given the deep commitment of Americans to ideas about progress, such ideological concerns are as likely to affect policy as any rational assessment of scientific or economic need."[89] Thus historians and the social sciences need to join the discussion about the human future in space.

The analogy of the 15th-century Chinese treasure fleet, commanded by Zheng He, has often been used as a lesson to be learned for those who would withdraw from the Space Age to seek shorter term goals on Earth. It is a matter of historical fact that, from 1405 to 1433, China sent seven massive expeditions into the Indian Ocean and perhaps beyond; the first expedition alone may have included 62 "junks" three or four times larger than Columbus's flagship, 225 support vessels, and 27,000 men. It is also well known that following a maritime tradition stretching back to the 11th century, these ships plied the seas of southeast Asia, and then they sailed to India, the Persian Gulf, the Red Sea, and down the east coast of Africa. And the sudden end of this distant voyaging is indisputable: with changing internal political conditions and the external threat of the Mongols, the fleet was withdrawn in 1433 and its records burned. The subsequent inward turn, it is argued, set China back centuries.[90]

The interpretation begins with the effect of this inward turn. There is no doubt that, although Chinese state revenues were probably 100 times Portugal's, after the 1430s the Ming emperors had other priorities, leaving the Portuguese

---

88. William Goetzmann, *New Lands, New Men: America and the Second Great Age of Discovery* (New York, NY: Penguin Books, 1987).
89. Taylor E. Dark III, "Reclaiming the Future: Space Advocacy and the Idea of Progress," in *Societal Impact of Spaceflight*, ed. Dick and Launius, pp. 555–571. See also Linda Billings, "Overview: Ideology, Advocacy, and Spaceflight—Evolution of a Cultural Narrative," in the same volume, pp. 483–499. On the importance of the idea of progress in Western civilization, see Robert Nisbet, *History of the Idea of Progress* (New York, NY: Basic Books, 1980), and J. B. Bury, *The Idea of Progress: An Inquiry into Its Origin and Growth* (New York, NY: Dover Publications, 1932).
90. Louise Levanthes, *When China Ruled the Seas: The Treasure Fleet of the Dragon Throne, 1405–1433* (Oxford, U.K.: Oxford University Press, 1994); Edward L. Dreyer, *Zheng He: China and the Oceans in the Early Ming Dynasty, 1405–1433* (New York, NY: Pearson Longman, 2007). The voyages of Zheng He have received increasing attention during the 600th anniversary of the voyages and because of the controversial thesis in Gavin Menzies, *1421: The Year China Discovered America* (New York, NY: William Morrow, 2002). Menzies's thesis that the Chinese discovered America seven decades before Columbus is plausible but unproven. For the role of the voyages in the history of exploration, see Felipe Fernandez-Armesto, *Pathfinders: A Global History of Exploration* (New York, NY: W. W. Norton, 2006), pp. 109–117.

and other European countries to lead the way in exploration. As Librarian of Congress and historian Daniel Boorstin noted, "When Europeans were sailing out with enthusiasm and high hopes, landbound China was sealing her borders. Within her physical and intellectual Great Wall, she avoided encounter with the unexpected . . . . Fully equipped with the technology, the intelligence, and the national resources to become discoverers, the Chinese doomed themselves to be discovered." Historians J. R. McNeill and William McNeill came to the same conclusions, and historians in general (even Chinese historians) tend to agree that the Chinese chose poorly in the mid-15th century. By the 1470s, the McNeills wrote, even the skills needed to build great ships were lost; some would draw a parallel to the Saturn V rockets, the last three of which found their rest in museum settings rather than in exploration. Boorstin called the withdrawal of the Chinese into their own borders, symbolized by the Great Wall of China that took its current form at that time, "catastrophic . . . with consequences we still see today."[91]

The lesson of 15th-century China is perhaps not quite so simple, because history is driven by complex factors. Nevertheless, China's maritime withdrawal is certainly one element in its well-documented demise, and it is an undisputed fact that the Chinese are now building a massive reproduction of one of the treasure ships in the ancient Ming shipyard at Nanjing, and they are using it to shape perceptions of China's rise to global prominence after 600 years.[92] It is also an undisputed fact that the Chinese now have a human space program and that they have ambitions to land on the Moon. The question goes to the geopolitical impact mentioned earlier: whether or not the United States decides to return humans to the Moon, the Chinese or another nation will ultimately do so, with real consequences for the global balance of power. History shows that the United States will likely wait until the Chinese act before committing resources to the same end. The ISS notwithstanding, the past 50 years demonstrate that, for the United States, competition trumps cooperation as a national modus operandi for space. The result would again be a Moon race, perhaps this time the key to the rest of the universe. If so, it will be yet another case of not learning the lessons of history.

---

91. Daniel Boorstin, *The Discoverers* (New York, NY: Random House, 1983), esp. pp. 186–201; J. R. McNeill and William H. McNeill, *The Human Web: A Bird's-Eye View of World History* (New York, NY, and London, U.K.: W. W. Norton and Co., 2003), p. 166.
92. Mara Hvistendahl, "Rebuilding a Treasure Ship," *Archaeology* (March/April 2008): 40–45.

Skeptics of the benefits of exploration might well point to the fate of Portugal and Spain, the leaders of the Age of Discovery who eventually lost their leadership. As one historian has pointed out, "the rewards of national strength and wealth proved elusive. Portugal never achieved true great power status. Its population was too small, its commitments too many and its new-found overseas wealth flowed too quickly into foreign hands."[93] Portugal came under the rule of Philip II of Spain in 1581. Spain itself came to dominate Western Europe during the late 16th and early 17th centuries, but the treasures from the New World also proved ephemeral. In a scenario tempting to compare to the present case for the United States, Spain also overcommitted itself and, by the mid-17th century, weakened by the Thirty Years' War, lost its status as a world power.

But the ultimate lesson is not that exploration lacks geopolitical impact. As Norman Augustine, Chair of the Augustine Committee, argued in his report, "Rising Above the Gathering Storm," in the *Report of the Advisory Committee on the Future of the U.S. Space Program,* leadership among nations is not a birthright; it must be earned and reearned.[94] The report showed how already, in 1990, American leadership in science and technology had begun to erode, and it argued that the federal government must urgently address this situation. Surely exploration is an important part of that picture and an important part of national leadership. Each of the ages of exploration in the past was the product of specific decisions of certain cultures: the Europeans (and briefly the Chinese) for the first age, the Europeans and Americans for the second age, and the Soviet Union—soon joined by the United States, then Europe, and other countries—for the third age. As historian Stephen J. Pyne has argued, "Exploration is a specific invention of specific civilizations conducted at specific historical times. It is not . . . a universal property of all human societies. Not

93. Fritze, *New Worlds*, p. 240.
94. "Americans, with only 5% of the world's population but with nearly 30% of the world's wealth, tend to believe that scientific and technological leadership and the high standard of living it underpins is somehow the natural state of affairs. But such good fortune is *not* a birthright. If we wish our children and grandchildren to enjoy the standard of living most Americans have come to expect, there is only one answer: We must get out and *compete*" (Norman Augustine, "Rising Above the Gathering Storm, Energizing and Employing America for a Brighter Economic Future," statement before U.S. House of Representatives Committee on Science, 20 October 2005). The report was published by the National Academy of Sciences in 2007 and is available at *http://history.nasa.gov/augustine/racfup1.htm.* The report's Executive Summary is at *http://history.nasa.gov/augustine/racfup2.htm,* and its main recommendations are summarized at *http://history.nasa.gov/augustine/racfup6.htm.*

all cultures have explored or even traveled widely. Some have been content to exist in xenophobic isolation."[95]

In the end, what does history offer in this great debate? It was the arch-Darwinian T. H. Huxley who said that the great end in life is not knowledge, but action. The importance of our knowledge of history is that it empowers us to act wisely, if cautiously. Not without reason does there exist a National Archives in the United States with the words "What is Past is Prologue" scrolled along the top of its impressive façade, a building whose function is duplicated in all civilized countries of the world. Not without reason did the Columbia Accident Investigation Board devote an entire chapter to history in its official report and conclude that "history is not just a backdrop or a scene-setter, history is cause."[96] Not without reason does the Smithsonian Institution strive to display thoughtful commentary in its exhibits, despite criticism from its wide variety of audiences, each with its own interpretations of history. And not without reason does every high school, college, and university teach history. As Hermann Wouk said in the context of his novel *War and Remembrance*, "the beginning of the end of War lies in Remembrance."[97] For the United States, the beginning and end of the exploration of space lie in remembrance, remembrance of what happens to cultures that have turned too much inward. It would be ironic if, having led the world in space exploration during its first 50 years, the United States squandered that lead during the next 50. Put in a more ecumenical sense, it may be better to cooperate than compete, and it would be an extraordinary lost opportunity if the United States did not lead the international cooperation of space, as it has in the ISS, whose most important product may be a model of cooperation, difficult though it has been at times.

Unfortunately one of the great lessons of history is that we do not learn the lessons of history. As a recent author put it in while contemplating Herodotus's ancient message about intercultural understanding, "it goes unheeded, as it always has and as it always will, because history teaches us that we do not learn from history, that we fight the same wars against the same enemies for the same reasons in different eras, as though time really stood still and history

---

95. Stephen J. Pyne, "The Third Great Age of Discovery," in *Space: Discovery and Exploration*, ed. Martin J. Collins and Sylvia K. Kraemer (Southport, CT: Hugh Lauter Levin Associates, 1994).
96. Columbia Accident Investigation Board, *Report*, vol. 1, p. 195.
97. Herman Wouk, preface to *War and Remembrance*, 1st ed. (New York, NY: Little, Brown and Co., 1978).

itself as moving narrative was nothing but artful illusion."[98] Even in an optimistic frame of mind, in a world in which we might apply lessons learned if only we paid attention, the problem is determining exactly what those lessons are. To give only one recent example, confusion in the political world between any attempt at "negotiation" and Chamberlainian "appeasement" does not inspire confidence in lessons learned, especially where ideologies are at stake. Realizing the difficulties and ambiguities of the task, in closing, I nevertheless offer six macrolessons that should be learned from the first 50 years of the Space Age:

1. Absent an Asimovian "psychohistory" that would allow us to foresee the statistical probabilities of the future, history is not predictive, and it cannot guarantee that exploration (human or robotic) will result in a more creative society.[99] Numerous factors regulate society, which, after all, is composed of individuals more unpredictable than the gas molecules of statistical mechanics. But history, nevertheless, suggests that robust exploration, undertaken by a nation that continually looks forward to the new, enhances its chances of survival as a vibrant society.
2. It is always tempting to sacrifice long-term goals for perceived short-term needs. And it is almost always a bad idea, unless survival is at stake and there is no long term. This is one lesson that the U.S. Congress could particularly take to heart.[100]
3. Long-term goals need to be better understood in the political process. If this were true, we might not throw away a $25 billion investment on launch technology, as the United States did with Apollo, with consequences we are still suffering more than three decades later. As space policy analyst John Logsdon has memorably put it, NASA at 50 is still suffering from NASA at 12.
4. There is never enough funding to do everything. Painful priorities must be set. This seems to be common sense. But NASA has often not set priorities, tried to do too much, and failed to achieve major goals as a consequence.

---

98. Justin Marozzi, *The Way of Herodotus: Travels with the Man Who Invented History* (Cambridge, MA: Da Capo Press, 2008), p. 95. The entire book is a meditation on the lessons of current history through the eyes of Herodotus; see esp. pp. 72–77.
99. In his famous fictional Foundation series beginning in the 1950s, Isaac Asimov postulated "psychohistory," a discipline that used statistics to assess probabilities of future events. While it seems far-fetched in some ways, the 2008 Nobel Laureate in Economics, Paul Krugman, confessed to being influenced by it in his work on economics. See, for example *http://www.technovelgy.com/ct/Science-Fiction-News.asp?NewsNum=1925* (accessed 3 December 2008).
100. Kay, *Democracies.*

5. Human spaceflight will not, and should not, go away. Robotic spaceflight will not, and should not, go away. It is always a question of balancing resources, but in the end, each needs the other, and they should exist in a synergistic relationship. The Hubble Space Telescope servicing missions are the role model here. If in the long term, humans become intelligent robots, the problem of this false dichotomy will disappear.[101]
6. Risk and exploration have always gone hand-in-hand, and they will forever go hand in hand. Safety is a priority, but it is the number-two priority. The number-one priority is to go, to get off the launchpad. Otherwise no explorer would ever have left the ports of Palos, Lisbon, and Sanlúcar de Barrameda. And no rocket would ever have left its launchpad. NASA understands this; the astronauts understand it; but the public does not. Thousands are killed each year on highways, but no one calls for an end to automobiles. A forward-looking nation must take risks.

As we stand at NASA's 50th anniversary and on the verge of a presidential transition—always a perilous time for government agencies—we and our leaders need to remember that (rhetoric notwithstanding) exploration is not a destiny, but a choice. It is a choice that any society must make in the midst of many other priorities. History hints, at least, that those societies that make the wrong choice will suffer the consequences. At the 100th anniversary of NASA in 2058, our descendants will be looking back at the choices we made as the leading agency for exploration in the world, as well as the choices made by the other nations of the world. The choice to explore or not to explore, in the midst of a world perpetually swamped by more pressing problems, is the ultimate challenge to NASA, the nation, and nation states constituting planet Earth. That choice is the proper context embodying the meaning and the essence of the Space Age. The universe awaits the nation, or consortium of nations, willing to take the risks and meet the challenge.

---

101. This is not as far-fetched as it may seem; see Launius and McCurdy, *Robots in Space*.

About the Authors

Laurence Bergreen is an award-winning biographer, historian, and chronicler of exploration. His books have been translated into 18 languages worldwide. In October 2007, Alfred A. Knopf published *Marco Polo: From Venice to Xanadu*, a groundbreaking biography of the iconic traveler. Warner Brothers is developing a feature film based on this book starring Matt Damon and written by William Monahan, who won an Oscar for *The Departed*. His previous work, *Over the Edge of the World: Magellan's Terrifying Circumnavigation of the Globe*, was published to international acclaim by William Morrow/HarperCollins in October 2003. A *New York Times* "Notable Book" for 2003, it is also in development as a motion picture and is now in its 20th printing. In addition, Bergreen is the author of *Voyage to Mars: NASA's Search for Life Beyond Earth*, a narrative of NASA's exploration of Mars, published in November 2000 by Penguin Putnam. Dramatic rights were acquired by TNT. In 1997, Bantam Doubleday Dell published *Louis Armstrong: An Extravagant Life*, a comprehensive biography drawing on unpublished manuscripts and exclusive interviews with Armstrong's colleagues and friends. It appeared on many "Best Books of 1997" lists, including those of the *San Francisco Chronicle*, the *Philadelphia Inquirer*, and *Publishers Weekly*, and it has been published in Germany, Finland, and Great Britain. In 1994, Simon & Schuster published his *Capone: The Man and the Era*. A Book-of-the-Month Club selection, it has been published in numerous foreign languages, was optioned by Miramax, and was a *New York Times* "Notable Book." His biography, *As Thousands Cheer: The Life of Irving Berlin*, appeared in 1990. This book won the Ralph J. Gleason Music Book Award and the ASCAP-Deems Taylor award, and it received front-page reviews in major American and British newspapers and appeared on bestseller lists; it was also a *New York Times* "Notable Book" for 1990. His previous biography, *James Agee: A Life*, was also critically acclaimed and was a *New York Times* "Notable Book" for 1984. His first book was *Look Now, Pay Later: The Rise of Network Broadcasting*, published by Doubleday in 1980. He has written for many national publications including *Esquire, Newsweek, TV Guide, Details, Prologue, the Chicago Tribune*, and *Military History Quarterly*. He has taught at the New School for Social Research and served as an assistant to the president of the Museum of Television and Radio in New York City, New York. In 1995, he served as a judge for the National Book Awards and, in 1991, as a judge for the PEN/Albrand Nonfiction Award. A frequent lecturer at major universities and symposiums, he also serves as a Featured Historian for the History Channel. Mr. Bergreen graduated from Harvard University in 1972. He is a member of PEN American Center, the Explorers Club, the Authors Guild, and the board of the New York Society Library. He lives in New York, New York.

Linda Billings is a communication researcher and policy analyst based in Washington, DC. She currently serves as coordinator of communications for NASA's Astrobiology Program in the Science Mission Directorate, under an Intergovernmental Personnel Agreement with the SETI Institute of Mountain View, California. In this position, Dr. Billings is responsible for managing communications, education, and outreach activities for the astrobiology program. From September 2002 through December 2006, Dr. Billings conducted science and risk communication research for NASA's Planetary Protection Office. From September 1999 through August 2002, she was the director of communications for SPACEHAB Inc., a builder of space habitats. Dr. Billings has three decades of experience in Washington, DC, as a researcher, analyst, and journalist. She was the founding editor of *Space Business News* (1983–85) and the first senior editor for space at *Air & Space/Smithsonian* magazine (1986–88). She was a contributing author for *First Contact: The Search for Extraterrestrial Intelligence* (New York, NY: New American Library, 1990). Her freelance articles have been published in outlets such as the *Chicago Tribune, Washington Post Magazine*, and *Space News*. Dr. Billings was a member of the staff for the National Commission on Space (1985 to 1986). Dr. Billings's expertise is in mass communication, science communication, risk communication, rhetorical analysis, journalism studies, and social studies of science. Her research has focused on the role that journalists play in constructing the cultural authority of scientists and the rhetorical strategies that scientists and journalists employ in communicating about science. She earned her B.A. in social sciences from the State University of New York at Binghamton, her M.A. in international transactions from George Mason University, and her Ph.D. in mass communication from Indiana University's School of Journalism. Dr. Billings served as an officer of Women in Aerospace for 15 years and was president of Women in Aerospace in 2003.

Andrew Butrica earned a Ph.D. in history of technology and science from Iowa State University in 1986. He was a Chercheur Associé at the Center for Research in the History of Science and Technology (Centre de Recherches en Histoire des Sciences et Techniques) at the Cité des Sciences et de l'Industrie in Paris, France, before embarking on a career as a research historian. Among other books, he has written *To See the Unseen*, which won the Leopold Prize of the Organization of American Historians, and *Single Stage to Orbit: Politics, Space Technology, and the Quest for Reusable Rocketry*, a history of the Reagan Revolution and the DC-X experimental rocket, which won the 2005 Michael C. Robinson Prize of the National Council on Public History. More recently, he has written monographs on NASA's role in the manufacture of integrated circuits during the Apollo era and the Agency's contributions to the early history of microelectromechanical systems; he currently is writing a history of NASA's deep space navigation.

Erik Conway is the historian at JPL. His duties include research and writing, conducting oral histories, and contributing to the Lab's historical collections. Before JPL, he worked as a contract historian at LaRC. Conway enjoys studying the historical interaction between national politics, scientific research, and technological change. For his current research in robotic Mars exploration, he analyzes the effects of changing policies on project management and planetary science. His *History of Atmospheric Science at NASA* was published by Johns Hopkins University Press in 2008. He also is a coauthor on two articles on the history of climate science published this year: Naomi Oreskes, Erik M. Conway, and Matthew Shindell, "From Chicken Little to Dr. Pangloss: William Nierenberg, Global Warming, and the Social Deconstruction of Scientific Knowledge," *Historical Studies in the Natural Sciences* 38, no. 1 (February 2008): 113–156; Oreskes and Conway, "Challenging Knowledge: How Climate Science Became a Victim of the Cold War," in *Agnotology: The Making and Unmaking of Ignorance*, ed. Robert N. Proctor and Londa Schiebinger (Palo Alto, CA: Stanford University Press, 2008), pp. 55–89.

David H. DeVorkin is the senior curator for the history of astronomy and the space sciences at NASM. DeVorkin's major research interests are in the origins and development of modern astrophysics during the 20th century and the origins and development of the space sciences from the V-2 to the present. He is the author, editor, or compiler of 9 books and over 100 scholarly and popular articles, most recently, with Robert Smith and Elizabeth Kessler, *The Hubble Space Telescope: New Views of the Universe* (Washington, DC: National Geographic Books, 2004). Earlier works include *Beyond Earth: Mapping the Universe* (Washington, DC: National Geographic Books, 2002); *Henry Norris Russell: Dean of American Astronomers* (Princeton, NJ: Princeton University Press, 2000); *The American Astronomical Society's First Century* (Melville, NY: American Institute of Physics, 1999); *Science with a Vengeance: How the Military Created the U.S. Space Sciences After World War II* (New York, NY: Springer-Verlag, 1992); and *Race to the Stratosphere: Manned Scientific Ballooning in America* (New York, NY: Springer-Verlag, 1989). DeVorkin holds a Ph.D. in the history of astronomy from the University of Leicester (1978) and an M.Phil. in astronomy from Yale University (1970). DeVorkin has curated "Stars" (1983–97); "V-2: The World's First Ballistic Missile System" (1990–); and most recently, "Explore the Universe" (2001–).

Steven J. Dick served as the Chief Historian for NASA and Director of the NASA History Division from 2003 to 2009. He obtained his B.S. in astrophysics (1971) and M.A. and Ph.D. (1977) in history and philosophy of science from Indiana University. He worked as an astronomer and historian of science at the U.S. Naval Observatory in Washington, DC, for 24 years, including 3 years on a mountaintop in New Zealand, before coming to NASA Headquarters in 2003. Among his books are *Plurality of Worlds: The Origins of the Extraterrestrial Life Debate from Democritus to Kant* (Cambridge, U.K.: Cambridge University Press, 1982) (translated into French), *The Biological Universe: The Twentieth Century Extraterrestrial Life Debate and the Limits of Science* (Cambridge, U.K.: Cambridge University Press, 1996), and *Life on Other Worlds: The 20th-Century Extraterrestrial Life Debate* (Cambridge, U.K.: Cambridge University Press, 1998), the latter translated into Chinese, Italian, Czech, Greek, and Polish. His most recent books are (with James Strick) *The Living Universe: NASA and the Development of Astrobiology* (Piscataway, NJ: Rutgers University Press, 2004) and a comprehensive history of the U.S. Naval Observatory, *Sky and Ocean Joined: The U.S. Naval Observatory, 1830–2000* (Cambridge, U.K.: Cambridge University Press, 2003). The latter received the Pendleton Prize of the Society for History in the Federal Government. He also is the editor of *Many Worlds: The New Universe, Extraterrestrial Life, and the Theological Implications* (Radnor, PA: Templeton Foundation Press, 2000) and (with Keith Cowing) *Risk and*

Exploration: Earth, Sea and the Stars (Washington, DC: NASA SP-2005-4701, 2005). His latest works are edited volumes (with Roger Launius) on *Critical Issues in the History of Spaceflight* (Washington, DC: NASA SP-2006-4702, 2006) and *Societal Impact of Spaceflight* (Washington, DC: NASA SP-2007-4801, 2007). Dr. Dick is the recipient of the Navy Meritorious Civilian Service Medal, the NASA Exceptional Service Medal, the NASA Group Achievement Award for his role in NASA's multidisciplinary program in astrobiology, and the 2006 LeRoy E. Doggett Prize for Historical Astronomy of the American Astronomical Society. He has served as chairman of the Historical Astronomy Division of the American Astronomical Society, as president of the History of Astronomy Commission of the International Astronomical Union, and as president of the Philosophical Society of Washington. He is a corresponding member of the International Academy of Astronautics. Minor planet 6544 Stevendick is named in his honor.

Robert Ferguson is a Washington, DC-based historian of technology. On behalf of the NASA History Division, he is completing a manuscript on the history of NASA's aeronautics research. Dr. Ferguson is a graduate of the University of Minnesota Program in the History of Science and Technology. He taught at the Hong Kong University of Science and Technology and was awarded the university-wide medal for distinguished teaching. He was the 2003 Ramsey Fellow in Naval Aviation History at NASM. His other principal area of research is World War II American aircraft manufacturing.

James R. Fleming is a professor of science, technology, and society at Colby College. He earned a Ph.D. in history from Princeton University, with earlier degrees in astronomy (B.S., Pennsylvania State University) and atmospheric science (M.S., Colorado State University). He worked in climate modeling and airborne observations, held a fellowship with the Joseph Henry Papers at the Smithsonian Institution, and served as the historian of the American Meteorological Society. Since coming to Colby College, Professor Fleming has held National Endowment for the Humanities, National Science Foundation, Smithsonian (Lindbergh), and American Association for the Advancement of Science (Revelle) fellowships, and he was appointed a visiting scholar at MIT (1992), Harvard (1999), NASM (2005), and the Woodrow Wilson International Center for Scholars (2006). He was elected a fellow of the American Association for the Advancement of Science "for pioneering studies on the history of meteorology and climate change and for the advancement of historical work within meteorological societies" and is the founder and first president of the International

Commission on History of Meteorology. Professor Fleming recently served on the National Research Council's Committee on Scientific Accomplishments of Earth Observations from Space and as chair of the American Association for the Advancement of Science section on history and philosophy of science. He has written or edited more than a dozen books, including *Meteorology in America, 1800–1870* (Baltimore, MD: Johns Hopkins University Press, 1990), *Historical Perspectives on Climate Change* (New York, NY: Oxford University Press, 1998), and *The Callendar Effect: the Life and Work of Guy Stewart Callendar* (Boston, MA: American Meteorological Society, 2007).

Edward Goldstein served as the lead writer at NASA Headquarters. For over six years, he has written speeches, testimony, and other communications for NASA's senior leaders. He also was project manager for the commemorative publication *NASA: 50 Years of Exploration and Discovery* and for organizing the NASA speaker panels at the 2008 Smithsonian Folklife Festival. Dr. Goldstein received his Ph.D. in public administration from the George Washington University in January 2007 for the dissertation, "NASA's Earth Science Program: The Bureaucratic Struggles of the Space Agency's Mission to Planet Earth." From 2007, he has served as an adjunct faculty member teaching about the environment, politics, and public policy at American University, the Catholic University of America, and Georgetown University. In the administration of President George H. W. Bush, Dr. Goldstein was the deputy associate director for energy, environmental, and natural resources policy in the White House Office of Domestic and Economic Policy.

Michael Griffin served as the NASA Administrator from 2005 until 20 January 2009. Prior to being nominated as NASA Administrator, he served as the Space Department Head at the Johns Hopkins University's Applied Physics Laboratory in Laurel, Maryland. He was previously president and chief operating officer of In-Q-Tel, Inc., and also served in several positions within Orbital Sciences Corporation, Dulles, Virginia, including as the CEO of Orbital's Magellan Systems division and general manager of the Space Systems Group. Earlier in his career, Griffin served as Chief Engineer and as Associate Administrator for exploration at NASA, and as deputy for technology at the Strategic Defense Initiative Organization. He has been an adjunct professor at the University of Maryland, the Johns Hopkins University, and the George Washington University, where he taught courses in spacecraft design, applied mathematics, guidance and navigation, compressible flow, CFD, spacecraft attitude control, astrodynamics, and introductory aerospace engineering. He is the lead author of more than two dozen technical papers, as well as the textbook *Space Vehicle Design*.

A registered professional engineer in Maryland and California, Dr. Griffin is a member of the National Academy of Engineering and the International Academy of Astronautics, an honorary fellow of AIAA, a fellow of the American Astronautical Society, and a senior member of the Institute of Electrical and Electronic Engineers. He is a recipient of the NASA Exceptional Achievement Medal, the AIAA Space Systems Medal, and the DOD Distinguished Public Service Medal, the highest award given to a nongovernment employee. Griffin received a bachelor's degree in physics from the Johns Hopkins University, a master's degree in aerospace science from the Catholic University of America, a Ph.D. in aerospace engineering from the University of Maryland, a master's degree in electrical engineering from the University of Southern California, a master's degree in applied physics from the Johns Hopkins University, a master's degree in business administration from Loyola College, and a master's degree in civil engineering from the George Washington University. He is a certified flight instructor with instrument and multiengine ratings.

Richard P. Hallion is the 2007–08 Alfred Verville Fellow, Department of Aeronautics, NASM, Smithsonian Institution, Washington, DC. He was formerly Senior Adviser for Air and Space Issues, Directorate for Security, Counterintelligence and Special Programs Oversight, Office of the Secretary of the Air Force, the Pentagon, Washington, DC, before retiring in November 2006. Dr. Hallion has broad experience in science and technology policy, research, development, and management analysis, and he has served as a consultant to various professional organizations. He has flown as a mission observer (not pilot) in a wide range of high-performance military and civilian fixed- and rotary-wing aircraft. Dr. Hallion is the author and editor of numerous books, articles, and essays on aerospace technology and military operations, and he consults, teaches, and lectures widely.

J. D. Hunley, a former NASA Headquarters and Air Force historian, retired in 2001 as the first chief historian at DRFC. From 2001 to 2002, he was the Ramsey Fellow at NASM. He has written or edited a large number of books and articles about various aspects of aerospace history. Most recently, he has published three books about the history of rocketry: *The Development of Propulsion Technology for U.S. Space-Launch Vehicles, 1926–1991* (College Station, TX: Texas A&M University Press, 2007); *Preludes to U.S. Space-Launch Vehicle Technology: Goddard Rockets to Minuteman III* (Gainesville, FL: University Press of Florida, 2008); and *U.S. Space-Launch Vehicle Technology: Viking to Space Shuttle* (Gainesville, FL: University Press of Florida, 2008). The last two volumes received the AIAA 2010 Gardner-Lasser Aerospace History Literature Award.

Stephen Johnson is an associate research professor with the Institute for Science and Space Studies at the University of Colorado at Colorado Springs, and he also is a health management systems engineer for the Advanced Sensors and System Health Management Branch of MSFC. He was a faculty member at the University of North Dakota Department of Space Studies from 1997 to 2005, teaching military space, space history, and management and economics of space endeavors. He is the author of *The United States Air Force and the Culture of Innovation, 1945–1965* and *The Secret of Apollo: Systems Management in American and European Space Programs*, both published in 2002. He also was the editor of *Quest: The History of Spaceflight Quarterly* from 1998 to 2005, and he is currently the general editor for a two-volume encyclopedia of space history to be published in 2009 by ABC-CLIO, *Space Exploration and Humanity: A Historical Encyclopedia*. He currently works on the Ares I Crew Launch Vehicle project of the Constellation program. His current research involves dependable space system design and operations, space industry management and economics, the history of space science and technology, and the history of cognitive psychology and artificial intelligence. He received his bachelor's degree in physics from Whitman College in 1981 and his doctorate in the history of science and technology from the University of Minnesota in 1997, where he was also the associate director of the Babbage Institute for the History of Computing. Prior to 1997, he worked for Northrop and Martin Marietta, and he was co-owner of his own small business, managing computer simulation laboratories, designing space probes, and developing engineering processes.

John Krige is the Kranzberg Professor in the School of History, Technology, and Society at the Georgia Institute of Technology in Atlanta, Georgia. His main area of research concerns the place of science and technology in U.S. foreign policy in Western Europe after World War II. He is currently the Principal Investigator on a NASA-funded project to write the history of NASA's international relations in space.

W. Henry "Harry" Lambright is a professor of public administration and political science at the Maxwell School of Syracuse University. He is the author or editor of seven books and over 275 articles, papers, and reports. His books include a biography, *Powering Apollo: James E. Webb of NASA* (Baltimore, MD: Johns Hopkins University Press, 1995), and *Space Policy in the 21st Century* (Baltimore, MD: Johns Hopkins University Press, 2003). A longstanding student of leadership and change in government, he has received support from IBM for monographs on other NASA Administrators: *Transforming NASA: Dan Goldin and the Remaking of NASA* (Washington, DC: IBM, 2001), *Executive Response to Changing Fortune: Sean O'Keefe as NASA Administrator* (Washington, DC: IBM, 2005), and *Launching a New Mission: Michael Griffin and NASA's Return to the Moon* (Washington, DC: IBM, 2009). IBM is supporting research on NASA Administrator Michael Griffin and his effort to implement the Moon-Mars decision of 2004. Dr. Lambright is also currently researching a book under NASA sponsorship on Mars exploration.

Dr. Lambright has served as a guest scholar at the Brookings Institution, the director of the Science and Technology Policy Center at the Syracuse Research Corporation, and the director of the Center for Environmental Policy and Administration at the Maxwell School of Syracuse University. He also has served as an adjunct professor in the Graduate Program of Environmental Science in the College of Environmental Science and Forestry at the State University of New York. Dr. Lambright's interest in NASA and space policy goes back to his years as a graduate student. Early in his career, he served as a special assistant at NASA, working in the Office of University Affairs, and also writing speeches for NASA Administrator Tom Paine. The recipient of a range of grants from federal and private organizations for his research in science and technology policy, he is frequently cited in the media. He teaches courses in science, technology, public policy, and energy and environmental policy. He holds a bachelor's degree from the Johns Hopkins University, and he holds master's and Ph.D. degrees from Columbia University.

John M. Logsdon was, until mid-2008, the director of the Space Policy Institute at the George Washington University's Elliott School of International Affairs, where he is also Professor Emeritus of Political Science and International Affairs. In September 2008, Dr. Logsdon became the Lindbergh Chair in Aerospace History for 2008–09 at NASM. He holds a B.S. in physics from Xavier University (1960) and a Ph.D. in political science from New York University (1970). Dr. Logsdon's research interests focus on the policy and historical aspects of U.S. and international space activities. Dr. Logsdon is the author of *The Decision to Go to the Moon: Project Apollo and the National Interest* (Chicago, IL: University of Chicago Press, 1976) and is general editor of the eight-volume series *Exploring the Unknown: Selected Documents in the History of the U.S.*

Civil Space Program (Washington, DC: NASA SP-4407 series). He has written numerous articles and reports on space policy and history. He is frequently consulted by electronic and print media for his views on space issues.

Dr. Logsdon is a member of the NASA Advisory Council and of the Commercial Space Transportation Advisory Committee of the Department of Transportation. In 2003, he served as a member of the Columbia Accident Investigation Board. He is a recipient of the NASA Distinguished Public Service and Public Service Medals, the 2005 John F. Kennedy Award from the American Astronautical Society, and the 2006 Barry Goldwater Space Educator Award of the AIAA. He is a Fellow of the AIAA and the American Association for the Advancement of Science. He is a member of the International Academy of Astronautics.

Maura Mackowski is currently working on a history of NASA life sciences research from 1980 to 2005, under contract to the NASA Headquarters History Division. She is the author of *Testing the Limits: Aviation Medicine and the Origins of Manned Space Flight* (College Station, TX: Texas A&M University Press, 2006), which won an Honorable Mention for the 2006 Eugene Emme Astronautical Literature Award from the American Astronautical Society. Before earning her Ph.D. in modern U.S. history at Arizona State University, she was a freelance writer covering science, technology, medicine, and high-tech business for publications in the United States and Europe. She currently resides in Gilbert, Arizona.

Robert R. MacGregor is currently a graduate student in the history of science program at Princeton University. Before coming to Princeton, he studied at Rice University in Houston, Texas, where he received a B.S. in chemical physics and a B.A. in history. Robert also has studied at Moscow State University in Russia where he studied Russian language, history, and culture. His current work focuses on the processes in the U.S. government that led to the formation of NASA between the launch of Sputnik in October 1957 and the signing into law of the National Air and Space Act in July 1958. In the future, he plans to delve into the history of the Soviet space program and the early amateur rocket societies in Germany, the United States, and the Soviet Union.

Howard McCurdy is a professor in the School of Public Affairs at American University in Washington, DC, and chair of its Department of Public Administration and Policy. Author or coauthor of seven books on the U.S. space program, he is known for *Space and the American Imagination* (Washington, DC: Smithsonian Institution, 1999), winner of the Eugene M. Emme Astronautical

Literature Award, and *Inside NASA: High Technology and Organizational Change in the U.S. Space Program* (Baltimore, MD: Johns Hopkins University Press, 1994), a study of NASA's organizational culture that received the Henry Adams Prize for that year's best history on the federal government. He recently authored *Faster, Better, Cheaper: Low-Cost Innovation in the U.S. Space Program* (Baltimore, MD: Johns Hopkins University Press, 2003), a critical analysis of cost-cutting initiatives in the U.S. space program, and he just finished *Robots in Space: Technology, Evolution, and Interplanetary Travel* (Baltimore, MD: Johns Hopkins University Press, 2008), coauthored with Roger Launius, which examines the continuing controversy between advocates of human and robotic spaceflight. Dr. McCurdy is often consulted by the media on space policy issues and has appeared on national news outlets such as the Jim Lehrer News Hour, National Public Radio, and NBC Nightly News. He received his bachelor's and master's degrees from the University of Washington and his doctorate from Cornell University.

Michael Meltzer is an environmental scientist who has been writing about science and technology for 30 years. His books and articles have investigated topics that include NASA expeditions to Jupiter and Saturn, planetary environmental protection, solar house design, industrial pollution prevention, and the history of U.S. commercial fishing. He also has published two science fiction stories with environmental themes. Michael worked for 15 years at Lawrence Livermore National Laboratory, where he helped start a pollution prevention program. He lives in Oakland, California, with his wife, Naisa, and seven-year-old daughter, Jordana.

Michael J. Neufeld is the chair of the Space History Division of NASM in Washington, DC. Born in Canada, he received history degrees from the University of Calgary and the University of British Columbia before getting a Ph.D. in modern European history from the Johns Hopkins University in 1984. Before Dr. Neufeld came to NASM in 1988 as the A. Verville Fellow, he taught at various universities in upstate New York. From 1989 to 1990, he held Smithsonian and National Science Foundation fellowships at NASM. In 1990, he was hired as a museum curator in the Aeronautics Division, where he remained until early 1999. After transferring to the Space History Division, he took over the collection of German World War II missiles, and from 2003 to 2007, he ran the collection of Mercury and Gemini spacecraft and components. In fall 2001, he was a senior lecturer at the Johns Hopkins University. He was named chair of Space History in January 2007.

In addition to authoring numerous scholarly articles, Dr. Neufeld has written three books: *The Skilled Metalworkers of Nuremberg: Craft and Class in the Industrial Revolution* (Piscataway, NJ: Rutgers University Press, 1989); *The Rocket and the Reich: Peenemünde and the Coming of the Ballistic Missile Era* (Cambridge, MA: Harvard University Press, 1995), which won two book prizes; and *Von Braun: Dreamer of Space, Engineer of War* (New York, NY: Alfred A. Knopf, 2007). He has also edited Yves Béon's memoir, *Planet Dora* (1997), and he is the coeditor of *The Bombing of Auschwitz: Should the Allies Have Attempted It?* (New York, NY: St. Martin's Press, 2000).

Anthony M. Springer is the lead for communications and education in aeronautics research at NASA Headquarters. At NASA, he is responsible for all communications and education activities related to aeronautics for the Agency. Previously he served as the alliance development manager for the Office of Aerospace Technology, where he was responsible for the development and coordination of strategic alliances between NASA and its stakeholders including industry, academia, other agencies, and the public. He has also served as the director for all NASA Centennial of Flight activities and on the National Centennial of Flight Commission History and Education Panel. He has been the NASA resident manager for the X-34 project at Orbital Sciences Corporation in Dulles, Virginia, where he was responsible for the day-to-day activities and contractor interfaces with the contractor designing and fabricating the X-34 for NASA. Mr. Springer has also held positions as a project and test engineer at MSFC in Huntsville, Alabama. At MSFC, he conducted wind tunnel tests and developed the preliminary aerodynamics for numerous future launch vehicle concepts. During that time, Mr. Springer did research on the application of rapid prototyping methods to high-speed wind tunnel testing and other topics related to advanced applications for NASA's wind tunnels. In addition, Mr. Springer served as the lab lead engineer on the Bantam Launch Vehicle program, a test engineer for solid rocket motor airflow testing, and a diver in MSFC's Neutral Buoyancy Simulator, among other tasks. His started his career with NASA as a cooperative education student at MSFC in the late 1980s. Mr. Springer has published many technical and historical publications and received numerous NASA awards over his career. He is a graduate of the University of Illinois with a B.S. in aeronautical and astronautical engineering, and he holds an A.A. from Kankakee Community College. He is a fellow of the AIAA, and he is married and lives outside Washington, DC, with his wife, Emily, and son, Oliver.

Joseph N. Tatarewicz is an associate professor in the Department of History and the director of the Human Context of Science and Technology Certificate program at the University of Maryland, Baltimore County. He holds M.A. degrees in philosophy (Catholic University of America, 1976) and history and philosophy of science (Indiana University, 1980), as well as a Ph.D. in history and philosophy of science (Indiana University, 1984). He is the author of *Space Technology and Planetary Astronomy* (Bloomington, IN: Indiana University Press, 1990), a contributor to *The Space Telescope: A Study of NASA, Science, Technology, and Politics* (New York, NY: Cambridge University Press, 1989 and 1993), and has authored numerous articles and reviews for professional journals and publications in the history of science and technology. His book-length, NASA-sponsored scholarly history, *Exploring the Solar System: the Planetary Sciences Since Galileo*, was awarded the AIAA History Manuscript Award and will be published by Johns Hopkins University Press. He began his career in space history as a NASA History Office intern (1980), followed by predoctoral and postdoctoral Guggenheim fellowships at the Smithsonian Institution's NASM, where he later served in curatorial and administrative positions. He has served as the associate director and curator of the Center for the History of Electrical Engineering and as a consulting historian for various government agencies.

Acronyms and Abbreviations

AAAF	Association Aéronautique et Astronautique de France
AAF	Army Air Forces
AAOE	Airborne Antarctic Ozone Experiment
AAS	Advanced Airways System
ABL	Allegany Ballistics Laboratory
ABMA	Army Ballistic Missile Agency
ACE	Advanced Composition Explorer
ACEE	Aircraft Energy Efficiency
ACS	Advanced Camera for Surveys
AEC	Atomic Energy Commission
AFCRC	Air Force Cambridge Research Center
AFFDL	Air Force Flight Dynamics Laboratory
AFFTC	Air Force Flight Test Center
AFHRA	Air Force Historical Research Agency
AGARD	Advisory Group for Aerospace Research and Development
AIAA	American Institute of Aeronautics and Astronautics
ALSEP	Apollo Lunar Surface Experiments Package
AMSR-E	Advanced Microwave Scanning Radiometer-EOS
APL	Applied Physics Laboratory
ARC	Ames Research Center
ARPA	Advanced Research Projects Agency
ARRMD	Affordable Rapid Response Missile Demonstrator
ARTCC	Air Route Traffic Control Center
ASCENDS	Active Sensing of CO2 Emissions over Nights, Days, and Seasons

ASIC	application-specific integrated circuit
ASSET	Aerothermo-Structural Systems-Environmental Tests
ASTP	Apollo-Soyuz Test Project
ATC	air traffic control
ATMOS	Atmospheric Trace Molecules Spectroscopy
ATS	Applications Technology Satellite

BESS	Biomedical Experiments Scientific Satellite
BHT	Bell Helicopter Textron
BST	Behavioral Science Technology

Caltech	California Institute of Technology
CASETI	Cultural Aspects of SETI
CASTL	Cassini Technical Library
CCD	charge-coupled device
CDC	Center for Disease Control and Prevention
CEES	Committee on Earth and Environmental Sciences
CELSS	Closed Environmental Life Support Systems
CEO	chief executive officer
CES	Committee on Earth Sciences
CfA	Center for Astrophysics
CFD	Computational Fluid Dynamics
CFES	Continuous Flow Electrophoresis System
CIA	Central Intelligence Agency
CIS	Commonwealth of Independent States
CLARREO	Climate Absolute Radiance and Refractivity Observatory
CO	carbon monoxide
COBE	Cosmic Background Explorer
COPUOS	Committee on the Peaceful Uses of Outer Space
COSPAR	Committee on Space Research
CSSC	Communications Support Services Center
CTAS	Center TRACON Automation System

ΔDOR	delta-Differenced One-Way Range
DA	Descent Advisor
DESDynI	Deformation, Ecosystem Structure, and Dynamics of Ice

DFRC	Dryden Flight Research Center
DOD	Department of Defense
DOT	Department of Transportation
DOR	Differenced One-Way Range
DRVID	Differenced Range Versus Integrated Doppler

EELV	Evolved Expendable Launch Vehicle
ELVs	Expendable Launch Vehicles
ENSO	El Niño Southern Oscillation
EOS	Earth Observing System
EOSDIS	Earth Observing System Data and Information System
ERBS	Earth Radiation Budget Satellite
ESA	European Space Agency
ESO	European Southern Observatory
EUI	European University Institute
EVA	extravehicular activity

FAA	Federal Aviation Administration
FACET	Future Air traffic management Concepts Evaluation Tool
FAST	Final Approach Spacing Tool
FIRE	Flight Investigation Reentry Environment
FLEXSTAB	Flexible Airplane Analysis Computer System
FRC	Flight Research Center
FRR	Flight Readiness Review
FUSE	Far Ultraviolet Spectroscopic Explorer

GACM	Global Atmospheric Composition Mission
GALCIT	Guggenheim Aeronautical Laboratory at the California Institute of Technology
GALEX	Galaxy Evolution Explorer
GAO	General Accounting Office
GASL	General Applied Sciences Laboratory
GD/A	General Dynamics/Astronautics
GEO-CAPE	Geostationary Coastal and Air Pollution Events
GISS	Goddard Institute of Space Studies
GOES	Geostationary Operational Environmental Satellites

GPO	Government Printing Office
GPS	Global Positioning System
GRACE	Gravity Recovery and Climate Experiment
GSBC	Gordon S. Black Corporation
GSFC	Goddard Space Flight Center
GTO	geosynchronous transfer orbit
GWE	Global Weather Experiment

HEAO	High-Energy Astronomical Observatory
HEIC	Hubble European Space Agency Information Centre
HMSO	Her Majesty's Stationery Office
HROs	high-reliability organizations
HSFS	High-Speed Flight Station
HTPB	hydroxyl-terminated polybutadiene
HUA	Harvard University Archives
HyspIRI	Hyperspectral Infrared Imager
HYWARDS	Hypersonic Weapon and Research and Development System

---

IAA	International Academy of Astronautics
IAS	Institute of the Aeronautical Sciences
ICASE	Institute for Computer Applications to Science and Engineering
ICBM	intercontinental ballistic missile
ICESat	Ice, Cloud, and land Elevation Satellite
IGY	International Geophysical Year
IMBP	Institute for Medical and Biological Problems
INMARSAT	International Maritime Satellite Organization
IOD	Industrial Operations Directorate
IRAS	Infrared Astronomical Satellite
ISF	Industrial Space Facility
ISS	International Space Station
ITA	Independent Technical Authority
ITAR	International Traffic in Arms Regulations
IU	instrument unit
IUE	International Ultraviolet Explorer

JPL	Jet Propulsion Laboratory
JSC	Johnson Space Center
KSC	Kennedy Space Center
LaRC	Langley Research Center
L/D	lift-to-drag ratio
LGM 1	Little Green Men 1
LIST	Lidar Surface Topography
LO	low observable
MAWD	Mars Atmospheric Water Detector
MERs	Mars Exploration Rovers
MESSENGER	MErcury Surface, Space ENvironment, GEochemistry, and Ranging
MISR	Multi-angle Imaging SpectroRadiometer
MISS	Man-in-Space-Soonest
MIT	Massachusetts Institute of Technology
MOC	Mars Orbiter Camera
MODIS	Moderate Resolution Imaging Spectroradiometer
MOL	Manned Orbital Laboratory
MOLA	Mars Orbiter Laser Altimeter
MOPITT	Measurement of Pollution in the Troposphere
MPH	miles per hour
MSC	MacNeal-Schwedier Corporation
MSFC	Marshall Space Flight Center
MTPE	Mission to Planet Earth
NAA	North American Aviation
NACA	National Advisory Committee for Aeronautics
NASA	National Aeronautics and Space Administration
NASM	National Air and Space Museum
NASP	National Aero-Space Plane Program
NASTRAN	NASA Structural Analysis

NATO	North Atlantic Treaty Organization
NavTechMisEu	Naval Technical Mission to Europe
NEAR	Near Earth Asteroid Rendezvous
NIH	National Institutes of Health
NOAA	National Oceanic and Atmospheric Administration
NOZE	National Ozone Experiment
NPD	NASA Policy Directive
NPOESS	National Polar-orbiting Operational Environmental Satellite System
NRAO	National Radio Astronomy Observatory
NRL	Naval Research Laboratory
NRO	National Reconnaissance Office
NSB	National Science Board
NSBRI	National Space Biomedical Research Institute
NSC	National Space Council
NSCORTs	National Specialized Centers of Research and Training
NSF	National Science Foundation
NSPD	National Security Presidential Directive
NSSDC	National Space Science Data Center

OAO	Orbiting Astronomical Observatory
OCO	Orbiting Carbon Observatory
OECD	Organization for Economic Cooperation and Development
OHI	oral history interview
OMB	Office of Management and Budget
OMSF	Office of Manned Space Flight
ONIPS	Optical Navigation Image Processing System
ONP	Optical Navigation Program
ONR	Office of Naval Research
OPEC	Organization of Petroleum Exporting Countries
OSO	Orbiting Solar Observatory
OSSA	Office of Space Science and Applications

PAM	Payload Assist Module
PAO	Public Affairs Office
PATH	Precipitation and All-weather Temperature and Humidity
PBMA	Process-Based Mission Assurance

PETA	People for the Ethical Treatment of Animals
P.L.	Public Law
PRIME	Precision Recovery Including Maneuvering Entry
PSAC	President's Scientific Advisory Committee

QuikSCAT	Quick Scatterometer

R&D	research and development
RAF	Royal Air Force
RAPP	Research Airplane Projects Panel
RAS	Royal Astronomical Society
RAVE	Research on Volcanic Eruptions
RCS	radar cross-section
RFPs	Requests for Proposals
RM	Research Memorandum
RMS	Remote Manipulator System
ROSAT	Röntgen Satellite
ROTC	Reserve Officer Training Corps
RTG	radioisotope thermoelectric generator
RTOP	Research and Technology Objectives and Plans

SAB	Scientific Advisory Board
SAC	Science Advisory Committee
SAE	Society of Automotive Engineers
SAG	Scientific Advisory Group
SAGE	Semi-Automatic Ground Equipment
SAL	Space Astronomy Laboratory
SAO	Smithsonian Astrophysical Observatory
SAOHP	Space Astronomy Oral History Project
SAO/SIA	Smithsonian Astrophysical Observatory, Smithsonian Institution Archives
SCLP	Snow and Cold Land Processes
SCRJ	Supersonic Combustion Ram-Jet
SCW	supercritical wing
SDI	Space Defense Initiative
SeaWiFS	Sea-viewing Wide Field-of-view Sensor

SEI	Space Exploration Initiative
SETI	Search for Extraterrestrial Intelligence
SHMA/AIP	Sources for History of Modern Astronomy, American Institute of Physics
SITE	Satellite Instructional Television Experiment
SMAP	Soil Moisture Active and Passive
SMC/HO	Space and Missile Systems Center History Office
SMS	Surface Management System
SOHO	Solar and Heliospheric Observatory
SPI	Space Policy Institute
SPOT	Satellite Pour l'Observation de la Terre
SRBs	solid rocket boosters
SRMs	solid rocket motors
SSB	Space Science Board
SSBE	Shaped Sonic Boom Experiment
SSC	Stennis Space Center
SSME	Space Shuttle main engine
SSTO	single stage to orbit
STG	Space Task Group
STOL	short takeoff and landing
STS	Space Transportation System
STScI	Space Telescope Science Institute
SWOT	Surface Water Ocean Topography

TACT	Transonic Aircraft Technology
TCP	Technological Capabilities Panel
TIROS	Television Infrared Observation Satellite
TMA	Traffic Management Advisor
TOMS	Total Ozone Mapping Spectrometer
TOPEX	Topography Experiment
TQM	Total Quality Management
TR	Technical Report
TRACON	Terminal Radar Approach Control
TRMM	Tropical Rainfall Measuring Mission
TSTO	two stage to orbit
TVA	Tennessee Valley Authority

UARS	Upper Atmosphere Research Satellite
UCAC	U.S. Naval Observatory CCD Astrograph Catalog
UNESCO	United Nations Educational, Scientific and Cultural Organization
USCCRI	U.S. Climate Change Research Initiative
USGCRP	U.S. Global Change Research Program
USNC	United States National Committee
USSR	Union of Soviet Socialist Republics
USSRC	U.S. Space and Rocket Center

VAB	Vehicle Assembly Building
VDT	Variable Density Tunnel
VLBI	Very Long Baseline Interferometry
VSTOL	vertical short takeoff and landing
VTOL	vertical takeoff and landing

WASPs	Women's Airforce Service Pilots
WEP	Wisconsin Experiment Package
WMAP	Wilkinson Microwave Anisotropy Probe

XMM	X-ray Multi-Mirror Mission

The NASA History Series

REFERENCE WORKS, NASA SP-4000:

Grimwood, James M. *Project Mercury: A Chronology*. NASA SP-4001, 1963.

Grimwood, James M., and Barton C. Hacker, with Peter J. Vorzimmer. *Project Gemini Technology and Operations: A Chronology*. NASA SP-4002, 1969.

Link, Mae Mills. *Space Medicine in Project Mercury*. NASA SP-4003, 1965.

Astronautics and Aeronautics, 1963: Chronology of Science, Technology, and Policy. NASA SP-4004, 1964.

Astronautics and Aeronautics, 1964: Chronology of Science, Technology, and Policy. NASA SP-4005, 1965.

Astronautics and Aeronautics, 1965: Chronology of Science, Technology, and Policy. NASA SP-4006, 1966.

Astronautics and Aeronautics, 1966: Chronology of Science, Technology, and Policy. NASA SP-4007, 1967.

Astronautics and Aeronautics, 1967: Chronology of Science, Technology, and Policy. NASA SP-4008, 1968.

Ertel, Ivan D., and Mary Louise Morse. *The Apollo Spacecraft: A Chronology, Volume I, Through November 7, 1962*. NASA SP-4009, 1969.

Morse, Mary Louise, and Jean Kernahan Bays. *The Apollo Spacecraft: A Chronology, Volume II, November 8, 1962–September 30, 1964*. NASA SP-4009, 1973.

Brooks, Courtney G., and Ivan D. Ertel. *The Apollo Spacecraft: A Chronology, Volume III, October 1, 1964–January 20, 1966*. NASA SP-4009, 1973.

Ertel, Ivan D., and Roland W. Newkirk, with Courtney G. Brooks. *The Apollo Spacecraft: A Chronology, Volume IV, January 21, 1966–July 13, 1974*. NASA SP-4009, 1978.

Astronautics and Aeronautics, 1968: Chronology of Science, Technology, and Policy. NASA SP-4010, 1969.

Newkirk, Roland W., and Ivan D. Ertel, with Courtney G. Brooks. *Skylab: A Chronology*. NASA SP-4011, 1977.

Van Nimmen, Jane, and Leonard C. Bruno, with Robert L. Rosholt. *NASA Historical Data Book, Volume I: NASA Resources, 1958–1968*. NASA SP-4012, 1976, rep. ed. 1988.

Ezell, Linda Neuman. *NASA Historical Data Book, Volume II: Programs and Projects, 1958–1968*. NASA SP-4012, 1988.

Ezell, Linda Neuman. *NASA Historical Data Book, Volume III: Programs and Projects, 1969–1978*. NASA SP-4012, 1988.

Gawdiak, Ihor, with Helen Fedor. *NASA Historical Data Book, Volume IV: NASA Resources, 1969–1978*. NASA SP-4012, 1994.

Rumerman, Judy A. *NASA Historical Data Book, Volume V: NASA Launch Systems, Space Transportation, Human Spaceflight, and Space Science, 1979–1988*. NASA SP-4012, 1999.

Rumerman, Judy A. *NASA Historical Data Book, Volume VI: NASA Space Applications, Aeronautics and Space Research and Technology, Tracking and Data Acquisition/Support Operations, Commercial Programs, and Resources, 1979–1988*. NASA SP-4012, 1999.

Rumerman, Judy A. *NASA Historical Data Book, Volume VII: NASA Launch Systems, Space Transportation, Human Spaceflight, and Space Science, 1989–1998*. NASA SP-2009-4012, 2009.

Astronautics and Aeronautics, 1969: Chronology of Science, Technology, and Policy. NASA SP-4014, 1970.

Astronautics and Aeronautics, 1970: Chronology of Science, Technology, and Policy. NASA SP-4015, 1972.

Astronautics and Aeronautics, 1971: Chronology of Science, Technology, and Policy. NASA SP-4016, 1972.

Astronautics and Aeronautics, 1972: Chronology of Science, Technology, and Policy. NASA SP-4017, 1974.

Astronautics and Aeronautics, 1973: Chronology of Science, Technology, and Policy. NASA SP-4018, 1975.

Astronautics and Aeronautics, 1974: Chronology of Science, Technology, and Policy. NASA SP-4019, 1977.

Astronautics and Aeronautics, 1975: Chronology of Science, Technology, and Policy. NASA SP-4020, 1979.

Astronautics and Aeronautics, 1976: Chronology of Science, Technology, and Policy. NASA SP-4021, 1984.

Astronautics and Aeronautics, 1977: Chronology of Science, Technology, and Policy. NASA SP-4022, 1986.

Astronautics and Aeronautics, 1978: Chronology of Science, Technology, and Policy. NASA SP-4023, 1986.

Astronautics and Aeronautics, 1979–1984: Chronology of Science, Technology, and Policy. NASA SP-4024, 1988.

Astronautics and Aeronautics, 1985: Chronology of Science, Technology, and Policy. NASA SP-4025, 1990.

Noordung, Hermann. *The Problem of Space Travel: The Rocket Motor*. Edited by Ernst Stuhlinger and J. D. Hunley, with Jennifer Garland. NASA SP-4026, 1995.

Astronautics and Aeronautics, 1986–1990: A Chronology. NASA SP-4027, 1997.

Astronautics and Aeronautics, 1991–1995: A Chronology. NASA SP-2000-4028, 2000.

Orloff, Richard W. *Apollo by the Numbers: A Statistical Reference*. NASA SP-2000-4029, 2000.

Lewis, Marieke, and Ryan Swanson. *Astronautics and Aeronautics: A Chronology, 1996–2000*. NASA SP-2009-4030, 2009.

Ivey, William Noel, and Ryan Swanson. *Astronautics and Aeronautics: A Chronology, 2001–2005*. NASA SP-2010-4031, 2010.

MANAGEMENT HISTORIES, NASA SP-4100:

Rosholt, Robert L. *An Administrative History of NASA, 1958–1963*. NASA SP-4101, 1966.

Levine, Arnold S. *Managing NASA in the Apollo Era*. NASA SP-4102, 1982.

Roland, Alex. *Model Research: The National Advisory Committee for Aeronautics, 1915–1958*. NASA SP-4103, 1985.

Fries, Sylvia D. *NASA Engineers and the Age of Apollo*. NASA SP-4104, 1992.

Glennan, T. Keith. *The Birth of NASA: The Diary of T. Keith Glennan*. Edited by J. D. Hunley. NASA SP-4105, 1993.

Seamans, Robert C. *Aiming at Targets: The Autobiography of Robert C. Seamans*. NASA SP-4106, 1996.

Garber, Stephen J., ed. *Looking Backward, Looking Forward: Forty Years of Human Spaceflight Symposium*. NASA SP-2002-4107, 2002.

Mallick, Donald L., with Peter W. Merlin. *The Smell of Kerosene: A Test Pilot's Odyssey*. NASA SP-4108, 2003.

Iliff, Kenneth W., and Curtis L. Peebles. *From Runway to Orbit: Reflections of a NASA Engineer*. NASA SP-2004-4109, 2004.

Chertok, Boris. *Rockets and People, Volume I*. NASA SP-2005-4110, 2005.

Chertok, Boris. *Rockets and People: Creating a Rocket Industry, Volume II*. NASA SP-2006-4110, 2006.

Chertok, Boris. *Rockets and People: Hot Days of the Cold War, Volume III*. NASA SP-2009-4110, 2009.

Laufer, Alexander, Todd Post, and Edward Hoffman. *Shared Voyage: Learning and Unlearning from Remarkable Projects*. NASA SP-2005-4111, 2005.

Dawson, Virginia P., and Mark D. Bowles. *Realizing the Dream of Flight: Biographical Essays in Honor of the Centennial of Flight, 1903–2003*. NASA SP-2005-4112, 2005.

Mudgway, Douglas J. *William H. Pickering: America's Deep Space Pioneer*. NASA SP-2008-4113.

PROJECT HISTORIES, NASA SP-4200:

Swenson, Loyd S., Jr., James M. Grimwood, and Charles C. Alexander. *This New Ocean: A History of Project Mercury*. NASA SP-4201, 1966; rep. ed. 1999.

Green, Constance McLaughlin, and Milton Lomask. *Vanguard: A History*. NASA SP-4202, 1970; rep. ed. Smithsonian Institution Press, 1971.

Hacker, Barton C., and James M. Grimwood. *On the Shoulders of Titans: A History of Project Gemini*. NASA SP-4203, 1977; rep. ed. 2002.

Benson, Charles D., and William Barnaby Faherty. *Moonport: A History of Apollo Launch Facilities and Operations*. NASA SP-4204, 1978.

Brooks, Courtney G., James M. Grimwood, and Loyd S. Swenson, Jr. *Chariots for Apollo: A History of Manned Lunar Spacecraft*. NASA SP-4205, 1979.

Bilstein, Roger E. *Stages to Saturn: A Technological History of the Apollo/Saturn Launch Vehicles*. NASA SP-4206, 1980 and 1996.

Compton, W. David, and Charles D. Benson. *Living and Working in Space: A History of Skylab*. NASA SP-4208, 1983.

Ezell, Edward Clinton, and Linda Neuman Ezell. *The Partnership: A History of the Apollo-Soyuz Test Project*. NASA SP-4209, 1978.

Hall, R. Cargill. *Lunar Impact: A History of Project Ranger*. NASA SP-4210, 1977.

Newell, Homer E. *Beyond the Atmosphere: Early Years of Space Science*. NASA SP-4211, 1980.

Ezell, Edward Clinton, and Linda Neuman Ezell. *On Mars: Exploration of the Red Planet, 1958–1978*. NASA SP-4212, 1984.

Pitts, John A. *The Human Factor: Biomedicine in the Manned Space Program to 1980*. NASA SP-4213, 1985.

Compton, W. David. *Where No Man Has Gone Before: A History of Apollo Lunar Exploration Missions*. NASA SP-4214, 1989.

Naugle, John E. *First Among Equals: The Selection of NASA Space Science Experiments*. NASA SP-4215, 1991.

Wallace, Lane E. *Airborne Trailblazer: Two Decades with NASA Langley's 737 Flying Laboratory*. NASA SP-4216, 1994.

Butrica, Andrew J., ed. *Beyond the Ionosphere: Fifty Years of Satellite Communications*. NASA SP-4217, 1997.

Butrica, Andrew J. *To See the Unseen: A History of Planetary Radar Astronomy*. NASA SP-4218, 1996.

Mack, Pamela E., ed. *From Engineering Science to Big Science: The NACA and NASA Collier Trophy Research Project Winners*. NASA SP-4219, 1998.

Reed, R. Dale. *Wingless Flight: The Lifting Body Story*. NASA SP-4220, 1998.

Heppenheimer, T. A. *The Space Shuttle Decision: NASA's Search for a Reusable Space Vehicle*. NASA SP-4221, 1999.

Hunley, J. D., ed. *Toward Mach 2: The Douglas D-558 Program*. NASA SP-4222, 1999.

Swanson, Glen E., ed. *"Before This Decade Is Out . . ." Personal Reflections on the Apollo Program*. NASA SP-4223, 1999.

Tomayko, James E. *Computers Take Flight: A History of NASA's Pioneering Digital Fly-By-Wire Project*. NASA SP-4224, 2000.

Morgan, Clay. *Shuttle-Mir: The United States and Russia Share History's Highest Stage*. NASA SP-2001-4225.

Leary, William M. *"We Freeze to Please:" A History of NASA's Icing Research Tunnel and the Quest for Safety*. NASA SP-2002-4226, 2002.

Mudgway, Douglas J. *Uplink-Downlink: A History of the Deep Space Network, 1957–1997*. NASA SP-2001-4227.

Dawson, Virginia P., and Mark D. Bowles. *Taming Liquid Hydrogen: The Centaur Upper Stage Rocket, 1958–2002*. NASA SP-2004-4230.

Meltzer, Michael. *Mission to Jupiter: A History of the Galileo Project*. NASA SP-2007-4231.

Heppenheimer, T. A. *Facing the Heat Barrier: A History of Hypersonics*. NASA SP-2007-4232.

Tsiao, Sunny. *"Read You Loud and Clear!" The Story of NASA's Spaceflight Tracking and Data Network*. NASA SP-2007-4233.

CENTER HISTORIES, NASA SP-4300:

Rosenthal, Alfred. *Venture into Space: Early Years of Goddard Space Flight Center*. NASA SP-4301, 1985.

Hartman, Edwin P. *Adventures in Research: A History of Ames Research Center, 1940–1965*. NASA SP-4302, 1970.

Hallion, Richard P. *On the Frontier: Flight Research at Dryden, 1946–1981*. NASA SP-4303, 1984.

Muenger, Elizabeth A. *Searching the Horizon: A History of Ames Research Center, 1940–1976*. NASA SP-4304, 1985.

Hansen, James R. *Engineer in Charge: A History of the Langley Aeronautical Laboratory, 1917–1958*. NASA SP-4305, 1987.

Dawson, Virginia P. *Engines and Innovation: Lewis Laboratory and American Propulsion Technology*. NASA SP-4306, 1991.

Dethloff, Henry C. *"Suddenly Tomorrow Came . . .": A History of the Johnson Space Center, 1957–1990*. NASA SP-4307, 1993.

Hansen, James R. *Spaceflight Revolution: NASA Langley Research Center from Sputnik to Apollo*. NASA SP-4308, 1995.

Wallace, Lane E. *Flights of Discovery: An Illustrated History of the Dryden Flight Research Center*. NASA SP-4309, 1996.

Herring, Mack R. *Way Station to Space: A History of the John C. Stennis Space Center*. NASA SP-4310, 1997.

Wallace, Harold D., Jr. *Wallops Station and the Creation of an American Space Program*. NASA SP-4311, 1997.

Wallace, Lane E. *Dreams, Hopes, Realities. NASA's Goddard Space Flight Center: The First Forty Years*. NASA SP-4312, 1999.

Dunar, Andrew J., and Stephen P. Waring. *Power to Explore: A History of Marshall Space Flight Center, 1960–1990*. NASA SP-4313, 1999.

Bugos, Glenn E. *Atmosphere of Freedom: Sixty Years at the NASA Ames Research Center*. NASA SP-2000-4314, 2000.

Schultz, James. *Crafting Flight: Aircraft Pioneers and the Contributions of the Men and Women of NASA Langley Research Center*. NASA SP-2003-4316, 2003.

Bowles, Mark D. *Science in Flux: NASA's Nuclear Program at Plum Brook Station, 1955–2005*. NASA SP-2006-4317, 2006.

Wallace, Lane E. *Flights of Discovery: An Illustrated History of the Dryden Flight Research Center*. NASA SP-2007-4318, 2007. Revised version of NASA SP-4309.

GENERAL HISTORIES, NASA SP-4400:

Corliss, William R. *NASA Sounding Rockets, 1958–1968: A Historical Summary*. NASA SP-4401, 1971.

Wells, Helen T., Susan H. Whiteley, and Carrie Karegeannes. *Origins of NASA Names*. NASA SP-4402, 1976.

Anderson, Frank W., Jr. *Orders of Magnitude: A History of NACA and NASA, 1915–1980*. NASA SP-4403, 1981.

Sloop, John L. *Liquid Hydrogen as a Propulsion Fuel, 1945–1959*. NASA SP-4404, 1978.

Roland, Alex. *A Spacefaring People: Perspectives on Early Spaceflight*. NASA SP-4405, 1985.

Bilstein, Roger E. *Orders of Magnitude: A History of the NACA and NASA, 1915–1990*. NASA SP-4406, 1989.

Logsdon, John M., ed., with Linda J. Lear, Jannelle Warren Findley, Ray A. Williamson, and Dwayne A. Day. *Exploring the Unknown: Selected Documents in the History of the U.S. Civil Space Program, Volume I: Organizing for Exploration*. NASA SP-4407, 1995.

Logsdon, John M., ed., with Dwayne A. Day and Roger D. Launius. *Exploring the Unknown: Selected Documents in the History of the U.S. Civil Space Program, Volume II: External Relationships*. NASA SP-4407, 1996.

Logsdon, John M., ed., with Roger D. Launius, David H. Onkst, and Stephen J. Garber. *Exploring the Unknown: Selected Documents in the History of the U.S. Civil Space Program, Volume III: Using Space*. NASA SP-4407, 1998.

Logsdon, John M., ed., with Ray A. Williamson, Roger D. Launius, Russell J. Acker, Stephen J. Garber, and Jonathan L. Friedman. *Exploring the Unknown: Selected Documents in the History of the U.S. Civil Space Program, Volume IV: Accessing Space*. NASA SP-4407, 1999.

Logsdon, John M., ed., with Amy Paige Snyder, Roger D. Launius, Stephen J. Garber, and Regan Anne Newport. *Exploring the Unknown: Selected Documents in the History of the U.S. Civil Space Program, Volume V: Exploring the Cosmos*. NASA SP-2001-4407, 2001.

Logsdon, John M., ed., with Stephen J. Garber, Roger D. Launius, and Ray A. Williamson. *Exploring the Unknown: Selected Documents in the History of the U.S. Civil Space Program, Volume VI: Space and Earth Science*. NASA SP-2004-4407, 2004.

Logsdon, John M., ed., with Roger D. Launius. *Exploring the Unknown: Selected Documents in the History of the U.S. Civil Space Program, Volume VII: Human Spaceflight: Projects Mercury, Gemini, and Apollo*. NASA SP-2008-4407, 2008.

Siddiqi, Asif A., *Challenge to Apollo: The Soviet Union and the Space Race, 1945–1974*. NASA SP-2000-4408, 2000.

Hansen, James R., ed. *The Wind and Beyond: Journey into the History of Aerodynamics in America, Volume 1: The Ascent of the Airplane*. NASA SP-2003-4409, 2003.

Hansen, James R., ed. *The Wind and Beyond: Journey into the History of Aerodynamics in America, Volume 2: Reinventing the Airplane*. NASA SP-2007-4409, 2007.

Hogan, Thor. *Mars Wars: The Rise and Fall of the Space Exploration Initiative*. NASA SP-2007-4410, 2007.

MONOGRAPHS IN AEROSPACE HISTORY, NASA SP-4500:

Launius, Roger D., and Aaron K. Gillette, comps. *Toward a History of the Space Shuttle: An Annotated Bibliography*. Monographs in Aerospace History, No. 1, 1992.

Launius, Roger D., and J. D. Hunley, comps. *An Annotated Bibliography of the Apollo Program*. Monographs in Aerospace History, No. 2, 1994.

Launius, Roger D. *Apollo: A Retrospective Analysis*. Monographs in Aerospace History, No. 3, 1994.

Hansen, James R. *Enchanted Rendezvous: John C. Houbolt and the Genesis of the Lunar-Orbit Rendezvous Concept*. Monographs in Aerospace History, No. 4, 1995.

Gorn, Michael H. *Hugh L. Dryden's Career in Aviation and Space*. Monographs in Aerospace History, No. 5, 1996.

Powers, Sheryll Goecke. *Women in Flight Research at NASA Dryden Flight Research Center from 1946 to 1995*. Monographs in Aerospace History, No. 6, 1997.

Portree, David S. F., and Robert C. Trevino. *Walking to Olympus: An EVA Chronology*. Monographs in Aerospace History, No. 7, 1997.

Logsdon, John M., moderator. *Legislative Origins of the National Aeronautics and Space Act of 1958: Proceedings of an Oral History Workshop*. Monographs in Aerospace History, No. 8, 1998.

Rumerman, Judy A., comp. *U.S. Human Spaceflight: A Record of Achievement, 1961–1998*. Monographs in Aerospace History, No. 9, 1998.

Portree, David S. F. *NASA's Origins and the Dawn of the Space Age*. Monographs in Aerospace History, No. 10, 1998.

Logsdon, John M. *Together in Orbit: The Origins of International Cooperation in the Space Station*. Monographs in Aerospace History, No. 11, 1998.

Phillips, W. Hewitt. *Journey in Aeronautical Research: A Career at NASA Langley Research Center*. Monographs in Aerospace History, No. 12, 1998.

Braslow, Albert L. *A History of Suction-Type Laminar-Flow Control with Emphasis on Flight Research*. Monographs in Aerospace History, No. 13, 1999.

Logsdon, John M., moderator. *Managing the Moon Program: Lessons Learned from Apollo*. Monographs in Aerospace History, No. 14, 1999.

Perminov, V. G. *The Difficult Road to Mars: A Brief History of Mars Exploration in the Soviet Union*. Monographs in Aerospace History, No. 15, 1999.

Tucker, Tom. *Touchdown: The Development of Propulsion Controlled Aircraft at NASA Dryden*. Monographs in Aerospace History, No. 16, 1999.

Maisel, Martin, Demo J. Giulanetti, and Daniel C. Dugan. *The History of the XV-15 Tilt Rotor Research Aircraft: From Concept to Flight*. Monographs in Aerospace History, No. 17, 2000. NASA SP-2000-4517.

Jenkins, Dennis R. *Hypersonics Before the Shuttle: A Concise History of the X-15 Research Airplane*. Monographs in Aerospace History, No. 18, 2000. NASA SP-2000-4518.

Chambers, Joseph R. *Partners in Freedom: Contributions of the Langley Research Center to U.S. Military Aircraft of the 1990s*. Monographs in Aerospace History, No. 19, 2000. NASA SP-2000-4519.

Waltman, Gene L. *Black Magic and Gremlins: Analog Flight Simulations at NASA's Flight Research Center*. Monographs in Aerospace History, No. 20, 2000. NASA SP-2000-4520.

Portree, David S. F. *Humans to Mars: Fifty Years of Mission Planning, 1950–2000*. Monographs in Aerospace History, No. 21, 2001. NASA SP-2001-4521.

Thompson, Milton O., with J. D. Hunley. *Flight Research: Problems Encountered and What They Should Teach Us*. Monographs in Aerospace History, No. 22, 2001. NASA SP-2001-4522.

Tucker, Tom. *The Eclipse Project*. Monographs in Aerospace History, No. 23, 2001. NASA SP-2001-4523.

Siddiqi, Asif A. *Deep Space Chronicle: A Chronology of Deep Space and Planetary Probes, 1958–2000*. Monographs in Aerospace History, No. 24, 2002. NASA SP-2002-4524.

Merlin, Peter W. *Mach 3+: NASA/USAF YF-12 Flight Research, 1969–1979*. Monographs in Aerospace History, No. 25, 2001. NASA SP-2001-4525.

Anderson, Seth B. *Memoirs of an Aeronautical Engineer: Flight Tests at Ames Research Center: 1940–1970*. Monographs in Aerospace History, No. 26, 2002. NASA SP-2002-4526.

Renstrom, Arthur G. *Wilbur and Orville Wright: A Bibliography Commemorating the One-Hundredth Anniversary of the First Powered Flight on December 17, 1903*. Monographs in Aerospace History, No. 27, 2002. NASA SP-2002-4527.

No monograph 28.

Chambers, Joseph R. *Concept to Reality: Contributions of the NASA Langley Research Center to U.S. Civil Aircraft of the 1990s*. Monographs in Aerospace History, No. 29, 2003. NASA SP-2003-4529.

Peebles, Curtis, ed. *The Spoken Word: Recollections of Dryden History, The Early Years*. Monographs in Aerospace History, No. 30, 2003. NASA SP-2003-4530.

Jenkins, Dennis R., Tony Landis, and Jay Miller. *American X-Vehicles: An Inventory—X-1 to X-50*. Monographs in Aerospace History, No. 31, 2003. NASA SP-2003-4531.

Renstrom, Arthur G. *Wilbur and Orville Wright: A Chronology Commemorating the One-Hundredth Anniversary of the First Powered Flight on December 17, 1903*. Monographs in Aerospace History, No. 32, 2003. NASA SP-2003-4532.

Bowles, Mark D., and Robert S. Arrighi. *NASA's Nuclear Frontier: The Plum Brook Research Reactor*. Monographs in Aerospace History, No. 33, 2004. NASA SP-2004-4533.

Wallace, Lane, and Christian Gelzer. *Nose Up: High Angle-of-Attack and Thrust Vectoring Research at NASA Dryden, 1979–2001*. Monographs in Aerospace History, No. 34, 2009. NASA SP-2009-4534.

Matranga, Gene J., C. Wayne Ottinger, Calvin R. Jarvis, and D. Christian Gelzer. *Unconventional, Contrary, and Ugly: The Lunar Landing Research Vehicle*. Monographs in Aerospace History, No. 35, 2006. NASA SP-2004-4535.

McCurdy, Howard E. *Low-Cost Innovation in Spaceflight: The History of the Near Earth Asteroid Rendezvous (NEAR) Mission*. Monographs in Aerospace History, No. 36, 2005. NASA SP-2005-4536.

Seamans, Robert C., Jr. *Project Apollo: The Tough Decisions*. Monographs in Aerospace History, No. 37, 2005. NASA SP-2005-4537.

Lambright, W. Henry. *NASA and the Environment: The Case of Ozone Depletion*. Monographs in Aerospace History, No. 38, 2005. NASA SP-2005-4538.

Chambers, Joseph R. *Innovation in Flight: Research of the NASA Langley Research Center on Revolutionary Advanced Concepts for Aeronautics*. Monographs in Aerospace History, No. 39, 2005. NASA SP-2005-4539.

Phillips, W. Hewitt. *Journey into Space Research: Continuation of a Career at NASA Langley Research Center*. Monographs in Aerospace History, No. 40, 2005. NASA SP-2005-4540.

Rumerman, Judy A., Chris Gamble, and Gabriel Okolski, comps. *U.S. Human Spaceflight: A Record of Achievement, 1961–2006*. Monographs in Aerospace History, No. 41, 2007. NASA SP-2007-4541.

Dick, Steven J., Stephen J. Garber, and Jane H. Odom. *Research in NASA History*. Monographs in Aerospace History, No. 43, 2009. NASA SP-2009-4543.

Merlin, Peter W. *Ikhana: Unmanned Aircraft System Western States Fire Missions*. Monographs in Aerospace History, No. 44, 2009. NASA SP-2009-4544.

Fisher, Steven C., and Shamim A. Rahman. *Remembering the Giants: Apollo Rocket Propulsion Development*. Monographs in Aerospace History, No. 45, 2009. NASA SP-2009-4545.

ELECTRONIC MEDIA, NASA SP-4600:

Remembering Apollo 11: The 30th Anniversary Data Archive CD-ROM. NASA SP-4601, 1999.

Remembering Apollo 11: The 35th Anniversary Data Archive CD-ROM. NASA SP-2004-4601, 2004. This is an update of the 1999 edition.

The Mission Transcript Collection: U.S. Human Spaceflight Missions from Mercury Redstone 3 to Apollo 17. NASA SP-2000-4602, 2001.

Shuttle-Mir: The United States and Russia Share History's Highest Stage. NASA SP-2001-4603, 2002.

U.S. Centennial of Flight Commission Presents Born of Dreams—Inspired by Freedom. NASA SP-2004-4604, 2004.

Of Ashes and Atoms: A Documentary on the NASA Plum Brook Reactor Facility. NASA SP-2005-4605, 2005.

Taming Liquid Hydrogen: The Centaur Upper Stage Rocket Interactive CD-ROM. NASA SP-2004-4606, 2004.

Fueling Space Exploration: The History of NASA's Rocket Engine Test Facility DVD. NASA SP-2005-4607, 2005.

Altitude Wind Tunnel at NASA Glenn Research Center: An Interactive History CD-ROM. NASA SP-2008-4608, 2008.

CONFERENCE PROCEEDINGS, NASA SP-4700:

Dick, Steven J., and Keith Cowing, eds. *Risk and Exploration: Earth, Sea and the Stars*. NASA SP-2005-4701, 2005.

Dick, Steven J., and Roger D. Launius. *Critical Issues in the History of Spaceflight*. NASA SP-2006-4702, 2006.

Dick, Steven J., ed. *Remembering the Space Age: Proceedings of the 50th Anniversary Conference*. NASA SP-2008-4703, 2008.

SOCIETAL IMPACT, NASA SP-4800:

Dick, Steven J., and Roger D. Launius. *Societal Impact of Spaceflight*. NASA SP-2007-4801, 2007.

Dick, Steven J., and Mark L. Lupisella. *Cosmos and Culture: Cultural Evolution in a Cosmic Context*. NASA SP-2009-4802, 2009.

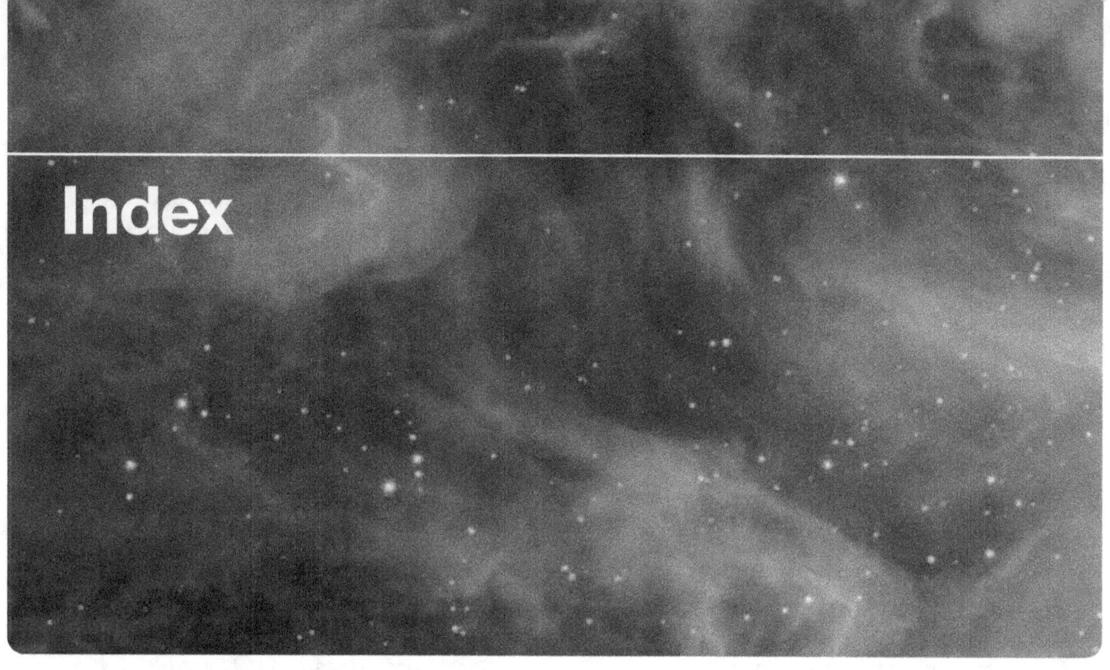

Index

A

B

C

D

E

G

H

J

K

M

N

O

P

Q

R

S

T

U

W

X

Z

www.ingramcontent.com/pod-product-compliance
Lightning Source LLC
Chambersburg PA
CBHW081424170526
45166CB00008B/2104

* 9 7 8 1 4 7 0 0 2 4 7 5 8 *